INDUSTRIAL ELECTRICITY

5th Edition

John M. Nadon
Bert J. Gelmine
Edward D. McLaughlin

NOTICE TO THE READER

Publisher does not warrant or guarantee any of the products described herein or perform any independent analysis in connection with any of the product information contained herein. Publisher does not assume, and expressly disclaims, any obligation to obtain and include information other than that provided to it by the manufacturer.

The reader is expressly warned to consider and adopt all safety precautions that might be indicated by the activities described herein and to avoid all potential hazards. By following the instructions contained herein, the reader willingly assumes all risks in connection with such instructions.

The publisher makes no representations or warranties of any kind, including but not limited to, the warranties of fitness for particular purpose or merchantability, nor are any such representations implied with respect to the material set forth herein, and the publisher takes no responsibility with respect to such material. The publisher shall not be liable for any special, consequential, or exemplary damages resulting, in whole or in part, from the readers' use of, or reliance upon, this material.

Cover Photo Courtesy of Niagara Mohawk

Cover Design By Cheri Plasse

Delmar Staff

Senior Executive Editor: Mark Muth
Assistant Editor: Nancy Belser
Senior Project Editor: Laura Gulotty
Production Coordinator: Karen Smith
Art/Design Supervisor: Cheri Plasse

For information, address Delmar Publishers Inc.
3 Columbia Circle, Box 15-015
Albany, NY 12212-5015

1 2 3 4 5 6 7 8 9 10 X X X 00 99 98 97 96 95 94

Printed in the United States of America
Published simultaneously in Canada
by Nelson Canada,
A Division of The Thomson Corporation

Library of Congress Cataloging in Publication Data

Nadon, John M.
Industrial electricity / John M. Nadon, Bert J. Gelmine, Edward D. McLaughlin. —5th ed.
p. cm.
Includes index.
ISBN 0-8273-6074-6
1. Electric engineering. I. Gelmine, Bert J. II. McLaughlin, Edward D. III. Title
TK146.N23 1994 621.31—dc20 93–20198
ISBN: 0-8273-6074-6 text CIP
0-8273-6075-4 I6

CINCINNATI STATE
WITHDRAWN

CONTENTS

PREFACE

The continued growth of the electrical industry and technological advancements place increasing demands upon electrical personnel. Every electrical worker needs general knowledge of electricity and electronics to prepare for advanced studies as well as specialization in a specific area.

A thorough knowledge of electrical theory is not enough to be successful in today's business. Prospective electrical personnel must also learn to apply this knowledge to present-day equipment. Since INDUSTRIAL ELECTRICITY was first published, it has proved to be a valuable tool for beginners in electrical trades and technology. The text covers the theory of electricity and focuses on the installation, maintenance, and industrial application of electrical equipment and controls.

This well-illustrated text first introduces basic electrical symbols and theory. The material is presented from the simpler to the more complex topics in industrial electricity. All terms are defined as they are introduced, and a glossary of important terms is included. Important metric units of measurement, which are valuable for a thorough understanding of theory and application, are introduced.

Educational objectives are stated at the beginning of each chapter to clearly outline what the student is expected to learn. Reviews at the end of each chapter help to determine if the student has achieved the goals set by the objectives. The reviews contain a variety of multiple-choice questions, essay questions, and mathematical problems, as appropriate to the material covered.

Technological progress in the field of electricity has necessitated this revision of INDUSTRIAL ELECTRICITY. The authors have maintained the extent of prior editions of the text while revising those areas that required technical updating.

A comprehensive *Instructor's Guide*, detailing answers to the reviews and providing some metric examples of mathematical problems, is available to course instructors.

Edward D. McLaughlin has taught vocational electrical courses at the high school and adult education levels and has lectured at educational conferences. He has many years' experience in electrical construction and maintenance and has provided consulting services in these areas. He has teaching credentials in the areas of occupational education and special education. In addition, Mr. McLaughlin is the author of Electricity for Electricians and coauthor of Electrical Wiring Methods (Hickok Teaching Systems).

Bert J. Gelmine earned B.S. and M.S. degrees in electrical engineering and taught at Ford Motor Company for thirty-nine years. After his retirement in 1968, Mr. Gelmine taught part-time at Henry Ford Community College in Dearborn, Michigan.

John M. Nadon is a retired supervisor of electrical engineering instruction at the Apprentice School of Ford Motor Company.

ACKNOWLEDGMENTS

The authors are grateful for the important suggestions made by the reviewers of this edition:

Jack Bohannon, Long Beach City College
Vaughn Treadway, Southern Technical College
Don Hartshorn, Columbus State Community College

The authors wish to express their appreciation to the many individuals and organizations who supplied the photographs and information necessary for revision of the text. Special appreciation is extended to the following, whose extra efforts have enhanced the quality of the book: Niagara Mohawk Power Corp.; W.S. Cahill, C & I Publicity Division, General Electric Company, Nela Park, Cleveland, OH; Paul Sackenheim, advertising manager, IC Division, Square D Company, Milwaukee, WI; Barbara Duggan, assistant to the advertising manager, Wiremold Company, West Hartford, CT; H.F. van der Voort, vice president, Planning and Development, Carlon—An Indian Head Company, Cleveland, OH; Professor Samuel Noodleman, University of Arizona at Tucson; June M. Lloyd, Inland Motor Division, R&D Group, Radford, VA; Frank W. Laltham, director of vocational education, Central Florida Community College; and Michael Finocchiaro, electrical instructor, Central Florida Community College.

The authors also thank Alan Levy and George Tashjian of Medford Public Schools, Medford, MA, for their expert photography; and the students and faculty at Medford Vocational-Technical High School for supplying the material and assisting Mr. Levy and Mr. Tashjian. Acknowledgment is also due to the editorial staff at Delmar Publishers for their invaluable skills and guidance.

1

Language of Electricity

Objectives

After studying this chapter, the student will be able to:

- Explain the purpose of electrical symbols and drawings.
- List three main categories of electrical symbols and drawings.
- Identify and draw from memory some common electrical symbols.
- Use electrical symbols to construct electrical drawings.

One of the greatest accomplishments of humanity is the development of systems of symbols for communication. The modern English alphabet is a set of 26 symbols, called letters, which convey information when combined in the proper order. Our number system is a set of 10 symbols, which can be used individually or grouped to represent small quantities or large quantities. Symbols used in arithmetic indicate the operation to be performed. For example, the symbol + indicates addition and the symbol × indicates multiplication. Another set of symbols known as shorthand enables stenographers to write words at a rapid pace.

ELECTRICAL SYMBOLS

Electrical symbols are used throughout the electrical industry. The symbols allow engineers and other professionals to convey information regarding type of equipment, location, and proper electrical connections in a simple and concise manner.

In order that electrical personnel throughout the continent can understand the symbols, a uniform set of electrical symbols has been developed. These symbols are divided into three categories: schematic, architectural, and pictorial.

Schematic Symbols

Schematic symbols are used in diagrams that indicate the electrical connections but not necessarily the location of the components. Schematic symbols are also frequently incorporated

TABLE 1-1: Schematic Symbols

METER INSTRUMENTS		ROTATING MACHINERY		MISCELLANEOUS	
A	AMMETER	G	GENERATOR, GENERAL	BELL	GROUND
AH	AMPERE-HOUR METER	G	GENERATOR, DC	BUZZER	CONTACTS, NORMALLY CLOSED
DB	DECIBEL METER	G	GENERATOR, AC	BATTERY, MULTICELL	CONTACTS, NORMALLY OPEN
F	FREQUENCY METER	GS	GENERATOR, SYNCHRONOUS	CELL	MOMENTARY PUSH BUTTON, NORMALLY OPEN
GD	GROUND DETECTOR	M	MOTOR, GENERAL	PUSH BUTTON	MOMENTARY PUSH BUTTON, NORMALLY CLOSED
OHM	OHMMETER	M	MOTOR, DC	CIRCUIT BREAKER	MOMENTARY PUSH BUTTON, TWO-CONTACT NORMALLY CLOSED NORMALLY OPEN (TWO-CIRCUIT)
PF	POWER FACTOR METER	M	MOTOR, AC	FUSE	TRANSFORMER, GENERAL
SY	SYNCHROSCOPE	MS	MOTOR, SYNCHRONOUS	THREE-WAY SWITCH	NONSATURATING TRANSFORMER (MAGNETIC CORE)
V	VOLTMETER		COMPENSATION WINDING	FOUR-WAY SWITCH	AUTOTRANSFORMER
VA	VOLT-AMMETER		SERIES FIELD WINDING	INCANDESCENT LAMP	SINGLE-POLE, SINGLE-THROW SWITCH, FULLY OPEN
VAR	VARMETER		SHUNT FIELD WINDING	PILOT LIGHT	DOUBLE-POLE, SINGLE-THROW SWITCH, FULLY OPEN
W	WATTMETER		ARMATURE	RESISTOR, FIXED	THREE-POLE, SINGLE-THROW SWITCH, FULLY OPEN
WH	WATTHOUR METER			RESISTOR, TAPPED	THERMAL OVERLOAD RELAY
				RESISTOR, ADJUSTABLE	MAGNETIC RELAY
				CAPACITOR, FIXED	
				WIRES CROSSING, NOT CONNECTED	
				WIRES CONNECTED	
				SPLICE	
				HEATER	

into other types of diagrams. Some common schematic symbols are listed in Table 1-1.

Architectural Symbols

Architectural symbols, Table 1-2, are used in conjunction with architectural drawings (for buildings and structures). These symbols indicate the type of device or equipment and its location, as

TABLE 1-2: Architectural Electrical Symbols

Symbol	Description
	SURFACE OR PENDANT INCANDESCENT OR SIMILAR FIXTURE
	WALL FIXTURE, INCANDESCENT OR SIMILAR
R	RECESSED INCANDESCENT OR SIMILAR CEILING FIXTURE
R	RECESSED INCANDESCENT OR SIMILAR WALL FIXTURE
O	SURFACE OR PENDANT FLUORESCENT CEILING FIXTURE
O	SURFACE FLUORESCENT WALL FIXTURE
O	SURFACE OR PENDANT FLUORESCENT CEILING FIXTURE: CONTINUOUS ROW
OR	RECESSED FLUORESCENT CEILING FIXTURE: CONTINUOUS ROW
WP	WEATHERPROOF WALL FIXTURE
X	SURFACE OR PENDANT CEILING EXIT LIGHT
X	WALL EXIT LIGHT
RX	WALL EXIT LIGHT: RECESSED
B	CEILING BLANKED OUTLET
B	WALL BLANKED OUTLET
	GROUNDED SINGLE RECEPTACLE OUTLET
UNG	UNGROUNDED SINGLE RECEPTACLE OUTLET
	GROUNDED DUPLEX RECEPTACLE OUTLET
UNG	UNGROUNDED DUPLEX RECEPTACLE OUTLET
	GROUNDED DUPLEX RECEPTACLE OUTLET, SPLIT WIRED
	GROUNDED SINGLE SPECIAL PURPOSE OUTLET
R	GROUNDED RANGE OUTLET
DW	GROUNDED DISHWASHER OUTLET
	(The above symbol is used for all special-purpose outlets. Subscript letters are changed to indicate function, e.g., CD — clothes dryer, WH — water heater, etc.)
C	GROUNDED CLOCK-HANGER RECEPTACLE
F	GROUNDED FAN-HANGER RECEPTACLE
	GROUNDED FLOOR SINGLE RECEPTACLE OUTLET
	GROUNDED FLOOR DUPLEX RECEPTACLE OUTLET
	PUBLIC TELEPHONE OUTLET
	PRIVATE TELEPHONE OUTLET
S	SINGLE-POLE SWITCH
S_2	DOUBLE-POLE SWITCH
S_3	THREE-WAY SWITCH
S_4	FOUR-WAY SWITCH
S_K	KEY-OPERATED SWITCH
S_P	SWITCH AND PILOT LIGHT
S_L	SWITCH FOR LOW-VOLTAGE SWITCHING SYSTEM
S_{LM}	MASTER SWITCH FOR LOW-VOLTAGE SWITCHING SYSTEM
S_D	DOOR SWITCH
S_T	TIME SWITCH
	PUSH BUTTON
	BUZZER
	BELL
	COMBINATION BELL/BUZZER
CH	CHIME
D	ELECTRIC DOOR OPENER
BT	BELL TRANSFORMER
TV	TELEVISION OUTLET
MC	MOTOR CONTROL
	CONDUIT OR CABLE RUN. SLASHES INDICATE THE NUMBER OF CONDUCTORS
	SUPPLY FROM PANEL. ARROWS INDICATE THE NUMBER OF CIRCUITS
	CONDUIT OR CABLE. THE CURVED DASHED LINE INDICATES THE APPARATUS IS SWITCHED
	POWER PANEL
	LIGHTING PANEL

TABLE 1-3: Pictorial Symbols

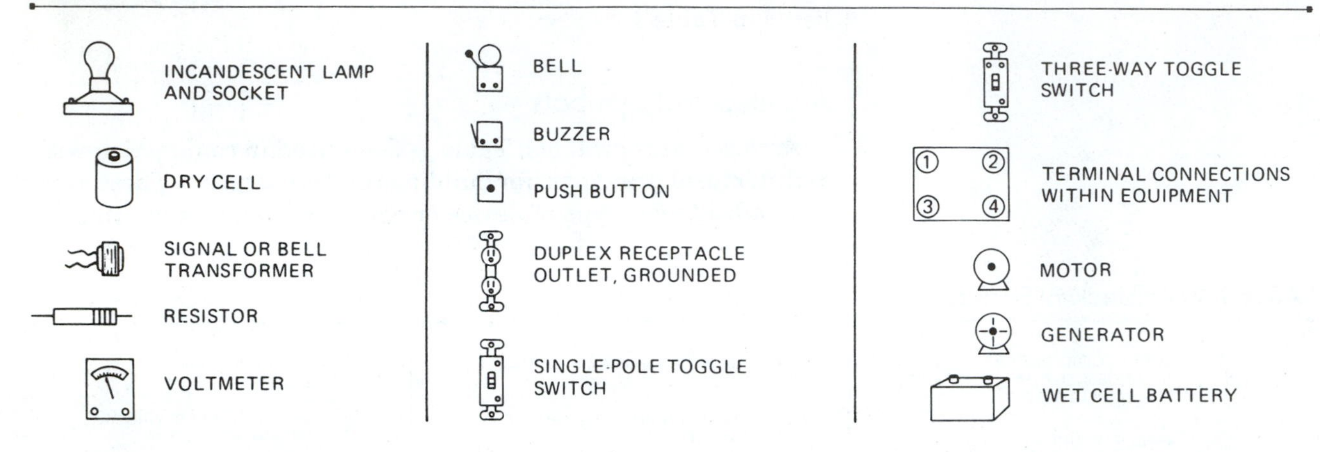

NOTE: Pictorial symbols are not necessarily standardized and may vary according to the engineer's preference.

well as the number of electrical conductors necessary for proper operation.

Pictorial Symbols

Pictorial symbols designate the component, its physical location with respect to other components, and the electrical connections. These symbols are frequently used in panel diagrams (which illustrate the location and connections of electrical and mechanical components within a panel). Pictorial diagrams frequently incorporate schematic symbols within the diagram. Some common pictorial symbols are listed in Table 1-3.

ELECTRICAL DRAWINGS

Electrical drawings and *diagrams* convey information to the people who construct, assemble, install, and maintain electrical equipment. The electrical worker must be able to interpret these drawings and apply the information to the actual job.

Schematic Diagrams

Schematic diagrams supply detailed information about electrical connections and operation of equipment, in a concise and simplified manner. These diagrams are grouped into two general categories: elementary wiring diagrams and circuit or control diagrams.

The elementary wiring diagram usually consists of two vertical lines, which represent the supply conductors with the components connected between them. Elementary wiring diagrams are sometimes referred to as ladder diagrams because the horizontal lines resemble the rungs of a ladder.

Frequently, if the circuit is complex, the diagram is separated into two or more parts, Figure 1-1A. Figure 1-1B shows the same circuit in a more complicated drawing.

Architectural Drawings

Architectural drawings are used in conjunction with the plans of buildings and structures. One type of architectural drawing is the floor plan. The floor plan is a sort of bird's-eye view of the building or a section of the building. It notes the location of structural elements such as walls, doors, windows, rooms, and stairs. The electrical floor plan also indicates the types and locations of the electrical equipment, Figure 1-2.

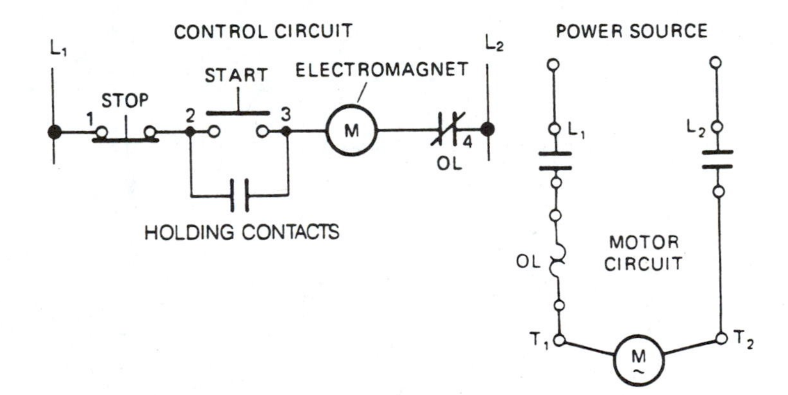

Figure 1-1A
Motor and control circuit (in two views)

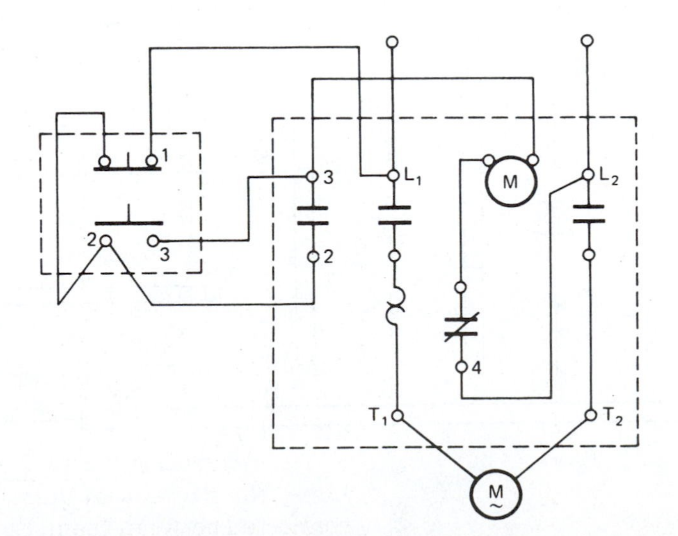

Figure 1-1B
Motor and control circuit with the electrical components shown in their actual physical locations

Figure 1-2
Typical electrical floor plan of a small assembly plant

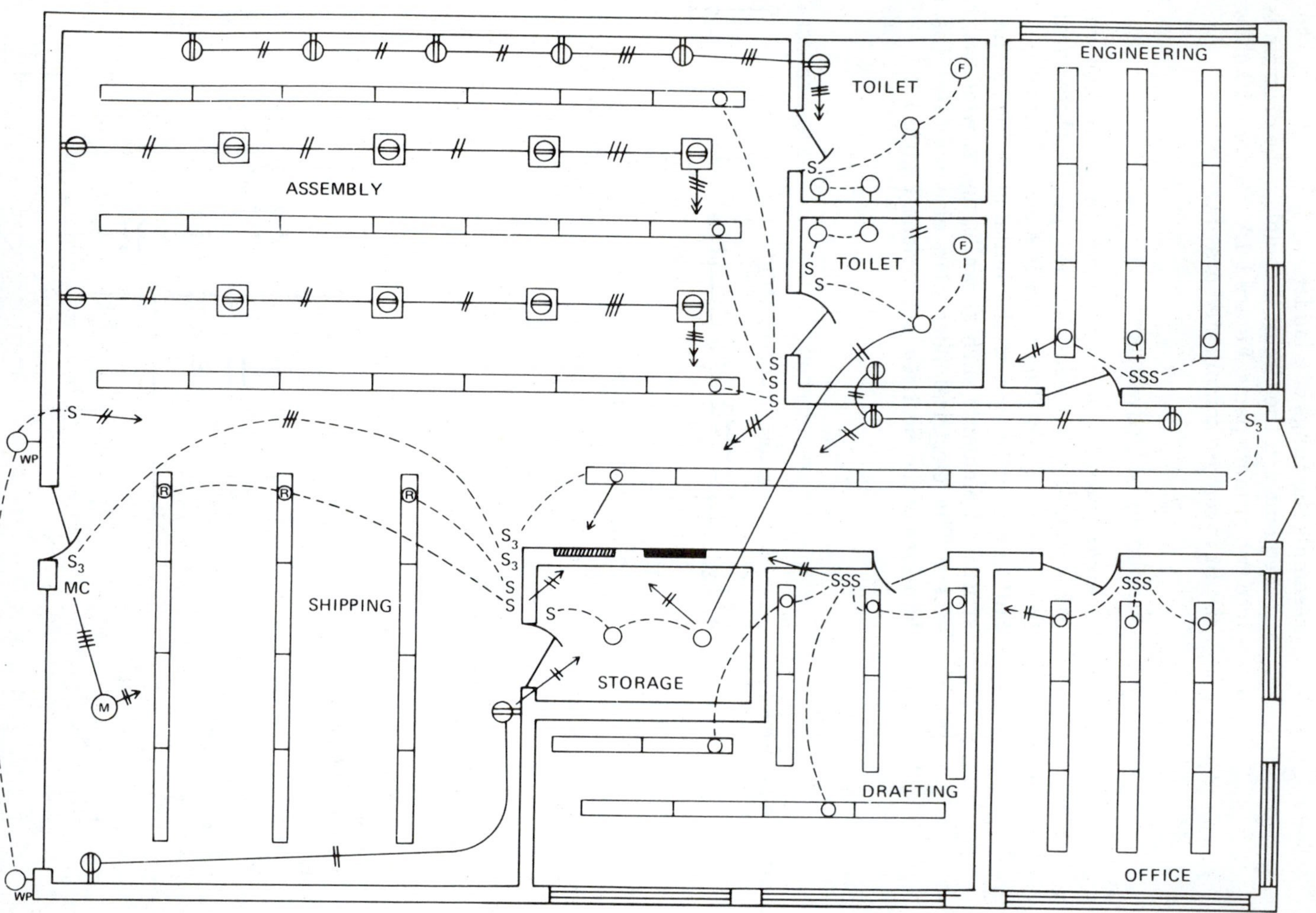

Pictorial Diagrams

Pictorial diagrams, as the name implies, are drawings that give an abbreviated picture of the installation, circuit, panel, and components. There are two basic types of pictorial diagrams: *line drawings* and *orthographic projections*.

The line drawing is often used to provide information relative to the circuitry and location of the electrical and mechanical components within a panel or control device. For example, Figure 1-3 is a pictorial line drawing of a Honeywell protectorelay as it relates to a residential heating system.

An orthographic projection, or drawing, is required when detailed information is necessary to assemble and/or install certain equipment. As many views are drawn as are necessary for clarity. In some cases actual photographs are used. Figure 1-4 shows a view of the Honeywell protectorelay. A cutaway section of the component containing the pyrostat contacts is shown in Figure 1-5.

In Figure 1-6, an orthographic drawing of a relay, two views are shown. If a relay is to be assembled in the field, the electrician will need detailed drawings to ensure correct assembly.

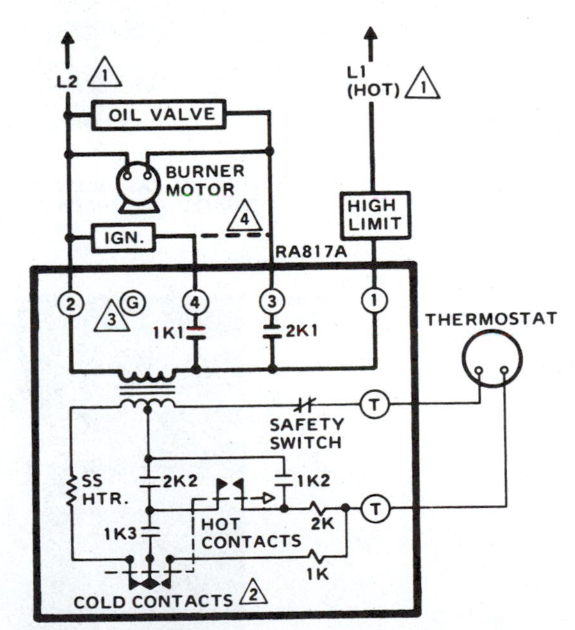

Figure 1-3
Line drawing of a protectorelay controlling a residential heating system (*Courtesy of Honeywell, Inc., Minneapolis, MN*)

Figure 1-4
Protectorelay—master control for a residential heating system *(Courtesy of Honeywell, Inc., Minneapolis, MN)*

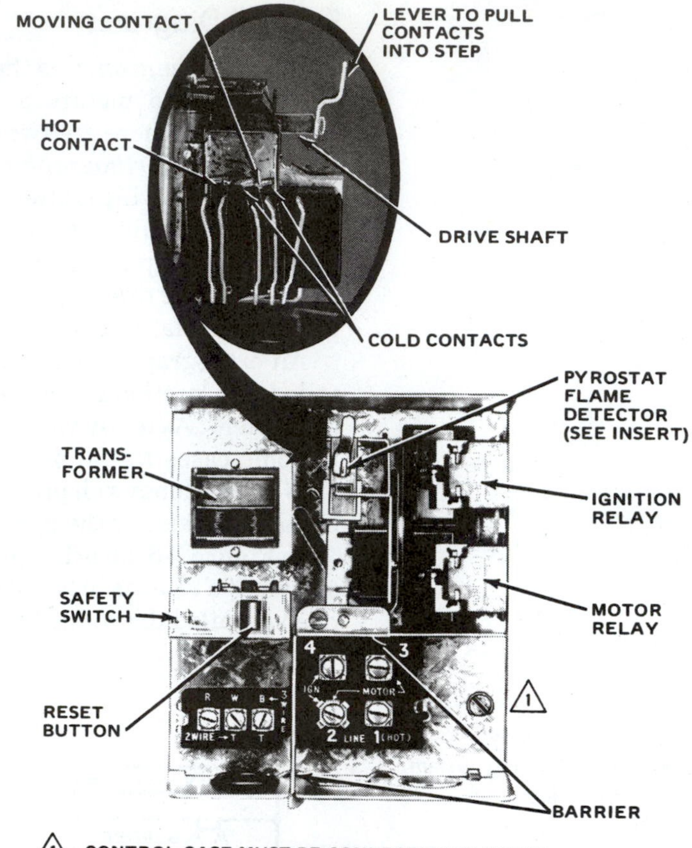

Figure 1-5
Cutaway section of pyrostat contacts (*Courtesy of Honeywell, Inc., Minneapolis, MN)*

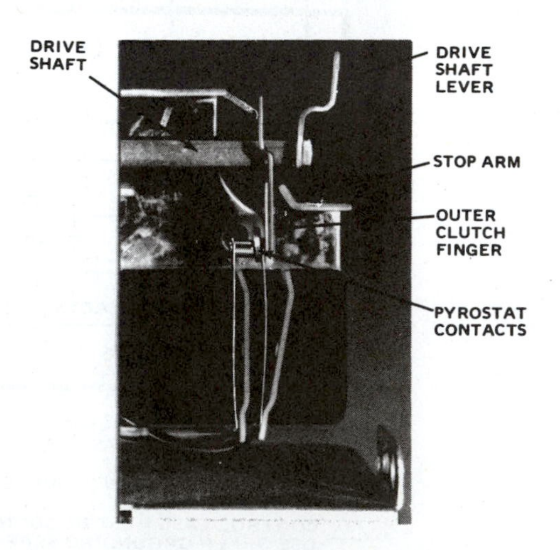

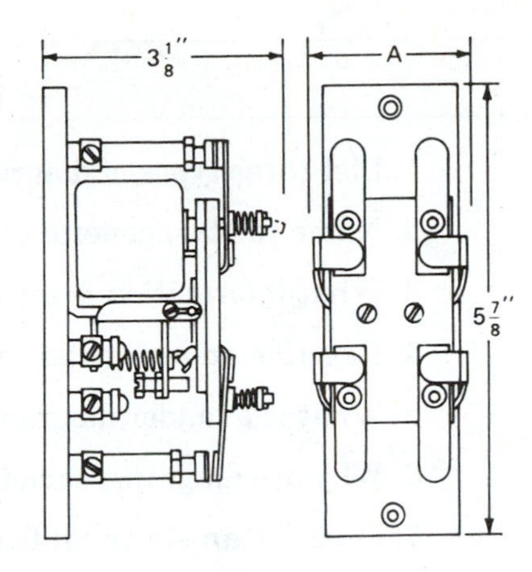

Figure 1-6
Orthographic drawing showing two views of a common relay

REVIEW

A. Multiple choice.
Select the best answer.

1. Electrical symbols enable engineers and other personnel to convey information regarding the
   a. quality of equipment.
   b. cost of equipment.
   c. type of equipment.
   d. rating of equipment.
2. Electrical symbols are divided into
   a. two categories.
   b. three categories.
   c. four categories.
   d. six categories.
3. Schematic symbols are used primarily to indicate the
   a. electrical connections.
   b. specific location of devices.
   c. rating of equipment.
   d. quality of equipment.
4. Architectural electrical symbols are used in conjunction with
   a. building plans.
   b. panel diagrams.
   c. motor control circuit diagrams.
   d. schematic diagrams.
5. Pictorial electrical symbols are used in conjunction with
   a. building plans.
   b. panel diagrams.
   c. internal circuits of motors.
   d. schematic diagrams.

REVIEW *(continued)*

B. Give complete answers.

1. List three types of diagrams.
2. Name the two general categories of schematic diagrams.
3. What information is conveyed in schematic diagrams?
4. Describe an elementary wiring diagram.
5. What is a ladder diagram?
6. Why are diagrams sometimes separated into several parts?
7. Describe an electrical floor plan.
8. What is a pictorial diagram?
9. Describe the two types of pictorial diagrams.
10. What is an orthographic drawing?
11. Draw the architectural electrical symbols for the following:
    a. Incandescent ceiling outlet
    b. Surface-type fluorescent ceiling fixture
    c. Weatherproof wall fixture
    d. Grounded duplex receptacle outlet
    e. Grounded range outlet
12. Draw the schematic symbols for the following:
    a. DC motor
    b. AC generator
    c. Ammeter
    d. Ohmmeter
    e. Circuit breaker
13. Draw the pictorial symbols for the following:
    a. Motor
    b. Single-pole toggle switch
    c. Bell
    d. Bell transformer
    e. Terminal connection points within equipment

REVIEW *(continued)*

B. Give complete answers (continued).

14. Name these architectural electrical symbols.

a. 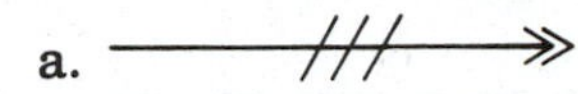

b.

c. s

d. 

e.

15. This is a pictorial diagram of a single pole toggle switch controlling an incandescent lamp. Draw a schematic diagram of the same circuit.

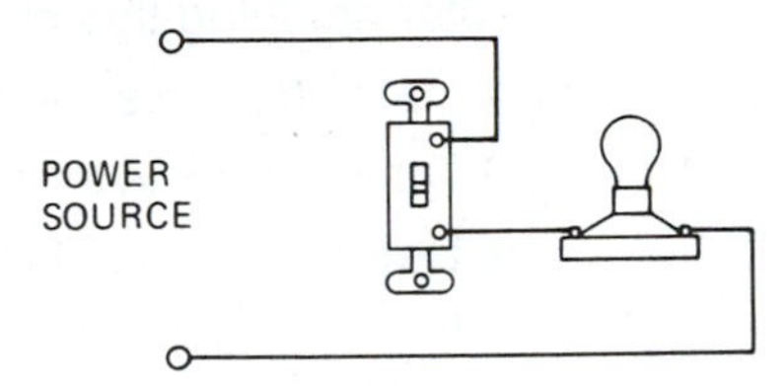

2

ELECTRICAL FUNDAMENTALS

Objectives

After studying this chapter, the student will be able to:

- Describe the structure of matter.
- Define *static electricity* and explain its effect.
- Define the unit of electricity and describe current flow.
- List and define the three electrical quantities that are present in all energized electrical circuits.
- Describe six methods used to produce electricity.
- State and apply Ohm's Law.

STRUCTURE OF MATTER

It is necessary to study the structure of matter in order to develop a thorough understanding of electricity. *Matter* is anything that occupies space and has weight. Some examples of matter are wood, air, metal, and water. The smallest particle of matter that retains the same chemical properties is the *molecule*. A molecule can be divided into smaller parts called *atoms*. Dividing a molecule into atoms creates a chemical change. For example, a molecule of water undergoes a chemical change to become two parts of hydrogen (two hydrogen atoms) and one part oxygen (one oxygen atom).

Currently we know of more than 100 kinds of atoms, or elements. *Elements* are frequently referred to as the "building blocks" of nature. Singly or in combination, they are the materials that constitute all matter. Each element is composed solely of one type

Figure 2-1
Atom containing protons and neutrons in the nucleus, with electrons orbiting around the nucleus

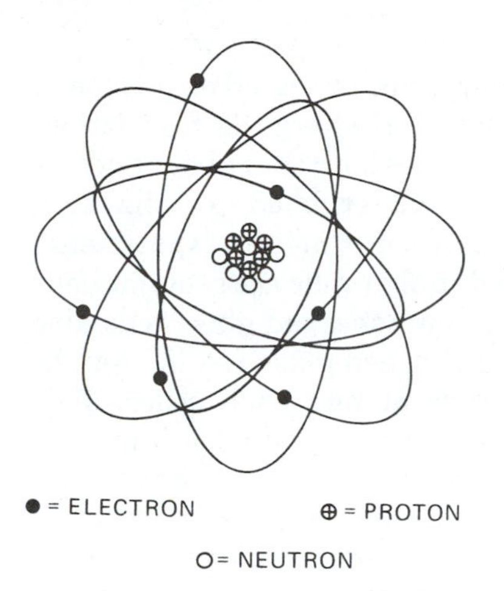

Figure 2-2
Atom of hydrogen

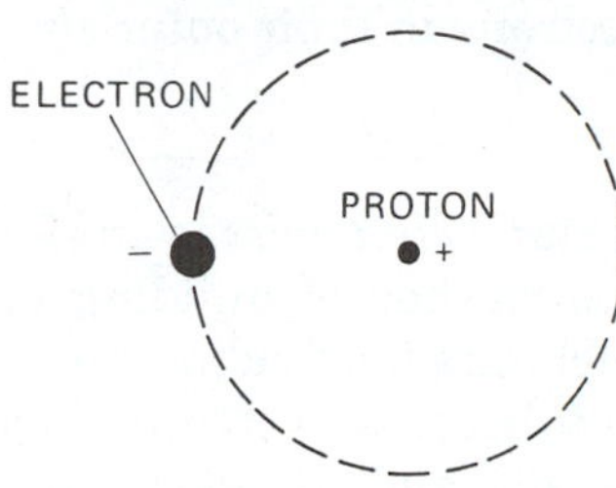

Figure 2-3
Atom of aluminum

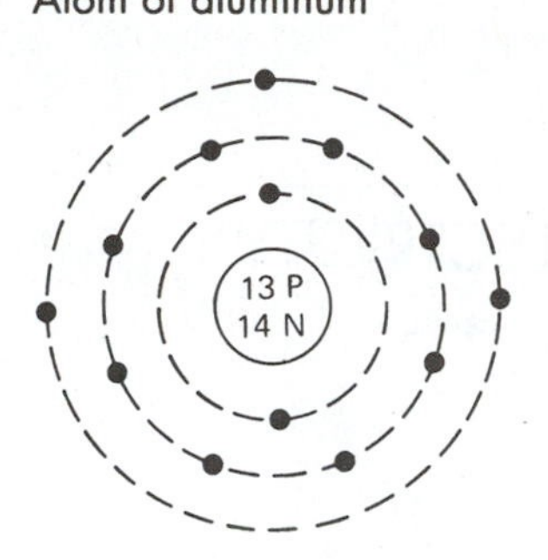

of atom. Some examples of elements are iron, copper, aluminum, lead, carbon, and hydrogen.

The structure of an atom can be compared to the sun's planetary system. The nucleus, like the sun, is at the center, with tiny particles called *electrons* orbiting around it, Figure 2-1. Electrons have a negative electrical charge. The nucleus of the atom consists of *protons*, which have a positive electrical charge, and *neutrons*, which have no electrical charge.

The hydrogen atom, Figure 2-2, is the simplest of all atoms. The hydrogen atom contains 1 proton in the nucleus, with 1 electron revolving around the nucleus. An atom of aluminum has 13 protons and 14 neutrons in the nucleus, with 13 electrons orbiting around the nucleus, Figure 2-3.

Atoms in their natural state contain an equal number of electrons and protons. In both the aluminum atom and the copper atom, Figure 2-4, as in all atoms, the electrons orbit in arranged rings around the nucleus. These rings are called *shells*. The maximum number of electrons in any shell is determined by the "number" of the shell. The first shell is the one nearest the nucleus, and the numbers of the shells increase consecutively as their distance from the nucleus increases. The maximum number of electrons contained in any shell of any atom is as follows: the first shell, 2 electrons; the second shell, 8; the third shell, 18; the fourth shell, 32; the fifth and sixth shells, 18 each; and the seventh shell, 2 electrons.

Atoms that have from 1 to 4 electrons in their last, or outer, shell are generally good conductors of electricity. For example, aluminum and copper are good conductors. (The aluminum atom has 3 electrons in its last shell; the copper atom has 1 electron in its last

Figure 2-4
Atom of copper

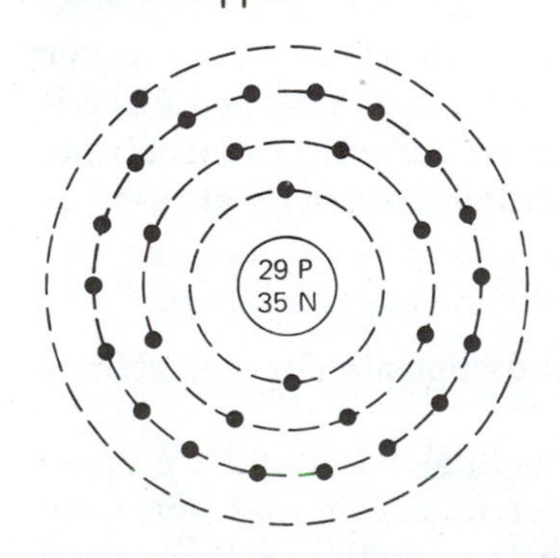

shell.) Atoms with 5, 6, or 7 electrons in their outer shell are classified as nonmetals and are poor conductors. Atoms with 8 electrons in their outer shell are insulators.

STATIC ELECTRICITY

Static electricity is an electric charge accumulated on an object. One method of building up an electrical charge is by friction. Rubbing a hard rubber rod with fur causes both the rubber and the fur to become electrified. When a glass rod is rubbed with silk, both the silk and the glass become electrified. Further experiments demonstrate that when two charged rubber rods are brought near each other, a repelling force exists. Two charged glass rods also repel each other. However, when a charged glass rod is brought near a charged rubber rod, the rods attract each other. This attraction and repulsion indicate that there must be a difference between the charge on the rubber and the charge on the glass.

An object in its natural state contains an equal number of electrons and protons. Therefore, it is *uncharged*, or *neutral*. In order to be in a state of *charge*, the object must contain more electrons than protons or more protons then electrons. Removing electrons from an object causes it to take on a positive charge. Adding electrons to an object causes it to become negatively charged. Note that electrons are the particles that are added to or removed from the object. The protons are held firm in the nucleus, but the electrons, being in the outer portion of the atom, are more easily removed from one object and deposited on another.

When a rubber rod is rubbed with fur, electrons are transferred from the fur to the rod. This transfer causes the rod to become negatively charged and the fur to become positively charged. Rubbing glass with silk removes electrons from the glass and deposits them on the silk. Two charged glass rods repel each other because they contain like charges. If a charged rubber rod is placed near a charged glass rod, the rods are attracted to each other because they contain opposite charges. Objects charged electrically by friction retain their charge for an indefinite period of time. This storage of an electrical charge is called "electricity at rest," or static electricity.

Two rules for electric charges are:

1. Like charges repel each other, and opposite charges attract each other.
2. The strength of the attraction or repulsion is directly proportional to the strength of the electric charge and inversely proportional to the square of the distance between the charged objects.

Two unlike charged objects are attracted to each other because of the nature of electrical balance. If the two objects touch, the electrons in the negatively charged object will move into the positively charged object until there is a balance between them. This is nature's way of restoring objects to their natural state. If the difference in potential (the surplus of electrons as opposed to the deficit of electrons) is great enough, it is not necessary for the objects to touch. They need only come close to each other and the electrons will move through the air to the object containing the positive charge. This discharge through the air is called a static discharge. Lightning is a common form of static discharge.

ELECTRIC CURRENT

An *electric current* is electricity in motion. It is the movement of electrons from a negatively charged object to a positively charged object. The direction of current flow is the same as the direction of electron flow.

In nonmetallic materials such as rubber and glass, the electrons are held firmly to their parent atom, and it is difficult to cause them to move through the material. In metals, however, some electrons are free to flow. These are called *free electrons*. Materials that contain free electrons are called *conductors*. Materials that do not contain free electrons are called *insulators*. Rubber and glass are good insulators; copper and aluminum are good conductors. If a copper wire is connected between two unlike charged bodies, electrons will flow freely through the wire, attempting to restore balance.

Figure 2-5
Battery with terminals connected by a length of copper wire. The arrows indicate the direction of electron flow.

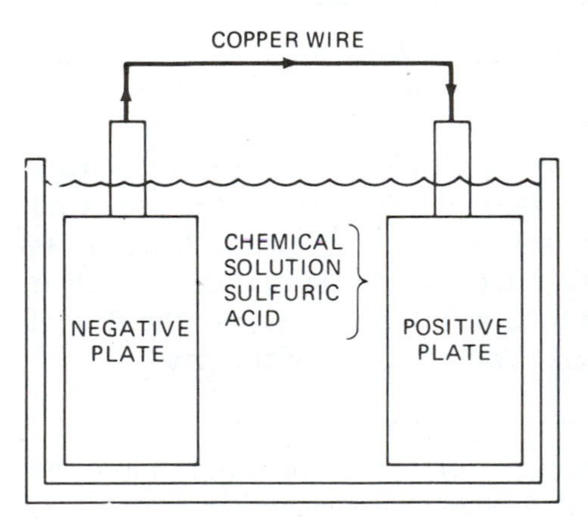

A *battery* is an apparatus that uses chemicals to cause an electron unbalance between its terminals. If a copper wire is connected to the terminals of a battery, the free electrons are attracted toward the positive terminal and repelled from the negative terminal, Figure 2-5. The difference in potential between the battery terminals causes this electron flow. As electrons arrive at the positive side of the battery, the chemical action forces them through the battery to the other terminal, thus maintaining an electrical unbalance. This movement of electrons continues as long as the battery chemicals are active. The flow of electrons through a circuit constitutes an electric current. It can be said, then, that current flows from the negative terminal of the battery, through the circuit, and back to the positive terminal.

CURRENT MEASUREMENT

It is necessary to measure the rate of flow of electrons in order to determine circuit characteristics. The current (flow of electrons) through a 100-watt (W) lamp is approximately 6×10^{18} electrons per second. Because working with such large numbers is cumbersome, a unit much larger than the electron has been established. This unit is the *coulomb* (C). One coulomb is equal to 628×10^{16} electrons. A motor that permits 5 coulombs per second to flow through its windings is rated at 5 amperes. The word *ampere* (A) means coulombs per second (C/s). A 10-ampere heater permits 10 coulombs per second to flow through it. The relationship between the coulomb and the ampere can be expressed mathematically as follows:

$$Q = It \qquad \text{(Eq. 2.1)}$$

where Q = quantity of coulombs (C)
I = current, in amperes (A)
t = time, in second(s)

Example 1

A battery forces a current of 10 A through a circuit for 1 hour (h). (a) How many coulombs will flow through the circuit? (b) How many electrons per second will flow through the circuit?

(a) $Q = It$
$Q = 10\text{ A} \times 3600\text{ s}$
$Q = 36{,}000\text{ C}$
(b) $10 \times 628 \times 10^{16} = 628 \times 10^{17}$ electrons/s

Figure 2-6
Ammeter connected to measure the current flowing through a lamp

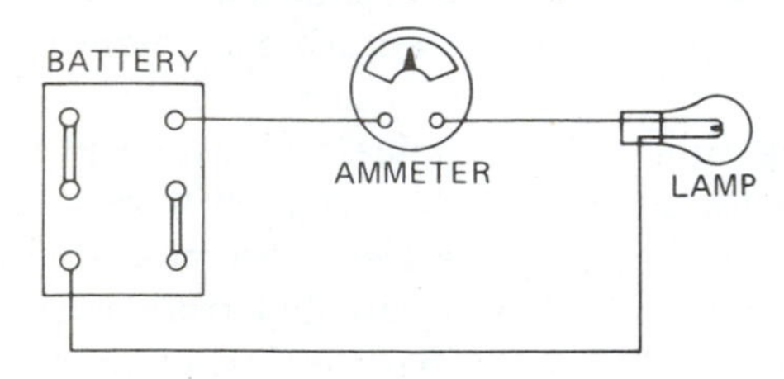

To measure current flow through a circuit, a *current meter* must be inserted into the circuit. It must be connected so that all the current that flows in the circuit will flow through the meter. A current meter is called an *ammeter* because it measures the rate of current, or coulombs per second flow. Figure 2-6 shows an ammeter connected to measure the current flowing through the lamp. Because the ammeter must be inserted into the circuit, it must have a very low resistance so it will not hinder the flow of current and thereby alter the characteristics of the circuit.

VOLTAGE

The flow of electrons through a circuit can be compared to the flow of water through a pipe. A good comparison is a closed-loop system similar to that used on a forced hot water heating system. As shown in Figure 2-7A, the pump blades cause the water to enter the pump at point D at a low pressure and leave at point A at a high pressure. The section of pipe between A and B causes very little friction. Therefore, the pressure at B is only slightly less than at A. Section BC causes considerable friction, so the pressure at point C is much less than at point B. The section of pipe from C to D causes the same amount of friction as from A to B. Therefore, the pressure drop is the same as from A to B.

Figure 2-7A
Closed-loop hydraulic system

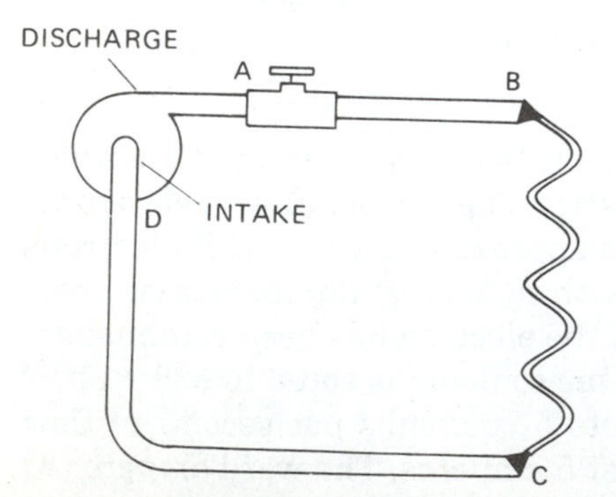

Pressure is measured in *pounds per square inch* (psi). If the drop in pressure from A to B is 5 pounds per square inch (0.35 kilogram per square centimeter), from B to C is 100 pounds per square inch (7.0 kilograms per square centimeter), and from C to D is 5 pounds per square inch (0.35 kilogram per square centimeter), then the

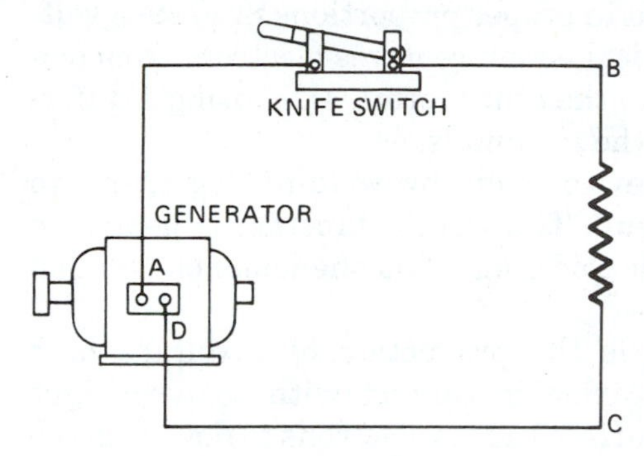

Figure 2-7B
Closed-circuit electrical system

total drop is 5 + 100 + 5 = 110 pounds per square inch (7.7 kilograms per square centimeter). This means that the pump must develop a pressure of 110 pounds per square inch (7.7 kilograms per square centimeter) between the intake at D and the discharge at A.

A similar condition exists in an electrical circuit, but the pressure drop is expressed in units called *volts* (V). Volt is used in an electric circuit in the same manner as the pounds per square inch unit is used in a hydraulic circuit. Figure 2-7B illustrates an electric circuit comparable to the hydraulic circuit in Figure 2-7A. The section of the wire from A to B has low resistance (electrical friction) and the pressure drop is low. Section BC has high resistance, causing a large drop in pressure. Section CD has the same resistance as AB and therefore has the same pressure drop. If the drop in pressure from A to B is 5 volts, B to C is 100 volts, and C to D is 5 volts, the total drop is 5 + 100 + 5 = 110 volts. This means that the generator, an electric pump, must develop a pressure of 110 volts between its terminals.

In an electric circuit, the term *electromotive force* (emf) is used to indicate the voltage generated by the source, such as a battery or generator. The term *voltage drop* is used to denote the drop in pressure in a particular part of a circuit. In circuits like that in Figure 2-7B, the sum of the voltage drops is equal to the emf.

VOLTAGE SOURCES

The methods used to produce voltage include the following:

- Mechanical magnetic (induction)
- Chemical
- Thermoelectric
- Photovoltaic
- Piezoelectric
- Friction

Generators are used to produce a voltage by induction. There are two types of generators: the direct current (dc) generator and the alternating current (ac) generator. The *direct current generator* produces a voltage that acts in one direction only. In other words, it produces an emf that forces the current to flow in one direction only. The *alternating current generator* (sometimes called an *alternator*) produces a voltage that alternates back and forth. That is, it produces a voltage first in one direction and then in the opposite direction. The current flowing as a result of this alternating emf flows first in one direction and then in the opposite direction. This type of current is called *alternating current* (ac).

Generators produce a voltage by rotating coils of wire through a *magnetic field*, an invisible force produced by a magnet. As the coils move through the magnetic field, a voltage develops. It is not necessary to move the coils if the magnetic field is moved across the coils. Either method produces a voltage.

Various chemicals mixed in proper proportions produce a voltage in batteries. The chemical action removes electrons from one terminal and forces them to the other terminal, causing a difference in potential between the terminals.

A thermocouple is a device made by welding together two dissimilar metals at one end. The welded junction is heated to produce a voltage across the cold ends. This phenomenon is called the *thermoelectric effect*.

The *photovoltaic effect* is the production of a voltage in a material, such as copper oxide in contact with copper. Light striking the copper-oxide surface causes electrons to flow from the copper to the copper oxide, thus developing a potential difference between the two metals. The amount of emf produced is proportional to the intensity of the light.

A voltage can be produced by applying pressure to certain crystals, such as Rochelle salts. When pressure is applied, a small emf is produced. The amount of voltage varies with the pressure. This phenomenon is known as the *piezoelectric effect*.

Electrostatic generators are used to develop a voltage by friction. The *electrostatic generator* consists of a latex belt, which revolves over a metal pulley. The friction of the belt moving over the pulley builds up an electrostatic charge. A hollow metal ball surrounding part of the belt attracts the electrons. An emf of many thousands of volts can be produced with this machine.

VOLTAGE MEASUREMENTS

The amount of water flowing in a pipe depends upon the difference in pressure between the two ends of the pipe. Figure 2-8A shows one method used to measure this difference. A pressure gauge is tapped (connected) into each end of the pipe at points X

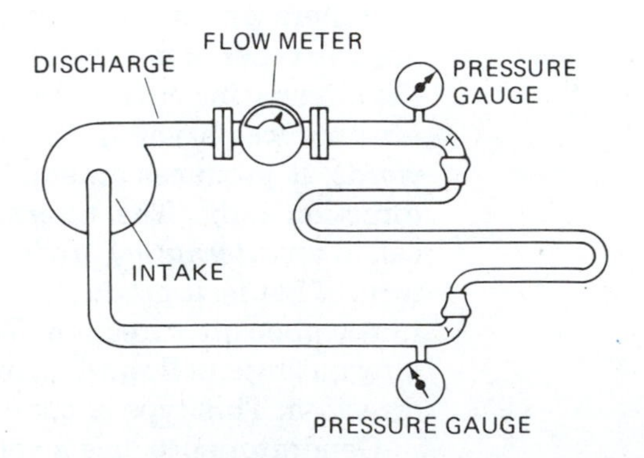

Figure 2-8A
Pressure gauges tapped into a pipe indicate water pressure at points X and Y.

Figure 2-8B
A voltmeter connected across points X and Y indicates the voltage between the two points.

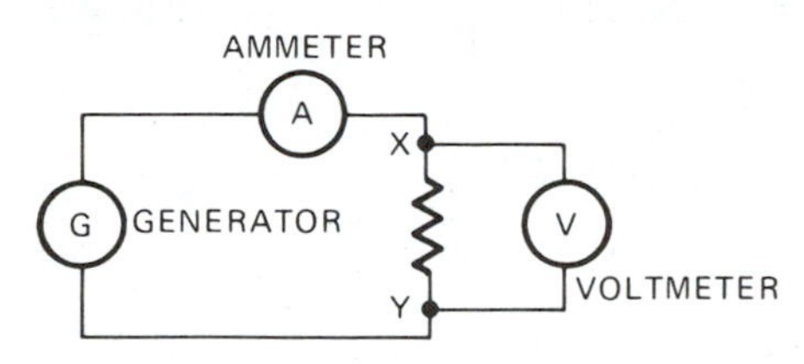

and Y. The difference between the two readings gives the amount of pressure necessary to force the water through the pipe.

The amount of current flowing through a conductor depends upon the difference in pressure between the two ends of the conductor. This difference in pressure is called the *potential difference* or *VOLTAGE*. To measure the potential difference between the two ends X and Y, Figure 2-8B, the voltmeter leads must be tapped to the ends of the wire. The voltmeter is an instrument that measures the difference in electrical pressure between two points. It will, therefore, indicate the voltage across the wire XY.

RESISTANCE

Figure 2-9
Ohmmeter connected to measure circuit resistance

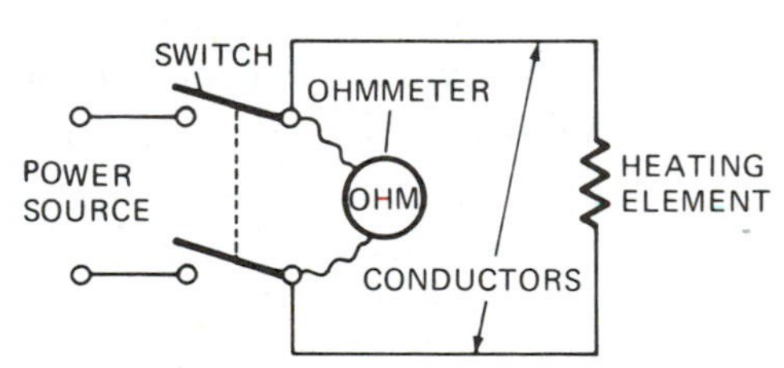

Some materials have many free electrons. These materials require only a comparatively low voltage to force the electrons to flow from atom to atom, which establishes a high current flow. These materials are good conductors. Other materials have only a few free electrons, thus permitting only a low current flow when the same voltage is applied. These materials are poor conductors.

All materials have some opposition to current flow. This opposition is caused by the type of material and by friction produced when the electrons move through the material. Opposition to the flow of electrons is called *resistance*. The unit of measurement for resistance is the *ohm* (Ω). The instrument for measuring resistance is called the *ohmmeter*, Figure 2-9.

> **CAUTION:** Never connect an ohmmeter to a circuit until the circuit has been disconnected from the power source. The ohmmeter has its own source of power, and connecting it across another power source can cause serious damage.

OHM'S LAW

The amount of current flowing through a circuit is directly proportional to the electromotive force (voltage) and inversely proportional to the resistance of the circuit. In other words, if the voltage applied to a circuit is increased and the resistance remains the same, the current will increase in proportion to the increase in voltage. If the voltage remains the same but the resistance is increased, the current will decrease in proportion to the increase in resistance. This relationship between current, voltage, and resistance is called *Ohm's Law*. The mathematical formula for Ohm's Law is

$$I = \frac{E}{R} \ (\text{or } E = IR, \text{ or } R = \frac{E}{I}) \qquad \text{(Eq. 2.2)}$$

Where I = current, in amperes (A)
E = electromotive force, in volts (V)
R = resistance, in ohms (Ω)

Example 2

How much current flows through a 30-Ω heating element when it is connected to a 120-V source?

$$I = \frac{E}{R}$$
$$I = \frac{120}{30}$$
$$I = 4 \text{ A}$$

Example 3

What is the resistance of an electric iron when it is connected to a 120-V source, and a current of 3 A flows?

$$R = \frac{E}{I}$$
$$R = \frac{120}{3}$$
$$R = 40 \ \Omega$$

Example 4

What is the emf necessary to force 6 A through a circuit having a resistance of 20 Ω?

$$E = IR$$
$$E = 6 \times 20$$
$$E = 120 \text{ V}$$

Ohm's Law may be applied to an entire circuit or any part of the circuit. This may be stated as follows:

- The total current flowing in a circuit is equal to the total voltage applied to the circuit divided by the total resistance of the circuit.
- The current flowing in any part of a circuit is equal to the voltage across that part of the circuit divided by the resistance of that part of the circuit.

REVIEW

A. Multiple choice.
Select the best answer.

1. The nucleus of an atom is surrounded by
   a. neutrons.
   b. electrons.
   c. protons.
   d. positrons.

2. A positively charged body has a
   a. deficiency of electrons.
   b. surplus of electrons.
   c. surplus of protons.
   d. deficiency of protons.

3. Electrical pressure is measured with
   a. a voltmeter.
   b. an ammeter.
   c. an ohmmeter.
   d. a wattmeter.

4. Electric current is the movement of
   a. positrons.
   b. protons.
   c. neutrons.
   d. electrons.

5. Good conductors contain many free
   a. protons.
   b. neutrons.
   c. electrons.
   d. positrons.

6. Without changing it chemically, the smallest particle of matter is
   a. an element.
   b. a molecule.
   c. a proton.
   d. an atom.

7. Static electricity is generally produced by
   a. friction.
   b. induction.
   c. heat.
   d. light.

REVIEW *(continued)*

A. Multiple choice. Select the best answer (continued).

8. One coulomb is equal to
   a. 628 electrons.
   b. 628×10^{8} electrons.
   c. 628×10^{16} electrons.
   d. 628×10^{18} electrons.
9. Voltage is the
   a. movement of electrons through a circuit.
   b. opposition to current flow.
   c. electrical pressure that causes current flow.
   d. electrical resistance.
10. Resistance is
    a. the movement of protons through a circuit.
    b. the opposition to current flow.
    c. an immovable force.
    d. the movement of electrons through a circuit.

B. Give complete answers.

1. What elements are present in a molecule of water?
2. Describe the structure of an atom.
3. What is static electricity?
4. Write two rules related to static electricity.
5. What is electric current?
6. Define *resistance* and identify the cause of resistance in electric circuits.
7. Describe six methods used to produce electricity.
8. What instrument is used to measure electrical current?
9. How is a voltmeter connected in a circuit? (Use a diagram if necessary.)
10. Write Ohm's Law.

REVIEW *(continued)*

C. Solve each problem, showing the method used to arrive at the solution.

1. It takes 360,000 C to charge a certain storage battery. If the charging rate is 15 A, how long will it take to fully charge the battery?
2. A voltmeter, which has a resistance of 10,000 Ω, indicates 120 V. How much current is flowing through it?
3. A current of 2.5 A is flowing through the heating element of a soldering iron. If the iron is connected to a 120-V line, what is the resistance of the element?
4. What is the voltage drop across a resistor if its resistance is 100 Ω and 0.2 A flows through it?
5. How much current flows through a 144-Ω lamp connected to a 120-V line?
6. A circuit has a resistance of 480 ohms and is connected across 120 volts. How much current flows through the circuit?
7. If the circuit in problem 6 is connected across 240 volts, how much current will flow?
8. If the resistance of the circuit in problem 6 is reduced to 240 ohms, how much current will flow?
9. If the circuit in problem 6 requires 1 ampere, what voltage must be applied to the circuit?
10. What is the resistance of a circuit if an ammeter in the line indicates 4 amperes and the circuit is connected across 480 volts?

3

ELECTRIC CIRCUITS

Objectives

After studying this chapter, the student will be able to:

- Describe an electric circuit.
- Describe and construct series circuits, parallel circuits, and combination circuits.
- Write from memory and apply the rules for current, voltage, and resistance, relative to the above circuits.
- State Kirchhoff's Laws for current and voltage.
- Apply Ohm's Law and Kirchhoff's Laws to the above circuits.

An *electric circuit* is the path over which the electrons flow. There are two basic types of circuits: series circuits and parallel circuits. A combination circuit is the complete arrangement whereby components are connected into a network so that some are joined in series and others are joined in parallel.

SERIES CIRCUITS

A *series circuit* is a circuit that has only one path through which the current can flow, Figure 3-1. Because there is only one path for the current, the same value of current flows through each circuit component. Therefore, in order to connect electrical apparatuses in series, they must all be able to carry the same amount of current. The resistance and/or voltage ratings may differ, but the current rating must always be the same. A break in any part of the circuit will render all circuit components inoperable.

For example, if the lamp in Figure 3-1 should burn out and cause the filament to disintegrate, the current will be interrupted to all of the circuit components.

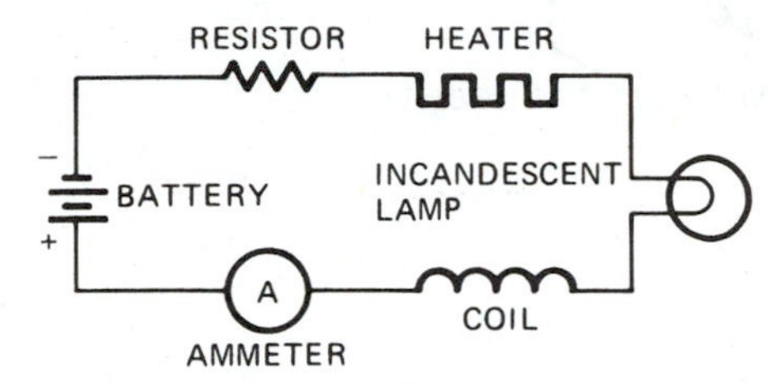

Figure 3-1
Series circuit

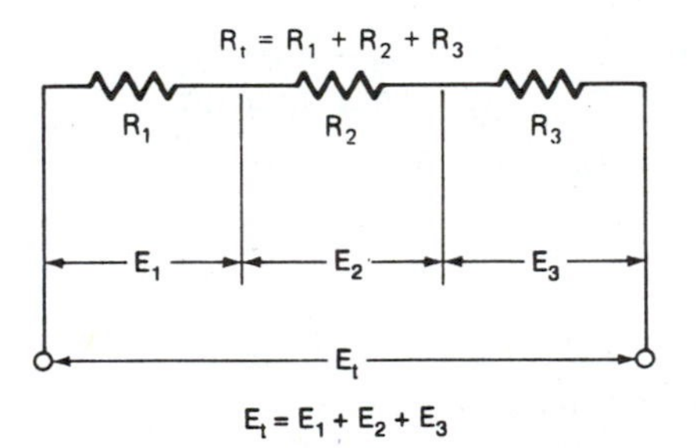

Figure 3-2
Three resistors connected in series

Rules for Series Circuits

The following rules apply to all series circuits:

1. The value of current flowing in a series circuit is the same through all parts of the circuit.
2. The total voltage of a series circuit is equal to the sum of the voltages across each part of the circuit.
3. The total resistance of a series circuit is equal to the sum of the resistances of each part of the circuit.

These rules for series circuits may be expressed mathematically as follows:

$$(1)\ I_t = I_1 = I_2 = I_3 = \ldots \quad \text{(Eq. 3.1)}$$
$$(2)\ E_t = E_1 + E_2 + E_3 \ldots \quad \text{(Eq. 3.2)}$$
$$(3)\ R_t = R_1 + R_2 + R_3 + \ldots \quad \text{(Eq. 3.3)}$$

The subscript number indicates the circuit component. The subscript letter t indicates the value for the entire circuit.

When applying Ohm's Law to an entire circuit, be sure to use the values for the entire circuit. When applying Ohm's Law to a part of a circuit, use only the values that apply to that part of the circuit. Figure 3-2 shows three resistors connected to form a series circuit. The Ohm's Law equation (Equation 2.2) for the various parts may be written as

$$E_1 = IR_1;\ \text{or}\ E_2 = IR_2;\ \text{or}\ E_3 = IR_3$$

The subscript number for I has been omitted because the current is the same value for all parts of the circuit. The Ohm's Law equation for the entire circuit is $E_t = IR_t$.

Example 1

What value of voltage must be supplied by a generator to force 2 A through a series circuit consisting of an 80-Ω lamp, a 10-Ω rheostat, and a 30-Ω soldering iron?

(a) $R_t = R_1 + R_2 + R_3$
$R_t = 80 + 10 + 30$
$R_t = 120\ \Omega$

(b) $E_t = IR_t$
$E_t = 2 \times 120$
$E_t = 240\ \text{V}$

Figure 3-3
Two incandescent lamps and two current-limiting resistors connected in series

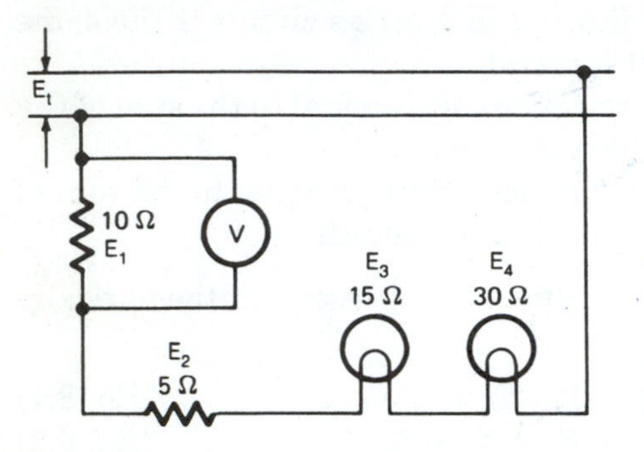

Figure 3-4A
Rheostat. A sliding contact connects the terminal to the uninsulated section of the resistance wire.

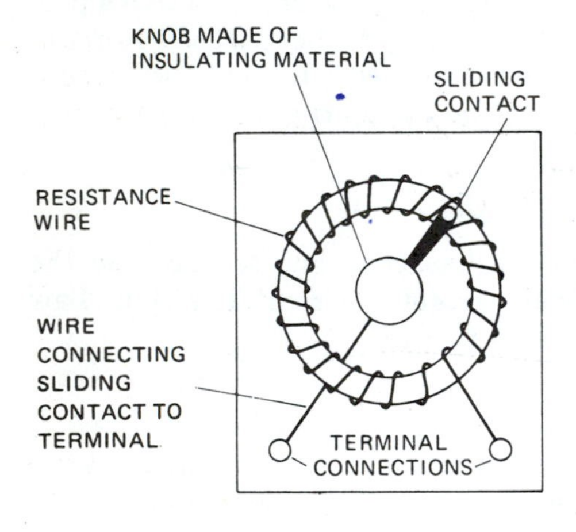

Figure 3-4B
Rheostat controlling the current to two incandescent lamps

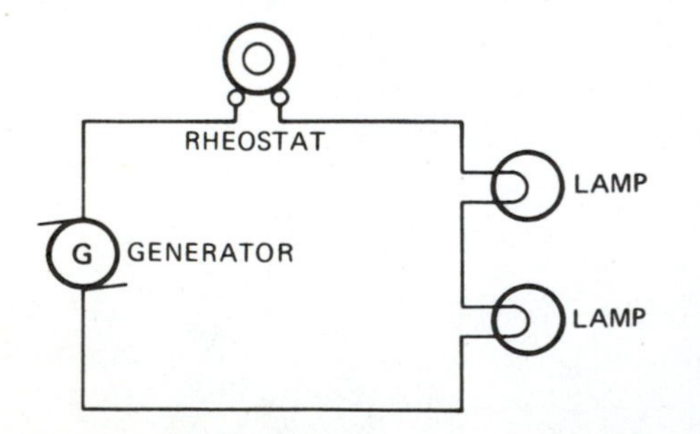

Example 2

Determine the line voltage (V_t) in Figure 3-3, with the voltmeter indicating 20 V.

(a) Calculate the current through the resistor.

$$I_1 = \frac{E_1}{R_1}$$

$$I_1 = \frac{20}{10}$$

$$I_1 = 2\ \text{A}$$

The current is the same value for all parts of a series circuit. Therefore, the current through each component is 2 A.

(b) $E_2 = IR_2$
$E_2 = 2 \times 5$
$E_2 = 10\ \text{V}$

(c) $E_3 = IR_3$
$E_3 = 2 \times 15$
$E_3 = 30\ \text{V}$

(d) $E_4 = IR_4$
$E_4 = 2 \times 30$
$E_4 = 60\ \text{V}$

(e) $E_t = E_1 + E_2 + E_3 + E_4$
$E_t = 20 + 10 + 30 + 60$
$E_t = 120\ \text{V}$

Example 3

What value of resistance must be connected in series with a heating element to limit the current to 3 A? The (hot) resistance of the element is 85 Ω and it is to be connected to a 600-V supply.

(a) $R_t = \frac{E_t}{I}$

$$R_t = \frac{600}{3}$$

$$R_t = 200\ \Omega$$

(b) The total resistance of a series circuit is equal to the sum of the resistances of the separate parts. In this series circuit, the value of the second resistance is equal to the total resistance (200 Ω) minus the resistance of the heating element (85 Ω), so that 200 Ω – 85 Ω = 115 Ω.

A resistance of 115 Ω must be connected in series with the heating element.

A common way to control the amount of current supplied to an apparatus is to vary the circuit resistance. This may be accomplished without opening the circuit by using a rheostat. A *rheostat* is a variable resistor made up of a wire-wound resistor and a sliding contact, Figure 3-4A. Figure 3-4B illustrates a rheostat controlling the current to two incandescent lamps. When the contact is in the position shown, all the resistance is in series with the lamps, and the current is low. (The lamps glow dimly.) Rotating the knob in a clockwise direction moves the sliding contact along the wire-wound resistor and removes some of the resistance from the circuit. This allows the current to increase and the lamp to glow brighter.

PARALLEL CIRCUITS

A *parallel circuit* is a circuit that has more than one path through which the current can flow, Figure 3-5. Note that each component is connected directly across the supply voltage. Therefore, the voltage rating of each component must be of the same value as the supply voltage. The value of current varies according to the resistance of the component.

Each component forms its own path, from the source through the individual branch and back to the source. Therefore, a break in one branch of the circuit will not affect the other branches. For example, if the lamp in Figure 3-5 burns out, current will continue to flow through the other branches. Notice that all the components are connected directly across the battery. Thus, the voltmeter will indicate the battery voltage, which is equal to the voltage across any one of the branches.

The more paths that are available for the current, the less opposition there is to the flow. Every time another path is added to a parallel circuit, the circuit resistance decreases. Adding paths to a parallel circuit decreases the circuit's resistance and increases its conductance. *Conductance* is a measurement of the ability to conduct an electric current. The mathematical symbol for conductance is G, and its unit of measurement is siemens. Conductance is the reciprocal of resistance. Therefore, $G = 1/R$, where G is the conductance, in siemens, and R is the resistance, in ohms.

Rules for Parallel Circuits

The following rules apply to all parallel circuits:

1. The total current supplied to a parallel circuit is equal to the sum of the currents through the branches.
2. The voltage across any branch of a parallel circuit is equal to the supply voltage.
3. The total conductance of a parallel circuit is equal to the sum of the conductance of all of the branches.

Figure 3-5
Parallel circuit

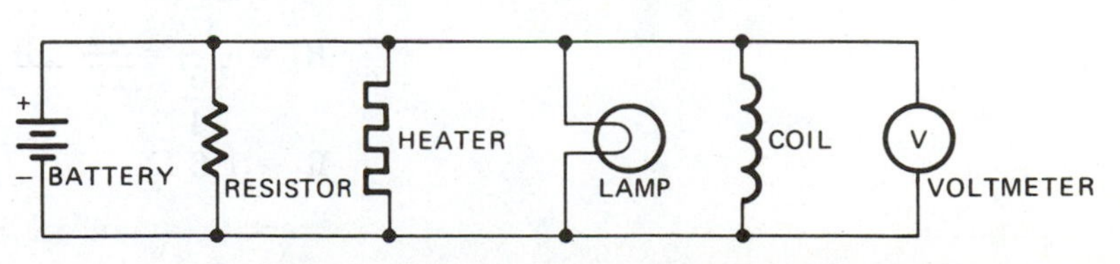

4. As the conductance of a circuit increases, its resistance decreases.
5. The combined or equivalent resistance of a parallel circuit is always smaller than the resistance of any of the branches.

These rules for parallel circuits may be expressed mathematically as follows:

$$(1)\ I_t = I_1 + I_2 + I_3 + \ldots \qquad \text{(Eq. 3.4)}$$

$$(2)\ E_t = E_1 = E_2 = E_3 = \ldots \qquad \text{(Eq. 3.5)}$$

$$(3)\ G_t = G_1 + G_2 + G_3 + \ldots \qquad \text{(Eq. 3.6)}$$

$$(4)\ R_t = \frac{1}{G_1 + G_2 + G_3 + \ldots};\ \text{or}\ R_t = \frac{1}{G_t} \qquad \text{(Eq. 3.7)}$$

Example 4

If the resistor in Figure 3-5 has a resistance of 12 Ω, the heater 3 Ω, the lamp 24 Ω, and the coil 6 Ω, find (a) The conductance of each component, (b) the total conductance, and (c) the combined resistance of the circuit. If the emf of the battery is 12 V, (d) what is the voltage across each component? (e) how much current flows through each component? and (f) what is the total current flowing in the circuit?

(a) The conductance is the reciprocal of the resistance. The reciprocal of a number is 1 divided by the number. For example, the reciprocal of 5 is 1/5. Therefore, the conductance of each of the components in Figure 3-5 is as follows:
Resistor: 12-Ω resistance, 1/12 siemens conductance
Heater: 3-Ω resistance, 1/3 siemens conductance
Lamp: 24-Ω resistance, 1/24 siemens conductance
Coil: 6-Ω resistance, 1/6 siemens conductance

(b) $G_t = G_1 + G_2 + G_3 + G_4$

$$G_t = \frac{1}{12} + \frac{1}{3} + \frac{1}{24} + \frac{1}{6}$$

$$G_t = \frac{2}{24} + \frac{8}{24} + \frac{1}{24} + \frac{4}{24}$$

$$G_t = \frac{15}{24} = 0.625 \text{ siemens}$$

(c)

$$R_t = \frac{1}{G_t} \qquad R_t = \frac{1}{G_t}$$

$$R_t = \frac{1}{\frac{15}{24}} = \frac{24}{15} \quad \text{or} \quad R_t = \frac{1}{0.625}$$

$$R_t = 1.6\ \Omega \qquad R_t = 1.6\ \Omega$$

(d) The voltage across a parallel combination is the same value as the voltage across each branch. Therefore, the voltage across each component is 12 V.

(e)

Resistor	Heater	Lamp	Coil
$I_1 = \frac{E}{R_1}$	$I_2 = \frac{E}{R_2}$	$I_3 = \frac{E}{R_3}$	$I_4 = \frac{E}{R_4}$
$I_1 = \frac{12}{12}$	$I_2 = \frac{12}{3}$	$I_3 = \frac{12}{24}$	$I_4 = \frac{12}{6}$
$I_1 = 1\text{ A}$	$I_2 = 4\text{ A}$	$I_3 = 0.5\text{ A}$	$I_4 = 2\text{ A}$

(f) $I_t = I_1 + I_2 + I_3 + I_4$ $\qquad$ $I_t = \frac{E_t}{R_t}$

$I_t = 1 + 4 + 0.5 + 2$ $\qquad$ *or* $\qquad$ $It = \frac{12}{1.6}$

$I_t = 7.5\text{ A}$ $\qquad$ $It = 7.5\text{ A}$

Example 5

A heater with a resistance of 5 Ω, a bell with a resistance of 20 Ω, and a lamp with a resistance of 50 Ω are connected in parallel across 110 V. Calculate (a) the circuit conductance, (b) the combined resistance, and (c) the total current flowing through the circuit.

(a) $G_t = G_1 + G_2 + G_3$

$$G_t = \frac{1}{5} + \frac{1}{20} + \frac{1}{50}$$

$$G_t = \frac{20}{100} + \frac{5}{100} + \frac{2}{100}$$

$$G_t = \frac{27}{100}$$

$G_t = 0.27$ siemens

(b) $R_t = \frac{1}{G_t}$

$$R_t = \frac{1}{0.27}$$

$R_t = 3.704\ \Omega$

(c) $I_t = \frac{E}{R_t}$

$$I_t = \frac{110}{3.704}$$

$I_t = 29.698\text{ A}$

Another method frequently used to solve for the combined resistance if there are only two resistances in parallel, or if it is desired to solve only two at a time, is to divide their product by their sum. The formula is written as follows:

$$R_t = \frac{R_1 R_2}{R_1 + R_2}$$

Example 6

Two loads are connected in parallel. Load A has a resistance of 5 Ω; load B has a resistance of 10 Ω. Calculate the combined resistance of the circuit.

$$R_t = \frac{R_1 R_2}{R_1 + R_2}$$

$$R_t = \frac{5 \times 10}{5 + 10}$$

$$R_t = \frac{50}{15}$$

$$R_t = 3.333\ \Omega$$

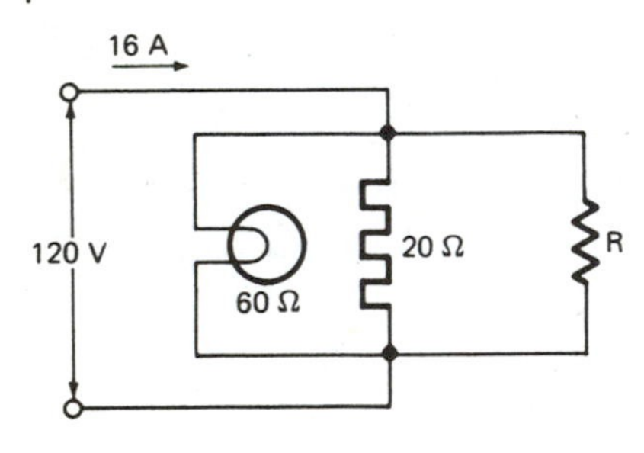

Figure 3-6
Three devices connected to form a parallel circuit

Example 7

Find the resistance of R in Figure 3-6. The voltage across R is 120 V. Before the resistance of R can be calculated, the current through R must be known. In order to determine the current through R, it is necessary to find the current through the other two branches.

$$I_1 = \frac{E}{R_1} \qquad I_2 = \frac{E}{R_2}$$

$$I_1 = \frac{120}{60} \qquad I_2 = \frac{120}{20}$$

$$I_1 = 2\ A \qquad I_2 = 6\ A$$

The current through R is found in this way:

$$I_3 = I_t - (I_1 + I_2)$$

$$I_3 = 16 - (6 + 2)$$

$$I_3 = 16 - 8$$

$$I_3 = 8\ A$$

then

$$R_3 = \frac{E}{I_3}$$

$$R_3 = \frac{120}{8}$$

$$R_3 = 15\ \Omega$$

COMBINATION CIRCUITS

Figure 3-7A
Combination circuit

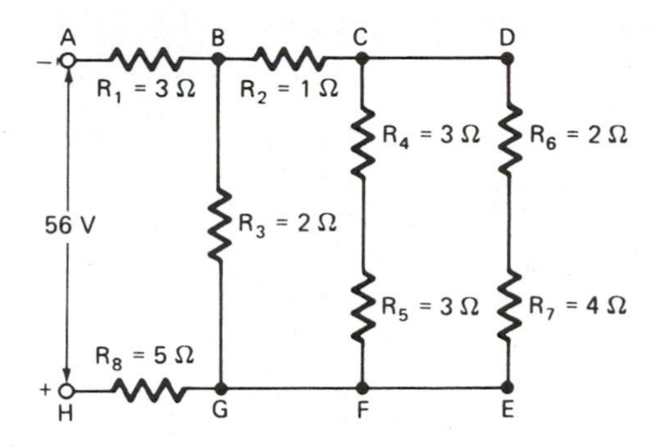

Figure 3-7B
Equivalent circuit to Figure 3-7A. R_6 and R_7 are replaced with one resistance (R_a) of 6 Ω.

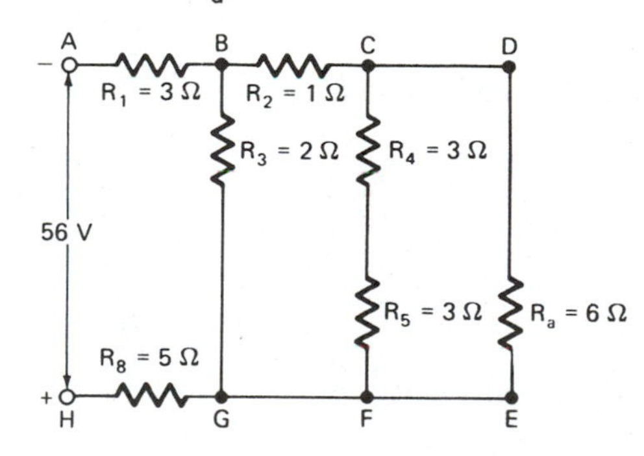

Figure 3-7C
Equivalent circuit to Figure 3-7A. R_4 and R_5 are replaced with one resistance (R_b) of 6 Ω.

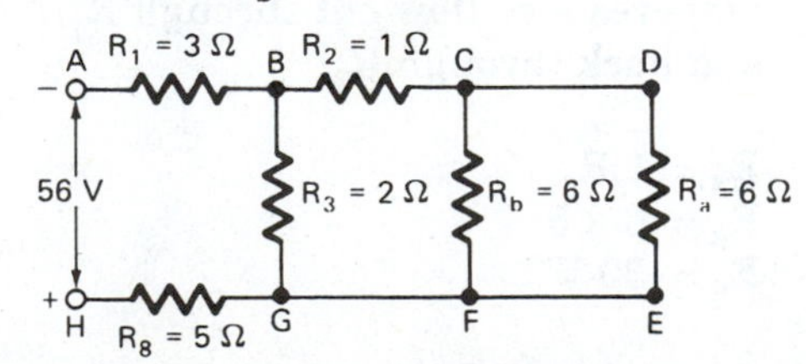

Combination circuits of series and parallel connections are often used to achieve the desired operation. The arrangement may appear to be very complex. However, a simple solution is to break down the circuit into series and/or parallel groups.

To solve problems involving combination circuits, each group may be dealt with individually. Each series group may be replaced by one resistance having a value equal to the sum of all the resistances in the group. Each parallel group may be replaced by one resistance having a value equivalent to the combined resistance of that group.

To determine the values of current, voltage, and resistance for each component, it is often necessary to construct equivalent circuits.

Example 8

Determine the combined resistance of the circuit shown in Figure 3-7A.

(1) Combine R_6 and R_7.
$R_a = R_6 + R_7$
$R_a = 2 + 4$
$R_a = 6\ \Omega$

(2) Draw and equivalent circuit. See Figure 3-7B.

(3) Combine R_4 and R_5.
$R_b = R_4 + R_5$.
$R_b = 3 + 3$
$R_b = 6\ \Omega$

(4) Draw and equivalent circuit. See Figure 3-7C.

(5) Combine R_a and R_b.

Note: The combined resistance of two or more *equal* resistances in parallel is *equal* to the resistance of one resistor divided by the number of resistors in parallel. Stated mathematically, the equation is

$$R_c = \frac{R}{N}$$

where R_c = combined resistance, in ohms (Ω)
R = value of one resistor, in ohms (Ω)
N = number of equal resistors in parallel

$R_c = \frac{R}{N}$

$R_c = \frac{6}{2}$

$R_c = 3\ \Omega$

Figure 3-7D
Equivalent circuit to Figure 3-7A.
R_4, R_5, R_6 and R_7 are replaced with one resistance (R_c) of 3 Ω.

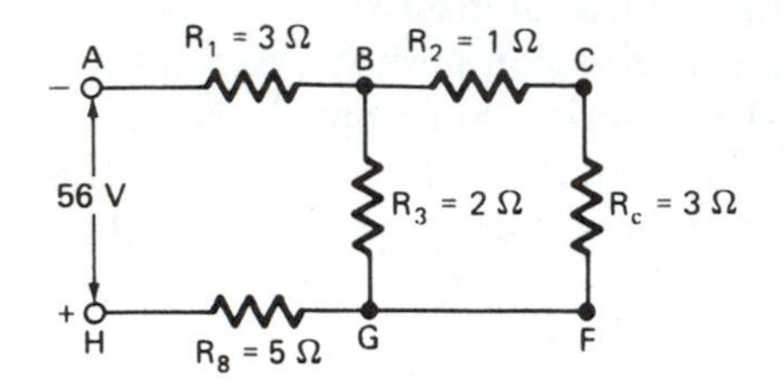

(6) Draw and equivalent circuit. See Figure 3-7D.

(7) Combine R_2 and R_c.
$R_d = R_2 + R_c$
$R_d = 1 + 3$
$R_d = 4\ \Omega$

(8) Draw and equivalent circuit. See Figure 3-7E.

(9) Combine R_3 and R_d.

$$R_e = \frac{1}{\frac{1}{R_3} + \frac{1}{R_d}}$$

$$R_e = \frac{1}{\frac{1}{2} + \frac{1}{4}}$$

$$R_e = \frac{1}{\frac{2}{4} + \frac{1}{4}}$$

$$R_e = \frac{1}{\frac{3}{4}}$$

$$R_e = \frac{4}{3}$$

$$R_e = 1\frac{1}{3}\ \Omega$$

Figure 3-7E
Equivalent circuit to Figure 3-7A.
R_2, R_4, R_5, R_6, and R_7 are replaced with one resistance (R_d) of 4 Ω.

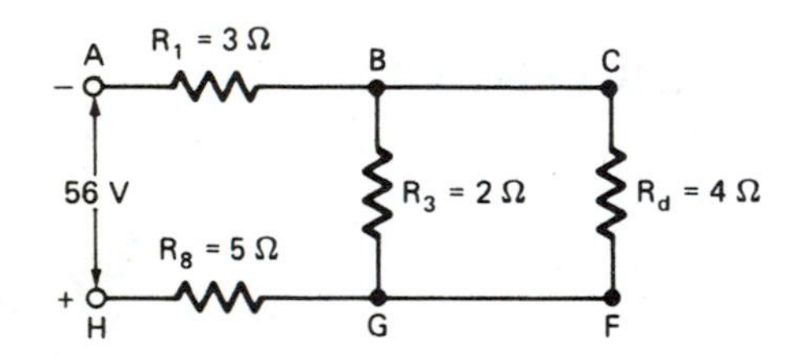

(10) Draw an equivalent circuit. See Figure 3-7F.

(11) Combine R_1, R_e, and R_8.
$R_t = R_1 + R_e + R_8$

$$R_t = 3 + 1\frac{1}{3} + 5$$

$$R_t = 9\frac{1}{3}\ \Omega$$

The combined resistance of the circuit is $9\frac{1}{3}$ Ω.

Example 9

Calculate the current through, and the voltage drop across, each resistance in Figure 3-7A.

(1) $I_t = \frac{E_t}{R_t}$

$$I_t = \frac{56}{9\frac{1}{3}}$$

$I_t = 6$ A

(2) Six amperes flow from point A to point B in Figure 3-7A. Six amperes also flow out through R_1 and back through R_8.

(3) $E_1 = I_1R_1$
$E_1 = 6 \times 3$
$E_1 = 18$ V

(4) $E_8 = I_8R_8$
$E_8 = 6 \times 5$
$E_8 = 30$ V

Figure 3-7F
Equivalent circuit to Figure 3-7A.
R_2, R_3, R_4, R_5, R_6, and R_7 are replaced with one resistance (R_e) of 1 1/3 Ω. The total circuit resistance equals 9 1/3 Ω.

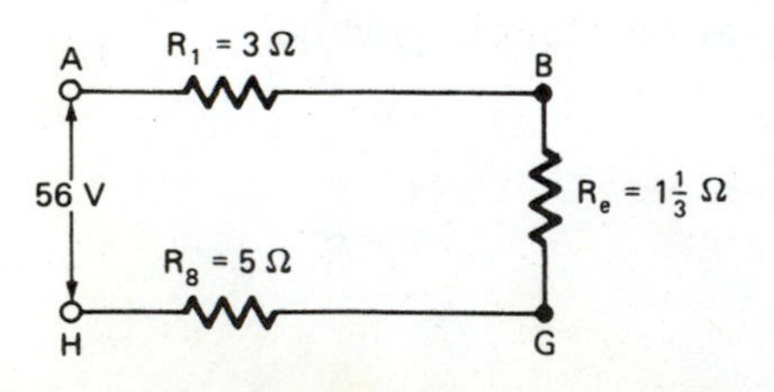

(5) $E_3 = E_t - (E_1 + E_8)$
$E_3 = 56 - (18 + 30)$
$E_3 = 8\ V$

(6) $I_3 = \frac{E_3}{R_3}$
$I_3 = \frac{8}{2}$
$I_3 = 4\ A$

(7) $I_2 = I_t - I_3$
$I_2 = 6 - 4$
$I_2 = 2\ A$

(8) $E_2 = I_2R_2$
$E_2 = 2 \times 1$
$E_2 = 2\ V$

(9) $E_{cf} = E_3 - E_2$
$E_{cf} = 8 - 2$
$E_{cf} = 6\ V$

(10) $I_{cf} = \frac{E_{cf}}{R_{cf}}$
$I_{cf} = \frac{6}{6}$
$I_{cf} = 1A$
$I_4 = 1A$
$I_5 = 1\ A$

(11) $I_{de} = \frac{E_{de}}{R_{de}}$
$I_{de} = \frac{6}{6}$
$I_{de} = 1\ A$
$I_6 = 1\ A$
$I_7 = 1\ A$

(12) $E_4 = I_4R_4$
$E_4 = 1 \times 3$
$E_4 = 3\ V$

(13) $E_5 = I_5R_5$
$E_5 = 1 \times 3$
$E_5 = 3\ V$

(14) $E_6 = I_6R_6$
$E_6 = 1 \times 2$
$E_6 = 2\ V$

(15) $E_7 = I_7R_7$
$E_7 = 1 \times 4$
$E_7 = 4\ V$

Summary of Answers ($R = \frac{E}{I}$)	Voltage (E) in volts	Current (I) in amperes	Resistance (R) in ohms
R_1	18	6	3
R_2	2	2	1
R_3	8	4	2
R_4	3	1	3
R_5	3	1	3
R_6	2	1	2
R_7	4	1	4
R_8	30	6	5

THREE-WIRE CIRCUITS

Most utility companies supply their customers with three-wire systems. The available voltages may vary slightly according to the

Figure 3-8
Two-wire circuit

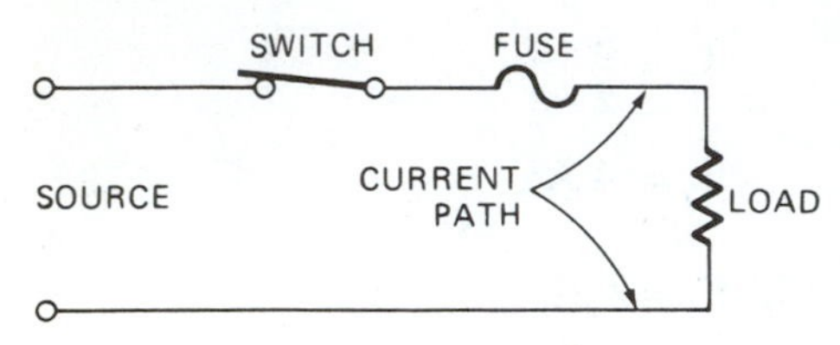

Figure 3-9A
Two two-wire circuits

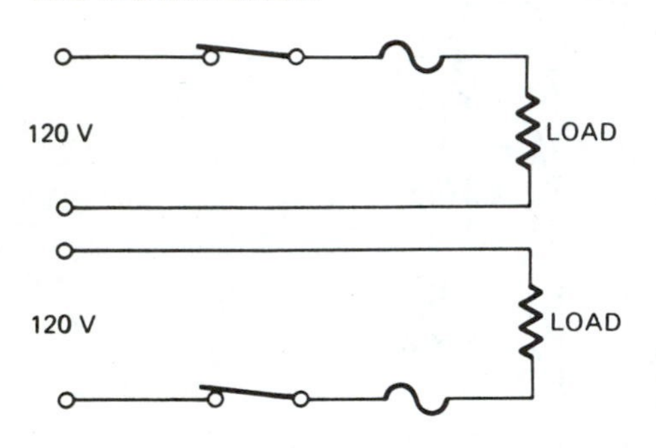

Figure 3-9B
Three-wire circuit

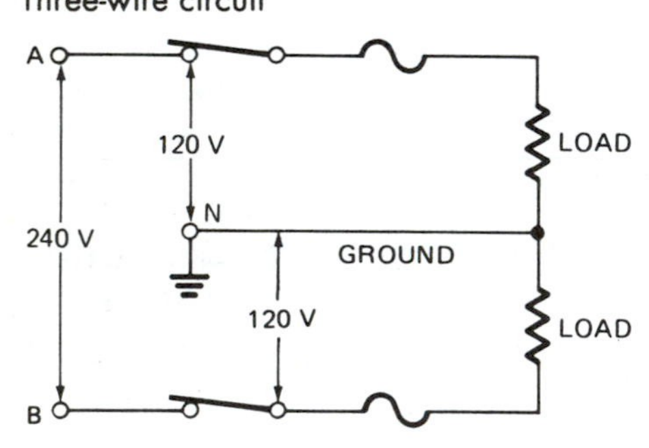

characteristics of the equipment, but the standard values are usually 115/230 volts or 120/240 volts. This type of system gives the customer a choice of two values of voltage as well as the advantage of using both two-wire and three-wire circuits.

A two-wire circuit is an electrical circuit that utilizes two wires to form a current path from the source to the load and back to the source. (Refer to Figure 3-8.) A three-wire circuit can be used in place of two two-wire circuits. This saves one wire. (See Figures 3-9A and 3-9B.) Three-wire circuits are frequently used to supply household electric ranges and clothes dryers. These appliances may contain some components that require 120 volts and other components that require 240 volts.

Figure 3-9A shows two ordinary two-wire circuits. If these two circuits are combined to form one three-wire circuit, Figure 3-9B, the cost of one length of wire is saved. It is necessary to analyze the circuit structure in order to better understand this arrangement.

The National Electrical Code® (*NEC®*)* definition of a *multiwire branch circuit* states that the circuit must consist of three or more conductors, one of which is grounded. There must be a voltage between the ungrounded conductors. An equal voltage must exist between each unground conductor and the grounded conductor. The grounded conductor must be identified and connected to the neutral conductor of the system.

The circuit in Figure 3-9B meets all these requirements. There is a voltage between A and B. The voltage between A and N is equal to the voltage between B and N. N is connected to the neutral conductor of the system at the panelboard, and the neutral is connected to the system ground. The final requirement is that N be identified. This is accomplished during the installation by using white or natural gray insulation on N.

All multiwire circuits must conform to the definition set down by the *NEC®* in order to maintain minimum safety standards.

There are two types of three-wire circuits: *balanced* and *unbalanced*, Figures 3-10A and 3-10B. In a balanced three-wire circuit the load between A and N is equal to the load between B and N. In an unbalanced three-wire circuit the loads are not equal. The ideal arrangement is the balanced circuit.

A mathematical analysis of the circuits in Figures 3-10A and 3-10B helps one to understand the characteristics of three-wire circuits. Figure 3-10A is a balanced three-wire circuit because the load between A and N is equal to the load between B and N. This can be determined by Ohm's Law. The voltage across load X is 120 volts. Therefore, by the equation $I = E/R$, it can be determined that the current through X is 24 amperes. Load Y has the

**National Electrical Code®* and *NEC®* are Registered Trademarks of the National Fire Protection Association, Inc., Quincy, MA.

Figure 3-10A
Balanced three-wire circuit

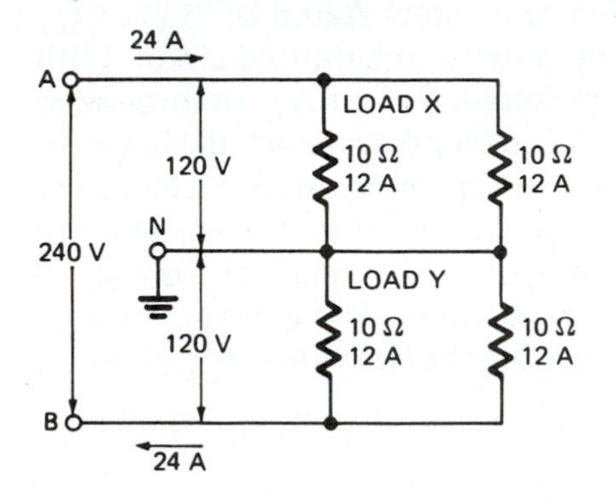

Figure 3-10B
Unbalanaced three-wire circuit

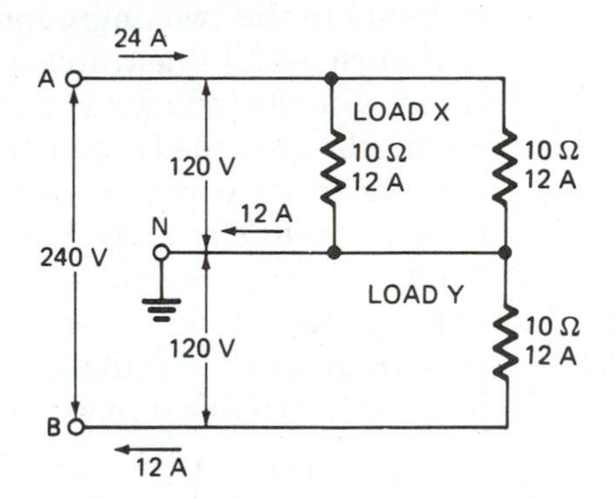

Figure 3-11A
Balanced three-wire circuit equivalent to Figure 3-10A

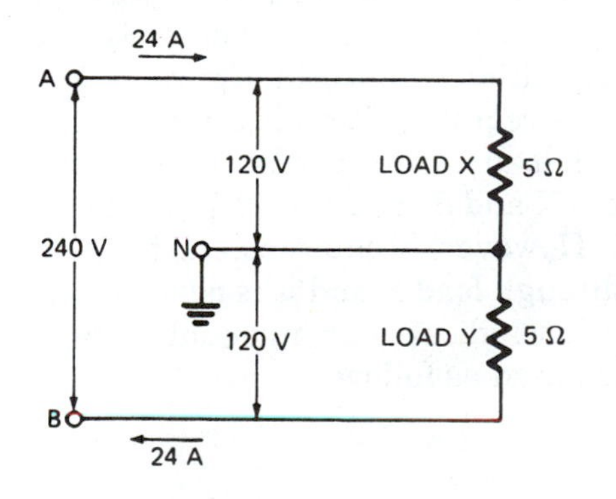

Figure 3-11B
Balanced three-wire circuit with conductor N disconnected

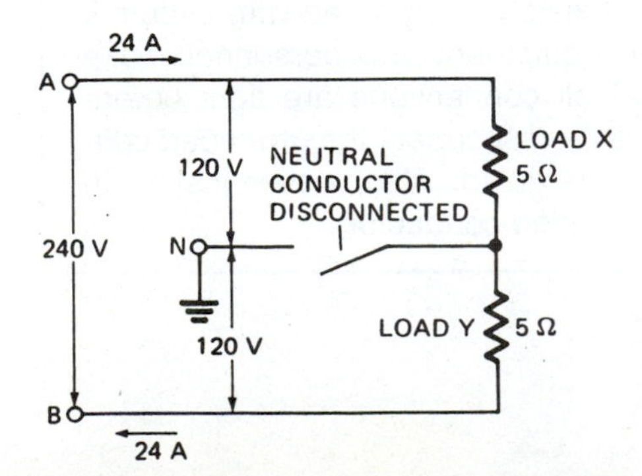

same resistance as load X and has a 120-volt input. Therefore, 24 amperes flow through load Y. Under these conditions, no current flows in conductor N.

Figure 3-11A illustrates a circuit equivalent to the circuit in Figure 3-10A. This equivalent circuit consists of two 5-ohm resistances, each supplied with 120 volts. The current through each resistance is 24 amperes.

Figure 3-11B shows the same circuit with conductor N disconnected. This arrangement now becomes a series circuit with two 5-ohm resistances connected in series across 240 volts. From Ohm's Law, the current through each resistance is 24 amperes. The current flow in both circuits is as indicated by the arrows, and no current flows in conductor N.

Why not omit conductor N, thereby achieving an additional cost savings? If load X is disconnected while conductor N is connected, as in Figure 3-11A, load Y will not be affected. The current will flow out conductor N, through load Y, and back through conductor B. If, however, load X is disconnected while conductor N is open, Figure 3-11B, the entire circuit will be open and load Y will not operate.

Unbalanced three-wire circuits present a somewhat different situation. The circuit in Figure 3-12A is identical to the circuit in Figure 3-10A, with one exception: one 12-ampere load has been removed from between conductors B and N. Removing this load causes an unbalanced condition and changes the circuit characteristics.

The combined load resistance between conductors A and N is 5 ohms; between B and N the load is 10 ohms. Using Ohm's Law, the current through conductor A is 24 amperes. Twelve amperes will flow through each 10-ohm load. There is only one 10-ohm path between conductors N and B, which allows only 12 amperes to flow back through B. The other 12 amperes must flow back through conductor N. Notice the amount and direction of current indicated

Figure 3-12A
Unbalanced three-wire circuit

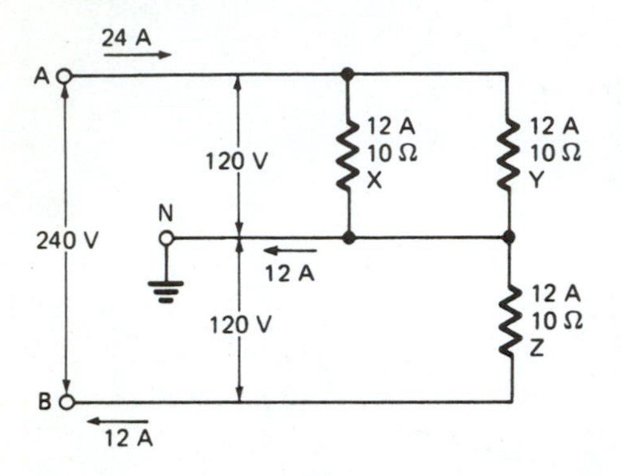

Figure 3-12B
Unbalanced three-wire circuit with conductor N disconnected

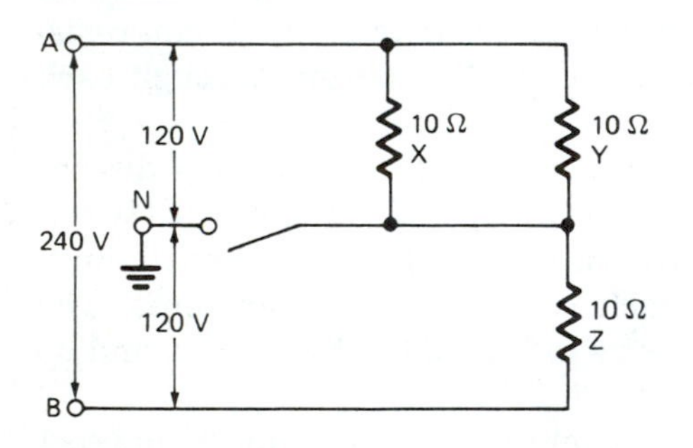

Figure 3-12C
Unbalanced three-wire circuit equivalent to Figure 3-12B

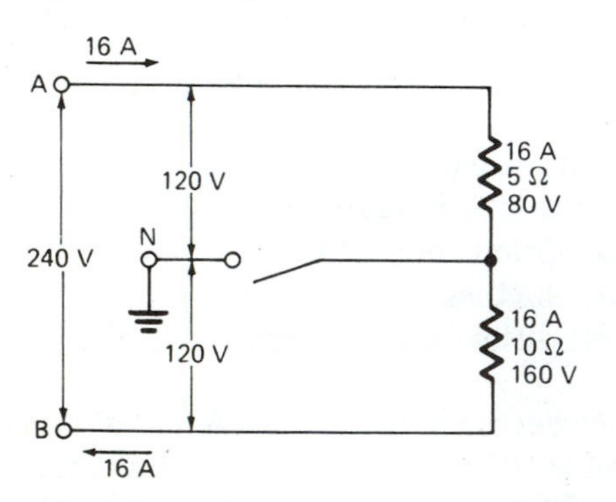

in Figure 3-12A. It can be observed that the grounded conductor N of the unbalanced circuit will carry the difference between the currents in the two ungrounded conductors A and B.

Figure 3-12B illustrates a three-wire unbalanced circuit with the grounded conductor N disconnected. Normally, in a three-wire circuit the grounded conductor N is never disconnected. However, this discussion explains what would happen if it were accidentally disconnected. The equivalent to this circuit is illustrated in Figure 3-12C, which shows two resistances, one of 5 ohms and one of 10 ohms, connected in series across 240 volts. The current through this circuit can be calculated by adding the two values of resistance and applying Ohm's Law:

$$5\ \Omega + 10\ \Omega = 15\ \Omega$$

$$I = \frac{E}{R}$$

$$I = \frac{240}{15}$$

$$I = 16\ A$$

Refer again to Figure 3-12B. It can be seen that with N disconnected, the current in conductor A (16 amperes) is not adequate. Under the normal conditions shown in Figure 3-12A, with N connected 24 amperes are required. With N disconnected only 8 amperes will flow through loads X and Y. Also, the 10-ohm resistance Z between conductors N and B requires 12 amperes for safe operation (Figure 3-12A). However, because conductor N is open, 16 amperes are forced through load Z and it is overloaded.

The above current distribution indicates an unequal voltage distribution, which can be calculated as follows:

Loads X and Y	**Load Z**
$E = IR$	$E = IR$
$E = 16 \times 5$	$E = 16 \times 10$
$E = 80$ volts	$E = 160$ volts

CAUTION: If this condition exists on any three-wire circuit, it may be hazardous to both equipment and personnel. Take precautions to be sure that all connections are tight before energizing the circuit, and never disconnect the grounded conductor when the circuit is energized. Also, never install an overcurrent device in the grounded conductor.

KIRCHHOFF'S LAWS

Figure 3-13
Combination circuit: Kirchhoff's Current Law

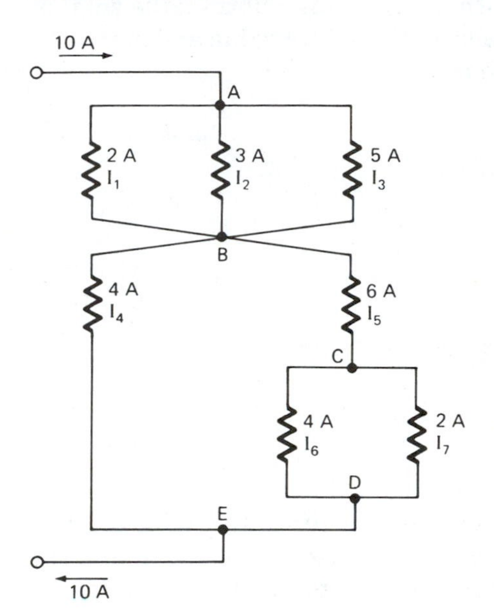

Figure 3-14
Thermometer

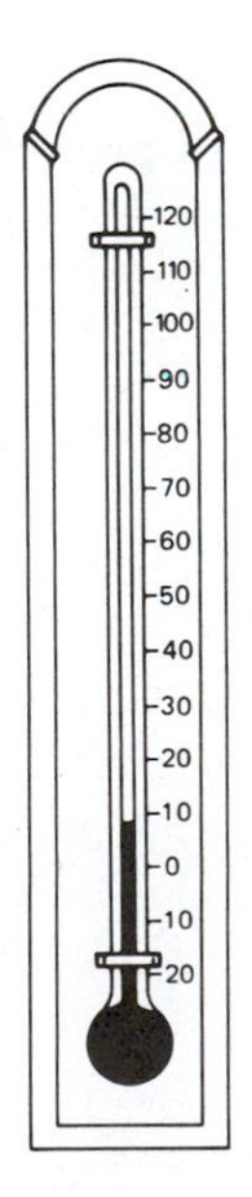

Some complex circuits are more easily solved through the application of *Kirchhoff's Laws*:

- *Current Law*. The sum of the currents entering a junction of a circuit is equal to the sum of the currents leaving the junction.
- *Voltage Law*. The algebraic sum of the voltages around any closed loop is equal to zero.

Kirchhoff's Current Law states that there can be neither an accumulation nor a loss of current at any junction. This can be more clearly understood by analyzing Figure 3-13. Current flow from the supply to junction A is 10 amperes. Leaving junction A, 2 amperes flow through branch I_1, 3 amperes flow through I_2, and 5 amperes flow through I_3. The current entering junction A is 10 amperes; the current leaving junction A is 2 amperes + 3 amperes + 5 amperes = 10 amperes. The current entering junction B is 2 amperes + 3 amperes + 5 amperes = 10 amperes. The current leaving junction B is 6 amperes + 4 amperes = 10 amperes. The current entering junction C is 6 amperes; the current leaving C is 4 amperes + 2 amperes = 6 amperes. The current entering junction D is 4 amperes + 2 amperes = 6 amperes; the current leaving D is 6 amperes. The current entering junction E is 4 amperes + 6 amperes = 10 amperes; the current leaving E is 10 amperes.

Kirchhoff's Voltage Law involves using positive and negative numbers. In arithmetic, the plus sign (+) and minus sign (-) are symbols of operation. In algebra, their meaning is extended to include signs of opposition.

The scale on a thermometer extends in both directions from zero, Figure 3-14. The numbers above zero are called positive numbers; those below zero are called negative numbers. For example, 7 degrees above zero is written +7°; 7 degrees below zero is written -7°. Frequently, the positive numbers are written without the sign of opposition. Therefore, 7 degrees above zero may be written +7° or 7°.

When the symbol + is used as a sign of opposition, it is called a positive sign. When it is used as a sign of operation, it is called a plus sign or a sign of addition. When the symbol – is used as a sign of opposition, it is called a negative sign. When it is used as a sign of operation, it is called a minus sign or a sign of subtraction. When the negative sign is used as a sign of opposition, it is frequently enclosed with the number within parentheses. For example, (–5) + (–4) means to add –5 and –4.

There are two important rules that must be followed when adding positive and negative numbers:

- To add two numbers with like signs, add the numbers as in arithmetic and prefix the common sign to the result.
- To add two numbers with unlike signs, subtract the smaller absolute value from the larger absolute value and prefix the sign of the larger to the result.

Example 10

$(+5) + (+6) = (+11)$ *or* $5 + 6 = 11$

Example 11

$(-5) + (-6) = (-11)$

Example 12

$(+5) + (-6) = (-1)$

Example 13

$(-5) + (+6) = (+1)$

If it is necessary to subtract two quantities, the following rule must be applied. To subtract one number from another, change the sign of the number to be subtracted and proceed as in addition.

Example 14

To subtract +2 from +6, change the sign preceding the 2 and follow the rules for addition.

$(+6) - (+2) =$
$(+6) + (-2) = +4$

Example 15

Subtract −2 from +6.

$(+6) - (-2) =$
$(+6) + (+2) = +8$

Example 16

Subtract +6 from −2.

$(-2) - (+6) =$
$(-2) + (-6) = -8$

Example 17

Subtract −2 from −6.

$(-6) - (-2) =$
$(-6) + (+2) = -4$

The following rules apply to multiplication and division of positive and negative numbers.

- When multiplying or dividing numbers having the same sign, the answer will always be a positive value.
- When multiplying or dividing numbers having different signs, the answer will always be a negative value.

Example 18

Multiply +3 by +5.

$(+3) \times (+5) = +15$

Example 19

Multiply –3 by –5.

$(-3) \times (-5) = +15$

Example 20

Multiply +3 by –5.

$(+3) \times (-5) = -15$

Example 21

Multiply –3 by +5.

$(-3) \times (+5) = -15$

Example 22

Divide +15 by +3.

$(+15) \div (+3) = +5$

Example 23

Divide –15 by –3

$(-15) \div (-3) = +5$

Example 24

Divide –15 by +3.

$(-15) \div (+3) = -5$

Example 25

Divide +15 by –3.

$(+15) \div (-3) = -5$

When applying Kirchhoff's Voltage Law, it is necessary to determine which sign of opposition to prefix to the number representing the value of voltage. One rule is to trace around the circuit in the direction of current flow. All voltages supplied to the circuit are given positive values, and all voltage drops are given negative values. If it is necessary to trace around the circuit opposite to the flow of current, then the voltages supplied to the circuit are given negative values and the voltage drops are given positive values. Example 26 is helpful in understanding the application of Kirchhoff's Laws.

Figure 3-15
Bridge circuit: Use of Kirchhoff's Laws

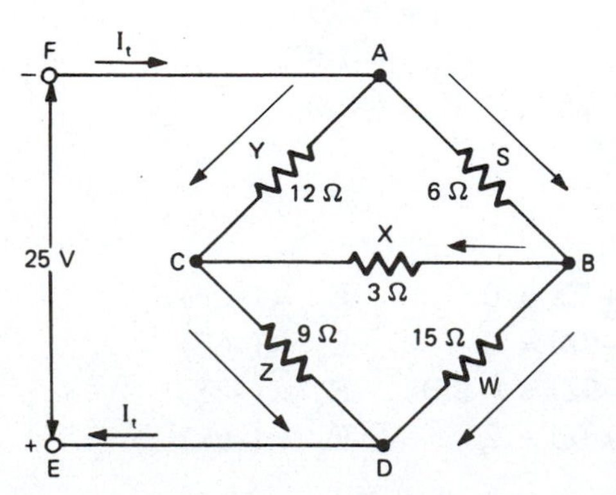

Example 26

Determine the values of current and voltage for each resistance and the total current supplied to the circuit in Figure 3–15.

The values of current for each resistance are represented by the letters S, W, X, Y, and Z. The total current is represented by the symbol I_t. Beginning at junction A and continuing through junction D, develop Kirchhoff's current equation for each junction. Remember that the current entering the junction must be equal to the current leaving the junction.

Junction A	Junction B	Junction C	Junction D
$I_t = S + Y$	$S = W + X$	$Z = X + Y$	$I_t = Z + W$
$S = I_t - Y$	$W = S - X$	$X = Z - Y$	$Z = I_t - W$
$Y = I_t - S$	$X = S - W$	$Y = Z - X$	$W = I_t - Z$

Now, develop Kirchhoff's voltage equations. Remember that the algebraic sum of the voltages around a closed loop is equal to zero.

When tracing around loop EFABDE, the equation is

$$E_t + E_s + E_w = 0$$

$E_t = 25$ V
$E_s = R_sS$ *and* $R_s = 6\ \Omega$

Therefore, $E_s = 6S$

$E_w = R_wW$ *and* $R_w = 15\ \Omega$

Therefore, $E_w = 15W$

Refer back to the original equation. Recognize that E_s and E_w are voltage drops and thus carry a negative sign, and substitute the correct values for E_t, E_s and E_w,

$$E_t + E_s + E_w = 0$$
$$(+25) + (-6S) + (-15W) = 0$$

which, after algebraic manipulation, reduces to

$$(+6S) + (+15W) = 25$$

so that $6S + 15W = 25$ is the first voltage equation.

List the equation for future reference and continue tracing around the various loops to develop several voltage equations.

Loop EFABDE

$6S + 15W = 25$

Loop EFACDE

$$E_t + E_y + E_z = 0 \qquad E_t = (+25)$$
$$(+25) + (-12Y) + (-9Z) = 0 \qquad E_y = (-12Y)$$
$$(-12Y) + (-9Z) = (-25) \qquad E_z = (-9Z)$$
$$12Y + 9Z = 25$$

Loop EFABCDE

$$E_t + E_s + E_x + Ez = 0 \qquad E_t = (+25)$$
$$(+25) + (-6S) + (-3X) + (-9Z) = 0 \qquad E_s = (-6S)$$
$$(-6S) + (-3X) + (-9Z) = (-25) \qquad E_x = (-3X)$$
$$6S + 3X + 9Z = 25 \qquad E_z = (-9Z)$$

Loop ABDCA

$$E_s + E_w + E_z + E_y = 0$$
$$(-6S) + (-15W) + (+9Z) + (+12Y) = 0$$

$E_s = (-6S)$
$E_w = (-15W)$
$E_z = (+9Z)$
$E_y = (+12Y)$

After development of the voltage equations, it is now necessary to solve these equations through the processes of substitution and combination of equations. The goal is to eliminate all the unknown values except for one in any single equation, then solve for that one unknown value. In order to do this, group all the equations together for easy reference.

Current Equations

$I_t = S + Y$	$S = W + X$	$Z = X + Y$	$I_t = Z + X$
$S = I_t - Y$	$W = S - X$	$X = Z - Y$	$Z = It - W$
$Y = I_t - S$	$X = S - W$	$Y = Z - X$	$W = It - Z$

Voltage Equations

$6S + 15W = 25$ $\qquad 12Y + 9Z = 25$ $\qquad 6S + 3X + 9Z = 25$

$$(-6S) + (-15W) + (+9Z) + (+12Y) = 0$$

Step 1. Beginning with the equation derived from loop ABDCA, eliminate the unknown values W and Z by substituting their values from the current equations.

$$(-6S) + (-15W) + (+9Z) + (+12Y) = 0$$
$$(-6S) + [-15(S - X)] + 9(X + Y) + (12Y) = 0$$
$$(-6S) + (-15S) + 15X + 9X + 9Y + 12Y = 0$$
$$(-21S) + 24X + 21Y = 0$$

The equation now has one fewer unknown quantity.

Step 2. Using the equation from loop EFABCDE and the current equation $Z = X + Y$, substitute for Z.

$$6S + 3X + 9Z = 25$$
$$6S + 3X + 9(X + Y) = 25$$
$$6S + 3X + 9X + 9Y = 25$$
$$6S + 12X + 9Y = 25$$

Two equations have now been developed that have the same unknown values S, X, and Y. Write the two equations as shown.

$$6S + 12X + 9Y = 25$$
$$-21S + 24X + 21Y = 0$$

Step 3. Of the two equations just given, the top equation has half the number of X units as the bottom equation. In order to remove the X terms, multiply the top equation by (–2).

$$-2(6S + 12X + 9Y = 25)$$
$$-12S - 24X - 18Y = -50$$

Step 4. Combine this result with the equation derived in Step 1 by adding the two equations.

$$-21S + 24X + 21Y = 0$$
$$\underline{-12S - 24X - 18Y = -50}$$
$$-33S \quad + 3Y = -50 \text{ or } (-33S) + (3Y) = (-50)$$

Step 5. Using the equation derived from loop EFACDE and the current equation Z = X + Y, substitute as follows:

$$12Y + 9Z = 25$$
$$12Y + 9(X + Y) = 25$$
$$12Y + 9X + 9Y = 25$$
$$9X + 21Y = 25$$

Step 6. Refer again to the equations 6S + 12X + 9Y = 25 (Step 2) and 9X + 21Y = 25 (Step 5). The X terms can be eliminated by (a) multiplying the first equation by 3, (b) multiplying the second equation by –4, and (c) adding them.

(a) $6S + 12X + 9Y = 25$

$3(6S + 12X + 9Y = 25)$ (Multiply by 3.)

$18S + 36X + 27Y = 75$

(b) $9X + 21Y = 25$

$-4(9X + 21Y = 25)$ (Multiply by –4.)

$-36X - 84Y = -100$

(c) Combine the equations.

$$18S + 36X + 27Y = 75$$
$$\underline{-36X - 84Y = -100}$$
$$18S \quad - 57Y = -25 \textit{ or } 18S - 57Y = -25$$

Step 7. The same approach is continued. Referring again to the equation derived in Step 4, write the following:

$$-33S + 3Y = -50$$
$$6(-33S + 3Y = -50) \text{ (Multiply by 6.)}$$
$$-198S + 18Y = -300$$

$$18S - 57Y = -25$$
$$11(18S - 57Y = -25) \text{ (Multiply by 11.)}$$
$$198S - 627Y = -275$$

Step 8. Combine the two results obtained in Step 7.

$$\begin{array}{r} 198S - 627Y = -275 \\ -198S + 18Y = -300 \\ \hline -609Y = -575 \\ Y = 0.944A \end{array}$$

Step 9. Using the equation arrived at in Step 5, solve for X.

$$\begin{aligned} 9X + 21Y &= 25 \\ 9X + 21(0.944) &= 25 \\ 9X + 19.824 &= 25 \\ 9X &= 25 - 19.824 \\ 9X &= 5.176 \\ X &= 0.575A \end{aligned}$$

Step 10. Referring back to the equation arrived at in Step 2, solve for S.

$$\begin{aligned} 6S + 12X + 9Y &= 25 \\ 6S + 12(0.575) + 9(0.944) &= 25 \\ 6S + 6.9 + 8.496 &= 25 \\ 6S + 15.396 &= 25 \\ 6S &= 25 - 15.396 \\ 6S &= 9.604 \\ S &= 1.601\ A \end{aligned}$$

Step 11. Referring to the current equations, solve the following.

$I_t = S + Y$
$I_t = 1.601 + 0.944$
$I_t = 2.545\ A$

$W = S - X$
$W = 1.601 - 0.575$
$W = 1.026\ A$

$Z = X + Y$
$Z = 0.575 + 0.944$
$Z = 1.519\ A$

Summary of Currents

$I_t = 2.545\ A$	$X = 0.575\ A$
$S = 1.601\ A$	$Y = 0.944\ A$
$W = 1.026\ A$	$Z = 1.519\ A$

Check the work by using the equation derived from loop ABDCA.

$$\begin{aligned} (-6S) + (-15W) + (+9Z) + (+12Y) &= 0 \\ -6(1.601) + [-15(1.026)] + [9(1.519)] + [12(0.944)] &= 0 \\ (-9.606) + (-15.39) + (13.671) + (11.328) &= 0 \\ (-24.996) + (24.999) &= 0.003 \end{aligned}$$

Rounding off numbers to the nearest 1/1000 accounts for the discrepancy of 3/1000 ampere (0.003 A).

To solve for the voltage drops, Ohm's Law may be used.

$E_s = I_s R_s$
$E_s = 1.601 \times 6$
$E_s = 9.606V$

$E_z = I_z R_z$
$E_z = 1.519 \times 9$
$E_z = 13.671\ V$

$E_w = I_w R_w$
$E_w = 1.026 \times 15$
$E_w = 15.39\ V$

$E_y = I_y R_y$
$E_y = 0.944 \times 12$
$E_y = 11.328\ V$

$E_x = I_x R_x$
$E_x = 0.575 \times 3$
$E_x = 1.725\ V$

REVIEW

A. Multiple choice.
Select the best answer.

1. An electric circuit is the
   a. flow of electrons.
   b. path over which electrons flow.
   c. force that causes electrons to flow.
   d. opposition to electron flow.
2. A series circuit is a circuit that
   a. has only one path over which the current flows.
   b. has more than one path over which the current flows.
   c. has three current paths.
   d. contains many devices.
3. The value of current flowing in a series circuit is
   a. equal to the sum of the currents through the various branches.
   b. the same value through all parts of the circuit.
   c. equal to the resistance divided by the voltage.
   d. equal to the applied voltage.
4. The total resistance in a series circuit is
   a. equal to the sum of the resistances of each part of the circuit.
   b. inversly proportional to the voltage.
   c. always less than any resistance in the circuit.
   d. equal to the current divided by the voltage.

REVIEW *(continued)*

A. Multiple choice.
Select the best answer (continued).

5. A parallel circuit is a circuit that
   a. has only one path over which electrons can flow.
   b. has more than one path through which current can flow.
   c. contains many devices.
   d. has very low resistance.
6. The value of current flowing in a parallel circuit is
   a. equal to the sum of the currents through the various branches.
   b. the same value through all parts of the circuit.
   c. equal to the resistance divided by the voltage.
   d. equal to the applied voltage.
7. Conductance is
   a. opposition to current flow.
   b. a measurement of resistance.
   c. a measurement of electrical pressure.
   d. a measurement of the ability to conduct electric current.
8. The mathematical symbol for conductance is
   a. G.
   b. V.
   c. C.
   d. A.
9. The voltage across any branch of a parallel circuit is
   a. determined by the resistance of the branch.
   b. greater if the current is greater.
   c. equal to the supply voltage.
   d. equal to the current divided by the resistance.
10. The reciprocal of resistance is
    a. voltage.
    b. conductance.
    c. current.
    d. coulomb.

B. Give complete answers.

1. What is a combination circuit?
2. What is an equivalent circuit?
3. What is a two-wire circuit?
4. What is a three-wire circuit?'
5. Name at least one advantage of a three-wire circuit.

REVIEW *(continued)*

B. Give complete answers (continued).

6. Name at least one disadvantage of a three-wire circuit.
7. Describe a multiwire branch circuit.
8. Is a three-wire circuit a multiwire circuit?
9. Describe the difference between a balanced and an unbalanced three-wire circuit.
10. Write Kirchhoff's Current Law.

C. Solve each problem, showing the method used to arrive at the solution (continued).

1. What voltage must be supplied by a battery to force 0.4A through a series circuit consisting of a 7-Ω lamp, a 10-Ω rheostat, and a 12-Ω soldering iron?
2. What is the voltage drop across each component in Problem 1?
3. Three components are connected in series; the total resistance is 200 Ω. If the combined resistance of two of the components is 150 Ω, what is the resistance of the third component?
4. What value of current will flow through this series circuit?

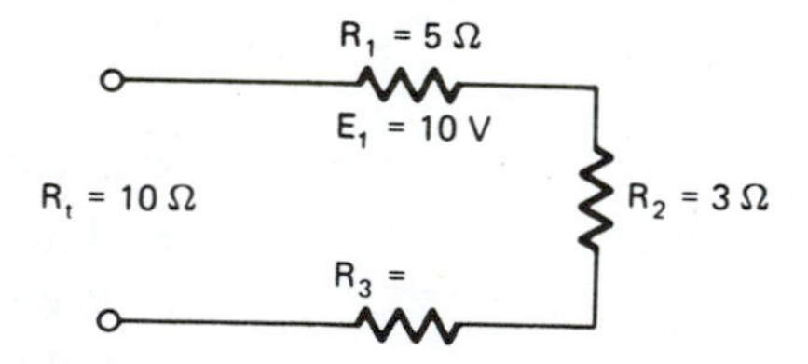

5. Calculate the resistance of R_3 in the circuit in Problem 4.
6. Calculate the supply voltage and the voltage drops across R_2 and R_3 in the circuit in Problem 4.
7. Using Ohm's Law, calculate the current through R_1, R_2, and R_3 in the circuit in Problem 4.
8. What value of resistance must be connected in series with an incandescent lamp to limit the current to 1 A? The (hot) resistance of the lamp is 200 Ω and it is connected to a 240-V supply.

REVIEW *(continued)*

C. Solve each problem, showing the method used to arrive at the solution (continued).

9. Determine the total voltage, current, and resistance for this four-element series circuit.

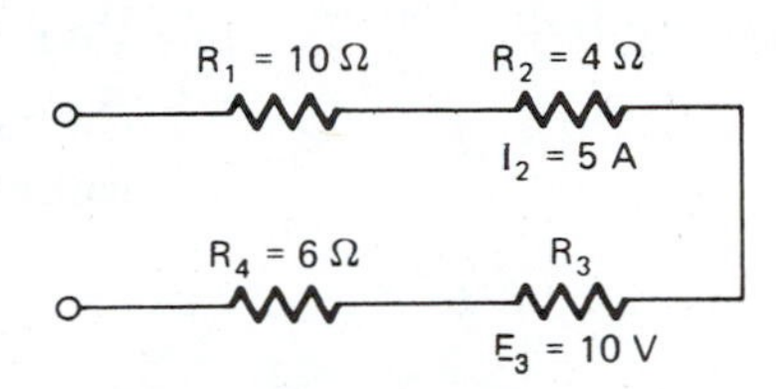

10. Prove that the current through R4 in Problem 9 is 5 A.
11. Three heating elements, each having a resistance of 30 Ω, are connected in parallel across 100 V. Calculate the combined resistance and the current through each heater.
12. Five lamps of equal resistance are connected in parallel across a 120-V line. If the total current is 3 A, what is the resistance of each lamp?
13. Three heating elements having resistances of 30 Ω, 60 Ω, and 80 Ω are connected in parallel. If a current of 8 A flows through the 30-Ω branch, determine the current through each of the other two branches.
14. An unknown resistance, a resistance of 15 Ω, and a resistance of 20 Ω are connected in parallel across a 120-V generator. If the generator supplies 22 A, determine the value of the unknown resistance.
15. What is the combined resistance of a 0.3-Ω and a 0.75-Ω cast-iron grid resister connected in parallel?
16. The combined resistance of two lamps in parallel is 80 Ω. If the resistance of one is 200 Ω, what is the resistance of the other?
17. What voltage will be required to force a current of 10 A through a parallel circuit consisting of a 16-Ω resistor and a 20-Ω resistor?
18. The current through a 4-Ω coil and an 8-Ω coil in parallel is 10 A. Determine:
    (a) The voltage across the coils.
    (b) The current through each coil
19. Three lamps having resistances of 60 Ω, 80 Ω, and 100 Ω are connected in parallel. If the current through the combination is 8 A, determine the current through each.

REVIEW *(continued)*

C. Solve each problem, showing the method used to arrive at the solution (continued).

20. Three equal resistances connected in series have a combined resistance of 25 Ω. What would be their combined resistance if they were reconnected in parallel?

21. Calculate the combined resistance of this three-element combination circuit.

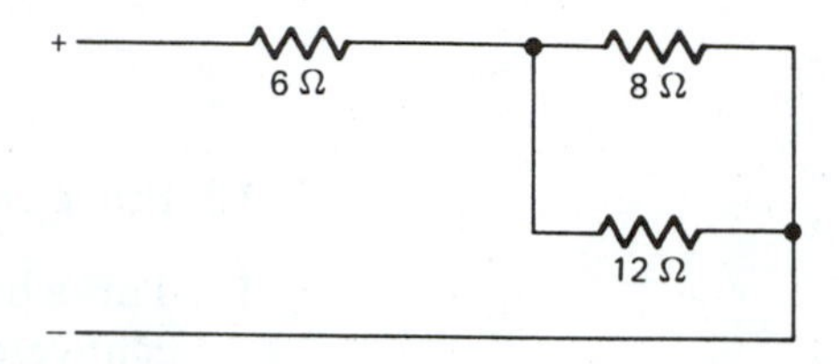

22. Calculate the combined resistance of this five-element combination circuit.

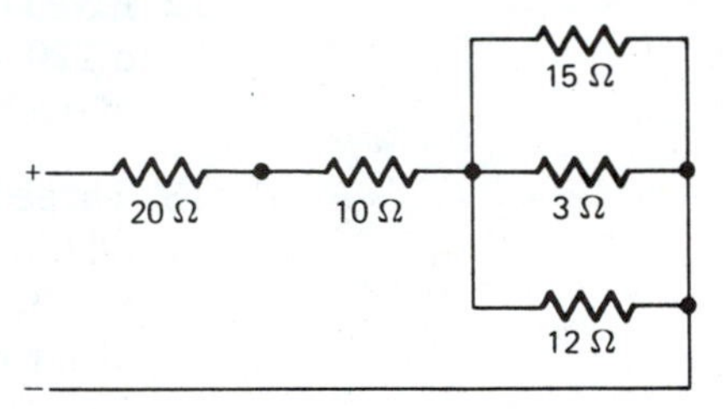

23. Calculate the combined resistance of this four-element combination circuit.

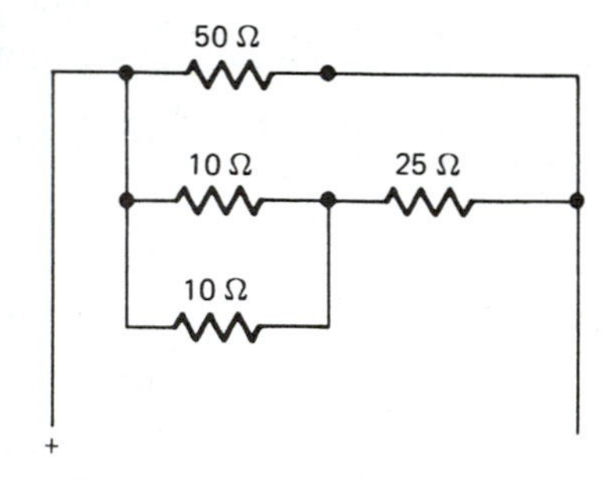

REVIEW *(continued)*

C. Solve each problem, showing the method used to arrive at the solution (continued).

24. Calculate the combined resistance of this five-element combination circuit.

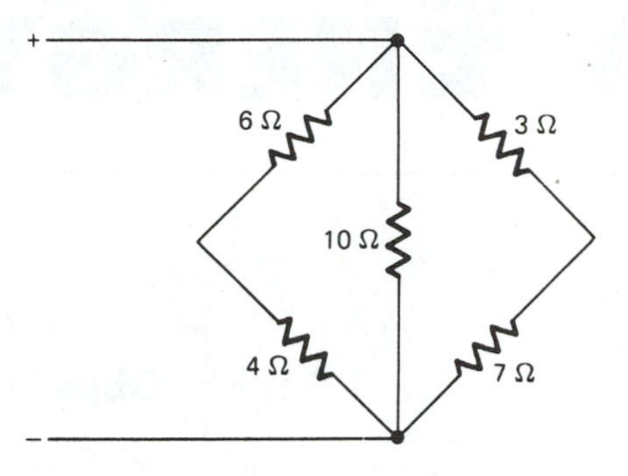

25. Calculate the current through each resistance and the voltage drop across each resistor in this bridge circuit.

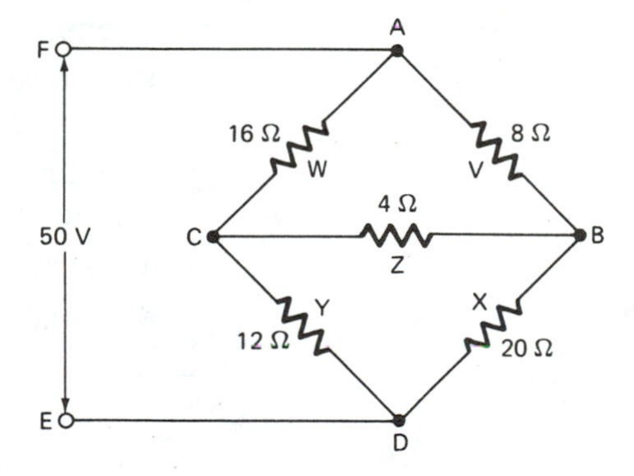

4

Electrical Power and Energy

Objectives

After studying this chapter, the student will be able to:

- Define the terms relating to electrical power and energy.
- Calculate the power necessary to perform various electrical jobs.
- Calculate the amount of electrical energy required to perform various electrical jobs.
- Determine the horsepower necessary to drive various machines.
- Calculate the efficiency of electrical equipment.
- Calculate the cost of operating electrical equipment.

WORK

Work may be defined as the overcoming of resistance through a distance. The unit of measurement of work is the *foot-pound* (kilogram-meter), abbreviated ft·lb, which is the work performed when a force of 1 pound (0.4536 kilogram) acts through a distance of 1 foot (0.3048 meter). The unit of measurement of work in the metric system is the *joule*. One joule (J) is equal to 0.7376 foot-pound (0.102 kilogram-meter).

POWER

Machines were invented to assist in doing work. The amount of work a machine can do in a specific time determines its power. *Power* is a measurement of the rate of doing work. The amount of work can be calculated by the formula Work = Force × Distance, W = FD, where the force is measured in pounds (kilograms) and the distance is measured in feet (meters).

To determine the power a machine must deliver, it is necessary to know the speed at which the work must be accomplished. The formula for power is Power = Work/Time, P = W/T, where work is measured in foot-pounds and time is measured in minutes (min).

Horsepower

Horsepower (hp) is the common unit of measurement of mechanical power. In early times, water was pumped from mines by horses pulling and turning a wheel on the end of a shaft to drive a pump. Later, horses were replaced by the steam engine. Steam engines are rated as having the power of a certain number of horses (horsepower). It was determined that an average horse could do work at the rate of 33,000 foot-pounds (44,740 joules) per minute. This means that the average horse can move 1000 pounds (453.6 kilograms) 33 feet (10.058 meters) in one minute. The mathematical formula for horsepower is

$$hp = \frac{ft{\cdot}lb/min}{33{,}000} \qquad \text{(Eq. 4.1)}$$

Example 1

A machine can place 50,000 lb (22,680 kg) of scrap metal onto a truck 10 ft (3.048 m) high in 5 min. What horsepower is the machine capable of delivering?

$$(50{,}000 \times 10) \div 5 = 100{,}000 \text{ ft}{\cdot}\text{lb/min}$$

$$hp = \frac{ft{\cdot}lb/min}{33{,}000}$$

$$hp = \frac{100{,}000}{33{,}000}$$

$$hp = 3$$

Electric Power

Electric power is measured in *watts* (W) or *kilowatts* (kW). One thousand watts equal one kilowatt. It is relatively simple to convert from mechanical power to electrical power; 1 horsepower is equal to 746 watts. (One watt per second equals one joule.)

One watt is developed when one volt forces one ampere through a circuit. This can be expressed mathematically by the formula

$$P = IE \qquad \text{(Eq. 4.2)}$$

where P = power, in watts (W)
I = current, in amperes (A)
E = electrical pressure, in volts (V)

Sometimes it is necessary to solve for power when the current or the voltage is unknown. If the resistance is known, the problem can be solved as in the following Examples 2 and 3.

Example 2

How much electrical power is required to operate a water heater that is rated at 240 V, if its resistance (R) is 24 Ω?

Solution A

$$I = \frac{E}{R} \qquad P = IE$$

$$I = \frac{240}{24} \qquad P = 10 \times 240$$

$$I = 10\text{ A} \qquad P = 2400\text{ W, } \textit{or } 2.4\text{ kW}$$

Solution B

$P = IE$ *and* $I = E/R$

$P = \frac{E}{R} \times E$ (Substitute $\frac{E}{R}$ for I.)

$$P = \frac{E^2}{R}$$

$$P = \frac{57{,}600}{24}$$

$P = 2400$ W, *or* 2.4 kW

Therefore, the problem can be solved by the formula

$$P = \frac{E^2}{R} \qquad \text{(Eq. 4.3)}$$

Example 3

How much power is dissipated by an electric oven if 30 A flow through a resistance of 10 Ω?

Solution A

$$E = IR \qquad P = IE$$

$$E = 30 \times 10 \qquad P = 30 \times 300$$

$$E = 300\text{ V} \qquad P = 9000\text{ W, } \textit{or } 9\text{ kW}$$

Solution B

$P = IE$ *and* $E = IR$

$P = IIR$ (Substitute IR for E.)

$P = I^2R$

$P = 900 \times 10$

$P = 9000$ W, *or* 9 kW

The problem can be solved by the formula

$$P = I^2R \qquad \text{(Eq. 4.4)}$$

ENERGY

Energy is the ability to do work. Energy cannot be destroyed or consumed. However, it can be converted from one form to another. Some forms of energy are heat, light, mechanical, electrical, and chemical energy.

Many methods for converting energy from one form to another have been developed through scientific research. Energy conversion can be used to do work. For example, the electric motor converts electrical energy into mechanical energy. In the process the motor shaft rotates to drive a machine.

The term *energy loss*, which is frequently used in the study of electricity, can be misleading. As just stated, energy cannot be destroyed or consumed. Energy loss means that during the process of energy conversion, some energy is converted into a form other then that which is desired. This so-called loss takes place in the electric motor. During the process of converting electrical energy into mechanical energy, some of the electrical energy is converted into heat energy. Because the heat energy is not desired and does not serve a useful purpose, it is considered a loss.

Mechanical Energy

Mechanical Energy is energy that causes motion. It is measured in the same units as work (foot-pounds or joules).

Heat Energy

Heat Energy is energy that causes substances to rise in temperature. It may be expressed in either the British unit or the metric unit. In the British system, heat is measured in *British thermal units* (Btu). One Btu is the amount of heat required to raise the temperature of one pound of water one degree Fahrenheit (F). In the metric system, heat energy is measured in units called *calories* (cal). One calorie is the amount of heat required to raise the temperature one gram (g) of water one degree Celsius (C).

Electrical Energy

Electrical energy is the product of power and time. Therefore, its unit of measurement is the *watthour* (Wh), or the *kilowatt-hour* (kW·h). A 100-watt incandescent lamp will utilize 1 kilowatt-hour of energy if it is operated for 10 hours (100 watts X 10 hours = 1000 watthours = 1 kilowatt-hour).

Electrical consumers are billed by the utility company according to the amount of energy utilized. In other words, they are billed for the amount of power expended in a specific period of time. Billing is usually based on a sliding scale according to the number of kilowatt-hours indicated for the billing period.

Example 4

The customer, an industrial plant, expends 5000 kW·h of electrical energy each working day. The plant is in operation 5 days per week. If the utility company charges $0.03 per kW·h for the first 50,000 kW·h and $0.015 per kW·h for all energy above 50,000 kW·h, how much does the customer pay for a 10-week period?

5000 kW·h/day × 5 days/week × 10 weeks = 250,000 kW·h
50,000 kW·h × $0.03/kW·h = $1500.00
250,000 kW·h – 50,000 kW·h = 200,000 kW·h
200,000 kW·h × $0.015/kW·h = $3000.00
$1500.00 + $3000.00 = $4500.00 paid for the 10-week period

EFFICIENCY

Efficiency is the ratio of the useful power output to the total power input. Efficiency is generally stated in percent (%). No machine is 100 percent efficient. Not all the energy delivered to a machine serves the purpose for which it was intended. A motor is designed to produce motion (mechanical energy). However, some of the electrical energy received by the motor produces heat. The energy that is converted into heat is considered a loss. In order to operate machines at minimum cost, it is necessary to keep all losses to a minimum. A machine with very low energy loss is considered to be very efficient. The nearer the efficiency is to 100 percent, the more efficient is the machine. Efficiency can be calculated by using one of the following formulas:

$$\%\ \text{Eff} = \frac{\text{Useful energy output}}{\text{Total energy input}} \times 100 \qquad \text{(Eq. 4.5)}$$

or

$$\%\ \text{Eff} = \frac{\text{Power output}}{\text{Power input}} \times 100 \qquad \text{(Eq. 4.6)}$$

MECHANICAL TRANSMISSION OF POWER

Machines may be divided into two classes, the *driving machine* and the *driven machine*. The driving machine delivers the power to the machine that is being driven. Some types of driving machines are gasoline engines, steam turbines, and electric motors. Some examples of driven machines are presses, lathes, elevators, pumps, and saws.

DRIVES

The usual types of connections between driving machines and driven machines are belts on pulleys, chains on sprockets, gear assemblies, and direct drives. It is very important that the proper drive be used to meet the needs of the job. Consideration must be given to such factors as speed requirements, direction of rotation, starting torque (twisting effort when the machine is starting), full-load torque, and starting current.

Speed Requirements

If the speeds of both machines are the same, it may be possible to use a direct drive. There are two methods used to obtain direct drive. The machine and the motor may be mounted on the same shaft. This method is frequently used with motor-generator sets. A second method is to use a shaft coupling, a device that fastens two shafts together. This causes the entire assembly to operate as one machine.

If the speed of the driven machine differs from the speed of the motor, pulleys or gears are usually used. When pulleys are used, the speed of the machine is determined by the sizes of the pulleys. The speed relationship of the two machines is inversely proportional to the diameter relationship of their pulleys. The formula for calculating pulley sizes is

$$\frac{N_1}{N_2} = \frac{D_2}{D_1} \qquad \text{(Eq. 4.7)}$$

where N_1 = speed of the motor, in revolutions per minute (r/min)
N_2 = speed of the driven machine, in revolutions per minute (r/min)
D_1 = diameter of the motor pulley, in inches (in.) or centimeters (cm)
D_2 = diameter of the driven machine pulley, in inches (in.) or centimeters (cm)

Example 5

The nameplate of a motor indicates that it rotates at a speed of 1700 r/min. The diameter of the pulley on the motor shaft is 6 in. (15.24 cm). If the machine must be driven at a speed of 850 r/min, what size of pulley must be used on the machine?

$$\frac{N_1}{N_2} = \frac{D_2}{D_1}$$

$$\frac{1700}{850} = \frac{D_2}{6}$$

$$85D_2 = 1020$$

$$D_2 = 12 \text{ in.}$$

Example 6

A motor operates at a speed of 3250 r/min. The machine it is driving requires a speed of 650 r/min. If the machine pulley is 20 in. (50.8 cm), what size pulley must be installed on the motor?

$$\frac{N_1}{N_2} = \frac{D_2}{D_1}$$

$$\frac{3250}{650} = \frac{20}{D_1}$$

$$5D_1 = 20$$

$$D_1 = 4 \text{ in.}$$

Example 7

A motor operates at a speed of 500 r/min. The pulley attached to its shaft has a diameter of 20 in. (50.8 cm). The machine that the motor is driving has a pulley that is 8 in. (20.32 cm) in diameter. At what speed will the machine operate?

$$\frac{N_1}{N_2} = \frac{D_2}{D_1}$$

$$\frac{500}{N_2} = \frac{8}{20}$$

$$8N_2 = 10{,}000$$

$$N_2 = 1250 \text{ r/min}$$

A gear drive is generally used where slippage is a factor. Although the gear drive is much more expensive than belts and pulleys, it is a positive drive and will not slip.

When using gears, the speed ratio is determined by the number of teeth in each gear. If both gears have the same number of teeth, the machine and the motor will operate at the same speed. The speed relationship is inversely proportional to the relationship of the number of teeth. With only two gears, the maximum speed ratio is generally 10 to 1. More than one set of gears is usually required to obtain greater speed ratios.

The formula for calculating speed and/or the number of teeth is

$$\frac{N_1}{N_2} = \frac{T_2}{T_1} \qquad \text{(Eq. 4.8)}$$

where N_1 = speed of the driving gear, in revolutions per minute (r/min)

N_2 = speed of the driven gear, in revolutions per minute (r/min)

T_1 = number of teeth in the driving gear

T_2 = number of teeth in the driven gear

Example 8

The driving gear on a motor has 60 teeth and the driven gear has 90 teeth. If the driven gear revolves at 100 r/min, what is the speed of the driving gear?

$$\frac{N_1}{N_2} = \frac{T_2}{T_1}$$

$$\frac{N_1}{100} = \frac{90}{60}$$

$$6N_1 = 900$$

$$N_1 = 150 \text{ r/min}$$

Example 9

The driving gear on a motor revolves at a speed of 1150 r/min and has 25 teeth. The speed of the machine must be 115 r/min. How many teeth must there be on the driven gear?

$$\frac{N_1}{N_2} = \frac{T_2}{T_1}$$

$$\frac{1150}{115} = \frac{T_2}{25}$$

$$T_2 = 250 \text{ teeth}$$

Example 10

Drive gear A on a motor has 20 teeth and revolves at a speed of 2250 r/min. Gear A drives a second gear, B, which has 30 teeth and meshes with gear C, which has 40 teeth. What is the speed of gear C?

Solution

$$\frac{N_A}{N_C} = \frac{T_C}{T_A}$$

$$\frac{2250}{N_C} = \frac{40}{20}$$

$$N_C = 1125 \text{ r/min}$$

Proof

$$\frac{N_A}{N_B} = \frac{T_B}{T_A}$$

$$\frac{2250}{N_B} = \frac{30}{20}$$

$$3N_B = 4500$$

$$N_B = 1500 \text{ r/min}$$

$$\frac{N_B}{N_C} = \frac{T_C}{T_B}$$

$$\frac{1500}{N_C} = \frac{40}{30}$$

$$4N_C = 4500$$

$$N_C = 1125 \text{ r/min}$$

Direction of Rotation

The direction of rotation of the motor and the machine must be considered when selecting the type of drive. When a direct drive, belt drive (Figure 4-1), or chain drive is used, the machine rotor (revolving part) will rotate in the same direction as the motor rotor. Reversing the direction of rotation of the driven pulley necessitates reversing the direction of rotation of the motor rotor or crossing the belt, Figure 4-2. Crossing the belt is not generally recommended because of the extra wear on the belt.

Figure 4-1
Motor using a belt drive

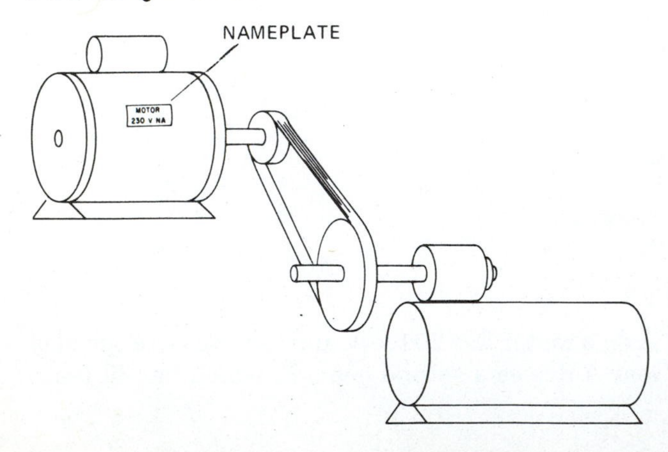

Figure 4-2
Crossed belt drive for reversing the direction of the driven machine rotor

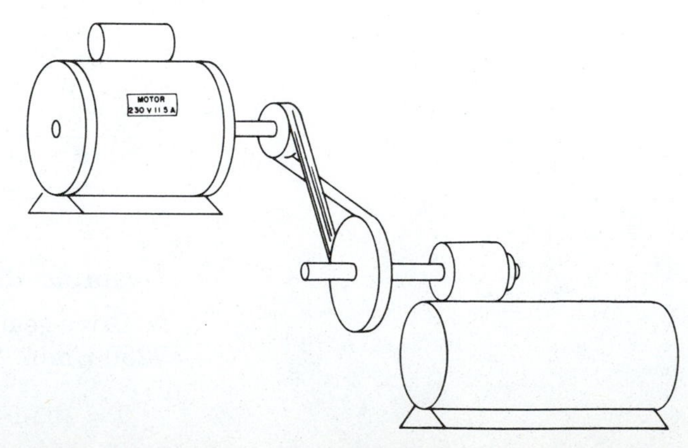

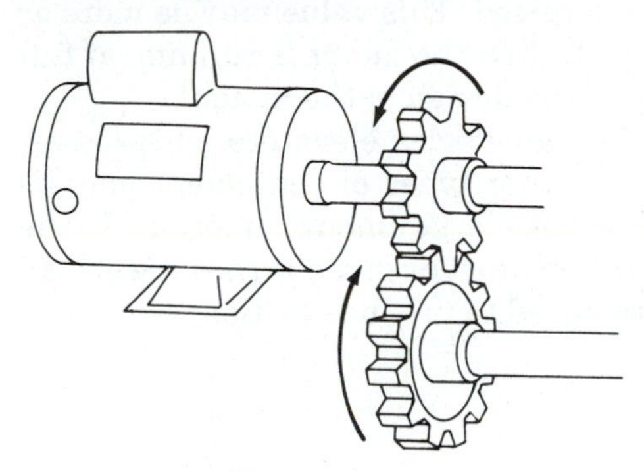

Figure 4-3
Two-gear drive. The arrows indicate the direction of gear rotation.

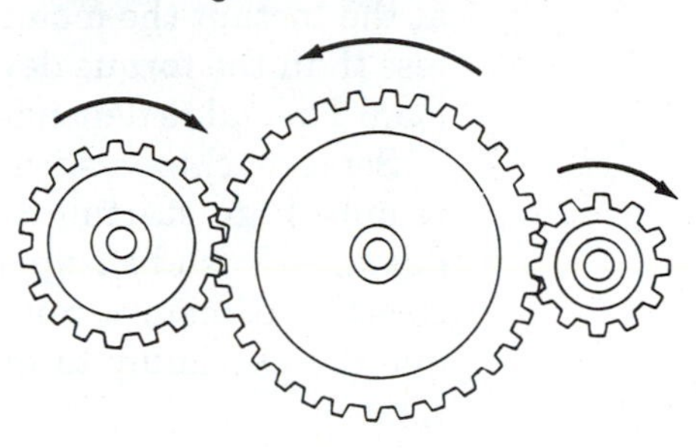

Figure 4-4
Three-gear drive. The arrows indicate the direction of gear rotation.

When a two-gear drive is used, the driven gear will rotate in the opposite direction from the driving gear, Figure 4-3. Three gears must be used in order to have the driven gear rotate in the same direction as the driving gear, Figure 4-4.

A motor can deliver only the amount of horsepower indicated on the nameplate. No arrangement of gears and/or pulleys will change the horsepower delivered to the machine.

Torque

It can be observed that horsepower is a combination of *torque* (a rotating force) and speed. At a specific horsepower the torque is inversely proportional to the speed. Any change in speed will cause a change in torque. An increase in speed will result in a decrease in torque, and vice versa. One must ensure adequate torque to drive the machine at the required speed. The value of torque can be calculated for various speeds by the formula

$$\text{hp} = \frac{TN}{5252} \qquad \text{(Eq. 4.9)}$$

where hp = horsepower (hp)
T = torque, in pound-feet (lb·ft)
N = speed, in revolutions per minute (r/min)

Example 11

How much torque is required to drive a machine rated at 10 hp? The speed of the machine is 500 r/min.

$$T = \frac{5252\ \text{hp}}{N}$$

$$T = \frac{5252 \times 10}{500}$$

$$T = 105.04\ \text{lb·ft}$$

Starting Torque

The *starting torque* of a motor is the amount of torque developed at the instant the motor is energized. This value may be more or less than the torque developed when the motor is running at full load. The value depends upon the design of the motor.

Some machines, such as electric vehicles, elevators, and presses, require high starting torque. Other types of equipment may be damaged by sudden application of a large amount of torque. In the selection of motors, consideration must be given to the type of load and the maximum torque required to produce motion.

Starting Current

Starting current is the value of current required by the motor from the instant it is energized until it reaches its rated speed. The largest amount of current flows at the instant the motor is energized. A motor that produces a high starting torque requires a high starting current.

The value of current available is limited by the size of the circuit conductors and the overcurrent devices protecting the circuit. With some motors the starting current may be great enough to cause the overcurrent devices to open. To eliminate this problem, special motor starting equipment is connected into the circuit. This equipment limits the maximum current to a specific value during the starting cycle. Many sophisticated devices have been designed for this purpose. A very simple device to use, however, is the rheostat. A rheostat connected in series with the motor will both limit and control the starting current.

Other Factors in Selecting Drive

Other factors that must be considered in the selection of the type of drive include the following:

- Safety requirements
- Space requirements
- Size and type of shafts
- Horizontal or vertical shafts
- Distance from the motor to the machine

SIZING MOTORS

The size of a motor for driving a machine depends upon the speed of the machine, the torque necessary to drive the machine at the rated speed, and the efficiency of the machine. For the most efficient operating conditions, motors should be sized as accurately as possible. Oversizing results in wasted power and high operating costs. Undersized motors cause excessive heating, insulation breakdown, and frequent operation of the overcurrent protection devices. The power required to drive a machine depends upon the rate that the machine does work and the efficiency of the machine. The power may be calculated by using variations of the following formula:

$$hp = \frac{Wh}{33{,}000 \cdot Eff} \qquad \text{(Eq. 4.10)}$$

Where hp = mechanical power, in horsepower (hp)
W = weight lifted, in pounds (lb)
h = height lifted, in feet per minute (ft/min)
Eff = efficiency, in percent (%)

Example 12

What size motor is required to drive a hoist if it must lift 1000 lb (453.6 kg) to a height of 30 ft (9.144 m) in one minute? The hoist is 90% efficient.

$$hp = \frac{Wh}{33{,}000 \cdot Eff}$$

$$hp = \frac{1000 \times 30}{33{,}000 \times 0.9}$$

$$hp = \frac{30{,}000}{29{,}700}$$

$$hp = 1$$

Manufacturers' literature and machine reference books provide information to calculate the horsepower of the driving motor for other types of driven machines, such as pumps, compressors, fans, and presses. Basically, the equations used are adaptations of Equation 4.10.

REVIEW

A. Multiple choice.
Select the best answer.

1. Energy is the
   a. overcoming of resistance through a specific distance.
   b. rate of doing work.
   c. ability to do work.
   d. power.
2. The unit of measurement of work is the
   a. foot-pound.
   b. watt.
   c. kilowatt-hour.
   d. horsepower.
3. A common unit of measurement of mechanical power is the
   a. watt.
   b. horsepower.
   c. kilowatt-hour.
   d. volt.
4. One kilowatt is equal to
   a. 0.001 watt.
   b. 1000 watts
   c. 100 watts
   d. 1,000,000 watts.
5. One horsepower is equal to
   a. 746 watts.
   b. 986 watts.
   c. 1000 watts.
   d. 0.5 watt.
6. An incandescent lamp converts electrical energy into
   a. light energy.
   b. heat energy.
   c. both a and b.
   d. chemical energy.
7. Energy loss means that
   a. energy has been consumed.
   b. energy has been destroyed.
   c. one form of energy has been converted into an unwanted form of energy.
   d. energy has dissipated.

REVIEW *(continued)*

A. Multiple choice.
Select the best answer (continued).

8. The electric motor converts
   a. light energy into mechanical energy.
   b. electrical energy into mechanical and heat energy.
   c. mechanical energy into electrical and heat energy.
   d. chemical energy into mechanical energy.
9. Horsepower is a measurement of
   a. mechanical power.
   b. electrical power.
   c. useful power.
   d. atomic power.
10. The percent efficiency of a motor is the percent of
   a. power lost.
   b. useful power.
   c. total power that is performing useful work.
   d. both a and b.

B. Give complete answers.

1. Define *power*.
2. What is meant by the term *watthour*?
3. Write three formulas, in equation form, for calculating electrical power.
4. What is the unit of measurement of electrical energy?
5. Explain how to produce 1 watt of electrical power.
6. Define *work*.
7. Write the formula, in equation form, for calculating work.
8. Name the most common unit of measurement of mechanical power.
9. How many watts are there in one kilowatt?
10. List five forms of energy.
11. Name the two classes of machines.
12. List four types of machine drives.
13. Identify two methods used to obtain direct drive.
14. What method of drive is used if the speed of the driven machine differs from the speed of the motor?

REVIEW *(continued)*

B. Give complete answers (continued).

15. List two factors that determine the speed of a machine that is belt driven.
16. If the direction of rotation of the machine rotor is opposite to the direction of the rotation of the motor rotor, what type of drive is preferred?
17. Name one disadvantage of chain drives.
18. Name on advantage and one disadvantage of belt drives.
19. Write the formula (in equation form) for horsepower when the torque and speed are known.
20. List seven factors that must be considered when selecting the type of drive to be used on a machine.

C. Solve each problem, showing the method used to arrive at the solution.

1. A person raised 450 lb of shingles 100 ft in 20 min. Calculate the average horsepower expended.
2. A current of 20 A flows through a heating element when a pressure of 110 V is applied. What power does it dissipate?
3. A current of 50 A flows through a circuit having a resistance of 10 Ω. What power is dissipated by the circuit?
4. A bank of heating lamps requires 230 V for full heat. If the lamps' (hot) resistance is 5 Ω, what power do they require?
5. What horsepower is developed by the heating element in Problem 2?
6. How much electrical energy is utilized in 1 week by the circuit in Problem 3 if it is in operation for 8 hours a day, 6 days per week?
7. How much will it cost to operate the lamps in Problem 4 if they are to be used for 10 hours? The cost of energy is $0.05 per kW·h.
8. A machine motor requires 20 A at 120 V. If the motor's output is 2300 W, what is the efficiency of the motor?
9. Motors are usually rated in terms of their horsepower output. In order to calculate the efficiency, the power (output and input) must be in the same unit of measurement. An electric motor on a forklift is rated at 10 hp. The input current to the motor is 70 A at 120 V. What is the efficiency of the motor?

REVIEW *(continued)*

C. Solve each problem, showing the method used to arrive at the solution.

10. A compressor motor delivers 20 hp to the pump. If the motor is 75% efficient, what is the kW input to the motor?
11. The driving gear on a motor has 30 teeth. If it revolves at a speed of 1150 r/min, how many teeth must the driven gear have if it is to revolve at a speed of 575 r/min?
12. A machine must be driven by a gear that rotates at a speed of 875 r/min and has 100 teeth. If the rotor of the driving motor rotates at a speed of 1750 r/min, how many teeth must the driving gear contain?
13. The revolving component of a machine rotates at a speed of 100 r/min and is driven by a gear containing 100 teeth. If the driving gear on the motor has 25 teeth, what is the speed of the motor?
14. A motor shaft contains a gear that has 10 teeth. The shaft rotates at a speed of 2350 r/min. If the machine that is being driven by the motor has a gear containing 200 teeth, what is the speed of the machine?
15. A machine rotor is belt driven and must revolve at a speed of 345 r/min. If the driving motor shaft rotates at a speed of 1725 r/min, determine the size of pulleys for both the motor and the machine.
16. The nameplate of a motor indicates that it rotates at a speed of 3550 r/min. The diameter of the pulley on the motor shaft is 4 in. (10.16 cm). If the machine must be driven at a speed of 710 r/min, what size pulley must be used on the machine?
17. A motor operates at a speed of 2950 r/min. The machine it is driving requires a speed of 295 r/min. If the machine has a pulley 30 in. (76.2 cm) in diameter, what size pulley must be installed on the motor?
18. A motor operates at a speed of 250 r/min. The pulley attached to its shaft has a diameter of 30 in. (76.2 cm). The machine the motor is driving has a pulley that is 4 in. (10.16 cm) in diameter. What is the speed of the machine?
19. How much torque is required to drive a machine rated at 50 hp? The machine must operate at a speed of 250 r/min.
20. What size motor is required to operate an elevator if its maximum capacity is 3000 lb (1360.8 kg)? The elevator travels at a speed of 10 ft (3.048 m) per minute. The motor is 80% efficient.

5

ELECTRICAL MEASURING INSTRUMENTS

Objectives

After studying this chapter, the student will be able to:

- Explain how to use the more common electrical measuring instruments.
- Describe the function of various meters.
- Explain the basic principle of operation of the more common electrical measuring instruments.
- Describe the care and maintenance procedures for the instruments discussed in this chapter.
- List the safety precautions to follow when using various electrical instruments.

Electrical personnel use many different types of measuring instruments. Some jobs require very accurate measurements; other jobs need only rough estimates. Some instruments are used solely to determine whether or not a circuit is complete. The most common measuring and testing instruments are voltage testers, voltmeters, ammeters, ohmmeters, continuity testers, megohmmeters, wattmeters, and watthour meters.

All meters used for measuring electrical values are basically *current meters*. They measure or compare the values of current flowing through them. The meters are calibrated and the scale is designed to read the value of the desired unit.

SAFETY PRECAUTIONS

Correct meter connections are very important for the safety of the user and for proper maintenance of the meters. A basic knowledge of the construction and operation of meters will aid the user in making proper connections and maintaining them in safe working order.

Many instruments are designed to be used on dc or ac only; others can be used interchangeably. *Note*: It is very important to use each meter only with the type of current for which the meter is designed. Using a meter with an incorrect type of current can result in damage to the meter and may cause injury to the user.

Some meters are constructed to measure very low values. Other meters can measure extremely high values.

CAUTION: Never allow a meter to exceed its rated maximum limit. The importance of never allowing the actual value to exceed the maximum value indicated on the meter cannot be overemphasized. Exceeding maximum values can damage the indicating needle, interfere with proper calibration, and in some instances may cause the meter to explode, resulting in injury to the user. Some meters are equipped with overcurrent protection. However, a current many times greater than the instrument's design limit may still be hazardous.

BASIC METER CONSTRUCTION AND OPERATION

Many meters operate on the principle of *electromagnetic interaction*. This interaction is caused by an electric current flowing through a conductor that is placed between the poles of a permanent magnet. This type of meter is especially suitable for direct current.

Whenever an electric current flows through a conductor, a magnetic force is developed around the conductor. The magnetic force caused by the electric current reacts with the force of the permanent magnet. This causes the indicating needle to move. The larger the amount of current, the farther the needle will move.

The conductor is formed into a coil, which is placed on a pivot between the poles of the permanent magnet. The coil is connected to the terminals of the instrument through two spiral springs. These springs supply a reacting force proportional to the deflec-

Figure 5-1A
Permanent-magnet meter

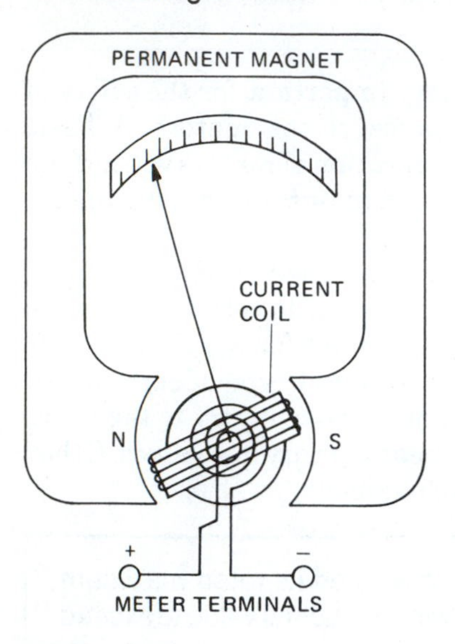

Figure 5-1B
Permanent-magnet meter with external shunt

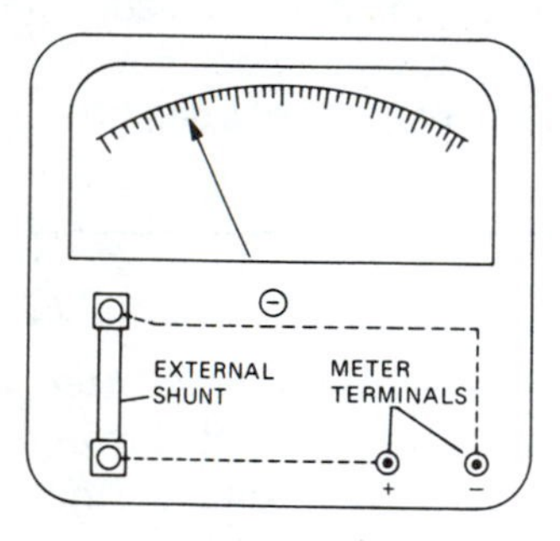

Figure 5-2
Voltmeter-type electrodynamometer instrument

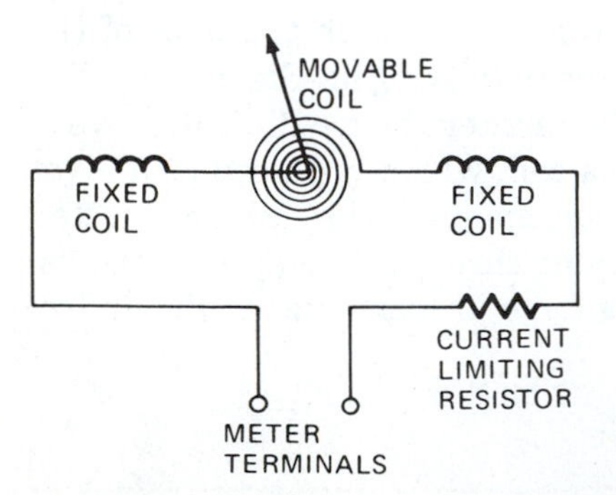

tion. When no current is flowing, the springs cause the needle to return to zero.

The meter scale is designed to indicate the amount of current being measured. The movement of the coil (and thus the movement of the indicating needle) is proportional to the amount of current flowing in the coil. If it is necessary to measure larger currents than the coil can safely carry, a *bypass circuit*, or *shunt*, is included. The shunt may be contained within the meter housing or connected externally.

Example 1

A meter is constructed to measure 10 A on a maximum scale. The coil can safely carry 0.001 A. The shunt must be designed to carry 9.999 A. The meter is designed to indicate 10 A when 0.001 A flows in the coil.

Figure 5-1A illustrates a *permanent-magnet meter*. Figure 5-1B shows an external shunt connected across the meter terminals. The permanent-magnet meter can be used as an ammeter or a voltmeter. When the scale is designed to indicate current and the internal resistance is kept to a minimum, the meter functions as an ammeter. When the scale is designed to indicate voltage, the internal resistance will be relatively high, depending upon the value of voltage for which the meter is designed. *Note*: Regardless of the design, the distance the needle moves is determined by the amount of current flowing in the coil.

A slight change must be made in the design in order to use this type of meter on ac. A *rectifier* is a device that changes ac to dc. It must be incorporated into the meter and the scale must be drawn to indicate the correct value of ac voltage. Rectifier-type ac meters cannot be used on dc and are generally designed as voltage meters.

The *electrodynamometer*, Figure 5-2, is another design for both ammeters and voltmeters that can be used on alternating current. This instrument consists of two stationary coils and one movable coil. The three coils are connected in series with each other through two spiral springs. The springs act upon the movable coil. When current flows through the coils, the movable coil moves in a clockwise direction.

In the electrodynamometer the scale is not divided uniformly, as it is in permanent-magnet meters. The force on the movable coil varies with the square of the current flowing through the coils. This requires that the divisions near the beginning of the scale be made closer together than those near the end. The greater the distance between the divisions, the more accurately one can read the meter. It is important to strive for an accurate reading.

The *moving-vane meter* is another type of construction for ac meters. Current flowing in the coil causes two iron strips (the vanes) to become magnetized. One vane is movable; the other is stationary. The magnetic reaction between the two vanes causes the movable one to turn. The amount of movement depends upon the value of current flowing in the coil.

CAUTION: All of the instruments described depend upon magnetism for their operation, so it is important that they not be placed near other magnets. The magnetic force from another magnet may damage the meter and/or cause incorrect measurements.

USE OF MEASURING INSTRUMENTS

A *voltmeter* is designed to measure the electrical pressure applied to a circuit and/or the voltage drop across a component. Voltmeters must always be connected in parallel with the circuit or the component being measured.

Voltage Tester

The ac-dc *voltage tester* is a rather crude but useful instrument for the electrician. This instrument is designed to indicate approximate values of voltage. The more common types indicate the following values of voltage: ac, 110, 220, 440, and 550 volts; dc, 125, 250, and 600 volts. Many of these instruments also indicate the "polarity" of dc; i.e., which conductor of the circuit is positively or negatively charged.

The voltage tester is used to check common voltages, to identify the grounded conductor, to check for blown fuses, and to distinguish between ac and dc. The voltage tester is small and rugged, making it easier to carry and store than the average voltmeter. Figures 5-3 and 5-4 depict methods for testing fuses with a voltage tester.

To determine which conductor of a circuit or a system is grounded, connect the tester between one conductor and a well-established ground. If the tester indicates a voltage, the conductor is not grounded. Continue this procedure with each conductor until zero voltage is indicated (see Figure 5-5).

To determine the approximate voltage between any two conductors, connect the tester between the two conductors.

CAUTION: Always read and follow the instructions that are supplied with the voltage tester.

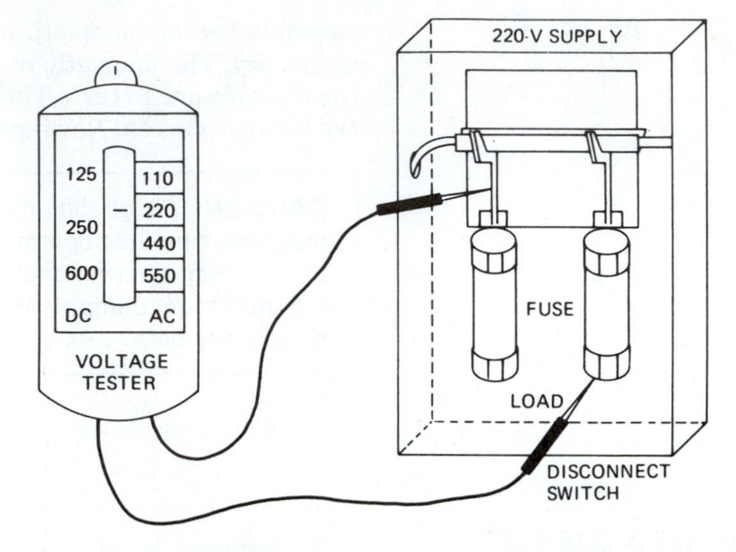

Figure 5-3
Testing cartridge fuses with a voltage tester. The tester indicates 220 V ac. The right-hand fuse is good. If the fuse is blown, the tester will indicate zero voltage.

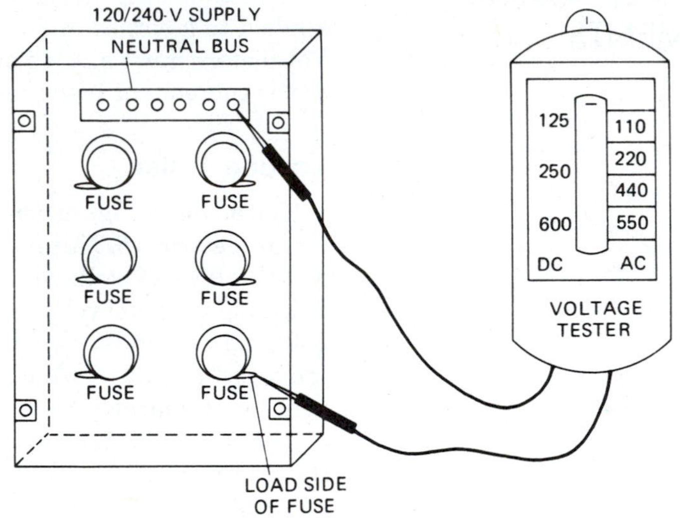

Figure 5-4
Testing plug fuses with a voltage tester. Tester value of zero volt indicates that the fuse is blown.

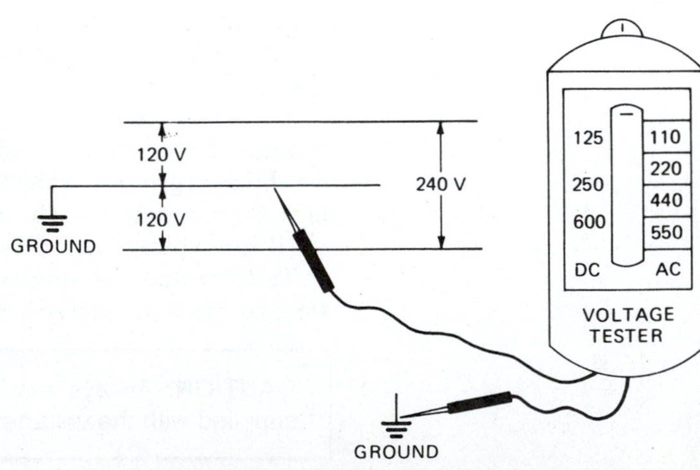

Figure 5-5
Testing to locate the grounded conductor. If the tester indicates zero, the conductor is grounded.

Voltmeter

The voltmeter is a much more accurate measuring instrument than the voltage tester. Because voltmeters are connected in parallel with the circuit or the component being considered, it is necessary that they have relatively high resistance. The internal resistance keeps the current through the meter to a minimum. The lower the value of current through the meter, the less effect it has on the electrical characteristics of the circuit.

The sensitivity (and therefore the accuracy) of the meter is stated in ohms per volt (Ω/V). The higher the ohms per volt, the better the quality of the meter. High values of ohms per volt minimize any change in circuit characteristics.

The average meter used by the electrician is generally between 95 percent and 98 percent accurate. This range of accuracy is satisfactory for most applications. It is very important, however, that the electrical worker strive to obtain the most accurate reading possible. An accurate reading can be obtained by standing directly in front of the meter face and looking directly at it. If the meter has a mirror behind the scale, adjust the angle of sight until there is no reflection of the indicating needle in the mirror. For extreme accuracy, a digital meter may be used, Figure 5-6.

Voltmeters can be used for the same applications as voltage testers. Voltmeters are much more accurate than voltage testers. Therefore, much more information can be obtained. For example,

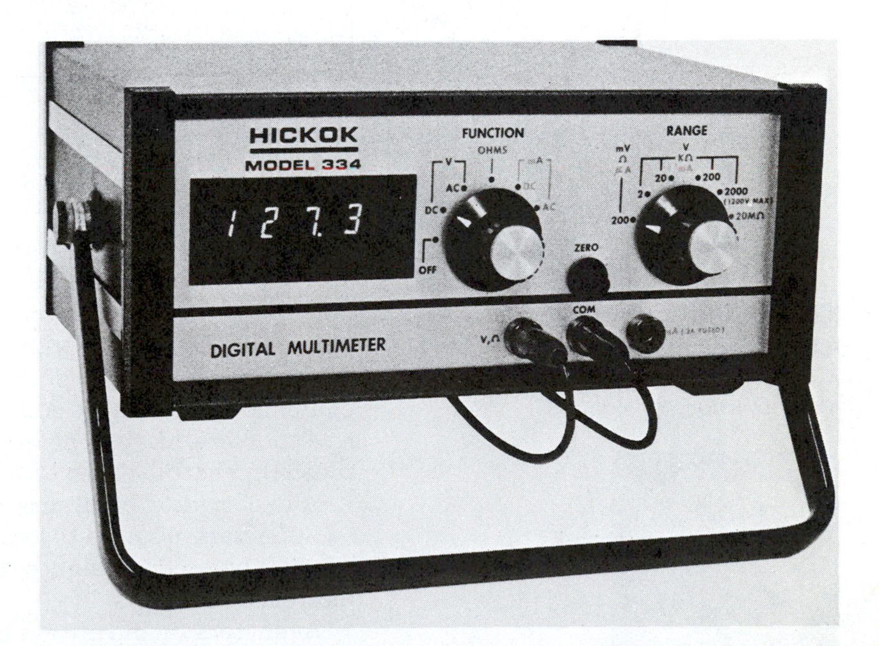

Figure 5-6
Digital multimeter set to indicate voltage. *(Courtesy of Hickok Teaching Systems, Inc., Peabody, MA)*

Figure 5-7
Multirange voltmeter *(Courtesy of Hickok Teaching Systems, Inc., Peabody, MA)*

Figure 5-8A
Ammeter measuring current through a lamp bank

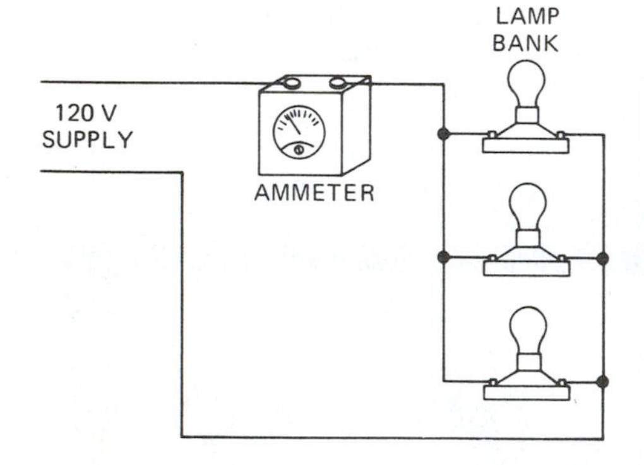

Figure 5-8B
Alternating current ammeter
(Courtesy of Westinghouse Electric Corp., Meter Division, Raleigh, NC)

if the supply voltage to a building is slightly below normal, the voltmeter can indicate this problem. The voltmeter can also be used to determine the amount of voltage drop on feeder and branch circuit conductors.

Voltmeters sometimes have more than one scale. It is very important to select the scale that will provide the most accurate measurement. A range selector switch is provided for this purpose. *Note*: It is advisable to begin with a high scale and work down to the lowest scale so as not to exceed the range limit of any scale. Setting the selector switch on the lowest usable scale provides the most accurate reading.

Before using the meter, check to be sure that the indicating needle is pointing to zero. An adjustment screw is provided just below the face of the meter. A very slight turn will cause the needle to move. The needle can be aligned with the zero line on the scale by turning the screw. Figure 5-7 shows a voltmeter with an indicating needle.

When voltmeters are used on dc, it is very important to maintain proper polarity. Most dc power supplies and meters are color coded to indicate the polarity. Red indicates the positive terminal; black indicates the negative terminal. If the polarity of the circuit or component is unknown, touch the leads to the terminals while observing the indicating needle. If the indicating needle attempts to move backward, the meter lead connections must be reversed.

CAUTION: Do not leave a meter connected with the polarity reversed.

Ammeter

Ammeters are designed to measure the amount of current flowing in a circuit or part of a circuit. They are always connected in series with the circuit or circuit component being considered. The resistance of the meter must be extremely low so it does not restrict the flow of current through the circuit. In the measurement of current flowing through very sensitive equipment, even a slight change in current caused by the ammeter may cause the equipment to malfunction.

Ammeters, like voltmeters, have an adjustment screw to set the indicating needle to zero. Many meters have mirrors to assist the user in obtaining an accurate reading.

Ammeters are used to locate overloads and open circuits. They can also be used to balance the loads on multiwire circuits and to locate malfunctions.

Ammeters should always be connected in series with the circuit or component under consideration. If direct current is being used, always check the polarity. Figure 5-8A shows an ammeter measuring the current through a circuit. Figure 5-8B shows and ac ammeter.

Figure 5-9
Ohmmeter

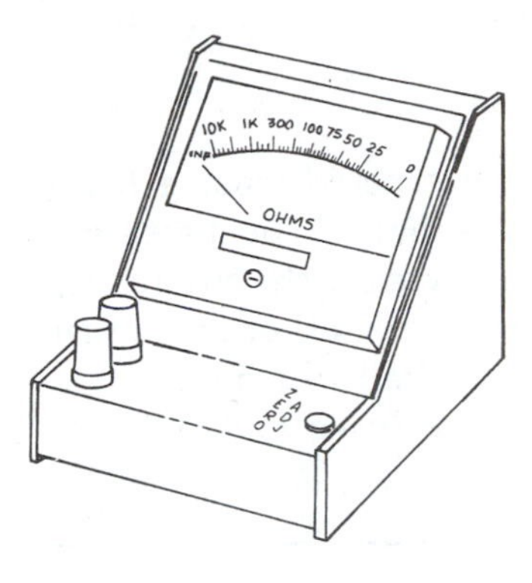

Ohmmeter

An *ohmmeter* is used to measure resistance. Batteries located in the meter case furnish the power for its operation.

> **CAUTION:** It is very important to be sure that the circuit or component is disconnected from its regular power source before connecting the meter. Connecting an ohmmeter to a circuit that has not been deenergized can result in damage to the meter and possible injury to the user.

The ohmmeter scale is designed to be read in the direction opposite to other meters. When the meter circuit is open, the indicating needle should point to infinity. The needle can be aligned with the infinity mark by turning the adjustment screw, Figure 5-9.

Most ohmmeters have several ranges. The range selector switch must be set on the scale that will provide the most accurate measurement. The ranges are generally indicated as follows: R × 1, R × 100, and R × 10,000. If the selector switch is set on R × 1, the value indicated on the scale is the actual value. If the selector switch is set on R × 100, the value indicated on the scale must be multiplied by 100. For R × 10,000, the value must be multiplied by 10,000, Figure 5-10.

Figure 5-10
Multirange ohmmeter *(Courtesy of Hickok Teaching Systems, Inc., Peabody, MA)*

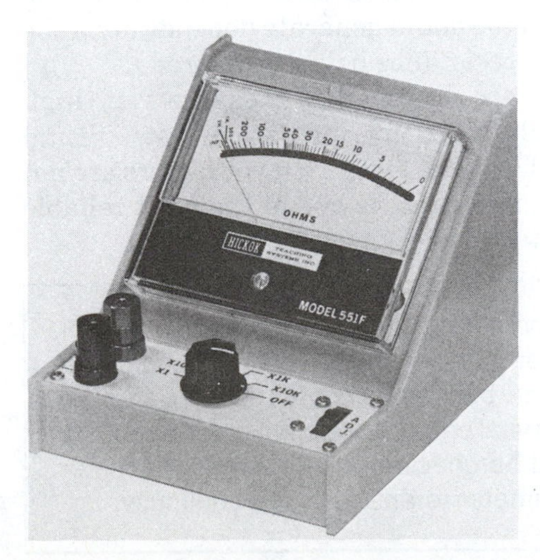

Before a resistance is measured, it is important to check the zero adjustment. The procedure is as follows:

1. Set the range selector switch on the desired range.
2. Connect the test leads together. The indicating needle should point to zero.
3. If the needle is not aligned with the zero line, turn the zero adjustment until the needle and the zero line are aligned. *Note*: Do not leave the test leads connected together, because under zero resistance the batteries will deteriorate rapidly.
4. **CAUTION**: Be sure the circuit is deenergized.
5. Connect the test leads across (in parallel with) the component or circuit to be measured. The needle will indicate the resistance value.

The ohmmeter is an excellent continuity tester. Connect the ohmmeter across the circuit to be tested. If the circuit is complete, the needle will indicate the resistance of the circuit. A reading of zero generally indicates a short circuit. A reading of infinity indicates an open circuit.

Continuity Tester

A simple *continuity tester* can be constructed from a 6-volt battery and bell, Figure 5-11. One terminal of the battery is connected to the bell, and the other is connected to the circuit. The second terminal on the bell is connected to the other side of the circuit. If the current path is complete, the bell will ring.

Continuity testers can be used to check fuses and switches, Figures 5-12 and 5-13.

Figure 5-11
Continuity tester consisting of a 6-V bell and a 6-V battery

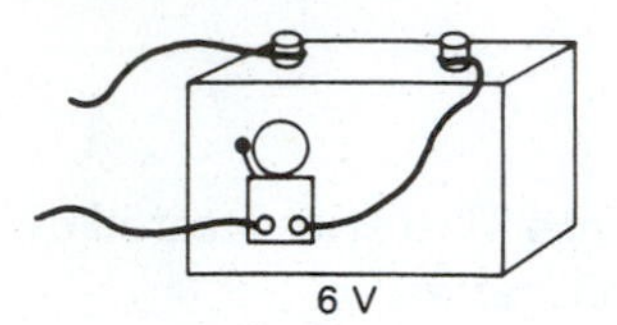

Figure 5-12
Continuity tester consisting of a 6-V bell and a 6-V battery. The tester is connected to a switch. When the circuit is complete (switch contacts closed), the bell will ring.

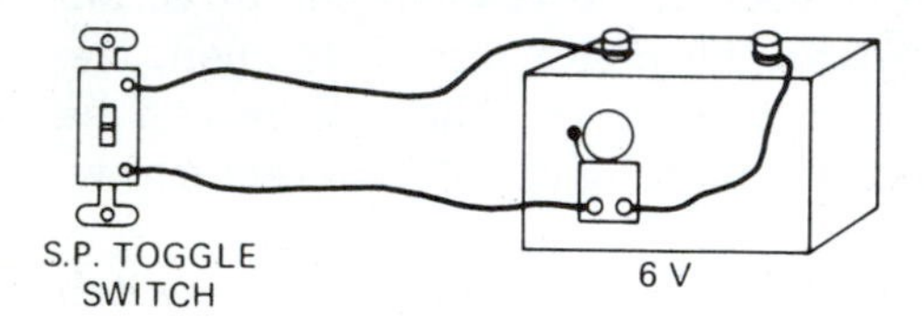

CAUTION: Never attach a continuity tester to a circuit that is energized.

Megohmmeter (MEGGER®)*

A *megohmmeter*, commonly known by the trade name MEGGER®, is an instrument used to measure very high values of resistance. For example, it is used to measure the resistance of the insulation on circuit conductors and motor windings, Figure 5-14. A megohmmeter is designed to measure the resistance in megohms; one *megohm* (MΩ) is equal to one million ohms.

A small generator called a *magneto* is contained within the megohmmeter housing. The magneto furnishes the power for the instrument, just as batteries do for an ohmmeter. The magneto can be hand powered or driven by batteries or other sources of electricity. Megohmmeters have many different voltage ratings. Some of the most common are designed to operate on one of the following values: 500 volts, 1000 volts, and 10,000 volts. The amount of voltage the magneto should generate depends upon the ohmic value and the type of resistance being measured.

Because megohmmeters are designed to measure very high resistances, they are usually used for insulation tests. Visual inspection of insulation and leakage tests with voltmeters are not always reliable. A megohmmeter test is one of the most reliable tests available to the maintenance electrician.

Figure 5-13
Continuity tester connected across a 100-A fuse. If the fuse is good, the bell will ring.

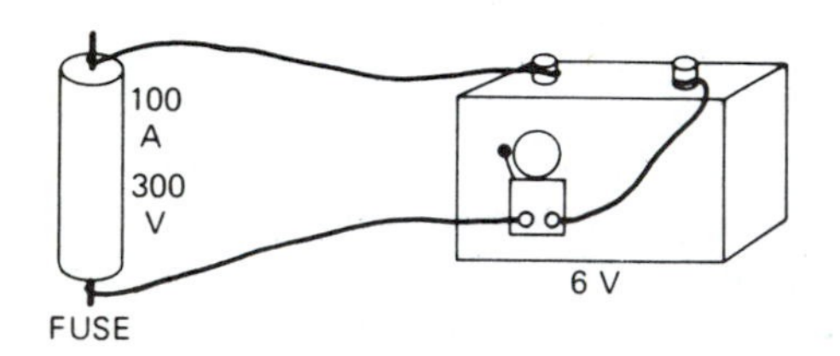

CAUTION: Before a megohmmeter is connected to a conductor or a circuit, the circuit must be deenergized. The testing of the insulation is generally done between each conductor and the ground. A good ground is a vital part of the testing procedure. The ground connection should be checked with the megohmmeter and with a low-range ohmmeter to ensure good continuity.

*MEGGER® is a trade name used by James G. Biddle Co. for its megohmmeter design.

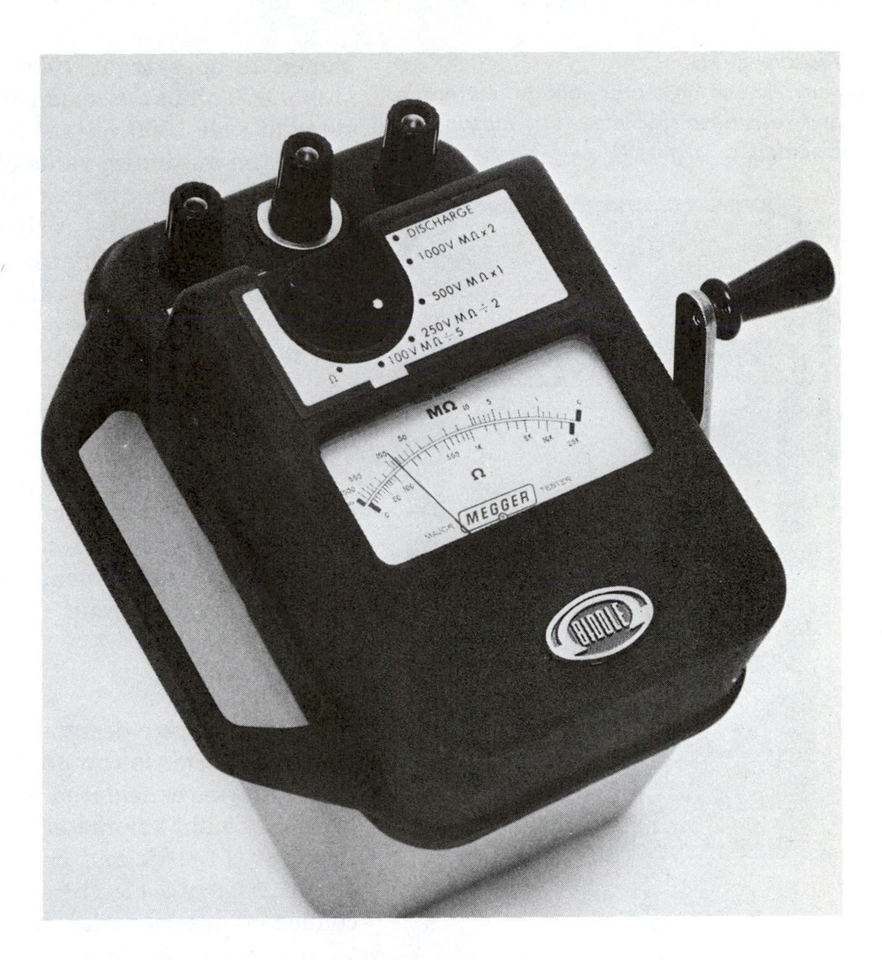

Figure 5-14
Insulation tester *(Courtesy of James G. Biddle Co., Plymouth Meeting, PA)*

Insulation tests should be made at the time of the installation and periodically thereafter. For circuits and equipment rated at 600 volts or less, the 1000-volt magneto can be used. A log (record) should be maintained, noting dates of tests, time of day, temperature, humidity, and resistance values.

Because atmospheric conditions affect the resistance of insulation, many different readings may be obtained over a period of time. This is because the resistance of insulation varies with temperature, humidity, and air quality.

The common enemies of insulation are moisture, dirt, oil, and chemicals. It is important to keep the equipment and conductors as clean and dry as possible. Good maintenance practices and periodic testing of the insulation should be the rule rather than the exception. Megohmmeter scales are usually drawn to indicate a minimum of 10,000 ohms and a maximum of 200 megohms (200,000,000 ohms). (Megohm is abbreviated MΩ.) Conductor insulation designed for 600 volts should indicate a resistance of 600,000 ohms or higher. For motors, generators, transformers, and similar equipment, the minimum resistance for those de-

Figure 5-15
A multimeter measures voltage, current, and resistance. *(Courtesy of Simpson Electric Co., Elgin, IL)*

signed to operate at 1000 volts or less should be 1 megohm (1,000,000 ohms). A good rule of thumb for wires and equipment rated at over 1000 volts is to divide the voltage rating by 1000 to obtain the minimum value of insulation resistance in megohms.

Periodic testing of insulation should take place at least once every 2 months. The resistance values will vary according to temperature and other atmospheric conditions. However, a steady downward trend in the value of the resistance longer than 1 year to 18 months indicates trouble, and the circuit or equipment should be checked out.

Multimeter

Multimeters are designed to measure more than one unit. For example, the volt-ohm-milliammeter, Figure 5-15, measures dc and ac voltages, dc current, and resistance. The advantage of this type of meter is that measurements can be taken with or without deenergizing the circuit.

Wattmeter

Wattmeters are designed to measure electrical power. Because electrical power is the product of current and voltage, the wattmeter may be considered a combination ammeter-voltmeter. It is generally a dynamometer-type instrument constructed as shown in Figures 5-16A and 5-16B. The coils are connected into the circuit, Figure 5-17. The stationary coils, called *current coils*, are connected in series with the load. The movable coil is connected in

Figure 5-16A
Basic construction of a dynamometer-type wattmeter

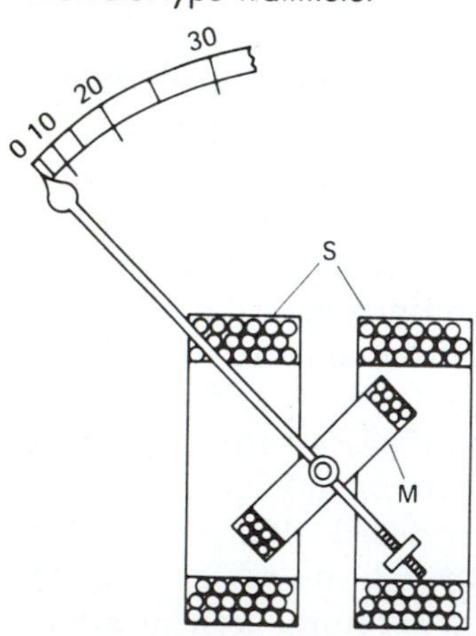

Figure 5-16B
Dynamometer-type wattmeter showing an aluminum vane

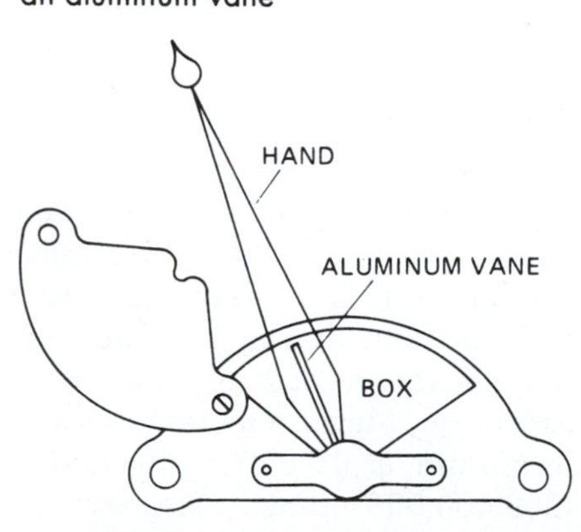

Figure 5-17
Dynamometer type wattmeter showing coil connections

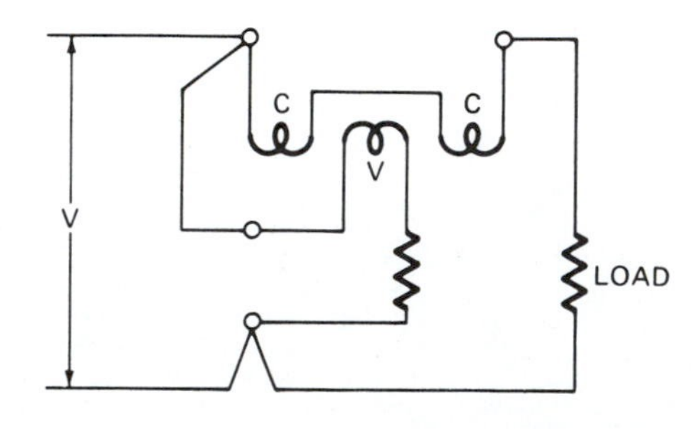

series with a high resistance. This series combination is connected in parallel with the load.

The movable coil is known as the *voltage coil*. The magnetic reaction between the two coils causes the voltage coil to move. The indicating needle is attached to the voltage coil and moves according to the amount of current flowing through both coils. The scale is designed to indicate the power in watts. Figure 5-18 shows a standard wattmeter.

Because of the two circuits in the wattmeter, either coil can burn out under overload. For this reason, current and voltage ratings are marked on the instrument. *Note*: Care must be taken to not exceed these limits.

Figure 5-18
Dynamometer-type wattmeter *(Courtesy of Triplett Corp., Bluffton, OH)*

Watthour Meter

Electrical energy is the product of power and time. The *watthour meter* measures the amount of power expended in a specific amount of time. In a dc watthour meter, the speed is directly proportional to the power. It is a meter that registers the number of watthours or kilowatt-hours delivered to the customer. Because most customers require a large amount of energy, the standard meter is designed to indicate kilowatt-hours. (One kilowatt-hour is equal to one thousand watthours.)

The ac kilowatt-hour meter works on the principle of induction. Moving magnetic fields cause currents to flow in an aluminum disk. These currents, called *eddy currents*, produce magnetic fields that interact with the moving magnetic fields, causing the disk to rotate. The rotating disk drives a gear chain, which in turn drives the indicating needles.

Kilowatt-hour meters have either four or five dials. Each dial has an indicating needle and a scale from 0 to 9. The dials are read from right to left. Beginning with the right hand dial and working to the left, the dials indicate units, tens, hundreds, thousands, and ten thousands. Figure 5-19 shows a five-dial kilowatt-hour meter. In Figure 5-20, the dials of a four-dial meter are shown. The dials indicate 1238 kilowatt-hours. If the indicating needle is between two numbers, the smaller of the two numbers is always read.

Figure 5-19
Kilowatt-hour meter *(Courtesy of General Electric Co., Meter Department, Somersworth, NH)*

Figure 5-20
Kilowatt-hour meter dials. The dials indicate 1238 kW·h of electrical energy.

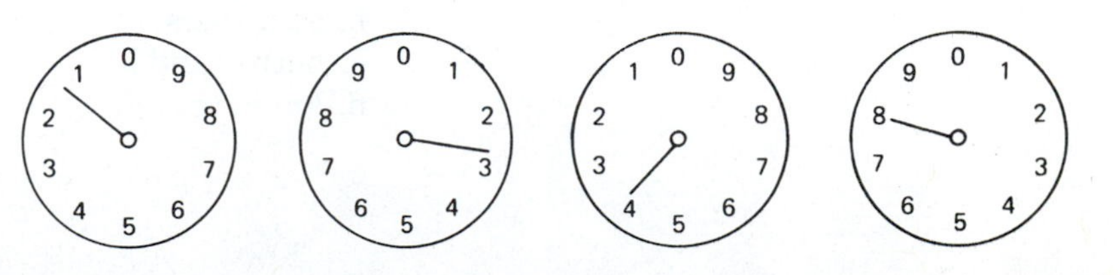

REVIEW

A. Multiple choice.
Select the best answer.

1. One important fact to remember about all meters designed for measuring electrical values is that they are all basically
   a. current meters.
   b. voltage meters.
   c. energy meters.
   d. power meters.

2. Exceeding the maximum range of the meter can damage the
   a. meter case.
   b. meter scale.
   c. terminal connections.
   d. indicating needle.

3. Electromagnetic interaction is caused by
   a. two permanent magnets placed next to one another.
   b. the earth's magnetic force.
   c. an electric current flowing through a conductor placed between the poles of a magnet.
   d. an electric current flowing through a permanent magnet.

4. The indicating needle of an electric meter is attached to the
   a. permanent magnet.
   b. spiral spring.
   c. meter scale.
   d. meter case.

5. A meter shunt is
   a. a special type of ammeter.
   b. a meter bypass circuit.
   c. an overcurrent device.
   d. an overload device.

6. An ammeter has
   a. a very low resistance.
   b. a very high resistance.
   c. no resistance.
   d. a very low conductance.

7. A rectifier is used in
   a. dc meters.
   b. ac meters.
   c. both a and b.
   d. ammeter only.

REVIEW *(continued)*

A. Multiple choice.
Select the best answer (continued).

8. A rectifier unit changes
   a. ac to dc.
   b. low voltage to high voltage.
   c. dc to ac.
   d. high voltage to low voltage.

9. An electrodynamometer-type instrument is a type of
   a. dc meter.
   b. ac meter.
   c. both a and b.
   d. energy meter.

10. A voltage tester is a very
    a. accurate voltmeter.
    b. rugged voltmeter.
    c. sensitive voltmeter.
    d. rare voltmeter.

11. Voltmeters are always connected
    a. in parallel with the component under consideration.
    b. in series with the component under consideration.
    c. next to the component under consideration.
    d. across the line.

12. The sensitivity of a voltmeter is determined by the number of
    a. ohms per watt.
    b. ohms per ampere.
    c. ohms per volt.
    d. amperes per volt.

13. The average meter used by an electrician is
    a. 90 percent accurate.
    b. 80 percent accurate.
    c. 96 percent accurate.
    d. 85 percent accurate.

14. A mirror is used on a meter to
    a. take measurements behind obstacles.
    b. align one's sight for accurate readings.
    c. magnify the numbers.
    d. reflect the numbers.

REVIEW *(continued)*

A. Multiple choice.
Select the best answer (continued).

15. Before using a meter, one should check to be sure that the indicating needle is
    a. pointing to zero.
    b. moving up the scale.
    c. stable.
    d. indicating full-scale deflection.

16. Most dc meters are marked to indicate polarity. The color red indicates the
    a. positive terminal.
    b. negative terminal.
    c. common terminal.
    d. both a and c.

17. A voltmeter has
    a. high resistance.
    b. high conductance.
    c. low resistance.
    d. variable resistance.

18. Ammeters are always connected
    a. in series with the component under consideration.
    b. in parallel with the component under consideration.
    c. to the positive terminal.
    d. across the line.

19. The adjustment screw on a meter is used to
    a. position the indicating needle.
    b. select the range.
    c. adjust the magnetic force.
    d. control the current.

20. An ohmmeter scale reads from
    a. left to right.
    b. right to left.
    c. bottom to top.
    d. top to bottom.

B. Give complete answers.

1. List the common ranges for an ohmmeter.
2. Describe how to check the zero adjustment on an ohmmeter.
3. Explain how to read an ohmmeter.
4. List three uses for the ohmmeter.
5. What is a continuity tester?

REVIEW *(continued)*

B. Give complete answers (continued).

6. Describe how to construct a continuity tester.
7. What is a megohmmeter?
8. Describe the operation of a megohmmeter.
9. List two uses for the megohmmeter.
10. List three uses for a voltage tester.
11. Describe the volt-ohm-milliammeter.
12. What kind of meter is used to measure electrical power?
13. List three factors that affect the resistance of the insulation on conductors.
14. How many coils are there in a wattmeter?
15. What kind of meter is used to measure electrical energy?
16. How many ohms are equal to a megohm?
17. How many watts are equal to a kilowatt?
18. What is the purpose of the adjustment screw on an ohmmeter?
19. List three uses for an ammeter.
20. List five precautions to observe when using an ammeter.
21. Explain how to use the meters listed below in terms of connections, adjustments, range selections, reading the scale, and precautions. Use diagrams if necessary.
    a. Ammeter
    b. Voltmeter
    c. Ohmmeter
    d. Megohmmeter
    e. Wattmeter
22. Explain the basic principle of operation of these meters:
    a. Ammeter
    b. Megohmmeter
    c. Kilowatt-hour meter
23. List three important rules to follow in the care and maintenance of meters.
24. List the most important safety rule to follow when using meters.
25. What is the function of the kilowatt-hour meter?

6

Conductor Types and Sizes

Objectives

After studying this chapter, the student will be able to:

- Define the units of measurement of electrical conductors and calculate their area.
- Describe the tools and methods used to determine the size of electrical conductors.
- Define the term *resistivity* and calculate the resistance of electrical conductors.
- Explain the effect of temperature on the resistance of conductors and calculate any change in resistance.
- Determine the ampacity of conductors under various conditions of use.
- Describe a variety of types of conductors and insulations used in the electrical industry.
- Define and calculate *line drop* and *line loss*.
- Describe and know how to make various types of splices and connections used in the electrical industry.

A *conductor* is any material that offers little opposition to the flow of an electric current. Although a conductor may be in the form of a solid, liquid, or gas, most conductors used in the electrical industry are solids. Copper and aluminum are the most common materials used. Other good conductors are silver, zinc, brass, platinum, iron, nickel, tin, and lead. Electrical conductors used for general transmission and distribution are manufactured in circular shapes known as *wires* and/or *cables* and in rectangular shapes called *bus bars*. Wires, cables, and bus bars are used extensively in the electrical industry.

UNITS OF MEASUREMENT

The size of electrical conductors varies widely according to their use and the amount of current they carry. Therefore, special units of measurement have been established for conductors.

Circular Mils

Many applications require very small wires. A measurement much smaller than the inch is needed. The unit *mil* has been selected as the basis for measuring electrical conductors. One mil is equal to one thousandth of an inch (1/1000 inch or 0.001 inch). Rather than stating that a conductor's diameter is 0.050 inch, it is stated that it measures 50 mils.

In the metric system, there are 1,000,000 *micrometers* (μ) in 1 meter (m). Because there is 0.0254 meter in 1 inch, and 1 mil is equal to 0.001 inch, it can be calculated that 1 mil is equal to 25.4 micrometers. *Note:* Micrometers are sometimes called *microns*.

Conductor sizes are usually expressed as unit area rather than diameter. To measure circular conductors (wires and cables), the unit *circular mil* is used. To measure rectangular conductors (bus bars), the unit *square mil* is used.

One circular mil is the area of a circle one mil in diameter. One square mil is the area of a square in which each side measures one mil. The linear measurement in mils in denoted by the abbreviation "m." The area in circular mils is abbreviated "cm," and square mils is abbreviated "sq m."

> *Note:* In the metric system of measurement, the symbol "m" denotes meter, and the symbol "cm" denotes centimeter. When dealing with the size of conductors, the reader is cautioned to recognize the abbreviations "cm" and "sq m" as circular mil and square mil.

To determine the circular mil area of a wire, the following formula may be used:

$$A = m^2 \qquad \text{(Eq. 6.1)}$$

where A = area, in circular mils (cm)
m = diameter, in mils (m)

Example 1

Calculate the circular mil area of a wire that measures 0.025 in. in diameter.

Customary System

$0.025 \times 1000 = 25$ mils (m)

$A = m^2$

$A = 25^2$

$A = 625$ cm

The circular mil (English units) area of a conductor may be changed to circular millimeters (metric units, abbreviated c mm) by this formula:

$$\text{c mm} = \frac{\text{cm}}{1550} \qquad \text{(Eq. 6.2)}$$

Example 2

Calculate the circular millimeter value of the conductor in Example 1.

$$\text{c mm} = \frac{\text{cm}}{1550}$$

$$\text{c mm} = \frac{625}{1550}$$

$$\text{c mm} = 0.403$$

Example 3

What is the diameter of a wire in inches if it has an area of 5184 cm?

$$m = \sqrt{\text{cm}}$$

$$m = \sqrt{5184}$$

$$m = 72 \text{ mils}$$

$$72 \div 1000 = 0.072 \text{ in.}$$

Example 4

What is the diameter of the wire in Example 3, in millimeters (mm)?

$$1 \text{ in.} = 25.4 \text{ mm}$$

$$0.072 \times 25.4 = 1.829 \text{ mm}$$

Square Mils

Areas of bus bars are more conveniently expressed in square mils. The square mil area of a conductor may be determined by multiplying the thickness in mils by the width in mils. Use the formula

$$A = TW \quad \text{(Eq. 6.3)}$$

where A = area, in square mils (sq m)
T = thickness, in mils (m)
W = width, in mils (m)

Example 5

Calculate the square mil area of a bus bar 1/4 in. thick and 2 in. wide.

(1) Change inches to mils

$$\frac{1}{4} \text{ in.} = 0.25 \text{ in.}$$

$$0.25 \times 1000 = 250 \text{ m}$$

$$2 \times 1000 = 2000 \text{ m}$$

(2) $A = TW$

$$A = 250 \times 2000$$

$$A = 500{,}000 \text{ sq m}$$

Example 6

Determine the number of square mils in a bus bar that measures 1/2 in. by 1 in.

$$\frac{1}{2} \text{ in.} = 0.5 \text{ in.}$$

$$0.5 \times 1000 = 500 \text{ m}$$

$$1 \times 1000 = 1000 \text{ m}$$

$$A = TW$$

$$A = 500 \times 1000$$

$$A = 500{,}000 \text{ sq m}$$

The square mil area of a conductor may be changed to circular mils by the formula

$$\text{cm} = \frac{\text{sq m}}{0.7854} \quad \text{(Eq. 6.4)}$$

Example 7

What is the circular mil area of the conductor in Example 6?

$$\text{cm} = \frac{\text{sq m}}{0.7854}$$

$$\text{cm} = \frac{500{,}000}{0.7854}$$

$$\text{cm} = 636{,}618$$

The square mil area of a conductor may be changed to square millimeters (sq mm) by the formula:

$$\text{sq mm} = \frac{\text{sq m}}{1550} \quad \text{(Eq. 6.5)}$$

Example 8

Calculate the square millimeter area of the conductor in Example 6.

$$\text{sq mm} = \frac{\text{sq m}}{1550}$$

$$\text{sq mm} = \frac{500{,}000}{1550}$$

$$\text{sq mm} = 322.58$$

Example 9

What is the circular mil area of a bus bar that is 1/8 in. (3.175 mm) thick and 1 in. (25.4 mm) wide?

$$1/8 \text{ in.} = 0.125 \text{ in.}$$

$$0.125 \text{ in} \times 1000 = 125 \text{ m}$$

$$1 \times 1000 = 1000 \text{ m}$$

$$A = TW$$

$$A = 125 \times 1000$$

$$A = 125{,}000 \text{ sq m}$$

$$\text{cm} = \frac{\text{sq m}}{0.7854}$$

$$\text{cm} = \frac{125{,}000}{0.7854}$$

$$\text{cm} = 159{,}154$$

Figure 6-1
Gauge for measuring wire sizes *(Courtesy of the L. S. Starrett Co., Athol, MA)*

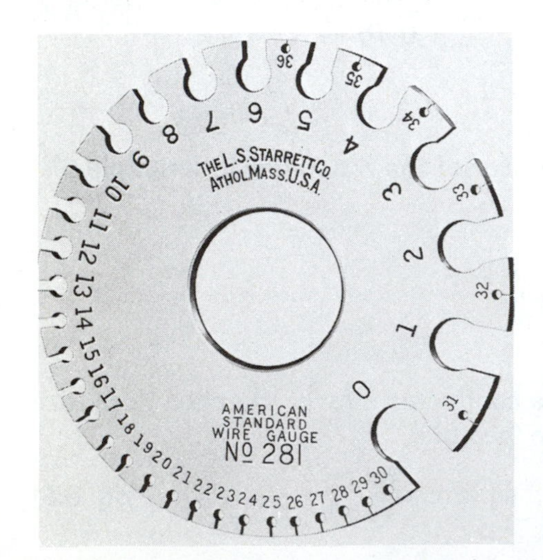

American Wire Gauge

A special scale has been established for the more common wires used in the electrical industry. The common name for this scale is the *American Wire Gauge* (AWG). The wire sizes range from No. 50 AWG, which is the smallest standard size, to No. 4/0 AWG, which is the largest standard size for this scale. Figure 6-1 illustrates the instrument used to measure the gauge of a wire. All wire sizes larger and smaller than the AWG scale are expressed in circular mils.

The *NEC*® lists standard wire sizes from No. 18 AWG to No. 0000 (4/0, or four aught) AWG. Also listed are the standard sizes from 250,000 circular mils (250 MCM) to 2,000,000 circular mils (2000 MCM). (MCM means thousands of circular mils.) Number 18 wire has an area of 1620 circular mils and 4/0 has an area of 211,600 circular mils. Table 6-1 in this text is taken from *Chapter 9, Table 8* of the 1993 *NEC*®.

TABLE 6-1: Properties of conductors *(Source: Chapter 9, Table 8, 1993 National Electrical Code®)*

Table 8. Conductor Properties

Size AWG/ kcmil	Area Cir. Mills	Conductors				DC Resistance at 75°C (167°F)		
		Stranding		Overall		Copper		Aluminum
		Quan-tity	Diam. In.	Diam. In.	Area In.^{2}	Uncoated ohm/kFT	Coated ohm/kFT	ohm/ kFT
18	1620	1	—	0.040	0.001	7.77	8.08	12.8
18	1620	7	0.015	0.046	0.002	7.95	8.45	13.1
16	2580	1	—	0.051	0.002	4.89	5.08	8.05
16	2580	7	0.019	0.058	0.003	4.99	5.29	8.21
14	4110	1	—	0.064	0.003	3.07	3.19	5.06
14	4110	7	0.024	0.073	0.004	3.14	3.26	5.17
12	6530	1	—	0.081	0.005	1.93	2.01	3.18
12	6530	7	0.030	0.092	0.006	1.98	2.05	3.25
10	10380	1	—	0.102	0.008	1.21	1.26	2.00
10	10380	7	0.038	0.116	0.011	1.24	1.29	2.04
8	16510	1	—	0.128	0.013	0.764	0.786	1.26
8	16510	7	0.049	0.146	0.017	0.778	0.809	1.28
6	26240	7	0.061	0.184	0.027	0.491	0.510	0.808
4	41740	7	0.077	0.232	0.042	0.308	0.321	0.508
3	52620	7	0.087	0.260	0.053	0.245	0.254	0.403
2	66360	7	0.097	0.292	0.067	0.194	0.201	0.319
1	83690	19	0.066	0.332	0.087	0.154	0.160	0.253
1/0	105600	19	0.074	0.373	0.109	0.122	0.127	0.201
2/0	133100	19	0.084	0.419	0.138	0.0967	0.101	0.159
3/0	167800	19	0.094	0.470	0.173	0.0766	0.0797	0.126
4/0	211600	19	0.106	0.528	0.219	0.0608	0.0626	0.100
250	—	37	0.082	0.575	0.260	0.0515	0.0535	0.0847
300	—	37	0.090	0.630	0.312	0.0429	0.0446	0.0707
350	—	37	0.097	0.681	0.364	0.0367	0.0382	0.0605
400	—	37	0.104	0.728	0.416	0.0321	0.0331	0.0529
500	—	37	0.116	0.813	0.519	0.0258	0.0265	0.0424
600	—	61	0.099	0.893	0.626	0.0214	0.0223	0.0353
700	—	61	0.107	0.964	0.730	0.0184	0.0189	0.0303
750	—	61	0.111	0.998	0.782	0.0171	0.0176	0.0282
800	—	61	0.114	1.03	0.834	0.0161	0.0166	0.0265
900	—	61	0.122	1.09	0.940	0.0143	0.0147	0.0235
1000	—	61	0.128	1.15	1.04	0.0129	0.0132	0.0212
1250	—	91	0.117	1.29	1.30	0.0103	0.0106	0.0169
1500	—	91	0.128	1.41	1.57	0.00858	0.00883	0.0141
1750	—	127	0.117	1.52	1.83	0.00735	0.00756	0.0121
2000	—	127	0.126	1.63	2.09	0.00643	0.00662	0.0106

These resistance values are valid ONLY for the parameters as given. Using conductors having coated strands, different stranding type, and, especially, other temperatures changes the resistance.

Formula for temperature change: $R_2 = R_1 [1 + \alpha(T_2 - 75)]$ where: $\alpha_{cu} = 0.00323$, $\alpha_{AL} = 0.00330$.

Conductors with compact and compressed stranding have about 9 percent and 3 percent, respectively, smaller bare conductor diameters than those shown. See Table 5A for actual compact cable dimensions.

The IACS conductivities used: bare copper = 100%, aluminum = 61%.

Class B stranding is listed as well as solid for some sizes. Its overall diameter and area is that of its circumscribing circle.

(FPN): The construction information is per NEMA WC8-1976 (Rev 5=1980). The resistance is calculated per National Bureau of Standards Handbook 100, dated 1966, and Handbook 109, dated 1972.

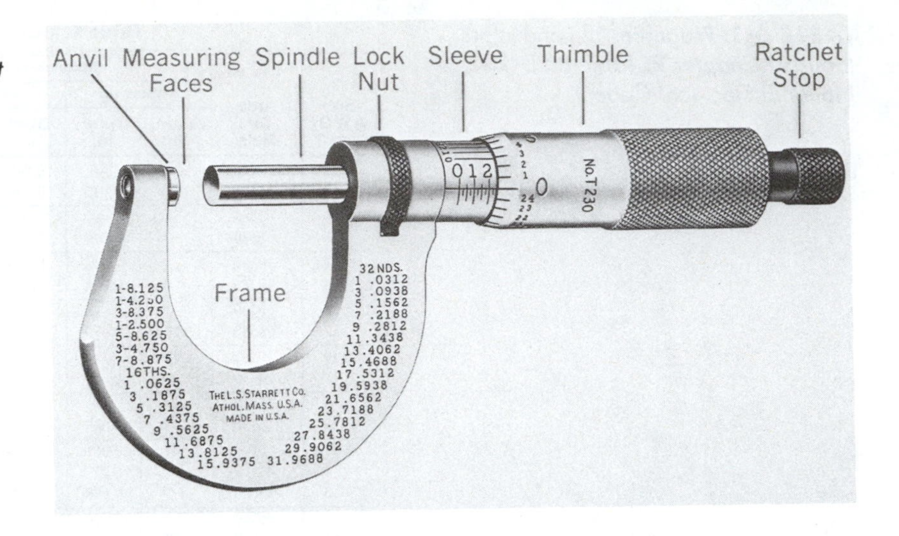

Figure 6-2
Micrometer *(Courtesy of the L. S. Starrett Co., Athol, MA)*

Rectangular conductors are measured with a micrometer, Figure 6-2. Micrometers are very accurate measuring instruments and are calibrated in 1/1000-inch increments. Determining conductor size is relatively easy because 1 mil is equal to 1/1000 inch. Therefore, each 1/1000 inch on the micrometer is equal to 1 mil. (*Note:* Do not confuse the instrument "micrometer" with the unit of measurement "micrometer.")

To measure a rectangular conductor, measure the width and thickness and then multiply the values to obtain the square mil area. Micrometers can also be used to measure circular conductors. Measure the diameter and then square the number to obtain the circular mil area.

All electrical personnel should be able to use a micrometer and an American Wire Gauge.

MIL-FOOT WIRE

A wire one mil in diameter and one foot long is called a *mil-foot of wire*, Figure 6-3. One foot of any wire may be considered to be composed of a number of mil-foot wires in parallel. In other words, a foot of wire with a cross-sectional area of 5 circular mils equals five 1 mil-foot wires.

Figure 6-3
One mil foot of wire

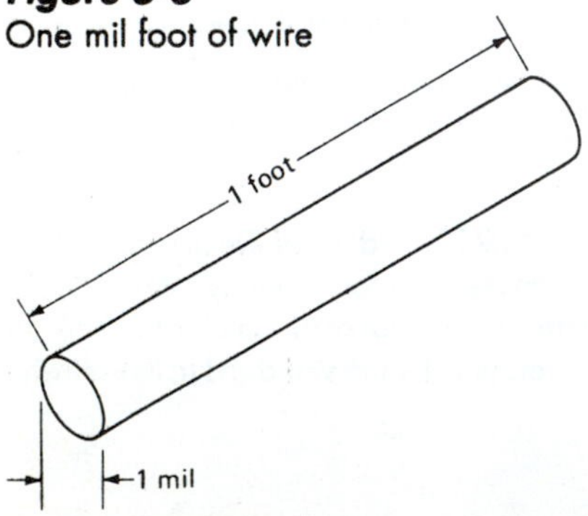

The resistance of any number of identical wires in parallel is equal to the resistance of 1 wire divided by the total number of wires. For example, the resistance of 1 mil-foot of copper building wire is 10.4 ohms. A wire of the same material and length, but 10 circular mils in cross section, is 10.4 ohms ÷ 10 circular mils, or 1.04 ohms. The resistance of wire is directly proportional to its length and inversely proportional to its cross-sectional area. If the 10 circular-mils wire is 1000 feet long, its resistance is 1.04 ohms × 1000 feet, or 1040 ohms. (See Equation 6.6.)

RESISTIVITY

The *resistivity* (K) of a conducting material is the resistance per mil-foot at a temperature of 68° Fahrenheit (F), or 20° Celsius (C). The resistance of the conductor depends upon the material from which it is made, its cross-sectional area, its length, and the operating temperature. (The operating temperature is the ambient temperature plus the increase in temperature caused by the current flow.)

Table 6-2 lists the resistivity of various conducting materials. To calculate the resistance of any conductor at 68°F (20°C), divide the resistance of 1 mil-foot by the cross-sectional area (in circular mils) and multiply by its length in feet. From this information the following equation can be constructed:

$$R = \frac{K\ell}{A} \qquad \text{(Eq. 6.6)}$$

where R = resistance of the conductor, in ohms (Ω)
K = resistivity
ℓ = length, in feet (ft)
A = area, in circular mils (cm)

Example 10

Determine the resistance of a copper wire 0.15 in. in diameter and 1 mile long.

$$0.15 \times 1000 = 150 \text{ m}$$
$$150 \times 150 = 22{,}500 \text{ cm}$$
$$R = \frac{K\ell}{A}$$
$$R = \frac{10.4 \times 5280}{22{,}500}$$
$$R = 2.44\ \Omega$$

Example 11

Calculate the length of a copper wire 0.025 in. in diameter, if it has a resistance of 5 Ω.

$$0.025 \times 1000 = 25 \text{ m}$$
$$25 \times 25 = 625 \text{ cm}$$
$$R = \frac{K\ell}{A}$$
$$R = \frac{10.4\ell}{625}$$
$$5 = \frac{10.4\ell}{625}$$
$$10.4\ell = 3125$$
$$\ell = 300 \text{ ft}$$

TABLE 6-2: Resistivity and temperature coefficients of conducting materials

Materials	Resistivity (K) at 20°C or 68°F	Temperature Coefficient (a) per Degree C.*
Aluminum	17.7	0.0043
Carbon	20,000	−0.0005
Constantan	296.	0.
Copper	10.4	0.0043
German Silver Wire	200	0.0004
Iron Wire	60	0.006
Iron (cast)	500	0.0008
Manganin	266	0.00002
Nichrome	660	0.0002
Nickel	60	0.006
Silver	9.5	0.004
Steel (soft)	90	0.0044
Steel (hard)	275	0.0016
Tungsten (annealed)	26	0.005
Tungsten (hard drawn)	33	0.005

*Average of values between 0° and 100° centigrade.

Example 12

What is the resistance of a feeder consisting of two cooper bus bars, each 250 ft long? Each bar is 1/4 in. thick and 2 in. wide.

$$0.25 \times 1000 = 250 \text{ m}$$
$$2 \times 1000 = 2000 \text{ m}$$
$$250 \times 2000 = 500{,}000 \text{ sq m}$$

$$\text{cm} = \frac{\text{sq m}}{0.7854}$$

$$\text{cm} = \frac{500{,}000}{0.7854}$$

$$\text{cm} = 636{,}618$$

$$R = \frac{K\ell}{A}$$

$$R = \frac{10.4 \times (2 \times 250)}{636{,}618}$$

$$R = 0.008\ \Omega$$

THERMAL EFFECT

The average temperature coefficients of various materials between 0°C and 100°C (32°F and 212°F) are listed in Table 6-2. The temperature coefficient is the amount by which the resistance of a material changes for each degree of change in temperature, for each ohm of resistance.

The resistance of pure metals such as silver, copper, and aluminum increases as the temperature increases. The resistance of some metal alloys, such as constantan and manganin, remains almost unaffected by the temperature. Other materials, such as carbon and most electrolytes, react in an opposite way to pure metals in that their resistance decreases as the temperature increases.

To calculate the resistance of a material at a specific temperature, use the following formula:

$$R_o = R + Rat_o \textit{ or } R_o = R(1 + at_o) \qquad \text{(Eq. 6.7)}$$

where Ro = resistance at the operating temperature
R = resistance at zero degrees Celsius (0°C)
a = temperature coefficient at zero degrees Celsius (°C)
t_o = operating temperature, in degrees Celsius (°C)

Example 13

The resistance of a nichrome heating element is 50 Ω at 0°C. What is its resistance at 1000°C?

$R_o = R(1 + at_o)$
$R_o = 50[1 + (0.0002 \times 1000)]$
$R_o = 50(1 + 0.2)$
$R_o = 50 \times 1.2$
$R_o = 60\ \Omega$

Example 14

A carbon electrode has a resistance of 0.02 Ω at 0°C. What is its resistance at 1500°C?

$R_o = R(1 + at_o)$
$R_o = 0.02[1 + (-0.0005 \times 1500)]$
$R_o = 0.02[1 + (-0.75)]$
$R_o = 0.02 \times 0.25$
$R_o = 0.005\ \Omega$

Example 15

The resistance of a coil wound with copper wire is 100 Ω at 0°C. What is the resistance at 60°C?

$R_o = R(1 + at_o)$
$R_o = 100[1 + (0.0043 \times 60)]$
$R_o = 100(1 + 0.258)$
$R_o = 100 \times 1.258$
$R_o = 125.8\ \Omega$

If the operating temperature is expressed in degrees Fahrenheit, the temperature can be converted to degrees Celsius by using a temperature conversion chart or the formula

$$°C = \frac{5}{9}(°F-32) \qquad \text{(Eq. 6.8)}$$

It is frequently necessary to determine the operating resistance when a change in temperature occurs from a value other than zero degrees Celsius. This may be accomplished as follows:

$$R = \frac{R_i}{1 + at_i} \qquad \text{(Eq. 6.9)}$$

where R = resistance at zero degrees Celsius (0°C)
R_i = resistance at the initial temperature
a = temperature coefficient at zero degrees Celsius (0°C)
t_i = initial temperature, in degrees Celsius (°C)

Example 16

The resistance of a coil wound with iron wire is 200 Ω at 20°C. What is the resistance at 80°C?

$$R = \frac{R_i}{1 + at_i}$$

$$R = \frac{200}{1 + (0.006 \times 20)}$$

$$R = \frac{200}{1.12}$$

$$R = 178.571\ \Omega \text{ at } 0°C$$

$$R_o = R(1 + at_o)$$
$$R_o = 178.571[1 + (0.006 \times 80)]$$
$$R_o = 178.571[1 + 0.48]$$
$$R_o = 178.571 \times 1.48$$
$$R_o = 264.3\ \Omega \text{ at } 80°C$$

It is not necessary to first find the resistance at zero degrees Celsius if the two formulas are combined into one. The new formula becomes

$$R_o = \frac{R_i(1 + at_o)}{1 + at_i} \qquad \text{(Eq. 6.10)}$$

Example 17

Solve the problem in Example 16, using Equation 6.10.

$$R_o = \frac{R_i(1 + at_o)}{1 + at_i}$$

$$R_o = \frac{R_i(1 + 0.006t_o)}{(1 + 0.006t_i)}$$

Example 17 continues on following page

$$R_o = \frac{200(1 + 0.006 \times 80)}{1 + 0.006 \times 20}$$

$$R_o = \frac{200 \times 1.48}{1.12}$$

$$R_o = 264.3\ \Omega$$

Example 18

The copper winding of a motor has a resistance of 10 Ω when the machine is started. The resistance increases to 12 Ω after it has run several hours at full load. If the room temperature is 20°C, what is the temperature rise of the winding?

$$R_o = \frac{R_i(1 + at_o)}{1 + at_i}$$

$$R_o = \frac{R_i(1 + 0.0043t_o)}{(1 + 0.0043t_i)}$$

$$12 = \frac{10(1 + 0.0043t_o)}{(1 + 0.0043 \times 20)}$$

$$10(1 + 0.0043t_o) = 12(1 + 0.0043 \times 20)$$

$$(1 + 0.0043t_o) = \frac{12(1 + 0.0043 \times 20)}{10}$$

$$1 + 0.0043t_o = \frac{13.032}{10}$$

$$1 + 0.0043t_o = 1.3032$$

$$0.0043t_o = 0.3032$$

$$t_o = 70.5°C$$

The temperature rise is 70.5°C – 20°C = 50.5°C.

Chapter 9, Table 8, of the 1993 *NEC*® indicates the resistance of copper and aluminum wires at 75°C. From this information, the resistance at any other temperature can be calculated by using Equation 6.10.

INSULATION AND AMPACITY OF CONDUCTORS

Heat is generated whenever current flows through a conductor. The amount of heat depends upon the value of current and the resistance of the conductor. As the heat is generated, the operating temperature of the conductor increases. This rise in temperature continues until the rate at which heat leaves the conductor equals the rate at which heat is produced. When the two are equal, the temperature remains constant.

The amount of current a conductor can safely carry is called its *ampacity*. The ampacity of a conductor is limited by the temperature that its insulation can tolerate without deteriorating and/or losing its insulating quality. When considering the ampacity of a conductor, it is also necessary to consider its operating temperature. *Table 310-17* of the *NEC®*. lists the allowable ampacity of insulated conductors having an insulation voltage rating from zero volt to 2000 volts, and installed as a single conductor in free air (not enclosed). It is based on an ambient temperature of 30°C. The maximum allowable operating temperature is listed at the top of each column. This value is determined by the type of insulation used. The operating temperatures range from 60°C to 90°C.

Tables 6-3 and 6-4 are taken from *NEC® Tables 310-16* and *310-17*. These tables list the allowable ampacities of the same types of conductors. Table 6-3 is for conductors installed in a raceway, a cable, or earth. Table 6-4 applies to single insulated conductors in free air. The values in Table 6-3 apply only when there are not more than three conductors in the raceway, cable, or earth. Ambient and operating temperatures are the same for both tables. The ampacity ratings in Table 6-3 are lower because heat cannot be dissipated as rapidly when the conductors are grouped and/or placed in an enclosure.

Correction Factors

The operating temperature of a conductor depends upon the ambient temperature and the amount of current flowing in the conductor. For example, Table 6-3 indicates that insulation types TW and UF have a maximum operating temperature of 60°C (140°F). Insulations that are subjected to temperatures higher than their maximum operating temperatures will deteriorate.

The values indicated in Tables 6-3 and 6-4 are for use where the ambient temperature does not exceed 30°C (86°F). For conductors installed in areas where the ambient temperature is above 30°C (86°F), the ampacity is determined by applying the correction factor indicated at the bottom of the table.

Example 19

A No. 12 AWG copper conductor is covered with RH insulation. If this conductor is installed in an area where the ambient temperature is 122°F (50°C), what is the ampacity of the conductor?

Table 6-3 indicates that the load current rating of No. 12 RH copper wire is 20 A, (see footnote) provided that not more than three conductors are installed in the same raceway and the maximum ambient temperature does not exceed 86°F (30°C). The correction factor for 122°F (50°C) is 0.75. Therefore,

$$20 \times 0.75 = 15 \text{ A}$$

The safe ampacity for the conductor in this example is 15 A.

TABLE 6-3: Allowable ampacities of insulated conductors rated 0 to 2000 volts, 60°C to 90°C *(Source: Article 310, Table 310-16, 1993 National Electrical Code®)*

Table 310-16. Allowable Ampacities of Insulated Conductors Rated 0-2000 Volts, 60° to 90°C (140° to 194°F) Not More Than Three Conductors in Raceway or Cable or Earth (Directly Buried), Based on Ambient Temperature of 30°C (86°F)

Size	Temperature Rating of Conductor. See Table 310-13.						Size
	60°C (140°F)	75°C (167°F)	90°C (194°F)	60°C (140°F)	75°C (167°F)	90°C (194°F)	
AWG kcmil	TYPES TW†, UF†	TYPES FEPW†, RH†, RHW†, THHW†, THW†, THWN†, XHHW† USE†, ZW†	TYPES TA, TBS, SA SIS, FEP†, FEPB†, MI RHH†, RHW-2, THHN†, THHW†, THW-2, THWN-2, USE-2, XHH, XHHW† XHHW-2, ZW-2	TYPES TW†, UF†	TYPES RH†, RHW†, THHW†, THW†, THWN†, XHHW†, USE†	TYPES TA, TBS, SA, SIS, THHN†, THHW†, THW-2, THWN-2, RHH†, RHW-2 USE-2 XHH, XHHW XHHW-2, ZW-2	AWG kcmil
	COPPER			ALUMINUM OR COPPER-CLAD ALUMINUM			
18	....	....	14	....	....	....	....
16	....	....	18	....	....	....	....
14	20†	20†	25†	....	....	....	....
12	25†	25†	30†	20†	20†	25†	12
10	30	35†	40†	25	30†	35†	10
8	40	50	55	30	40	45	8
6	55	65	75	40	50	60	6
4	70	85	95	55	65	75	4
3	85	100	110	65	75	85	3
2	95	115	130	75	90	100	2
1	110	130	150	85	100	115	1
1/0	125	150	170	100	120	135	1/0
2/0	145	175	195	115	135	150	2/0
3/0	165	200	225	130	155	175	3/0
4/0	195	230	260	150	180	205	4/0
250	215	255	290	170	205	230	250
300	240	285	320	190	230	255	300
350	260	310	350	210	250	280	350
400	280	335	380	225	270	305	400
500	320	380	430	260	310	350	500
600	355	420	475	285	340	385	600
700	385	460	520	310	375	420	700
750	400	475	535	320	385	435	750
800	410	490	555	330	395	450	800
900	435	520	585	355	425	480	900
1000	455	545	615	375	445	500	1000
1250	495	590	665	405	485	545	1250
1500	520	625	705	435	520	585	1500
1750	545	650	735	455	545	615	1750
2000	560	665	750	470	560	630	2000

CORRECTION FACTORS

Ambient Temp. °C	For ambient temperatures other than 30°C (86°F), multiply the allowable ampacities shown above by the appropriate factor shown below.						Ambient Temp. °F
21-25	1.08	1.05	1.04	1.08	1.05	1.04	70-77
26-30	1.00	1.00	1.00	1.00	1.00	1.00	78-86
31-35	.91	.94	.96	.91	.94	.96	87-95
36-40	.82	.88	.91	.82	.88	.91	96-104
41-45	.71	.82	.87	.71	.82	.87	105-113
46-50	.58	.75	.82	.58	.75	.82	114-122
51-55	.41	.67	.76	.41	.67	.76	123-131
56-60	....	.58	.71	....	.58	.71	132-140
61-70	....	.33	.58	....	.33	.58	141-158
71-80	....	....	.41	....	....	.41	159-176

†Unless otherwise specifically permitted elsewhere in this Code, the overcurrent protection for conductor types marked with an obelisk (†) shall not exceed 15 amperes for No. 14, 20 amperes for No. 12, and 30 amperes for No. 10 copper; or 15 amperes for No. 12 and 25 amperes for No. 10 aluminum and copper-clad aluminum after any correction factors for ambient temperature and number of conductors have been applied.

TABLE 6-4: Allowable ampacities of insulated conductors rated 0 through 2000 volts, 60°C to 90°C *(Source: Article 310, Table 310-17, 1993 National Electrical Code®)*

Table 310-17. Allowable Ampacities of Single Insulated Conductors, Rated 0 through 2000 Volts, in Free Air Based on Ambient Air Temperature of 30°C (86°F)

Size	Temperature Rating of Conductor. See Table 310-13.						Size
	60°C (140°F)	75°C (167°F)	90°C (194°F)	60°C (140°F)	75°C (167°F)	90°C (194°F)	
AWG kcmil	TYPES TW†, UF†	TYPES FEPW†, RH†, RHW†, THHW†, THW†, THWN†, XHHW† ZW†	TYPES TA, TBS, SA SIS, FEP†, FEPB†, MI, RHH†, RHW-2, THHN†, THHW†, THW-2, THWN-2, USE-2, XHH, XHHW†, XHHW-2, ZW-2	TYPES TW†, UF†	TYPES RH†, RHW†, THHW†, THW†, THWN†, XHHW†	TYPES TA, TBS, SA, SIS, THHN†, THHW†, THW-2, THWN-2, RHH†, RHW-2, USE-2, XHH, XHHW†, XHHW-2, ZW-2	AWG kcmil
	COPPER			ALUMINUM OR COPPER-CLAD ALUMINUM			
18	....	....	18	....	....	....	....
16	....	....	24	....	....	....	....
14	25†	30†	35†	....	....	....	....
12	30†	35†	40†	25†	30†	35†	12
10	40†	50†	55†	35†	40†	40†	10
8	60	70	80	45	55	60	8
6	80	95	105	60	75	80	6
4	105	125	140	80	100	110	4
3	120	145	165	95	115	130	3
2	140	170	190	110	135	150	2
1	165	195	220	130	155	175	1
1/0	195	230	260	150	180	205	1/0
2/0	225	265	300	175	210	235	2/0
3/0	260	310	350	200	240	275	3/0
4/0	300	360	405	235	280	315	4/0
250	340	405	455	265	315	355	250
300	375	445	505	290	350	395	300
350	420	505	570	330	395	445	350
400	455	545	615	355	425	480	400
500	515	620	700	405	485	545	500
600	575	690	780	455	540	615	600
700	630	755	855	500	595	675	700
750	655	785	885	515	620	700	750
800	680	815	920	535	645	725	800
900	730	870	985	580	700	785	900
1000	780	935	1055	625	750	845	1000
1250	890	1065	1200	710	855	960	1250
1500	980	1175	1325	795	950	1075	1500
1750	1070	1280	1445	875	1050	1185	1750
2000	1155	1385	1560	960	1150	1335	2000

CORRECTION FACTORS

Ambient Temp. °C	For ambient temperatures other than 30°C (86°F), multiply the allowable ampacities shown above by the appropriate factor shown below.						Ambient Temp. °F
21-25	1.08	1.05	1.04	1.08	1.05	1.04	70-77
26-30	1.00	1.00	1.00	1.00	1.00	1.00	78-86
31-35	.91	.94	.96	.91	.94	.96	87-95
36-40	.82	.88	.91	.82	.88	.91	96-104
41-45	.71	.82	.87	.71	.82	.87	105-113
46-50	.58	.75	.82	.58	.75	.82	114-122
51-55	.41	.67	.76	.41	.67	.76	123-131
56-60	....	.58	.71	....	.58	.71	132-140
61-70	....	.33	.58	....	.33	.58	141-158
71-80	....	....	.41	....	....	.41	159-176

†Unless otherwise specifically permitted elsewhere in this Code, the overcurrent protection for conductor types marked with an obelisk (†) shall not exceed 15 amperes for No. 14, 20 amperes for No. 12, and 30 amperes for No. 10 copper; or 15 amperes for No. 12 and 25 amperes for No. 10 aluminum and copper-clad aluminum.

A correction factor must also be applied when there are more than three conductors in the raceway.

Example 20

If the raceway in Example 19 contains six conductors, what is the ampacity of the conductors?

The correction factor for more than three conductors is listed in the notes to *NEC® Tables 310-16* through *310-19*. The factor indicated for six conductors is 80% or 0.8, so that

$$20 \times 0.75 \times 0.8 = 12\text{ A}$$

The safe ampacity for this conductor is 12 A.

Types of Insulations

Plastic is by far the most common insulating material used on conductors. For general-purpose wiring in buildings, types TW, THW, and THWN are used more than any other insulations. Tables 6-5 and 6-6 list specifications for various types of wires.

Insulations such as TFN and MTW are designed for special purposes. Type TFN is designed for wiring lighting and similar fixtures. Type MTW is used for wiring machine tools and other applications exposed to harsh liquids. Tables 6-7 and 6-8 are specification tables for types TFN, TFFN, and MTW.

NEC® Table 310-13 lists the various insulations that meet the *Code* standards. This table provides information as to the maximum operating temperatures, types of installations (dry or wet), and maximum voltages. For installing electrical conductors, it is necessary to consider all factors that may affect the size and material of the conductors and the type of insulation used.

TABLE 6-5: Specifications for Type TW Wire *(Courtesy of Cyprus Wire and Cable Company, Rome Cable Corporation, Rome, NY)*

SPEC 2005
September 1,1981
Supersedes Issue
June 1, 1980

ROME TW

PVC Insulation, 600 Volts

APPLICATION: General purpose wiring for lighting and power — residential, commercial, industrial buildings in accordance with National Electrical Code, maximum conductor temperature of 60°C in wet or dry locations, for circuits not exceeding 600 volts.

STANDARDS:

1. Listed by Underwriters Laboratories as Type TW per UL Standard 83 for Thermoplastic-Insulated Wires.
2. All sizes carry the VW-1 designation.
3. Conforms to Federal Specification J-C-30A.
4. Cable complies with the requirements of OSHA when installed and used in accordance with the NEC.

CONSTRUCTION: Annealed uncoated copper conductor, PVC insulation, surface printed.

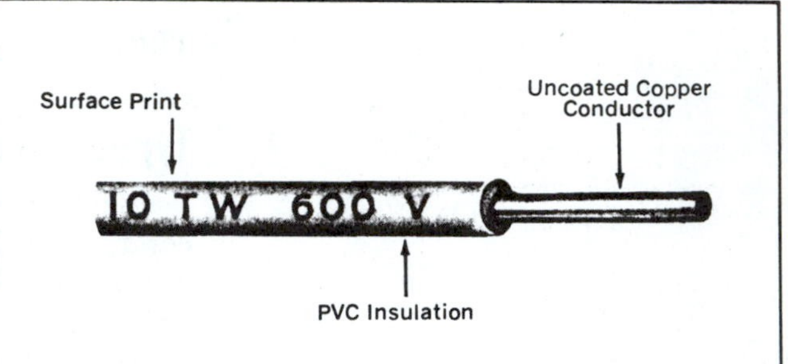

Size AWG	No. of Strands	Insulation Thickness Mils	Nom. Diam. Inches	NEC Ampacity*	Approx. Wt. Lb./1000 Ft.		Put-up	Stock Items①											
					Net	Shipping		1	2	3	4	5	6	7	8	9	10	11	12
Solid																			
14	Solid	30	.13	20†	19	19	500′ Carton 2500′ NR Reel	S	S	S	S	S	S	S	S			S	
12	Solid	30	.15	25†	26	27	500′ Carton 2500′ NR Reel	S	S	S	S	S	S	S	S	S		S	
10	Solid	30	.17	30†	39	41	500′ Carton 2500′ NR Reel	S	S	S	S	S							
Stranded																			
14	7	30	.14	20†	19	20	500′ Carton 2500′ NR Reel	S	S	S	S	S	S	S	S				
12	7	30	.16	25†	28	29	500′ Carton 2500′ NR Reel	S	S	S	S	S	S	S	S				
10	7	30	.18	30†	41	44	500′ Carton 2500′ NR Reel	S	S	S	S	S							
8	7	45	.24	40	67	70	500′ Carton 2500′ NR Reel	S	S	S	S								
6	7	60	.30	55	109	111	500′ Coil	S											
4	7	60	.35	70	162	164	500′ Coil												
2	7	60	.41	95	245	247	500′ Coil												

*Ampacity in accordance with NEC for not more than three conductors in raceway, 60°C conductor temperature and 30°C ambient in wet or dry locations.

†The load current rating and the over current protection shall not exceed 15 amperes for 14 AWG, 20 amperes for 12 AWG and 30 amperes for 10 AWG copper.

NOTES: ① Color Code: 1 black, 2 white, 3 red, 4 blue, 5 green, 6 yellow, 7 orange, 8 brown, 9 purple, 10 pink, 11 gray, 12 tan.

Information on this sheet subject to change without notice.

TABLE 6-6: Specifications for Types USE, RHW, and RHH Wire *(Courtesy of Cyprus Wire and Cable Company, Rome Cable Corporation, Rome, NY)*

Rome Cable
CORPORATION

SPEC 2150

September 1, 1981
Supersedes Issue
June 1, 1980

ROME USE or RHW or RHH

Rome-XLP Insulation, 600 Volts

APPLICATION:

A—Where NEC jurisdiction applies; as Type USE for direct burial, 75°C in wet locations; and as Type RHW, 75°C in wet locations, or Type RHH, 90°C in dry locations, installed in air, conduit, or other recognized raceway. Sizes 250 MCM and larger may be installed in cable tray per Article 318 of the NEC.

B—Otherwise, for general purpose applications at maximum conductor temperature of 90°C in wet or dry locations — including conduit, duct, tray, trough, direct burial, and aerial installations.

STANDARDS:

1. Listed by Underwriters Laboratories as Type USE per Standard 854 and as Types RHW or RHH per Standard 44.
2. Conforms to ICEA Pub. No. S-66-524 utilizing Column A thicknesses.
3. Conform to Federal Specification J-C-30A.
4. Sizes 12-4 AWG stranded copper approved under FAA Advisory Circular 150/5345-7D per Spec L-824 Airport Lighting Cable, Type C.
5. Cable complies with the requirements of OSHA when installed and used in accordance with the NEC.

CONSTRUCTION: Annealed copper or aluminum conductor, Rome-XLP thermosetting chemically crosslinked polyethylene insulation, surface printed.

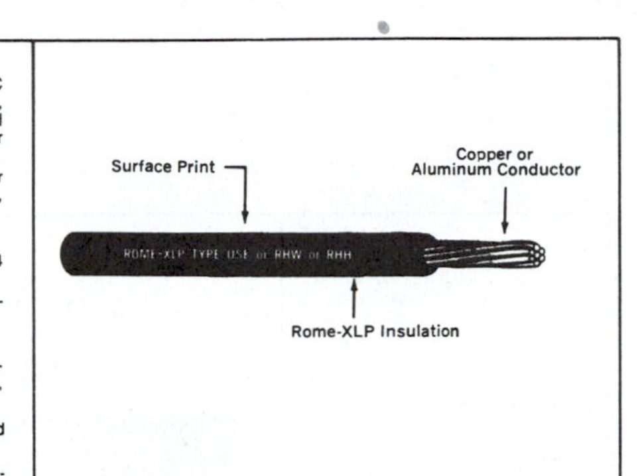

Size AWG or MCM	No. of Strands	Insulation Thickness Mils	Nom. Diam. Inches	Copper NEC Ampacity* 90°C RHH	Copper NEC Ampacity* 75°C RHW USE	Copper Approx. Wt. Lb./1000 Ft. Net	Copper Approx. Wt. Lb./1000 Ft. Shipping	Copper Put-up		Copper Stock Items	Aluminum NEC Ampacity* 90°C RHH	Aluminum NEC Ampacity* 75°C RHW USE	Aluminum Approx. Wt. Lb./1000 Ft. Net	Aluminum Approx. Wt. Lb./1000 Ft. Shipping	Aluminum Put-up	Aluminum Stock Items
Solid																
12	Solid	45	.18	30†	25†	31	33		NS NR reel	S	—	—	—	—	—	—
10	Solid	45	.20	40†	35†	44	46	500′ ctn	NS NR reel	S	—	—	—	—	—	—
8	Solid	60	.26	55	50	68	73	—	—	—	—	—	—	—	—	—
Stranded																
12	7	45	.19	30†	25†	31	33	500′ ctn	NS NR reel	S	—	—	—	—	—	—
10	7	45	.21	40†	35†	47	49	500′ ctn	NS NR reel	S	—	—	—	—	—	—
8	7	60	.27	55	50	71	79	500′ ctn	1000′ NR reel	S	—	—	—	—	—	—
6	7	60	.31	75	65	105	115	500′ coil	1000′ NR reel	S	60	50	50	58	—	—
4	7	60	.36	95④	85④	155	175	500′ coil	1000′ NR reel	S	75	65	68	87	—	—
2	7	60	.42	130④	115④	240	255	500′ coil	1000′ NR reel	S	100④	90④	97	115	—	—
1	19	80	.49	150④	130④	310	330	—	1000′ NR reel	S	115④	100④	130	150	—	—
1/0	19	80	.53	170④	150④	380	405	—	1000′ NR reel	S	135④	120④	155	175	—	—
2/0	19	80	.58	195④	175④	470	495	—	1000′ NR reel	S	150④	135④	190	210	—	—
3/0	19	80	.63	225	200	585	615	—	1000′ NR reel	S	175④	155④	230	260	—	—
4/0	19	80	.69	260	230	730	765	—	1000′ NR reel	S	205④	180④	275	315	—	—
250	37	95	.76	290	255	870	925	—	NS NR reel	S	230	205	335	370	—	—
300	37	95	.81	320	285	1030	1090	—	—	—	255	230	380	450	—	—
350	37	95	.86	350	310	1190	1250	—	NS NR reel	S	280	250	445	505	—	—
400	37	95	.91	380	335	1350	1410	—	—	—	305	270	500	555	—	—
500	37	95	.99	430	380	1670	1760	—	1000′ NR reel	S	350	310	610	695	—	—
600	61	110	1.10	475	420	2010	2110	—	—	—	385	340	715	815	—	—
750	61	110	1.20	535	475	2500	2600	—	NS NR reel	S	435	385	895	1000	—	—
1000	61	110	1.35	615	545	3290	3420	—	—	—	500	445	1160	1290	—	—
1250	91	125	1.51	665	590	4130	4310	—	—	—	545	485	1440	1620	—	—
1500	91	125	1.63	705	625	4930	5110	—	—	—	585	520	1700	1880	—	—
1750	127	125	1.74	735	650	5720	5990	—	—	—	615	545	1960	2230	—	—
2000	127	125	1.85	750	665	6510	6910	—	—	—	630	560	2220	2620	—	—

*AMPACITY in accordance with NEC for not more than three conductors in raceway or cable or direct burial. As RHH: 90°C conductor temperature and 30°C ambient in dry locations. As USE or RHW: 75°C conductor temperature and 30°C ambient in wet or dry locations.

†The load current rating and the over current protection shall not exceed 20 amperes for 12 AWG and 30 amperes for 10 AWG copper.

NOTES: ①Available in black only.

②On non-stock items, contact Rome Cable for minimum acceptable manufacturing quantities. All stocking items are available in long NS (non-standard lengths) on NR reels.

③Sizes #12-4 AWG stranded copper approved under FAA Advisory Circular 150/5345-7D per Spec L-824—Airport Lighting Cable, Type C.

④For three wire, single phase residential service, allowable ampacities are as follows:

Size AWG	Copper	Aluminum
4	100	—
2	125	100
1	150	110
1/0	175	125
2/0	200	150
3/0	—	175
4/0	—	200

Information on this sheet subject to change without notice.

TABLE 6-7: Specifications for Types TFN and TFFN Wire
(Courtesy of Cyprus Wire and Cable Company, Rome Cable Corporation, Rome, NY)

SPEC 5115
June 1, 1980
Initial Issue
March 1, 1975

ROME TFN or TFFN FIXTURE WIRE

PVC — Nylon, 600 Volts

Multi-Rated Wire

APPLICATION: For wiring of lighting and similar fixtures in accordance with the National Electrical Code, 90C maximum conductor temperature, 600 volts. Also used for wiring of machine tools (TFFN construction), appliances and control circuits not exceeding 600 volts.

STANDARDS: Listed by Underwriters Laboratories as follows:
(a) Type TFN — 18 and 16 AWG solid and 7-strand, 90C
(b) Type TFFN — 18 and 16 AWG flexible strand, 90C
(c) Gasoline and Oil Resistant II
(d) Type MTW — 90C Machine Tool Wire (TFFN construction only)
(e) AWM Style 1316—105C, 80C where exposed to oil.

CONSTRUCTION: Annealed uncoated copper conductor PVC insulation, surface printed, nylon jacket.

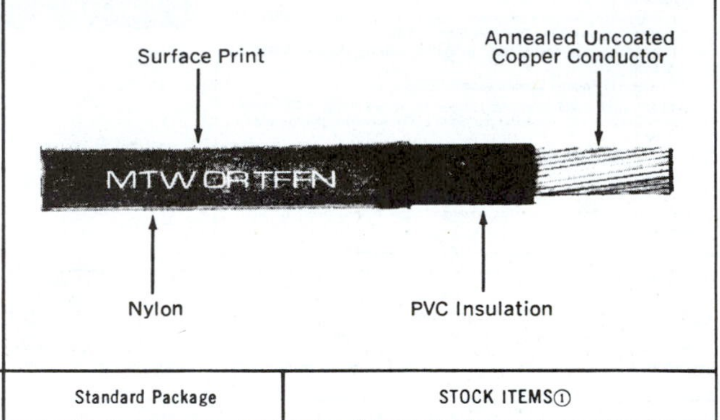

Size AWG	No. of Strands	Thickness in Mils		Nominal Diam. Inches	NEC Ampacity*	Approx. Net. Wt. Lb./1000 Ft.	Standard Package		STOCK ITEMS①											
		PVC Insulation	Nylon Jacket				Length	Put-up	1	2	3	4	5	6	7	8	9	10	11	12
Type TFN																				
18	solid	15	4	.09	6	7	1000′ spls.	2 per ctn.	S	S	S	S	S	S	S	S				
18	7	15	4	.10	6	8	1000′ spls.	2 per ctn.	S	S	S	S	S	S	S	S				
16	solid	15	4	.10	8	11	1000′ spls.	2 per ctn.	S	S	S	S	S	S	S	S				
16	7	15	4	.11	8	12	1000′ spls.	2 per ctn.	S	S	S	S	S	S	S	S				
Type MTW or TFFN																				
18	19	15	4	.10	6	8	1000′ spls.	2 per ctn.	S	S	S	S	S	S	S	S	S	S	S	
16	26	15	4	.11	8	12	1000′ spls.	2 per ctn.	S	S	S	S	S	S	S	S	S	S	S	

①Color Code: 1 Black, 2 White, 3 Red, 4 Blue, 5 Green, 6 Yellow, 7 Orange, 8 Brown, 9 Purple, 10 Pink, 11 Gray, 12 Tan.

*Ampacity in accordance with Article 402 of the NEC.

Information on this sheet subject to change without notice.

TABLE 6-8: Specifications for Type MTW Wire *(Courtesy of Cyprus Wire and Cable Company, Rome Cable Corporation, Rome, NY)*

SPEC 5160
June 1, 1980
Supersedes Issue
January 1, 1976

ROME MACHINE TOOL WIRE

PVC Insulation, 600 Volts
Multi-Rated Wire

APPLICATIONS: For wiring of machine tools, appliances and control circuits not exceeding 600 volts. Suitable for use as Machine Tool Wire at 90°C and lower temperatures in dry locations and at 60°C where exposed to moisture, oil or coolants — such as cutting oils.

STANDARDS:

1. Conforms to Underwriters Laboratories Standard 1063 for Machine Tool Wires and Cables.
2. Also multi-listed by Underwriters Laboratories as follows:
   (a) AWM — 105°C in air, 60°C in oil or wet locations UL Style No's 1230, 1231, 1232
   (b) AWM — 105°C in air, 60°C in oil UL Style No's 1015, 1028, 1283
   (c) AWM — 90°C in air, 60°C in oil UL Style No's 1013, 1024, 1026
   (d) Meets "VW-1" flame resistant requirements of UL.
3. Approved by Canadian Standards Association as 105°C Appliance Wiring Material rated 600 volts.
4. Conforms to Joint Industrial Council Electrical Standards for General Purpose Machine Tools.
5. Conforms to National Fire Protection Association Electrical Standard #79 for Metal Working Machine Tools.

CONSTRUCTION: Annealed uncoated copper conductor, PVC insulation, surface printed.

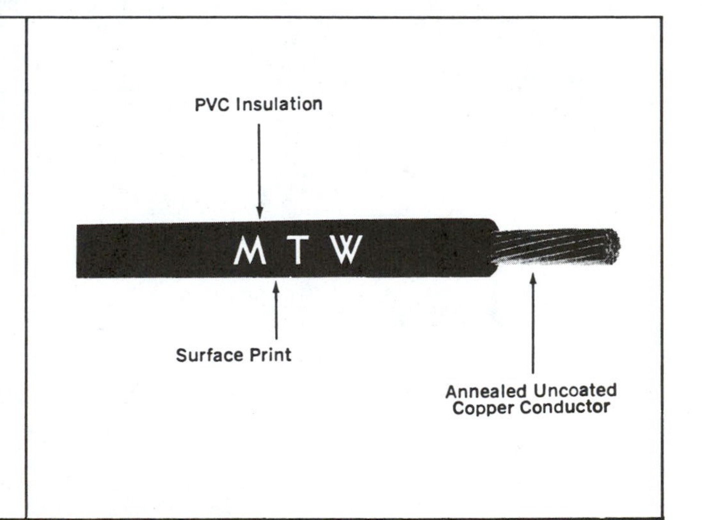

Size AWG	No. of Strands	Insulation Thickness Mils	Nominal Diam. Inches	Approx. Net Wt. Lb./1000 Ft.	UL AWM Style Numbers	Standard Package		Stock Items①											
						Length	Put-up	1	2	3	4	5	6	7	8	9	10	11	12
Multi-Listed as: MTW or 105°C AWM or 90°C AWM																			
14	19	30	.14	18	1230, 1015, 1013	500′ spls.	4 per ctn.	S	S	S	S	S	S	S	S				
12	19	30	.16	27	1230, 1015, 1013	500′ spls.	4 per ctn.	S	S	S	S	S	S	S	S				
10	19	30	.18	40	1230, 1015, 1013	500′ spls.	4 per ctn.	S	S	S	S	S							
8	19	45	.24	69	1231, 1028, 1024	500′ spl.	1 per ctn.	S	S	S									
Multi-Listed as: MTW or 105°C AWM or 90°C AWM																			
14	19	45	.17	22	1231, 1028, 1024	500′ spls.	4 per ctn.												
12	19	45	.19	31	1231, 1028, 1024	500′ spls.	4 per ctn.												
10	19	45	.21	45	1231, 1028, 1024	500′ spl.	1 per ctn.												
8	19	60	.27	75	1232, 1283, 1026	500′ spl.	1 per ctn.												
6	19	60	.31	110	1232, 1283, 1026	500′	coil	S	S	S									
4	19	60	.36	160	1232, 1283, 1026	500′	coil	S											
2	19	60	.42	245	1232, 1283, 1026	500′	coil	S	S										

① Color Code: 1 Black, 2 White, 3 Red, 4 Blue, 5 Green, 6 Yellow, 7 Orange, 8 Brown, 9 Purple, 10 Pink, 11 Gray, 12 Tan.

Information on this sheet subject to change without notice.

FLEXIBLE CORDS AND CABLES

Some of the most common types of cords and cables used in the electrical industry are lamp cord, tinsel cord, heater cord, vacuum cleaner cord, hard service cord, range and dryer cable, data processing cable, and elevator cable. The conductors in these cords and cables are made up of fine strands in order to obtain maximum flexibility.

Flexible cords and cables are permitted for use as pendants, fixture wiring, connections to portable lamps and/or appliances, elevators, cranes and hoists, and some stationary equipment.

They may not be used for fixed wiring of a building. *NEC® Article 400* covers the use of flexible cords. Various types of cords are shown in Figures 6-4A, 6-4B, and 6-4C.

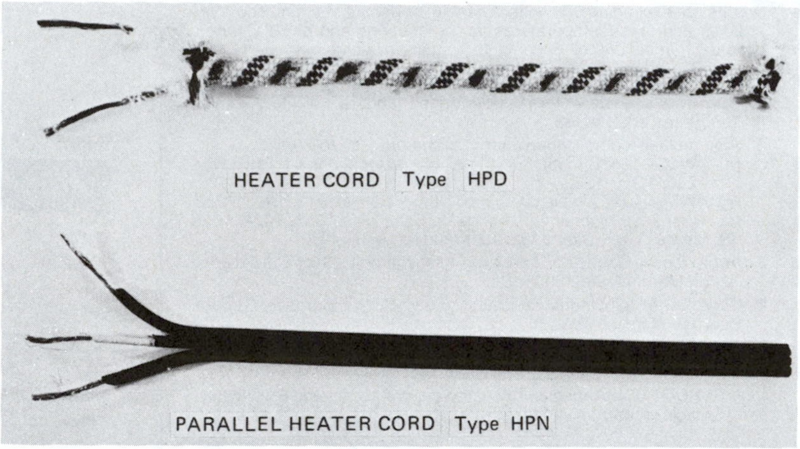

Figure 6-4A
Flexible heater cords

Figure 6-4B
Parallel cords

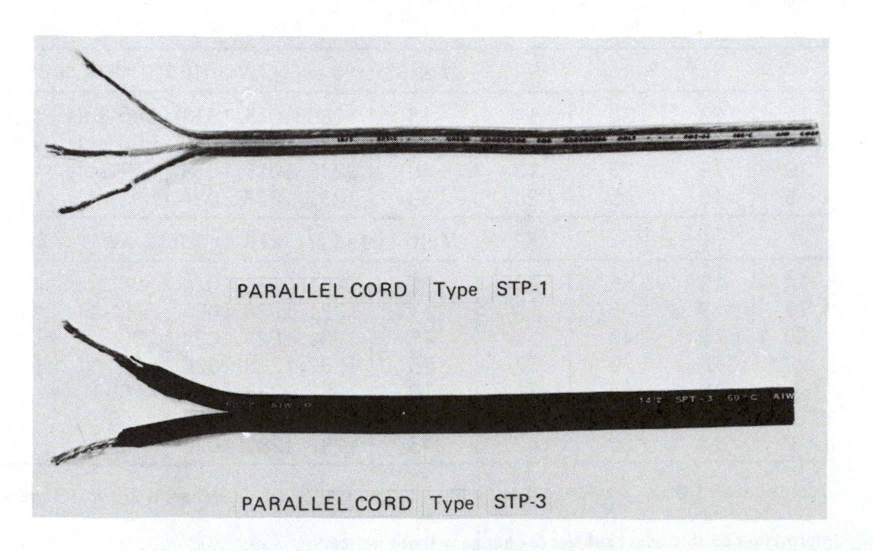

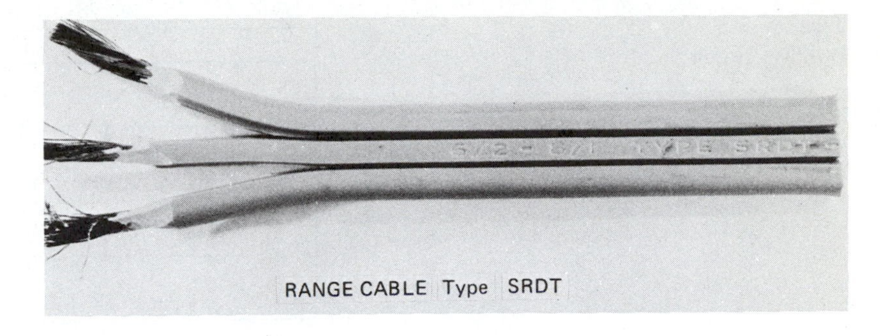

Figure 6-4C
Range cable

ELECTRICAL DISTRIBUTION

Conductors for distributing electricity to various parts of a building are classified as feeders and branch circuits. A *feeder circuit* consists of the conductors between the service equipment and the final branch circuit overcurrent device. A *branch circuit* consists of all the conductors between the final circuit overcurrent protective device and the load.

Transmitting electricity in large buildings frequently requires that conductors be installed over long distances. It is necessary to consider the distance between the supply and the load because the resistance of a conductor increases with its length.

Line Drop

The resistance of the conducting wires can become so great that it affects the operating characteristics of the equipment. An electrical pressure (voltage) is required to force the current through the resistance of the wires. This extra voltage must be supplied in addition to the rated voltage of the load. This extra pressure is called *line drop*, the voltage drop caused by current flowing though the resistance of the wires.

The *NEC*® allows a maximum voltage (line) drop of 3 percent for branch circuits and 5 percent for combined feeder and branch circuit conductors. To calculate the percentage of voltage drop, the voltage must be measured at the source and at the point farthest from the source while the circuit is fully loaded. The following equation is used:

$$\%\text{ voltage drop} = \frac{E_1 - E_2}{E_1} \times 100 \qquad \text{(Eq. 6.11)}$$

where E_1 = source voltage
E_2 = voltage at the load

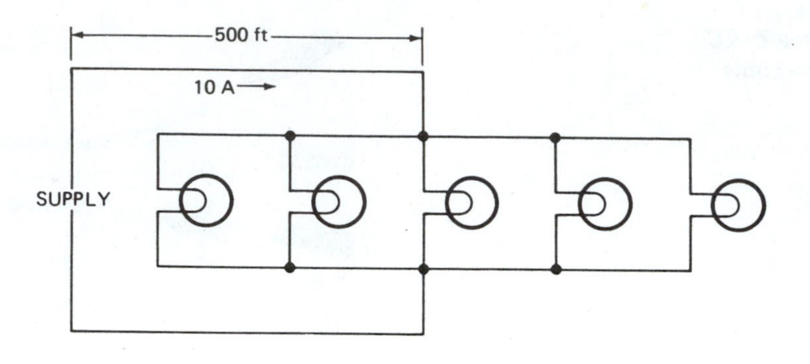

Figure 6-5
Lighting circuit

Example 21

The source voltage of a circuit is 120 V, and at the point farthest from the source, a voltmeter indicates 118 V at full load. Calculate the percent voltage drop caused by the circuit conductors.

$$\% \text{ voltage drop} = \frac{E_1 - E_2}{E_1} \times 100$$

$$\% \text{ voltage drop} = \frac{120 - 118}{120} \times 100$$

$$\% \text{ voltage drop} = 1.67\%$$

Example 22

A lighting load is located 500 ft (152.4 m) from the supply. If the circuit conductors are No. 12 AWG copper wire, how much voltage is necessary to force 10 A through the conductors? (See Figure 6-5.)

Five hundred feet (152.4 m) of No. 12 copper wire has a resistance of 0.965 Ω at 167°F (75°C). (See Table 6-1.)

Figure 6-6 illustrates an equivalent circuit to Figure 6-5. The following procedure may be used to calculate the voltage (line) drop.

$$E_{L1} = IR_{L1} \qquad E_{L2} = IR_{L2} \qquad E_{Lt} = E_{L1} + E_{L2}$$

$$E_{L1} = 10 \times 0.965 \qquad E_{L2} = 10 \times 0.965 \qquad E_{Lt} = 9.65 + 9.65$$

$$E_{L1} = 9.65 \text{ V} \qquad E_{L2} = 9.65 \text{ V} \qquad E_{Lt} = 19.3 \text{ V}$$

To force 10 A through the conductors, 19.3 V are necessary. Therefore, the voltage line drop is 19.3 V.

Example 23

If the supply voltage for the circuit in example 22 is 120 V, what is the voltage across the load?

$$120 - 19.3 = 100.7 \text{ V}$$

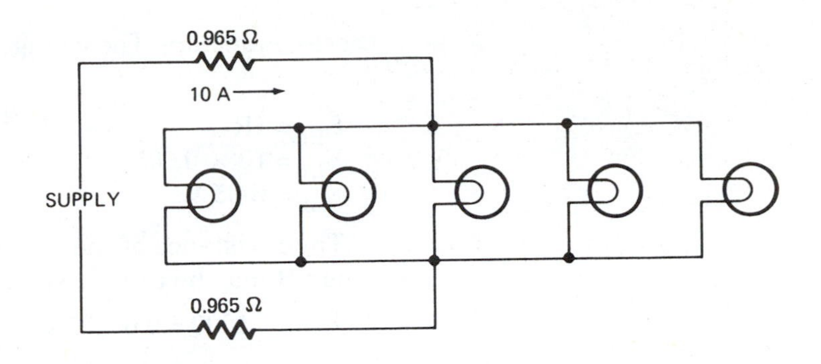

Figure 6-6
Equivalent circuit for Figure 6-5

Example 24

In Example 22, is the percent voltage drop within the maximum limits permitted by the *NEC®*?

$$\% \text{ voltage drop} = \frac{120 - 100.7}{120}$$

$$\% \text{ voltage drop} = 16\%$$

The maximum percent voltage drop permitted by the *NEC®* is 3%. Therefore, Example 22 does not conform to the standards.

Example 25

The circuit in Figure 6-7 is wired with No. 10 AWG copper wire. Each lamp requires 1 A and is rated at 115 V. What is the voltage drop from the source to group B? Are the lamps in group B delivering full light?

According to *Chapter 9, Table 8*, of the *NEC®*, No. 10 AWG copper wire has a resistance of 1.21 Ω per 1000 feet (304.8m) at a temperature of 167°F (75°C).

The resistance of the circuit conductors between the source and group A is 1.21 ÷ 2 = 0.605 Ω. Ten amperes are flowing through

Figure 6-7
Branch circuit for lighting load

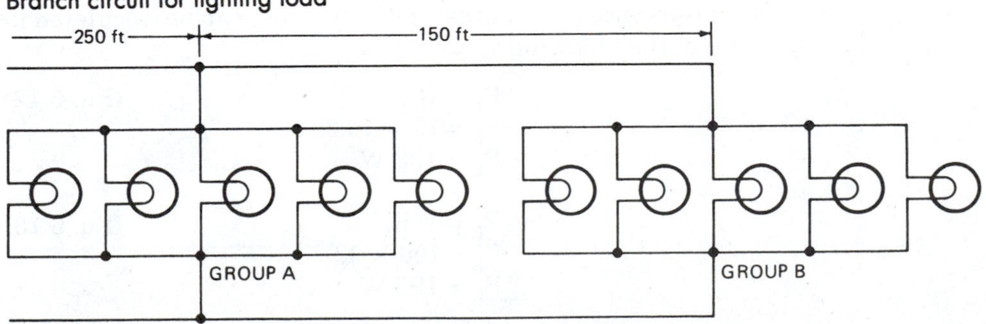

these conductors. The voltage drop between the source and group A is

$$E_{L1} = IR_{L1}$$
$$E_{L1} = 10 \times 0.605$$
$$E_{L1} = 6.05 \text{ V}$$

The resistance of the circuit conductors between group A and group B may be calculated as follows:

$$1.21 \div 1000 = 0.00121\ \Omega$$

The resistance of 1 ft of No. 10 AWG copper wire is 0.00121 Ω.

$$0.00121 \times 300 = 0.363\ \Omega$$

The resistance of the circuit conductors between group A and group B is 0.363 Ω. Five amperes flow in the conductors between group A and group B.

$$E_{L2} = IR_{L2}$$
$$E_{L2} = 5 \times 0.363$$
$$E_{L2} = 1.815 \text{ V}$$

The voltage drop between group A and group B is 1.815 V.

The total voltage (line) drop is

$$6.05 + 1.815 = 7.865 \text{ V}$$

The voltage across group B is

$$120 - 7.865 = 112.135 \text{ V}$$

Group B is glowing at less than full light because 115 V are required for the lamps to deliver their full light.

Line Loss

Line loss is the power dissipated as heat. It is caused by the current flowing through the feeder and branch circuit conductors. Large line losses result in very inefficient distribution systems and greater operating costs for the customer. Line losses should always be kept to a minimum.

With reference to Example 22, the line loss can be calculated in any one of the following ways:

$$P_L = IE_L \qquad \text{(Eq. 6.12)}$$
$$P_L = 10 \times 19.3$$
$$P_L = 193 \text{ W}$$

$$P_L = I^2 R_L \qquad \text{(Eq. 6.13)}$$
$$P_L = 100 \times 1.93$$
$$P_L = 193 \text{ W}$$

$$P_L = \frac{E^2{}_L}{R_L} \qquad \text{(Eq. 6.14)}$$

$$P_L = \frac{372.49}{1.93}$$

$$P_L = 193 \text{ W}$$

The power dissipated as heat is 193 W caused by current flowing through the circuit conductors. This loss of power can be reduced by increasing the size of the circuit conductors.

Example 26

If the conductors in Example 22 are increased in size to No. 4 AWG, what are the voltage drop and power loss in the circuit conductors?

Number 4 AWG copper wire has a resistance of 0.308 Ω per 1000 ft (1000 ft of wire are used for the circuit).

$$E_L = IR_L \qquad P_L = I^2 R_L$$
$$E_L = 10 \times 0.308 \qquad P_L = 100 \times 0.308$$
$$E_L = 3.08 \text{ V} \qquad P_L = 30.8 \text{ W}$$

The voltage drop is 3.08 V and the line loss is 30.8 W. This results in a considerable savings of power over the original circuit. The savings in operating costs over a short time will offset the increase in installation costs for the heavier wire.

Using No. 4 copper wire keeps the voltage drop within the *NEC*® standards.

When planning an electrical installation, it is advisable to locate the distribution panel close to the center of the area. This will minimize the distance between the panel and the loads.

The following two requirements must be met when selecting conductor sizes:

- The conductor must be large enough to carry the current without overheating.
- The conductor must be sized to keep the voltage drop within the maximum limits of the *NEC*®.

Electrical equipment is designed to operate best at its rated voltage. Efficiency and operating characteristics are adversely affected when equipment is operated at voltages other than rated values. With a lighting load, a small decrease in voltage causes a great decrease in light output. Motors subject to reduced voltage have poor starting torque and poor speed regulation. They also draw more current than they would at their rated voltage, and they are inefficient and tend to overheat.

TERMINAL CONNECTIONS AND SPLICES

It is frequently necessary to connect two or more wires together when installing electrical circuits. These connections are called *taps* or *splices*. There are several methods used for connecting wires together. The method selected depends upon the type of installation.

Before conductors are connected together, the insulation must be removed and the conductor thoroughly cleaned. During removal of the insulation, care must be taken to not nick the conductor. A nick in the conductor causes a weak point, which may result in a break when subject to vibration or tension. Insulation may be removed from the conductor with a knife or a wire stripper. When a knife is used, the insulation should be cut at about a 30-degree angle, Figure 6-8. Never cut through the insulation at right angles to the conductor. When using a wire stripper, Figure 6-9, always follow the manufacturer's instructions. The conductors may be cleaned by scraping with a knife or using very fine emery cloth.

Splices

A splice is a point in the wiring system where two or more wires are joined together. Splices should be avoided whenever possible because they are the weakest point in the wiring system. When it is necessary to make a splice, it must be mechanically and electrically secure.

New insulation must be applied after the splice is completed. This insulation must be equivalent to that which was removed. In other words, the new insulation must contain all the physical properties of the original insulation. If the original insulation was type T, it was a flame-retardant thermoplastic compound that can safely withstand operating temperatures up to 140°F (60°C) in dry locations. The replacement insulation must meet these standards. If the original insulation is rated at 600 volts, the replacement insulation must be thick enough to prevent current leakage at 600 volts.

Type MTW insulation is listed in the *NEC*® as being flame-retardant and moisture-, heat-, and oil-resistant thermoplastic.

Its maximum operating temperature is 140°F (60°C) for wet locations and 194°F (90°C) for dry locations. If a splice is made in a conductor having type MTW insulation, the replacement insulation must have the above characteristics. To determine the insulation characteristics, refer to the *NEC*® or the manufacturer's specifications.

Replacement insulation is generally supplied on a roll and is referred to as *tape*. The most common types are made of a thermoplastic compound.

Figure 6-8
Using a knife to remove insulation from a wire

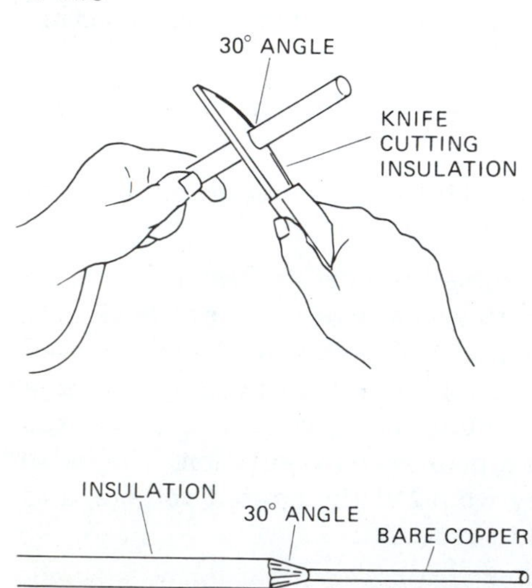

Figure 6-9
Two types of wire strippers *(Courtesy of Ideal Industries, Inc., Sycamore, IL)*

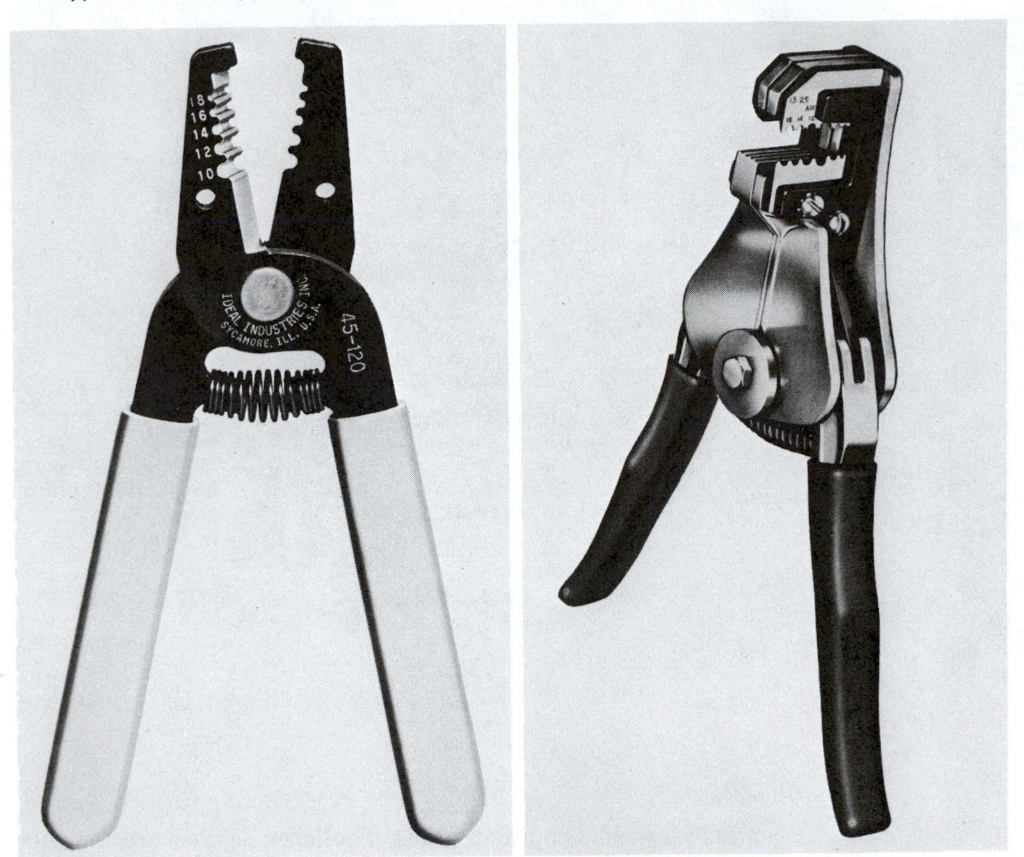

> **CAUTION:** Although various types may look alike, their characteristics may differ considerably. It is important to check the manufacturer's specifications to determine if the compound meets the requirements for the particular type of insulation.

When applying tape to a splice, allow each turn to overlap slightly, and also overlap the original insulation. Always keep the tape taut while it is being applied, and apply enough layers to provide electrical and mechanical protection equal to the original insulation.

Other types of replacement insulation are rubber, silicone rubber, varnish cambric, and silk.

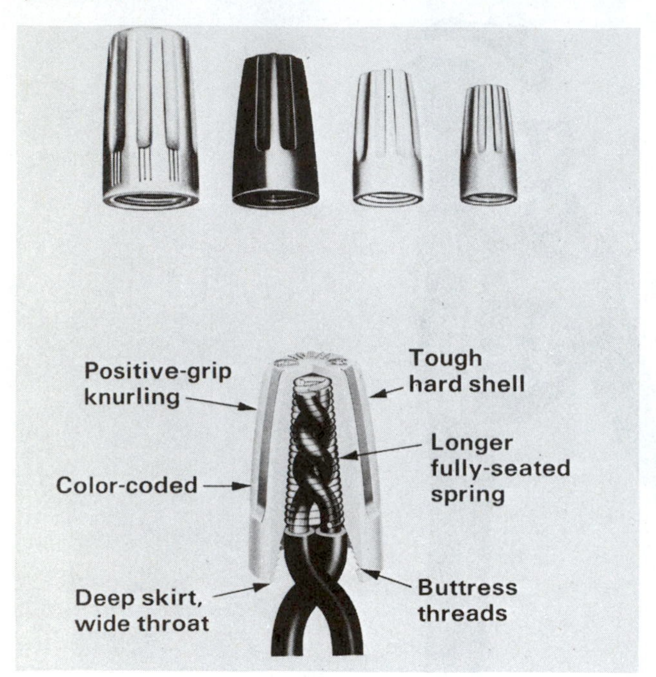

Figure 6-10A
Standard wire nuts *(Courtesy of Ideal Industries, Inc., Sycamore, IL)*

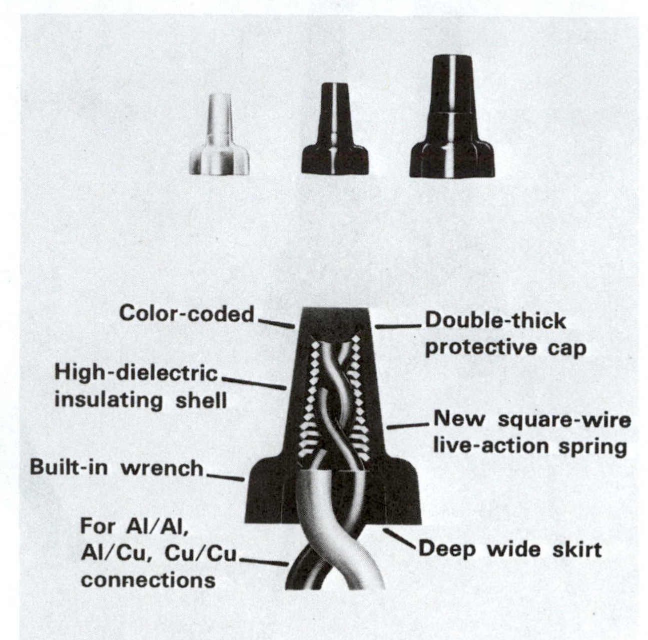

Figure 6-10B
Wire nut with built-in wrench *(Courtesy of Ideal Industries, Inc., Sycamore, IL)*

Splicing Devices

Some of the most common types of splicing devices are the *wire nut, sleeve connector, bolt connector,* and *clamp connector*. For any of these splices, the insulation is removed and the conductor is cleaned.

For small wires (No. 14 AWG to No. 10 AWG), the wire nut or sleeve-type connectors are generally used. The wire nut, Figures 6-10A and 6-10B, threads onto the wires. Sleeve connectors may be obtained in either of two styles. Figure 6-11A shows the crimp type, and Figure 6-11B shows the setscrew type.

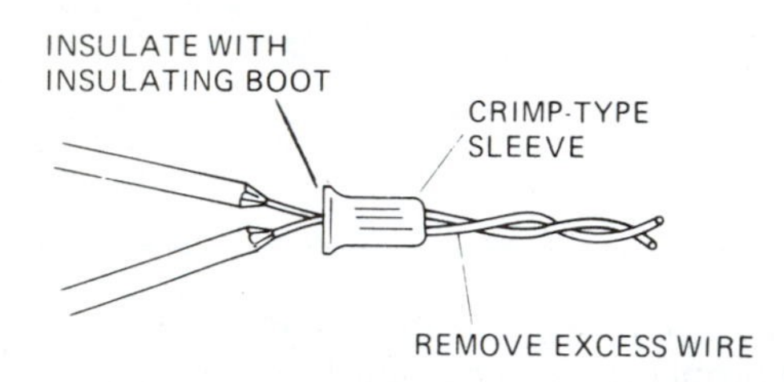

Figure 6-11A
Crimp-type sleeve connector

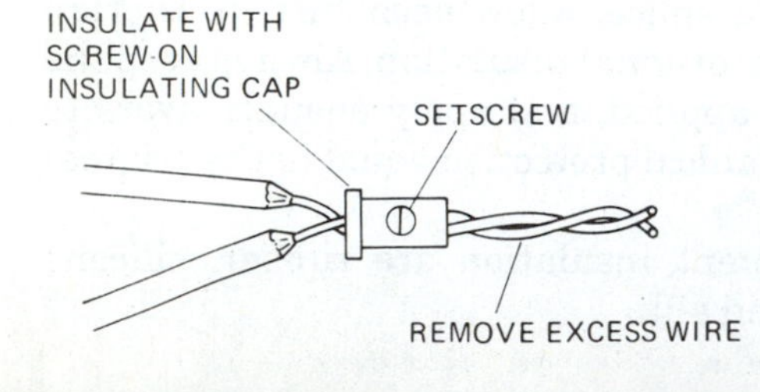

Figure 6-11B
Setscrew-type sleeve connector

The bolt connector is usually used for wires larger than No. 10 AWG, Figure 6-12. The clamp type, Figure 6-13, is generally used for ground connections.

When connecting electrical conductors together, always strive to use conductors of the same material. A chemical reaction between different materials can cause conductors to deteriorate rapidly, producing heat or other hazardous conditions.

Aluminum has become a popular conducting material because of its light weight and low cost. As a result, it is sometimes necessary to connect copper and aluminum conductors together. In order to make such a connection, special precautions must be

Figure 6-12
Bolt-type wire connector for splicing together two or more large conductors *(Courtesy of Burndy Corporation, Norwalk, CT)*

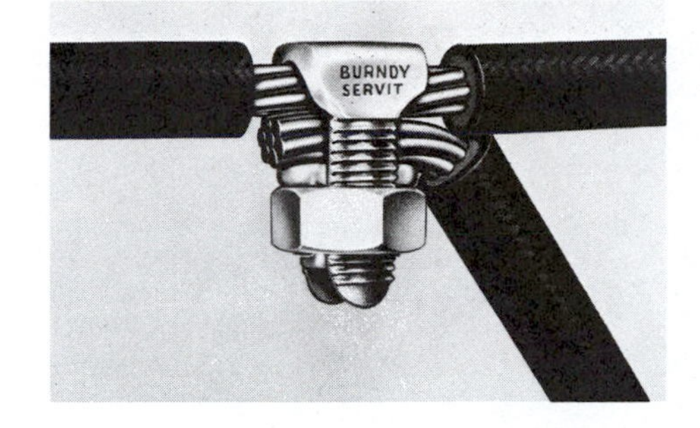

taken. Special chemical compounds have been developed that must be applied to the conducting materials.

> **CAUTION:** Follow the manufacturer's directions precisely, when applying the compound. Only by adhering to recommended procedures will the installation be safe.

Devices such as switches, receptacles, and wire connecting devices are rated for connections to copper, aluminum, or both. These devices are identified as shown in Figure 6-10B. Al indicates aluminum conductors, and Cu indicates copper. The wire nut in Figure 6-10B is manufactured for connections of aluminum to aluminum (Al/Al), aluminum to copper (Al/Cu), and copper to copper (Cu/Cu). (See *Article 110* of the *NEC®*.)

Terminal Connections

In most installations, the ends of conductors are connected to terminals of switchboards, panelboards, or devices. The most common method used to connect conductors to devices is shown in Figure 6-14. The device has an upturned lug and setscrew. The insulation is removed from the conductor, the conducting material is cleaned, and a loop is formed in the conductor, Figure 6-15. After the screw is tightened, the excess wire is removed and the screw is tightened again to ensure a good connection, Figure 6-16. Wires

Figure 6-13
Ground clamp for connecting system grounding conductor to grounding electrode

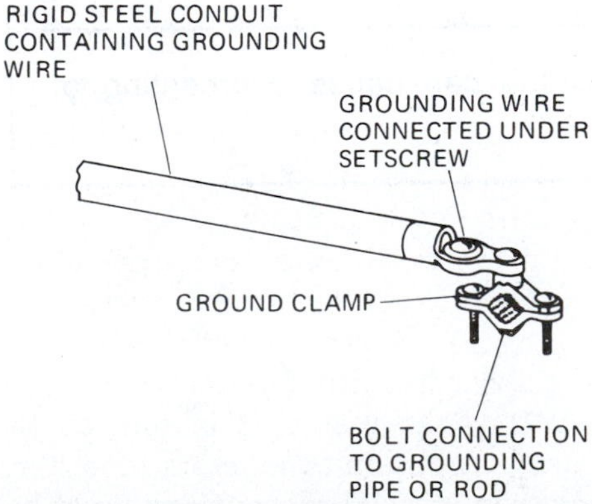

RIGID STEEL CONDUIT CONTAINING GROUNDING WIRE

GROUNDING WIRE CONNECTED UNDER SETSCREW

GROUND CLAMP

BOLT CONNECTION TO GROUNDING PIPE OR ROD

Figure 6-14
Device with upturned lugs for connection of wires

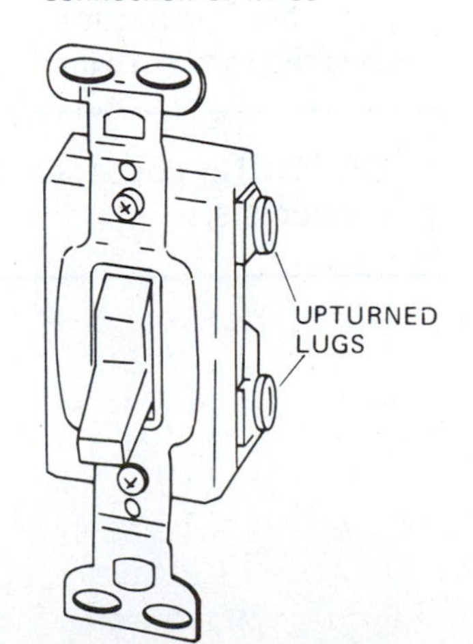

Figure 6-15
Conductor with loop for connecting to upturned lug. The loop closes around the screw in a clock-wise direction.

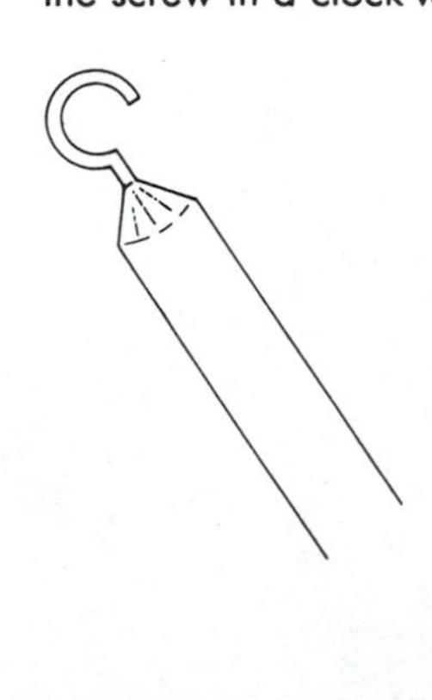

Figure 6-16
Connecting wires to upturned lug

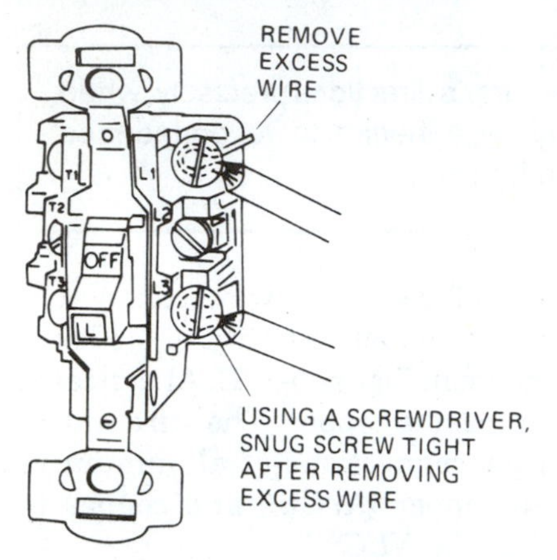

Figure 6-17
Conductor ends connected to terminal lugs on lighting contactor
(Courtesy of Square D Company, Milwaukee, WI)

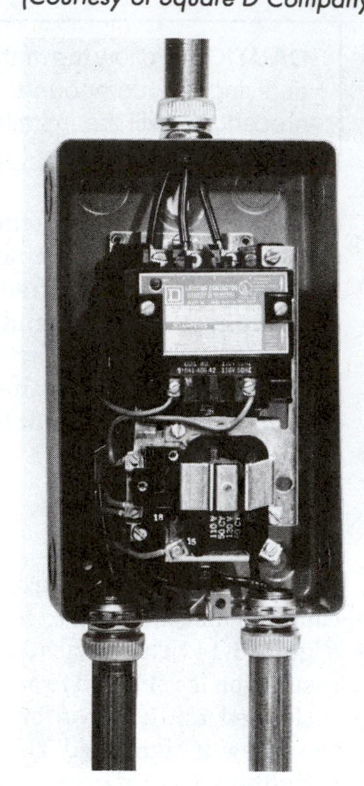

should always be wrapped around the screw in the direction the screw is tightened.

Conductor ends connected to terminal lugs are illustrated in Figure 6-17. All connections should be made tight and then rechecked to ensure mechanical security and electrical continuity.

CAUTION: Loose connections can cause overheating of conductors.

REVIEW

A. Multiple choice.
Select the best answer.

1. A conductor is
   a. any material that offers very little opposition to current flow.
   b. a material with an excess of electrons.
   c. a material with an excess of protons.
   d. a material with an excess of neutrons.
2. Most conductors used in the electrical industry are
   a. solids.
   b. liquids.
   c. gases.
   d. none of the above.
3. One mil is equal to
   a. 1/10 inch.
   b. 1/100 inch.
   c. 1/1000 inch.
   d. 1/10000 inch.
4. Circular conductors are usually measured in
   a. mils.
   b. circular mils.
   c. square mils.
   d. circular inches.
5. Areas of bus bars are usually expressed in
   a. circular mils.
   b. square mils.
   c. inches.
   d. mils.
6. The standard scale for measuring wires is called the
   a. American Wire Gauge.
   b. circular mil gauge.
   c. Universal Wire Gauge.
   d. International Wire Gauge.
7. An instrument used to measure rectangular conductors is the
   a. millimeter.
   b. Rectangular Wire Gauge.
   c. micrometer.
   d. American wire meter.

REVIEW *(continued)*

A. Multiple choice.
Select the best answer (continued).

8. A wire 1 mil in diameter and 1 foot long is called a
   a. foot-mil of wire.
   b. mil-foot of wire.
   c. circular mil-foot of wire.
   d. square mil-foot of wire.
9. The resistivity of copper building wire is
   a. 8.5 ohms.
   b. 6.2 ohms.
   c. 10.4 ohms.
   d. 12.3 ohms.
10. The resistance of a conductor depends upon the
    a. material from which it is made.
    b. operating temperature.
    c. room temperature.
    d. both a and b.
11. The temperature coefficient of a conductor is the
    a. increase in temperature from the ambient temperature to the operating temperature.
    b. amount by which the resistance of a material changes per degree change in temperature for each ohm of resistance.
    c. percentage of temperature increase.
    d. percentage of resistance change.
12. The amount of heat generated by current flowing in a conductor depends upon the
    a. amount of current flowing.
    b. resistance of the conductor.
    c. conductor insulation.
    d. both a and b.
13. The amount of current a conductor can safely carry is called its
    a. volume.
    b. ampacity.
    c. current ability.
    d. safety factor.
14. The ambient temperature of a conductor is
    a. the same as the operating temperature.
    b. lower than the operating temperature.
    c. higher than the operating temperature.
    d. none of the above.

REVIEW *(continued)*

A. Multiple choice. Select the best answer (continued).

15. The most common insulating material used on an electrical conductor is
    a. plastic.
    b. rubber.
    c. varnish.
    d. cotton.

16. A feeder is
    a. the conductors between the service equipment and the final branch-circuit overcurrent device.
    b. all the conductors between the final circuit overcurrent protective device and the load.
    c. the conductors entering the building.
    d. both a and b.

17. Voltage (line) drop is the
    a. current loss in the conductors.
    b. voltage loss in the conductors.
    c. power loss in the conductors.
    d. voltage applied to the circuit.

18. The *National Electrical Code®* allows a maximum of a
    a. 3 percent voltage drop for branch circuits.
    b. 3 percent voltage drop for feeders.
    c. 6 percent voltage drop for combined feeder and branch circuits.
    d. all of the above.

19. Line loss is the
    a. current loss in the conductors.
    b. voltage loss in the conductors.
    c. power loss in the conductors.
    d. none of the above.

20. The most common method used to connect conductors to devices is the use of the
    a. wire nut.
    b. upturned lug.
    c. screw connector.
    d. device connector.

REVIEW *(continued)*

B. Give complete answers.

1. Define the unit of measurement for electrical wires.
2. In what unit are bus bars commonly measured?
3. Describe an American Wire Gauge.
4. Describe the instrument called a micrometer.
5. Define the term *resistivity*.
6. Describe the effect of temperature change on the resistance of aluminum wire.
7. Does an increase in temperature affect the resistance of constantan?
8. How does an increase in temperature affect the resistance of carbon?
9. Name two of the most common materials used in building wiring.
10. Define the terms *voltage drop* and *line loss*.
11. What is an upturned lug?
12. What is the most common effect of loose connections?
13. List four types of connectors.
14. What is a splice?
15. Describe one method used to remove insulation from a wire.

C. Solve each problem, showing the method used to arrive at the solution.

1. Calculate the circular mil area of a wire that measures 1/4 in. in diameter.
2. What is the area of the wire in Problem 1, in circular millimeters?
3. A wire has an area of 250,000 circular mils. What is its diameter, in mils?
4. What is the diameter of the wire in Problem 3, in millimeters?
5. Calculate the square mil area of a bus bar 1/4 in. thick and 3/4 in. wide.
6. What is the circular mil area of the bus bar in Problem 5?
7. Calculate the square millimeter area of the conductor in Problem 5.

REVIEW *(continued)*

C. Solve each problem, showing the method used to arrive at the solution (continued).

8. Calculate the length of a copper building wire 1/4 in. in diameter, if it has a resistance of 0.25 Ω.
9. What is the resistance of 500 ft (152.4 m) of No. 12 AWG aluminum wire at a temperature of 20°C (68°F)?
10. What is the resistance of 500 ft (152.4 m) of No. 12 AWG copper wire at a temperature of 20°C (68°F)?
11. The resistance of a tungsten (annealed) filament is 10 Ω at 0°C. What is its resistance at 1500°C?
12. The resistance of a coil of aluminum wire is 20 Ω at 20°C. What is its resistance at 70°C?
13. If the coil in Problem 12 is wound with copper wire instead of aluminum, and all other facts remain the same, what is its resistance at 70°C?
14. The copper windings of a motor have a resistance of 10 Ω when the motor is operating at no load. The resistance is increased to 13 Ω when the motor is operating at full load. If the no-load temperature of the windings is 30°C, what is the temperature rise of windings?
15. Three No. 10 AWG aluminum wires covered with TW insulation are installed in a 1/2-in. conduit. If the ambient temperature is 100°F (37.8°C), what is the ampacity of the conductors?
16. If three more No. 10 AWG aluminum conductors are installed in the raceway in Problem 15, what is the ampacity of all the conductors in the raceway?
17. *Table 310-16* of the *NEC®* lists the ampacity of No. 6 AWG, type TW aluminum wire as 40 A. What is the voltage drop for 500 ft (152.4 m) of this wire if the wires are conducting at full capacity?
18. For the circuit in Problem 17, if the source voltage is 120 V, what is the percent voltage drop?
19. What is the line loss for the conductors in Problem 17?
20. The resistance of a coil wound with aluminum wire is 2 Ω at 20°C. What is its resistance at 80°C?

7
WIRING METHODS

Objectives

After studying this chapter, the student will be able to:

- Explain the purpose of electrical codes and of the National Fire Protection Association.
- List the type of installations covered by the *National Electrical Code®*.
- List and describe some common wiring methods used in the electrical industry.
- Describe the conditions of use of various wiring methods.

ELECTRICAL CODES

Electrical codes are rules and regulations governing the installation of electrical systems. In general, the purpose of an electrical code is to ensure safe installations.

There have been many electrical codes developed on local, state, and national levels. Most states have enacted laws setting forth minimum requirements for electrical safety. By far the most accepted standards are those developed by the National Fire Protection Association (NFPA). The primary goal of the NFPA is "to promote the science and improve the methods of fire protection."

The NFPA publishes materials pertaining to safe practices in the electrical industry. All electrical personnel should become familiar with this literature. The *National Electrical Code®* is the most frequently used publication. These are some other popular publications:

- *Recommended Practices for Electrical Equipment Maintenance*
- *Household Fire Warning Equipment*

- *Lightning Protection Code*
- *Central Station Protective Signaling Systems*
- *Electrical Metalworking Machine Tools*

The wiring methods and installations discussed in this text are covered by the *National Electrical Code*®. *Article 90* of the *NEC*® lists the types of installations covered by the *Code*. The types of installations include electrical equipment within or on public or private buildings and some other types of premises. Installations that are not covered by the *NEC*®, such as the wiring of ships, aircraft, underground mines, and automobiles, are covered by other codes designed specifically for the purpose.

In the design and installation of electrical systems, safety must be the top priority. Reference should be made to the *NEC*® and other safety regulations to determine the methods and materials to be used.

ARMORED CABLE

Armored cable, generally known as BX or AC cable, consists of rubber or plastic insulated wires surrounded by a flexible metal enclosure (armor). The armor protects the conductors from mechanical injury. Type AC cable may contain from two to four conductors in sizes from No. 14 AWG to No. 4 AWG. Unless there is a lead covering between the conductor insulation and the armor, the cable must contain an internal bonding strip. This strip must be made of either copper or aluminum, and it must be in contact with the armor for the entire length of the cable.

Together, the armor and the bonding strip serve as a grounding conductor. The purpose of the grounding conductor is to carry away stray currents caused by accidental grounds. If the stray current is great enough, it will cause the overcurrent protection device to open.

NEC® *Article 333* covers the conditions of use of armored cable. When a cable is terminated or spliced, an enclosure called a *box* must be installed. Figure 7-1 illustrates this method of splicing.

To install the cable in a box, 8 inches of armor must be removed from the cable. This is generally accomplished with a hacksaw. A blade with 32 teeth to the inch is used to cut through the armor. Care is taken so as not to damage the conductor insulation. The armor is then removed from the conductors, and an *antishort bushing* (plastic or fiber bushing) is installed between the conductors and the armor. The purpose of this bushing is to protect the conductor insulation from abrasion, Figure 7-2.

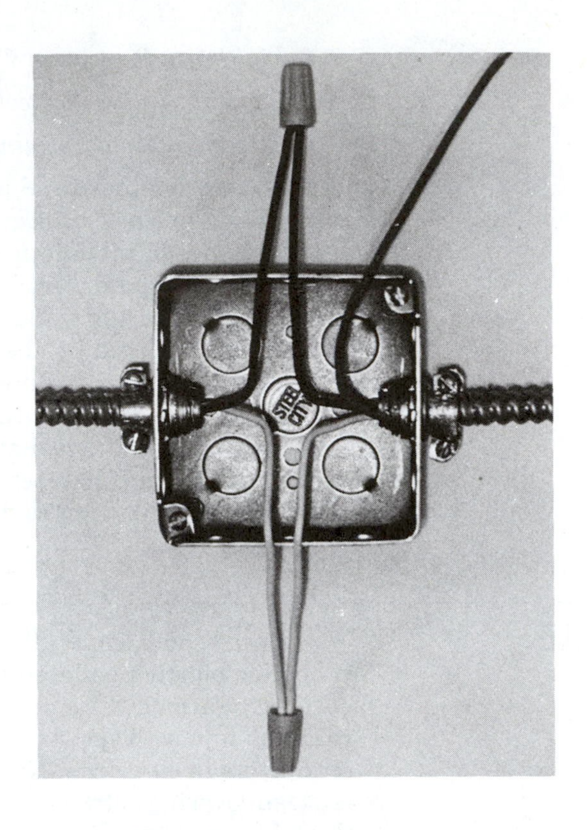

Figure 7-1
Conductors of armored cable spliced in a junction box

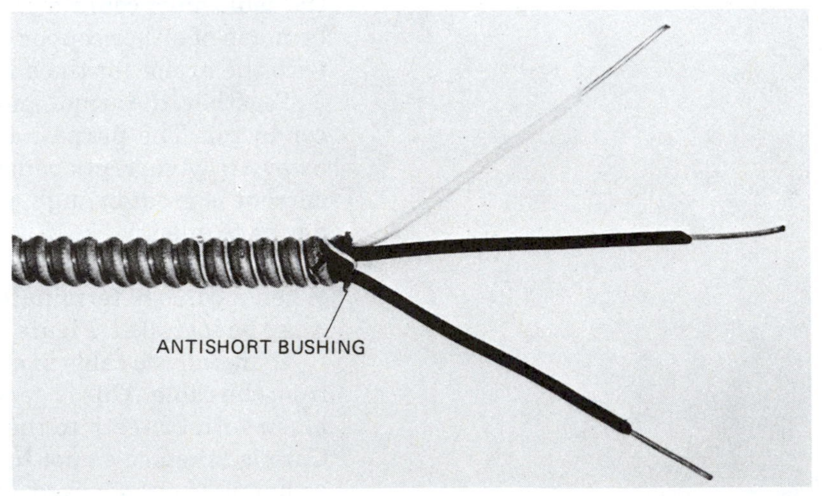

Figure 7-2
Armored cable with an antishort bushing installed to protect the insulation from abrasion

Figure 7-3
Various types of armored-cable box connectors
(Courtesy of Thomas & Betts Co., Raritan, NJ [A, B, C], and George Tashjian, Medford, MA [D])

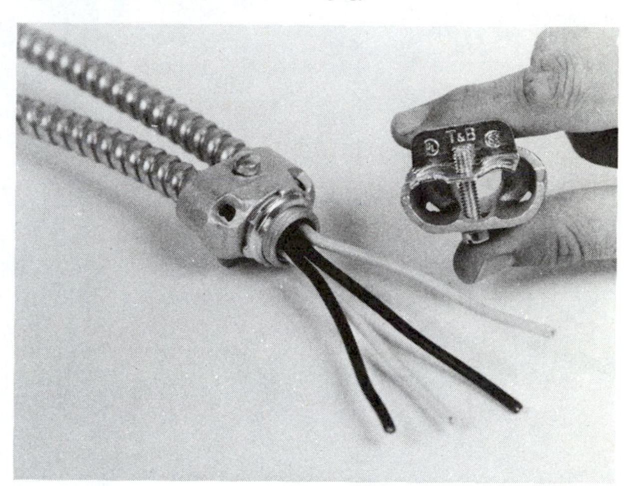

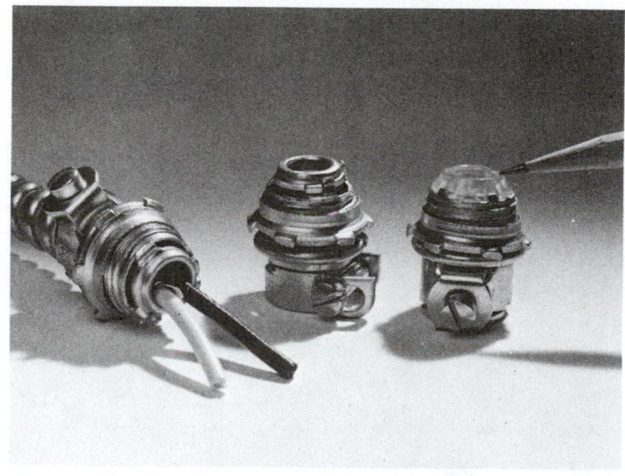

C

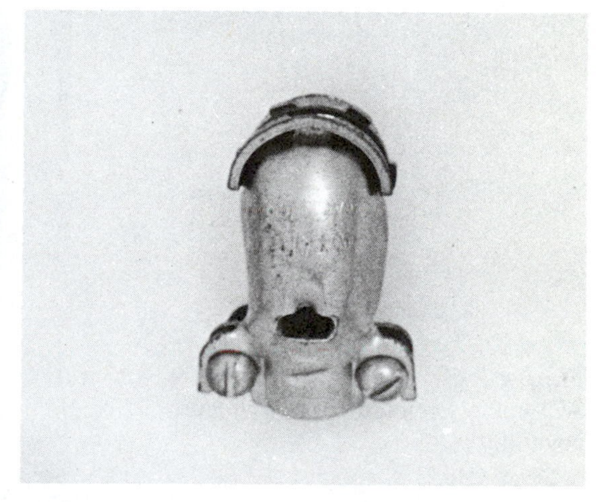

D

A *fitting* must be provided to connect the cable to the box. The fitting may be separate from the box or installed inside the box as a cable clamp. It must be designed so that the antishort bushing is visible after the installation has been completed. Figure 7-3 illustrates several types of BX connectors. Figure 7-4 shows a box containing cable clamps.

Type AC cable may be installed in dry locations and where not subject to physical damage. Type ACL (armored cable, lead covered) may be used where exposed to weather or continuous moisture, Figure 7-5.

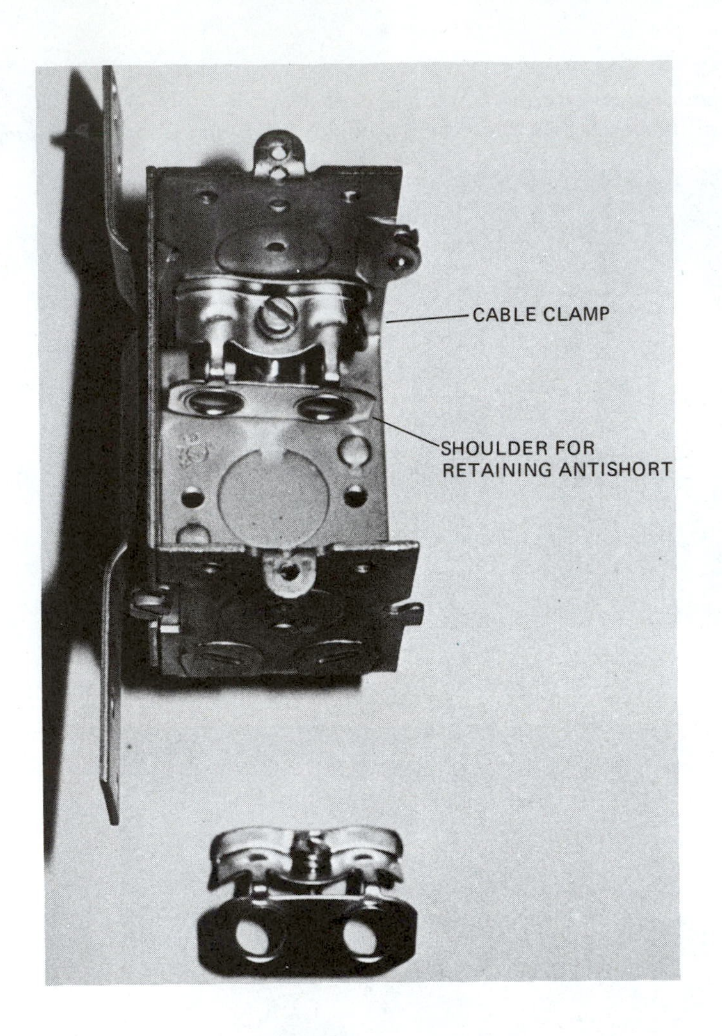

Figure 7-4
Box containing clamps for armored cable. Shoulder helps to hold antishort in place.

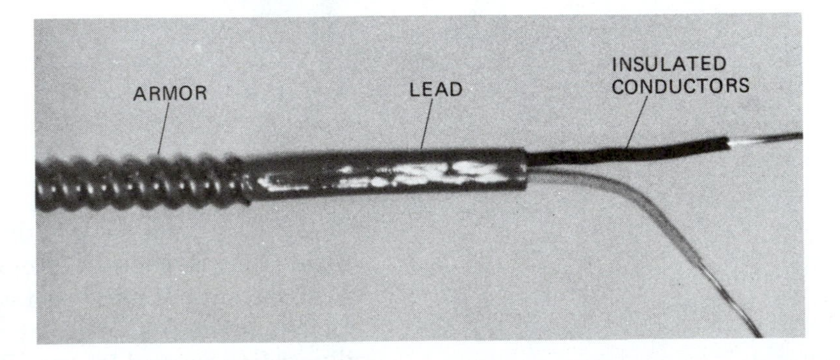

Figure 7-5
ACL cable (armored cable with lead covering)

NONMETALLIC-SHEATHED CABLE

Figure 7-6
Nonmetallic-sheathed cable, type NM *(Courtesy of Cyprus Wire and Cable Company, Rome Cable Corporation, Rome, NY)*

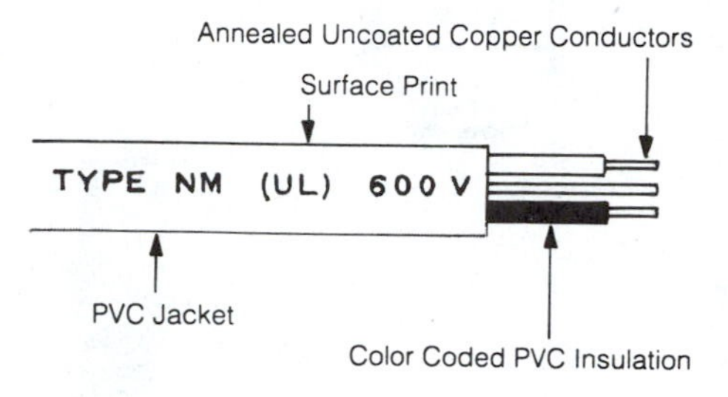

Figure 7-7
Type NMC cable with grounding wire

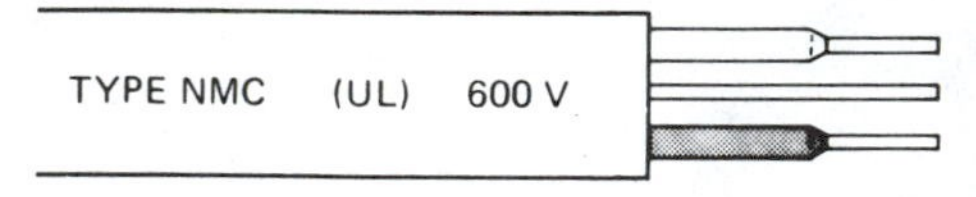

Figure 7-8
Nonmetallic-sheathed cable stripper *(Courtesy of VACO Products Co., Northbrook, IL)*

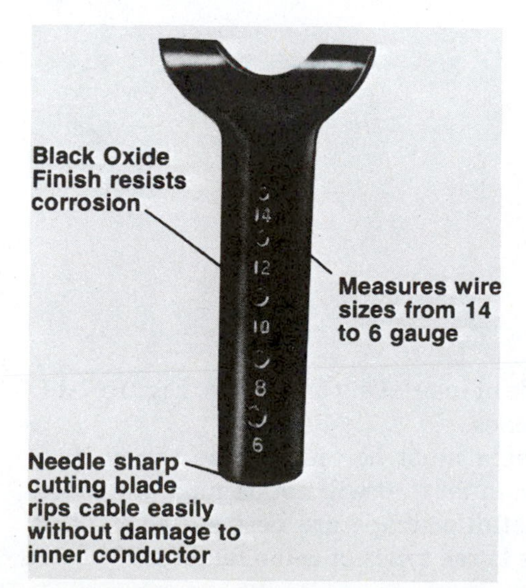

Nonmetallic-sheathed cable is known by several trade names such as Romex, Roflex, Cresflex, and Flexall. It consists of two or more insulated conductors enclased in a moisture-resistant, flame-retardant, nonmetallic material. The individual conductors are usually insulated with type T or TW insulation, and the outer covering is generally vinyl plastic.

The standard conductor sizes are No. 14 AWG through No. 2 AWG for copper wires, and No. 12 AWG through No. 2 AWG for aluminum wires. The cable is generally manufactured as a two- or three-conductor cable. If a grounding wire is required, an additional wire is included. This wire is made of the same material and size as the other conductors. The grounding conductor may be insulated with a green covering, or it may be bare. It should not be used for purposes other than grounding. Figure 7-6 illustrates a length of nonmetallic-sheathed cable with a grounding conductor.

Nonmetallic-sheathed cable is easy to install and requires no special tools. It is also less expensive than most other cables that are suitable for the same purpose. It is a popular wiring method for both residential and commercial buildings in metropolitan areas and for farm buildings.

There are two types of nonmetallic-sheathed cable—NM and NMC. *Article 336* of the *NEC®* covers both types. Type NM cable is used for exposed and concealed work in dry locations. Type NMC, Figure 7-7, may be used in dry or moist locations and where exposed to corrosive influences. For additional standards governing location, use, and methods of installation, refer to the *NEC®*.

In farm buildings such as barns and areas where moisture and other corrosive influences are present, a completely nonmetallic system is advised. A nonmetallic system is also advised in locations where it is difficult to obtain a good system ground.

CAUTION: Poor grounding of metal enclosures and parts can be hazardous.

When nonmetallic-sheathed cable is terminated or spliced, a box, fitting, cabinet, or similar enclosure must be installed. The enclosure may be constructed of metal or nonmetallic material. All noncurrent-carrying metal enclosures must be grounded. If metal enclosures are used, the cable must contain a grounding conductor.

To install nonmetallic-sheathed cable in a box, 8 inches of the sheath must be removed. This may be accomplished with a knife or with an NM cable stripper, Figure 7-8. A connector or cable clamp is used to secure the cable in place. Figure 7-9 illustrates several types of connectors. A box containing cable clamps is shown

Figure 7-9
Nonmetallic-sheathed cable connectors *(Courtesy of Thomas & Betts Co., Raritan, NJ)*

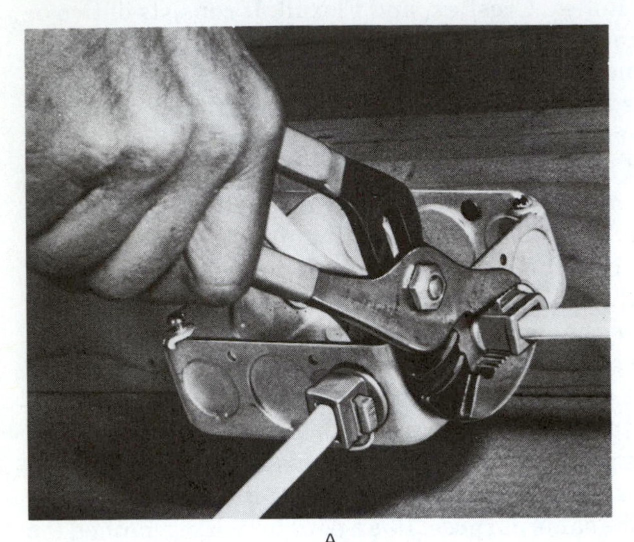

A

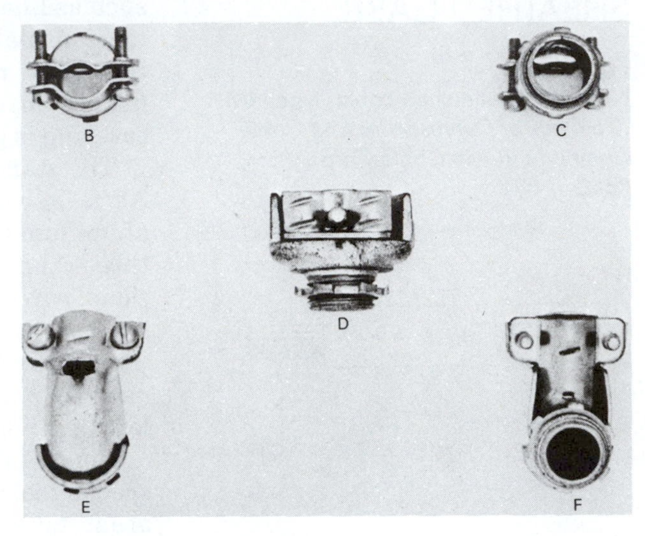

G

Figure 7-10
Switch box containing cable clamps

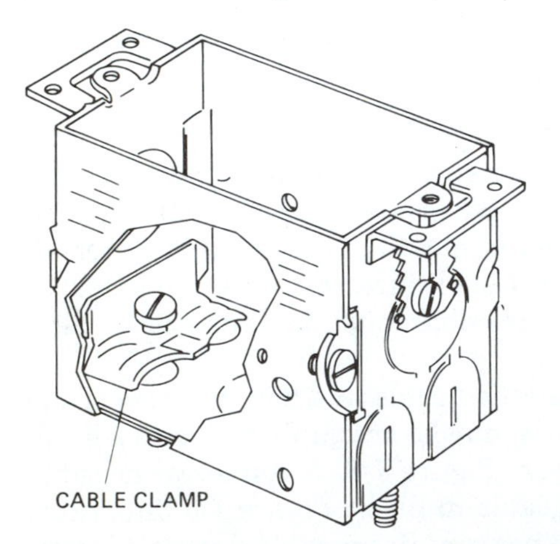

in Figure 7-10. Devices made of insulating materials, Figure 7-11, may be installed without boxes.

Nonmetallic-sheathed cable must be fastened to the surface with staples or straps in a manner that will not damage the cable. Insulated staples or nonmetallic straps are best suited for this purpose. Figure 7-12 shows three types of cable fasteners.

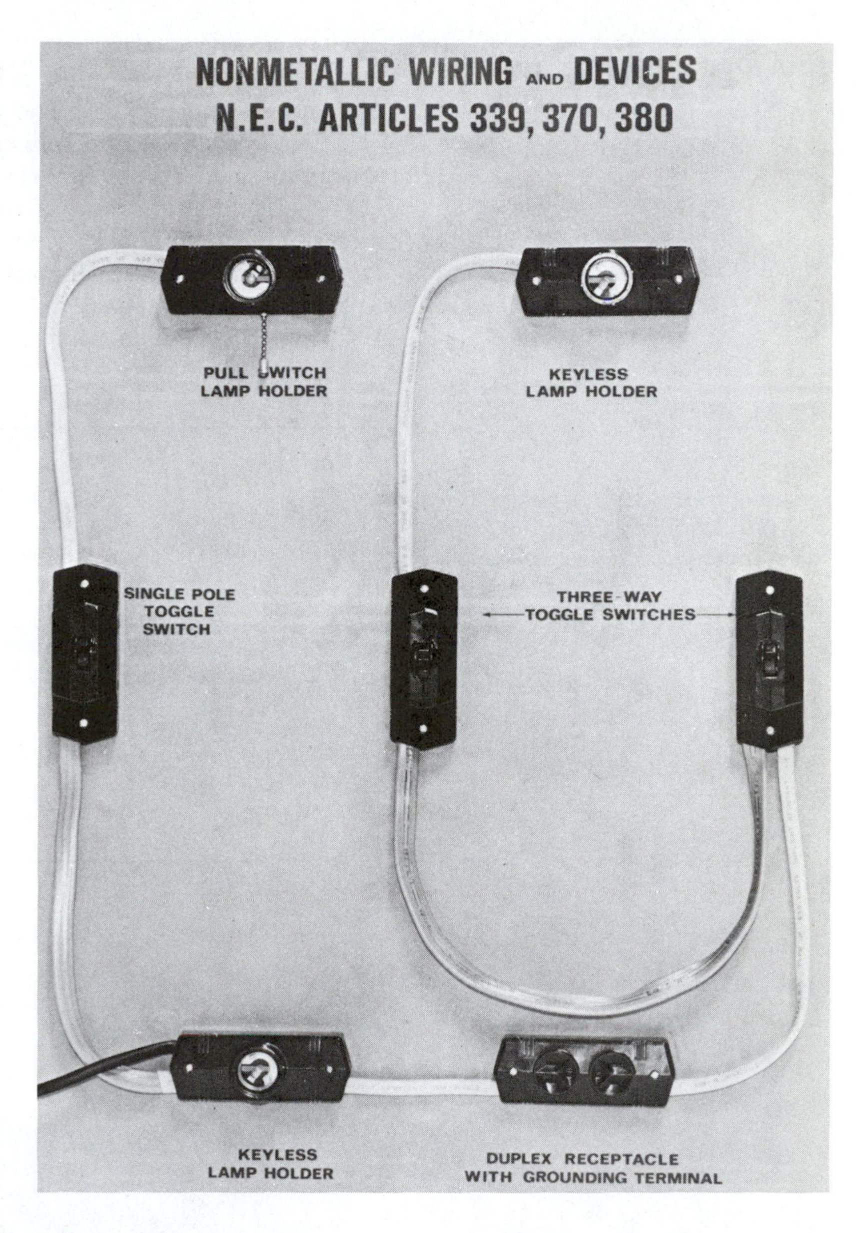

Figure 7-11
Devices of insulating material *(Courtesy of Hickok Teaching Systems, Peabody, MA)*

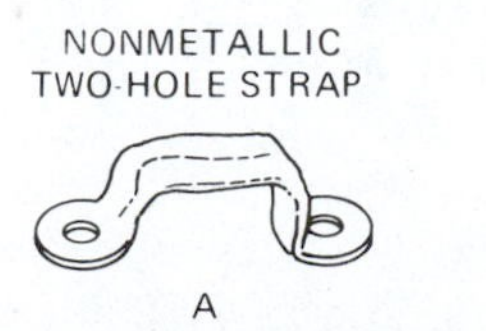

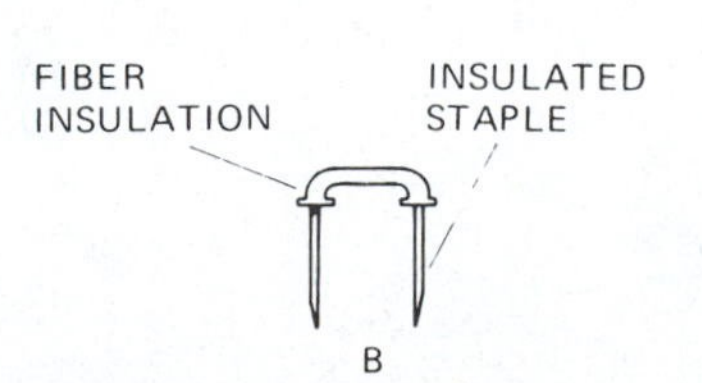

Figure 7-12
Nonmetallic-sheathed cable fasteners

SERVICE-ENTRANCE CABLE

Service-entrance cable consists of one or more conductors assembled into a cable. The cable is available in types SE and USE. Type SE cable is usually an assembly of two insulated conductors (THW or RHW) and one bare conductor. Surrounding these conductors is a flame-retardant, moisture-resistant, and sunlight-resistant covering, Figure 7-13. SE cable is also made with insulation covering all three conductors. When all the conductors are insulated, the cable may be used for interior wiring. Type USE cable is constructed to meet the *NEC*® standards for underground use.

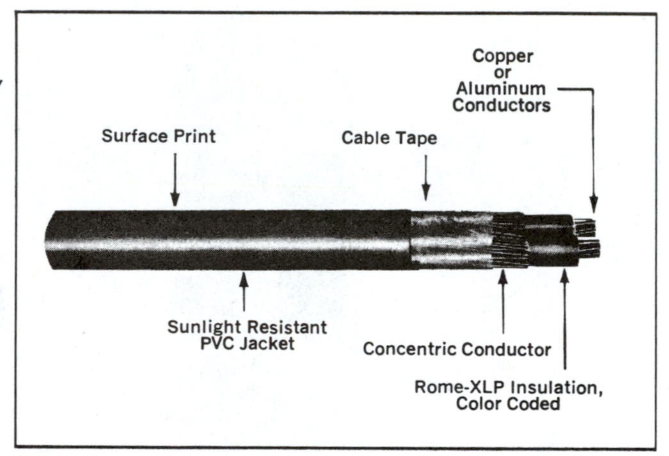

Figure 7-13
Service-entrance cable, type SE *(Courtesy of Cyprus Wire and Cable Company, Rome Cable Corporation, Rome, NY)*

Figure 7-15
Weathertight SE cable connector

When SE cable is used for overhead services, it may be continuous from the pole to the meter or it may terminate at the service head, Figure 7-14. A watertight connector must be installed where the cable enters the meter trough, Figure 7-15. At the point of entry to the building, a hole should be bored inward and upward to prevent the entrance of water. A *sill plate*, Figure 7-16, is then installed, and a waterproof, putty-type material called *duct seal* is packed around the cable. The cable is fastened to the surface with straps in a manner that will not injure the covering. Figure 7-17 shows a service wired with SE cable.

To install service-entrance cable in a box or fitting, it is necessary to remove the outer covering and install a cable connector. The covering may be removed with a knife (similar to removing the sheath from NM cable).

Service-entrance cable is made in standard sizes from No. 8 AWG through No. 4/0 AWG, in copper or aluminum.

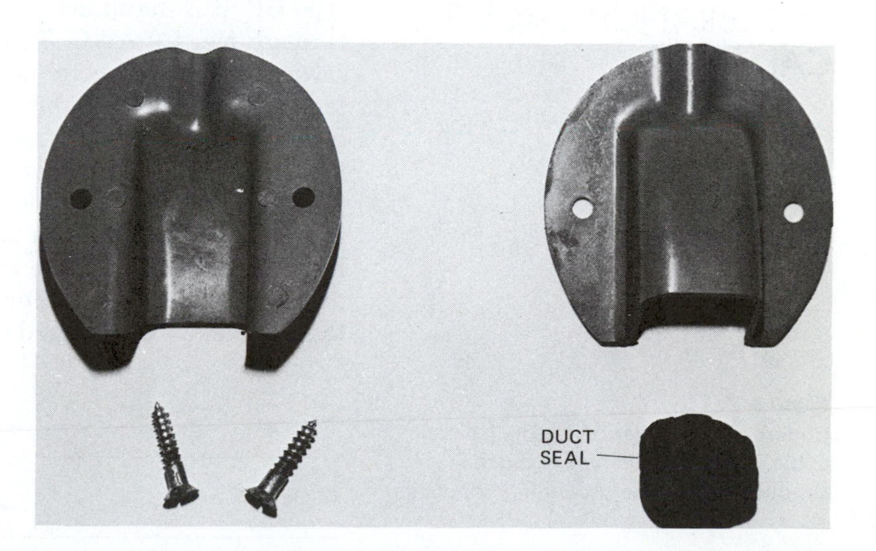

Figure 7-16
Sill plate for service entrance cable

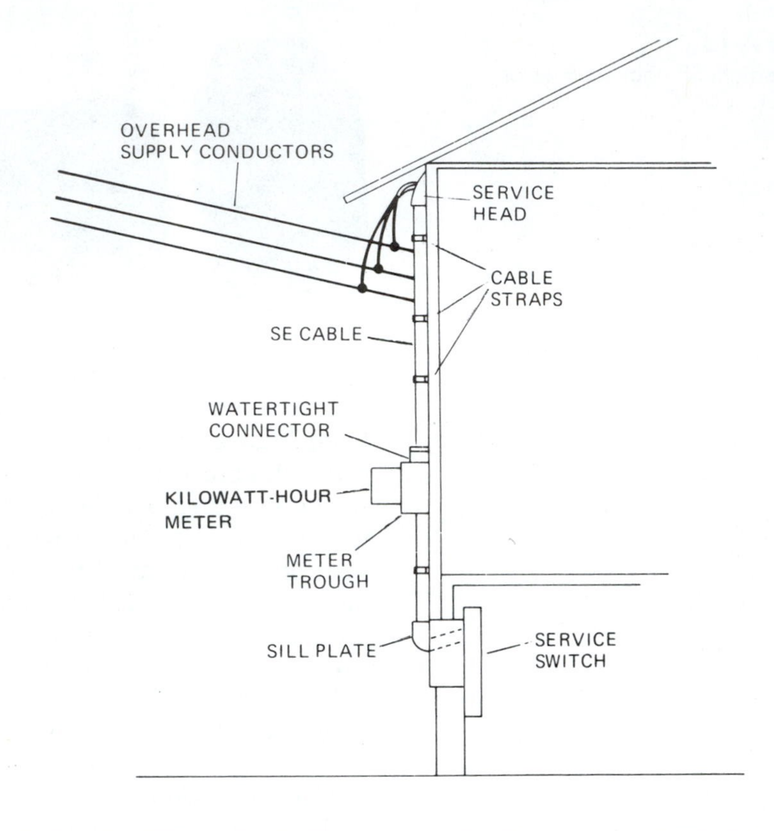

Figure 7-17
Type SE cable used for service-entrance conductors

UNDERGROUND FEEDER AND BRANCH-CIRCUIT CABLE

Underground feeder and branch-circuit cable is classified as type UF. It is manufactured in standard sizes from No. 14 AWG through No. 4/0 AWG. The conductors are made of copper and covered with a moisture-resistant insulation. They are assembled to form a cable. The cable is then encased in a flame-retardant, moisture-resistant, fungus-resistant, and corrosion-resistant material, which is suitable for direct burial in the earth, Figure 7-18.

UF cable may be installed in any area where NM or NMC cable is permitted. It may also be buried in the earth. It may not be installed where it will be exposed to the direct rays of the sun, unless it is specifically approved for this purpose. In areas where the cable may be damaged (for example, by construction or farm

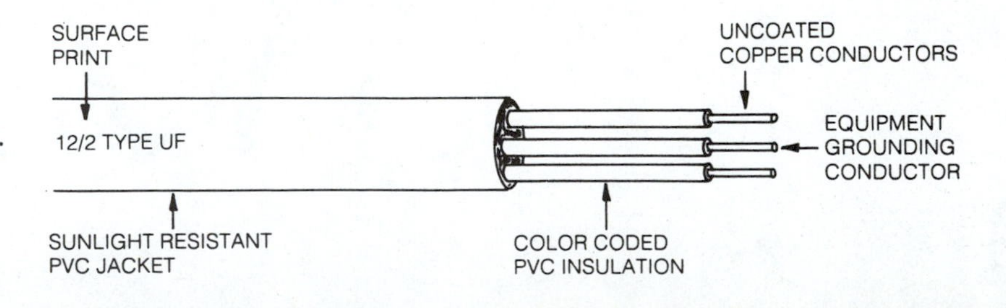

Figure 7-18
Underground feeder and branch-circuit cable, type UF. Contains two circuit conductors and one grounding conductor.

equipment), it should be adequately protected. Installing the cable in nonmetallic conduit is a method frequently used. When the cable is buried in the earth, it should always be at least 24 inches below the surface. Contact with sharp objects such as rocks or bricks should be avoided.

UF cable should be installed in the same manner as NM cable. *NEC® Article 339* applies to UF cable.

NONMETALLIC CONDUIT

Nonmetallic conduit is made of fiber, asbestos cement, soapstone, rigid polyvinyl chloride (PVC), high-density polyethylene, or rigid polyvinyl (PV). Nonmetallic conduit is resistant to moisture and chemical atmospheres. When installed above ground, it must be flame resistant as well as resistant to mechanical injury. It must also resist the effects of sunlight and low temperatures, and resist heat distortion. Figure 7-19A shows one type of nonmetallic conduit. Figure 7-19B shows some of the fittings used with this type of conduit.

Nonmetallic conduit is available in standard trade sizes from 1/2 inch through 6 inches. *Table 347-8* of the *NEC®* lists the standard sizes and the minimum spacing recommended between supports.

Nonmetallic conduit is recommended in areas where moisture and/or corrosive influences are present and where it is difficult to obtain a good ground.

In farm buildings such as barns, nonmetallic conduit provides better mechanical protection than NMC cable or UF cable. Other areas where the use of nonmetallic conduit is recommended include laundries, canneries, car washes, and locations where the walls are frequently washed. The entire system must be installed and equipped to prevent liquids from entering the boxes, fittings, and conduit.

Figure 7-19A
Nonmetallic conduit, type PVC
(Courtesy of Carlon, An Indian Head Company, Cleveland, OH)

Figure 7-19B
Fittings used with nonmetallic conduit, type PVC *(Courtesy of Carlon, An Indian Head Company, Cleveland, OH)*

METAL CONDUIT

Metal conduit is manufactured in both rigid and flexible types. The rigid type is available in three styles: heavy-wall, medium-wall, and thinwall.

Heavy-wall conduit, commonly referred to as rigid metal conduit, is discussed in *Article 346* of the *NEC®*. *Article 345* covers medium-wall conduit (Intermediate Metal Conduit), and *Article 348* pertains to thinwall conduit (Electrical Metallic Tubing).

Installing Wires in Conduits

With both metal and nonmetallic conduit systems, the conductors are pulled into and through the conduits. This must be done after the system has been completely installed. If the conduits are concealed or buried in masonry, plaster, or similar materials, the wires should not be installed until the material is dry. Following this procedures allows the wires to be withdrawn if the insulation becomes faulty.

Figure 7-20
Hook formed on the end of a fish tape

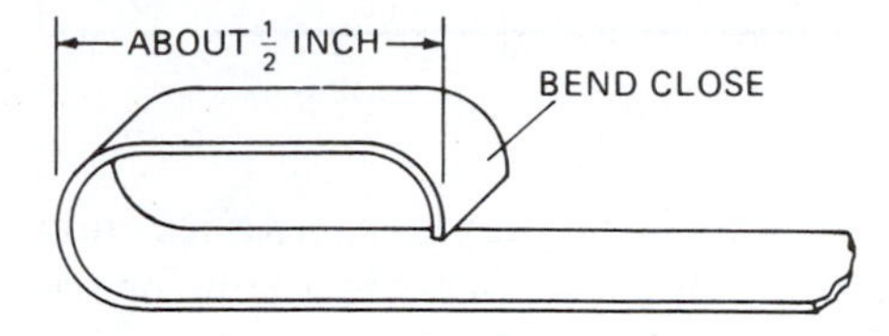

The most common method of installing conductors in conduits is to pull the conductors through the conduit with a tempered steel tape. This tape is called a *fish tape*. A hook is formed on one end of the fish tape, Figure 7-20. The hook serves two purposes: it permits the fish tape to slide through the conduit with ease, and it provides a means for attaching the conductors, Figures 7-21A and 7-21B.

If the run of conduit contains several bends, it may be difficult to insert the fish tape. The *NEC*® places restrictions on the radius of the bends as well as the number of bends between boxes and/or fittings.

If the fish tape cannot be pushed all the way through the conduit, a second tape may be pushed in from the other end. By careful manipulation the first tape can be hooked by the second, Figure 7-22. Another method frequently used is to place several loops of nylon line on the hook, as shown in Figure 7-23. The hook is tightly closed and the fish tape is pushed into the conduit. The fish tape with the nylon loop is manipulated until the first fish tape is caught. Power tools are also available to propel a nylon line through the conduit.

Figure 7-21A
Wires fastened to a fish tape

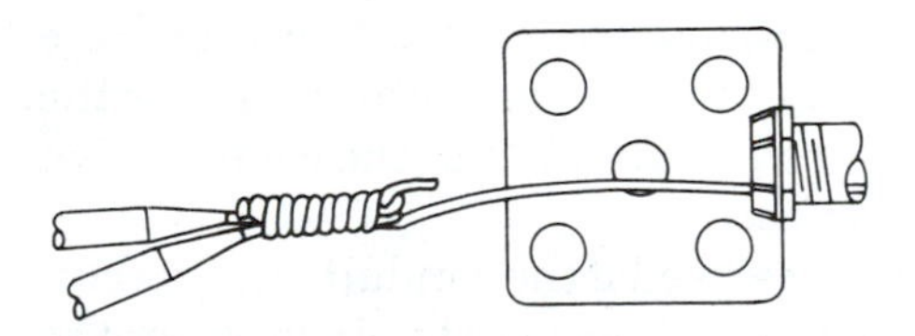

Installing the wires in the conduit usually requires two people. One person pulls the fish tape or line, and the other person "feeds" the wires. The person feeding the wires must be sure that they do not cross or kink. To reduce the friction it is common practice to apply an approved lubricant. Large conductors are frequently pulled through the conduit by machines designed for this purpose.

Figure 7-21B
Applying tape to make a smoother connection

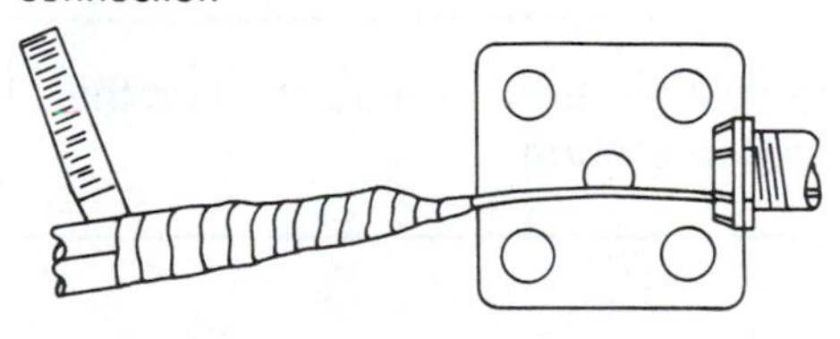

The maximum number of conductors in a conduit is determined by the size of the wire, the type of insulation, and the size of the conduit. *Tables 3A, 3B,* and *3C* of *Chapter 9* in the *NEC*® indicate the maximum number of conductors permitted in a conduit. The number of conductors permitted in boxes or fittings is governed by *NEC*® *Article 370*.

Figure 7-22
Catching one fish tape with another fish tape

Rigid Metal Conduit

Rigid metal conduit may be installed in most atmospheric conditions and is used extensively in industrial and commercial buildings. The pipes are manufactured in standard lengths of 10 feet, which include the coupling. Shorter lengths (12 inches and less) are also available. These short lengths are called nipples.

Figure 7-23
Using nylon line to catch a fish tape

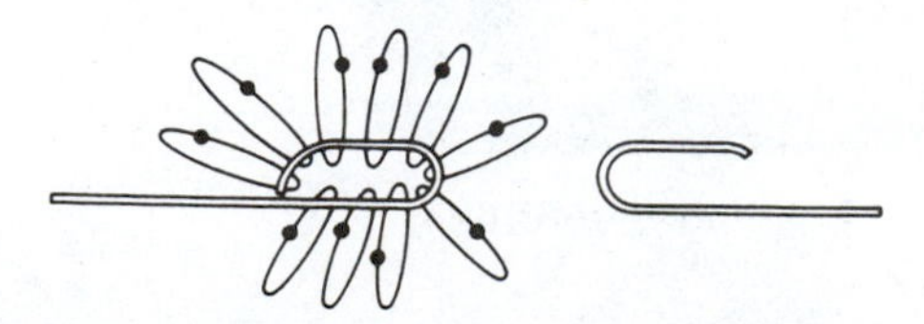

The standard lengths of conduit are threaded on both ends, and a coupling is installed on one end. For long runs, two or more lengths are coupled together. All connections must be made tight with a wrench. Locknuts and bushings are installed where a conduit enters a box through a punched hole. These connections must also be wrench tight. The metal conduit serves as an equipment ground (see *NEC*® *Article 100*).

Figure 7-24
Conduits entering a cabinet

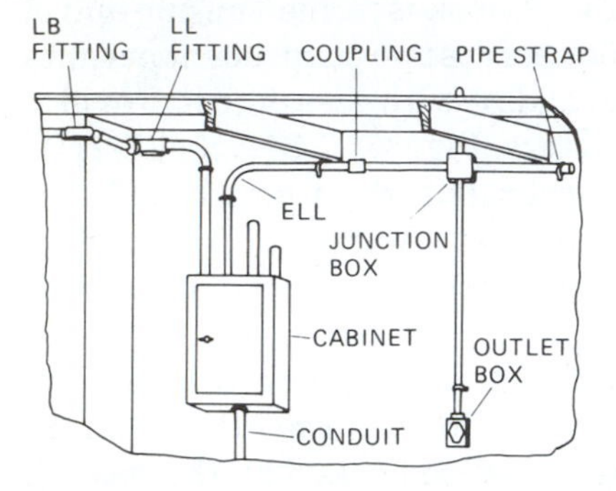

> **CAUTION:** Loose connections can cause hazardous conditions.

Installing a Conduit System

Conduit must always be fastened rigidly in place, with its ends connected to cabinets, boxes, or fittings. A *cabinet* is an enclosure for fuses or circuit breakers. Figure 7-24 illustrates a cabinet with conduits entering through punched holes.

Many types and styles of boxes are available for use with conduit. Boxes are installed at all outlets where splicing is necessary. They are also installed where it is necessary to pull the wires through the conduit. Figure 7-25 shows several types of boxes commonly used with conduit systems.

Some of the most common fittings used with conduit are shown in Figure 7-26. These fittings are manufactured with threads tapped into the hubs for threading onto the conduit, or they have compression-type hubs. The compression-type hub slides over the conduit (the conduit does not need a thread) and the compression fitting is tightened with a wrench.

A hacksaw or pipe cutter may be used if the conduit must be cut. The blade of a hacksaw should have 24 teeth to the inch. After the cut is made, the inside of the conduit is reamed or filed smooth to remove any sharp ridges caused by the cutting.

> **CAUTION:** Sharp ridges may cut through the insulation on the wires and cause a short circuit or ground.

Figure 7-25
Conduit boxes *(Courtesy of Appleton Electric Company, Chicago, IL)*

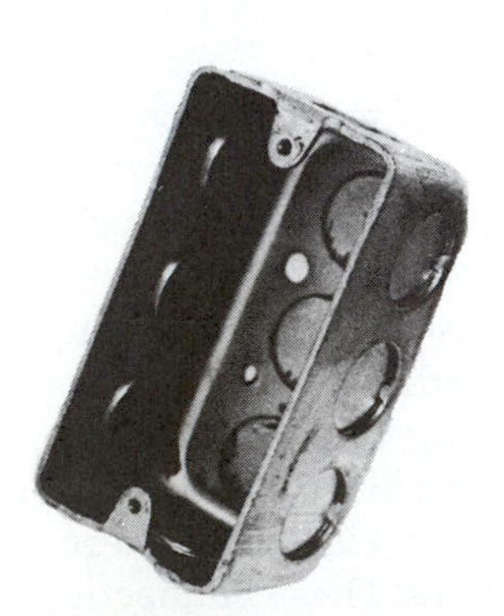
A. UTILITY BOX

B. 4-INCH OCTAGON BOX

C. 4-INCH SQUARE BOX

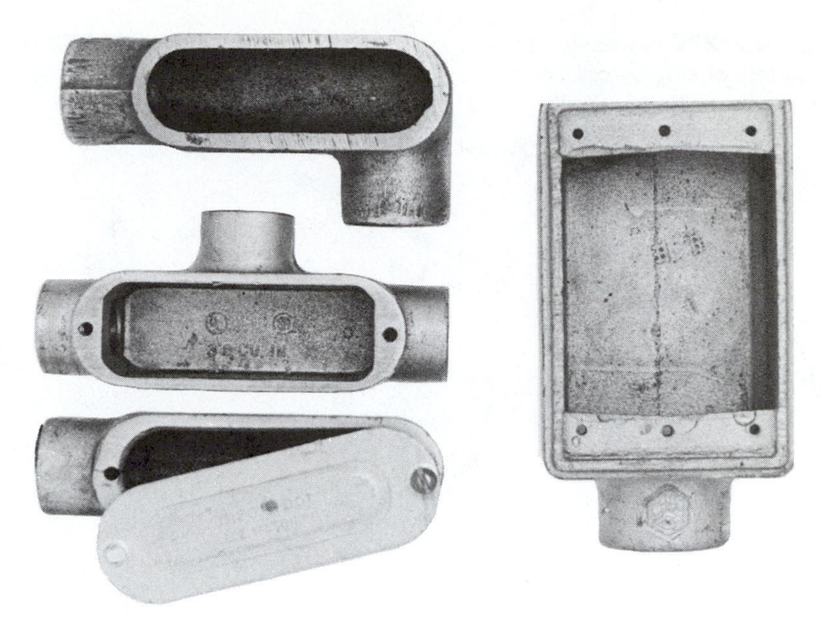

Figure 7-26
Assorted conduit fittings

After cutting and reaming, the conduit is then threaded with a standard industrial die. The length of the thread must be equal to the standard factory thread. Figures 7-27A through 7-27F illustrate some of the tools needed for this operation.

Many different types and angles of bends are required when installing conduit. These bends can be made with hand benders or power benders. When rigid metal conduit is to be bent, the hand

Figure 7-27A Chain vise for holding conduit
(Courtesy of Ridgid Tool Company, Elyria, OH)

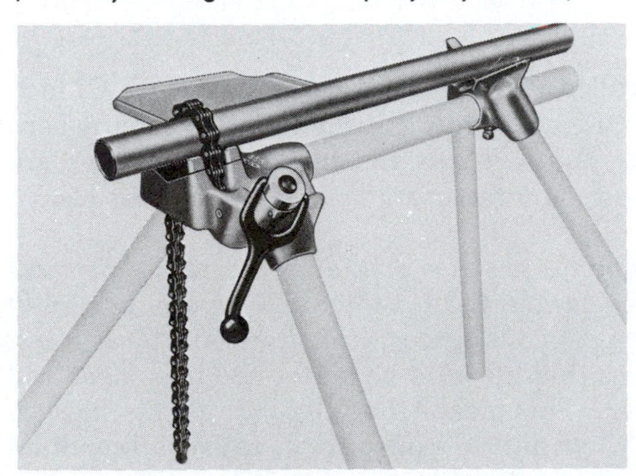

Figure 7-27B Adjustable die holder. It contains dies for cutting threads in conduit.
(Courtesy of Ridgid Tool Company, Elyria, OH)

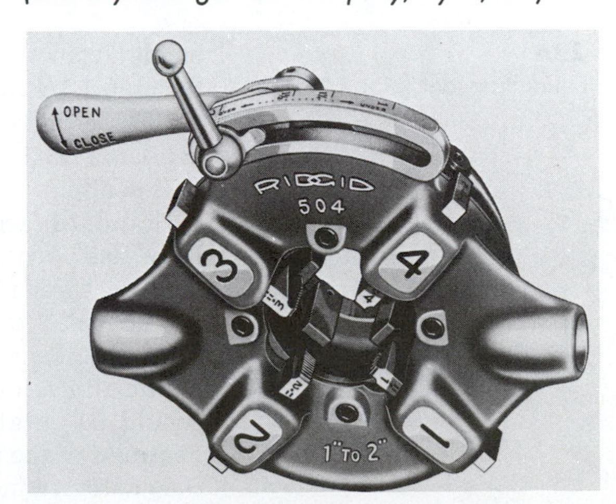

Figure 7-27C Power vise and die tool
(Courtesy of Ridgid Tool Company, Elyria, OH)

Figure 7-27D Conduit reamer
(Courtesy of Ridgid Tool Company, Elyria, OH)

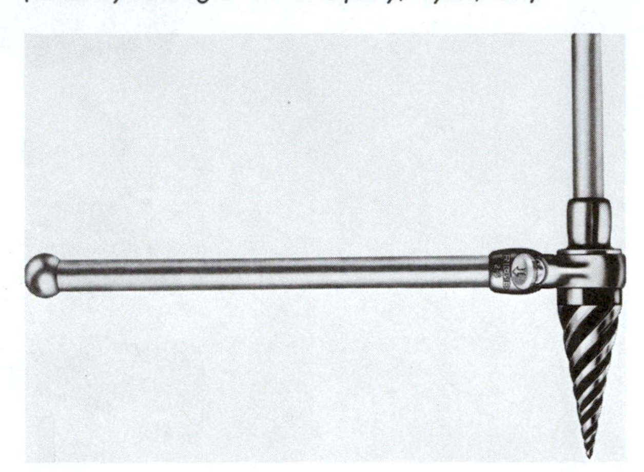

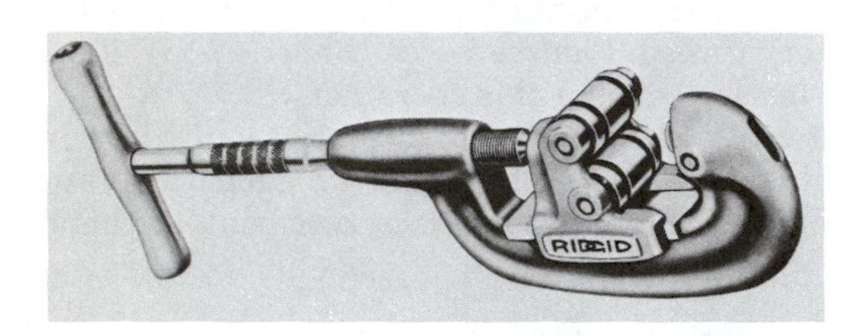

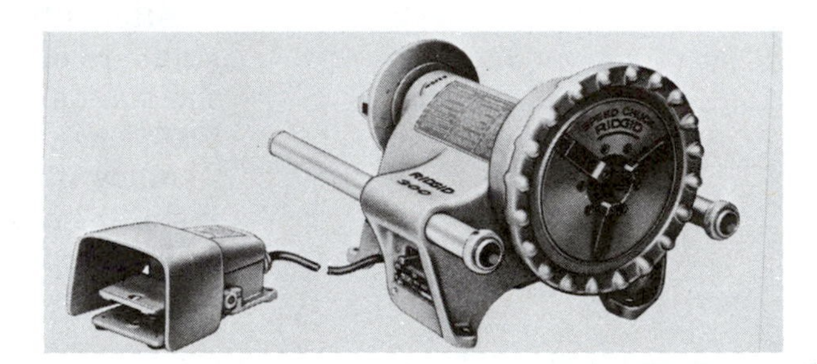

Figure 7-28A
Hand-type conduit bender (pipe hickey)

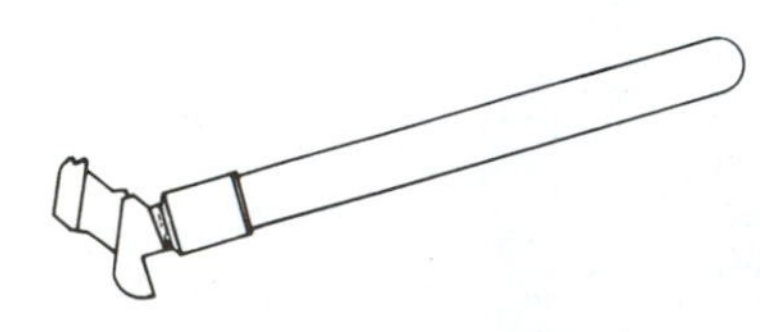

bender (sometimes called a pipe hickey) is used for sizes of 1/2 inch through 1 1/4 inches, Figure 7-28A. Larger sizes are usually bent with a power bender, Figure 7-28B. Instructions for bending are available from conduit manufacturers and manufacturers of conduit benders.

Standard factory bends are also available. These are particularly useful in sizes larger than 1 inch. They are generally manufactured for angles of 30 degrees, 45 degrees, and 90 degrees.

The *NEC*® requires that all work be performed in a "neat and workmanlike manner." This means that all conduit runs shall be plumb and/or level. When multiple lines are installed, all conduits should be parallel. Bends and fittings should be uniform. The neatness of the installation is frequently the yardstick by which the customer judges the quality of the work. Figure 7-29 shows multiple runs of conduit.

Figure 7-28B
Power-type conduit bender *(Courtesy of Greenlee Tool Company, Rockford, IL)*

Figure 7-29
Parallel runs of rigid metal conduit *(Courtesy of Republic Steel Corp., Cleveland, OH)*

Intermediate Metal Conduit

Intermediate metal conduit is a metal pipe designed to hold electrical conductors. It is lighter in weight and less rugged than rigid metal conduit. It can be used under most of the same conditions and is installed in the same manner as rigid metal conduit. (See *NEC® Article 345.*) The maximum size of intermediate metal conduit is 4 inches in diameter, electrical trade size. Rigid metal conduit is available in sizes of up to 6 inches in diameter.

Electrical Metallic Tubing

Electrical metallic tubing is a lightweight metal tube. It is manufactured in standard sizes from 1/2 inch to 4 inches in diameter, electrical trade size. It may be installed wherever rigid metal conduit is permitted, except where subject to severe mechanical damage or in hazardous locations. (See *NEC® Article 348.*)

Electrical metallic tubing is installed in much the same manner as rigid metal conduit except that it is not threaded. Special couplings, connectors, and fittings are designed for joining lengths together and for entering boxes or cabinets, Figure 7-30.

Electrical metallic tubing is less expensive than rigid metal conduit, and it is easier and faster to install. For this reason it is used extensively in residential, commercial, and industrial applications.

Flexible Metal Conduit

Flexible metal conduit consists of interlocking, spirally wound strips of steel. This type of conduit is much easier to install than the rigid types. It is easily cut with a hacksaw, and no bending tools are required. An installation using this wiring method is shown in Figure 7-31.

Flexible metal conduit is most commonly used with portable equipment and for wiring where excessive vibration may occur. It is frequently used for connecting motors to runs of rigid conduit. *NEC® Article 350* applies to flexible metal conduit.

Liquidtight Flexible Metal Conduit

Liquidtight flexible metal conduit is similar to flexible metal conduit, but has an outer jacket that is liquidtight, nonmetallic, and sunlight resistant. It is generally installed where both flexibility and protection from liquids are necessary. It is frequently used in machine shops and chemical plants. (See *NEC® Article 351.*)

Figure 7-30
Miscellaneous fittings for use with electrical metallic tubing

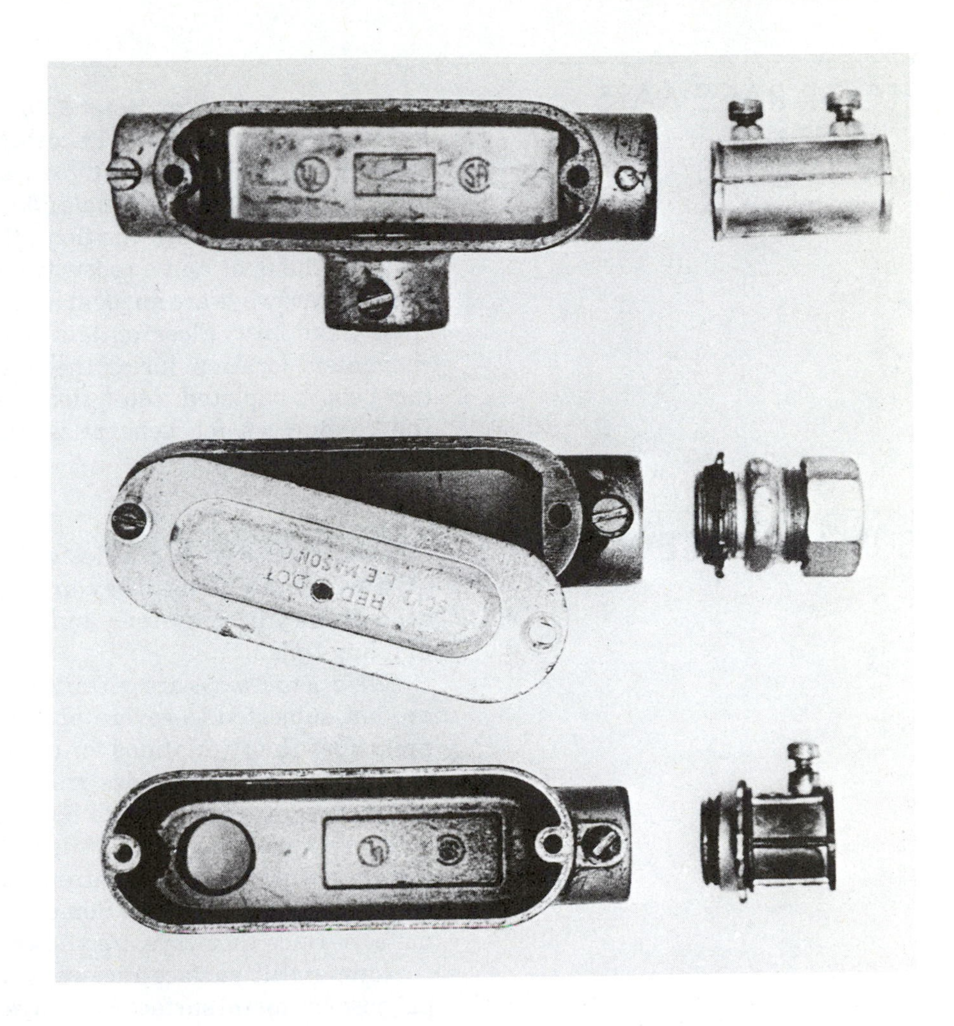

Figure 7-31
Installation of flexible metal conduit

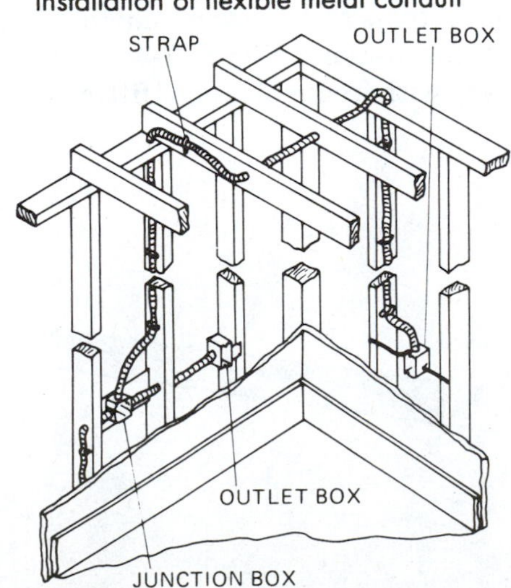

Flexible Metal Tubing

Flexible metal tubing is a circular, flexible liquidtight metal conduit. It does not have a nonmetallic covering, as on liquidtight flexible metal conduit. Flexible metal tubing must be used where it is not subject to mechanical damage. It is generally concealed in hung ceilings and similar areas.

FLOOR RACEWAYS

Floor raceways are designed in several styles. The more common styles are Underfloor Raceways (*NEC® Article 354*), Cellular Metal Floor Raceways (*NEC® Article 356*), and Cellular Concrete Floor Raceways (*NEC® Article 358*). These raceways are designed to be installed beneath the floor, flush with the finish floor, or to serve as the floor and a raceway. (See *NEC® Article 100*.)

Floor raceways are an ideal wiring method for offices because of their flexibility. Floor outlets may be installed at nearly any convenient location during the construction or after the building has been completed. Most floor raceways are designed to hold the conductors for general wiring as well as telephone and signal wires.

SURFACE RACEWAYS

Surface raceways are designed to be used for additions to existing installations. They are used where it is impractical to conceal the wiring system, and are designed to blend in with the building finish.

Surface raceways are permitted in dry locations and where they are not subjected to severe physical damage. *NEC® Article 352* prescribes the regulations for the use of surface raceways.

Surface raceways are designed in both metallic and nonmetallic types. They are intended to be installed on the finish surface of walls, ceilings, and floors. A length of metal surface raceway is shown in Figure 7-32A. Figure 7-32B shows a typical installation of metal surface raceway. Some of the fittings that are used with metal surface raceways are shown in Figure 7-33.

Nonmetallic surface raceways are used for most of the same purposes as metal surface raceways and, in general, are acceptable in similar locations. Good rules of thumb are to use the metal raceways when adding to metallic systems and to use nonmetallic raceways when adding to nonmetallic systems. Figure 7-34 illustrates a nonmetallic surface raceway with frequently used fittings.

Figure 7-32A
Length of metal surface raceway *(Courtesy of The Wiremold Co., West Hartford, CT)*

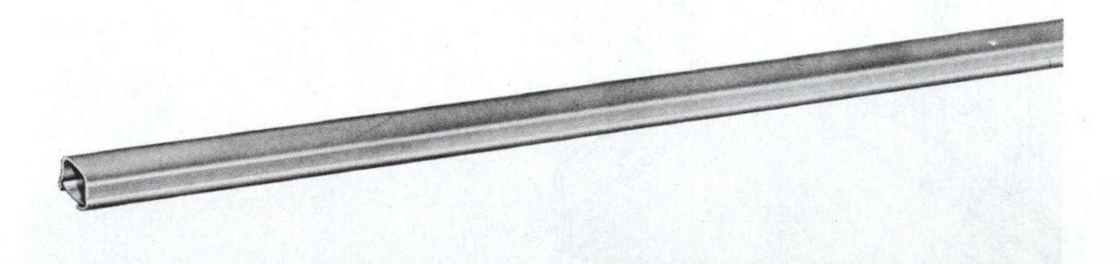

Figure 7-32B
Installation of metal surface raceway *(Courtesy of The Wiremold Co., West Hartford, CT)*

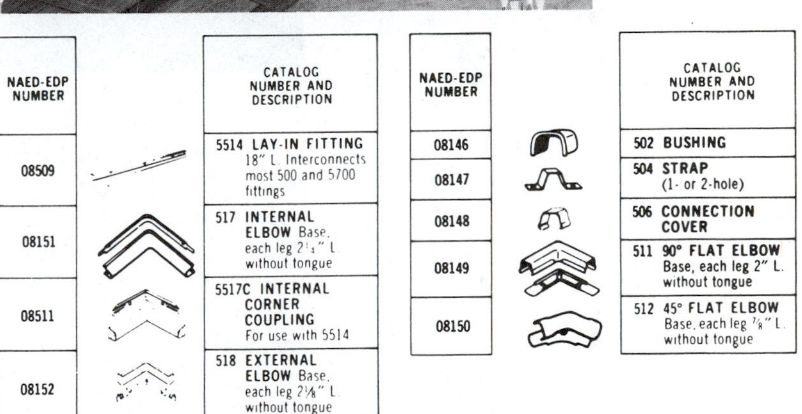

NAED-EDP NUMBER	CATALOG NUMBER AND DESCRIPTION
08509	5514 **LAY-IN FITTING** 18" L. Interconnects most 500 and 5700 fittings
08151	517 **INTERNAL ELBOW** Base, each leg 2¼" L without tongue
08511	5517C **INTERNAL CORNER COUPLING** For use with 5514
08152	518 **EXTERNAL ELBOW** Base, each leg 2⅛" L without tongue

NAED-EDP NUMBER	CATALOG NUMBER AND DESCRIPTION
08146	502 **BUSHING**
08147	504 **STRAP** (1- or 2-hole)
08148	506 **CONNECTION COVER**
08149	511 **90° FLAT ELBOW** Base, each leg 2" L. without tongue
08150	512 **45° FLAT ELBOW** Base, each leg ⅞" L. without tongue

Figure 7-33
Fittings used with metal surface raceway *(Courtesy of The Wiremold Co., West Hartford, CT)*

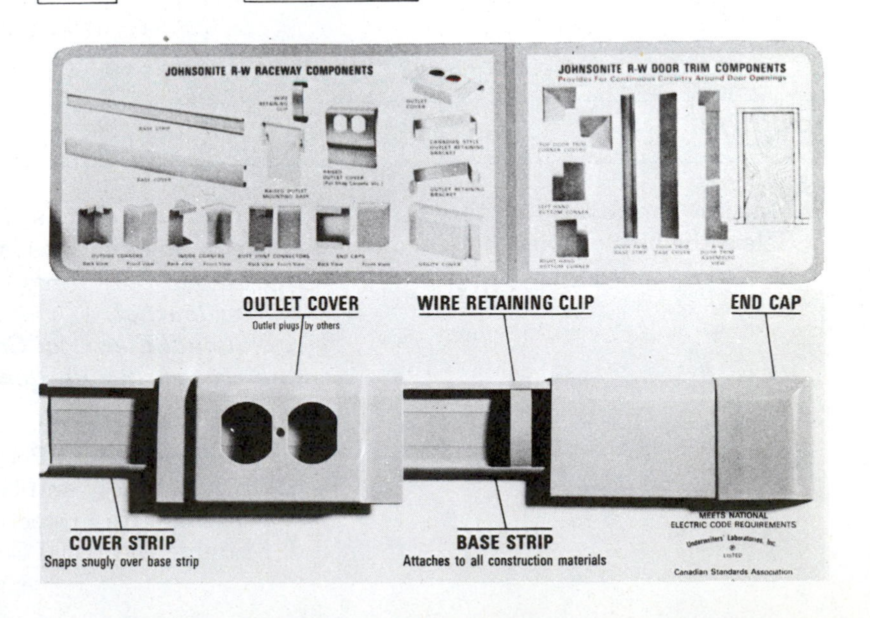

Figure 7-34
Nonmetallic surface raceway with fittings *(Courtesy of Carlon, An Indian Head Company, Cleveland, OH)*

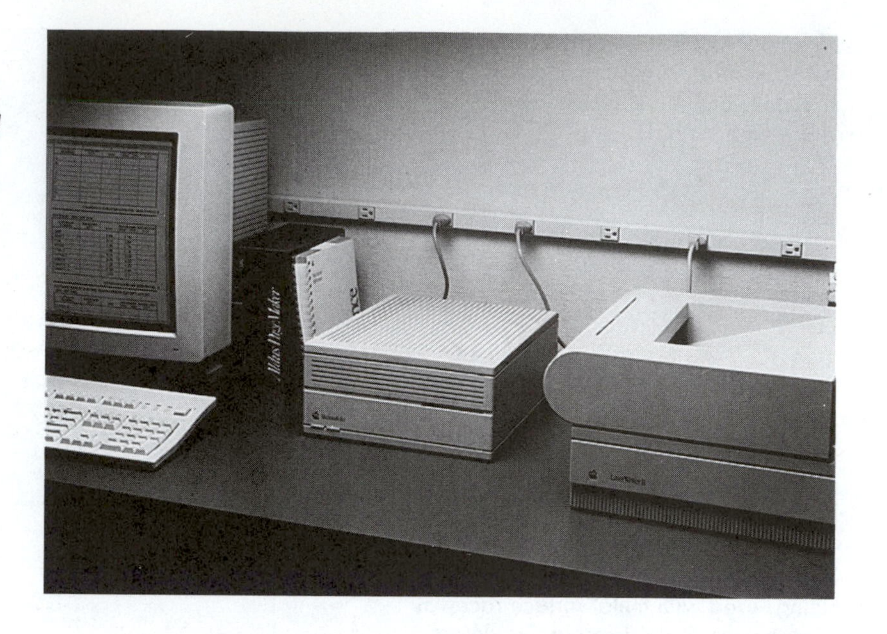

Figure 7-35
Multioutlet assembly—Ivory V2000 provides convenience receptacles on 6-inch centers *(Courtesy of The Wiremold Co., West Hartford, CT)*

A special type of surface raceway is called a *multioutlet assembly*. It is designed to hold connectors and receptacles that are assembled in the field or at the factory. A multioutlet assembly generally has several receptacles close to one another. It is frequently used where many small appliances are being used, such as assembly lines. Figure 7-35 shows and installation of this type of raceway.

REVIEW

A. Multiple choice.
Select the best answer.

1. The standard that is most frequently used to ensure safe electrical installations is the
   a. *National Standards.*
   b. *Fire Journal.*
   c. *National Electrical Code®.*
   d. *Electrical Installation Standards.*

2. The *National Electrical Code®* is published by the
   a. U.S. Government Printing Office.
   b. Office of Safety and Health Administration.
   c. National Fire Protection Association.
   d. National Electrical Contractors Association.

REVIEW *(continued)*

A. Multiple choice.
Select the best answer (continued).

3. Armored cable is a
   a. rigidly sheathed cable.
   b. flexible metal-covered cable.
   c. cable enclosed in copper armor.
   d. cable enclosed in rigid steel armor.
4. The purpose of the grounding conductor in armored cable is to
   a. carry away stray currents caused by accidental grounds.
   b. shield the circuit from lightening.
   c. decrease static interference.
   d. replace the neutral conductor.
5. In an installation of armored cable in a box, the minimum amount of armor to be removed is
   a. 4 inches.
   b. 6 inches.
   c. 8 inches.
   d. 12 inches.
6. An antishort is a
   a. fuse.
   b. circuit breaker.
   c. bushing.
   d. locknut.
7. The sheath on nonmetallic-sheathed cable must be
   a. moisture resistant.
   b. flame retardant.
   c. heat resistant.
   d. both a and b.
8. In comparison to other types of cables that are suitable for the same purpose, nonmetallic-sheathed cable is
   a. less expensive.
   b. more expensive.
   c. about the same price.
   d. less flexible.
9. When nonmetallic-sheathed cable is used in moist locations, it must be type
   a. MNA.
   b. NMC.
   c. NMM.
   d. MNW.

REVIEW *(continued)*

A. Multiple choice.
Select the best answer (continued).

10. Boxes and fittings used with nonmetallic-sheathed cable may be made of
    a. metal.
    b. nonmetallic materials.
    c. both a and b.
    d. none of the above.
11. Type SE cable may be used for
    a. the main service conductors.
    b. electric ranges and dryers.
    c. both a and b.
    d. underground service conductors.
12. Standard sizes of serive-entrance cables are
    a. No. 8 AWG through No. 4/0 AWG.
    b. No. 14 AWG through No. 2 AWG.
    c. No. 6 AWG through 1000 MCM.
    d. No. 10 AWG through No. 2 AWG.
13. Type UF cable may be used
    a. only indoors.
    b. only where subject to the weather.
    c. indoors, where subject to the weather, and underground.
    d. only underground.
14. Nonmetallic conduit is made of
    a. glass, wood, or hard rubber.
    b. fiber, soapstone, or plastic.
    c. nylon, latex, or leather.
    d. all of the above.
15. The most common method used to install conductors in conduits is to
    a. push them in.
    b. pull them in.
    c. slide them in as the conduit is being installed.
    d. all of the above.
16. Rigid metal conduit is manufactured in lengths of
    a. 6 feet, including one coupling.
    b. 10 feet, including one coupling.
    c. 8 feet (no couplings included).
    d. 12 feet, including two couplings.

REVIEW *(continued)*

A. Multiple choice. Select the best answer (continued).

17. Rigid metal conduit may be cut with a
    a. pipe cutter.
    b. torch.
    c. tap.
    d. die.
18. Rigid metal conduit is also cut with a hacksaw. The hacksaw blade should have
    a. 32 teeth to the inch.
    b. 30 teeth to the inch.
    c. 24 teeth to the inch.
    d. 18 teeth to the inch.
19. Floor raceways are designed to be installed
    a. on the surface of the floor.
    b. beneath the floor.
    c. above the floor.
    d. none of the above.
20. Surface raceways are designed to be installed
    a. within the walls.
    b. concealed within the hung ceiling.
    c. on finished walls, ceilings, and floors.
    d. underground.

B. Give complete answers.

1. Describe the general purpose of the *NEC*®.
2. What is the primary goal of the National Fire Protection Association?
3. What are some of the types of electrical installations covered by the *NEC*®?
4. List three types of electrical installations that are not covered by the *NEC*®.
5. Describe armored cable.
6. What is the purpose of the bonding strip in armored cable?
7. Describe how armored cable is installed in a box or cabinet.
8. Describe the locations in which it is permissible to install armored cable.
9. What is nonmetallic-sheathed cable?

REVIEW *(continued)*

B. Give complete answers (continued).

10. List at least three other names by which nonmetallic-sheathed cable is commonly known.
11. In what type of location is it permissible to install type NM cable?
12. What is the difference between NM cable and NMC cable?
13. Describe how nonmetallic-sheathed cable is installed in a cabinet.
14. What is the best-suited type of strap for fastening nonmetallic-sheathed cable to the surface?
15. Describe service-entrance cable.
16. Describe how service-entrance cable is brought into a house at the service.
17. Describe an underground feeder and branch-circuit cable.
18. In what type of installation would nonmetallic-sheathed cable best serve the purpose?
19. Name three styles of metal conduit.
20. How is electrical metallic tubing coupled together?
21. Why is it necessary to ream the inside of a conduit after it has been cut?
22. What is a pipe hickey?
23. Describe flexible metal conduit.
24. Describe liquidtight flexible metal conduit.
25. List three types of floor raceways.

8
WIRING APPLICATIONS

Objectives

After studying this chapter, the student will be able to:

- Apply the basic procedures used to calculate the size of various types of electrical services.
- Explain how the number and size of feeders and branch circuits are determined for various installations.
- Explain how to size the system grounding conductor and describe the various methods of grounding systems.
- List the general safety requirements for various electrical installations.
- Describe some common types of devices and equipment used in electrical installations and explain the reasons for using them.

There are certain basic procedures and methods that pertain to all electrical installations. When designing new installations or adding to existing ones, one must keep safety in mind as the prime consideration. Local and national codes must be observed. In a determination of the size of service, feeders, and branch circuits, allowances should be made for future needs. *NEC® Articles 215, 220, 225*, and *230* provide some guidelines for calculating the ampacity of conductors and equipment.

The local electric utility company should be contacted to determine the type of power available. There are several different values of voltage and types of systems that are in common use. Some questions that must be considered are:

- Does the system require ac, dc, or both?
- If the supply is ac, what type of system is required?

- In what part of the building will the service be located?
- Is the supply overhead or underground?
- What is the best method for grounding the system?

It is advisable to consult the local electrical inspector before the job is started. Most localities require that the electrical contractor file for a permit to do the work. Permits are obtained from the inspecting authority's office. Often, localities have special regulations that are peculiar to the area. The electrical inspector can make these regulations available beforehand.

Most states and/or municipalities require the licensing of electricians and/or electrical contractors. There are generally two types of licenses: *journeyman* and *master* (contractor). The journeyman's license is a certificate issued to electricians who work for contractors. These electricians are usually the workers who perform the actual installation. They are responsible for providing a complete installation according to all safety rules and regulations. The master's or contractor's license is a certificate issued to the person or company employing the electrical workers.

The requirements for obtaining a license vary for different areas. However, certain requirements are common to most areas. For a journeyman electrician's license, one must:

- Work a specific number of years under the direct supervision of a licensed journeyman electrician.
- Pass a written and practical test.

For a master's (contractor's) license, one must:

- Obtain a journeyman's license.
- Work a specific number of years as a licensed journeyman electrician.
- Pass a written test.

The tests required for the licenses are generally based on safety rules and regulations, usually as set forth in the *NEC®* and local codes. They are also based on practical knowledge and knowledge of electrical theory. The contractor's test may require more knowledge of design and maintenance.

RESIDENTIAL WIRING

The electrician or engineer should begin by referring to the basic floor plan. This plan provides the necessary information for calculating the general lighting load, the number and location of convenience outlets, and other electrical equipment. Other detailed information is provided in elevation and riser diagrams.

A simple and reliable method for determining the service size and the number of branch circuits is outlined in *NEC® Chapter 9*.

To determine the general lighting load, the "volt-amperes per square foot" method is used. The unit load for dwelling occupancies is 3 volt-amperes per square foot (VA/ft^2). In a calculation of the area of the building, the outside dimensions should be used. Unoccupied areas, open porches, and similar areas need not be included. The area of each floor should be determined and the total used for the above calculations.

After determination of the general lighting load, it is necessary to add all other electrical loads. For each dwelling unit, a load of not less than 3000 volt-amperes must be included for small appliances. Each laundry must be wired with a 1500-volt-ampere circuit for appliances. Other loads such as electric ranges, water heaters, air conditioners, dishwashers, electric heaters, and special lighting loads must be included.

> *Note*: In a calculation of ac loads, the term volt-ampere is used to indicate the product of the voltage and the current.

The *NEC*® does not require that the service be sized to carry the total connected load. In most residential occupancies, it is very unlikely that every light, appliance, and convenience outlet will be used at maximum capacity at the same time. For practical and economical purposes, the *NEC*® allows the application of a *demand factor*. A demand factor is the ratio of the maximum load used at any one time to the total connected load.

Applying the demand factor reduces the size of the service conductors and equipment. Caution and common sense must be used in order to avoid the installation of an inadequate service. It is always advisable to allow for future increases in the load. (See *NEC® Article 220*.)

The *NEC*® provides the minimum requirements for safety. Adhering to these rules ensures an installation that is reasonably safe, but not necessarily convenient. In addition, the *Code* cannot cover every unique condition that may arise.

It is assumed that the electrical worker will want to make an installation of high quality. Some rules that will aid in performing a flexible, convenient, trouble-free installation are as follows:

1. Install a 100-ampere, three-wire or larger service for a single-family dwelling.
2. Install a 200-ampere, three-wire or larger service for two- and three-family dwellings.
3. Install at least one double convenience outlet over each counter space in the kitchen.
4. For counter spaces longer than 7 feet, install one double convenience outlet for every 4 feet or fraction thereof.

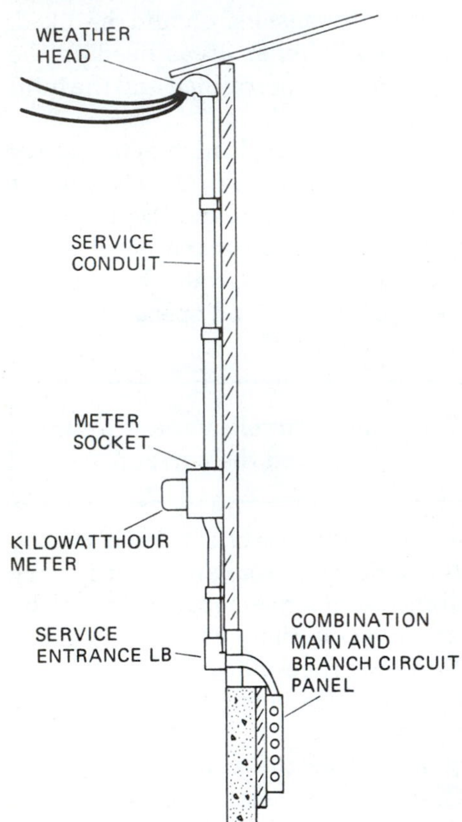

Figure 8-1
Typical service for a single-family residence

5. Do not install more than two double convenience outlets on any one appliance branch circuit.
6. Install no fewer than two 20-ampere branch circuits for the laundry.
7. Install ground fault interrupting devices for any convenience outlets that are within reach of water pipes or other grounded objects.
8. Do not load branch circuits to their maximum capacity; allow for future additions.
9. Always follow the standards prescribed by the *National Electrical Code®*.

Residential Service

A *residential service* consists of the service-entrance conductors, the kilowatt-hour meter, the main switch or circuit breaker, the overcurrent devices, and the common ground. Figure 8-1 shows a typical service for a single-family dwelling. Figure 8-2 shows two types of panels used for residential services.

NEC® Article 230 prescribes the regulations for services. All electrical personnel should be familiar with these rules.

The electrical service is the heart of the wiring system. It contains the safety devices (fuses and/or circuit breakers) that protect the system from excessive current flow. The common ground provides a path for stray currents, reducing the possibility of fires.

The main electrical service terminates in branch circuits, using fuses or circuit breakers to protect the circuits. Each branch circuit supplies power to certain areas or equipment. These circuits should be divided equally throughout the house, when practical. Special circuits that may be required by the *NEC®* and local codes must also be installed.

Fuses and circuit breakers are "safety valves." They disconnect the circuit when the current becomes too great. The operation of a fuse or circuit breaker indicates excessive current flow. The circuit should be checked and the fault eliminated before the fuse is replaced or the circuit breaker is reset.

CAUTION: It is very important to install the proper size of fuse or circuit breaker. Oversized fuses or circuit breakers will not give adequate protection against excessive current flow.

All ac systems are grounded at the service. The grounding conductor provides a low-resistance path for current caused by lightning and leakage currents. It should not carry current under normal conditions.

Figure 8-2A
Fuse panel used for a residential service

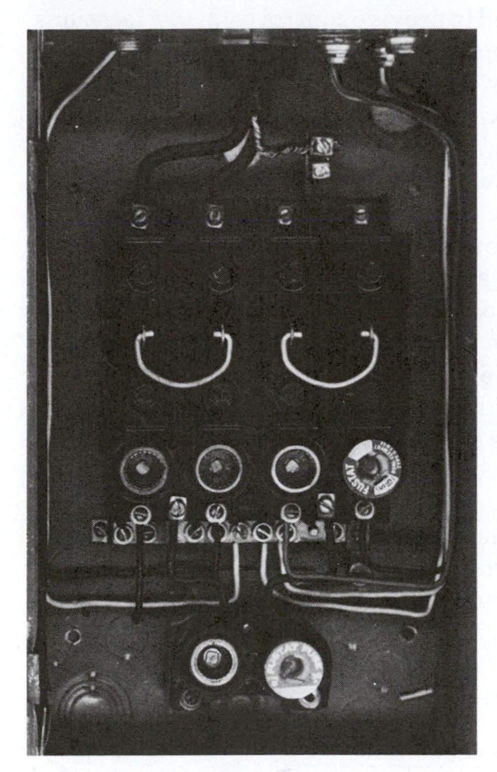

Figure 8-2B
Circuit-breaker for a single-family residence *(Courtesy of Square D Company, Lexington, KY)*

Calculating the Size of the Common Ground

The common grounding conductor must be large enough to conduct stray currents to ground without overheating. It is a very important part of the service. *NEC® Article 250* prescribes the rules for grounding. The common grounding conductor connects both the wiring system and the equipment to the grounding electrode.

The resistance of the common ground must be kept to a minimum. This can be done by ensuring tight connections and installing an adequately sized grounding conductor and electrode. *NEC® Table 250-94* lists the minimum size for the grounding conductor based on the size of service. *NEC® Section 250-H* sets the standards for the grounding electrode. The type of electrode to which the common grounding conductor is connected should be selected with care. A metal underground water-piping system generally provides an effective ground.

Ground-Fault Circuit Interrupter

A *ground-fault circuit interrupter* is a device that senses small currents to ground. The device opens the circuit when the current to ground exceeds a predetermined value, usually between 5 milliamperes and 30 milliamperes (mA).

Most circuits are protected against overcurrent by 15-ampere or larger fuses or circuit breakers. This protection is adequate against short circuits and overloads. Leakage currents to ground may be much less than 15 amperes and still be hazardous.

> **CAUTION:** The metal frame of electrical equipment will become energized if the ungrounded conductor (the hot wire) accidentally makes contact with it. A person touching the frame and the ground, or a grounded object, completes the circuit. Even a minute amount of current flowing through the heart, lungs, or brain can be fatal. It is therefore very important to take the necessary precautions to prevent leakage currents through the human body. The ground-fault circuit interrupter serves this purpose.

Lighting Circuits

Most lighting fixtures are controlled by switches. These switches are located in convenient places throughout the dwelling. They are usually installed in boxes set in the walls at a height of 42 inches to 48 inches.

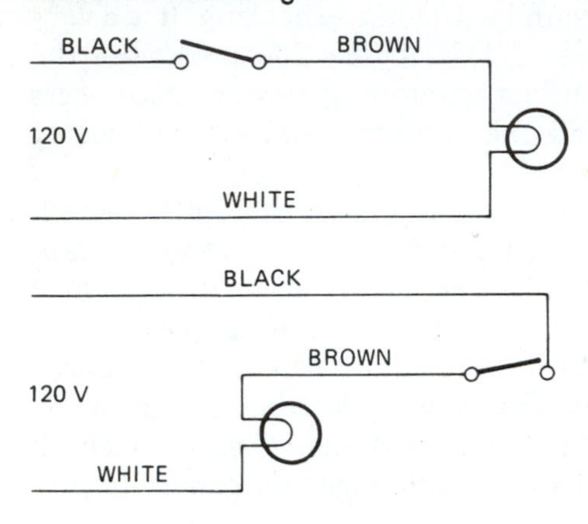

Figure 8-3
Two methods for connecting a single-pole switch to control a light

There are three common styles of switches used to control lights. The single-pole switch controls the light from one specific location. Three-way switches are used to provide control from two locations. If control is desired from more than two locations, a combination of three-way and four-way switches is used. Two methods for connecting a single-pole switch to control a light are depicted in Figure 8-3. Notice the color of the insulation for the wires. These colors are commonly used to identify the purpose of each conductor. The black is usually the ungrounded (hot) conductor from the supply to the switch. The brown wire is the switch leg (the wire that connects the switch to the lamp). The white wire is the grounded conductor, connecting the load to the grounded supply conductor.

Figure 8-4 shows three methods for connecting three-way switches to control a single light. Notice the color code for this arrangement. The black wire is the ungrounded (hot) conductor from the supply to one of the three-way switches. The orange wires are called the *travelers* (the conductors that connect the 2 three-way switches together). The white conductor is the conductor that connects the load to the grounded supply conductor.

The operation of this circuit can be better understood by tracing the path of current. Assume that the current is flowing out the

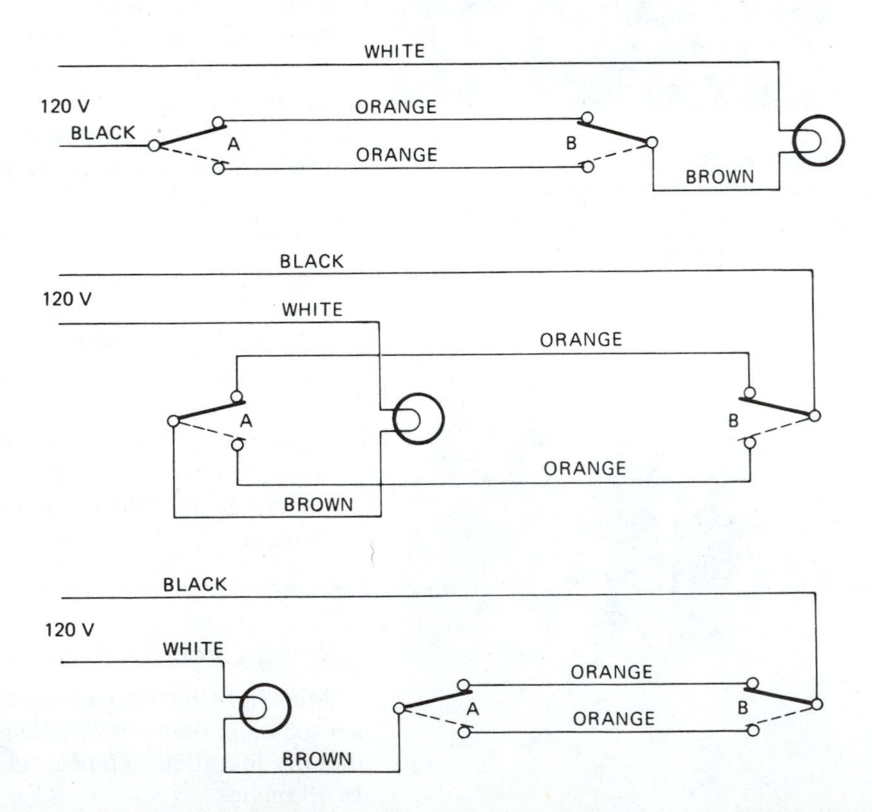

Figure 8-4
Connections for three-way switches

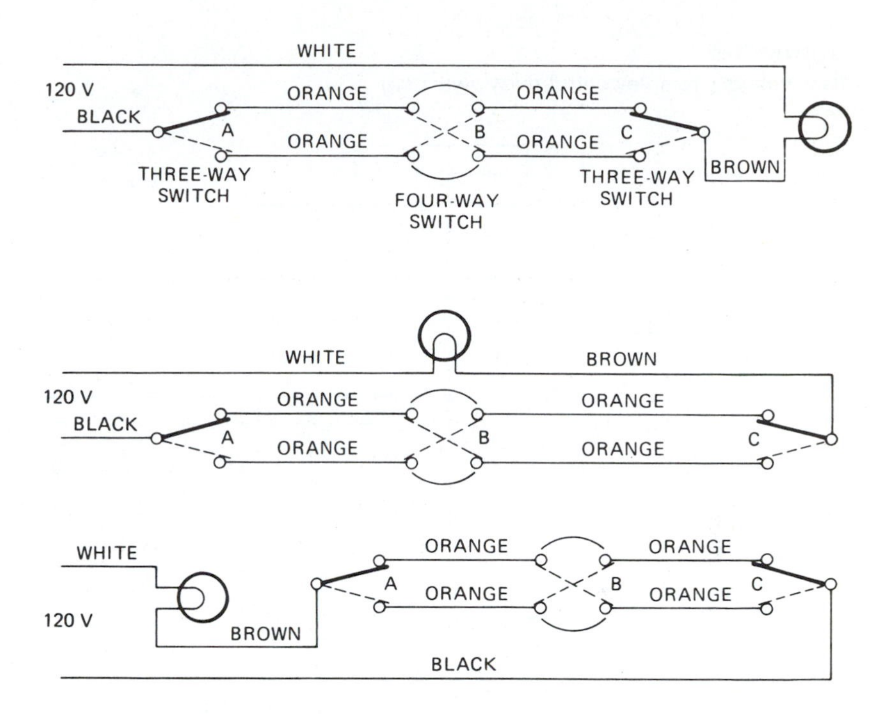

Figure 8-5
Connections for control of light fixture from more than two locations

black wire (Figure 8-4) to the switch. If the switch contacts are made in the solid position, the circuit is complete, and the lamp will glow. Moving switch A contacts to the dashed position breaks the circuit, and the lamp will be dark. Leaving the switch A contacts in the dashed position and moving the switch B contacts to the dashed position will again complete the circuit, and the lamp will glow. In this way, the lamp can be controlled from both switches.

Figure 8-5 shows a combination of three-way and four-way switches. The four-way switch changes the connections between the traveler conductors, thus changing the current path each time it is switched.

To trace the current path, refer to Figure 8-5 (top). Switches A and C are the three-way switches; switch B is a four-way switch. Assume that the current flows from the supply through the black wire to switch A and that all the contacts are in the solid position. The circuit is complete and the lamp will glow. If A is moved to the dashed position, the circuit is open and the lamp is dark. Moving switch B to the dashed position completes the circuit from the supply through the black wire to switch A, through the dashed path of switch A to the bottom traveler, through the traveler to switch B, through the dashed path of switch B to the top traveler, through the top traveler to switch C, through the solid contacts of switch C to the brown wire, through the brown wire to the lamp,

Figure 8-6
Low-voltage, remote-control relay switching

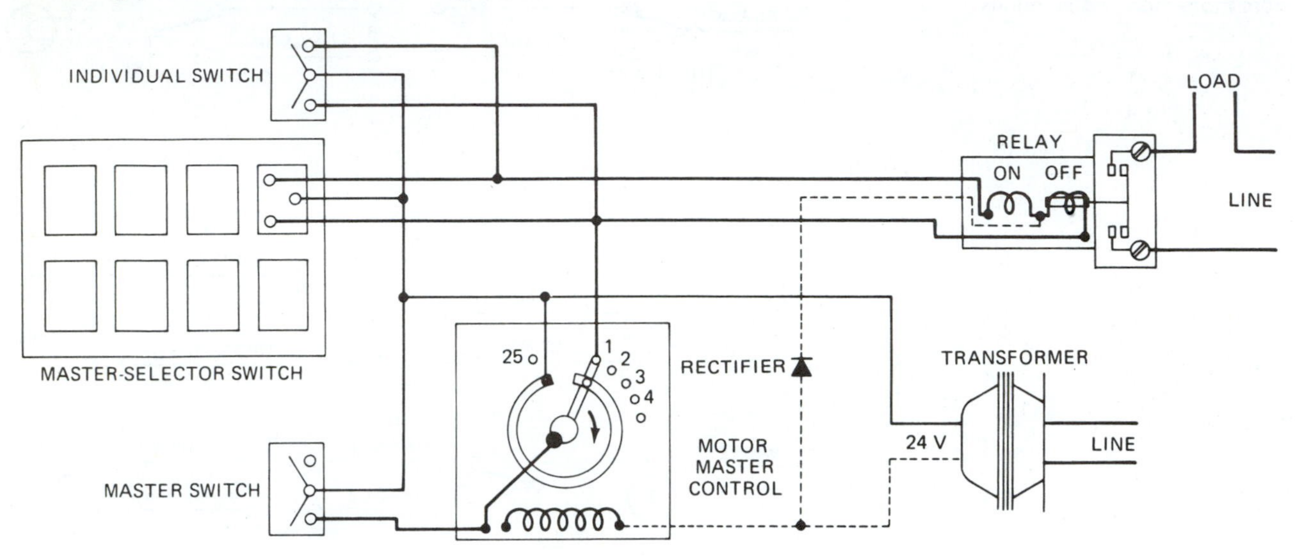

and back to the source by way of the white wire. The lamp is again lighted. Moving switch C to the dashed position again opens the circuit, and the lamp is dark. Continued operation of switches A, B, and C will alternately turn the lamp on and off.

NEC® Article 380 prescribes methods to install switches.

At times, lights are controlled by low-voltage switches and relays. This system is useful because it provides much more flexibility of control.

The individual switches of this system control a relay (an electromagnetic switch), which in turn controls the light. The relays may be located at the lighting fixtures or in one central location. The individual switches control the current to the low-voltage coil (usually 24 volts) in the relay. This eliminates the necessity of using three-way and four-way switches when control is desired from more than one location.

A master selector control may also be installed. This control permits all of the lights or specific groups of lights to be controlled from one location. The master control can also be wired with the burglar alarm system so that all the lights in the house will come on when the alarm is energized. Figure 8-6 illustrates a low-voltage, remote-controlled lighting system.

Alarm Systems

For the wiring of new houses or rewiring old ones, it is advisable to install burglar and fire alarm systems. There are many styles and types available. The selection of the type to use is a matter of personal preference. All systems should have a backup supply (usually batteries) in case of loss of power. When the main source of power is interrupted, a transfer switch automatically connects the batteries to supply power for the system. When the main power is restored, the transfer switch automatically disconnects the batteries and reconnects the main supply. The batteries should be checked periodically to ensure that they are adequately charged.

A fire alarm system may consist of smoke detectors, rate-of-rise heat detectors, and/or fixed-temperature heat detectors. A smoke detector reacts to smoke to start the alarm. A rate-of-rise heat detector energizes the alarm mechanism when the temperature increases very rapidly. A fixed-temperature heat detector starts the alarm when the temperature reaches a predetermined fixed value.

Burglar alarm systems fall into two major categories: those that signal when someone tries to enter a house (perimeter protection) and those that signal after someone enters the house (zone protection). Perimeter protection uses magnetic switches on doors and windows, which signal when a door or window is opened. They can be combined with other special devices, which signal when the glass is broken. Zone protection, on the other hand, signals only after someone has entered the house. It operates by detection of changes in light, heat, or ultrasonic sound waves. With either system, contacts can be placed under carpets to signal when someone steps on them. Sometimes, carpet contacts are used alone. Very often, perimeter protection devices and zone protection devices are used together to provide better security.

COMMERCIAL AND INDUSTRIAL WIRING

Commercial buildings are structures such as stores, offices, and warehouses. Commercial wiring is generally installed in rigid metal conduit, intermediate metal conduit, electrical metallic tubing, surface raceway, and cellular floor raceway. Circuit breakers and fuses are used for overcurrent protection.

Industrial installations consist of factories, foundries, machine shops, metal refineries, and similar structures. The wiring is generally installed in rigid metal conduit, wireways, and busways. Industrial establishments also use fuses and circuit breakers for overcurrent protection.

Industrial and commercial installations frequently use large amounts of power, thus requiring the installation of transformers on the premises. The power is supplied at a high voltage, such as 2300 volts. The transformers lower the voltage to 277/480 volts

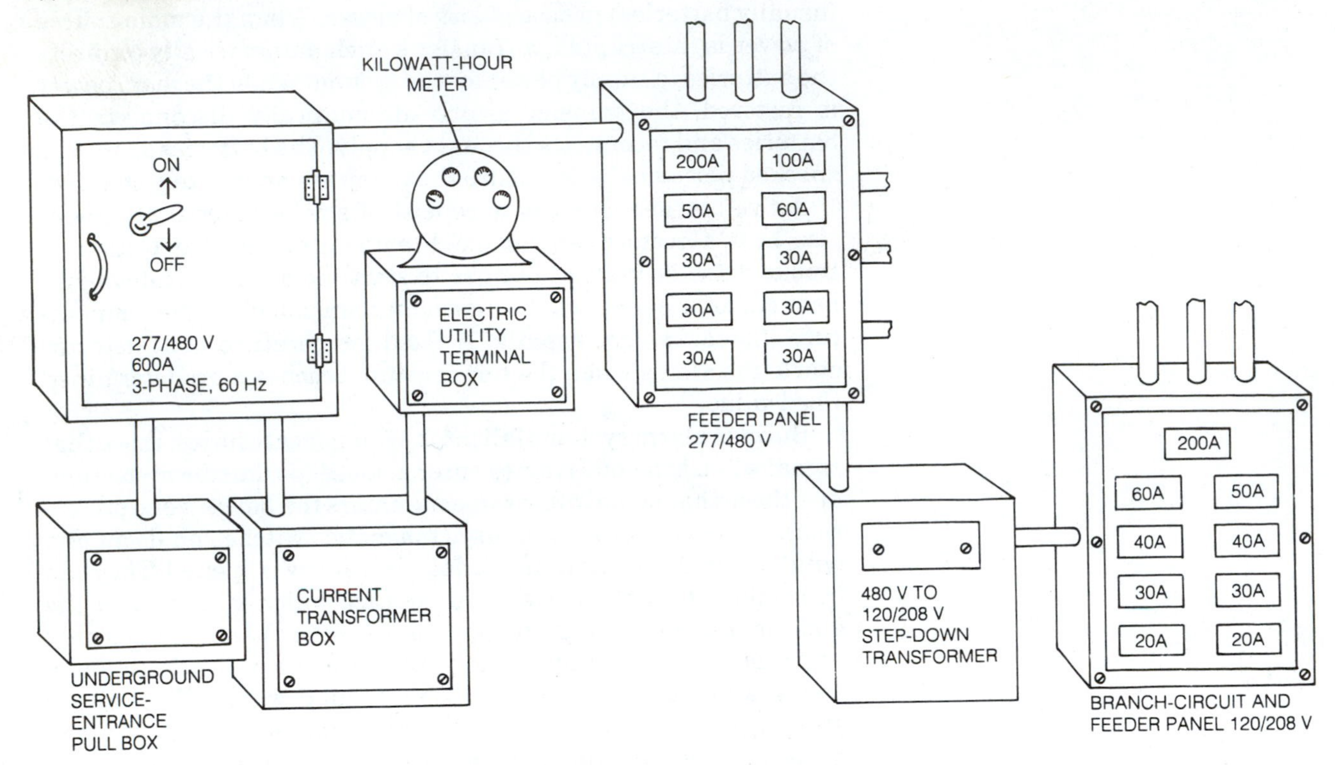

Figure 8-7
Typical commercial service

and/or 120/208 volts. Other voltages sometimes used are 120/240 volts and 480 volts. These values are standard voltages for alternating current systems, and are discussed in a later section.

Commercial and Industrial Services

The volt-amperes-per-square-foot method for calculating general lighting loads may be used for commercial and industrial buildings (See *Article 220* of the *NEC®*.) All other loads connected to the installation must be added. The service may be as small as 100 amperes or as large as several thousand amperes. The service, feeders, and branch circuits should be of adequate size to allow for future additions. Figure 8-7 shows a typical commercial service.

Calculating Loads

Determining the proper size of the service, feeders, and branch circuits requires accurate information and calculations. To make the calculations, it is common practice to start with the branch circuits and combine their loads at each panel.

To determine the feeder load, the combined branch-circuit loads are added to the load allowance for expansion, and the result is multiplied by the demand factor. (The demand factor varies with the type of installation.) After all the feeder loads have been calculated, they are combined to determine the size of the service. If the ampere rating of the service exceeds the maximum value for standard conductor sizes, multiple conductors must be installed in parallel. For example, if the ampacity of the service is 900 amperes, two type THW, 2,000,000 circular-mil cables must be installed for each conductor of the service. For determining conductor ampacity and for multiple connections, see *NEC® Article 310*.

Overcurrent Protection

A very large amount of current is usually used in commercial and industrial areas. Therefore, it is important to use extreme care in selecting *overcurrent protection devices*. The most common types of overcurrent protection devices are fuses and circuit breakers. Fuses and circuit breakers have two current ratings. One rating is the current value at which they will disconnect the circuit from the supply. The second rating is the current value at which they can operate without damage to the equipment to which they are connected and in which they are housed. If a fuse is rated at 15 amperes, the link will begin to melt when more than 15 amperes pass through the fuse. The length of time it will take to open the circuit depends upon the amount of current flowing. For example, if 16 amperes are flowing in a circuit protected by a 15-ampere fuse, it may take several minutes for the link to melt. If 1000 amperes are flowing in the circuit, however, the link will melt in less than a second. This time factor is very important to consider when selecting the correct overcurrent device.

A *short circuit* is a place in the wiring system where conductors of opposite polarity make contact. In theory, one might say that a short circuit is a circuit with zero resistance, but this is not true in actual practice. The conductors and equipment between the generator and the point of contact have resistance. The resistance, along with other variables, places some restriction on the amount of current that will flow under short-circuit conditions.

If the opposition to current flow is very low, the short-circuit current will build up to several thousand amperes in less than a second. A fuse rated at 15 amperes will start to melt when the current exceeds 15 amperes, and it will eventually open the circuit. The ordinary zinc link fuse is rated at 10,000-amperes interrupting current.

> **CAUTION:** If, under short-circuit conditions, the current should build to a value in excess of 10,000 amperes before the link melts, the fuse could explode. Such an explosion may cause damage to equipment and injure personnel. It is the responsibility of the electrical worker to determine the available short-circuit current at the installation. This information can usually be obtained from the utility company. Fuses and circuit breakers are available for various current-interrupting capacities.

Fuses having a current-interrupting capacity of other than 10,000 amperes have the rating marked on the fuse. Circuit breakers having a current-interrupting rating of other then 5000 amperes have the rating indicated on the body of the breaker. Figure 8-8 illustrates some of the most common fuses and circuit breakers.

Circuit breakers are mechanical devices that respond to current changes. These devices interrupt current flow, within rated values, without injury to themselves. There are three general types of circuit breakers:

- Thermal
- Magnetic
- Thermal-magnetic

The *thermal circuit breaker* responds to temperature. An increase in current produces an increase in the temperature of the

Figure 8-8
Various fuses and circuit breakers

sensing element. When the temperature reaches a predetermined value, the breaker opens the circuit.

The *magnetic circuit breaker* responds to a magnetic field produced by the current flowing through the breaker. As the value of current increases, the magnetic field becomes stronger. When the field strength reaches a predetermined value, the breaker disconnects the circuit from the supply.

The *thermal-magnetic circuit breaker* is a combination of both of the above types. It responds to both temperature and magnetism.

In some cases, the short-circuit current may exceed the interrupting rating of the circuit breaker or fuse. It then becomes necessary to install overcurrent devices with a greater interrupting capacity or to use current-limiting overcurrent devices. A *current-limiting overcurrent device* is a fuse or circuit breaker that opens the circuit much more quickly than the ordinary device does. Fast-acting overcurrent devices open the circuit before the current can build to an excessive value. Current-limiting devices are used extensively in commercial and industrial establishments.

The main overcurrent device for a commercial building may be rated at several thousand amperes. Extending from this device are feeders that are used to distribute the power throughout the building. These feeders are large conductors, which are protected by fuses or circuit breakers and which terminate in branch-circuit panels. The branch circuits are the final branches of the wiring system, with conductors smaller than the feeder conductors. Branch circuits are also protected by overcurrent devices rated according to the conductor sizes. Figure 8-9 illustrates a typical commercial installation.

Selection of Overcurrent Devices

The electrical system is the heart of a commercial or industrial establishment. The greatest operating expense is a shutdown. The major causes of shutdowns are short circuits, overloads, and burnouts. Careful selection of overcurrent devices can minimize shutdowns.

In the selection of overcurrent devices, the following factors should be considered:

- Maximum continuous current
- Maximum operating current
- Maximum interrupting current
- Frequency of the system
- Duty cycle
- Type of load

Figure 8-9
Typical commercial installation

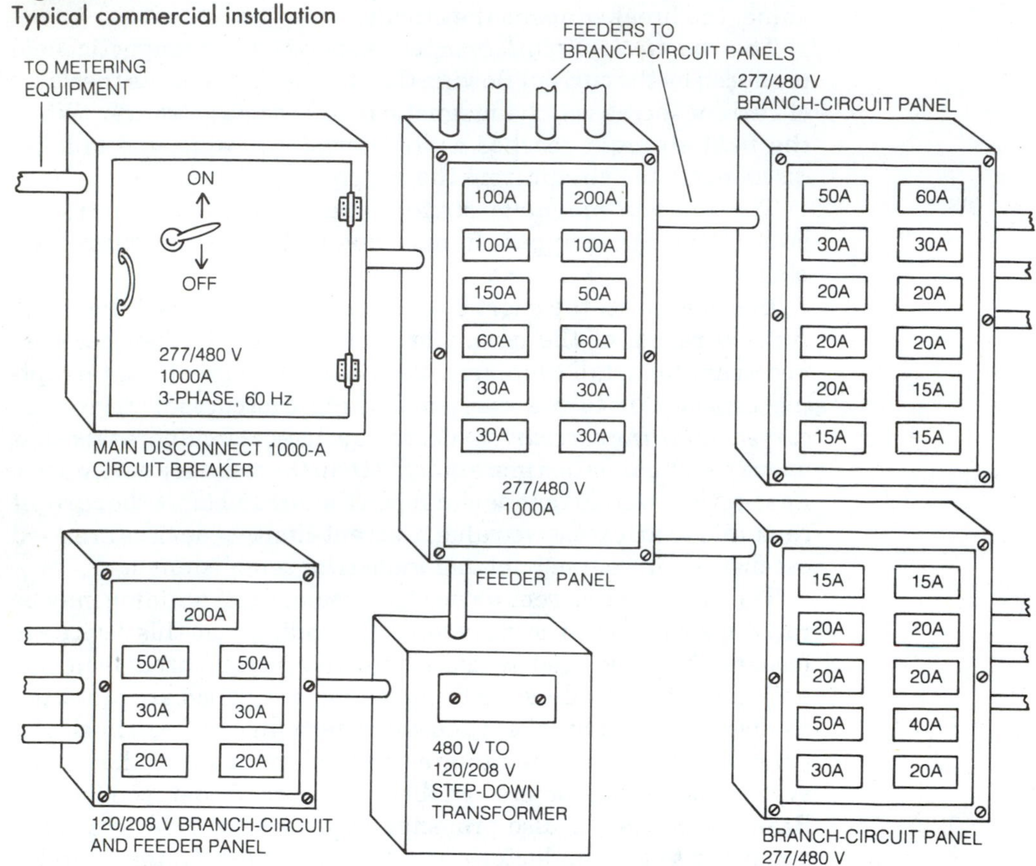

Because of their convenience, circuit breakers are often used in preference to fuses. Two of the more common types are the *molded-case circuit breaker* and the *air circuit breaker*. In the molded-case circuit breaker, Figure 8-10, the entire assembly is enclosed in a molded nonmetallic case. Molded-case circuit breakers are available in continuous current ratings up to and including 800 amperes at 600 volts or less. In the air circuit breaker, Figure 8-11, the operating mechanism is enclosed in a container, but it is accessible for inspection and maintenance.

The circuit breaker must be able to carry the continuous current without nuisance tripping. A common practice is to allow a breaker to carry only 80 percent of its continuous current rating. It must be able to open the circuit under short-circuit conditions without injury to itself.

A circuit breaker should always be rated at a voltage equal to or greater than the system voltage. The frequency of the system can also have an effect on the operation of a circuit breaker. If the frequency is other then 60 hertz, consult the breaker manufacturer.

Figure 8-10
Molded-case circuit breakers

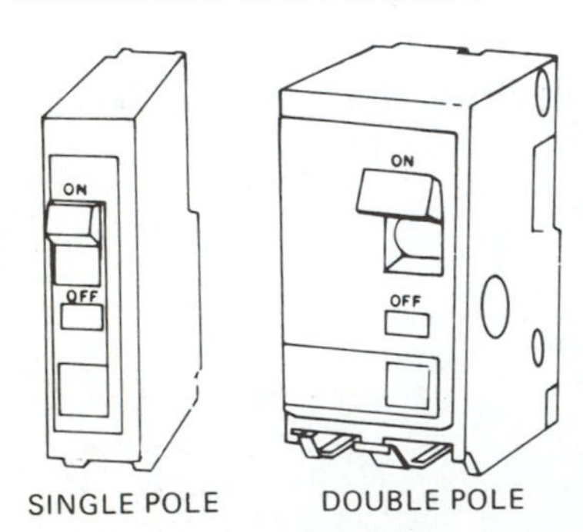

Frequency is the number of cycles of ac completed in one second. The unit of measurement of frequency is the *hertz* (Hz), which means cycles per second. Further discussion of this subject appears later in the text.

If a circuit breaker is protecting a load that is intermittent, such as frequent and periodic starting of motors or groups of flashing lights, a cumulative heating effect may occur. The ambient temperature and the temperature buildup caused by surges of current must always be considered when selecting circuit breakers. Time delay of operation is an important factor when considering motor starting currents.

In summary, the circuit breaker should never be subject to currents in excess of its interrupting capacity. It should be installed so as to be protected against overheating during normal operation. The breaker should never be subject to voltages in excess of its rating. The frequency of the system should always be considered when selecting a circuit breaker.

The maintenance engineer should have equipment available to test circuit breakers that have been subjected to short-circuit currents. Circuit breakers should be periodically switched off and on to ensure that the contacts do not corrode together.

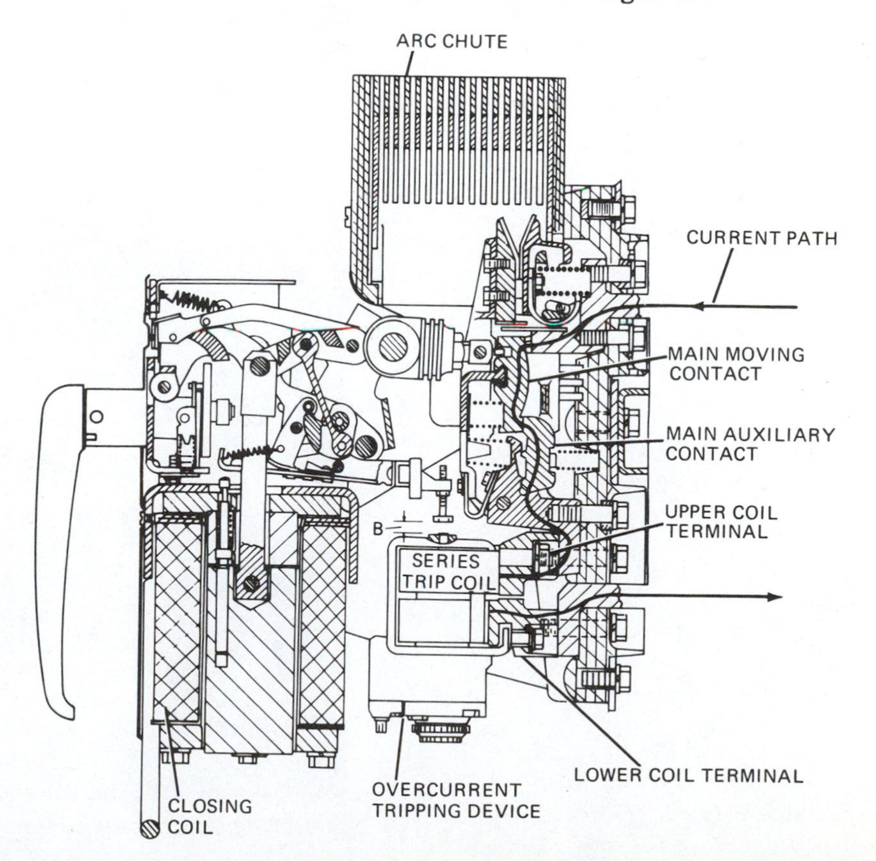

Figure 8-11
Internal diagram of an adjustable air circuit breaker

Air circuit breakers are available in continuous current ratings of from 15 amperes to 4000 amperes, and at voltages up to 600 volts. These breakers are suitable for lighting and power circuits as well as motor starting and running service. They are equipped with heavy-duty contacts, arc quenchers, and an operating mechanism, which may be either manual or automatic. The air circuit breaker has the advantages of being easy to inspect, test, and maintain. In these respects, the air circuit breaker is preferred to the molded-case breaker. The disadvantages are the initial cost, and in some instances, additional maintenance.

Air circuit breakers are adjustable for instantaneous, short-time, and/or time-delay operation. The standard time delays are 5 to 12 cycles, 12 to 20 cycles, and 20 to 30 cycles of time delay at 60 hertz. (In a 60-hertz circuit, a delay of 30 cycles is equivalent to a time delay of 1/2 second.)

Fuses are also used in industrial establishments and can provide maximum protection with minimum downtime. The two general types of fuses are the plug fuse and the cartridge fuse.

Plug fuses are available in ratings of up to 30 amperes, at 125 volts. There are three kinds of plug fuses: the standard Edison-base fuse, the Edison-base dual-element fuse, and the type S, duel-element tamperproof fuse. The dual-element fuse provides time delay for motor-starting currents and instantaneous short-circuit protection. Figure 8-12 shows the Edison-base and tamperproof fuses.

Figure 8-12
Various types of fuses

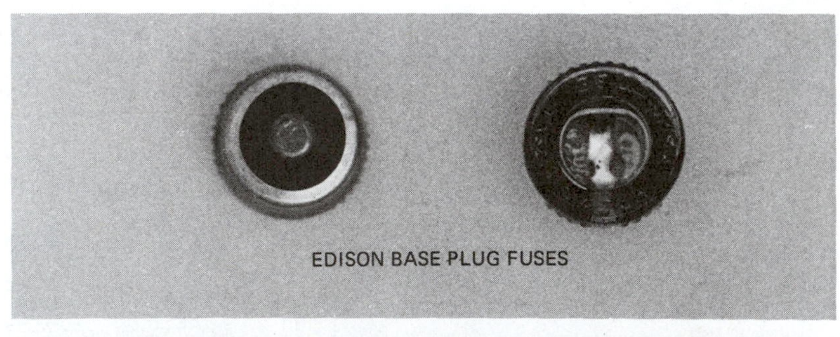

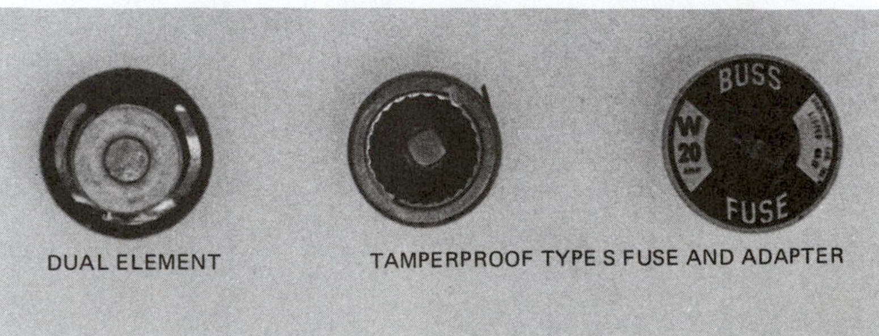

Figure 8-13A
One-time fuse (knife-blade type) *(Courtesy of The Chase-Shawmut Co., Newburyport, MA)*

Figure 8-13B
One-time fuse (ferrule type) *(Courtesy of The Chase-Shawmut Co., Newburyport, MA)*

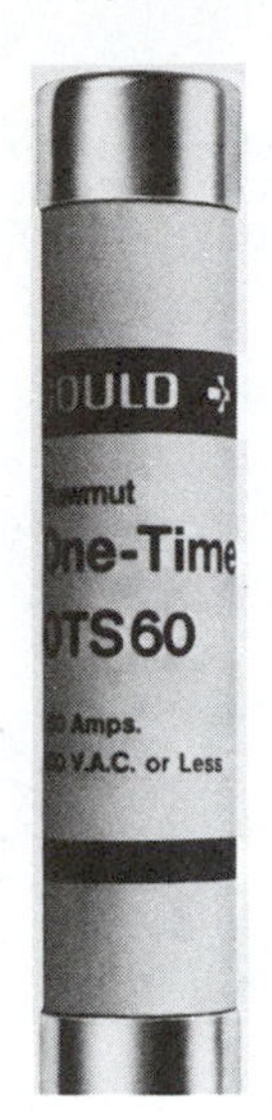

Figure 8-13C
Renewable ferrule type *(Courtesy of The Chase-Shawmut Co., Newburyport, MA)*

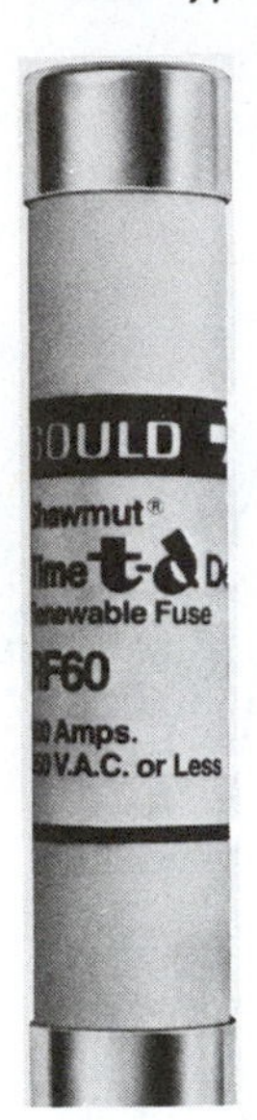

Figure 8-13D
Current-limiting fuse *(Courtesy of The Chase-Shawmut Co., Newburyport, MA)*

Cartridge fuses are the type most commonly used in industrial installations. They are available in single-element and dual-element styles as well as the renewable style. The renewable fuse has a replaceable link. The initial cost of the renewable fuse is higher than the onetime fuse. Using renewable links, however, is much less expensive then replacing the entire nonrenewable fuse.

Cartridge fuses are available in the ferrule and the knife-blade styles. The ferrule style is used for fuses rated at 60 amperes or less and up to 600 volts. The knife-blade style is used for fuses of more then 60 amperes and up to 600 volts. Figures 8-13A through 8-13D illustrate some types of cartridge fuses.

For circuits of over 600 volts, special overcurrent protective devices are used. The available short-circuit current in industrial areas frequently exceeds 10,000 amperes. The installation of standard fuses or circuit breakers backed up by current-limiting overcurrent device, maximizes protection. The current limiting overcurrent device is designed both to function as a short-circuit protective device and to allow for overload.

Current-limiting overcurrent protective devices have high current-interrupting capabilities. This feature and their fast action

Figure 8-14
Curves of a current-limiting fuse and a standard zinc link fuse *(Courtesy of The Chase-Shawmut Co., Newburyport, MA)*

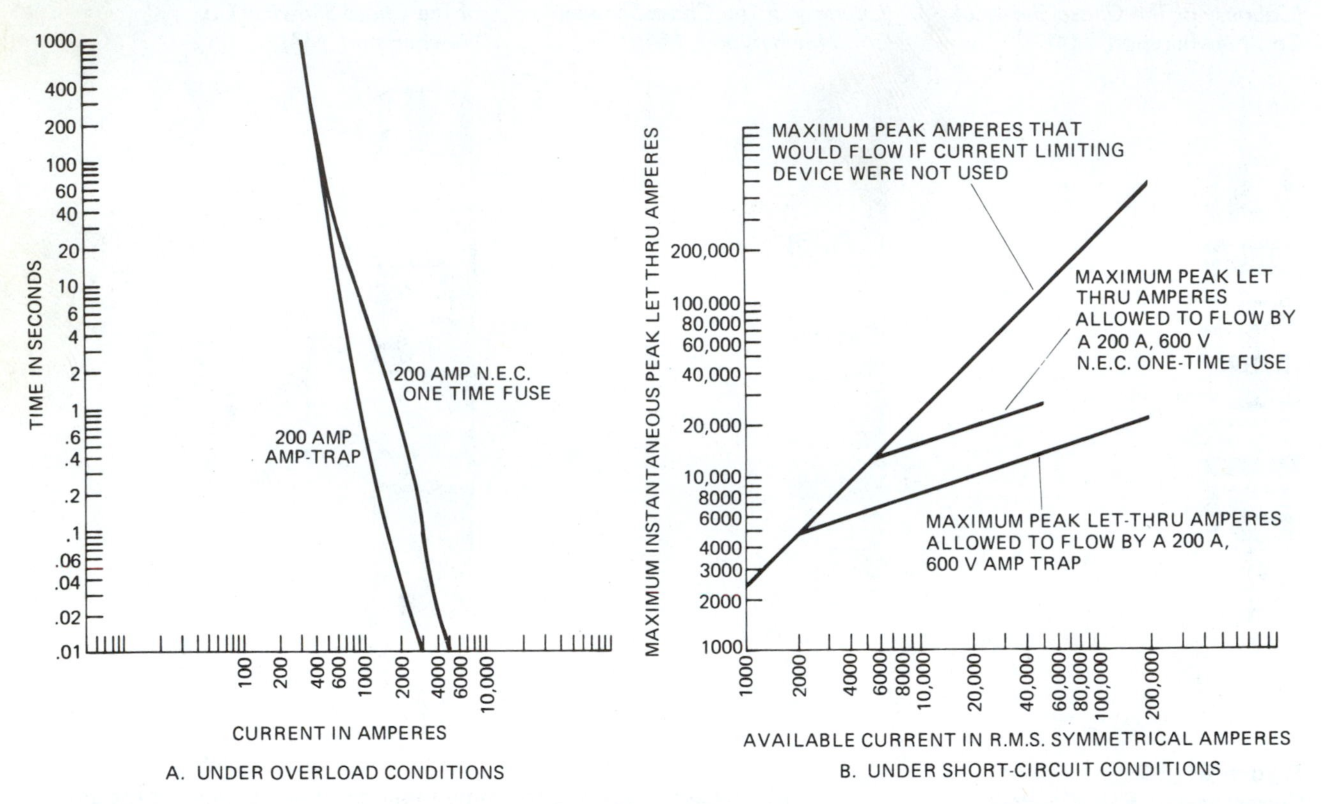

under short circuits make them a valuable device in industrial installations.

The key factor is the amount of time it takes to open the circuit under various values of current. A typical example is shown in Figures 8-14A and 8-14B. The curves show a 200-ampere current-limiting fuse compared to a standard zinc link fuse. Figure 8-14A shows the results of an overload, and Figure 8-14B shows the results of a short circuit. Under overload, the operating characteristics are basically the same up to 800 amperes. At higher currents, the current-limiting fuse operates much faster than the standard fuse. Under a short circuit, as in Figure 8-14B, the current-limiting fuse operates much faster than the standard fuse. This quick operation minimizes the value of short-circuit current.

Voltage (Line) Drop

Voltage drop (line drop) is another very important factor to consider when wiring an industrial establishment. Because such installations cover large areas and generally require high values of current, voltage drop can become a serious problem. Services should be installed in a central location to minimize the distance to the loads. In calculations of feeder and branch-circuit sizes, the length of the conductors and the ambient temperature must always be considered. Conductor sizes must be increased to compensate for high temperatures and long distances.

Grounding

Grounding of electrical systems and equipment is a very important safety factor. *Grounding* means to connect to the ground one wire of the electrical system and/or the metal enclosures of electrical conductors and equipment. Grounding of equipment is accomplished by connecting all the noncurrent-carrying metal parts together and then connecting them to the ground. All connections should be made wrench tight.

The main system ground for ac services is located on the line side of their service-disconnecting device. At this point, both the system ground and equipment ground are connected together to form a common ground. The grounding conductor connects to the common ground to the grounding electrode. All connections should be made tight to ensure good conductivity.

It may appear that it would be safer to keep all electrical systems ungrounded. If it were possible to have every installation perfect and to prevent any accidental grounds, it would be safer to have an ungrounded system. However, experience has shown that this is not possible. Oil damages insulation, and expansion of metal parts causes insulation breakdown. Accumulation of moisture and other impurities causes leakage to ground.

CAUTION: A loose connection or break in the grounding conductor can cause a serious hazard. For example, if the grounding conductor becomes disconnected and a ground fault occurs in the system, the current caused by the fault will seek other paths to ground. The alternative paths may have high resistance, thus producing heat and possibly causing a fire. Another hazard could occur if the ungrounded (hot) wire accidentally makes contact with a metal enclosure and there is no path to ground. The metal enclosure becomes energized with a voltage equal to that of the system voltage to ground. A person making contact with the energized enclosure and ground completes the circuit, and the resulting shock could be fatal.

A loose connection in the grounding conductor causes high resistance. Stray currents caused by lightning or by a ground fault flowing through the high resistance produce heat and possibly cause a fire.

NEC® Article 250 prescribes methods and standards for grounding.

Preventive Maintenance

Once an installation has been completed, a good preventive maintenance program should be developed. All conductors and equipment should be tested periodically for deterioration, malfunction, and safe operation. A log should be maintained, listing the times and dates of inspection, the condition of the equipment and insulation, and any other pertinent information. A good maintenance program minimizes breakdowns.

REVIEW

A. Multiple choice.
Select the best answer.

1. Appliance circuits must be included in the wiring plan for dwelling occupancies. The total minimum volt-amperes for these circuits must be
   a. 1000 volt-amperes.
   b. 1500 volt-amperes.
   c. 2000 volt-amperes.
   d. 3000 volt-amperes.
2. The power included for the laundry circuit must be
   a. 1000 volt-amperes.
   b. 1500 volt-amperes.
   c. 2000 volt-amperes.
   d. 3000 volt-amperes.
3. By applying the demand factor when calculating the service size, one is allowed to
   a. increase the service size.
   b. decrease the service size.
   c. install a demand meter.
   d. install larger-size fuses.

REVIEW *(continued)*

A. Multiple choice.
Select the best answer (continued).

4. The regulations for installing electrical services can be found in *NEC® Article*
   a. *220.*
   b. *230.*
   c. *240.*
   d. *250.*

5. Fuses and circuit breakers are
   a. safety valves.
   b. testing devices.
   c. current increasers.
   d. current reducers.

6. The common grounding conductor provides a path to ground for
   a. leakage currents.
   b. overcurrents.
   c. high voltages.
   d. short circuits.

7. The size of the common ground is based upon
   a. the voltage of the system.
   b. the size of the largest service conductor.
   c. the volt-amperes per square foot.
   d. the lighting load.

8. A ground-fault circuit interrupter is a device that
   a. opens the common grounding conductor when excessive current flows.
   b. senses small currents to ground and opens the circuit when they exceed a predetermined value.
   c. opens the circuit when there is a fault on the grounding conductor.
   d. disconnects the system when lightning occurs.

9. A current-limiting overcurrent device
   a. limits the current to a specific value.
   b. interrupts the current flow quickly when a short circuit occurs.
   c. limits the current to 100 amperes.
   d. limits the current to 5000 amperes.

REVIEW *(continued)*

A. Multiple choice.
Select the best answer (continued).

10. A circuit breaker is a
    a. mechanical device that interrupts current flow within its rated values, when short circuits or overloads occur, without injury to itself.
    b. device used instead of a fuse to prevent explosions.
    c. device that opens the circuit when the resistance increases.
    d. device used to open circuits from a remote area.

11. The demand factor for an industrial building is
    a. the same for all installations.
    b. different according to the type of installation.
    c. never used for industrial installations.
    d. limited to 20 percent of the total load.

12. A circuit breaker should always be rated at a voltage
    a. of 10,000 volts.
    b. equal to the system voltage.
    c. equal to or greater than the system voltage.
    d. equal to 150 percent of the system voltage.

13. Adjustable circuit breakers are
    a. air circuit breakers.
    b. molded-case circuit breakers.
    c. oil-immersed circuit breakers.
    d. all of the above.

14. One kind of plug fuse is the
    a. standard Edison-base type.
    b. prong type.
    c. ferrule type.
    d. renewable type.

15. One kind of cartridge fuse is the
    a. Edison-base type.
    b. knife-blade type.
    c. type S, dual-element tamperproof type.
    d. screw shell type.

16. The main ground for an ac service is located
    a. as near as possible to the grounding electrode.
    b. anywhere convenient in the building.
    c. on the line side of the service-disconnecting device.
    d. on the load side of the service-disconnecting means.

REVIEW *(continued)*

A. Multiple choice. Select the best answer (continued).

17. The common grounding wire
    a. connects the system ground to the grounding electrode.
    b. connects the equipment ground to the grounding electrode.
    c. both a and b.
    d. none of the above.

18. The local utility company
    a. installs the service switch.
    b. determines the type of power and value of voltage available.
    c. develops the laws governing the wiring of the building.
    d. installs the main switch.

19. Demand factor is the ratio of the
    a. maximum demand of the system to the total connected load.
    b. input to the output.
    c. available current to the total current.
    d. applied voltage to the voltage drop.

20. The minimum-size service for a single family house should be
    a. 50 amperes.
    b. 150 amperes.
    c. 100 amperes.
    d. 200 amperes.

B. Give complete answers.

1. List six questions that should be answered by the utility company before an electrical service is installed.
2. What information is found on the electrical floor plan?
3. Describe how to determine the general lighting load of a residence using the watts-per-square-foot method.
4. What is the minimum load that must be calculated for small appliances in a residence?
5. Define the term *demand factor*.
6. Why does the *NEC*® allow a demand factor?
7. List nine rules for residential wiring that will aid in carrying out a flexible, convenient, and trouble-free installation.

REVIEW *(continued)*

B. Give complete answers (continued).

8. List the components of a residential service.
9. What is the purpose of fuses and circuit breakers?
10. How can one ensure minimum resistance of the common ground.
11. Describe a ground-fault circuit interrupter.
12. List three common types of lighting switches.
13. Complete the following diagrams.

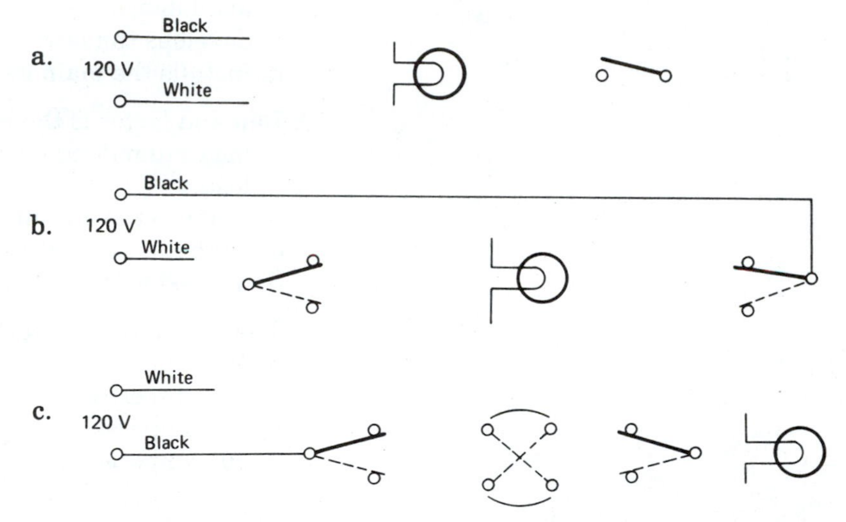

14. Explain the operation of low-voltage switching of lighting circuits.
15. List three types of detectors used with fire alarm systems.
16. What is the purpose of the backup supply used in conjunction with the fire and burglar alarm systems?
17. Describe a common type of residential burglar alarm system.
18. List five wiring methods used for commercial wiring.
19. What is the purpose of a power transformer in a commercial building?
20. Fuses and circuit breakers have two current ratings. What do these ratings indicate?
21. Using the *NEC*®, describe one method for calculating the size of a service for a single-family dwelling.

REVIEW *(continued)*

B. Give complete answers (continued).

22. Define the unit of measurement of frequency.
23. Name the two most common types of overcurrent protection devices.
24. List ten general safety rules associated with electrical wiring.
25. Explain how to calculate the size of service, the size and number of feeders, and the size and number of branch circuits for an industrial establishment.

9

MAGNETS AND MAGNETISM

Objectives

After studying this chapter, the student will be able to:

- Describe various types of magnets.
- Describe the nature of magnetic fields and forces.
- Explain the theories of magnetism.
- List the various uses for magnetism and methods of controlling magnetic forces.
- Explain the relationship between magnetism and electricity.

A good understanding of magnetism is of major importance in the study of electricity. Magnetism is involved in the operation of most electrical apparatuses. Some examples are motors, generators, transformers, controllers, relays, meters, and lifting magnets.

MAGNETS

A magnet is a material that attracts other metals. Magnets are divided into three classes: *natural, artificial,* and *electromagnets*. The first known magnets were stones found in Asia. These stones, composed of an iron ore called magnetite, had the unusual property of being able to attract small pieces of iron, steel, and other metals. It was later discovered that the stone, in an elongated form, would align itself in a nearly north and south direction, if freely suspended. This property of magnetite made it useful as a compass, which was the first practical use of a magnet. For this reason, the stone was called a leading stone, or lodestone.

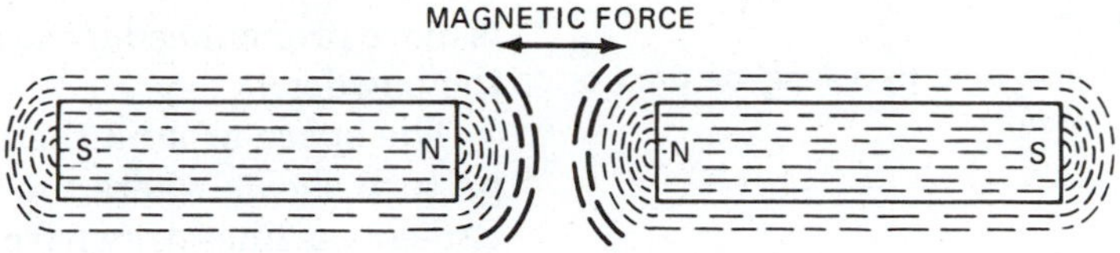

Figure 9-1
Like magnetic poles repel.

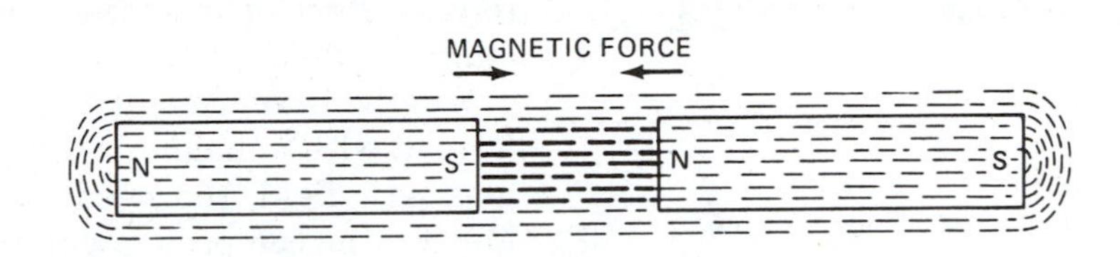

Figure 9-2
Unlike magnetic poles attract.

In 1600, Dr. William Gilbert showed that the Earth itself acts like a magnet, attracting the ends of a needle of lodestone. He called the end of the needle that pointed north the north-seeking pole, or north (N) pole. The end that pointed south he called the south-seeking pole, or south (S) pole.

Gilbert found that when two north poles of two magnets are brought close to each other and then released, they move apart. When two south poles are brought together, they also repel each other. Figure 9-1 illustrates this phenomenon. He also learned that if a north pole is placed near a south pole, they attract each other, Figure 9-2. From these observations, the following laws for magnetic attraction and repulsion were formulated:

1. Like magnetic poles repel each other.
2. Unlike magnetic poles attract each other.
3. The nearer the magnets are to each other, the greater the attraction or repulsion.

These rules are very important in the study of motors, generators, controllers, and solenoids.

Continued research led to the development of artificial magnets, which are stronger and more useful than natural magnets. Some materials used in the manufacture of artificial magnets are aluminum, nickel, cobalt, iron, and vanadium. When mixed in the proper proportions, these metals make strong permanent magnets.

MAGNETIC FIELDS AND FORCES

If a piece of paper is placed over a bar magnet and iron powder is sprinkled on the paper, the powder forms a pattern similar to that shown in Figure 9-3. The arrangement of the powdered iron is more pronounced if the paper is tapped gently while sprinkling the powder.

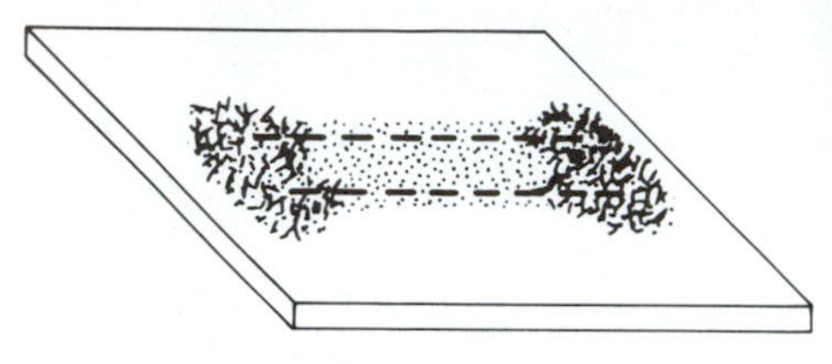

Figure 9-3
Powdered iron used to outline the area of magnetic force

The space around the magnet where the iron powder forms a pattern shows where the force exists. The pattern of the powder appears as lines drawn from one end of the bar to the other, Figure 9-4. Because of this pattern, the force is referred to as *magnetic lines of force*. The space in which this force exists is called the *magnetic field*.

Figure 9-4
Pattern of magnetic lines of force

Another experiment to identify the field of force can be performed with a magnetic compass. If the compass is placed in the magnetic field, the needle will line up with the lines of force. Moving the compass to different locations within the field shows that the lines have a definite direction. The lines appear to leave the magnet by the north pole and reenter at the south pole. Within the magnet, the lines of force continue from the south pole to the north pole.

MAGNETIC THEORIES

If a single bar magnet is cut in half, each half will be a magnet having north and south poles. If one should continue to separate each half into smaller and smaller particles, each particle will be a magnet. This separation of a single magnet into many smaller magnets has led to the belief that a magnetic substance is composed of molecular magnets (but not necessarily that each molecule is a magnet). This theory is called the *Molecular Theory of Magnetism*.

When a material is magnetized, the molecules are aligned as shown in Figure 9-5. When the material is not magnetized, the molecules are arranged as in Figure 9-6.

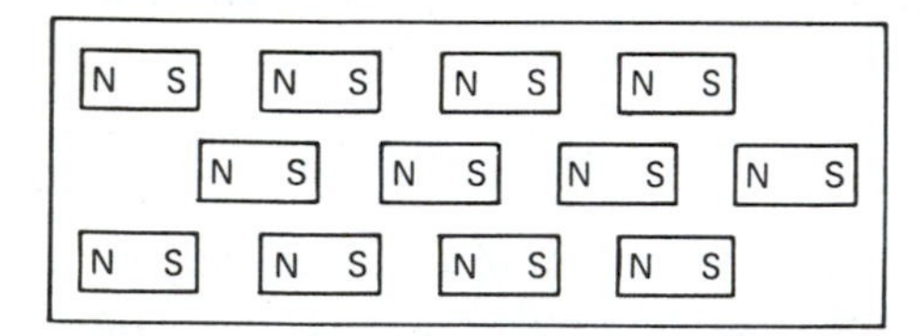

Figure 9-5
Magnetized material

Another theory that is widely accepted is called the *Electron Theory of Magnetism*. The electrons of each atom rotate in orbits about the nucleus and also spin, just as the Earth rotates on its axis. According to this theory, as the electrons spin, they produce a magnetic field. The direction of this field depends upon the direction of spin. In most atoms, there are an equal number of electrons spinning in opposite directions. These atoms are magnetically neutral because the magnetic fields of the electrons are equal and opposite.

Figure 9-6
Unmagnetized material

In atoms of magnetic materials, however, there are more electrons spinning in one direction than in the other direction. These atoms produce weak magnetic fields. When a large number of these magnetized atoms group together with their magnetic fields aligned, they form a *domain*. Each domain produces a magnetic field in a specific direction. In nonmagnetized materials, the arrangement of these domains is such that their magnetic effect is neutralized. When such material is placed under the influence of another magnetic force, the domains arrange themselves so that their north poles are in one direction and their south poles are in the opposite direction. The material thus becomes a magnet.

MAGNETIC MATERIALS

Artificial magnets can be classified as permanent or temporary. Their classification depends upon the materials from which they are made. *Permanent magnets* retain their magnetic properties for many years, possibly 100 years or more. *Temporary magnets* lose their magnetism almost as soon as they are removed from the magnetizing influence. Steel and alloys of steel are materials that make good permanent magnets. One of the most common types of permanent magnets is made of a material called *alnico*. Alnico is a mixture of aluminum, nickel, cobalt, and iron. Soft iron and iron alloys are used for temporary magnets.

Figure 9-7
Soft iron provides a path for magnetic lines of force.

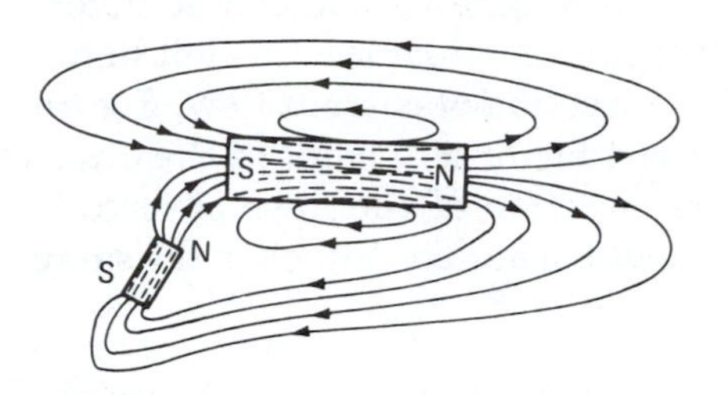

Magnetic Shields

Because all materials have some ability to conduct magnetic lines of force, it is not practical to make a magnetic insulator. Strong magnets placed near electrical instruments can cause permanent damage. Therefore, it is necessary to prevent the magnetic flux (lines of force) from passing through the instrument.

Magnetic shields can be constructed because magnetism passes more readily through some materials than others. When a piece of soft iron is placed near a magnet, the field is distorted and the lines tend to pass through the iron instead of the air, Figure 9-7. Iron is a better conductor of magnetism than air. This fact is utilized in constructing magnetic shields. Many expensive instruments that may be subjected to magnetic influences are surrounded by iron or iron alloys. The iron provides a magnetic path around the sensitive parts, Figure 9-8.

Figure 9-8
Soft iron provides a magnetic shield for expensive instruments.

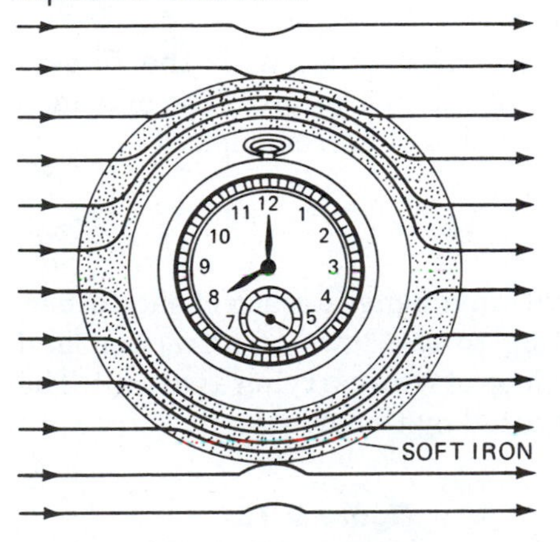

ELECTROMAGNETISM

Magnetism is produced by electron spin. Therefore, it follows that electricity and magnetism are closely related. An electric current (the movement of electrons along a conductor) produces a magnetic field. The field strength varies with the amount of current. The magnetic field produced by current flowing in a single conductor is illustrated in Figure 9-9.

Figure 9-9
Magnetic field produced by current flow

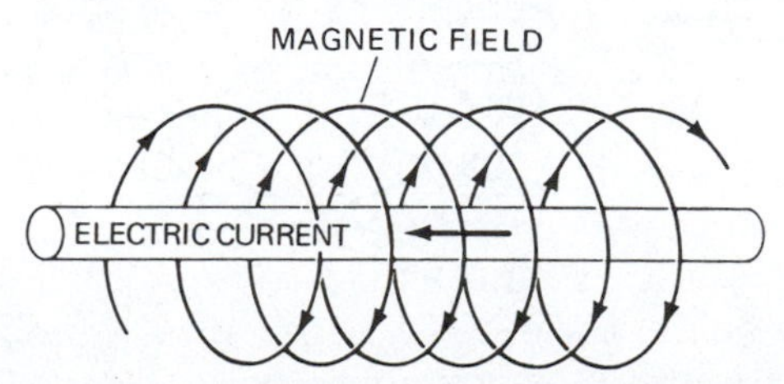

The magnetic effect of current flow in a conductor can be demonstrated by arranging a conductor and a piece of cardboard as shown in Figure 9-10. Using direct current, allow the current to flow while sprinkling powdered iron on the cardboard. The powder will take the shape of the magnetic field. An examination of the pattern shows that the field is circular, with the conductor in the

Figure 9-10
Iron powder takes the form of the magnetic field around a current-carrying conductor.

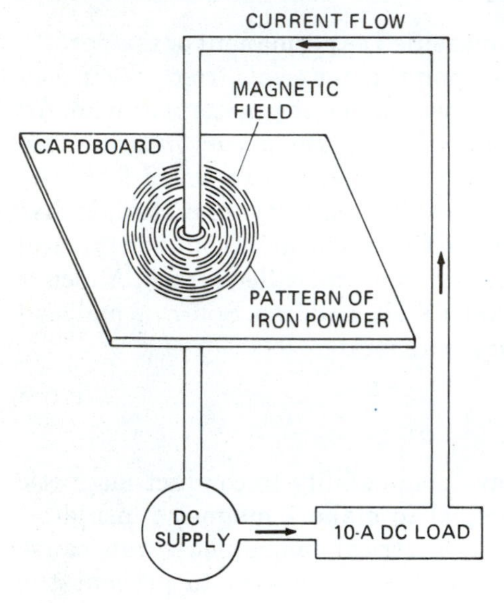

Figure 9-11
Compass needles line up with the magnetic field.

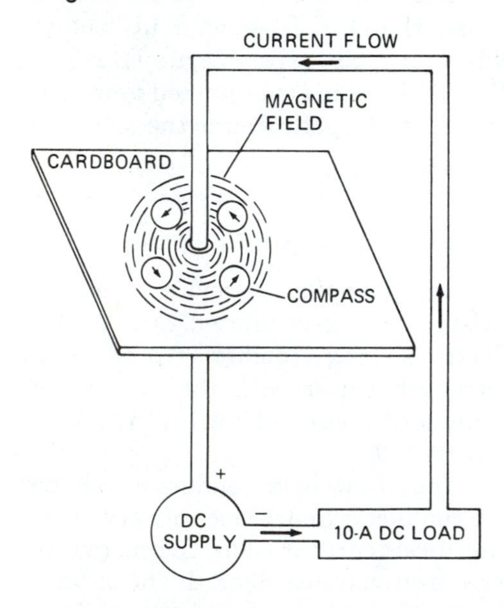

center. Another method used to see the magnetic reaction of current flow is to place compasses at various positions on the cardboard, Figure 9-11. When the current flows through the wire, the compass needles line up with the magnetic field. The needles indicate the direction of the lines of force.

Left-Hand Rule for a Single Conductor

The direction of current flow through a conductor and the direction of the lines of force of the magnetic field can be determined by the *left-hand rule*. Grasp the conductor in the left hand, with the thumb extended in the direction of current flow, Figure 9-12. The fingers encircling the conductor will point in the direction of the lines of force. If the direction of the lines of force is known, the direction of current flow can be determined in the same manner.

Rules

1. Grasp the conductor in the left hand, with the thumb pointing in the direction of current flow. The fingers will point in the direction of the magnetic force.
2. Grasp the conductor in the left hand, with the fingers pointing in the direction of the magnetic force. The extended thumb will point in the direction of current flow.

Magnetic Forces

If two magnetic forces are within magnetic reach of each other, their fields will react according to the laws of attraction and repulsion. Figure 9-13 shows a loop of wire carrying a current. The direction of the current is indicated by the arrows.

Figure 9-12
Left-hand rule. The thumb points in the direction of current flow. The finger encircles the conductor in the direction of the lines of force.

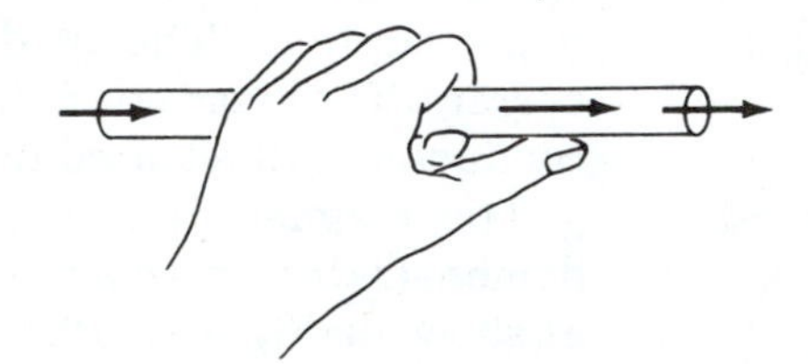

Figure 9-13
Loop of wire carrying current

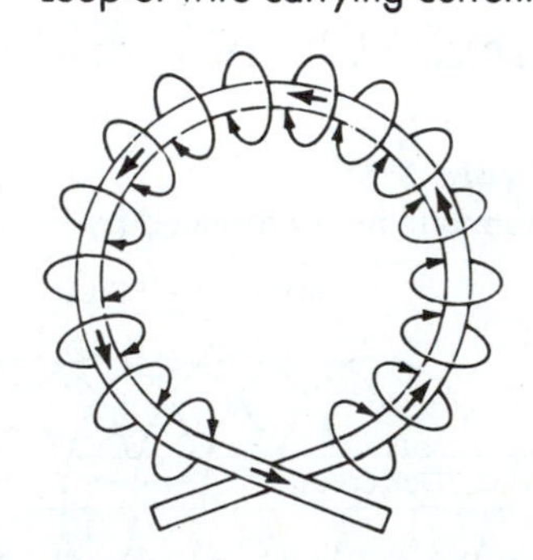

Figure 9-14
Conductors carrying current in opposite directions. The magnetic field forces the conductors apart.

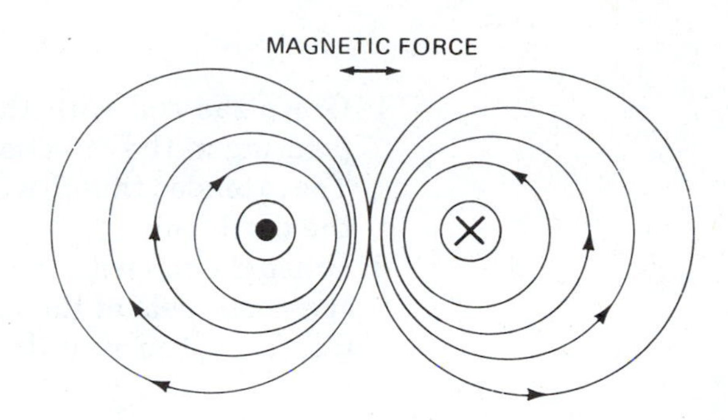

Figure 9-15
Two conductors carrying current in the same direction. Magnetic forces pull the conductors together.

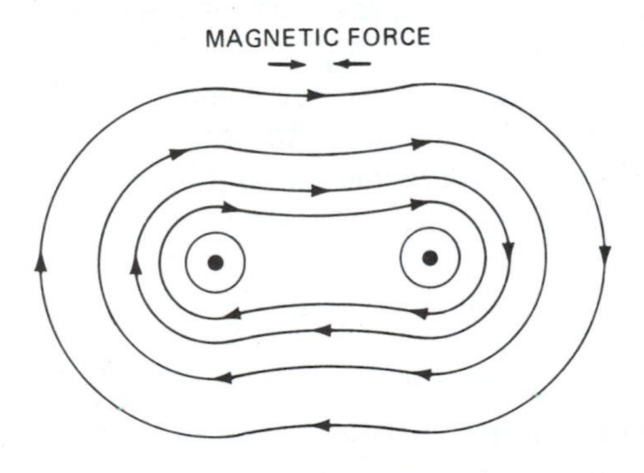

The lines of force are illustrated by the circles. Figure 9-14 shows only the end views of the same loop. The dot indicates that the current is coming out of the wire toward the viewer. The X indicates that the current is flowing into the wire, away from the viewer. Notice in Figure 9-14 that the two magnetic fields are within magnetic reach of each other and tend to force the loop apart.

If two conductors carrying current in the same direction are placed near each other, Figure 9-15, the magnetic forces tend to pull them together.

If several loops of wire are placed loosely together to form a coil, most of the lines of force will thread through the whole coil, Figure 9-16. If these loops are wound very closely together, nearly all the flux will thread through the coil, Figure 9-17. This produces a magnetic field with the same shape as a bar magnet.

Left-Hand Rule for a Coil

The magnetic polarity of a coil can be determined by grasping the coil in the left hand, with the fingers pointing in the direction of current flow. The extended thumb will point in the direction of the lines of force, toward the north pole of the magnet, Figure 9-18.

Figure 9-16
Loosely formed coil

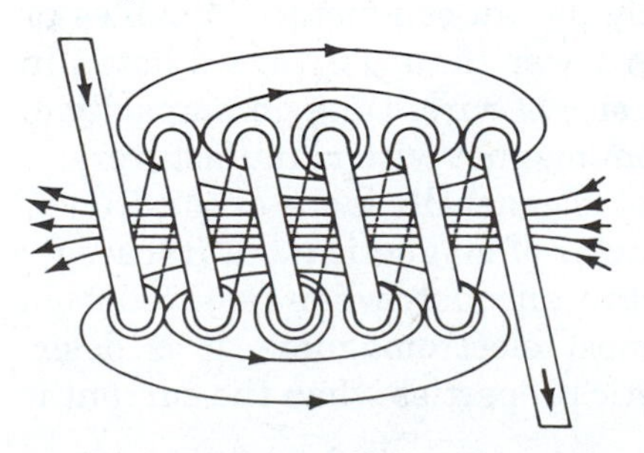

Figure 9-17
Lines of force through the entire coil

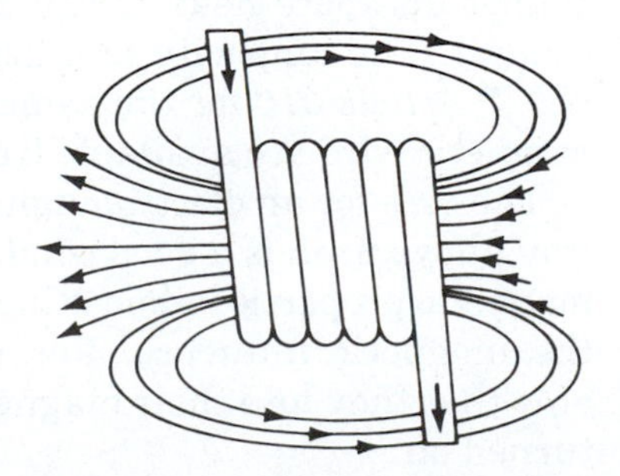

Figure 9-18
Left-hand rule for a coil

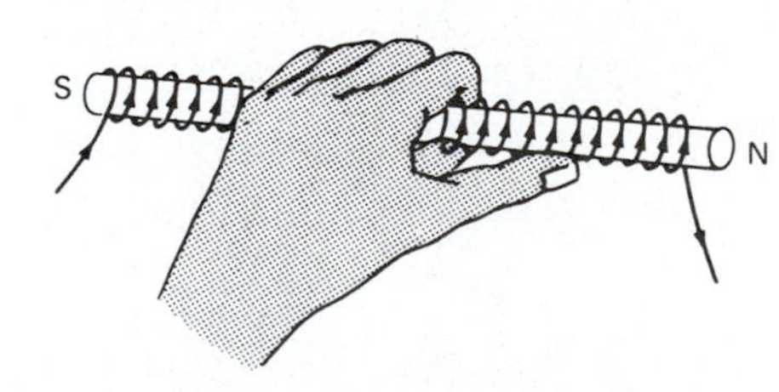

Rules

1. Grasp the coil with the left hand so that the fingers are pointing in the direction of current flow through the loops. The extended thumb will point in the direction of flux toward the north pole.
2. Grasp the coil with the left hand so that the thumb points to the north pole of the magnet. The fingers will point in the direction of current flow through the loops.

Electromagnets

A coil, as described previously, is known as an electromagnet. It has the same properties as a bar magnet. The advantage is that the magnetic force can be turned on and off with the current. For this reason, it is classified as a temporary magnet. Another advantage of an electromagnet is the ability to vary the field strength. An increase in current through the coil causes an increase in the strength of the magnetic field. Decreasing the current decreases the strength of the magnetic field.

Another way to increase the strength of an electromagnet is to insert an iron core into the coil. Iron is a better conductor of magnetism than air is and therefore increases the field strength.

In the construction of an electromagnet, there are several factors to consider:

- The number of loops of wire to be placed on the coil
- The amount of current the coil can carry without overheating
- The ability of the core to conduct magnetic lines of force

The first two items involve the electrical conductor used to form the coil. The material and size of the conductor determine the maximum amount of current it can carry. Most conductors for electromagnets are made of soft drawn copper, which has relatively low resistance and is easy to form into a coil. This type of wire is called *magnet wire*.

Magnet wire is formed into close-fitting loops. It therefore cannot dissipate heat as rapidly as can conductors in cables or conduit. The ampacity is much lower than the values listed in *NEC® Article 310* for the same size of wire. Data on ampacity of magnetic wire are available from magnet wire manufacturers.

The core for an electromagnet is generally made of soft iron or iron alloys. Iron is a good conductor of magnetism but it loses its magnetic properties almost instantaneously when removed from the magnetic influence. For most electromagnets, it is desirable that they lose their magnetic properties when the current is turned off.

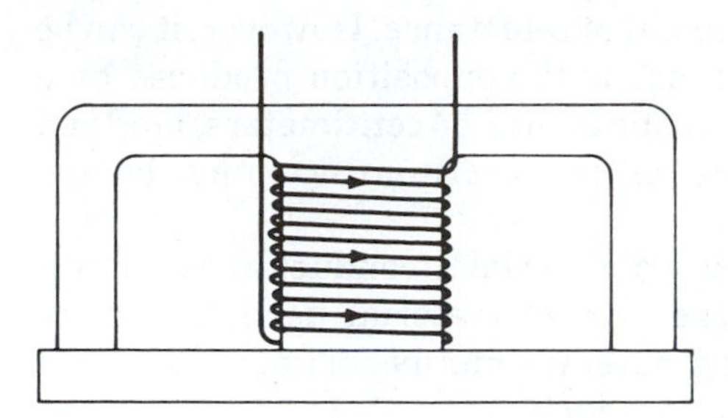

Figure 9-19A
Internal construction of a lifting magnet

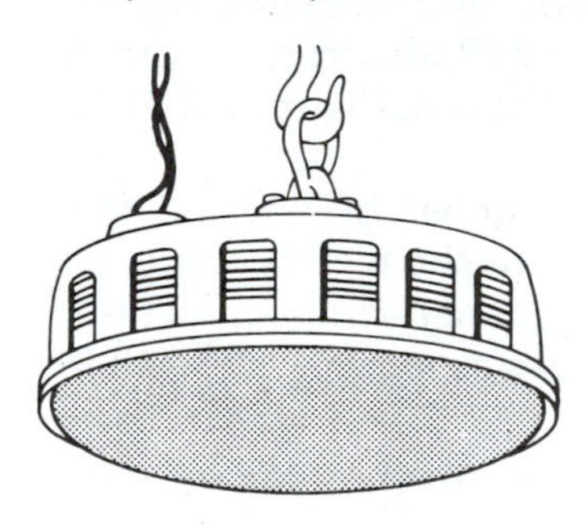

Figure 9-19B
Common lifting magnet used to move heavy metal objects

An electromagnet is sometimes referred to as a solenoid. Although the names are used interchangeably, the term *electromagnet* usually refers to coils with stationary iron cores. The term *solenoid* usually refers to coils with movable iron cores.

Application of Electromagnets

The lifting magnet is one of the common uses of electromagnets. Lifting magnets are used to move large amounts of iron and steel. These magnets can lift as much as 200 pounds per square inch (14 kilograms per square centimeter) of magnet surface. Figures 9-19A and 9-19B show a typical lifting magnet.

Many industrial machines use magnetic clutches to connect and disconnect the load from the driving source or to provide speed control. The amount of slip is varied by adjusting the distance between the driving magnet and the driven component. Another method is to vary the strength of the magnetic field.

Electromagnets are used in measuring instruments, as discussed in Chapter 5. The amount of current flowing through the instrument's coil determines the strength of the magnetic field. Permanent magnets within the instrument react with the electromagnetic field, causing the needle to move. The distance of travel depends upon the amount of current flowing in the electromagnet.

A solenoid can be used to open and close valves. Relays, circuit breakers, and door chimes are other examples of electromagnetism. Motors, generators, and transformers also depend upon electromagnetism for their operation.

MAGNETIC CIRCUITS AND MEASUREMENTS

The strength of an electromagnet depends upon its ability to conduct magnetism. In this respect, one may compare the path of the lines of force with the path of an electric current. They both form a complete path, Figures 9-20A and 9-20B.

The amount of flux in a magnetic circuit is determined by the number of lines of force and is measured in *maxwells* (Mx). One maxwell is equal to one magnetic line of force.

In an electromagnet, the amount of flux produced depends upon the *magnetomotive force* (mmf). The mmf is the product of the coil current (I) and the number of turns (T) of wire on the coil. Magnetomotive force is measured in *gilberts* (Gb). The unit gilbert is frequently used when working with permanent magnets. One gilbert is the mmf that will establish a flux of one maxwell in a magnetic circuit having a reluctance (rel) of one unit. Magnetomotive force is also measured in *ampere-turns*; 1 ampere-turn is equal to 1.257 gilberts.

Figure 9-20A
Horseshoe magnet illustrates the magnetic circuit (from the north pole, through the air, to the south pole, through the magnet and back to the north pole).

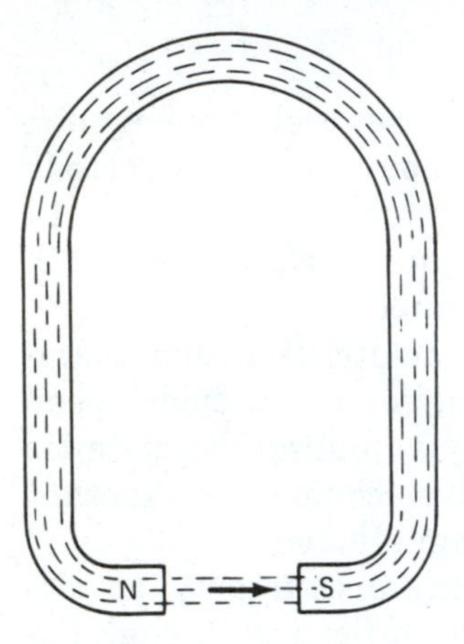

Figure 9-20B
Electric circuit (current flows from the negative terminal of the battery, through the load, to the positive terminal)

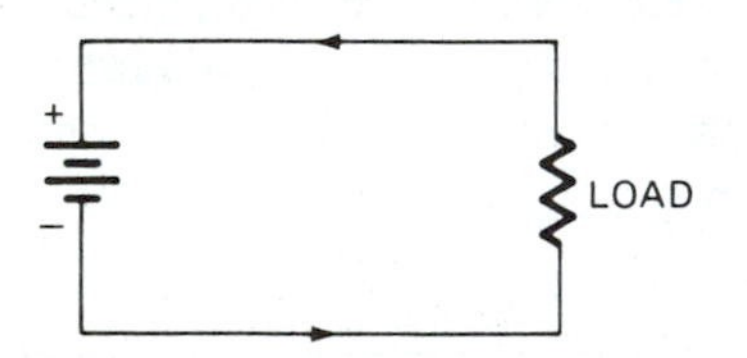

Figure 9-21A
Air core coil—weak electromagnet

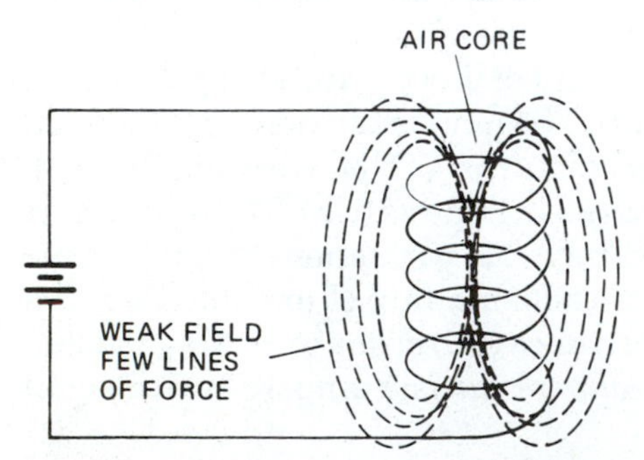

Reluctance is the opposition to the magnetic flux and can be compared to resistance in an electric circuit. No unit has been established for the measurement of reluctance. However, it can be said that one unit of reluctance is the opposition produced by a portion of the magnetic circuit one inch (2.54 centimeters) long and one square inch (6.451 square centimeters) in cross section, having unit permeability.

Permeability is the ability of a material to conduct lines of force. Permeability varies with the type of material used. The more permeable a material is, the greater the number of lines of force the material can conduct per square inch.

The strength of an electromagnet and/or solenoid can be varied by varying the mmf (ampere-turns). This is generally accomplished in the field by increasing or decreasing the current through the coil. The mmf can also be increased by increasing the flux density of the core. Figures 9-21A and 9-21B illustrate this condition. When an iron core is inserted into the coil, Figure 9-21B, the flux density is increased considerably.

Flux density is expressed in the number of lines of force (maxwells) per square inch. To convert to the metric value, one maxwell per square centimeter (0.155 square inch) is equal to one *gauss* (G). A flux density of 10,000 gauss or more is common for electrical machinery.

Other factors that are important to consider when designing magnets are:

- **Retentivity**, the ability of a material to retain its magnetism after being removed from the magnetizing influence
- **Residual magnetism**, the magnetism that remains in a material after being removed from the magnetizing influence
- **Magnetic saturation**, the point at which a material being magnetized by an electric current reaches saturation. Beyond this point, a large increase in current results in only a small increase in the magnetic strength.

These three factors depend upon the type, size, and length of the core. The greater the cross-sectional area of the core, the more permeable is the magnet. The longer the core, the more reluctance there is to the magnetic lines of force.

The residual magnetism and the retentivity vary with the material of the core. Soft iron is very permeable but has low retentivity. Hardened steel is not as permeable as soft iron, but it has high retentivity.

In the design of an electromagnet/solenoid, the most important factor to consider is the type of core. The selection of materials depends upon the use of the core. For example, a lifting magnet should have a core with good permeability and low retentivity. The

Figure 9-21B
Iron core coil—strong electromagnet

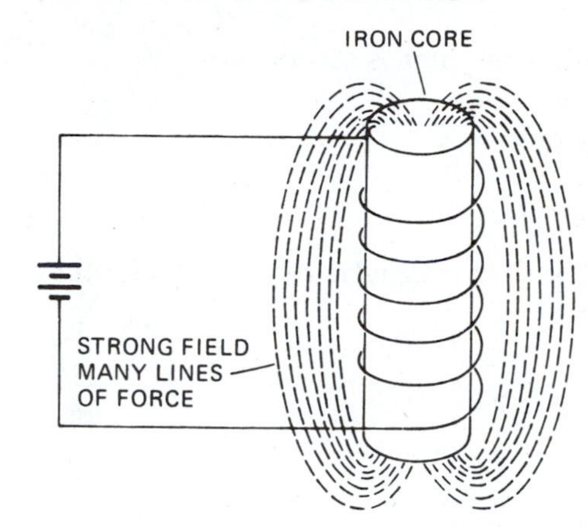

field cores of an electric generator should have good permeability and be able to retain magnetism for an indefinitely long period of time.

The magnetomotive force (mmf) and the core reluctance are the qualities that determine the amount of flux for any electromagnet. These values can be calculated by the following formula:

$$\Phi = \frac{f}{\mathfrak{R}} \qquad \text{(Eq. 9.1)}$$

where Φ = flux, in maxwells (Mx)
f = mmf, in gilberts (Gb)
$\mathfrak{R}$ = units of reluctance

SOLENOIDS

A *solenoid* is a type of electromagnet with a movable core. It is constructed by winding magnet wire on a hollow fiber or plastic form. The core is made of a material that is easily magnetized but does not retain its magnetism when the current no longer flows through the coil. The core, commonly called the *armature*, is arranged to move in and out of the coil.

When the coil is energized, the magnetic field established by the current pulls the core into the coil. When the coil is deenergized, the core returns to its original position.

Solenoids are commonly used in door chimes. When the coil is energized, the armature is pulled into the coil in such a way that it strikes a brass chime bar; as it returns to its original position, it strikes another chime bar.

Industrial uses for solenoids include such applications as opening and/or closing valves, operating clutches, and operating relays.

REVIEW

A. Multiple choice.
Select the best answer.

1. The magnetic force on a bar magnet is strongest
   a. in the center of the bar.
   b. at the north magnetic pole.
   c. at both ends.
   d. at the south magnetic pole.

2. Two unlike poles placed within magnetic reach of each other
   a. neutralize each other.
   b. attract each other.
   c. repel each other.
   d. none of the above.

3. Artificial magnets are
   a. stronger than natural magnets.
   b. less effective than natural magnets.
   c. weaker than natural magnets.
   d. the same as natural magnets.

4. The space around a magnet in which the magnetic force exists is called the magnetic
   a. range.
   b. field.
   c. area.
   d. path.

5. If a bar magnet is cut in half,
   a. one half will be the north pole and the other half will be the south pole.
   b. it will lose its magnetism.
   c. each half will be a magnet having a north pole and a south pole.
   d. each half will have better retentivity.

6. A nonmagnetized atom has
   a. most of the electrons spinning in one direction.
   b. half of the electrons spinning in one direction and half spinning in the opposite direction.
   c. none of the electrons spinning.
   d. no pattern to the spin.

7. An electric current produces
   a. a square magnetic field.
   b. an elongated magnetic field.
   c. a circular magnetic field.
   d. a rectangular magnetic field.

REVIEW *(continued)*

A. Multiple choice.
Select the best answer (continued).

8. Solenoids are used in
   a. transformers.
   b. relays.
   c. motors.
   d. generators.
9. Retentivity is the
   a. ability of a material to retain its magnetism after being removed from the magnetizing influence.
   b. magnetism that remains in a material after being removed from the magnetizing influence.
   c. state of a magnetic material being saturated.
   d. the ability of a material to be magnetized.
10. Residual magnetism is the magnetism
    a. produced by an electric current.
    b. that remains in a material after being removed from the magnetizing influence.
    c. produced by another magnet.
    d. that is lost when the magnet is removed from the magnetizing force.

B. Give complete answers.

1. Define the term *magnet*.
2. Name the three basic kinds of magnets.
3. What was the first practical use of a magnet?
4. Write the rules for magnetic attraction and repulsion.
5. List four materials used in making permanent magnets.
6. What are magnetic lines of force?
7. Define *magnetic flux*.
8. Define *magnetic field*.
9. Describe the experiment in which powdered iron and a bar magnet are used. What does the experiment prove?
10. How can it be proved that the lines of force leave from the north pole and return by the south pole.
11. Explain the Molecular Theory of Magnetism.
12. Explain the Electron Theory of Magnetism.

REVIEW

B. Give complete answers (continued).

13. Do the Electron and Molecular theories of magnetism conflict with each other? Explain.
14. Name at least two types of alloys used to make permanent magnets.
15. Explain the purpose of a magnetic shield, and describe how it works.
16. Describe how electromagnetism is produced.
17. Write the rule for determining the direction of the magnetic field around a current-carrying conductor.
18. Two wires carrying an electric current are placed next to each other. They are forced apart. Explain why this phenomenon takes place.
19. Why does a coil of wire carrying current produce a magnetic field similar to a bar magnet?
20. Write the rule for determining the polarity of an electromagnet.
21. What is the advantage of an electromagnet compared to a permanent magnet?

10

DC GENERATORS

Objectives

After studying this chapter, the student will be able to:

- Explain the theory of electromagnetic induction.
- Describe the basic operation of the ac generator.
- Describe the construction and operation of various types of dc generators.
- Describe the uses and operating characteristics of various types of dc generators.
- Discuss the methods of connecting dc generators and the basic troubleshooting procedures.

ELECTROMAGNETIC INDUCTION

In Chapter 9 we discussed the Electron Theory of Magnetism and explained that a magnetic field is produced whenever an electric current flows. Another relationship between electricity and magnetism is called *electromagnetic induction*. Electromagnetic induction takes place whenever a conductor moves across a magnetic field (cuts the lines of force) or when a magnetic field moves across a conductor. When magnetic lines of force are cut by a conductor, a voltage is induced into the conductor. This voltage is called *electromotive force* (emf) because it is the force that causes the current to flow in a circuit.

When magnetism is used to produce electricity, it makes no difference whether the conductor moves through the field or the magnetic field moves across the conductor. It is the interaction between the two that develops electrical pressure (emf), Figure 10-1.

The direction of the emf can be determined by the use of the left-hand rule for a generator. Using the left hand, extend the thumb, index finger, and middle finger so they form right angles to one

Figure 10-1A
Moving a conductor up through a magnetic field (electromagnetic induction)

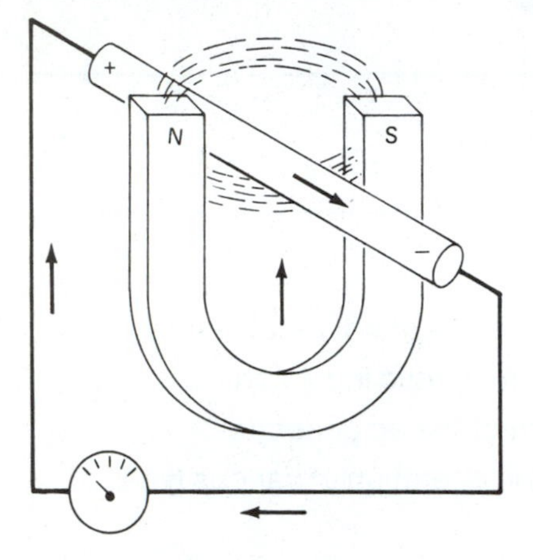

Figure 10-1B
Moving a magnetic field through a coil (electromagnetic induction)

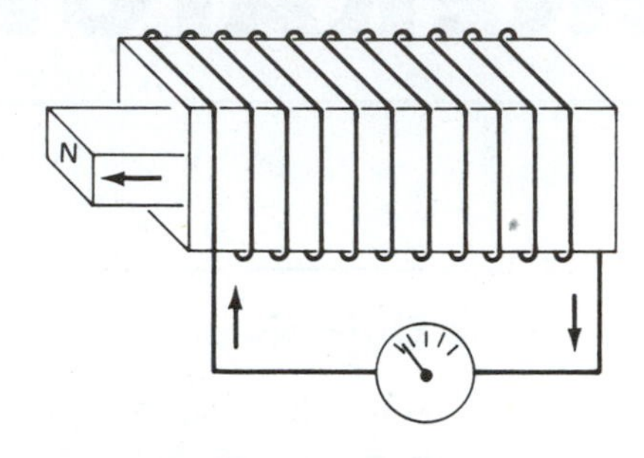

Figure 10-2
Left-hand rule for a generator

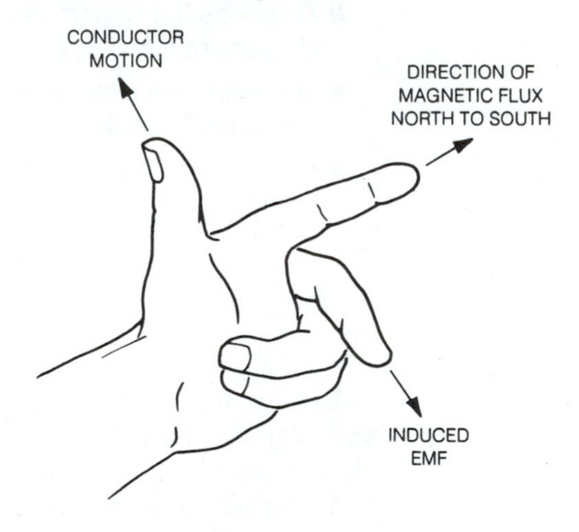

another. When the index finger points in the direction of the magnetic field (from north to south) and the thumb points in the direction of motion of the conductor, the middle finger will point in the direction of the induced emf, Figure 10-2.

GENERATOR CONSTRUCTION

It is frequently stated that a generator is a machine that changes mechanical energy into electrical energy. This statement, although true, is somewhat misleading. A generator actually converts both mechanical and magnetic energy into electrical energy.

There are many ways to drive generators. For example, in the northeastern United States, many generators are driven by steam turbines. The chemical energy of the fuel is converted into heat energy to make steam. The steam pressure drives the prime mover, which in turn drives the generator rotor. In large genera-

Figure 10-3
Alternating current generator in a typical generating station *(Courtesy of Boston Edison Co., Boston, MA)*

tors, the rotor contains the electromagnets. These magnets are rotated past coiled conductors wound on the *stator* (the stationary part of the generator). As the magnetic field moves across the stator coils, it induces an emf into them, producing electrical energy. Figure 10-3 shows a typical generating station.

Basic Generator

The simplest generator (the magneto) consists of a permanent magnet mounted on a frame (called the *yoke*). A coil of insulated wire mounted on a laminated iron core (called the *rotor* or *armature*) is positioned between the poles of the magnet. The armature is arranged so that it can rotate through the magnetic field, Figure 10-4. In small generators, such as those used in megohmmeters, the armature is rotated by hand.

The amount of emf produced by a generator depends upon the following factors:

- The strength of the main field flux (the number of lines of force per square inch)
- The number of loops of wire on the armature
- The angle at which the armature coils move across the lines of force
- The speed at which the coils rotate through the magnetic field.

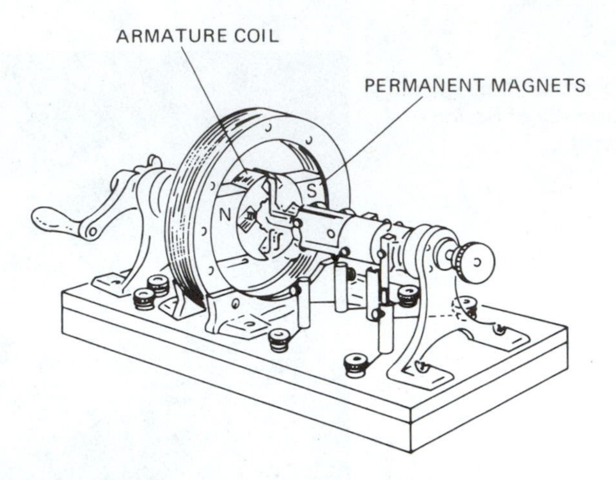

Figure 10-4
Hand-operated generator

A strong magnetic field produces a large emf. Many turns of wire on the armature also help to produce a large emf. When the coils move at right angles to the magnetic lines of force, they produce the most voltage. Coils moving parallel to the lines of force produce no voltage. The faster the coil rotates through the magnetic field, the greater the emf.

Because the armature revolves through the magnetic field, it does not produce a steady value of voltage. This can be understood by examining a single loop of wire as it rotates in a uniform magnetic field. Figure 10-5A illustrates the loop at the instant it is moving parallel with the lines of force. The two circles represent the ends of a single-loop coil. The meter indicates zero voltage. This is shown on the graph.

As the coil rotates and begins to cut across the field, a small emf is induced into the coil, and the meter indicates a low voltage, Figure 10-5B. Notice that the graph also indicates a low voltage. When the coil is at a 30-degree angle to the lines of force, Figure 10-5C, the emf is 5 volts. At a 60-degree angle, the emf is 8.67 volts, Figure 10-5D, and at 90 degrees the value is 10 volts, Figure 10-5E.

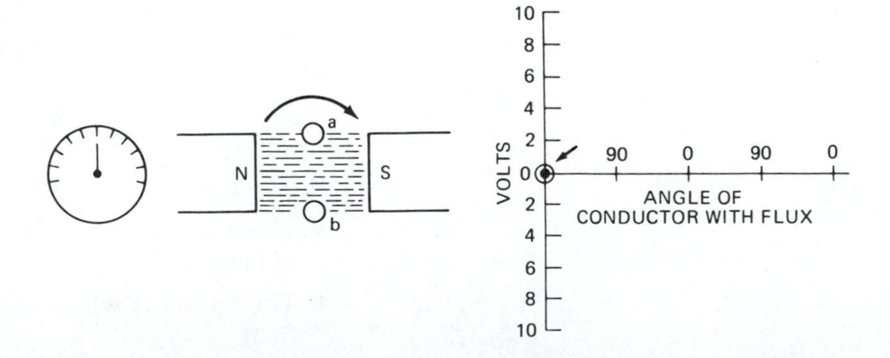

Figure 10-5A
Loop of wire moving parallel to magnetic flux. No voltage is induced into the loop.

Figure 10-5B
Loop of wire cutting lines of force at a slight angle. The induced emf is 2 V.

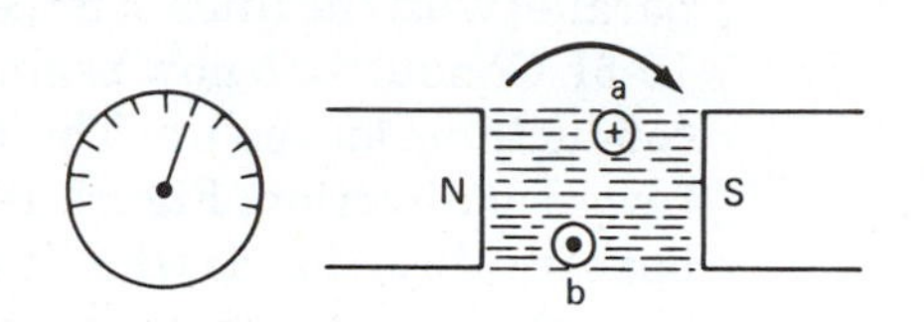

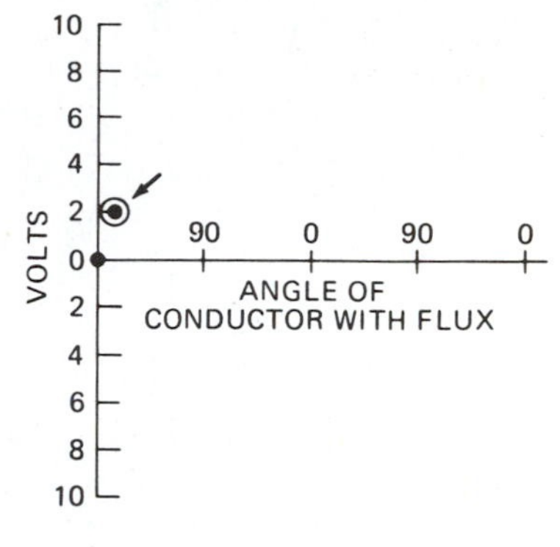

Figure 10-5C
Loop of wire cutting lines of force at a 30° angle. The induced emf is 5 V.

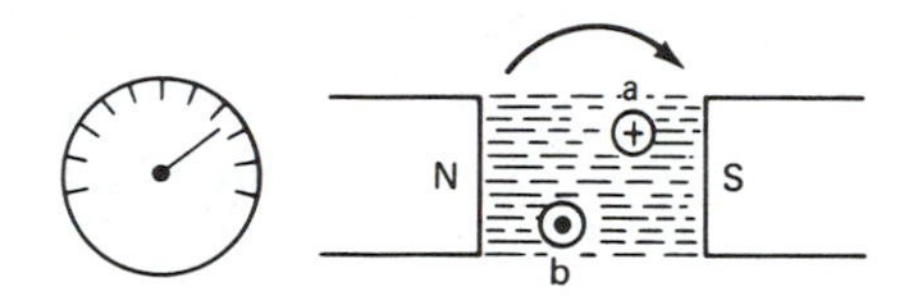

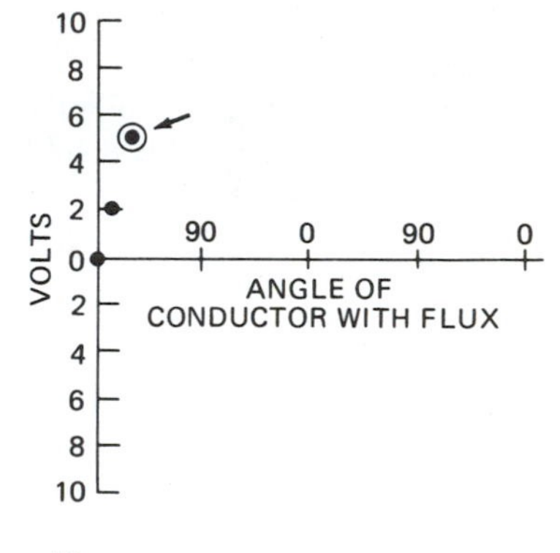

Figure 10-5D
Loop of wire cutting lines of force at a 60° angle. The induced emf is 8.67 V.

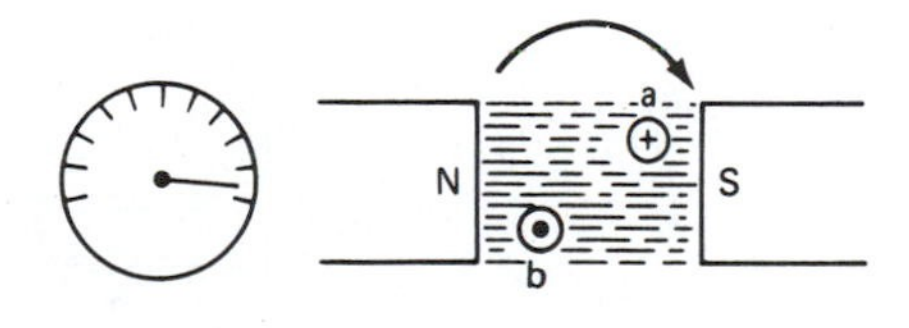

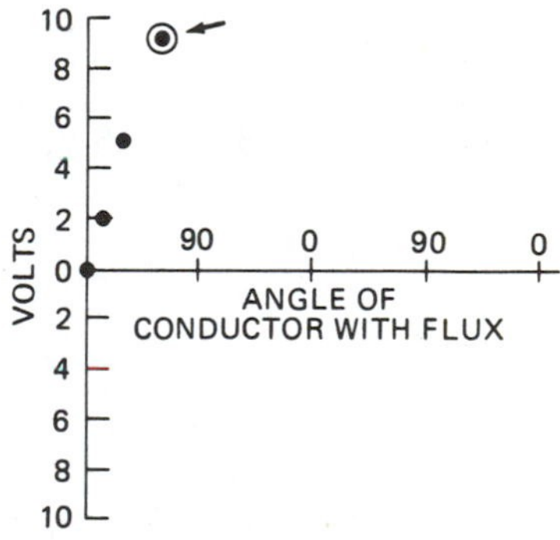

Figure 10-5E
Loop of wire cutting lines of force at a 90° angle. The induced emf is 10 V.

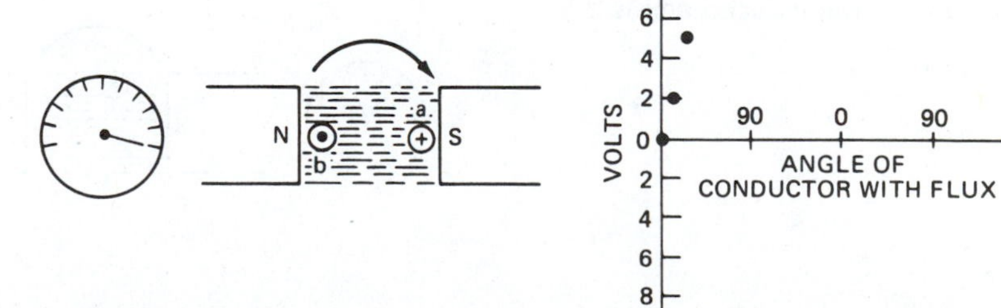

At this point, the angle begins to decrease and the voltage decreases, Figures 10-5F to 10-5H, until the coil is again moving parallel with the lines of force and the voltage is again zero, Figure 10-5I. Conductor *b* now begins to move down through the field and conductor *a* moves up. The meter indicates that the voltage has reversed direction, Figure 10-5J. The electrical pressure is in the opposite direction to what it was in Figures 10-5A through 10-5H. The graph indicates this change of direction by showing the values of voltage below the horizontal line.

In Figures 10-5I through 10-5P, it can be seen that the emf repeats the same values as before, but in the opposite direction.

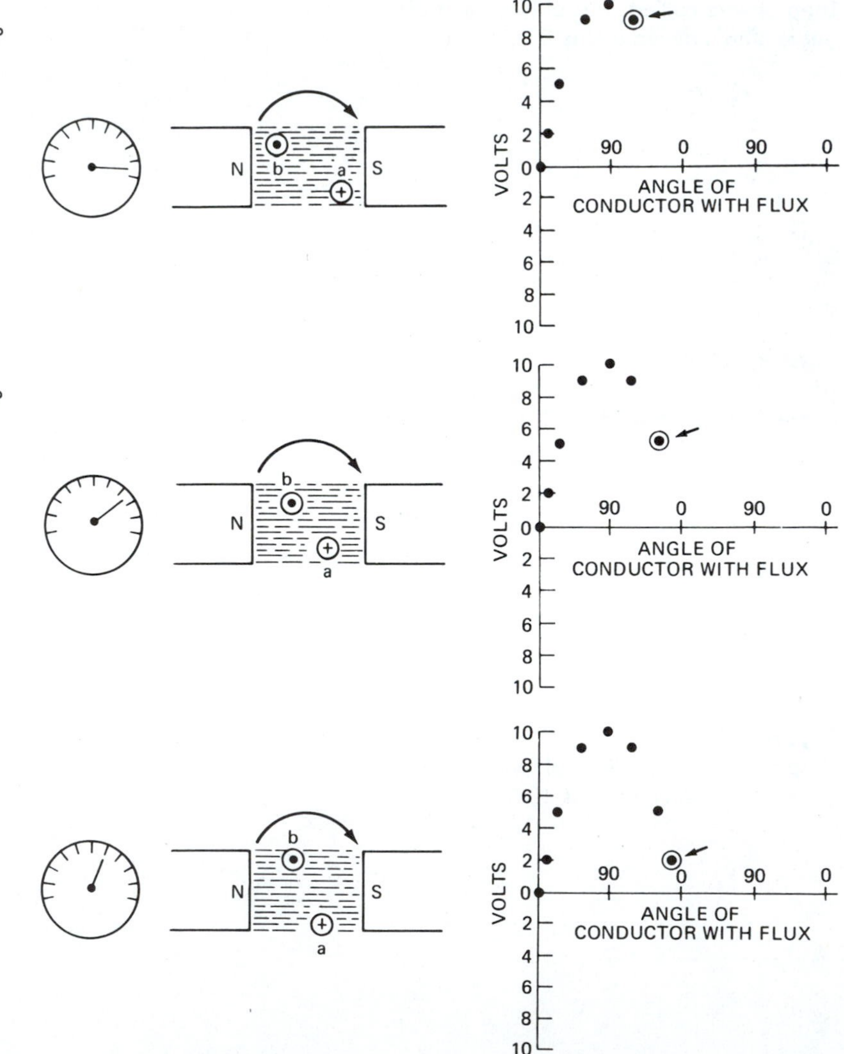

Figure 10-5F
Loop of wire cutting lines of force at a 60° angle. The induced emf is 8.67 V.

Figure 10-5G
Loop of wire cutting lines of force at a 30° angle. The induced emf is 5 V.

Figure 10-5H
Loop of wire cutting lines of force at a slight angle. The induced emf is 2 V.

Figure 10-5I
Loop of wire moving parallel to magnetic flux. No voltage is induced into the loop.

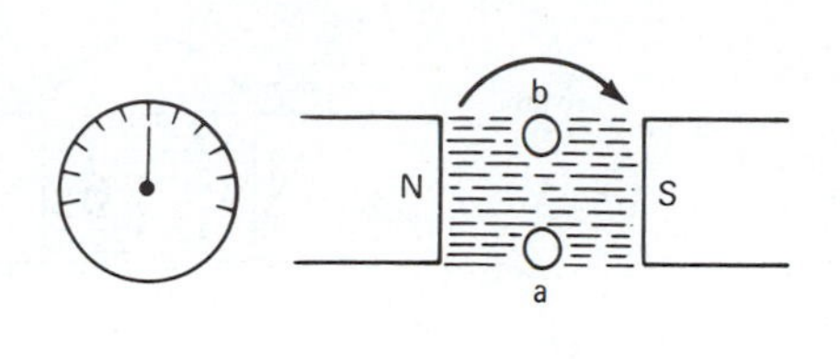

Figure 10-5J
Loop of wire cutting lines of force at a slight angle. The induced emf is 2 V.

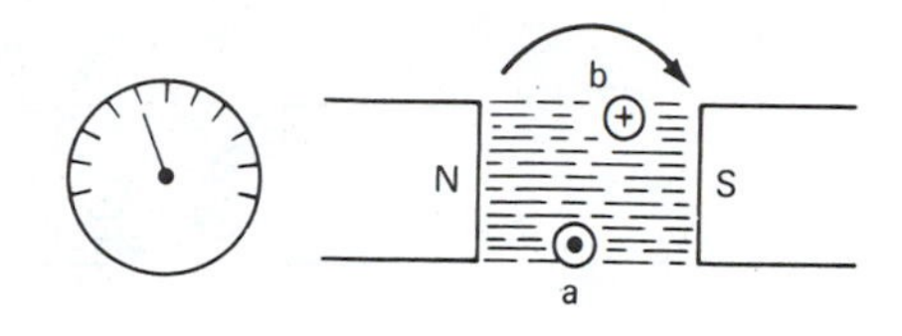

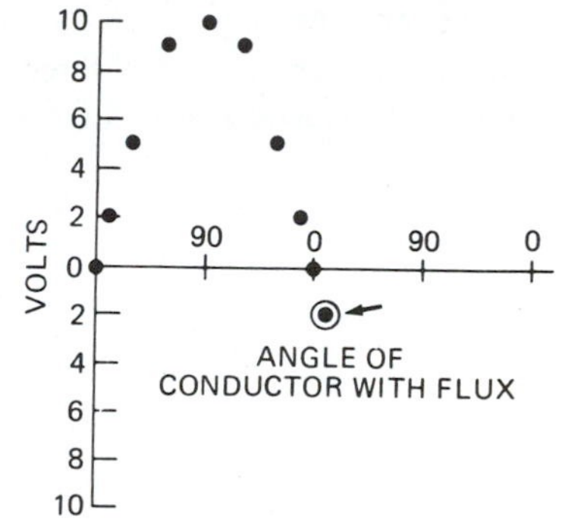

Figure 10-5K
Loop of wire cutting lines of force at a 30° angle. The induced emf is 5 V.

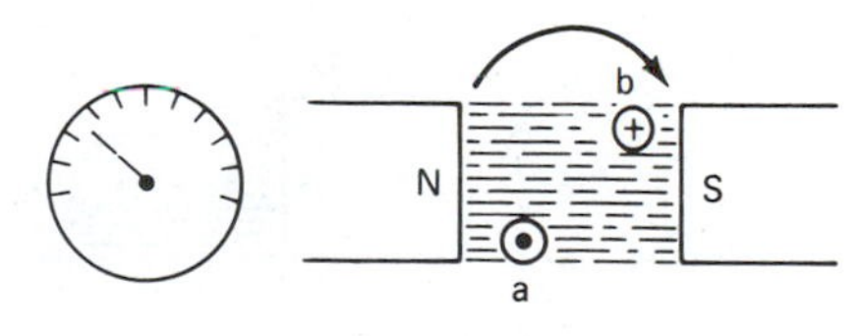

Figure 10-5L
Loop of wire cutting lines of force at a 60° angle. The induced emf is 8.67 V.

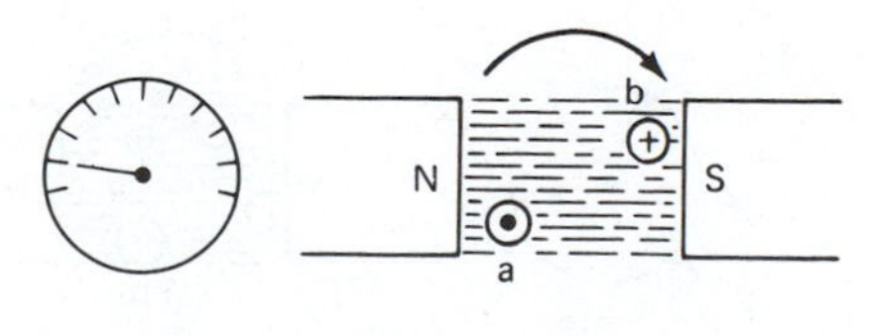

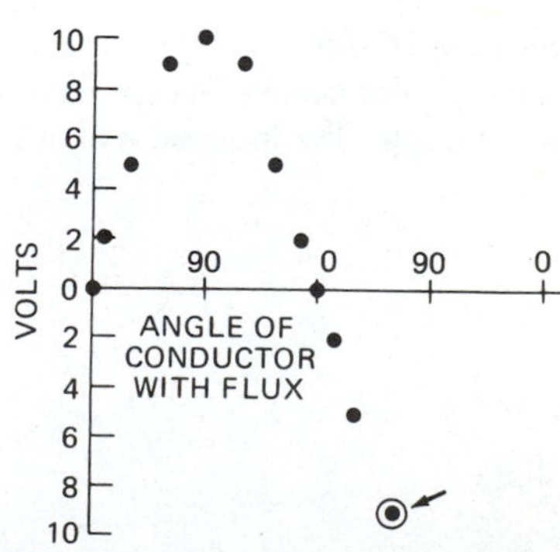

Figure 10-5M
Loop of wire cutting lines of force at a 90° angle. The induced emf is 10 V.

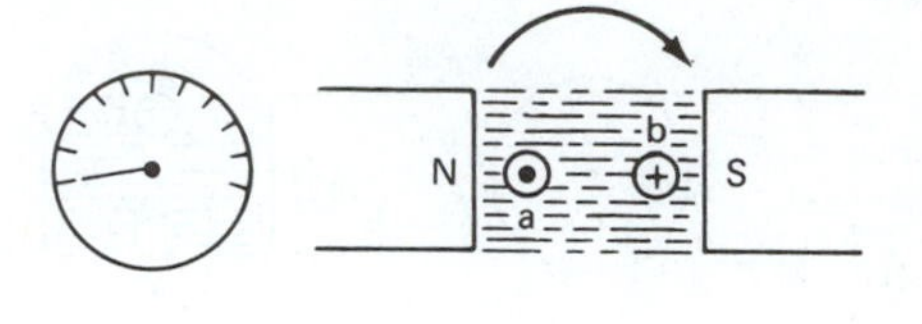

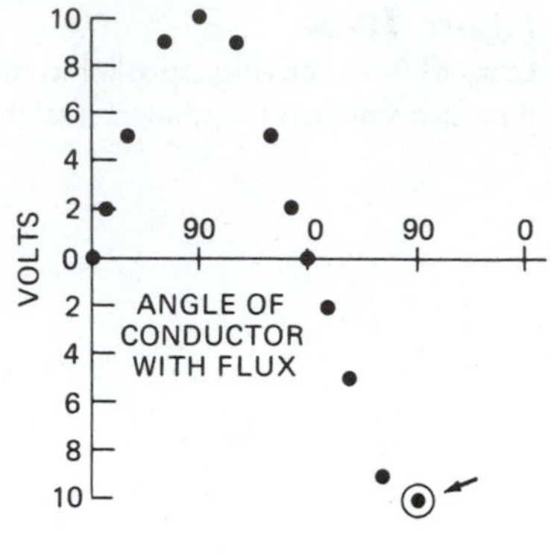

Figure 10-5N
Loop of wire cutting lines of force at a 60° angle. The induced emf is 8.67 V.

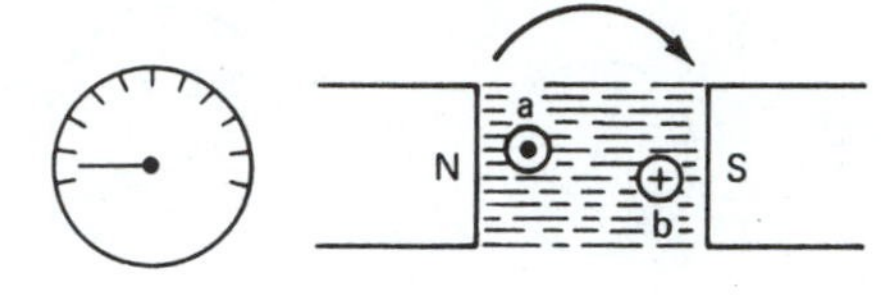

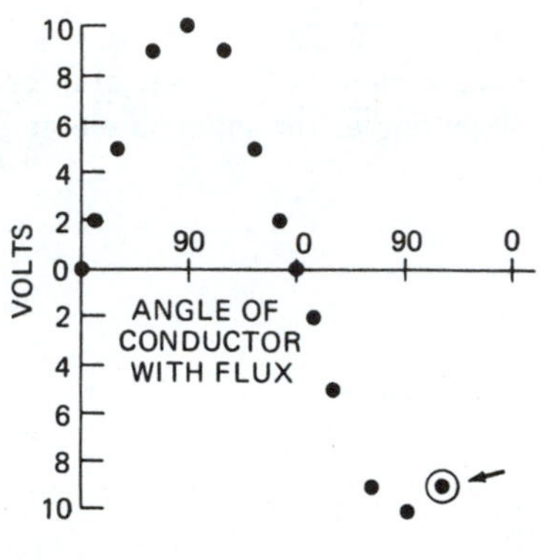

Figure 10-5O
Loop of wire cutting lines of force at a 30° angle. The induced emf is 5 V.

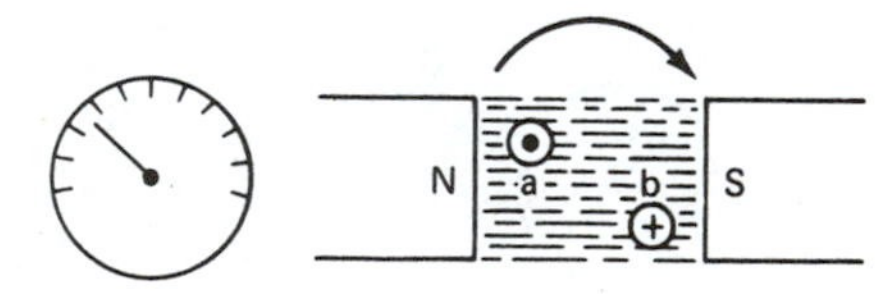

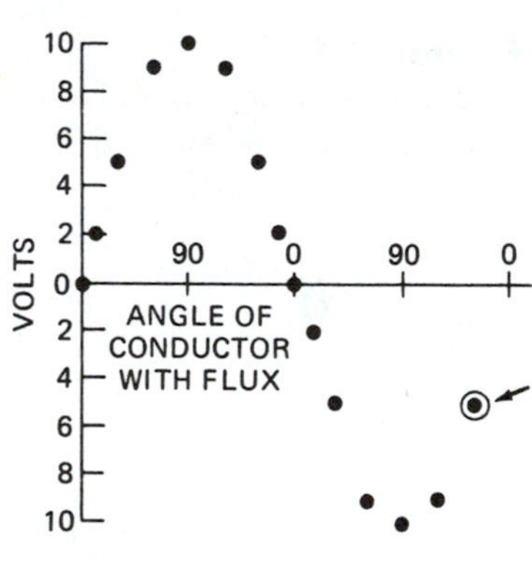

Figure 10-5P
Loop of wire cutting lines of force at a slight angle. The induced emf is 2 V.

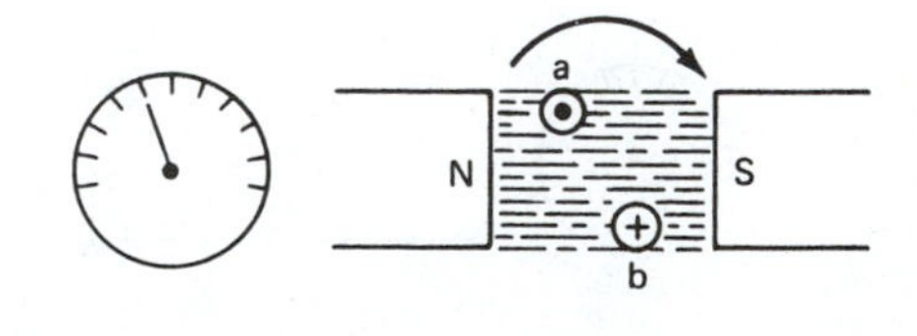

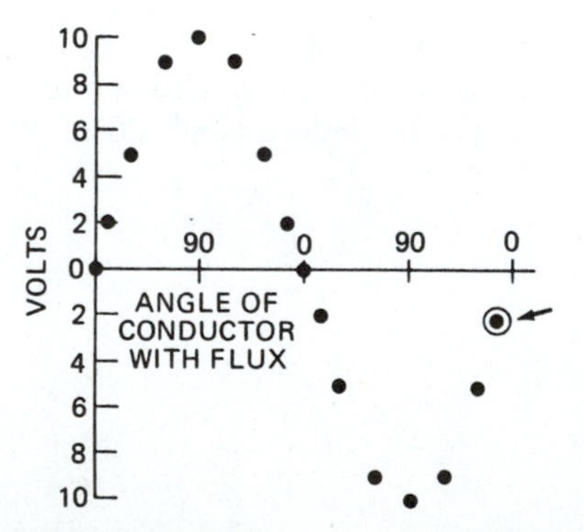

Figure 10-5Q
Typical graph of an alternating voltage

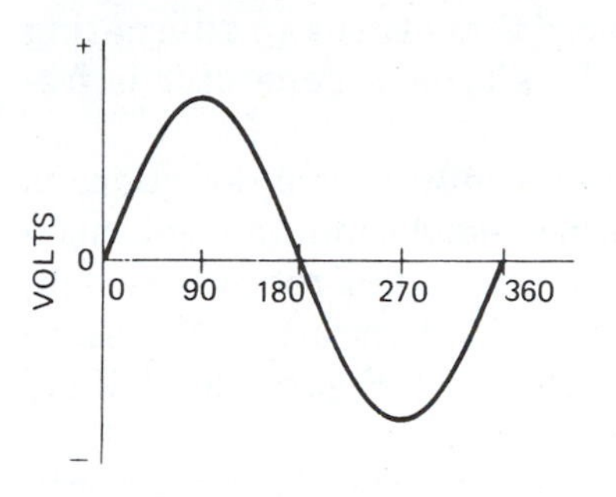

Because this force is first in one direction and then in the opposite direction, it is called an *alternating voltage*. The current that flows as a result of this voltage is called an *alternating current*. All rotating types of generators produce an alternating emf. Figure 10-5Q shows the typical graph of an alternating voltage.

It may be said that an alternating emf is an electrical pressure that reverses its direction of force periodically. An alternating current is the flow of electrons first in one direction and then in the opposite direction in equal periods of time.

In order to increase the amount of voltage produced by the above generator, it is necessary to add more loops to the coil. Each added loop produces an emf equal to that of the first loop. If three loops are used, they form a series circuit and their voltages add together to give a maximum value of 30 volts.

From the graph it can be seen that the values of emf vary with the angle at which the coil moves across the field. It has also been stated that the number of loops in series on the armature determines the value of voltage. If an electromagnet is used in place of the permanent field magnet, the field strength can be increased. This causes a further increase in the generated emf, Figure 10-6.

The type of current obtained from a rotating coil depends upon the method of connecting the load circuit to the generator (the takeoff system). Figure 10-7 shows a method used to obtain alternating current. The ends of the coil from the armature are connected to solid brass or copper rings. These rings, called *slip rings,* are mounted on and insulated from the rotor shaft. They

Figure 10-6
Conductor cutting lines of force at a 90° angle. The induced emf is 25 V.

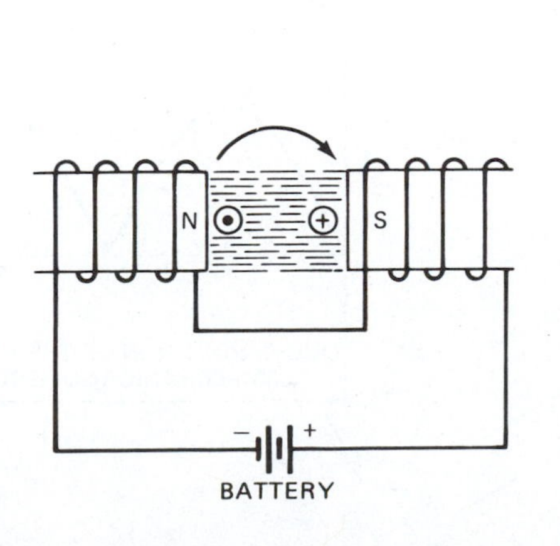

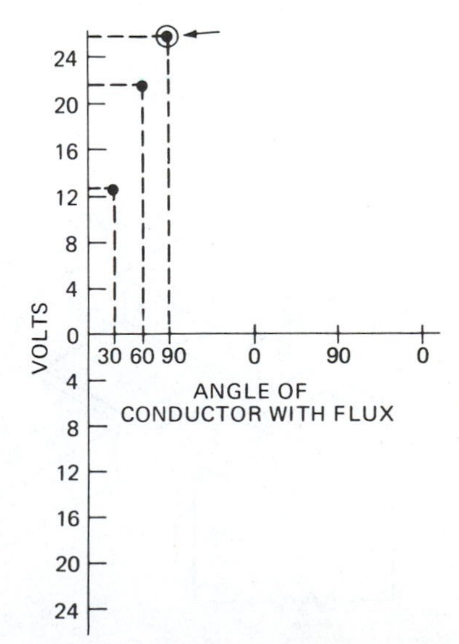

Figure 10-7
AC generator (alternator) takeoff system

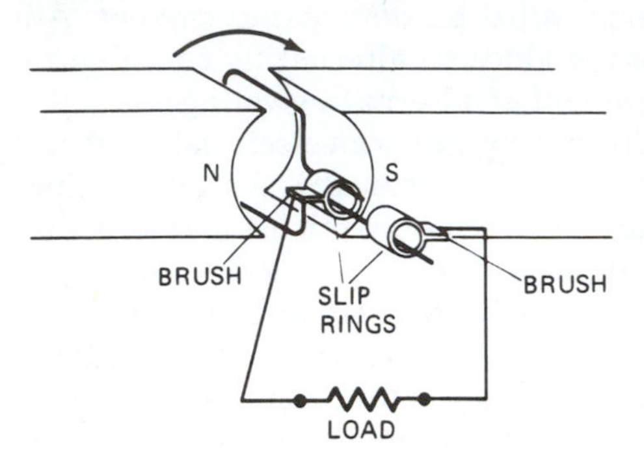

rotate with the armature. Stationary carbon brushes ride on the slip rings and make contact with the load circuit. As the armature rotates, an alternating emf is produced that causes an alternating current to flow through the load. This type of generator is frequently called an *alternator*.

If direct current (current that flows only in one direction) is desired, a different takeoff device is needed. A commutator is used in place of the slip rings. A *commutator* is a type of rotating switch, which changes the connections to the load circuit at the same instant as the emf reverses in the armature. Figure 10-8A illustrates the takeoff device for a dc generator.

Figure 10-8B is a graph of the emf produced by the generator in Figure 10-8A. Even though the voltage and current reverse direction in the armature, the current continues to flow in the same direction in the load circuit. Note that the pressure is always in the same direction, but it is not a steady pressure. For most dc apparatuses, it is necessary to maintain a reasonably constant value of voltage.

One method used to reduce the pulsation of voltage is to add more coils of wire to the armature and more segments to the commutator. For every coil added to the armature, two segments are added to the commutator. The ends of each coil are connected to a pair of commutator segments. The more coils and segments there are, the smoother the dc output will be. See Figures 10-9A, 10-9B, and 10-9C.

In order to maintain a steady direct voltage, the following conditions are necessary.

Figure 10-8A
DC generator takeoff system

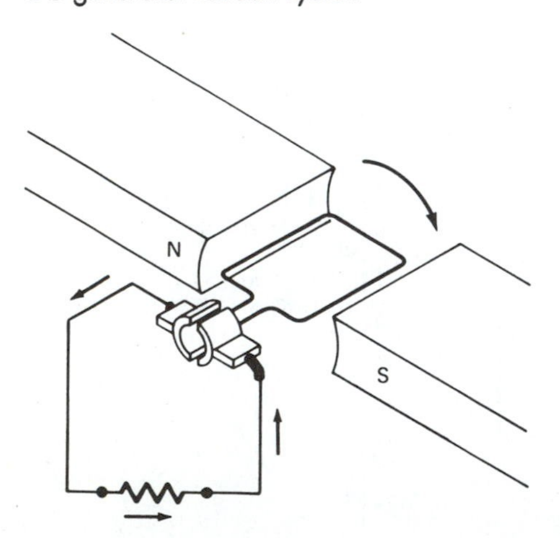

Figure 10-8B
Graph of output voltage of a single-loop dc generator

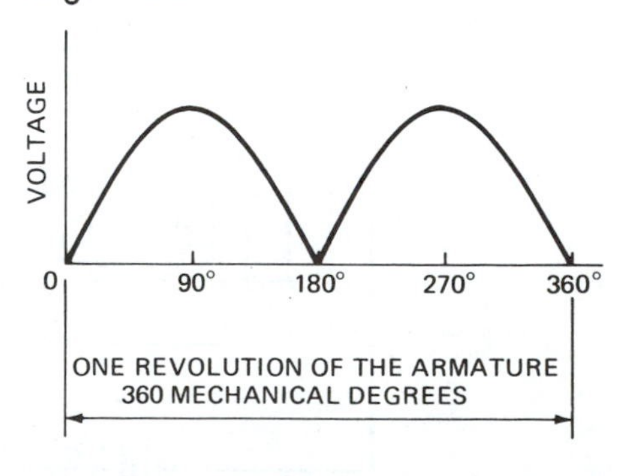

Figure 10-9A
Single-coil generator and graph of dc output

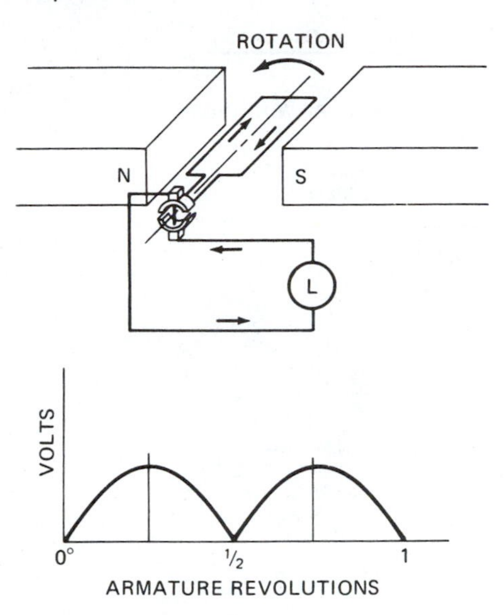

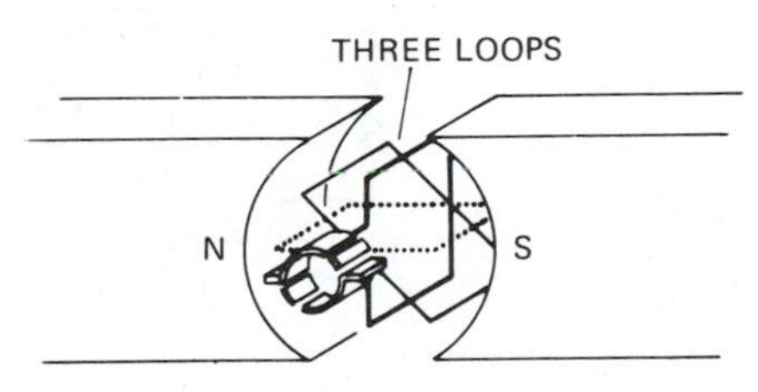

Figure 10-9B
Two-coil generator and graph of dc output

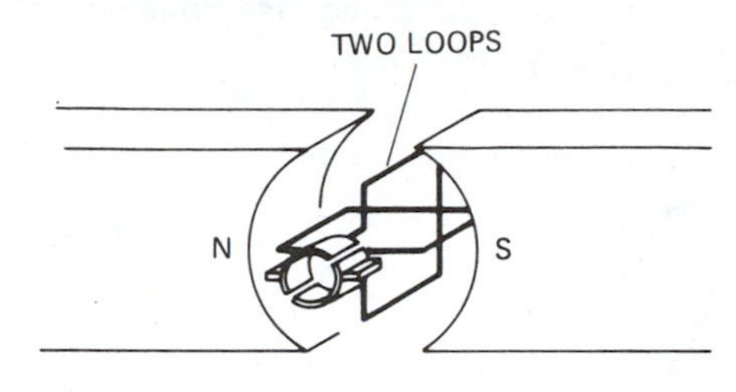

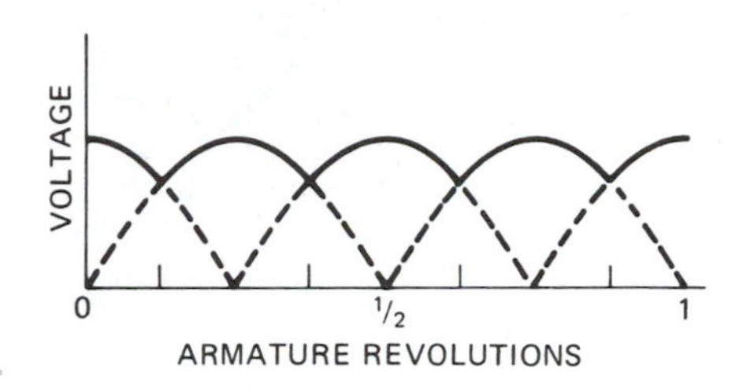

Figure 10-9C
Three-coil generator and graph of dc output

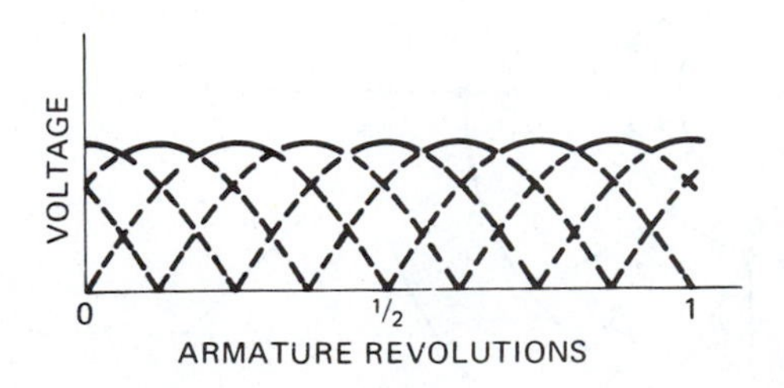

1. The main field flux must be uniform and steady.
2. The armature must rotate at a constant speed.
3. The armature must contain many coils; each coil must be connected to a pair of commutator segments.

Figure 10-10 shows a typical armature, including the commutator, for a dc generator.

Armature Construction

The armature coils of all rotating-type generators are wound on a core made of iron or mild steel. The core is mounted on a shaft set in bearings, which are inserted into the end bells. The entire assembly is arranged to rotate in the magnetic field of the generator. The core also serves as a path of low reluctance for the main field flux. Slots are cut lengthwise into the core and the windings are placed in the slots. The windings are arranged so that when one side of a coil is passing the north pole of the main field, the opposite side of the same coil is passing the south pole.

Two basic types of windings are in general use for dc generator armatures: the *lap winding* and the *wave winding.*

A lap winding is shown in Figure 10-11. Note that the ends of each winding are connected to adjacent commutator segments. The coils are arranged to form parallel paths, thereby allowing for more ampacity without increasing the voltage. On an armature of

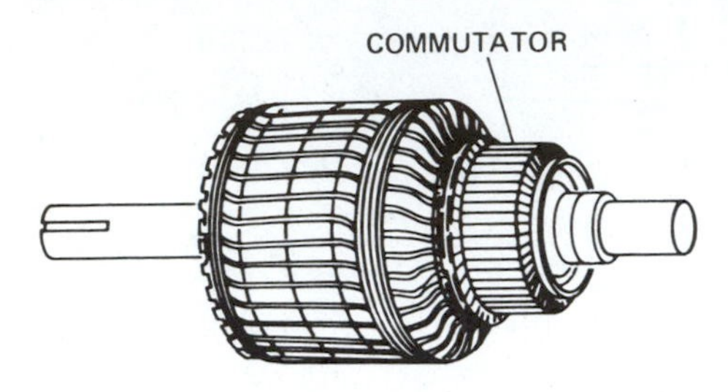

Figure 10-10
Armature with commutator for a dc generator

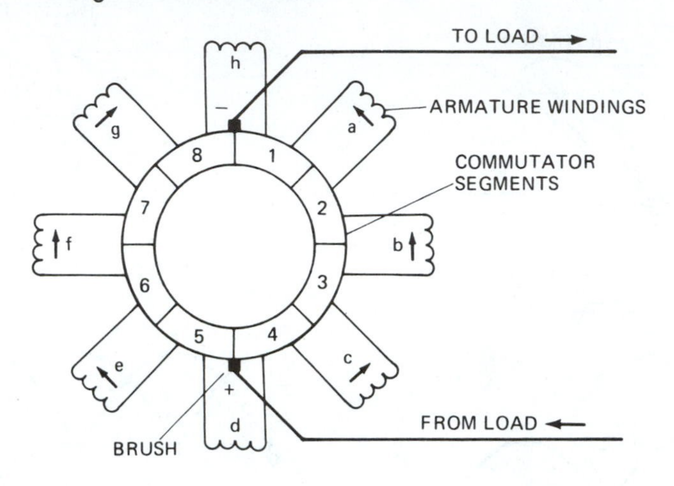

Figure 10-11
Lap winding. The arrows indicate the direction of current through the armature coils.

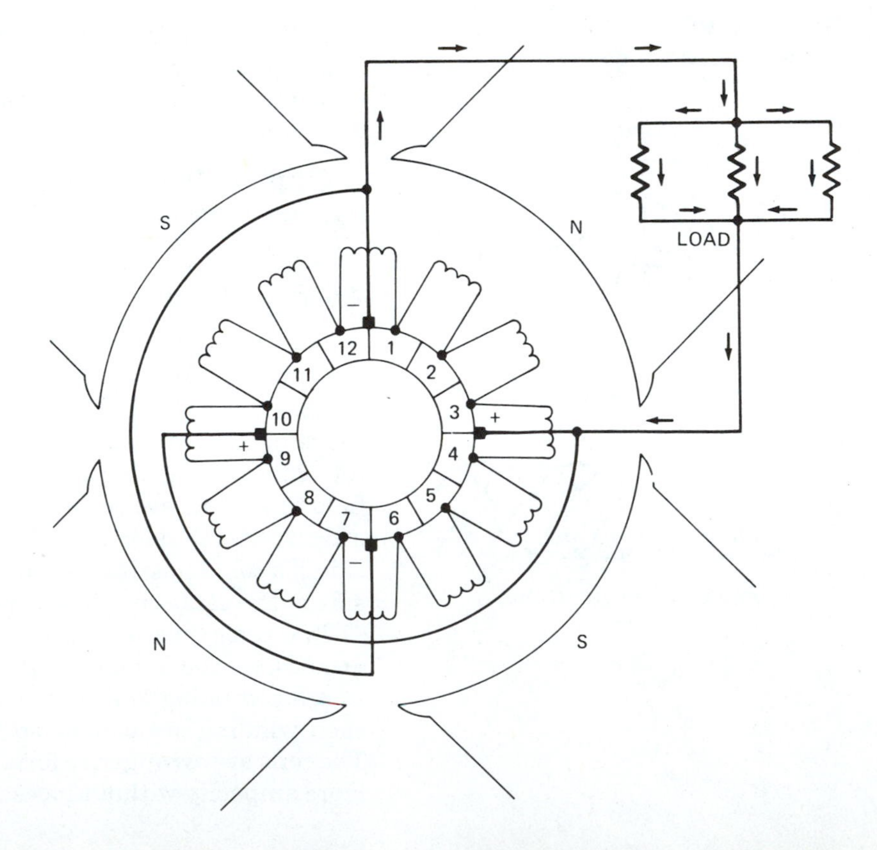

Figure 10-12
Lap-wound armature of a four-pole generator. Current in each path adds to equal load current.

this type, there are two paths for each set of field poles. In other words, a four-pole generator has four paths. The parallel paths are arranged in groups of two and connected to one set of brushes. A four-path generator contains four brushes. The voltage across any one set of brushes is equal to the voltage across any other set and is equal to the output voltage. The maximum current output of the generator is equal to the sum of the ampacity of each path. Figure 10-12 shows a lap-wound armature and the field poles of a four-pole generator.

The wave-wound armature contains only two paths. If it contained the same size, number, and type of coils as a lap-wound armature, it would produce twice as much voltage. However, the maximum current output would be one half that of the lap winding. The induced emf for each coil of a single path adds up to the total voltage of that path.

Notice in Figure 10-13 that the coil ends are not connected to adjacent segments. Their position on the commutator is determined by the number of field poles. The distance between the ends of each coil is equal to the distance between field poles of the same polarity.

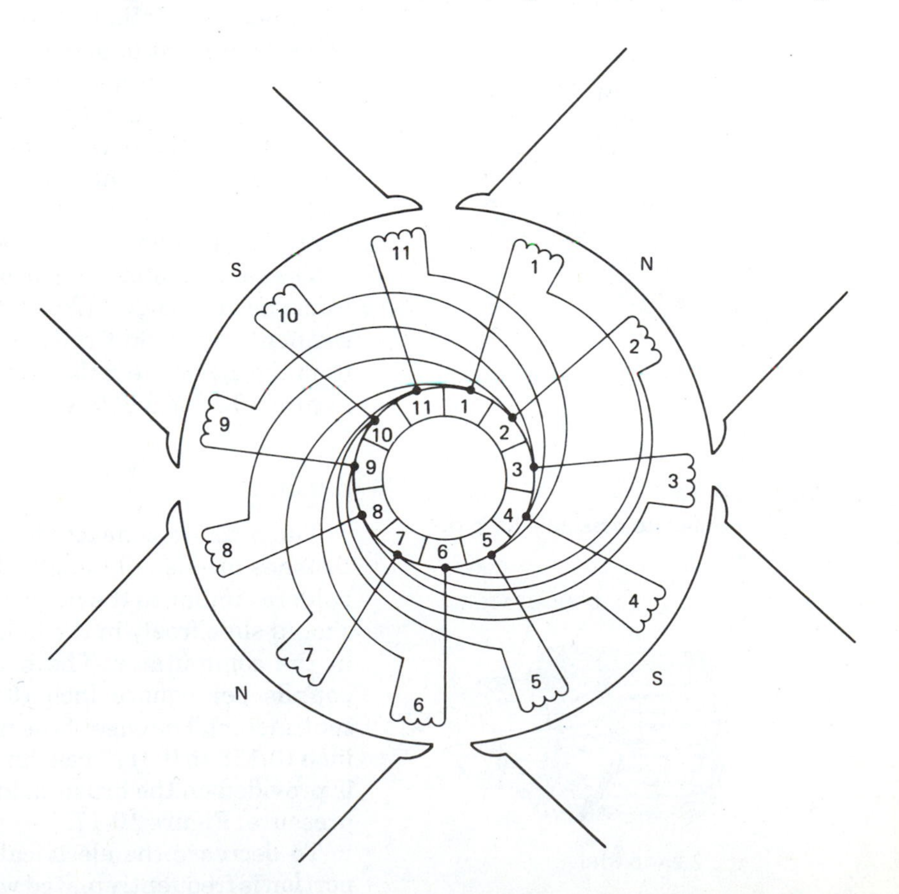

Figure 10-13
Wave-wound armature of a four-pole generator

Figure 10-14A
Typical disk (lamination) that makes up the rotor core

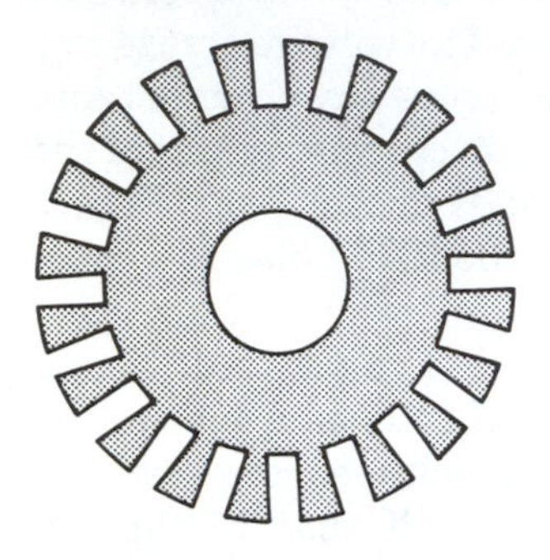

Figure 10-14B
Disks assembled and mounted on a shaft to form the armature core

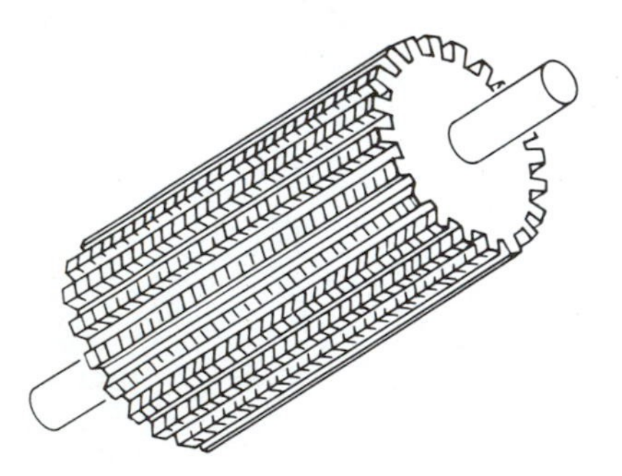

Figure 10-15
Complete armature assembly for a dc generator

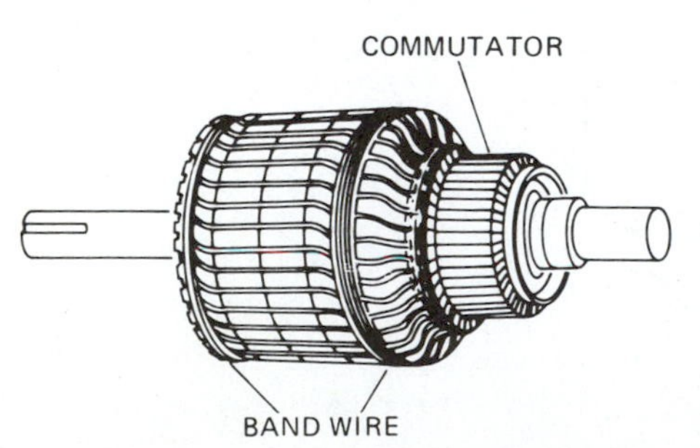

In summary, it can be said that a lap winding is used to obtain high current capacity, and a wave winding is used to obtain high voltage output. Various combinations of the two windings can be used to obtain the desired current and voltage combinations.

The armature core is made of soft iron or mild steel disks. Each disk is about 0.025 inch (0.0635 centimeter) thick. Figure 10-14A shows a typical disk. In Figure 10-14B, the disks are mounted on the rotor shaft. These disks are commonly called *laminations*.

Prior to assembly of the disks on the shaft, they are dipped into an insulating varnish. The varnish insulates them from one another and from the shaft. This type of construction is necessary because the core is a good conductor of electricity. If the core is solid, rotating it in a magnetic field will result in large currents that circulate within the metal. These currents, called *eddy currents*, cause the core to generate enough heat to melt the insulation on the windings, causing a short circuit within the armature.

The armature core is slotted to hold the windings. The slots are lined with an insulation called *fish paper*, which serves as an additional insulation between the core and the windings. It also helps to prevent damage to the insulation on the windings. The windings expand and contract as the load is increased and decreased. If it were not for the fish paper, this continued rubbing against the metal core would wear away the insulation. The windings are held in the slots with fiber or plastic wedges. The sections of the coils not in the slots are held in place by band wires. Figure 10-15 shows a complete armature assembly.

The commutator is made of copper segments that are insulated from one another, and from the supporting rings, with mica. These segments are held firmly in place by clamping rings. The leads from the armature coils are soldered to the commutator segments. Figure 10-16 is a cutaway view of the commutator.

Brushes

The *brushes* connect the commutator to the load conductors. Brushes are usually made of graphite and carbon and placed in holders similar to the holder shown in Figure 10-17. The brushes should slide freely in the holders so as to follow the irregularities in the commutator. The brush pressure should be 1.75 to 2.5 pounds per square inch (0.124 to 0.176 kilogram per square centimeter). The edge of the brush holder should be 0.0625 to 0.125 inch (0.158 to 0.3175 centimeter) from the commutator. A spring is provided on the brush holder and arranged to adjust the brush pressure, Figure 10-17.

To decrease the electrical resistance of the brush, the upper portion is frequently plated with copper. The brush and plating are

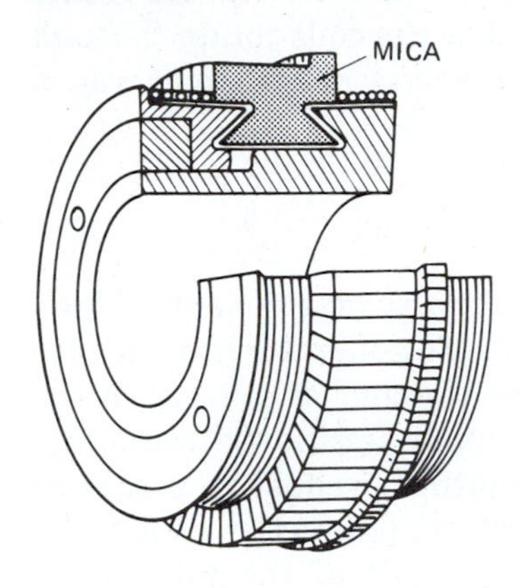

Figure 10-16
Cutaway view of a dc generator commutator

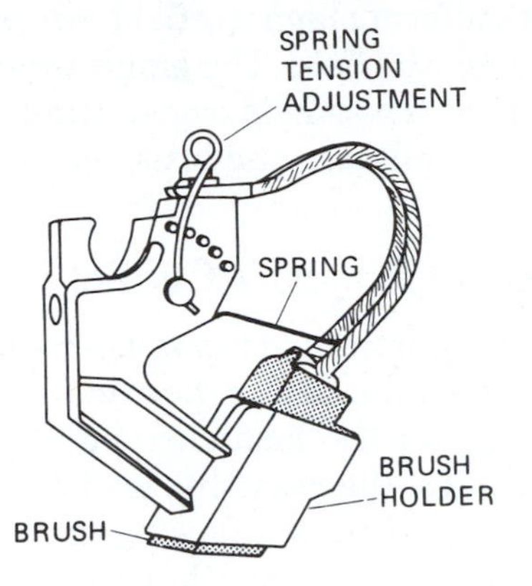

Figure 10-17
Brush holder for a dc generator

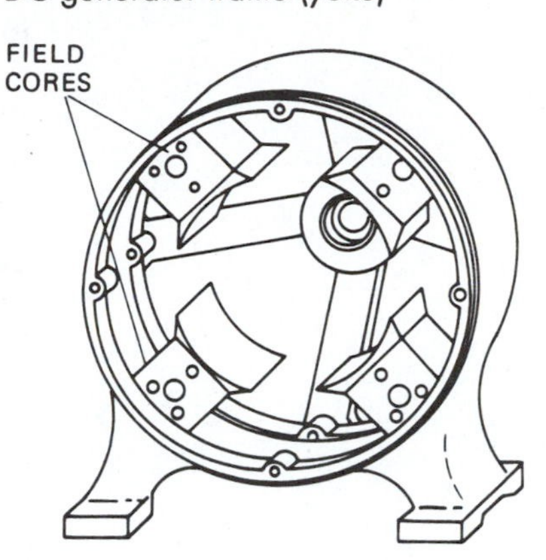

Figure 10-18
DC generator frame (yoke)

connected to the brush holder by a flexible copper wire. The brush holder is insulated from the frame of the machine.

Frame and Field Poles

The *frame*, or *yoke*, of a generator serves as the mechanical support for the machine and forms part of the magnetic circuit. The field cores are attached to the frame. They are made of steel laminations and have a rectangular shape. The ends near the armature are flared in order to hold the coils in place, Figure 10-18.

The field coils are usually wound with cotton-covered wire. After the coil is formed, it is wrapped with cotton tape and immersed in an insulating enamel. It is placed in an oven, and the enamel is baked to a hard finish. The coil is then installed on the field cores between the frame and the armature.

Field Excitation

In all generators, except the very small ones called magnetos, the field flux is produced by current flowing in coils placed on the pole pieces (field cores). The voltage for the field coils may be obtained from a separate source, such as batteries or another generator, or from the armature of the generator itself. When the field is excited from a separate source, the machine is called a *separately excited generator*. When the current is obtained from the machines's own armature, the machine is called a *self-excited generator*.

GENERATOR OPERATION

The basic dc generator consists of coils of insulated wire rotating in a uniform magnetic field. An emf is produced as the coils rotate through the field. The emf is produced as the coils rotate through the field. The emf is transmitted to the electrical system by way of the commutator and brushes.

Effect of Armature Current

A magnetic field is produced whenever an electric current flows. The emf induced into the armature coils of a generator provides the voltage for the load current. Because the emf originates in the armature, the resulting load current must also flow through the armature. Current flowing in the armature produces a magnetic field, which reacts with the main field flux. This action causes the main field to become distorted. The result of the interaction between the two fields is known as *armature reaction*, Figures 10-19A and 10-19B.

When there is no load on the generator, the main field flux forms a direct path from the north pole to the south pole, Figure 10-19A. As a load is applied, the armature flux causes the main field to bend, Figure 10-19B. The degree of bend depends upon the amount of current flowing in the armature. The more current flowing, the greater the degree of bend.

Figure 10-19A
Magnetic field of a two-pole generator when the armature current is zero

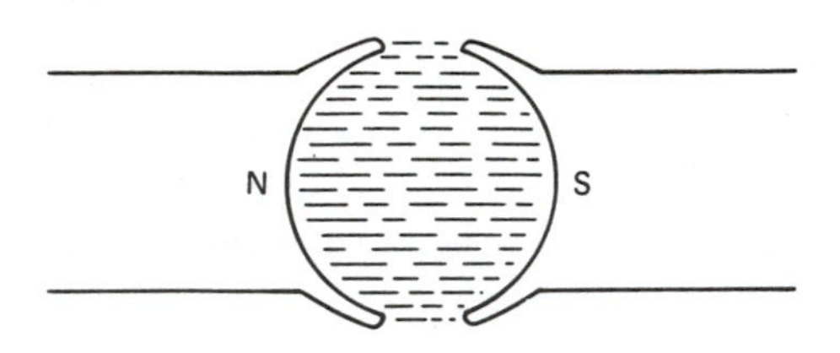

Figure 10-19B
Magnetic field of a two-pole generator with current flowing in the armature coils

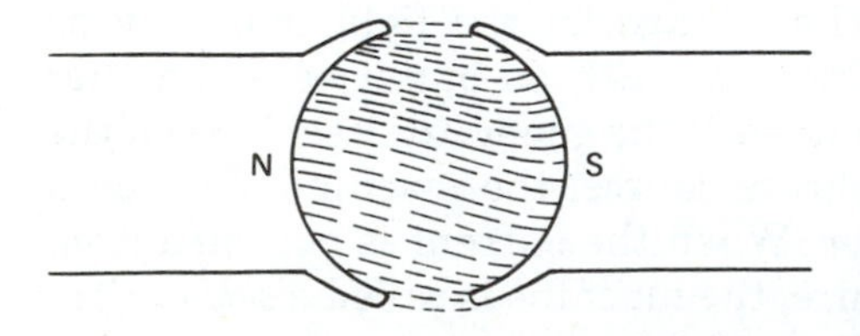

Neutral Plane

When the armature coils are moving parallel with the lines of force, no emf is induced within them. Figure 10-20A illustrates this phenomenon. The two circles (a and b) represent the ends of a single-loop coil. At the instant shown in Figure 10-20A, the coil is moving parallel with the main field flux. If a line is drawn connecting these two points, Figure 10-20B, it will indicate the area known as the *neutral plane* of the generator. Therefore, it can be said that when the armature coils are moving parallel with the lines of force, they are moving through the neutral plane.

The neutral plane is as indicated in Figure 10-20B when there is no load on the generator. However, when a load is applied and current flows in the armature coils, the main field flux becomes distorted. This distortion causes the neutral plane to shift. The shift is always in the direction in which the armature is rotating. The value of the current determines the degree of shift, Figure 10-20C.

Shifting of the neutral plane can be a serious problem if the generator is not designed correctly. The following situation illustrates the problems that may arise. Figure 10-21A represents a simple two-pole generator connected to a load. The armature

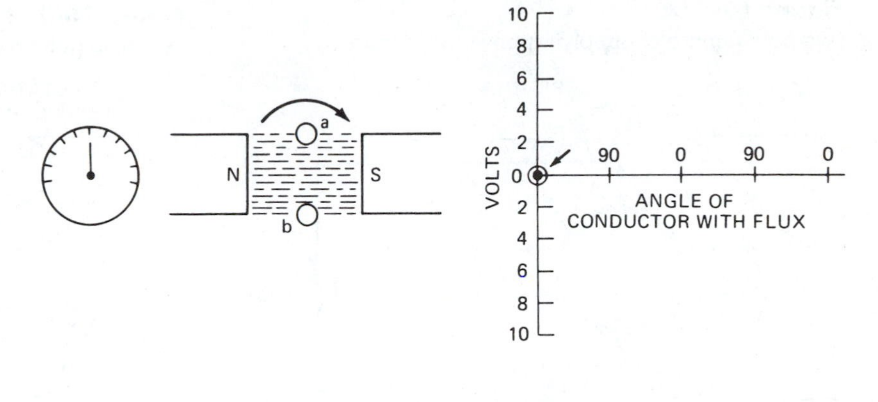

Figure 10-20A
Coil moving parallel with field flux. No emf is induced into the coil.

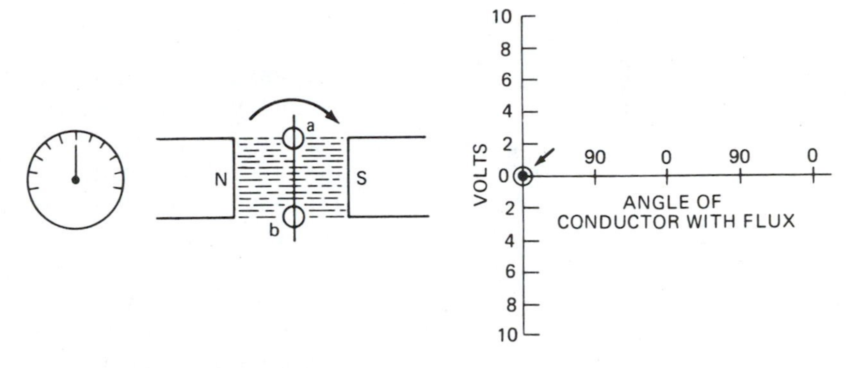

Figure 10-20B
Neutral plane of a dc generator. No current is flowing in the armature.

Figure 10-20C
Neutral plane shifts in the direction of armature rotation.

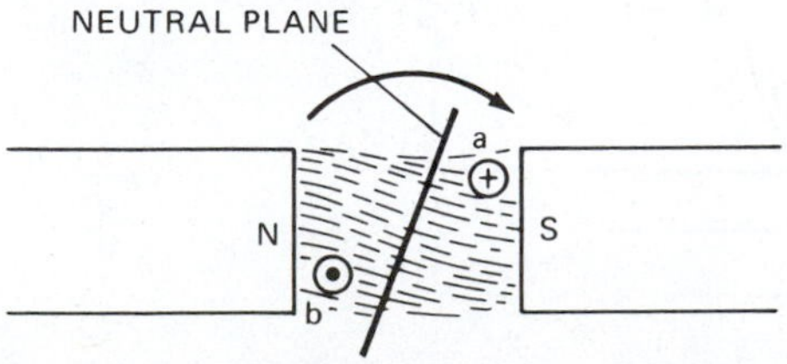

connections indicate two current paths.) If the rotation is in a clockwise direction, the current flow is as indicated by the arrows. (This can be determined by using the left-hand rule for a generator.) The current flows out brush A, through the load, and back to brush B. At brush B it divides equally, with 50 amperes flowing through the coils on the right and 50 amperes flowing through the coils on the left. Both currents return to brush A. At brush A, the two currents join and flow back out through the load.

It is important that the two paths in the armature be identical. The currents must divide equally, and the total emf induced into each path must be equal.

In Figure 10-21A, all coils are cutting the lines of force. Therefore, an emf is induced into all six windings. In Figure 10-21B, however, coils 3 and 6 are moving parallel with the lines of force, and no emf is produced. Coils 3 and 6 are short-circuited by the brushes. If an emf is induced into these coils, short-circuit currents will flow. The value of current depends upon the amount of emf and the resistance of the windings. If the current is large enough, overheating occurs.

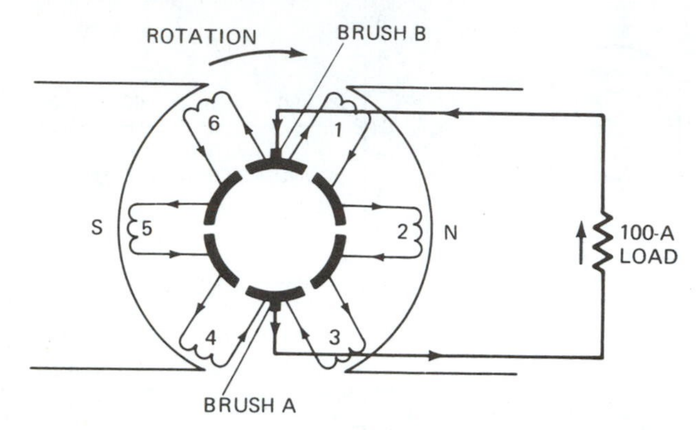

Figure 10-21A
Two-pole generator supplying power to a load

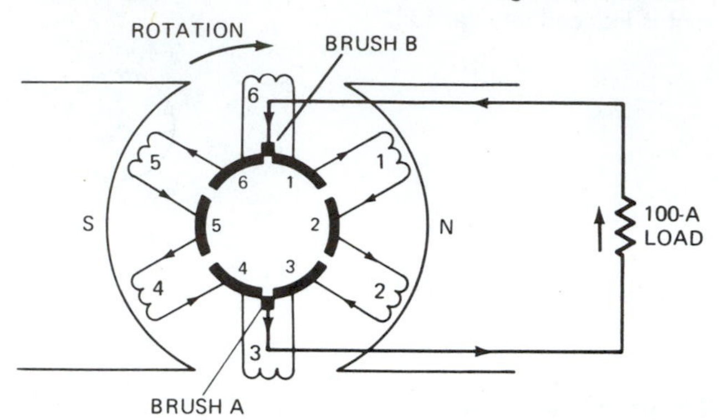

Figure 10-21B
Two-pole generator, brushes short-circuiting coils 3 and 6

Another important point to note is that each brush makes contact with two segments. Brush A is making contact with segments 3 and 4 while brush B is making contact with segments 1 and 6. As the armature rotates, the brushes will break contact with segments 1 and 4 and make contact with 3 and 6. If currents are flowing in coils 3 and 6 at the instant the contacts are broken, arcing will take place. The arcing will cause pitting of the commutator and excessive wear on the brushes. It is of extreme importance that the coils being short-circuited by the brushes always be in the neutral plane.

One way to eliminate this arcing is to move the brushes around the commutator into the new neutral plane. Figure 10-22. This method is successful only if the generator is supplying a constant load. The load on most generators increases and decreases according to the demand on the system. Under these conditions, a generator needs a means of shifting the brushes with every change in load. A way to eliminate the problem is discussed later in this chapter.

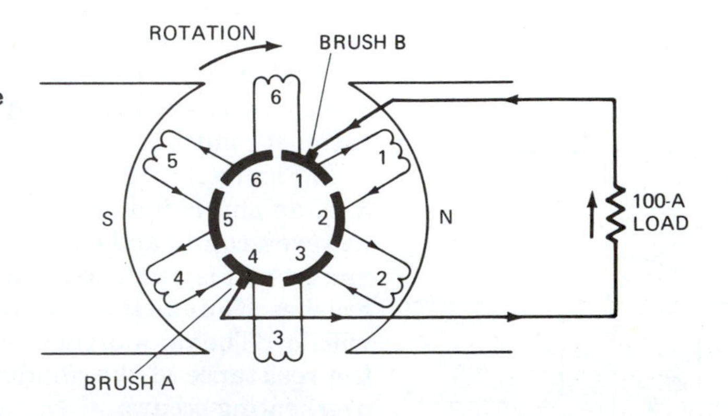

Figure 10-22
Two-pole generator supplying power to a load. The brushes have been shifted to the new neutral plane.

Figure 10-23A
Current flow through a straight wire and load

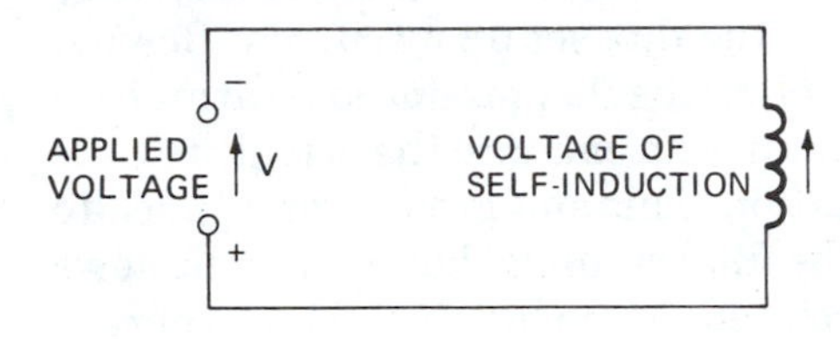

Figure 10-23B
Current flow through a coil

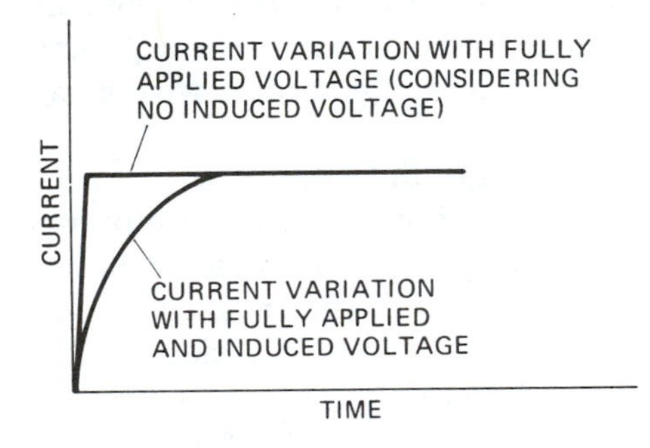

Figure 10-24
Neutral planes in a dc generator

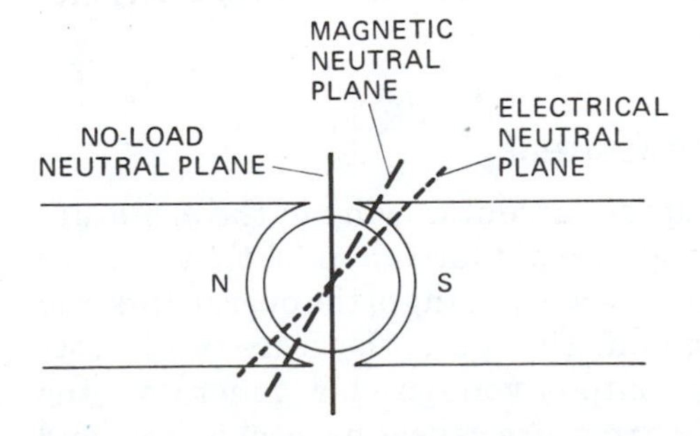

Armature Self-Induction

Self-induction is another phenomenon that takes place when current flows in a coil. When the current increases, the magnetic field caused by the current also increases. The field expands and moves across the loops on the coil. In other words, relative motion takes place between the conductors and the magnetic field. The coil does not move, but the magnetic field does. This relative motion induces an emf into the conductors of the coil. Therefore, two voltages exist across the coil:

1. The generated voltage, which is causing the current to flow
2. The voltage of self-induction (which is opposite to the applied voltage)

The end result is that it takes the current a longer period of time to reach a steady value, Figures 10-23A and 10-23B.

When the coil current decreases, the magnetic field collapses. This means that the field again moves across the conductors, but this time in the opposite direction. The induced emf caused by the collapsing magnetic field is in the same direction as the applied voltage. This tends to cause the current to flow even after the applied voltage has dropped to zero.

The current decreases as the armature coils of a generator move through the neutral plane. Theoretically, the current drops to zero. Because of self-induction, the current does not reach zero value until after the coil has moved beyond the neutral plane.

It can be said that a generator has three neutral planes, Figure 10-24. The no-load neutral plane, called the mechanical neutral plane, is midway between the main field poles. The magnetic neutral plane, caused by current flowing in the armature, is halfway between the mechanical neutral plane and the commutating plane. The electrical neutral plane, referred to as the commutating plane, is ahead of the magnetic plane a distance equal to the distance between the mechanical plane and the magnetic plane.

The brushes must be set in the commutating plane in order to obtain good commutation (to eliminate arcing). The commutating plane shifts with changes in the load. Therefore, this method of reducing arcing at the brushes is practical only for a constant load.

Interpoles (Commutating Poles)

Because most generators supply power to varying loads, some means other than shifting the brushes must be provided. One method is to install small poles midway between the main poles. These poles, called *interpoles* or *commutating poles*, are connected in series with the armature. As the load changes, the current

Figure 10-25
Four-pole dc generator with interpoles

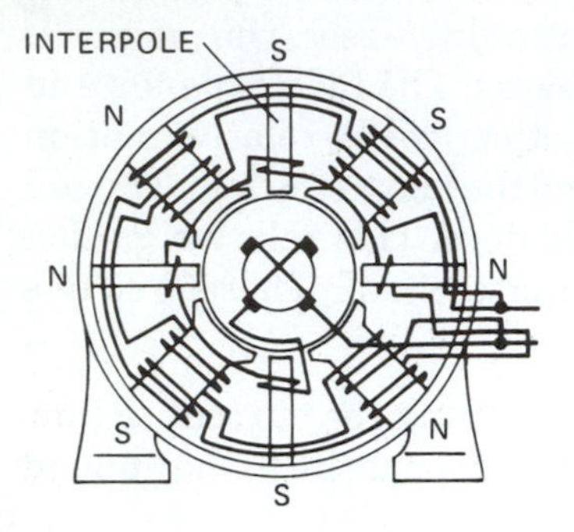

through the interpoles changes by the same amount. Figure 10-25 illustrates a four-pole generator with interpoles.

When interpoles are properly designed and installed, they produce a flux that neutralizes the flux set up by current flowing in the armature. Besides canceling the flux produced by armature reaction, the interpoles also induce an emf into the armature coils that are undergoing commutation. This emf is equal and opposite to the emf of self-induction. The final result is that no current flows in the coils that are short-circuited by the brushes. Thus, there is no arcing at the brushes.

Compensating for Armature Reaction

Interpoles provide good commutation, but they do not eliminate armature reaction. Another method (although more costly) is to install compensating windings. Compensating windings are placed in the main pole faces and are connected in series with the armature. They are arranged so that each conductor embedded in the field pole carries a current equal and opposite to the adjacent armature conductor. Figure 10-26 shows a generator with compensating windings.

Figure 10-26
DC generator with compensating windings

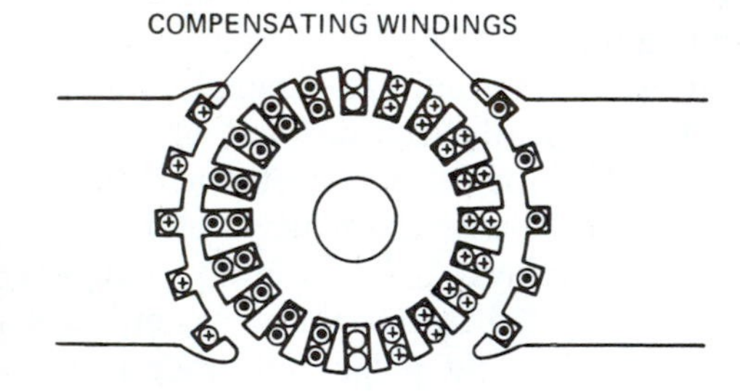

Although the compensating windings eliminate armature reaction, they do not solve the problem of self-induction. To eliminate arcing, the brushes must be set slightly ahead of the mechanical neutral plane (in the direction of rotation). If the brushes are set in the proper location, the coils undergoing commutation will be cutting enough flux in the opposite direction to generate a voltage equal and opposite to that of self-induction. It is almost impossible to eliminate arcing completely, but with careful engineering and proper brush placement, it can be reduced to a minimum.

Generators with interpoles are more common than machines with compensating windings. Interpole machines are much less expensive to build. Compensating windings are generally found on large generators that operate at high speeds and produce high voltages. Some machines that serve very heavy loads under wide load variations and that operate at high speeds use both interpoles and compensating windings.

Other Effects of Armature Current

The magnetic fields set up by current flowing in the armature distort the main field flux and also produce an mmf, that opposes the main field flux. The result is a weakening of the overall flux and a decrease in the generated emf. Therefore, there are two factors to consider that affect the output voltage of a generator: the resistance of the wire on the armature causes a voltage drop, and armature reaction reduces the generated emf. It can be said that

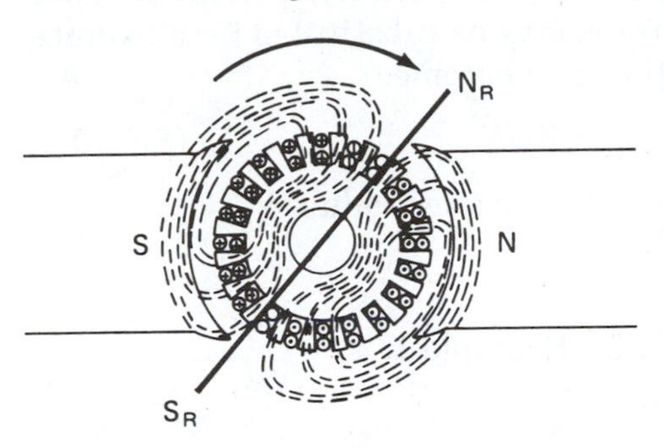

Figure 10-27
Motor action in a generator

as the load increases on a generator, the output voltage tends to decrease.

Reversed torque developed in the rotor is another result of armature reaction. In Figure 10-27, notice that the rotor is being driven in a clockwise direction. The polarity of the armature is such that the north pole of the armature is driven toward the north pole of the main field, and the south pole of the armature is driven toward the south pole of the main field. Because like poles repel, the prime mover (the machine driving the generator) must overcome this repelling force. Therefore, as the load is increased on the generator, the prime mover must develop more torque in order to maintain a constant rotor speed. This phenomenon is frequently called *motor action in a generator*.

GENERATOR VOLTAGE

The amount of emf induced into a conductor depends upon the speed at which it is cutting the flux. When a conductor cuts flux at the rate of 10^8 lines of force per second, 1 volt is induced. By utilizing this fact, we find it possible to derive an equation that will give the average emf produced by a generator. The equation is as follows:

$$E_G = \frac{PZ\Phi N}{10^8\,(60b)} \qquad \text{(Eq. 10.1)}$$

where E_G = generated emf, which at no load is the same value as the terminal voltage
P = number of poles in the main field
Φ = number of lines of force per pole
N = armature speed, in revolutions per minute (r/min)
b = number of parallel paths through the armature
Z = total number of conductors (inductors) on the armature. Because there are 2 inductors per turn, the total number of inductors is equal to 2 times the number of turns.

Example 1

Determine the average voltage generated in a 6-pole machine, running at 900 r/min, if the armature has 300 conductors cutting the field. The flux per pole is 5×10^6 magnetic lines. The armature has 6 parallel paths.

$$E_G = \frac{PZ\Phi N}{10^8\,(60b)}$$

$$E_G = \frac{6 \times 300 \times 5 \times 10^6 \times 900}{10^8 \times 60 \times 6}$$

$$E_G = 225\text{ V}$$

Figure 10-28A
Field current versus lines of force

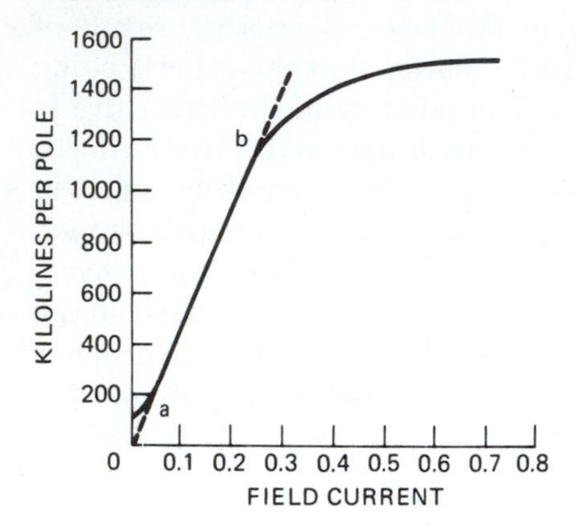

Figure 10-28B
Field current versus generated voltage

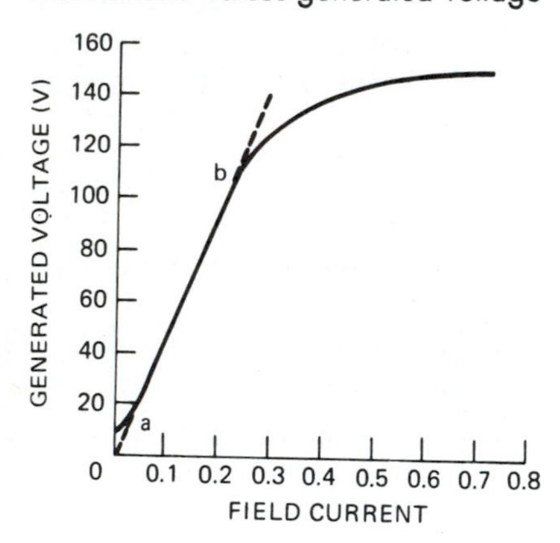

For a specific generator, all of the factors in the equation are fixed values with the exception of the speed (N) and the flux per pole (Φ). Therefore, the letter K may be substituted for all values except N and Φ. The equation now becomes

$$E_G = K\Phi N \qquad \text{(Eq. 10.2)}$$

where K = combined result of all the fixed values

Example 2

Determine the value of K for Example 1.

$$K = \frac{PZ}{10^8\,(60b)}$$

$$K = \frac{6 \times 300}{10^8 \times 60 \times 6}$$

$$K = 5 \times 10^{-8}$$

Proof:

$$E_G = K\Phi N$$
$$E_G = 5 \times 10^{-8} \times 5 \times 10^6 \times 900$$
$$E_G = 225 \text{ V}$$

The formula $E_G = K\Phi N$ can be useful for designing generators. In practice, however, generators usually operate at a fairly constant speed, and the only variable is the field flux (Φ). A rheostat or a similar device is frequently installed in the field circuit to provide a means of controlling the field current. Varying the field current will vary the flux and thus the generated emf.

Saturation Curve

The equation $E_G = K\Phi N$ indicates that the induced emf is proportional to the flux per pole and the revolutions per minute of the generator. At a constant speed, the generated emf is proportional to the field strength. In a given machine, the flux depends upon the field current. It is not, however, directly proportional to the field current, because beyond a certain number of ampere-turns, all electromagnets become saturated. Figure 10-28A is a graphic example of this feature. Because of the residual magnetism, the curved part at point *a* does not start at zero. Between points *a* and *b*, the curve is almost a straight line, indicating that the flux in this area is proportional to the field current. At point *b*, the line begins to curve sharply, indicating that the magnetic circuit is reaching saturation.

Because the generated emf varies directly with the field flux. the voltage curve is the same as the flux curve, Figure 10-28B.

Between points *a* and *b*, the voltage increases rapidly with a given change in field current. Beyond point *b*, a large increase in current causes only a slight increase in voltage. The voltage curve will vary somewhat with the design of the machine.

SELF-EXCITED GENERATOR

As discussed previously, a self-excited generator receives the current for the field from its own armature. This can be accomplished because of residual magnetism in the field cores.

There are three types of self-excited generators, depending upon how the armature and the field windings are connected. These types are the shunt generator, the series generator, and the compound generator. Figure 10-29 illustrates the connections for a self-excited shunt generator. It can be observed that there is no external source of power for the field. As the prime mover drives the generator, the armature rotates in a very weak magnetic field. This field is produced by residual magnetism in the field cores. The armature conductors, cutting the residual field, produce a small emf. Because the shunt field windings are connected directly across the armature, a small current flows through the shunt field. This current causes an increase in the field flux. A stronger field flux produces a larger emf, which forces even more current through the field coils. The flux density again increases, which induces a greater emf into the armature, again increasing the field current. This process continues until the field cores are saturated. At this point, the generator has reached its no-load voltage. The entire building-up process takes about 20 seconds to 40 seconds.

Figure 10-29
Schematic diagram of a shunt generator

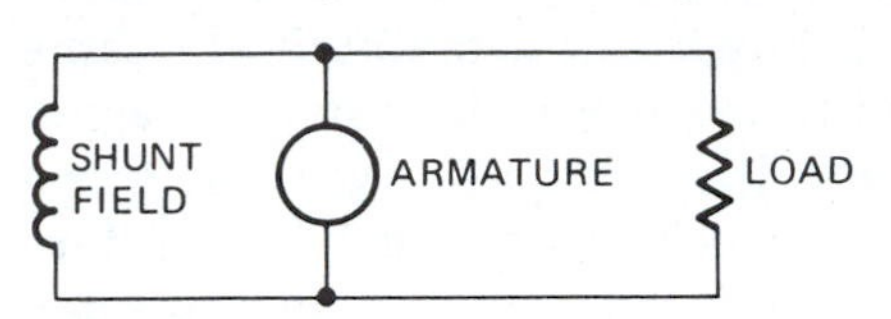

If a self-excited generator is being driven by the prime mover but fails to build up a voltage, there are several possible causes.

Causes

1. Loss of residual magnetism may be a cause. If the generator sits idle for a long period of time, it may lose its magnetism. If it has been moved, the jarring it received may have caused the loss of magnetism. Overloads and current surges could also be a cause.
2. If the voltage induced into the armature forces the current through the field in a direction that produces an mmf in opposition to the residual magnetism, it will not build up a voltage.
3. A break or opening in the field or armature circuit will prevent a voltage buildup.
4. Loose brush connections or contacts may be a cause.
5. A dirty or severely pitted commutator frequently prevents voltage buildup.

6. A short circuit in the armature or field may be the problem.
7. If the speed of rotation of the armature is too slow, the generator will not produce its rated voltage.
8. If the armature is rotating in the wrong direction, the magnetic fields will oppose each other.
9. If the field circuit resistance is too great, the voltage will not build up.

The problems listed can be eliminated as follows:

Solutions

1. Loss of residual magnetism. Disconnect the field from the armature, being sure to note the field polarity. Connect a dc source across the field. It is important to maintain the same polarity that the field had before it lost its residual magnetism. It is best to use a low-voltage dc for this purpose. For a 120-volt to 600-volt generator, a 12-volt automobile battery should be sufficient. Allow the dc source to remain connected for approximately 10 minutes.

   Disconnect the dc power and reconnect the armature and the field. Be sure to maintain correct polarity. The generator should now build up a voltage.

   > **CAUTION:** If a voltage greater than 0.1 times the rating of the generator is used, discharge resistors must be connected in parallel with the field. The purpose of these resistors is to absorb the power dissipated by the collapsing magnetic field when the dc supply is removed. The rating of this resistance bank depends upon the value of voltage used, the resistance of the field, and the inductive effect. Never use a voltage greater than the rating of the generator.

   If the polarity of the residual magnetism has been reversed, the polarity of the generated emf will be reversed. Reversal of polarity can damage some types of equipment.

   > **CAUTION:** It is important to check the polarity before restoring the generator to the line. If the residual magnetism is reversed, it will be necessary to recharge the residual field. After ensuring the correct polarity, apply the dc and allow it to remain connected for about 10 minutes.

2. The generator fails to build up a voltage because of reversed polarity on the armature. Reversing the armature connec-

Figure 10-30
Interchange the armature leads to change the polarity.

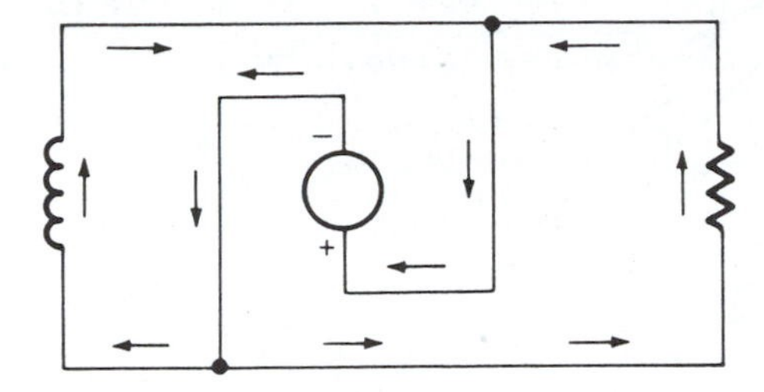

tions will correct this condition. Figure 10-30 illustrates the process.

3. Open circuits in the field or armature. This can be corrected by locating the open circuit and taking necessary steps to remedy it.
4. Poor brush contact or loose brush connections. Check the brushes for excessive wear, and replace them if necessary. Check the commutator for pitting. If necessary, "turn down" the commutator. This is accomplished by using a lathe to remove a small amount of copper, just enough to obtain a smooth surface. If necessary, undercut the mica so that it is slightly below the commutator segments. Always clean the commutator when poor brush contact is discovered. Check the brush tension and readjust it if necessary. Tighten any loose connections.
5. A dirty and/or pitted commutator. In this case, follow the same procedure as outlined in Step 4.
6. A short circuit in the armature or field. Short circuits should be located and cleared. A thorough inspection should be performed to determine the cause.
7. The armature is rotating too slowly. Increase the speed of the prime mover.
8. The armature rotation is reversed. The prime mover must be adjusted to drive it in the correct direction.
9. The field circuit resistance is too great. This can usually be corrected by adjusting the rheostat in the field circuit.

Shunt Generators

A *shunt generator* is a type of self-excited generator in which the field and armature are connected in parallel. (Figure 10-29 illustrates the connections for this type of generator.) The building-up process for a shunt generator is the same as described in the section on self-excited generators. When the magnetic circuit of the field has become saturated, the field coils will receive full armature voltage.

The field coils are constructed of many turns of small wire. The use of small wire and many turns produces a strong field while keeping the current to a minimum. This results in a more economical and compact construction, and it also improves the efficiency of the machine.

Because the armature supplies the emf for both the load and the field, all the current must flow though the armature. Figure 10-31 is a schematic drawing for a shunt generator. The arrows indicate the current flow.

The armature current increases as a load is added to a shunt generator. An increase in the armature current causes an increase

Figure 10-31
Shunt generator. The arrows show the direction of current flow.

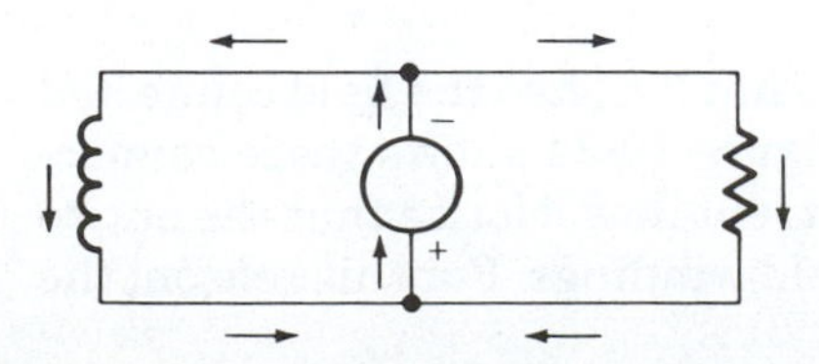

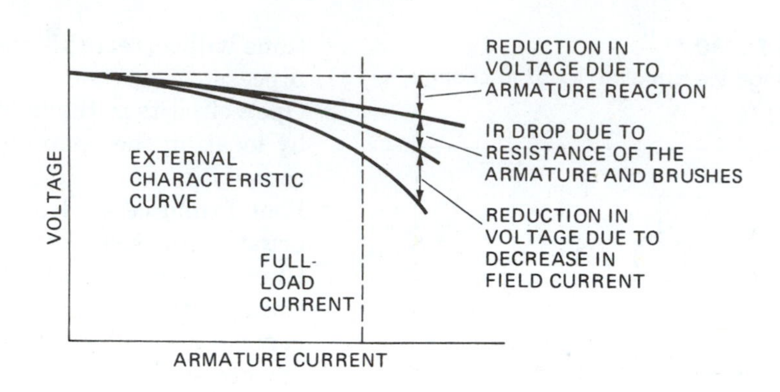

Figure 10-32
Load/voltage characteristics of a shunt generator

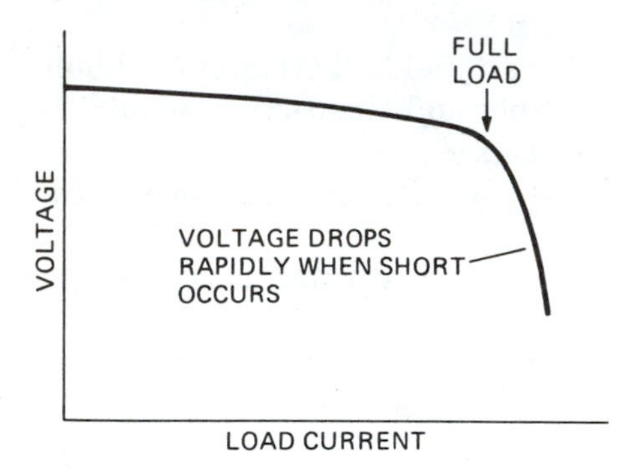

Figure 10-33
Voltage curve of a shunt generator when a short circuit occurs across the armature

in the effects of armature reaction. In other words, as the load current increases, the generated emf decreases. This decrease in the generated emf results in a lower output voltage. Another factor that affects the output voltage of a generator is the IR (voltage) drop in the armature.

Because the field receives its voltage from the armature, any decrease in terminal voltage results in less field current. The reduction in field current causes a weakening of the main field flux. The weaker field results in a further decrease in the output voltage.

It may appear from this information that the generated emf would quickly drop to zero. This does not occur, however, due to the internal design of the generator. Figure 10-32 is a graphic illustration of the load/voltage characteristics of a shunt generator.

It is generally desirable that the voltage across the load be a constant value. However, shunt generators still are used in industry because they are economical and provide efficient means for supplying power to constant loads. Well-designed generators do not have more than an 8 percent decrease in voltage from no load to full load. These machines are suitable when slight voltage fluctuations are not a problem.

One advantage of the load/voltage characteristics of the shunt generator is self-protection. A short circuit on the load will cause the output voltage to drop rapidly, Figure 10-33. This sharp drop in voltage reduces the current to a minimum and thus prevents overheating.

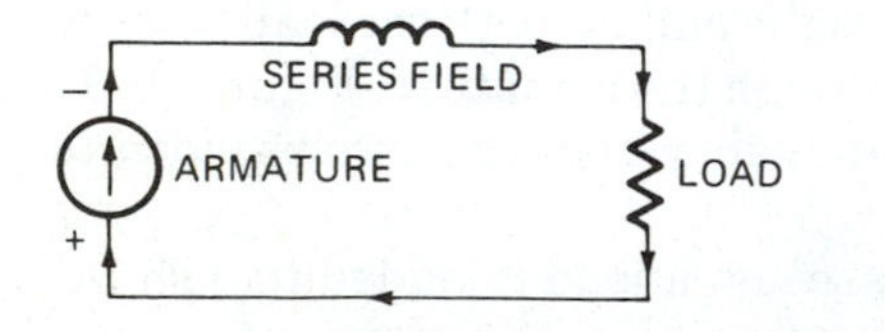

Figure 10-34
Schematic diagram of a series generator

Series Generators

A *series generator*, as the name implies, has the field connected in a series with the armature. Figure 10-34 shows these connections. The arrows indicate the current flow. Notice that the entire load current flows through the field windings. For this reason, the

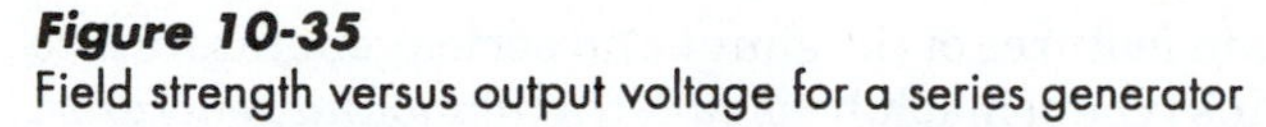

Figure 10-35
Field strength versus output voltage for a series generator

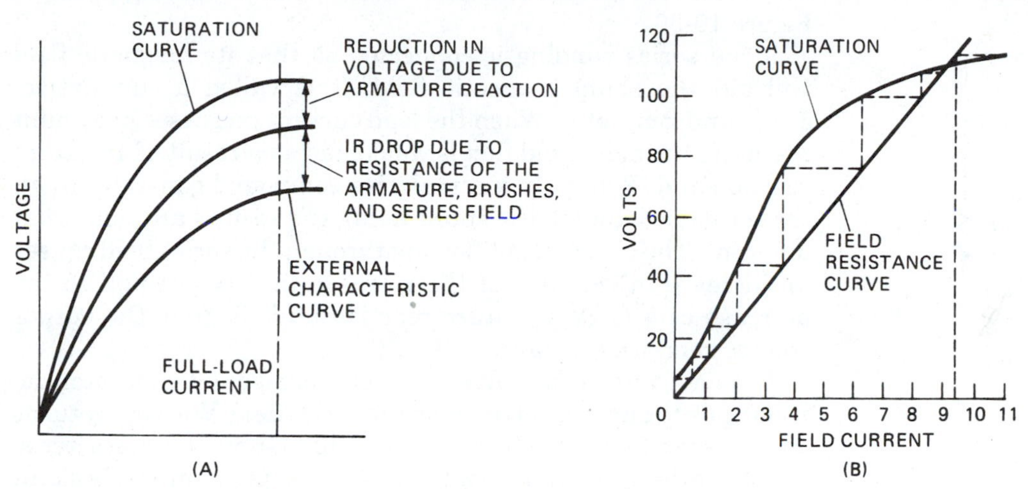

field coils must be wound with wire large enough to carry the full-load current of the generator.

In a series generator, an increase in load causes a similar increase in the field current. This results in both an increase in the generated emf and a higher output voltage. The increase in voltage with increases in load will continue until the magnetic circuit of the field is saturated. An increase in load beyond this point will cause the output voltage to decrease. The decrease is a result of IR drop and armature reaction. Figure 10-35 demonstrates the load/voltage characteristics of a series generator. The load/voltage curve in this figure can be compared to the saturation curve.

In the past, series generators were used extensively for series arc lighting, particularly for street lighting. Later, when arc lamps were replaced by series tungsten lamps, the series generator still supplied the power. Most tungsten lamps used for this purpose have been replaced by mercury vapor or high-pressure sodium lamps. These lamps require alternating current. Modern technology has practically phased out the series generator. However, it is sometimes found in remote areas or in applications where it is supplying constant loads.

Compound Generators

In industry, most loads require a constant value of voltage, but the load on the generator fluctuates considerably when a plant is in operation. This type of demand makes both the shunt and series generators undesirable for most conditions of use. However, it is possible to maintain a constant voltage under varying loads by

Figure 10-36
Schematic diagram of a compound generator

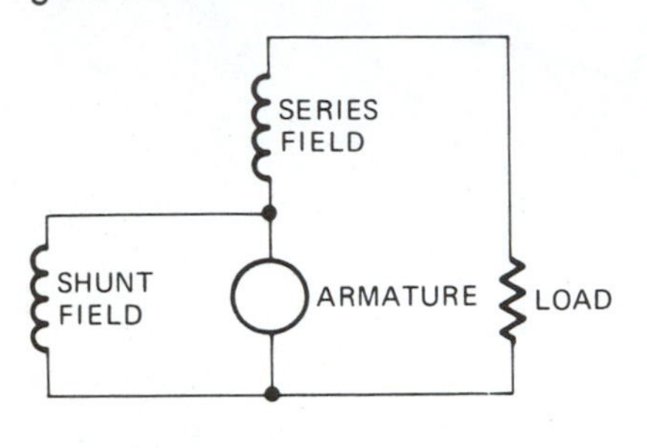

Figure 10-37
Compound generator with series field diverter

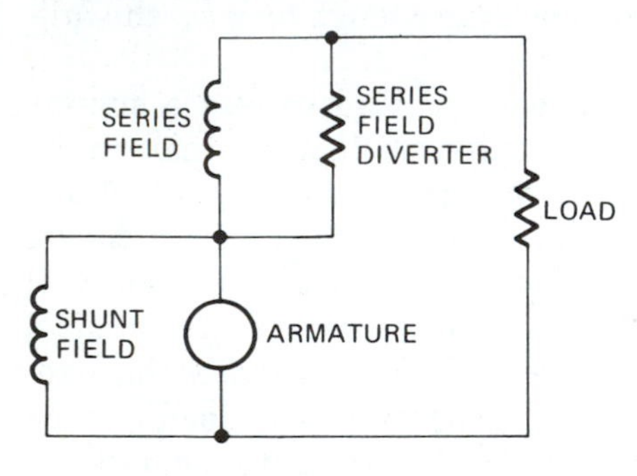

Figure 10-38
Load/voltage characteristics of compound generators

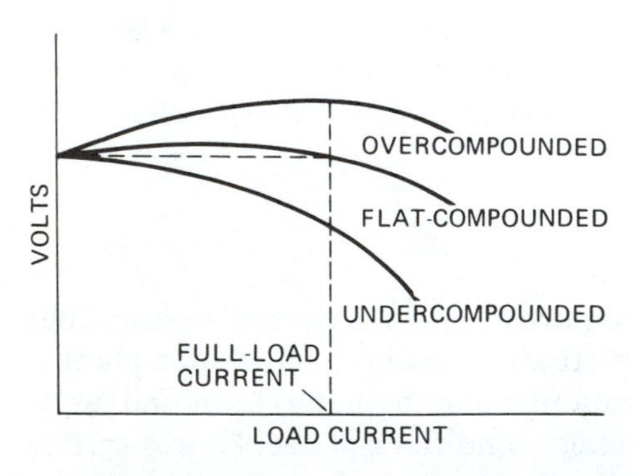

combining certain features of the shunt and series generators into one machine. This type of machine is called a *compound generator*, Figure 10-36.

If the series winding is connected so that its magnetic field will aid the shunt magnetic field, it is called a cumulative-compound generator. When the load current increases in a shunt machine, the main field flux is weakened as a result of armature reaction and IR drop. In a cumulative-compound generator (with the series field aiding the shunt field), the results are somewhat different. The load current flowing through the series field causes an increase in the flux. If the increase in flux is equal to the decrease caused by armature reaction and IR drop, the output voltage remains constant.

The load/voltage characteristics of a compound generator depend upon the number of turns on the series field. With many turns on the series field, the machine begins to assume the characteristics of a series generator. Few turns result in predominantly shunt characteristics.

The number of turns on the series field determines the amount of compounding. A machine with many turns on the series field is called an overcompounded generator. A generator with just enough series turns to maintain a steady voltage from no load to full load is a flat-compounded generator. One with fewer turns is said to be an undercompounded generator. Most compound generators are designed for overcompounding. The degree of compounding is determined by a *diverter* (a resistance of specific value) connected in parallel with the series field, Figure 10-37. The load/voltage curves for over-, under-, and flat-compounded generators are depicted in Figure 10-38. In practice, flat compounding does not actually produce a flat curve. The nearest approach is to adjust the machine so that the terminal voltage rises slightly and then drops again, reaching the same value at full load as at no load.

Flat-compounded generators are used when the load is located near the generator. If the load is located some distance away, an overcompounded machine is used. The overcompounding will compensate for the voltage drop in the line wires. The amount of overcompounding used depends upon the type of service. For generators in which the load is a great distance away, 10 percent overcompounding is common. Generators supplying power to street railway, monorail, and subway systems fall into this category.

The differential-compound generator is another type of compound generator. This machine is connected in such a way that the series field mmf opposes the shunt field mmf. In a machine of this type, the voltage drops off sharply as load is added. This machine is used most commonly in arc welding. When a heavy load occurs, the series field flux tends to neutralize much of the shunt field flux. The result is a decrease of the overall flux. The induced emf drops, reducing the armature current and preventing the armature from overheating.

SEPARATELY EXCITED GENERATOR

In a separately excited generator, the magnetization current for the field coils is supplied from a source outside of the generator. This supply may be another dc generator, batteries, or a rectifier. Figure 10-39 illustrates the connections for this type of generator.

Figure 10-39
Schematic diagram of a separately excited generator.

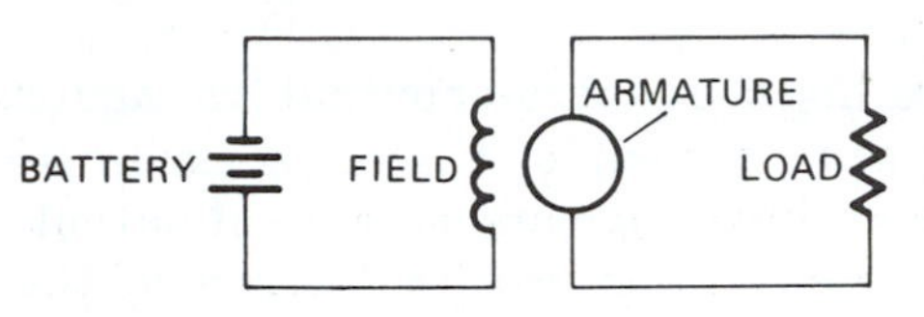

If a separately excited generator is operated at a constant speed and with a constant field voltage, the terminal voltage at no load will be equal to the generated emf. When a load is applied, terminal voltage will be less than the generated emf. This decrease in voltage is caused by armature reaction and IR drop. Figure 10-40 depicts the load/voltage characteristics of the separately excited generator.

Figure 10-40
Load/voltage characteristics of a separately excited generator

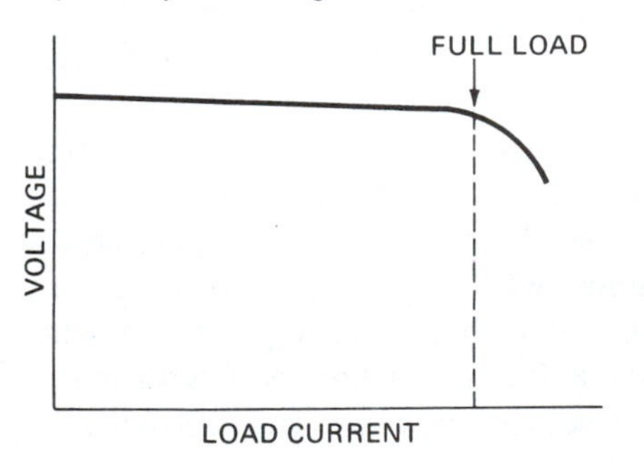

Because the field current is independent of the armature emf, the magnetic flux of the field is less affected by load changes than it is in the self-excited generator. One method for controlling the terminal voltage of the separately excited generator is to control the field current. This may be accomplished by installing a rheostat in the field circuit. If a constant voltage is required for all loads, the voltage source must be great enough to maintain the rated output voltage at full load.

Separately excited generators are used only for special installations because of the high cost of construction and their physical size.

VOLTAGE CONTROL VERSUS VOLTAGE REGULATION

The load/voltage curves for a generator illustrate the ability of the machine to regulate its output voltage with changes in load. A generator that maintains a nearly constant output voltage from no load to full load has excellent regulation. Voltage regulation, therefore, is determined by the design of the machine.

Voltage control takes place outside of the generator. It is generally accomplished by controlling the current through the shunt field. One method of achieving this is to install a rheostat in the field circuit. This method is used on separately excited generators, shunt generators, and compound generators. On series generators, a variable diverter is connected in parallel with the series field.

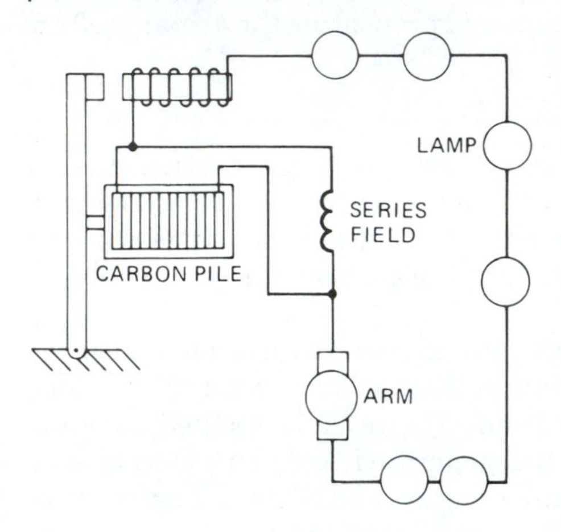

Figure 10-41
Automatic control of a series generator provides constant load current.

Hand control of generators is not satisfactory for all conditions of use. In most installations, an automatic control device is preferable.

Series generators that supply loads requiring a constant current employ an automatic variable diverter. In older installations, a *carbon pile* was used. This device consists of a special type of rheostat made of carbon and operated by a solenoid. Figure 10-41 illustrates this connection. When the current in the external circuit increases, the pull on the solenoid also increases and the carbon pile is compressed. This effectively reduces the resistance of the rheostat so that more current is diverted from the series field. The field flux is reduced, thus reducing the generated emf. The load current decreases until it reaches the value for which the solenoid was set.

Modern methods of controlling the generator output (voltage or current) are based on the same principle. A sensing device is installed to sense any changes in voltage and/or current output. This device actuates relays or electronic circuits that control the amount of current in the main field of the generator.

PARALLEL OPERATION OF GENERATORS

More than one generator is used when a large amount of dc power is required, such as in metal-refining plants, paper mills, electric railway systems, and marine work. In such applications, the generators are connected in parallel. Parallel operation provides a very efficient system. During reduced loading periods, some generators can be removed from the line. All generators perform at maximum efficiency when operating at full load. Preventive maintenance can be performed on the generators not in use. If a generator breaks down, it can be removed from the line and another one can be placed in service with minimum downtime.

Shunt Generators in Parallel

Certain procedures must be performed in order to connect generators in parallel. The following procedure is recommended for shunt generators, Figure 10-42. (Generator circuit breakers are omitted in order to simplify the drawing.) Generator A is supplying power to the load. It is operating at nearly full capacity. Additional equipment must be put into operation, requiring more dc power. This requires placing another generator on the line.

Power is supplied to the prime mover of generator B. When generator B has reached its rated speed, the output voltage is adjusted to the same value as generator A. A polarity test is performed to ensure that the positive terminal is connected to the positive bus and the negative terminal is conntected to the negative bus.

Figure 10-42
Shunt generators connected in parallel

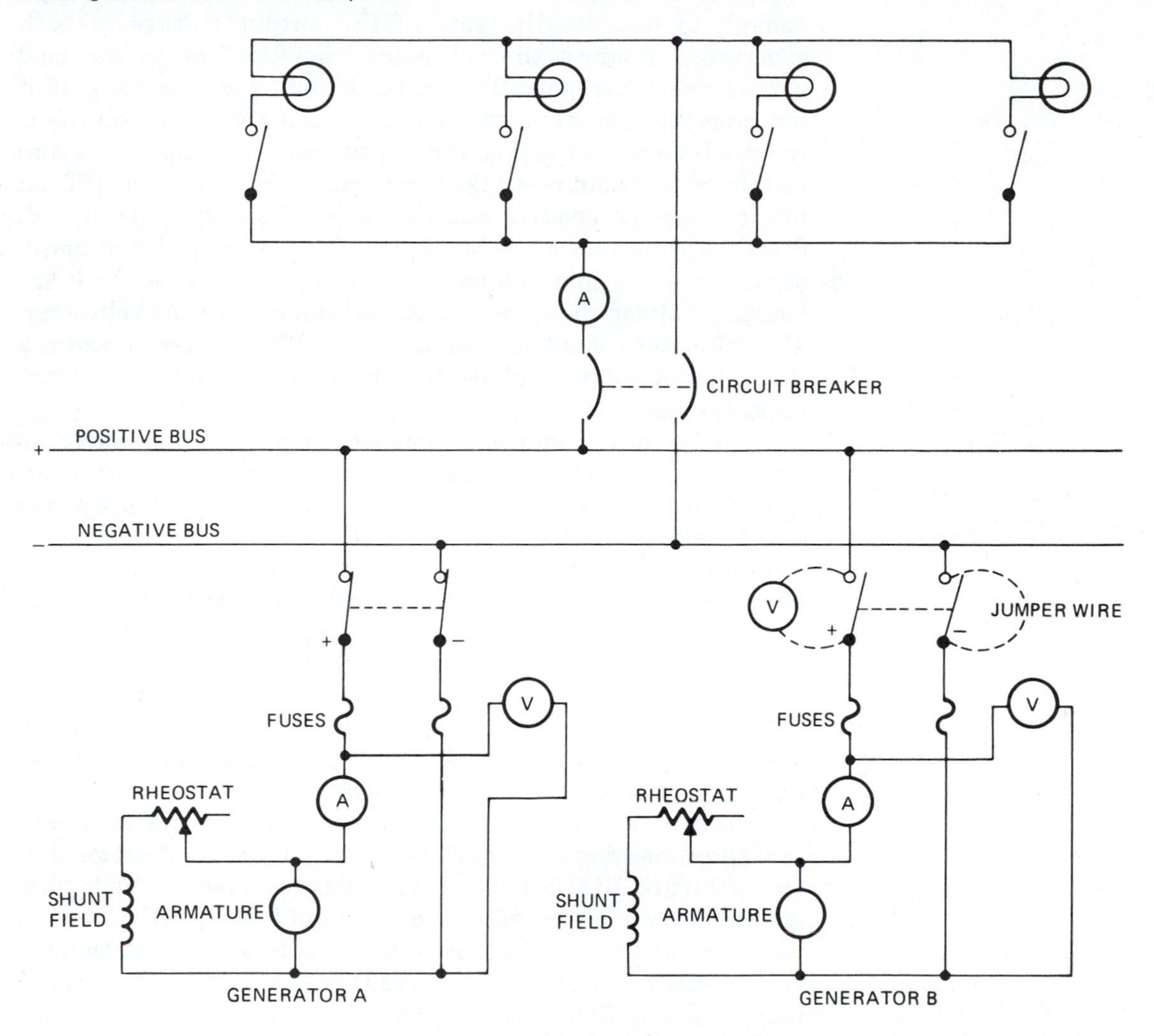

CAUTION: Reversed polarity can cause serious hazards. Therefore, a polarity test is very important.

A polarity test is accomplished as follows. With the switch for generator B open, connect a jumper wire from the line side to the load side of one pole.

CAUTION: Take great care in making this connection, because both the load and line sides of the switch are energized.

The line side is energized by generator B and the load side is energized by generator A. Select a voltmeter with a range high enough to indicate the sum of the output voltage of both generators. Connect the voltmeter from the line to the load side of the other pole. The meter should indicate zero. If it indicates the sum of the output voltages of A and B, the polarity is reversed. To correct this condition, remove the jumper wire and voltmeter, and shut down the prime mover for generator B. Then interchange the positive and negative lines from generator B. Start the prime mover and bring generator B up to its rated speed, adjusting the output voltage to the same value as A. With the values of voltage equal, reconnect the jumper wire and voltmeter. The voltmeter should now indicate zero. If the meter indicates a low value of voltage, adjust the rheostat of B until the meter indicates zero.

With the output voltage and polarity correct, close the switch to generator B. Disconnect the jumper wire and the voltmeter. Generator B is now floating on the line; it is not supplying power to the load.

In order to have generator B supply power to the load, it is necessary to increase the voltage of B while decreasing the voltage of A. During this procedure, the bus voltage must be kept at a constant value. This is accomplished by decreasing the voltage of A the same amount as B is increased, and at the same time. By way of this procedure, the load can be gradually shifted from A to B in any desired amounts. The additional load can thus be added to the line without overloading either generator.

To remove a generator from the line, first check the load to be sure the remaining generator(s) can carry it. Assume that generator A (Figure 10-42) is to be removed. Increase the voltage of generator B while decreasing the voltage of A. Be sure that the bus voltage remains constant. Check the ammeters to observe the load shift, and continue adjusting the voltages until the ammeter for A indicates zero. Generator A is now floating on the line, and the switch can be opened.

Compound Generators in Parallel

When connecting compound generators in parallel, one must give special consideration to the series field. Reaction of the series field to momentary changes in voltage and/or load currents can cause shifting of the load from one machine to the other. The load shifting becomes cumulative, and, eventually, one machine carries all the load, usually resulting in an overload.

In order to prevent the possibility of unwanted load shifting, connect the series field of each machine to a common equalizer bus. In effect, the series fields are connected in parallel with each other,

Figure 10-43. The result is that the series field currents divide proportionally between the generators.

> **CAUTION:** All the safety precautions for connecting shunt generators in parallel should be followed when connecting compound generators in parallel.

Figure 10-43
Compound generators connected in parallel

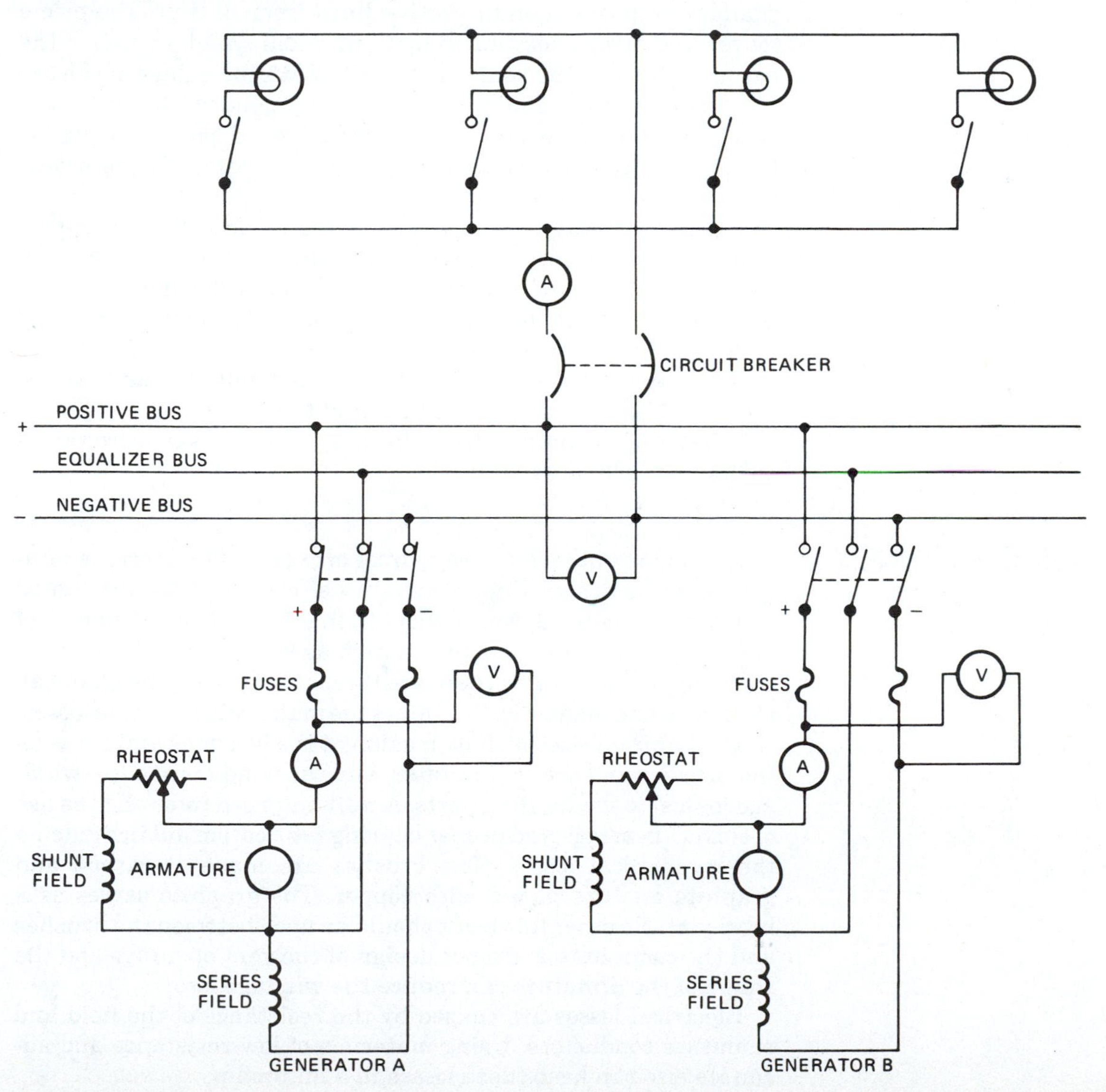

The general procedure is as follows. Generator A (Figure 10-43) is supplying power to the load. Bring generator B up to speed and adjust the output voltage to match the bus voltage. Connect a jumper wire across the positive side of the three-pole switch for generator B. Select a voltmeter with a range high enough to indicate the sum of the output voltages of both generators.

Connect the voltmeter across the negative side of the three-pole switch. The meter should indicate zero. If it indicates the sum of the output voltage of A and B, the polarity of generator B is reversed. To correct this condition, remove the jumper wire and voltmeter, and shut down the prime mover for B. Then interchange the positive and negative lines from B. Start the prime mover and bring generator B up to its rated speed, adjusting the output voltage to the same value as A. With the values of voltage equal, reconnect the jumper wire and voltmeter. The voltmeter should now indicate zero. If the voltmeter indicates a low value (less than the output voltage), adjust the rheostat of generator B until the voltmeter indicates zero.

With the voltage and polarity correct, close the three-pole switch for generator B. Remove the voltmeter and jumper wire. Generator B is now floating on the line. Following the same procedure outlined for the shunt generators divides the load between the generators as needed.

If the generators have interpoles, the equalizer bus must be connected between the series field and the interpoles.

To remove a generator from the line, follow the same procedure outlined for shunt generators.

GENERATOR EFFICIENCY

It is important to conserve sources of energy. Therefore, generators must be designed for maximum efficiency. A well-designed generator can be a very efficient machine. The efficiency of industrial generators is often as high as 90 percent.

Within the generator there are three major losses: mechanical, electrical, and magnetic. Friction is the major cause of these losses.

Mechanical losses include friction at the bearings and between the brushes and the commutator, and the wind resistance (windage losses) to the rotating parts. A well-balanced rotor and the use of correct bearings and proper bearing lubrication aid in reducing the mechanical losses. The brushes are made of carbon and graphite and are plated with copper. The graphite serves as a lubricant. No other lubricant should be used between the brushes and the commutator. Proper design of the vent openings and the shape of the armature can reduce the windage loss.

Electrical losses are caused by the resistance of the field and armature conductors. Using materials of low resistance and adequate size can keep these losses to a minimum.

Magnetic losses are a result of reluctance in the magnetic circuit. The two major losses are caused by eddy currents and hysteresis. The reluctance can be kept to a minimum by the use of good magnetic materials. It is also important to keep the air gap between the armature and the stator field to a minimum.

Eddy currents (currents set up in the armature and field cores) are caused by voltages induced into the cores. The best way to minimize eddy currents is to laminate the cores and insulate the laminations form one another.

Hysteresis losses are caused by fluctuating magnetic fields, which are produced by changes of current flowing in the coils. Hysteresis is a result of friction within the armature and field cores. Changes in magnetic strength and/or polarity cause a movement of the molecules and magnetic stresses within the cores.

Hysteresis can be reduced by selecting the proper materials for the core. Generally, materials that have good permeability produce less hysteresis loss.

REVIEW

A. Multiple choice.
Select the best answer.

1. Electromagnetic induction takes place
   a. whenever an electric current flows.
   b. when there is relative motion between a conductor and a magnetic field.
   c. when a steady direct current flows through a conductor.
   d. when a conductor is moved parallel with the lines of magnetic force.

2. In the northeastern United States, many generating stations depend upon
   a. water power.
   b. steam power.
   c. solar power.
   d. wind power.

3. Increasing the number of turns of wire on the armature coil causes the induced emf to
   a. decrease.
   b. increase.
   c. become more constant.
   d. none of the above.

4. When it is desired to take alternating current from a generator,
   a. pickup rings are used.
   b. a commutator is used.
   c. slip rings are used.
   d. compensating poles are used.

REVIEW *(continued)*

A. Multiple choice.
Select the best answer (continued).

5. One method used to reduce the pulsation of the output voltage of a dc generator is to
   a. add more coils of wire to the armature.
   b. add more coils of wire to the field.
   c. increase the speed of the rotor.
   d. decrease the speed of the rotor.
6. The armature of a dc generator contains a lap winding. This means that the ends of each winding are connected to
   a. every other segment.
   b. segments on opposite sides of the commutator.
   c. every third segment.
   d. adjacent segments.
7. An armature constructed with a wave winding has
   a. four current paths.
   b. three current paths.
   c. two current paths.
   d. one current path.
8. The insulation separating the segments of the commutator is made of
   a. mica.
   b. silk.
   c. PVC.
   d. rubber.
9. The brushes used on a generator are usually made of
   a. graphite and carbon.
   b. graphite and copper.
   c. aluminum.
   d. aluminum and copper.
10. The neutral plane of a generator is located
    a. in the armature windings.
    b. in the field winding.
    c. at a point where the coils cut the maximum number of lines of force.
    d. at a point where the armature coils do not cut lines of force.
11. When a load is placed on a generator, the neutral plane shifts
    a. in the direction of armature rotation.
    b. in the direction opposite the armature rotation.
    c. first in one direction and then in the other direction.
    d. 90 degrees.

REVIEW *(continued)*

A. Multiple choice.
Select the best answer (continued).

12. As the armature coils move through the neutral plane, the current
    a. increases.
    b. decreases.
    c. remains the same.
    d. fluctuates.

13. In order to obtain good commutation, it is necessary that the brushes be set in the
    a. current neutral plane.
    b. mechanical neutral plane.
    c. magnetic neutral plane.
    d. electrical neutral plane.

14. The purpose of interpoles is to
    a. improve commutation.
    b. reduce flux leakage.
    c. strengthen the main field flux.
    d. develop a greater emf.

15. Compensating windings compensate for
    a. motor action in the generator.
    b. armature reaction.
    c. flux losses.
    d. field reversals.

16. The generated emf varies directly with the
    a. armature current.
    b. strength of the field flux.
    c. resistance of the armature windings.
    d. resistance of the field.

17. A self-excited shunt generator has the field winding connected
    a. across the armature.
    b. to a separate dc source.
    c. in parallel with the interpoles.
    d. in series with the armature.

18. A shunt generator has the field connected
    a. in parallel with the armature.
    b. in series with the armature.
    c. to a separate source.
    d. in parallel with the interpoles.

REVIEW *(continued)*

A. Multiple choice. Select the best answer (continued).

19. A compound generator has
    a. two types of field windings.
    b. interpoles connected in series with the field.
    c. twin armatures.
    d. only one field.

20. Voltage regulation of a generator is
    a. using a rheostat to regulate the output voltage.
    b. varying the speed to regulate the output voltage.
    c. the ability of a generator to regulate its output voltage with changes in load.
    d. using a transformer to regulate the output voltage.

B. Give complete answers.

1. Describe electromagnetic induction.
2. What do the letters *emf* stand for? What is the unit of measurement of emf?
3. Write the left-hand rule for a generator.
4. What is a generator?
5. Describe the types of energy conversion that take place in a generating station.
6. List four factors that determine the amount of emf induced into a generator armature.
7. Why does a single-loop generator not produce a steady voltage?
8. What type of voltage is produced in the armature of all rotating-type generators?
9. What determines the type of current taken from a generator?
10. How does one obtain ac from a generator?
11. What is the purpose of brushes on a generator?
12. Define *commutator*, and explain its purpose on a generator.
13. What provisions are made in the construction of a generator in order to obtain a smooth dc?
14. Identify two types of windings that are used on a dc generator armature.
15. What is the least number of current paths in a dc generator armature?

REVIEW *(continued)*

B. Give complete answers (continued).

16. List the main parts of a dc generator, and explain the purpose of each part.
17. Explain how an alternating voltage is developed in the armature of a generator.
18. A generator is being constructed to produce a high voltage with minimum current. Will the plans call for a lap winding or a wave winding?
19. Describe the construction of the armature core of a dc generator.
20. Define *eddy currents*, and explain what causes them.
21. Why must eddy currents be kept to a minimum.
22. How are the armature and the field cores of generators constructed in order to keep eddy currents to a minimum?
23. Of what materials are brushes made, and why are these materials used?
24. Describe the yoke of a generator, and explain the purpose.
25. Describe the field coils, and explain how they are installed on a generator.
26. Explain the meaning of the phrase "field excitation."
27. Describe two methods commonly used to excite the field of a dc generator.
28. What is a prime mover?
29. What causes armature reaction in a generator?
30. Describe the effects of armature reaction in a generator.
31. Define the term *neutral plane.*
32. What causes the neutral plane to shift?
33. As the load is increased on the generator, in which direction does the neutral plane shift?
34. Describe the phenomenon known as self-induction.
35. Explain the purpose of interpoles, and describe how they are connected.
36. What are compensating windings, and what is their purpose in a generator.

REVIEW *(continued)*

B. Give complete answers (continued).

37. Describe motor action in a generator, and explain what causes it.
38. What is meant by the "saturation point" of a generator?
39. Draw a schematic diagram of a separately excited generator.
40. Draw a schematic diagram of a shunt generator.
41. Describe the building-up process of a shunt generator.
42. List nine reasons why a self-excited generator may fail to build up a voltage.
43. Draw a schematic diagram of a series generator.
44. Draw a schematic diagram of a compound generator.
45. What does the load/voltage curve of each of the following generators indicate?
    a. Shunt generator
    b. Series generator
    c. Flat-compounded generator
46. What is the most common type of dc generator in use today? Why?
47. Explain the procedures to be followed when connecting shunt generators in parallel.
48. Describe the procedure for connecting compound generators in parallel.
49. Explain the difference between overcompounded, flatcompounded, and undercompounded generators.
50. What is the difference between a cumulative-compound generator and a differential-compound generator?

C. Extended Study

1. In constructing a compound generator, how is the amount of compounding regulated?
2. Where are overcompounded generators usually used?
3. Name one use for a differential-compound generator, and explain why it is used for this purpose.
4. Define the term *voltage control*.

REVIEW *(continued)*

C. Extended Study (continued).

5. Define the term *voltage regulation*.
6. Describe one method used to control the output voltage of a shunt generator.
7. Describe one method used to control the output voltage of a series generator.
8. How is the output voltage of a compound generator controlled?
9. Define *hysteresis*.
10. List the three major losses that occur in a generator, and explain how to keep the losses to a minimum.

11
DC Motors

Objectives

After studying this chapter, the student will be able to:

- Explain the principles upon which dc motors operate.
- Describe the construction of dc motors.
- Discuss the different types of dc motors and their operating characteristics.
- Describe basic motor maintenance procedures.

The electric motor utilizes electrical energy and magnetic energy to produce mechanical energy. The purpose of a motor is to produce a rotating force (torque).

The basic construction of a dc motor is very similar to that of a dc generator. Each has a frame, end bells, field poles, an armature, and a commutator. There is, however, a vast difference in their use and operation.

A generator is driven by some type of mechanical machine (steam or water turbine, gasoline engine, or electric motor). It requires mechanical energy for its operation. The goal is to convert mechanical energy into electrical energy.

A motor requires electrical energy for its operation. An electric current flowing through the motor windings causes the armature to rotate. The goal is to produce mechanical motion, thus converting electrical energy into mechanical energy.

BASIC MOTOR OPERATION

The operation of a motor depends upon the interaction between two magnetic fields. One magnetic field is stationary and the other is free to rotate.

All current-carrying conductors produce a magnetic field. This magnetic field is circular in shape, and the direction of the flux depends upon the direction of current, Figure 11-1.

In Chapter 9, we described the action of two magnetic forces within magnetic reach of each other. Figure 11-2 depicts the results of this condition. This phenomenon can be used to demonstrate the theory of motor action.

If a current-carrying conductor is looped through a magnetic field, it will cause the main flux to become distorted, Figure 11-3.

In Figure 11-2, the crowding of the magnetic lines causes a pressure, which tends to force the conductors apart. A similar condition exits in Figure 11-3. The lines of force are crowded below the conductor on the left and above the conductor on the right. This produces an upward force on the left-hand conductor, and a downward force on the right-hand conductor. If the loop is free to rotate, it revolves in a clockwise direction until it is midway between the two poles. In this position the forces are equal and opposite to each other, Figure 11-4. If the loop is forced beyond this point, by either centrifugal force or other means, the magnetic action will be as shown in Figure 11-5. An oscillating force is

Figure 11-1
Magnetic field around a conductor carrying current

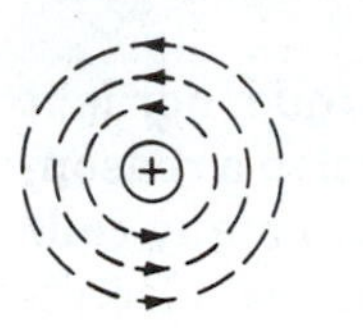
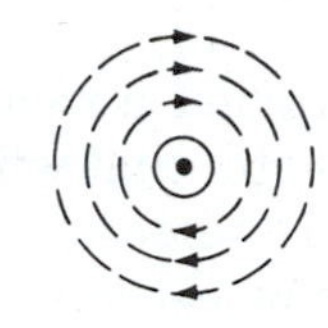

Figure 11-2
Two current-carrying conductors within magnetic reach of each other

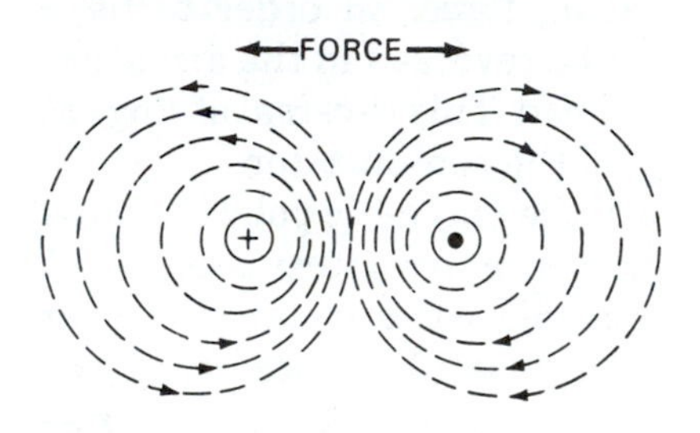

Figure 11-3
Current-carrying conductor looped through a magnetic field

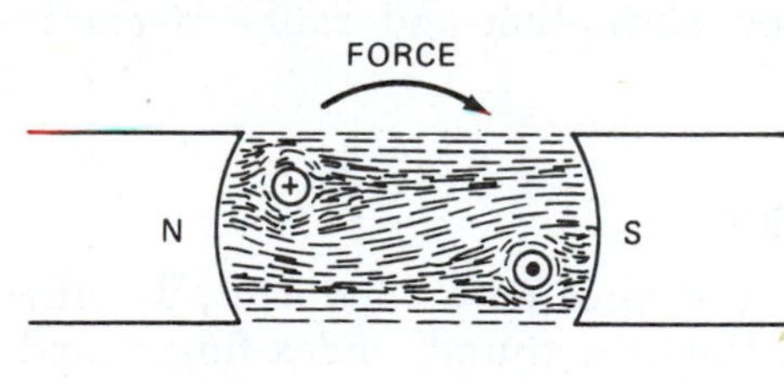

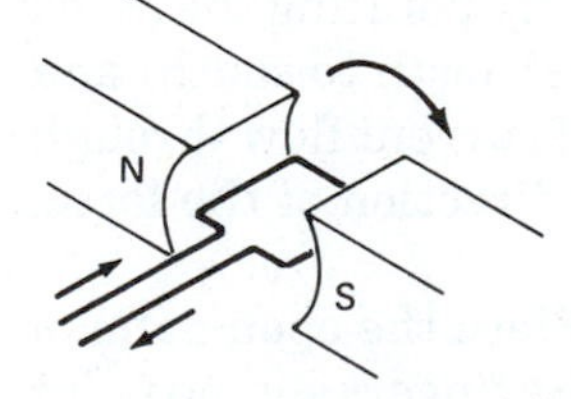

Figure 11-4
Current-carrying conductor in the neutral plane

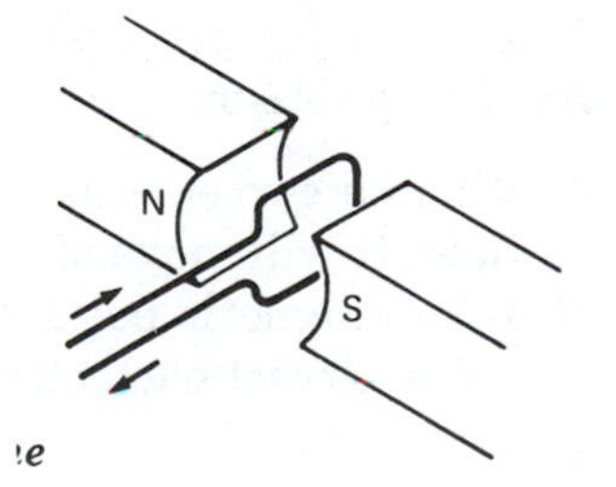

Figure 11-5
Magnetic reaction between two magnetic fields produces an oscillating force.

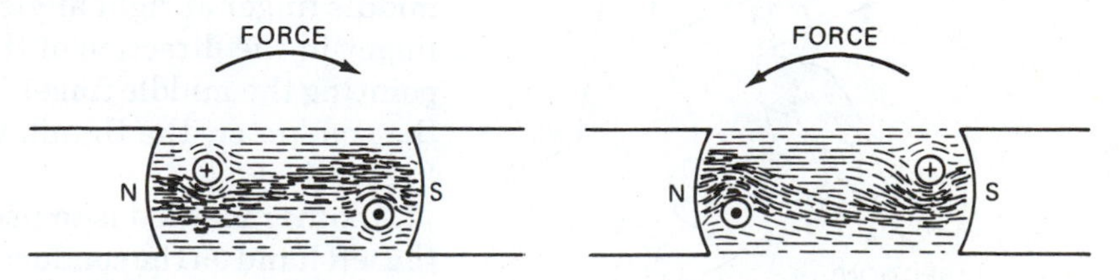

Figure 11-6
End view of motor armature. The wire is wound in the slots to form a coil.

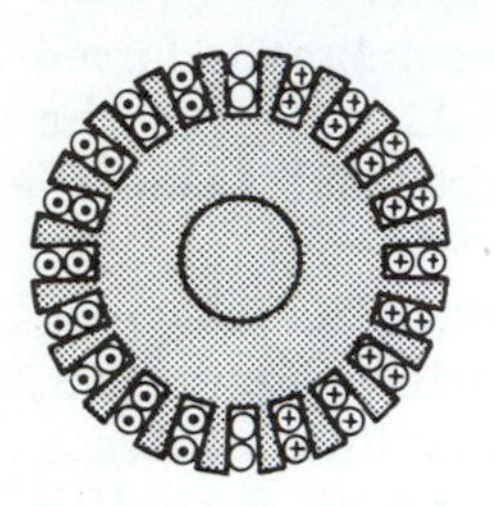

Figure 11-7
Magnetic reaction in a dc motor

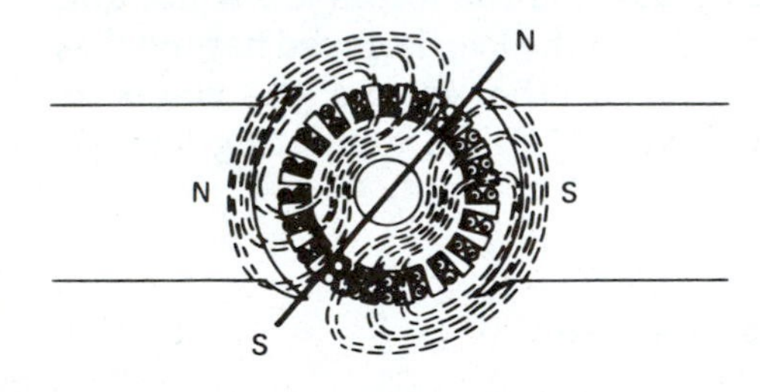

Figure 11-8
Right-hand rule for a motor

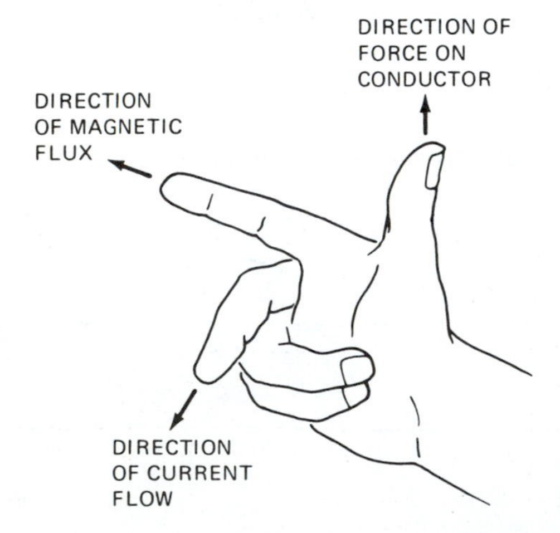

produced rather than a rotating force. In order to develop a rotating force, it is necessary to reverse the direction of current through the loop every time the loop passes the midpoint between the poles.

A single-loop motor does not produce a steady torque, nor is it practical. In order to construct a motor that will develop a reasonably steady and usable torque, it is necessary to have many coils on the armature. The armature now becomes a practical electromagnet with a north pole and a south pole A. cutaway end view of the armature is illustrated in Figure 11-6.

When a voltage is applied to the armature coils, a current will flow, causing the magnetic reaction shown in Figure 11-7. Notice that the north pole of the armature is attracted to the south pole of the main field, and the south pole of the armature is attracted to the north pole of the main field. When the opposite poles are aligned, there is no longer a rotating force. In order to have continuous torque, the current must be reversed in the armature at the instant the unlike poles are aligned. This reversal of current is accomplished through the action of the commutator.

When the current flow is reversed, the magnetic polarity of the armature reverses. The south pole of the armature is near the south pole of the main field; the north pole of the armature is near the north pole of the main field. Because like poles repel, the armature will continue to rotate. This action continues as long as the motor is in operation.

The operating principle of a motor depends upon the following laws of magnetism:

1. When a current-carrying conductor is placed in a magnetic field, it will move at right angles to the field.
2. Like magnetic poles repel each other and unlike magnetic poles attract each other.

Right-Hand Rule for a Motor

To determine the direction of force on a conductor, use the right-hand rule for a motor. Hold the thumb, index finger, and middle finger at right angles to one another. By pointing the index finger in the direction of the main field flux (north to south) and pointing the middle finger in the direction of current flow through the conductor, the thumb will point in the direction of the force, Figure 11-8.

Another method is to use the left hand. Place the open palm of the left hand on the conductor in question. The fingers should point in the direction of the main field flux (north to south), and the thumb should point in the direction of current flow. The force will be in the direction to cause the conductor to move away from the palm of the hand.

The direction of rotation of a dc motor may be reversed by interchanging either the armature or field connections. Reversing both connections will not change the direction of rotation. Because of the many different field connections, it is recommended that the armature connections be interchanged.

Force Exerted on a Conductor

The amount of force acting on a current-carrying conductor in a magnetic field is directly proportional to the following factors:

- The strength of the main field flux
- The length of the conductor within the field
- The amount of current flowing in the conductor

The following equation can be used to calculate this force:

$$F = \frac{8.85BI\ell}{10^8} \qquad \text{(Eq. 11.1)}$$

where F = force, in pounds (lb)
ℓ = length of the conductor within the field measured, in inches (in.)
B = flux density, in lines per square inch
I = amount of current flowing in the conductor, in amperes (A)

Figure 11-9A
Two-pole motor

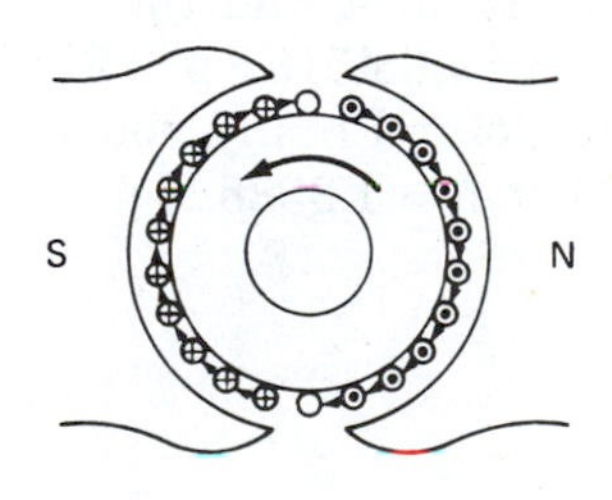

Example 1

Twelve in. of conductor are in a magnetic field. The current in the conductor is 50 A. If the main field flux has a density of 70,000 lines per square inch, how much force acts on the conductor?

$$F = \frac{8.85BI\ell}{10^8}$$

$$F = \frac{8.85 \times 7 \times 10^4 \times 50 \times 12}{10^8}$$

$$F = 3.717 \text{ lb}$$

Figure 11-9B
Four-pole motor

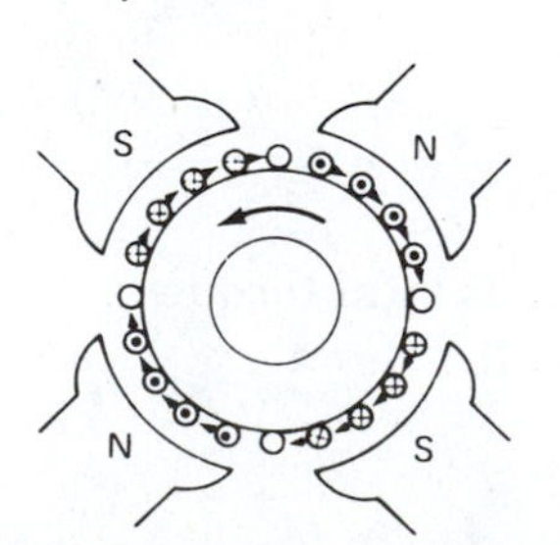

Torque and Power

Because the armature of a dc motor is wound much the same as a dc generator, the current from the supply must divide and flow through two or more paths. The currents through the conductors under a specific pole are in the same direction, Figures 11-9A and 11-9B.

The force acting on the conductors causes the armature to rotate in the direction indicated by the arrows. As the conductors move from the influence of one pole into the influence of the next, the current flow in that conductor is reversed. This reversal of current flow maintains a constant torque in one direction.

The force acting on each conductor adds to produce a total force. The effect of these forces in the armature depends upon their magnitude (total pounds exerted). The effect also depends upon the radial distance (the distance from the center of the rotor to the center of the conductors) through which they act. This effect is usually expressed as the product of the force and the radial distance. The unit of measurement is pound-feet (lb·ft). (The SI unit is newton-meters, or N·m.) The equation for torque is

$$T = FR \qquad \text{(Eq. 11.2)}$$

where T = torque, in pound-feet (lb·ft)
F = force, in pounds (lb)
R = radial distance, in feet (ft)

Example 2

The motor in Figure 11-9A has 20 conductors within the magnetic field. The length of each conductor within the field is 10 in. (25.4 cm) and the radial distance of the armature is 3 in. (7.62 cm). The current through each conductor is 20 A and the flux density is 60,000 lines per square inch. (1 in.^{2} = 6.4516 cm^2.) Find the torque, in pound-feet, developed by the motor. Finally, convert the pound-feet value to newton-meters. (1 lb·ft = 1.3588 N·m.)

(1) $F = \frac{8.85BI\ell}{10^8}$

$F = \frac{8.85 \times 6 \times 10^4 \times 20 \times 10}{10^8}$

$F = 1.062$ lb

(2) $T = FR$

$T = 1.062 \times \frac{3}{12}$

$T = 0.2655$ lb·ft on one conductor

(3) 0.2655 lb·ft × 20 conductors = 5.31 lb·ft total torque

(4) 1 lb·ft = 1.3588 N·m

(5) Therefore, 5.31 lb·ft = 7.215 N·m

Example 3

A dc motor has 100 conductors within the magnetic field. The length of each conductor within the field is 20 in. (50.8 cm) and the radial distance of the armature is 6 in. (15.24 cm). The current through each conductor is 30 A and the flux density is 60,000 lines per square inch. Find the torque developed by the motor. If the rated speed of the motor is 1000r/min, what is the horsepower output?

(1) $F = \dfrac{8.85BI\ell}{10^8}$

$F = \dfrac{8.85 \times 6 \times 10^4 \times 30 \times 20}{10^8}$

$F = 3.186 \text{ lb}$

(2) $T = FR$

$T = 3.186 \times \dfrac{6}{12}$

$T = 1.593 \text{ lb·ft}$

(3) 1.593 lb·ft × 100 conductors = 159.3 lb·ft = 216.5 N·m

The total torque is 159.3 lb·ft (216.5 N·m). From Equation 4.9,

$\text{hp} = \dfrac{TN}{5252}$

$\text{hp} = \dfrac{159.3 \times 1000}{5252}$

$\text{hp} = 30.33$

An instrument designed to measure torque is called a *prony brake*. There are several types available to industry. The most common types are the mechanical, the eddy current, and the electrodynameter. Other prony brakes available are the hysteresis, fluidics, and hydraulic types.

GENERATOR ACTION IN A MOTOR

In the motor, as in the generator, conductors rotate through a magnetic field. If an emf is induced into the armature conductors of a generator because they are cutting flux, the same must be true of motors.

Consider the motor in Figure 11-9A. If the armature is rotating in the direction indicated by the arrow, the emf is in a direction opposite to the current flow. This can be verified by using the left-hand rule for a generator (see Chapter 10).

From this information it can be seen that there are two voltages at work in a motor armature:

1. The applied voltage from the source that causes the current to flow
2. The induced emf that opposes the current flow

The induced emf is called *counter electromotive force* (cemf) because it acts against the applied voltage. The cemf serves a very useful purpose. Because it opposes the current flow, the armature circuit resistance may be kept to a minimum. The lower the resistance of the armature, the fewer the I^2R losses and the more efficient the motor.

A 5-horsepower, 230-volt motor can be used to illustrate this phenomenon. The resistance of the armature windings of this size of motor is about 0.1 ohm. According to Ohm's Law, the armature current would be $230 \div 0.1 = 2300$ amperes. Such a high current would cause overheating, and the insulation would burn on the armature windings. It may even burn the windings.

Because the induced voltage (cemf) opposes the applied voltage, the formula for current is

$$I_a = \frac{E_a - E_g}{R_a} \qquad \text{(Eq. 11.3)}$$

where I_a = armature current
E_a = voltage applied to the armature
E_g = voltage induced into the armature (cemf)
R_a = resistance of the armature windings

The cemf of the 5-horsepower, 230-volt motor is about 228 volts. Because of the cemf, the current flowing in the motor armature is only 20 amperes:

$$I_a = \frac{E_a - E_g}{R_a}$$

$$I_a = \frac{230 - 228}{0.1}$$

$$I_a = 20 \text{ A}$$

COMMUTATION

The armature coil current must periodically reverse its direction of flow in order to produce a constant torque. This is accomplished by the commutator, which is a type of rotating switch. The commutator must change the connections to the armature coils at the instant the coils pass through the neutral plane. At the same instant, the induced voltage in the armature also reverses direction.

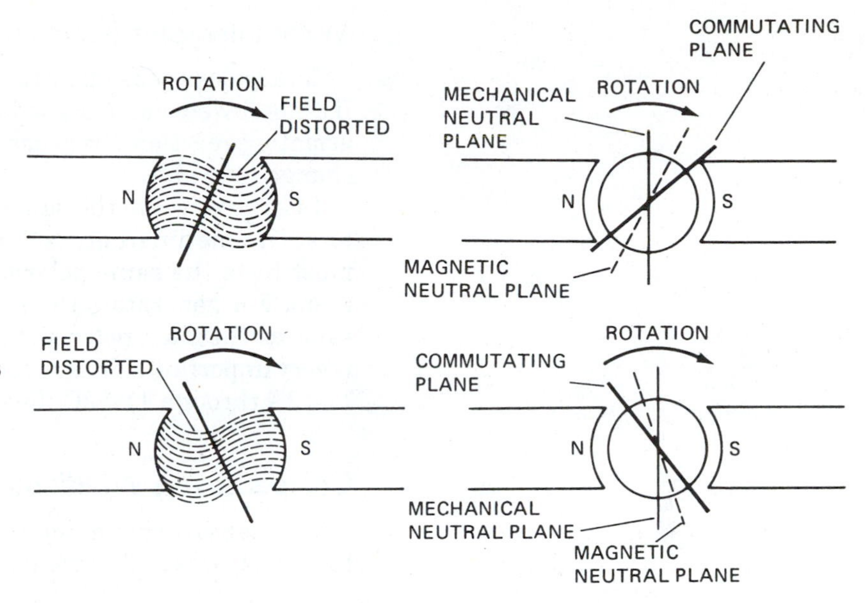

Figure 11-10A
Armature reaction in a generator. The result is a distorted field flux and a shifting of the neutral plane.

Figure 11-10B
Armature reaction in a motor. The result is a distorted field flux and a shifting of the neutral plane.

Figure 11-11A
No-load neutral plane

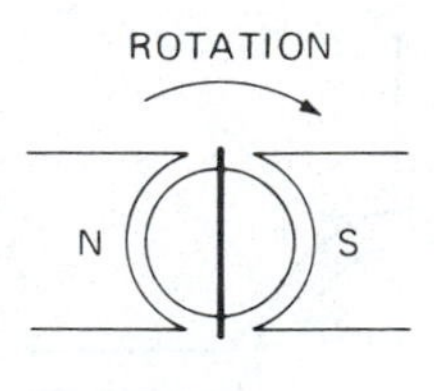

Figure 11-11B
Magnetic neutral plane

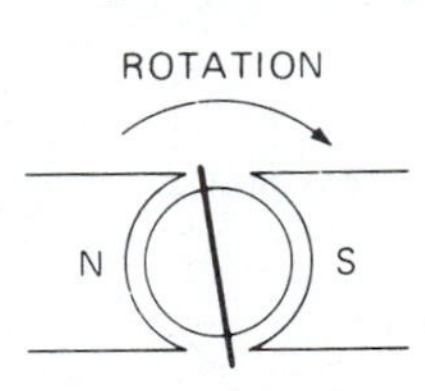

Figure 11-11C
Commutating plane

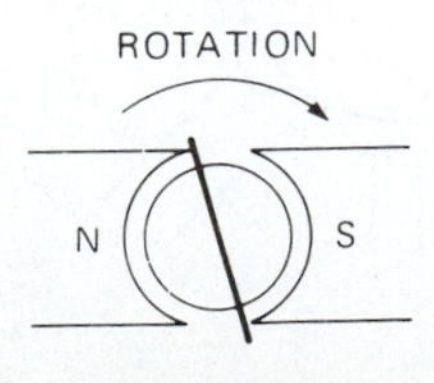

Armature Reaction in a Motor

In a generator, the armature current flows in the same direction as the induced emf. In a motor, the armature current flows in the direction opposite from the induced emf (cemf). Therefore, for the same direction of armature rotation, the magnetic polarity of the armature in a motor is the reverse of that in a generator. As a result, their neutral plane shift in opposite directions.

In a generator, the neutral plane shifts in the direction of the armature rotation. In a motor, the neutral plane shifts in the direction opposite to the direction of armature rotation. Figure 11-10A shows the field distortion and shifting of the neutral plane for a generator. Figure 11-10B shows the same conditions for a motor. It should be noted that the directions of rotation and field polarity are the same for both machines. The armature current, however, is in opposite directions. The direction of the current through the armature coils determines the pattern of the main field flux. (As a result, in a generator the brushes must be shifted in the direction of armature rotation. In a motor they must be shifted opposite to the direction of rotation.)

Self-induction takes place in a motor just as it does in a generator. It can be said that a motor also has three neutral planes:

1. The no-load neutral plane
2. The magnetic neutral plane
3. The commutating plane

Figures 11-11A, 11-11B, and 11-11C illustrate the neutral planes for a motor.

Motor Interpoles (Commutating Poles)

Interpoles are as important in motors as they are in generators. If a motor were not equipped with some means of compensating for armature reaction, the brushes would have to be shifted with every change in load.

Interpoles serve the same purpose in both motors and generators. The one difference is their polarity. For motors the interpoles must have the same polarity as the main poles directly behind them. For generators the polarity of the interpoles must be the same as the main poles just ahead of the them. Interpole polarity is very important. If it is reversed, severe arcing will occur. Figures 11-12A through 11-12D illustrate the correct polarity.

Compensating Windings in Motors

Compensating windings are seldom used in motors, because of their cost. When they are used, it is generally in conjunction with

Figure 11-12A
Interpoles in a two-pole generator

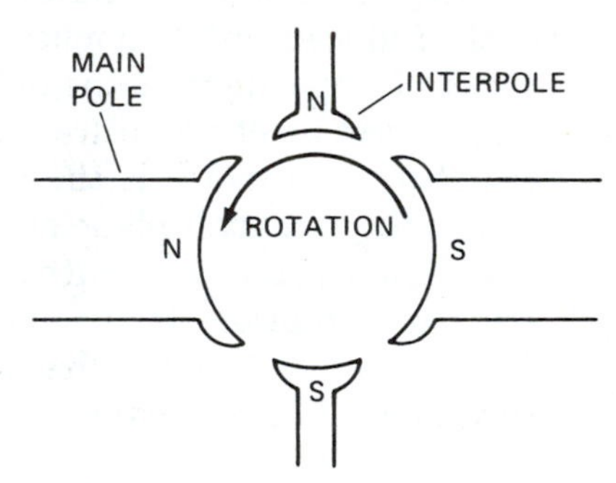

Figure 11-12B
Interpoles in a four-pole generator

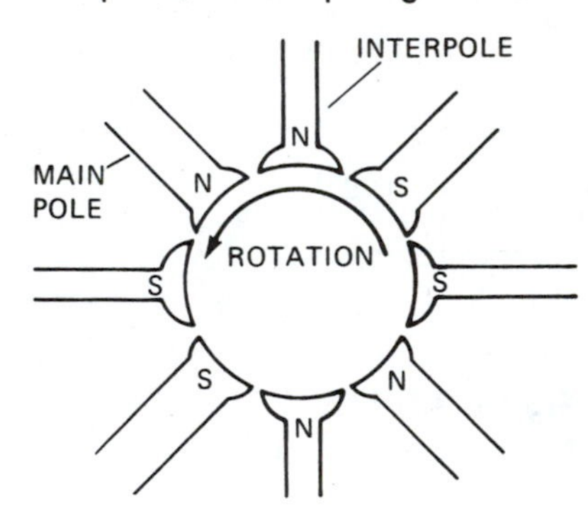

Figure 11-12C
Interpoles in a two-pole motor

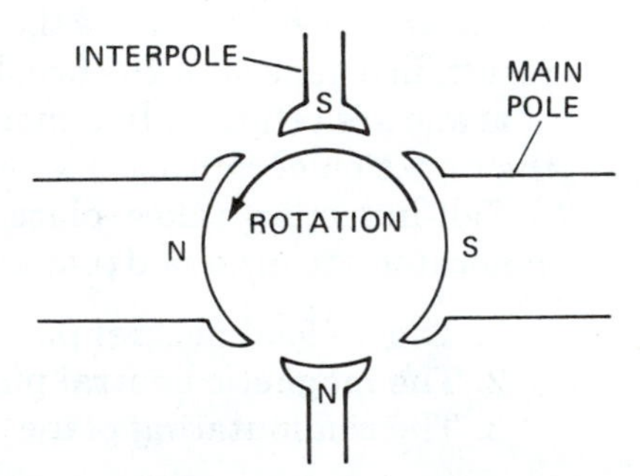

Figure 11-12D
Interpoles in a four-pole motor

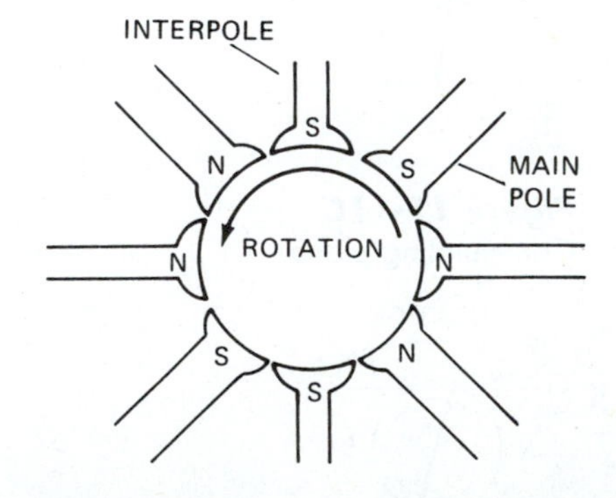

interpoles. Large machines that are subject to very heavy loads or extreme variations in load and speed may require both interpoles and compensating windings.

The compensating winding is embedded in the face of the main pole piece and then wound around the interpole. It has the same polarity as the interpole it surrounds.

MOTOR SPEED

The speed of a dc motor is proportional to the cemf. A weakening of the main field flux reduces the cemf. The lower cemf allows more current to flow in the armature circuit. This increase in the armature current provides a stronger magnetic field in the armature, which causes an increase in the armature speed. The speed increases until the cemf can limit the armature current to a new value. This value is determined by the main field strength. At this point the motor drives the load at a constant speed.

Decreasing the armature current also affects the motor speed. Assume that the motor is supplying a constant load. A decrease in the armature current results in a decrease in armature reaction. the decrease in armature reaction allows the main field flux to increase, and the armature slows down.

A motor is rated at its maximum horsepower and speed at a constant load. In other words, a 10-horsepower motor rated at 1200 revolutions per minute can deliver a maximum of 10 horsepower at a speed of 1200 revolutions per minute. From Equation 4.9, hp = TN/5252, it can be seen that at a specific horsepower, an increase in speed requires a similar decrease in torque in order to maintain the same horsepower. Therefore, to operate over a wide range of speeds a motor does not generally drive its maximum load. It should, however, operate as closely as possible to its maximum load. Motors are most efficient when operating at full load. Efficiency, as well as speed, should be of major concern when selecting motors, particularly for those requiring wide variations in speed.

Speed Regulation

Speed regulation refers to the manner in which a motor adjusts its speed to changes in load. It is the ratio of the loss in speed, between no load and full load, to the full-load speed. The formula for the percent of speed regulation is

$$\%\text{Reg} = \frac{N_1 - N_2}{N_2} \times 100 \qquad \text{(Eq. 11.4)}$$

where %Reg = percent of regulation
N_1 = no-load speed of the motor
N_2 = full-load speed of the motor

Example 4

A motor is rated at 1725 r/min. What is the percent of speed regulation if the no-load speed is 1775 r/min?

$$\%Reg = \frac{N_1 - N_2}{N_2} \times 100$$

$$\%Reg = \frac{1775 - 1725}{1725} \times 100$$

$$\%Reg = \frac{50}{1725} \times 100$$

$$\%Reg = 2.9\%$$

TYPES OF DC MOTORS

The three major types of dc motors are the series, shunt, and compound motors. In recent years the permanent-magnet motors have gained in popularity. They are extremely efficient, but at this time they are generally available only in smaller sizes, usually less than 1 1/2 horsepower.

Series Motors

Figure 11-13A
Series motor

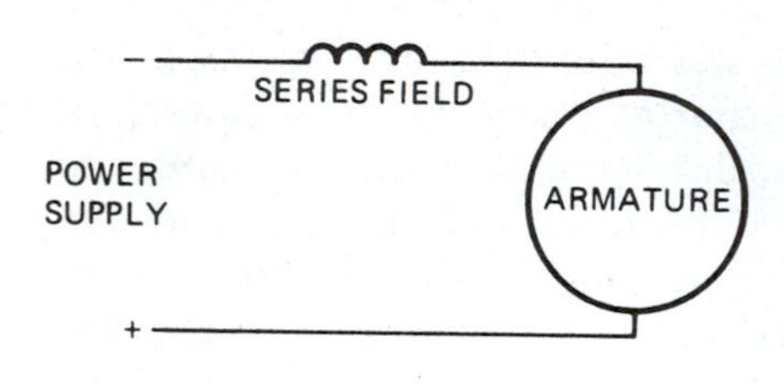

The *series motor,* like the series generator, has the field connected in series with the armature, Figure 11-13A. This method of connecting the field and armature has a very definite effect on the operating characteristics of the motor. Because all the current that flows through the armature must also flow through the field, the field strength varies with changes in the load.

A series motor has a very high starting torque. In some motors it may be as high as 5 times the full-load torque. The speed of the motor also varies with the load.

For this reason, series motors are always connected directly to the load. Belt drives are never used with series motors. As the load increases, the speed decreases. Figure 11-13B is a graph of the load/speed characteristics of a series motor. Figure 11-13C shows the effects of the load on the speed, torque, and efficiency.

Figure 11-13B
Load/speed characteristic curve for a series motor

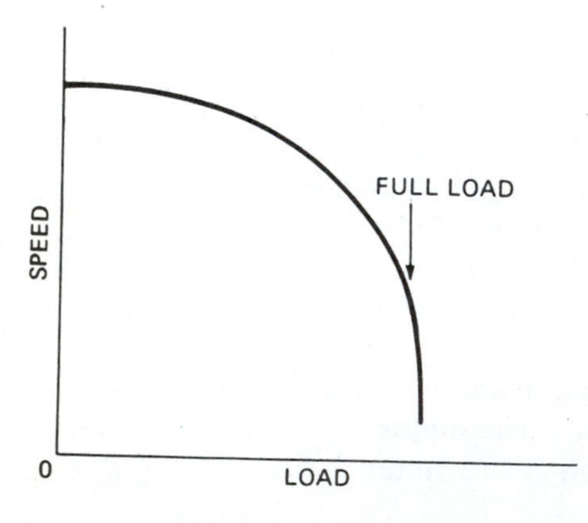

> **CAUTION:** This type of motor should never be operated without a load. The only field flux present at no load is that caused by residual magnetism; therefore, the field is very weak. Operating the motor without a load allows the rotor to reach such high speeds that the centrifugal force causes the windings to tear free.

Figure 11-13C demonstrates that when the armature current is 4 amperes, the motor develops a torque of 20 pound-feet (27.176

Figure 11-13C
Characteristics of a series motor

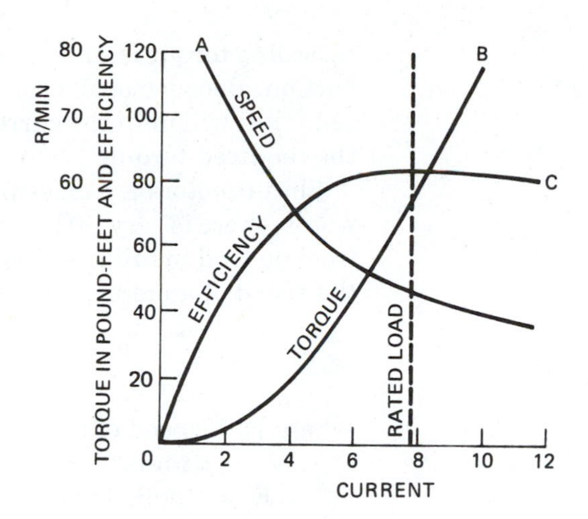

newton-meters). At 8 amperes, the torque is 80 pound feet (108.7 newton-meters). It can be seen that when the armature current doubles, the torque becomes 4 times as great. Thus, the torque increases rapidly near and above full load. Such characteristics make the use of series motors desirable when it is necessary to supply a large torque with a moderate increase in current.

Series motors are used chiefly for wide variations in load when extreme speed changes are not objectionable. They are used extensively for cranes, hoists, electric railway cars, and electric automobiles as well as for starting gasoline engines. In such applications, variations in speed with load are not objectionable. Occasionally the variations are desirable. For instance, when a crane is being used to lift a heavy load, it is generally desirable to proceed slowly. The enormous torque of a series motor makes it very suitable for work that demands frequent acceleration under heavy loads.

Series motors are also used extensively in portable tools because of their light weight in comparison to the horsepower they deliver. Because of their characteristics, the use of series motors is restricted to machines that require the presence of an operator.

CAUTION: It is most important to remember that a series motor should never be operated without a load.

Figure 11-14
Shunt motor

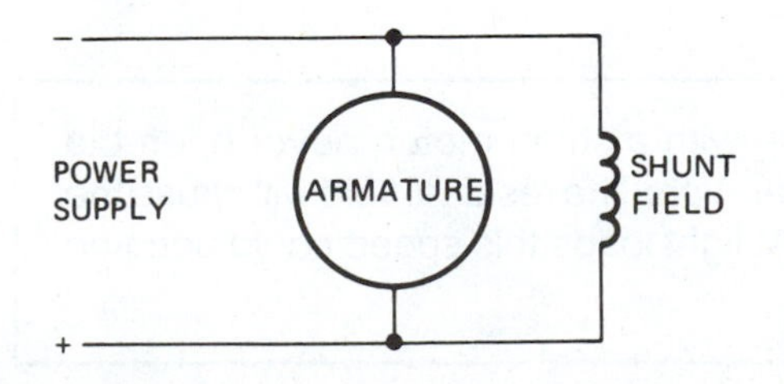

Shunt Motors

In a *shunt motor*, the field is connected directly across the line, Figure 11-14. As a result, the field current and the field flux are constant. When a shunt motor is operating without a load, the retarding torque is small since it is due only to the windage and the friction. Because of the constant field, the armature will develop cemf that will limit the current to the value needed to develop only the required torque.

Shunt motors are classified as constant speed motors. In other words, there is very little variation in the speed of the shunt motor from no load to full load. Equation 11.5 may be used to determine the speed of a motor at various loads.

$$N = \frac{E_a - IR}{K_1\Phi} \qquad \text{(Eq. 11.5)}$$

where N = speed of the armature, in revolutions per minute (r/min)
E_a = applied voltage
I = armature current at a specific load
R = armature resistance
Φ = lines of force per square centimeter
K_1 = a constant value for the specific motor

In a shunt motor, E_a, R, K_1, and Φ are practically constant values and I is the only variable. When no load is applied to the motor, the value of I is small because the speed and cemf are both at a maximum. In the equation for N, IR (the voltage drop in the armature coils) is negligible when compared to E_a. At full load, IR is generally 5 percent of E_a.The actual value depends upon the size and design of the motor. Consequently, at full load the speed is about 95 percent of the no-load value. This decrease in speed is reduced slightly by armature reaction, which causes a decrease in flux and a corresponding increase in speed. In some cases, armature reaction is sufficient to cause the speed to remain constant from no load to full load.

The starting torque of a shunt motor is about 2.75 times the full-load torque. The shunt motor does not have as high a starting torque as the series motor, but it has much better speed regulation.

The graph in Figure 11-15 illustrates the speed, torque, and efficiency of a shunt motor under varying loads.

The shunt motor is best suited for constant speed drives. It meets the requirements of many industrial applications. Some specific applications are machine tools, printing presses, blowers, and motor-generator sets.

CAUTION: When working with a shunt motor, never open the field circuit when it is in operation. The residual field will cause the rotor speed to increase. At light loads this speed could become dangerously high.

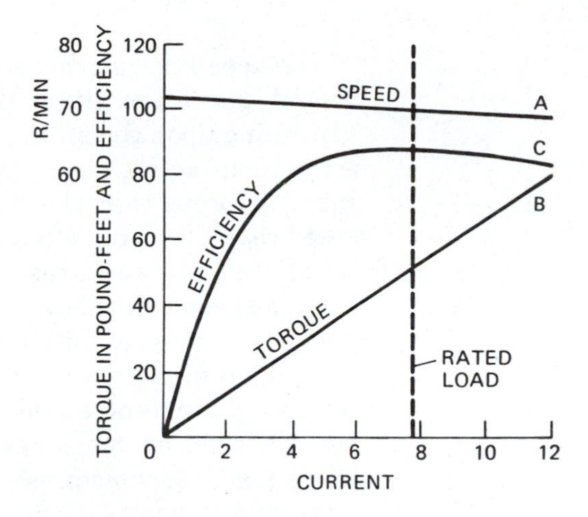

Figure 11-15
Speed, torque, and efficiency curves for a shunt motor

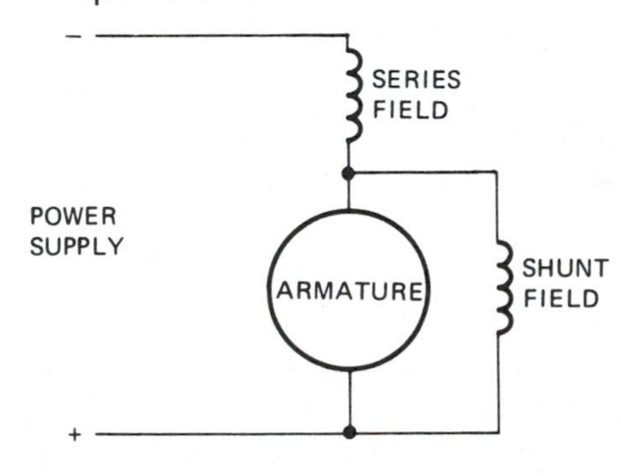

Figure 11-16
Compound motor

Compound Motors

The *compound motor* is a combination of the series motor and the shunt motor. It has two fields. One field is connected in parallel with the armature; the other field is connected in series with the armature, Figure 11-16.

Most compound motors are connected for cumulative compounding. The cumulative-compound motor has the combined characteristics of the shunt and the series motors, Figure 11-17. The cumulative-compound motor has a definite no-load speed and therefore may be operated without a load. As the load is increased, the speed decreases more rapidly than it does in shunt motors. This is caused by the increase in the series field flux as the armature current increases.

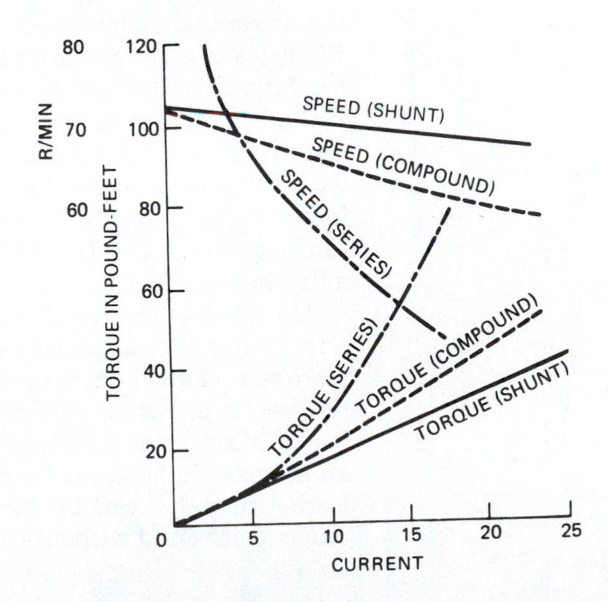

Figure 11-17
Load characteristics of series, shunt, and compound motors

The speed/torque characteristics of the compound motor may resemble those of either the series motor or the shunt motor, depending upon the strength of both fields. If the motor has only a few turns on the series field, it will have a better starting and running torque than the shunt motor. It will still retain the good speed regulation and efficient speed control of the shunt motor. If the motor has fewer turns on the shunt field (sufficient to limit the no-load speed to a safe value) and many turns on the series field, it will have most of the characteristics of the series motor.

Compound motors are used to drive machines that require a relatively constant speed under varying loads. They are frequently used on machines that require sudden application of heavy loads, such as presses, shears, compressors, reciprocating tools, and elevators. Compound motors are also used when it is desired to protect the motor by causing it to decrease in speed under heavy loads.

> **CAUTION:** Never open the shunt field of a compound motor when the motor is operating at light load.

Flywheel Effect

The power required by some machine tools is very irregular. For example, in a punch press or stamping machine, almost no power is required until the punch or die comes in contact with the material. If the moving parts in such a machine are not very heavy, the current taken by the driving motor will vary widely. Figure 11-18 shows this current curve. The motor selected to drive such a machine must be capable of carrying the greatest value of current without overheating or excessive arcing at the brushes. If a considerably overcompounded motor is used (about 30 percent) with a heavy flywheel, the armature current will vary less, as indicated by the dotted curve in Figure 11-18. From this curve it can be observed that the driving motor may be much smaller when a flywheel is used.

Figure 11-18
Current curve of a dc motor driving a punch press

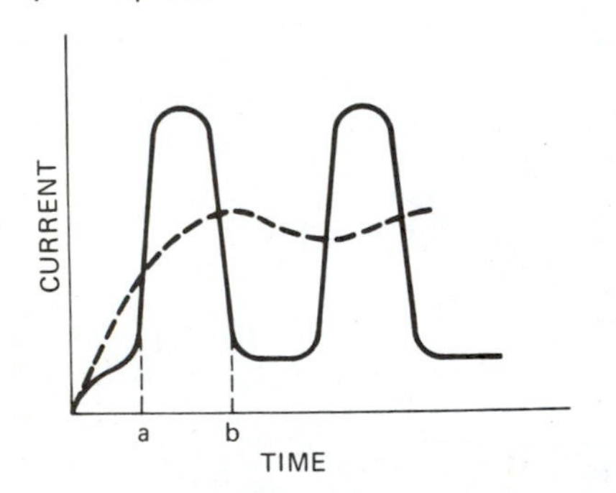

The effect of the flywheel can be explained as follows. At point *a* in Figure 11-18, the load on the motor and flywheel suddenly increases, causing them to slow down. The centrifugal force of the flywheel supplies the energy to help overcome the opposition of the load. This reduces the demand on the motor. In other words, with an increase in load, the flywheel assists the motor in driving the load. Although the motor current increases, it does not increase as much as it would without the flywheel.

Figure 11-19
Permanent-magnet servomotor

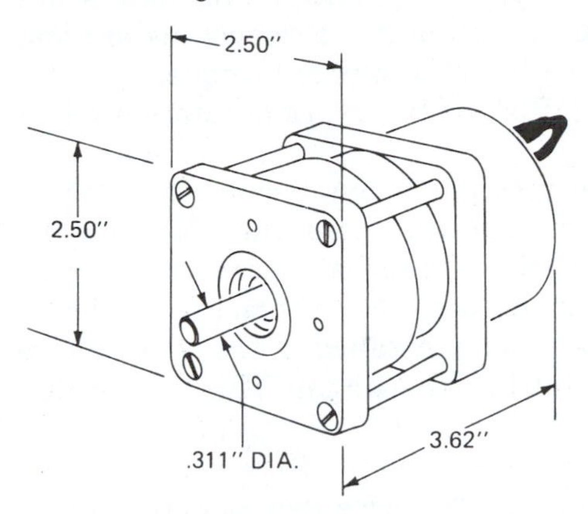

Figure 11-20A
Commutator and slip rings in the stator of a permanent-magnet motor

Figure 11-20B
Rolling contacts assembled for mounting on the rotor of a permanent-magnet motor

At point *b*, the power required to drive the load is less than the input to the motor. As a result, there is an increase in rotor speed. The increased speed produces a greater cemf, causing a decrease in current. This process continues as long as the machine is in operation.

Permanent-Magnet Motors

A *permanent-magnet motor* is a motor in which the main field flux is produced by permanent magnets. An electromagnet is used for the secondary field or armature flux.

Because of the constant field flux, the standard permanent-magnet motor has many of the same characteristics of the shunt motor. Variations in design, however, can change these characteristics considerably.

These motors are frequently used for servomotors, torque motors, and industrial drive motors. They are used on machines requiring exact positioning of an object or component, where high starting and operating torque are required, and where a constant torque is required. Some examples are the opening of a valve under pressure, precise positioning of dampers and three-way valves, and other specific operations in various control systems.

The stationary permanent-magnet motor consists of permanent magnets mounted on a frame with a rotating armature placed between them. Electrical energy is supplied to the rotor by way of a commutator and brushes. This is the conventual construction, and it is suitable for many uses. However, it is restricted in size because of its inability to dissipate heat rapidly. It is limited also by the strength of the permanent magnets. Figure 11-19 shows a motor of this type.

The revolving permanent-magnet and the wound-stator types of motors have several advantages over the wound-rotor type of motor. Because the windings are on the stator, they can dissipate heat more rapidly. In addition, there is less stress on the windings because of the lack of centrifugal force. For these two reasons the motors can be constructed in larger sizes, in some cases as large as 3 horsepower. Another advantage is easy access to the windings. This permits easier and more frequent insulation checks, monitoring of temperature, and use of static control.

The method used for commutation (reversing the current flow in the stator windings) is quite different from the methods for motors discussed previously. A specially designed commutator and slip rings are mounted inside the stator, Figure 11-20A. Rolling contacts are mounted on the rotor, Figure 11-20B. As the rotor revolves, one roller makes contact with a slip ring. Half of the

Figure 11-21A
Rotor of a permanent-magnet motor using salient-pole magnetic construction

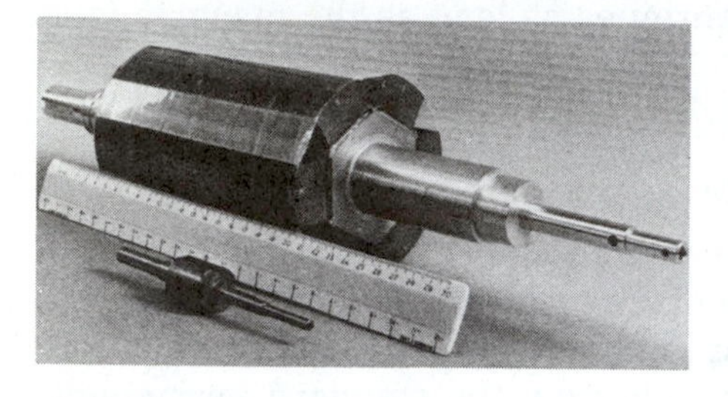

Figure 11-21B
Rotor of a permanent-magnet torque motor using cobalt-rare earth magnets

rollers contact one slip ring for one polarity; the other half contact the other slip ring for the opposite polarity. The rollers are positioned so as to energize the stator at the correct instant and with the proper polarity to maintain a constant torque.

Figures 11-21A and 11-21B show two types of rotors used on permanent-magnet motors.

The development of the rotating magnet motor has led to improvements in the magnetic materials used. One of the materials that has received wide acclaim is the cobalt-rate earth magnet. These magnets are made thin in the direction of magnetization to allow a large number of poles to be installed in the rotor, where they will funnel maximum flux into the air gap. This greater flux density results in a higher torque.

This type of construction yields a more efficient motor than those previously used. Figure 11-22 shows two industrial drive motors. The new design has approximately the same dimensions as the conventional one. However, for the same physical size, it provides 50 percent higher continuous rating. Its peak torque is 55 percent greater than that of the conventional design. Its accelerating ability is almost twice as good as that of the standard machine.

Stepping Motors

The stepping motor is similar in design to the permanent-magnet motor. Because they are designed to be used for jobs requiring minimum torque, their stators are connected directly to the power source.

Figure 11-23 is a simplified diagram of the magnetic polarity arrangement of the stator and rotor. The stator windings are

Figure 11-22
Industrial drive, permanent magnet motors. The top machine is a modern construction; the bottom machine is a conventional construction.

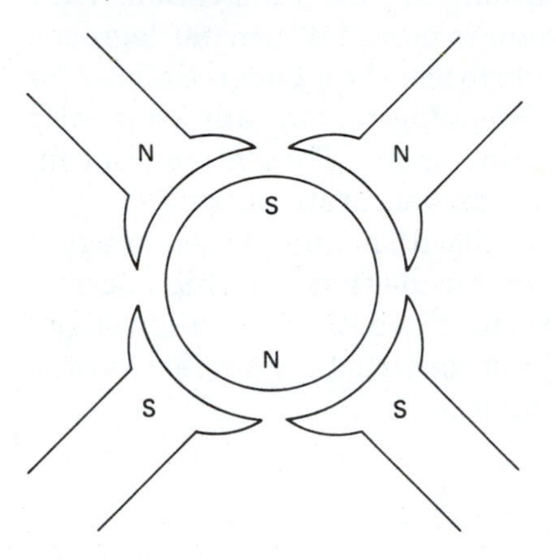

Figure 11-23
Simplified diagram of a stepping motor, stator, and rotor

arranged so that the direction of current through them can be reversed at the proper instant to obtain the desired direction of rotation. Also they are arranged to provide for controlling the number of degrees of rotation. Figure 11-24 illustrates the basic circuitry for the stator.

In actual practice, the design is more complex. The stator is designed to have a large number of poles, possibly as many as 40. The rotor is designed to have more poles than the stator. The relationship between the number of poles on the rotor and stator determines the amount of rotation *(step angle)*. The step angle is measured in the number of degrees of rotation.

An electric pulse fed into the stator winding sets up a magnetic field, causing the permanent magnets in the rotor to align themselves with the field of the stator. A series of pulses causes the stator field to rotate, and the rotor follows. A single pulse may advance the rotor as little as 0.5 degrees.

This can be better understood by referring to Figure 11-24. With switches A and B in the positions shown, the rotor field is in the vertical position. By moving switch A to connect with the top

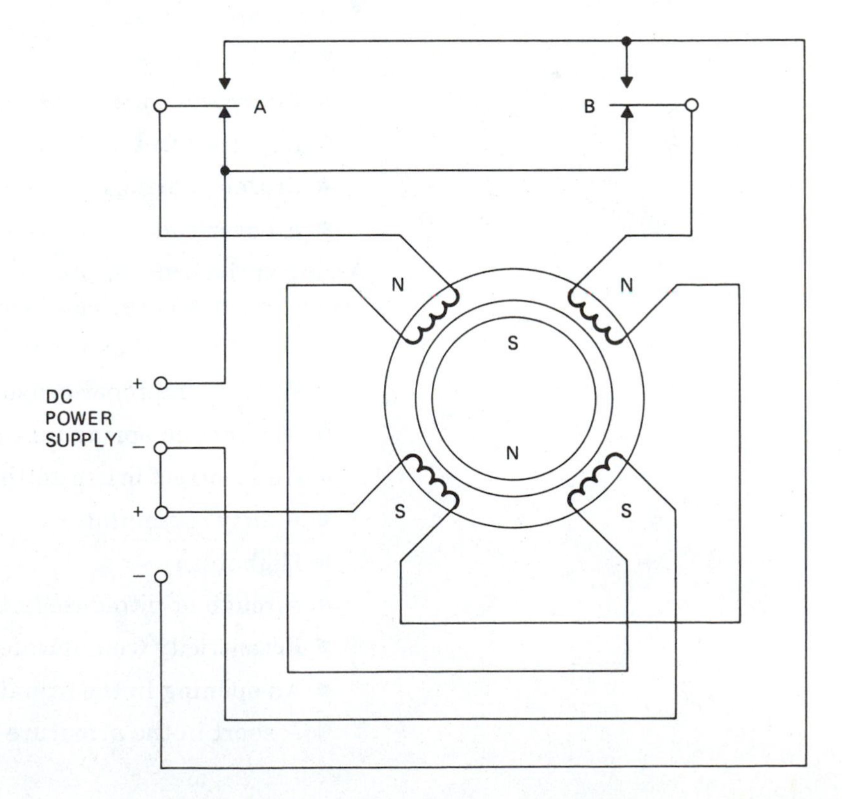

Figure 11-24
Basic circuitry for a permanent-magnet stepping motor

contact, the polarity reverses in one set of poles, and the rotor advances in a clockwise direction. Various switching sequences determine the direction of rotation. In this illustration, each switching operation causes the rotor to move 1/4 turn (90 degrees).

The ability to control the rotation to less than 1 degree allows for many applications in industry. Stepping motors are frequently used in printing shops and machine shops. They are often controlled by computers to obtain fast and accurate pulsing.

Some stepping motors are designed with nonpermanent magnet rotors. These rotors are made of materials that have high permeability. When the stator poles are energized, the rotor aligns itself with the magnetic field. This type of construction is referred to as a *variable reluctance stepping motor*.

MOTOR MAINTENANCE

A motor should start, deliver its rated load, and run within its speed ranges without excessive vibration, noise, heating, or arcing at the brushes. Failure to start may be a result of one or more of the following:

- A ground
- An open circuit
- A short circuit
- Incorrect connections
- Improper voltage
- Frozen bearings
- An overload

Arcing at the brushes may be traced to the brushes, commutator, armature, or an overload. Arcing may be the result of:

- Insufficient brush contact
- Worn or improper brushes
- Not enough spring tension
- An incorrect brush setting
- A dirty commutator
- High mica
- A rough or pitted commutator surface
- Eccentricity (commutator off center)
- An opening in the armature coil
- A short in the armature or field coils

Vibration and pounding noises may be caused by:

- Worn bearings
- Loose parts
- Rotating parts' hitting stationary parts
- Armature unbalance
- Improper alignment of the motor with the driven machine
- Loose coupling
- Insufficient end play
- The motor and/or driven machine loose on the base
- An intermittent load

Overheating is frequently caused by:

- Overload
- Arcing at the brushes
- A wet or shorted armature or field coils
- Too frequent starting or reversing
- Poor ventilation
- Incorrect voltage

Overheated bearings may be a result of:

- A lack of lubricant
- Too much grease
- A dirty lubricant
- Too tight a fitting at the bearings
- The oil ring's not rotating
- Too much belt tension or gear thrust
- Insufficient end play
- A rough or bent shaft
- A shaft out of round

To ensure good operating conditions, all motors should be checked periodically. Keep a permanent record of the results, replacement of parts, and general maintenance tasks. The log should indicate the date, time of day, ambient temperature, humidity, and other pertinent data that may affect the test results.

The interior and exterior of the motor should be kept clean and dry. Depending upon the atmospheric conditions, periodically disassemble the motor and clean the interior thoroughly. All loose dirt should be vacuumed away. Clean the commutator and con-

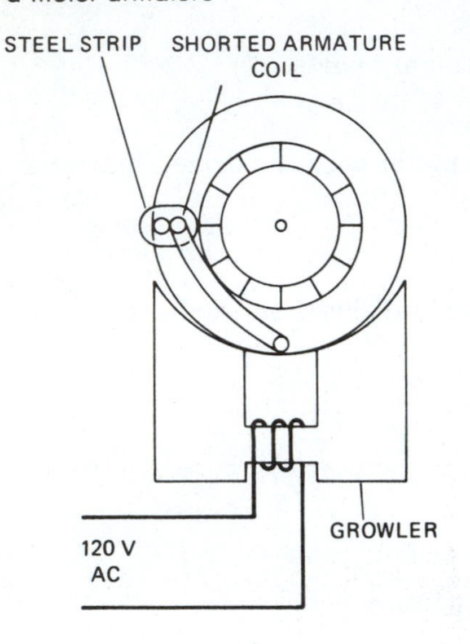

Figure 11-25
Using a growler to locate a shorted coil in a motor armature

tacts with a nontoxic, nonabrasive cleaner. Bearings should be lubricated as needed. (Do not overlubricate.) Replace worn bearings immediately.

Insulation tests should be performed on a regular basis. Check for grounds, shorts, open circuits, and leakage currents. A good insulation test can be performed with a megohmmeter (see Chapter 3). To test for shorted coils in the armature, *a growler* is frequently used, Figure 11-25. When the growler is energized with ac, an expanding and contracting flux is established. This moving flux induces a voltage into the armature windings. If the winding has no faults, little or no current will flow in the armature. If the coil is shorted, a current will flow in the short-circuited turns. To locate the faulty coil, a steel strip such as a hacksaw blade is passed over the slots containing the armature coils. It is arranged so that when a coil side lies under the strip, the other side of the same coil lies between the pole faces of the growler. When the steel strip is passed over the short-circuited coil, the flux caused by the current flow attracts the steel strip. If a part of the strip is held slightly away from the armature core, the strip will vibrate, making a noise and indicating a short in the coil. The growler gets its name from the noise it makes.

When cleaning or testing motors or doing both, always check the terminal connections to ensure that they are made tight. Any sign of overheating at the terminals is an indication of poor connections. If the connections appear to be tight, but there is an indication of overheating, disconnect the lead from the terminal. Clean all the surface area and reconnect it. Vibration or changes in temperature or both are frequently the cause of loose connections.

REVIEW

A. Multiple choice.
Select the best answer.

1. An electric motor is used to
   a. produce electrical energy.
   b. change mechanical energy into electrical energy.
   c. change electrical energy into mechanical energy.
   d. produce an alternating motion.

2. The operation of a motor depends upon
   a. static electricity.
   b. interaction between two magnetic fields.
   c. interaction between two electrostatic fields.
   d. interaction between an electrostatic field and a magnetic field.

3. When a current-carrying conductor is placed in a magnetic field, it will
   a. remain stationary.
   b. move parallel with the lines of force.
   c. move at right angles to the lines of force.
   d. rotate.

4. The direction of rotation of a dc motor armature may be reversed by
   a. interchanging either the armature or the field connections.
   b. interchanging both the armature and field connections.
   c. neither a nor b.
   d. interchanging the supply connections.

5. The amount of force acting on a current-carrying conductor in a magnetic field is proportional to
   a. the strength of the main field flux.
   b. the length of the conductor within the field.
   c. the amount of current flowing in the conductor.
   d. all of the above.

6. The effect of the forces in the armature of a motor depends upon
   a. the total force acting on the armature conductors.
   b. the radial distance through which the force acts.
   c. both a and b.
   d. neither a nor b.

7. An instrument designed to measure torque is called a
   a. torque meter.
   b. prony brake.
   c. torque scale.
   d. ohmmeter.

REVIEW *(continued)*

A. Multiple choice. Select the best answer (continued).

8. The voltage induced into a motor armature is called the
   a. voltage drop.
   b. applied voltage.
   c. counter electromotive force.
   d. IR drop.

9. The current in a motor armature flows in
   a. the same direction as the applied voltage.
   b. the opposite direction to the induced voltage.
   c. the same direction as the induced voltage.
   d. both a and b.

10. The polarity of an interpole in a dc motor is
    a. the same polarity as the main pole directly behind it.
    b. the same polarity as the main pole directly ahead of it.
    c. the opposite polarity of the main pole directly behind it.
    d. none of the above.

11. The speed of a dc motor is proportional to the
    a. applied voltage.
    b. counter electromotive force.
    c. armature reaction.
    d. IR drop in the armature.

12. Speed regulation in a motor refers to
    a. the manner in which a motor adjusts its speed to changes in load.
    b. a means of varying the speed of a motor.
    c. limiting the maximum speed of a motor.
    d. all of the above.

13. The series motor has
    a. low starting torque.
    b. high starting torque.
    c. low starting current.
    d. both b and c.

14. Series motors are generally used where
    a. good speed control is essential.
    b. they are subjected to wide variations in load.
    c. a constant speed is essential.
    d. a constant torque is needed.

REVIEW *(continued)*

A. Multiple choice. Select the best answer (continued).

15. A series motor
    a. is frequently operated without a load.
    b. should never be operated without a load.
    c. develops a constant speed at all loads.
    d. develops a constant torque at all loads.

16. Shunt motors are classified as
    a. variable speed motors.
    b. constant speed motors.
    c. constant torque motors.
    d. variable torque motors.

17. A shunt motor should never be operated
    a. without a load.
    b. with the field circuit open.
    c. with the armature circuit open.
    d. without an operator in attendance.

18. A compound motor has
    a. only a series field.
    b. only a shunt field.
    c. two fields.
    d. three fields.

19. Compound motors are used to drive machines that require a
    a. relatively constant speed under varying loads.
    b. relatively constant speed at a constant load.
    c. constant load and a constant speed.
    d. wide variations of loads and speed.

20. The permanent-magnet motor is manufactured in
    a. large sizes of 100 horsepower and larger.
    b. small sizes of usually only fractional horsepower.
    c. sizes up to 20 horsepower.
    d. sizes up to 5 horsepower.

B. Give complete answers.

1. Explain the principle of operation of a dc motor.
2. Write two laws of magnetism upon which the operation of a motor depends.
3. Describe the right-hand rule for a motor.
4. Describe a method used to reverse the direction of rotation of a dc motor armature.

REVIEW *(continued)*

B. Give complete answers (continued).

5. List three factors that determine the amount of force exerted on a conductor that is carrying current while in a magnetic field.
6. Define *torque*.
7. Identify the unit of measurement for torque.
8. List six types of prony brakes.
9. Define the term *counter electromotive force*.
10. List the factors that determine the value of current through the armature of a dc motor.
11. As the load is increased on a motor, does the neutral plane shift in the direction of rotation or opposite to the direction of rotation?
12. How are compensating windings installed in a motor?
13. Define *speed regulation* with regard to dc motors.
14. List three major types of dc motors and describe how their armatures and fields are connected.
15. What type of dc motor is used to drive machines with wide variations in load such as hoists, electric railway cars, and electric automobiles?
16. For what type of machines are shunt motors generally used?
17. What type of dc motor is usually used to drive machines that require relatively constant speeds and where sudden application of heavy loads occurs?
18. What is the advantage of a flywheel on a machine?
19. Name two types of permanent-magnet motors.
20. What is the advantage of using cobalt-rare earth magnets for the rotor of a permanent-magnet motor?
21. Give seven reasons why a dc motor may fail to start.
22. List ten problems that may cause arcing at the brushes on a dc motor.
23. List nine causes of motor vibration or pounding noises or both.
24. Give six reasons why a dc motor may overheat.
25. List nine causes of overheated bearings.

12
DC Motor Controls

Objectives

After studying this chapter, the student will be able to:

- Explain the purpose of dc motor starters.
- Show various methods of connecting dc motors to their appropriate starters, and explain their operation.
- Describe various means used to start a dc motor.
- Define the term *controller.*
- Describe various means used to control the speed and the direction of rotation of dc motor armatures.

DC MOTOR CONTROLLERS

A *controller* is any device that governs the electrical power to the apparatus to which it is connected. Switches, rheostats, potentiometers, and circuit breakers are controllers. *Manual controllers* are devices that require human intervention in order to operate, such as a switch on an electric drill and a hand-operated rheostat.

In Chapter 11, we stated that the amount of current flowing through the armature of a dc motor is dependent upon the resistance of the armature and the cemf. Because there is no cemf when the armature is stationary, the value of current is determined by the resistance alone. Therefore, the starting current of a dc motor will be extremely high unless a device is installed that will limit the current to a safe value.

One method used to limit the starting current is to connect resistance in series with the armature winding during the starting period. After the armature has begun to rotate and develop a cemf, the resistance must be removed. Some arrangements remove the resistance in steps. When the motor has reached approximately 80 percent of its rated full-load speed, all the resistance is removed.

Another method is to lower the voltage applied to the armature winding during the starting period. The voltage can be gradually increased as the armature begins to rotate. When the armature has reached the rated speed of the motor, full voltage is applied across its windings.

The method used depends upon the type of motor, the type of load, and the type of voltage (ac or dc) available from the utility company. Other factors such as space, cost of installation and maintenance, and efficiency of operation must also be considered.

Fractional horsepower motors can frequently be started without the use of current-limiting equipment. For motors rated at 1 horsepower and higher, one of the above methods must be used.

Many safety factors must be considered when one orders equipment and installs dc motors and controllers. For example, if there is a power loss or a serious voltage drop while the motor is in operation, provisions must be made to reconnect the current-limiting equipment into the circuit until the armature is again developing enough cemf to limit the current to a safe value.

A break in the field circuit may cause the armature to accelerate to a dangerously high speed. Provisions must be made to disconnect the motor from the supply should this occur.

These safety features can be obtained with various types of sensing devices. They may be incorporated into magnetic controllers, solid-state controllers, or hard wired in separately.

Figure 12-1
Magnetic switch

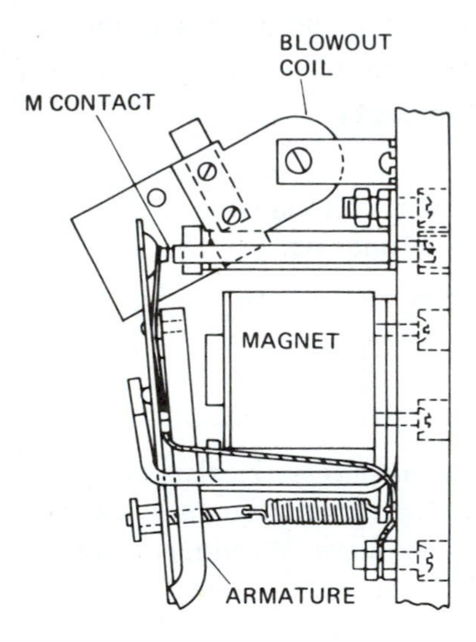

Figure 12-2
Magnetic blowout coil

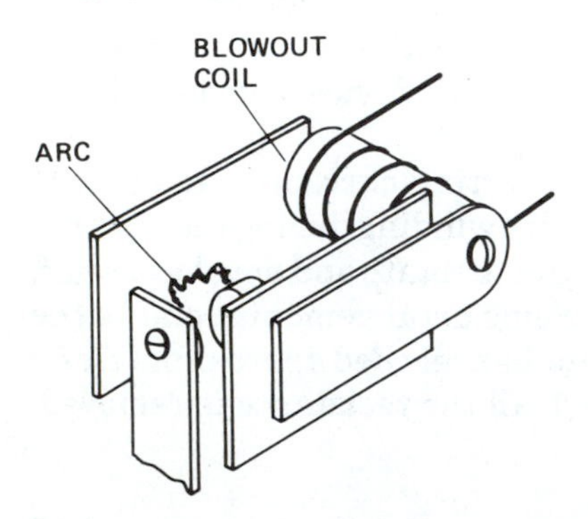

Magnetic Switching

A *magnetic switch* is a device that opens and closes a circuit or switches the current from one path to another through the use of magnetism, Figure 12-1. The essential parts of the magnetic switch are:

- The magnet, consisting of a coil of insulated wire wound on a metal core
- The armature (the moveable part of the switch)
- One or more movable contacts mounted on the armature and stationary contacts arranged to make or break with the movable contacts
- A spring arrange to hold the armature in the normal position. If the switch is constructed to break a large amount of current, a magnetic blowout coil may be incorporated into the device, Figure 12-2.

When the coil of the magnetic switch is energized, the magnet attracts the armature. The field is strong enough to overcome the spring tension, and the armature moves toward the magnet to

Figure 12-3A
Momentary push-button switch

Figure 12-3B
Push-button station wiring diagram

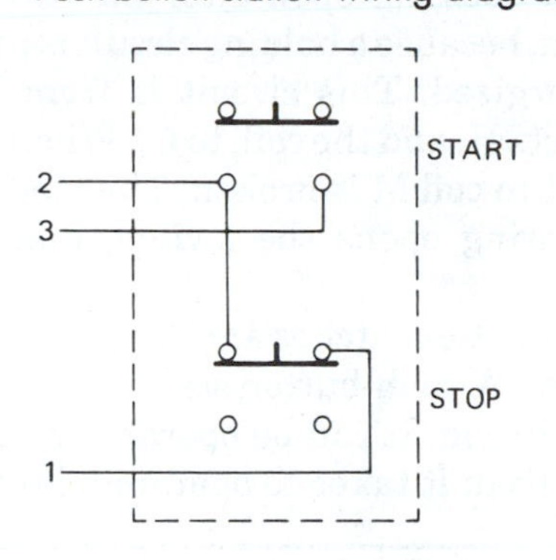

close and/or open the contacts. Deenergizing the coil releases the armature, and the spring causes it to return to the normal position.

Magnetic switches are frequently used in conjunction with *momentary start / stop buttons*. Figure 12-3A is a picture of a momentary start/stop button. Figure 12-3B is a schematic diagram of the same switch. They are called momentary buttons because the contacts are either made or broken only while the button is pressed.

Magnetic switches are commonly used in circuits similar to the circuit shown in Figure 12-4A. When the start button is pressed, the circuit is complete from L_1 through the stop button contacts, the start button contacts, and the coil of the magnetic switch, to L_2. Coil M becomes an electromagnet, attracting the armature. The armature moves toward the coil, closing contacts M and H.

Figure 12-4A
Start/stop buttons used with a magnetic relay to control the current to a load

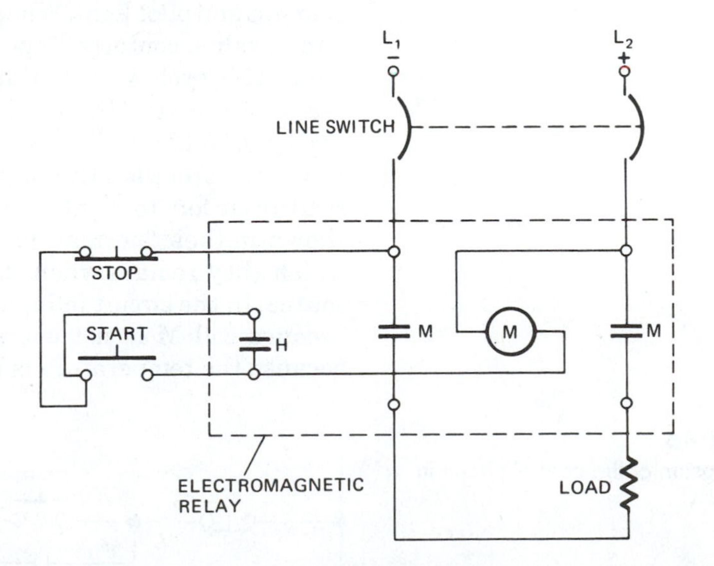

Current now flows from L_1 through the load to L_2. The magnetic switch does not open after the finger is removed from the start button, because a holding circuit is provided through which the coil is energized. This circuit is from L_1 through the stop button, contacts H, and the coil, to L_2. When the stop button is pressed, the circuit to coil M is broken. The electromagnet is deenergized and the spring opens the switch, disconnecting the load from the supply.

Magnetic switches are frequently used to control large values of current. A push-button station used with the switch provides ease of operation. It can be operated more conveniently and with less effort than it takes to operate a large manual switch.

> **CAUTION:** The operation of large current-carrying switches can be hazardous. It is much safer if the switch is located in a remote location, away from the operator.

Ladder Diagrams

Figure 12-4A is a schematic diagram illustrating the main circuit and the control circuit. In order to simplify drawings, *ladder diagrams* are frequently used. Figure 12-4B is a ladder diagram of the control circuit in Figure 12-4A. Figure 12-4C shows both the control circuit and the main circuit for an electric heater.

When the start button is pressed (Figure 12-4C), coil M is energized. The magnetic field produced by coil M causes contacts M to close. Closing contacts M completes the circuit to the heating element and pilot light. When the temperature reaches a predetermined value, contacts T open, disconnecting the heater and pilot light. This cycle will continue until the system is shut down by pressing the stop button. Pressing the stop button breaks the circuit to coil M. Coil M is deenergized and contacts M open.

To interpret a ladder diagram, one must begin at the top left and read from left to right working down the ladder. Contacts are shown in their "normal" position. Normal means the position in which they remain when they are not influenced by an outside source. In the circuit in Figure 12-4C, contacts M are shown open because coil M is not energized. Contacts T are shown closed because the temperature is below the setting of the thermostat.

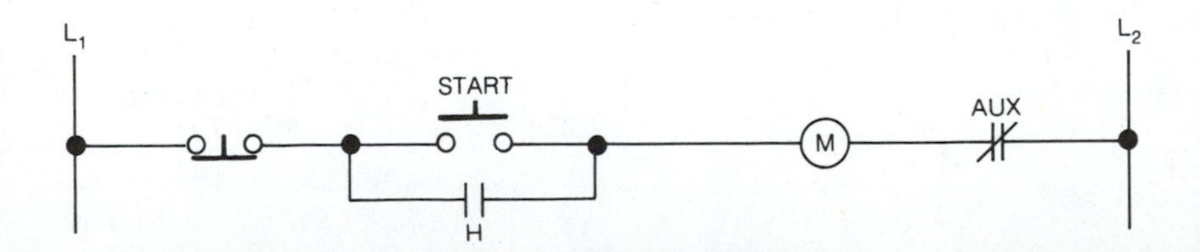

Figure 12-4B
Ladder diagram of the control circuit in Figure 12-4A

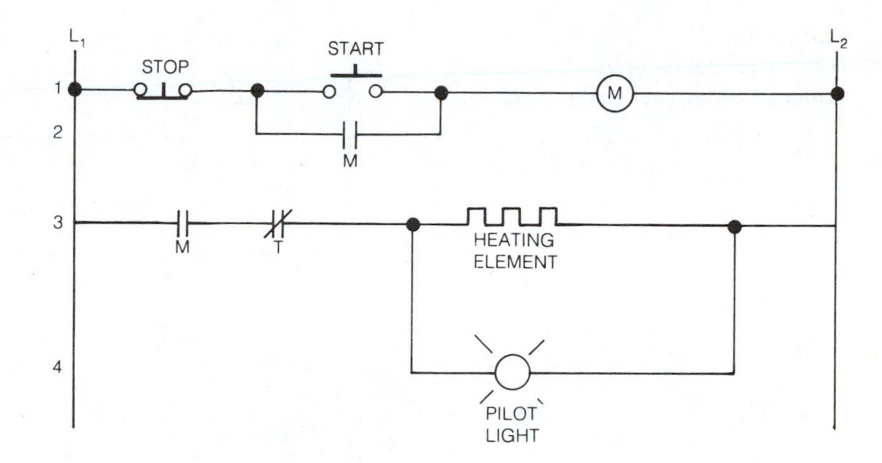

Figure 12-4C
Ladder diagram of a heating circuit

To trace the circuit in Figure 12-4C, begin at the top left. From L_1 the circuit is complete through the stop button. Pressing the start button completes the circuit through coil M to L_2. Coil M is now energized closing contacts M. The start button can be released and the coil remains energized through contacts M.

Beginning again at L_1, trace the circuit down to rung 3 of the ladder. Contacts M are closed because coil M is energized. Contacts T are closed; therefore the heating element and pilot light are energized.

DC MOTOR STARTERS

Figure 12-5 is a schematic diagram of motor connected to a stop/start station through an electromagnetic switch. This switch is commonly known by the following names: across-the-line starter, magnetic starter, and magnetic controller. Starting and speed controls have been omitted for simplicity.

Overload Protection

Overload protection is achieved by connecting a small electric heater in series with the motor. When the motor current exceeds a safe value, the current through the heater generates enough heat to activate the thermal device. The thermal device opens the normally closed overload contacts in the control circuit, as shown in Figure 12-5. When the control circuit is open, coil M is deenergized and the main contacts open.

The overload device is sensitive to heat and time. A small overload current takes longer to open the contacts than does a large one. It is important to realize that the overload device does not provide short-circuit protection. Protection against short circuits must be accomplished through the use of fuses, circuit breakers, or other devices.

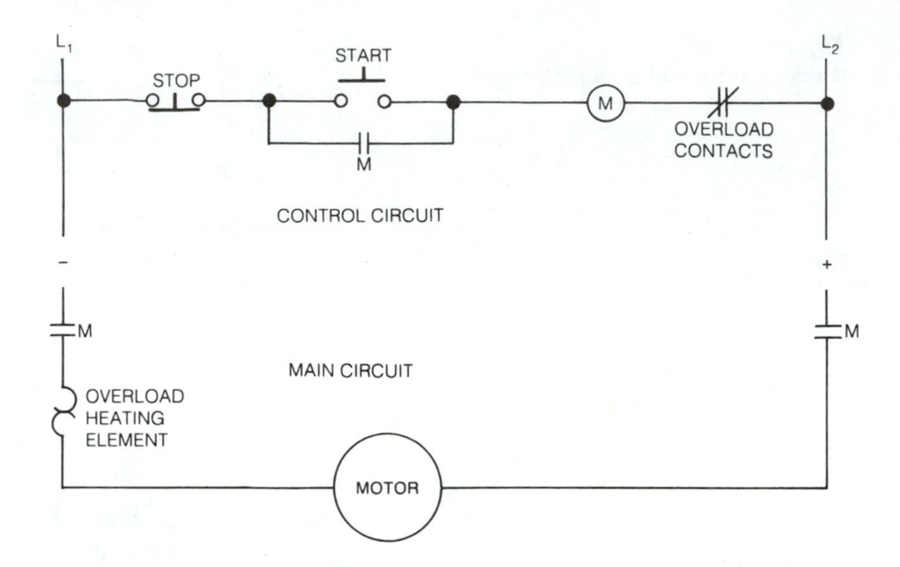

Figure 12-5
Push-button station used with magnetic starter provides overload protection.

The heating element has a specific rating. To provide adequate protection, the heater must be matched to the motor. The manufacturer specifies the rating according to the full-load running current of the motor. One must also refer to *Article 430* of the *NEC®* for the maximum safe setting.

No-Field Protection

If a break occurs in the shunt field circuit of a shunt or compound motor, the magnetic flux becomes very weak. As stated in Chapter 11, the speed of a dc motor increases as the strength of the field flux decreases. Depending upon the existing conditions, the speed could reach such high values that the centrifugal force would cause the windings to tear free from the armature. In order to eliminate this possibility, means must be provided to disconnect the armature circuit from the supply when the current through the shunt field drops below a specific value.

There are several methods used to provide no-field protection. One method is to use a relay that will react to the shunt field current. This relay is commonly called a field current relay (FCR) or a field circuit relay (FCR). The FCR consists of a coil connected in series with the shunt field. When the current drops below a specific value through the coil, the magnetic field becomes weak and allows the FCR contacts in the control circuit to open. This in turn deenergizes a relay that controls the armature circuit. The armature is then disconnected from the supply, (*See Figure 12-7*).

Another method is to use electronic circuits that monitor the field current and actuate a relay that opens the circuit to the armature if the field current drops below a predetermined value.

Time Delay Relays

Time delay relays are devices that open and/or close a circuit in a definite and predetermined period of time. There are many types of timing devices, such as clock timers, motor-driven timers, pneumatic timers, and electronic timers. Any of these can be used in conjunction with dc motors and starters. They may be wired separately or built into the starter.

Figure 12-6 illustrates a compound motor connected to a starter using time delay relays and an FCR. This arrangement is a

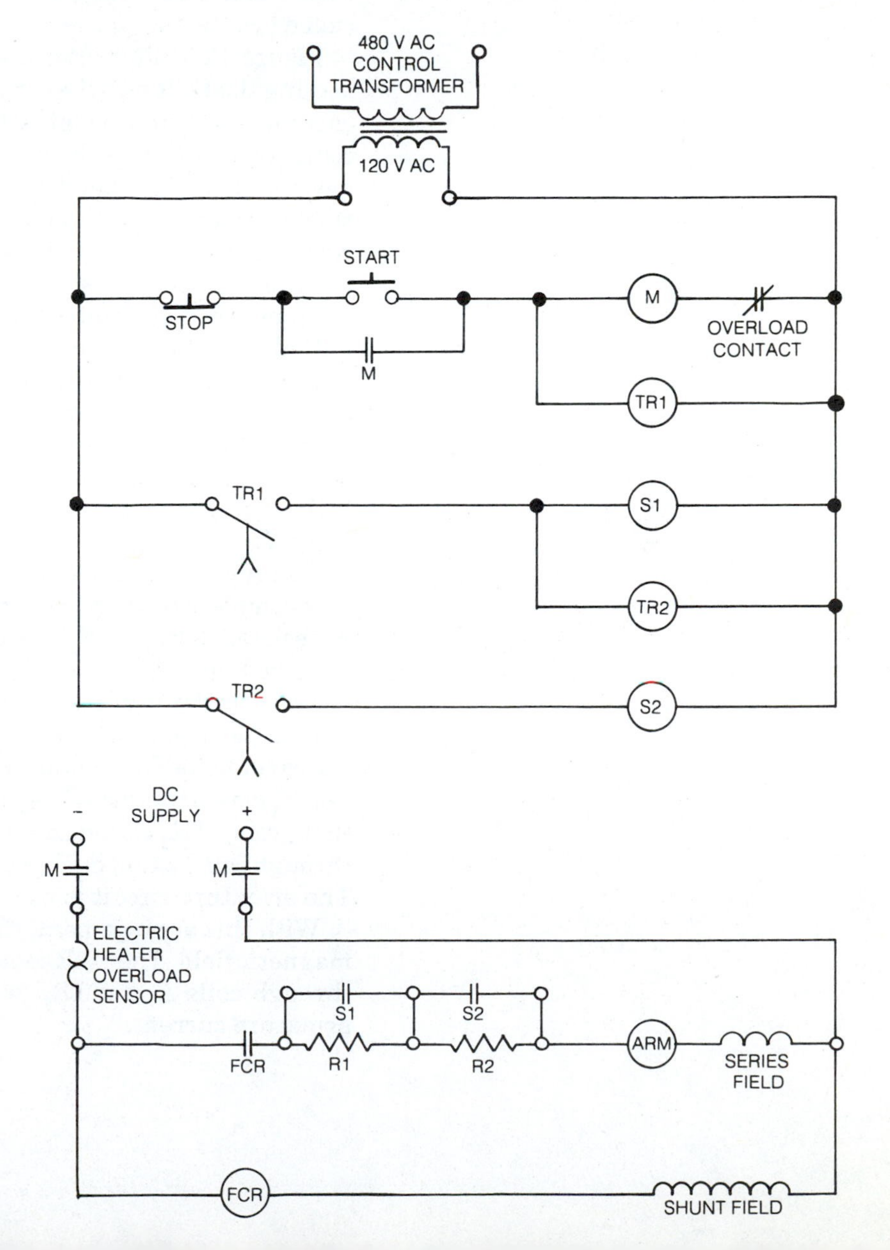

Figure 12-6
Motor starer using time delay relays and a field current relay

simplified method used to illustrate the basic operation of the FCR, but it is not practical for most installations. The FCR is connected in series with the shunt field, and therefore it must have a very low resistance. A high resistance would reduce the amount of current flowing in the shunt field winding, causing a weak field flux. The value of current flowing in the shunt field is low to begin with. Thus the current through the FCR is very limited.

The number of turns of wire on the FCR is also limited by physical restrictions. The result is a rather weak electromagnet. This weak electromagnet would not be capable of closing the rugged contacts required in the armature circuit.

Figure 12-7 illustrates a more practical arrangement for connecting the FCR into the circuit. When the start button is pressed, current flows from point X through the stop contacts, the start contacts, coil F, and the overload contacts. The control circuit is now complete from point X through the stop contacts, through the start contacts, and through the F contacts, which are in parallel with the start contacts, through coil F, and the overload contacts, to point Y. When the start button is released, the circuit remains complete through contacts F. Coil F remains energized.

Coil F also closes contacts F in the dc circuit. Current now flows from the negative line through the heater, contacts F, coil FCR, and the shunt field, to the positive line. Coil FCR cloases contacts FCR in the control circuit. Because coils F and FCR are both energized, the circuit is complete and current will flow from point X through contacts F and FCR and through coils A and TR1 to point Y.

Coil A closes contacts A in the armature circuit. The circuit is now complete from the negative line through the heater, contacts A, resistance R_1 and R_2, the armature, and the series field, to the positive line.

The armature will begin to rotate, developing a cemf in proportion to its speed. After a predetermined length of time, coil TR_1 closes contacts TR_1, completing the circuit through coils S_1 an TR_2. Coil S_1 closes contacts S_1, by passing resistance R_1. After a specific time, relay TR_2 closed contacts TR_2. The circuit is now complete through coil S_2. Coil S_2 closes contacts S_2, bypassing resistance R_2. The armature circuit is now connected across the full dc voltage.

With this arrangement, the FCR coil need develop only a weak magnetic field. The FCR contacts need interrupt current flow only through coils A and TR_1, which is much lower than the motor armature current.

Figure 12-7
Compound motor connected to a starting controller.

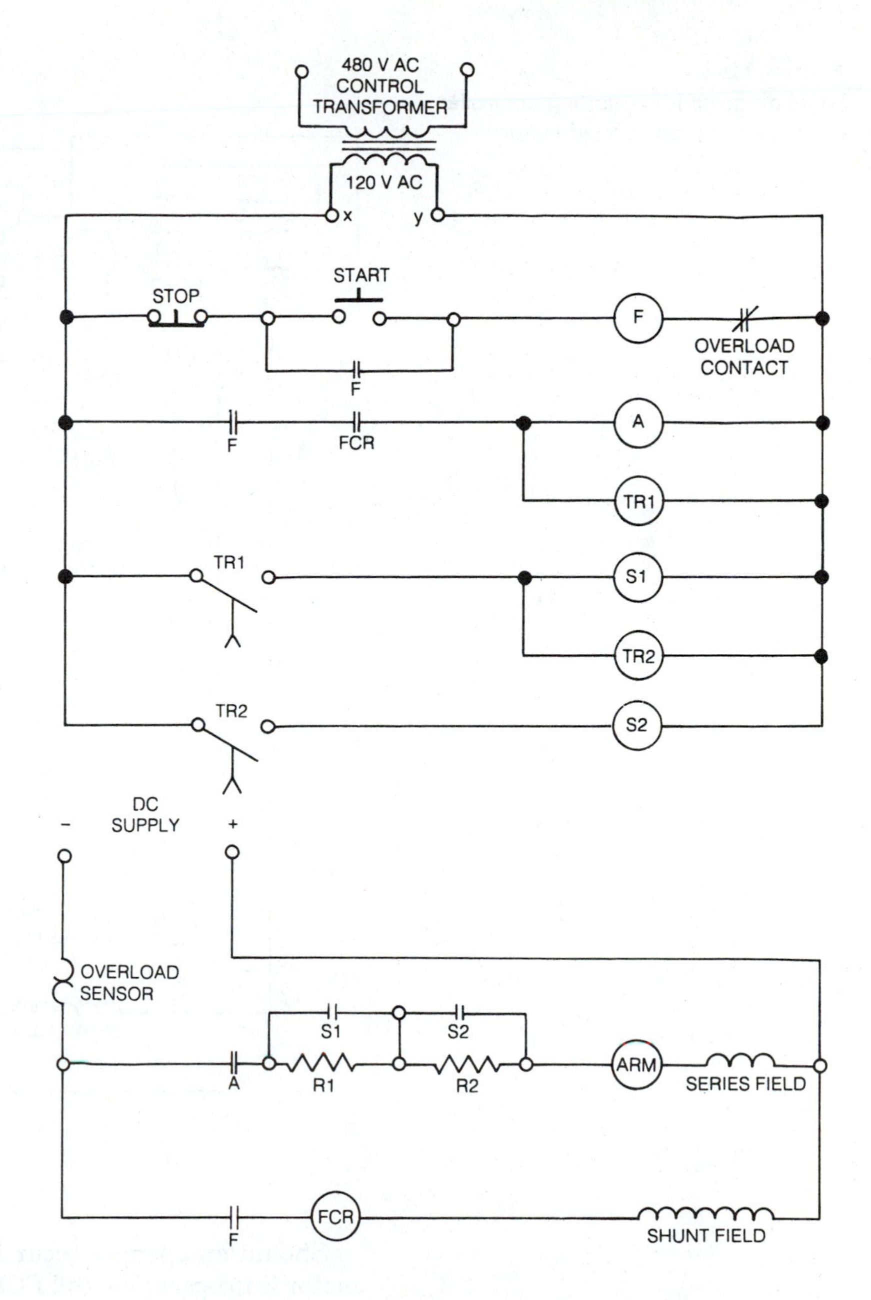

Figure 12-8
Panel diagram for a starting controller connected to a compound motor

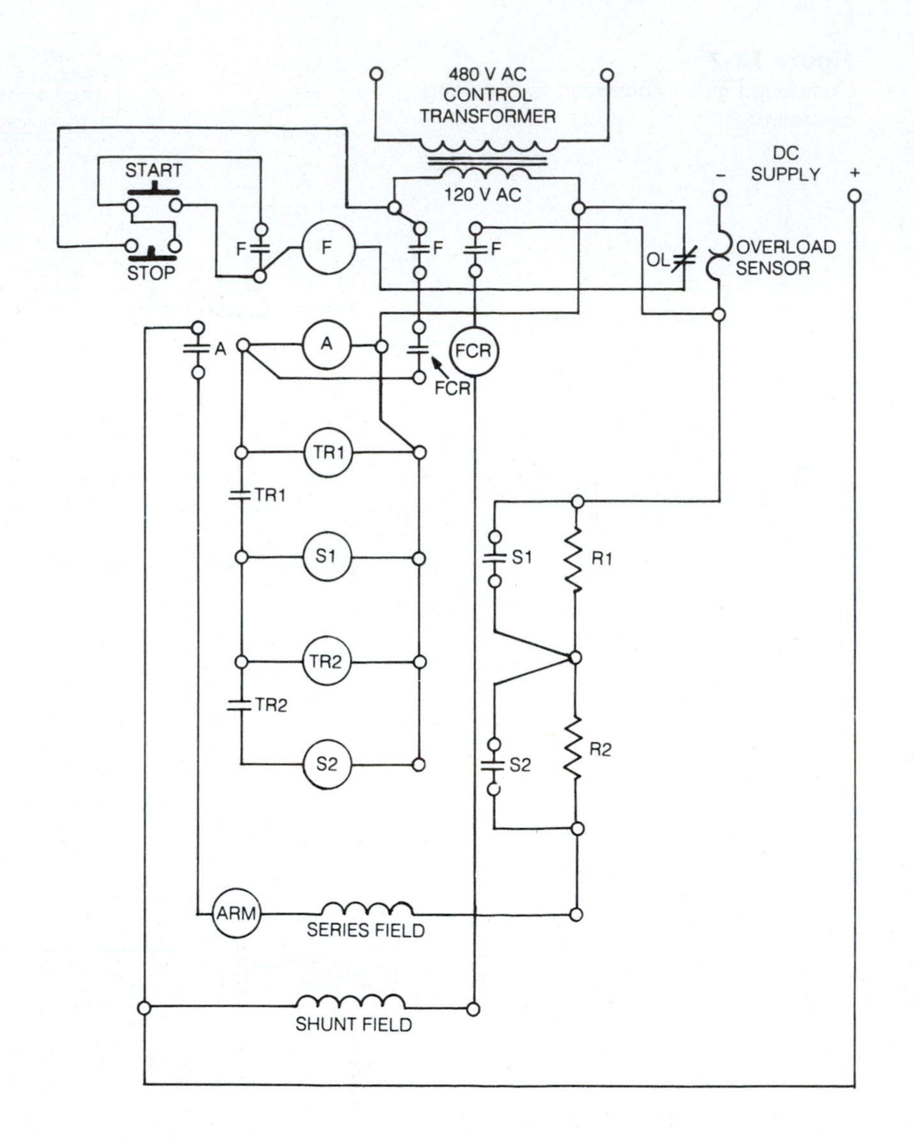

Should an opening occur in the shunt field circuit while the motor is in operation, coil FCR will be deenergized. Contacts FCR will open and coil A will be deenergized, causing contacts A in the armature circuit to open and removing the armature from the line.

Figure 12-8 is a panel diagram of the circuit in Figure 12-7.

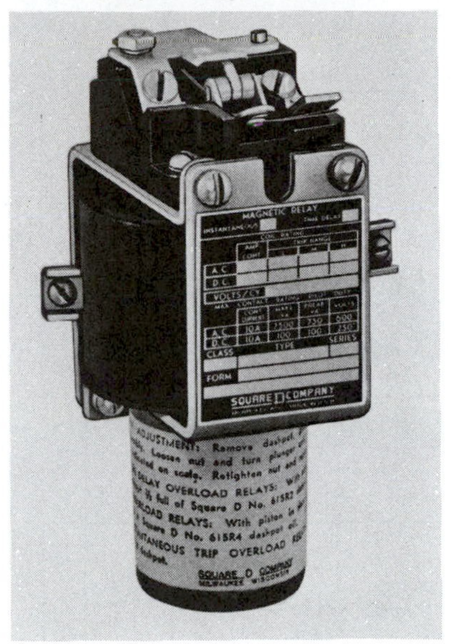

Figure 12-9
Fluid dashpot timing relay
(Courtesy of Square D Co.)

Control Transformers

Because utility companies usually supply high-voltage ac to industrial establishments, control transformers are frequently used to lower this voltage to a more practical value for the control circuit. Relays and other control devices are more readily available at 120 volts ac. Figures 12-6, 12-7, and 12-8 illustrate a control transformer used to lower the ac voltage from 480 volts to 120 volts.

The dc for the motor is usually supplied from a rectifier—a device or machine that converts ac to dc. This may be a motor-generator set or an electronic rectifier. The value of voltage form the rectifier depends upon the design of the motor. Many dc motors are supplied with two voltage sources—one for the armature and another for the field.

Relays may be used in either the ac or dc circuit. Where to install the relay depends upon the specific installation.

Dashpot Timing Starters

A dashpot timing relay is one type of time delay switch. The dashpot is a cylindrical container. A plunger inserted into the container acts as a timing device, Figure 12-9. The plunger is drawn up through air or liquid by a solenoid, and small holes in the plunger and/or an air inlet valve determine the speed at which the plunger travels. The time is usually set for the heaviest load under which the motor will operate. A bar of insulating material is attached to the plunger, Figure 12-10. As the plunger is drawn upward by the solenoid, copper contact points on the bar make contact with stationary fingers. As the fingers make contact in sequence, they bypass the starting resistance. When the plunger has traveled the length of the dashpot, all of the resistance has been bypassed.

Figure 12-10B is a ladder diagram of this circuit. Pressing the start button, Figure 12-10, completes the circuit through coil F, closing contacts F. Contacts F complete the dc circuit through the FCR coil and the shunt field. Coil FCR closes contacts FCR in the control circuit energizing coil A. Coil A closes contacts A in the dc circuit. Current now flows through the series field and armature and through resistances R_1, R_2, and R_3. It also closes contacts A in the control circuit energizing the solenoid coil. The magnetic effect of the solenoid causes the plunger to move upward. The bar attached to the plunger closes contacts S_1, S_2, and S_3 in sequence.

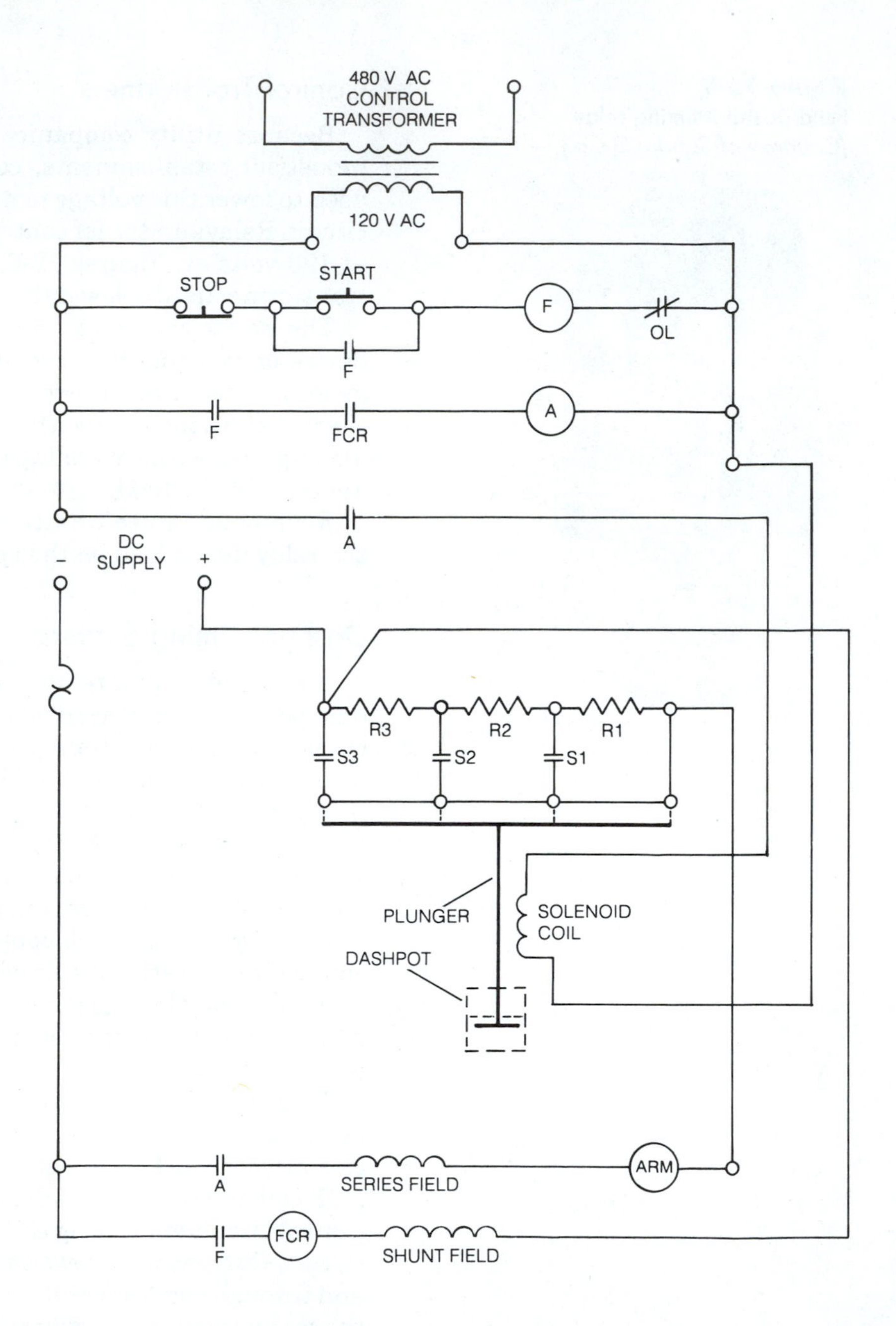

Figure 12-10A
Panel diagram of a dashpot dc motor starter

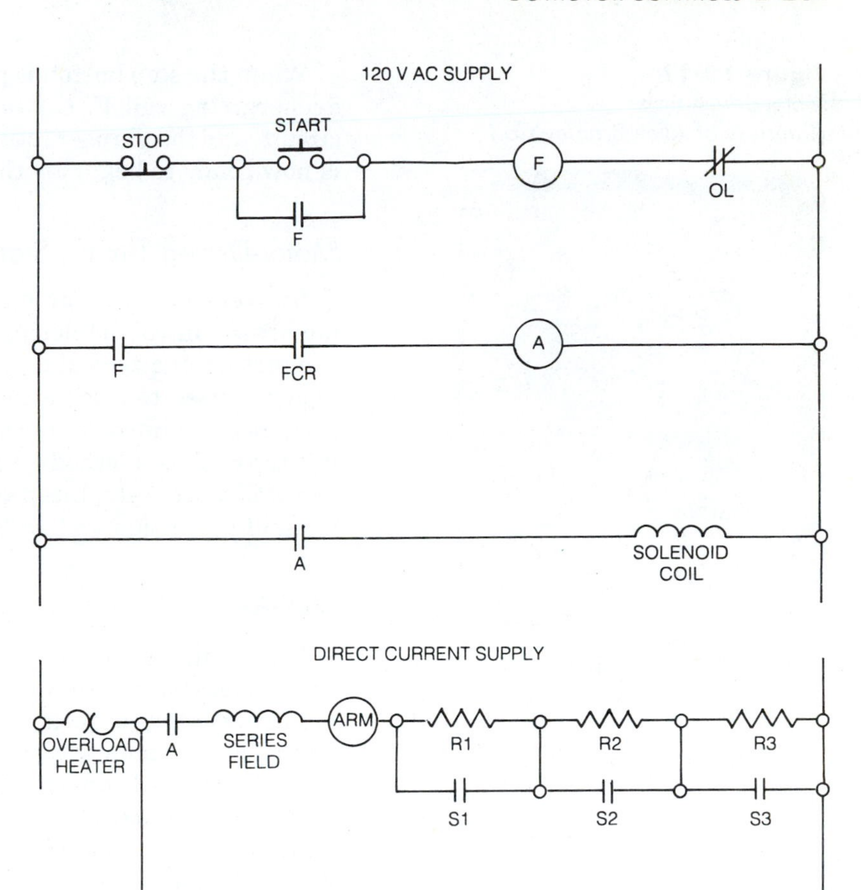

Figure 12-10B
Ladder diagram of a dashpot motor starter

Figure 12-11
Motor-driven timer
(Courtesy of Allen-Bradley Co.)

When the stop button is pressed, the circuit to coil F is opened, deenergizing coil F. Contacts F open, deenergizing the entire circuit, and the plunger returns to its original position. The starter is now ready to begin another starting sequence.

Motor-Driven Timing Starters

Starters for large dc motors sometimes use small ac timing motors to control the closing of acceleration contacts. Figure 12-11 is a picture of a typical motor-driven timer. A small synchronous motor is used to drive a cam assembly, which closes contacts in sequence. As the contacts close, they bypass the starting resistance in steps. This method provides a long timing period that is extremely accurate. Incorporated with this arrangement must be no-field protection and undervoltage protection.

Methods of Arc Quenching

Direct current contacts tend to arc when the circuit is broken. This causes excessive wear and results in frequent cleaning and replacing of the contacts. One method used to reduce the arc is to incorporate magnetic blowout coils. A blowout coil is an electromagnet installed and connected so that it produces a magnetic field across the contacts of the controller. It is generally arranged to be connected into the circuit an instant prior to the contacts' opening. When the contacts open, an arc is formed between them. An electric arc indicates current flow. The magnetic field produced by the arc reacts with the field from the blowout coil. This action causes the arc to bend. As a result, the length of the current path increases, the amount of current decreases, and the arc is extinguished quicker.

Frequently, arc chutes are also used. The blowout coils are contained within the arc chutes. The chutes confine and divide the arc, thereby extinguishing it rapidly. The use of both arc chutes and blowout coils greatly reduces wear on the contacts.

For some heavy loads, the described method alone is not adequate. In these cases, sets of contacts are connected in parallel, which divide the current between the number of contacts in parallel. Figure 12-12 illustrates this method. If arc chutes and blowout coils are incorporated with parallel contacts, then large currents can be interrupted quickly. The arc chutes separate each set of contacts, an a blowout coil is installed at each set of contacts.

Figure 12-12
Contacts connected in parallel

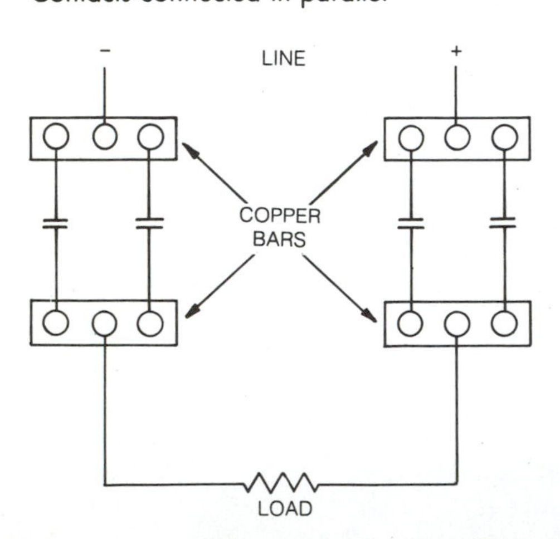

Electronic (Solid-State) Starters

All previously mentioned starters use resistance to limit the current to the armature of the motor during the starting period.

Electronic starters can be used in the same manner, but it is more economical to vary the voltage. When the armature is at a standstill, a very low voltage is applied. As it begins to rotate and produce a cemf, the applied voltage is increased. When the armature is rotating at the rated speed, full voltage is applied to the armature circuit.

Although it is desirable to apply a variable voltage to the armature, the field circuit must receive full voltage at the instant the motor is energized. For this reason, most electronic starters have two separate power sources. Because most industrial establishments are supplied with alternating current, electronic starters consist of a combination rectifier and starter. They consist of a circuit that converts alternating current into direct current and circuits that provide variable and steady dc voltages. Also incorporated within the starters can be no-field protection, undervoltage protection, and overload protection.

Electronic starters are generally less expensive than electromagnetic timing starters. They are also more flexible and more efficient. They require less maintenance and are easier to service.

For installations where there is much moisture, such as paper mills, the power to the shunt field can be left on indefinitely. Current flowing in the field circuit produces heat, which keeps the motor dry. Even on very large motors, the shunt field current is low. The energy utilized to keep the motor dry is far less expensive than it is to service motors subject to excessive moisture.

AUTOMATIC DC MOTOR CONTROLS FOR SPEED AND DIRECTION

Speed control of dc motors is obtained by varying the amount of current flowing through the armature or the shunt field. This can be accomplished either by inserting resistance into the armature or field circuit or by varying the voltage across the armature or shunt field circuits.

Electronic Speed Controllers

With some minor changes, an electronic starter can be used both to accelerate a motor to its rated speed and to vary the speed.

One device frequently used for this purpose is the silicon-controlled rectifier (SCR) speed controller. This controller consists of an SCR that can convert ac to dc as well as control the output voltage. Additional components determine when to increase or decrease the voltage and the rate and amount at which the voltage will change.

The SCR speed controller can be arranged to vary the armature and field voltages. Properly arranged, it can provide both above- and below-normal speed control. It must, however, be arranged so that the field is at full voltage when the armature voltage is decreased and so that the armature is at full voltage when the field voltage is decreased.

By varying the armature voltage, one can vary the speed of the armature from 0 to the rated speed at full load. By varying the voltage to the shunt field, one can increase the speed to approximately 25 percent above its rated full-load speed. For example, a motor rated at 1250 revolutions per minute could be operated from 0 to 1560 revolutions per minute at full load.

Safety features that are incorporated into the SCR speed controller are no-field control and current-limiting control. Although the shunt field current may be reduced enough to gain a 25 percent increase in speed, any greater decrease would disconnect the armature from the supply.

In order to ensure against excessive current to the armature during the starting period, a current-sensing device is installed, which will regulate the value of voltage applied to the armature.

The SCR speed controller can also be arrange to provide changes in the voltage applied to the armature in order to maintain a constant speed under varying loads.

Dynamic Braking and Reversing

Some industrial applications require that the direction of rotation of the armature be reversed periodically. To accomplish this, the motor must be stopped quickly and the current reversed through the armature or field. Stopping a motor quickly can be accomplished by *dynamic braking*. Figure 12-13 illustrates the connections for this operation. Note that the shunt field is connected directly across the line so that it will remain energized at all times.

While the motor is running, contacts M_1 and M_2 are closed and contacts M_3 are open. Thus, the resistance R is not in the circuit. Pressing the stop button breaks the circuit to coil M, causing contacts M_1 and M_2 to open and contacts M_3 to close. The armature is disconnected from the supply and is connected across resistor R. The armature is now rotating in a magnetic field produced by the shunt winding. A voltage is induced into the armature, causing the current to flow through the closed circuit. The current in the armature produces a magnetic field, which develops a torque in the direction opposite to the armature rotation. These steps take place very rapidly, and the armature stops quickly.

To reverse the direction of rotation, it is necessary to incorporate relays that will interchange the armature connections, thus reversing the direction of current through the armature. When a compound motor for this type of operation is installed, it is important to ensure that the current direction is reversed in the armature *only* and *not* the series field.

Figure 12-13
Simplified diagram of dynamic braking for a dc motor

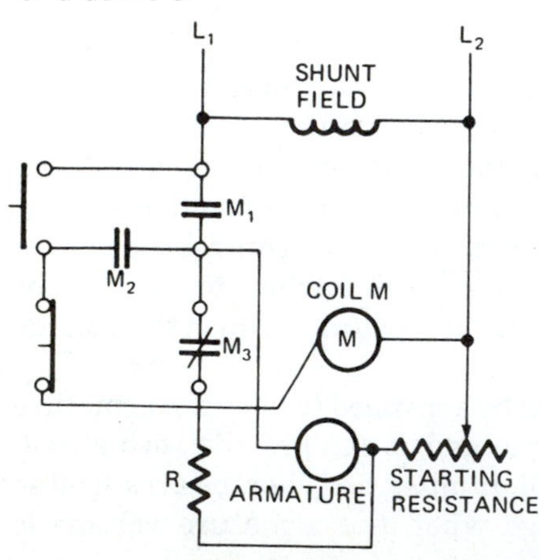

Figure 12-14
Dynamic braking and reversing

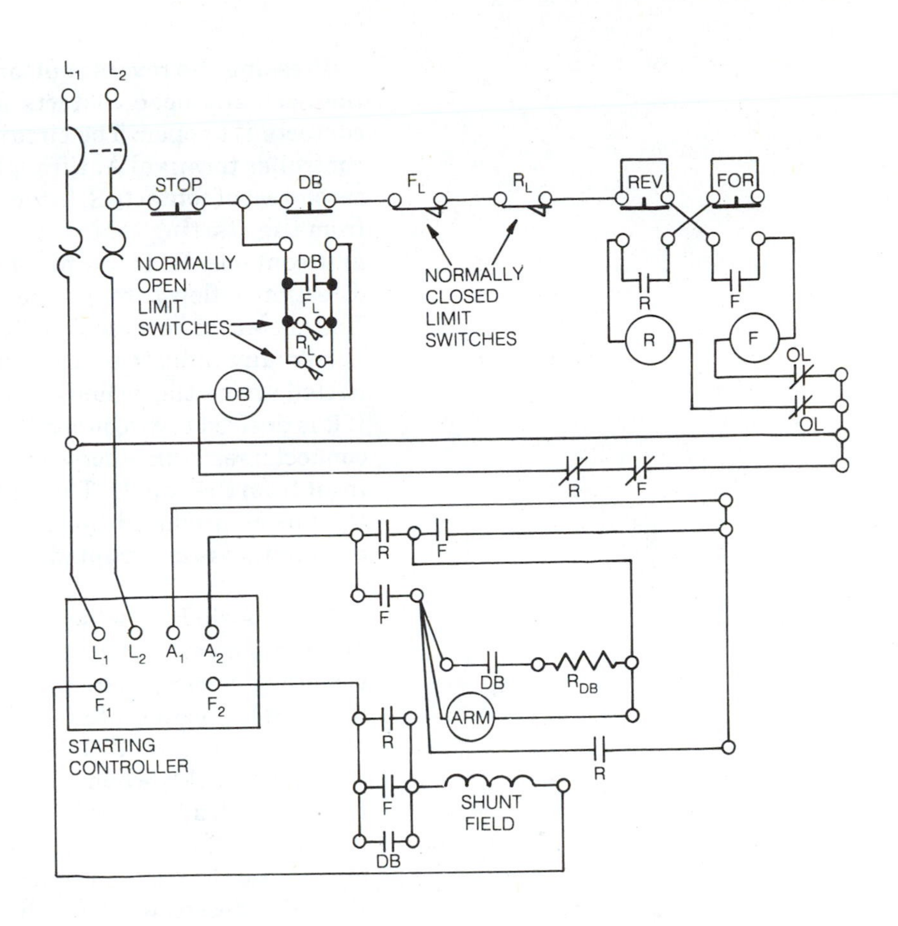

> **CAUTION:** Reversing the direction of current through the series field of a compound motor changes the operating characteristics of the motor and may result in a hazardous operation.

Figure 12-14 illustrates a shunt motor connected for forward and reverse operation. Pressing the forward button completes the circuit through coil F. When coil F is energized, the normally open contacts F close, and the normally closed contacts F open. This completes the circuit from the starting controller terminal A_1 through contacts F, the armature, and the second set of contacts F, to terminal A_2. The circuit is also complete from the starting controller terminal F_1 through the shunt field and contacts F, to terminal F_2. The motor is now operating in the forward direction.

Pressing the stop button deenergizes coil F, and contacts F return to their normal position. The motor will coast to a stop.

To stop the motor quickly, the dynamic brake (DB) button must be pressed.

Pressing the reverse button energizes the reverse coil, causing the normally open contacts R to close, and the normally closed contacts R to open. The circuit is now complete from the starting controller terminal A_1 through contacts R, the armature, and the second set of contacts R, to terminal A_2. The field circuit is complete from the starting controller terminal F_1 through the shunt field and contacts R, to terminal F_2. *Note*: the current reverses its direction of flow through the armature, but not through the field. The armature reverses its direction of rotation.

In many industrial installations, the shunt field remains connected across the dc line even when the motor is not in operation. If it is desired to disconnect the shunt field, it may be necessary to connect a resistance across the field an instant prior to disconnecting it from the supply. The purpose of the resistance is to dissipate the power produced by the collapsing magnetic field when the current flow is interrupted. This resistance is called field discharge resistance.

Limit switches, which determine the distance the movable mechanism can travel, are required by many machines such as elevators, cranes, slotters, and punch presses. These switches are frequently incorporated with reversing and dynamic braking operations. Figure 12-14 shows limit switches connected to stop the motor quickly when the mechanism has reached the maximum distance of travel in either direction.

A cam or finger arrangement is installed, which will momentarily close the normally open switches and momentarily open the normally closed switches. This action takes place an instant prior to reaching the maximum distance of travel. The normally closed switches are adjusted to open a fraction of a second prior to the closing of the normally open switches. This action deenergizes either coil F or R and energizes coil DB, causing the motor to stop quickly. Most limit switches can be adjusted to regulate the distance of travel.

REVIEW

A. Multiple choice.
Select the best answer.

1. A controller is a device that
   a. serves to govern the electric power to the apparatus to which it is connected.
   b. changes ac to dc.
   c. changes dc to ac.
   d. lowers the voltage to a practical value.

2. The starting current to the armature of a dc motor can be limited by
   a. installing fuses in the armature circuit.
   b. connecting resistance in series with the armature during the starting period.
   c. lowering the voltage applied to the armature circuit during the starting period.
   d. either b or c.

3. A break in the shunt field circuit of a dc motor may cause the
   a. armature to stall.
   b. armature to accelerate to dangerously high speeds.
   c. armature to reduce its speed but not stall.
   d. armature to increase its speed by 25 percent.

4. A magnetic switch is a
   a. switch that controls a magnet.
   b. switch that is magnetized.
   c. device that opens and closes a circuit through the use of electromagnetism.
   d. device that lowers the voltage through the use of electromagnetic induction.

5. An across-the-line starter is a
   a. type of magnetic switch.
   b. switch that reduces the voltage applied to the armature during the starting period.
   c. switch that shorts out resistance.
   d. device that controls the speed of a motor.

6. Overload protection for dc motors is achieved by connecting
   a. a fuse in the armature circuit
   b. a small electric heater in series with the motor.
   c. a fuse in the shunt field circuit.
   d. a coil in series with the motor.

7. Devices that open and/or close a circuit in a definite and predetermined period of time are called

REVIEW *(continued)*

A. Multiple choice.
Select the best answer (continued).

7. Devices that open and/or close a circuit in a definite and predetermined period of time are called
   a. definite time switches.
   b. predetermined time switches.
   c. time delay relays.
   d. set time switches.
8. A field current relay is used to
   a. limit the current through the shunt field.
   b. limit the current through the series field.
   c. disconnect the motor from the supply when the armature current drops below a specific value.
   d. disconnect the motor from the supply when the shunt field current drops below a specific value.
9. A control transformer is a device that
   a. lowers the supply voltage to a more practical value for the control circuit.
   b. raises the supply voltage to a more practical value for the control circuit.
   c. changes the ac voltage to dc voltage for the control circuit.
   d. reduces the current to a safe value for the control circuit.
10. A dashpot timing relay is a
    a. high-speed switch.
    b. time delay relay.
    c. motorized switch.
    d. dc switch.
11. A magnetic blowout coil is used to
    a. test for blown fuses.
    b. reduce arcing at contacts.
    c. blow the dust out of motors.
    d. test for open circuits.
12. Switch contacts are sometimes connected in parallel in order to
    a. enable the switch to handle heavy current loads without overheating.
    b. reduce arcing at the contacts when the circuit is broken.
    c. increase the power to the apparatus it is controlling.
    d. handle higher voltages.

REVIEW *(continued)*

A. Multiple choice.
Select the best answer (continued).

13. Electric motor starters generally
    a. use resistance to limit the current to the motor armature during the starting period.
    b. lower the voltage to the series field during the starting period.
    c. lower the voltage to the shunt field during the starting period.
    d. lower the voltage to the entire circuit during the starting period.

14. Below-normal speed control of a dc motor is obtained by varying the amount of current flowing through the
    a. armature winding.
    b. shunt field winding.
    c. series field winding.
    d. interpole winding.

15. Reversing the direction of current through the series field of a compound motor will
    a. change the magnetic polarity of the armature.
    b. change the direction of rotation of the armature.
    c. change the operating characteristics of the motor.
    d. have no effect on the operation of the motor.

B. Give complete answers.

1. What is a controller?
2. What determines the amount of current flowing through the armature coil of a dc motor?
3. List two methods used to limit the starting current of a dc motor.
4. What size dc motors require that the armature current be limited during the starting period?
5. List two safety requirements for dc motors.
6. What is a magnetic switch?
7. List four essential parts of a magnetic switch.
8. Describe the operation of a magnetic switch.
9. What is an across-the-line starter?
10. How is overload protection achieved with an across-the-line starter?

REVIEW *(continued)*

B. Give complete answers (continued).

11. How does a break in the shunt field circuit of a compound motor affect the operation of the motor?
12. What is a field current relay (FCR)?
13. Describe the operation of an FCR.
14. What is a time delay relay?
15. List four types of time delay relays.
16. What is a control transformer?
17. What is a rectifier?
18. What is a blowout coil?
19. What is an arc chute?
20. Why do most electronic starters have two separate power supplies?
21. How is speed control of dc motors accomplished?
22. Describe an electronic device frequently used to control the speed of a dc motor.
23. What is meant by the term *dynamic braking*?
24. How is armature rotation reversed for a dc shunt motor?
25. In many industrial installations, the shunt field of a dc motor remains connected across the dc line even when the motor is not in operation. What does this accomplish?

13

ALTERNATING CURRENT

Objectives

After studying this chapter, the student will be able to:

- Explain the difference between alternating current and direct current.
- Describe the basic principles of alternating current.
- List the advantages and disadvantages of alternating current.
- Describe the characteristics of alternating current with regard to resistance, inductance, and capacitance.
- Define *alternating-current power* and *power factor.*

BASIC AC THEORY

In Chapter 10, we stated that the voltage produced in the armature of a generator reverses direction as the coils rotate through the magnetic field. To obtain direct current for a system, a commutator must be used on the generator. No commutator is necessary for alternating current. The alternating voltage produced by the generator is impressed directly across the external circuit. This voltage not only reverses direction but also varies in strength.

Figure 13-1
Graph of an ac voltage

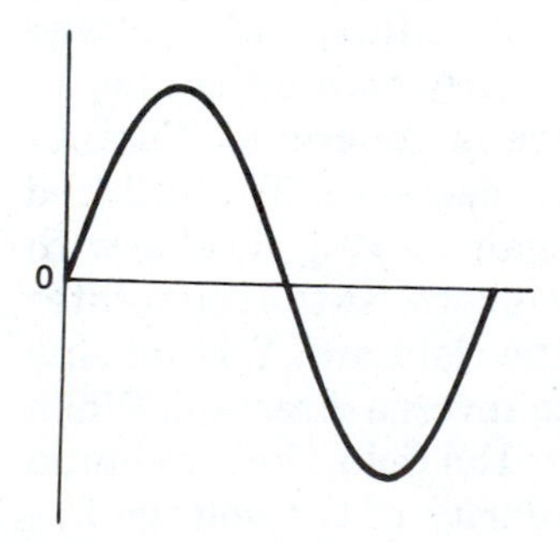

The standard ac generator (alternator) produces a voltage that, if plotted on a graph, forms a curve similar to that in Figure 13-1. This curve shows that the voltage increases from zero to a maximum value in one direction, drops to zero, increases to a maximum value in the opposite direction, and drops to zero again. When such a change in values has taken place, a cycle has been completed. From then on, the cycle is merely repeated.

The number of cycles completed in 1 second is called the *frequency*. The modern unit of measurement of frequency is the

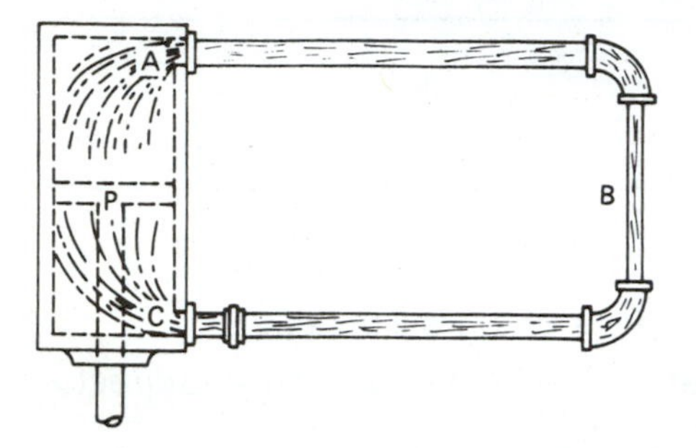

Figure 13-2
Closed system filled with water. The piston forces water through the pipe, producing an alternating water flow similar to ac flow in an electric circuit.

hertz (Hz). Hertz means cycles per second. The most common frequency in the United States is 60 hertz. Many European and Asian countries operate on a frequency of 50 hertz.

Hydraulic Analogy of Alternating Current

The flow of alternating current can be compared to the flow of water in a closed system, Figure 13-2. When the piston, P, is moved toward A, the water is forced out at A and drawn in at C, flowing through the pipe in the direction ABC. When the piston reaches the end of the stroke it moves back toward C. The water is then forced out at C and drawn in at A. The flow is reversed, flowing through the pipe in the direction CBA. As the piston is moved back and forth, it sets up an alternating pressure, causing the current of water to change direction at the end of each stroke. Thus the current of water alternates in direction, and the rate of flow varies widely.

Generation of a Voltage Curve

The amount of voltage produced by a coil rotating in a magnetic field depends upon the following factors:

- The strength of the magnetic field
- The speed at which the coil moves through the field
- The number of turns of wire in series on the coil
- The angle at which the coil moves through the field

The direction of the induced voltage depends upon the direction of motion of the coil. Figure 13-3 portrays various positions of the rotor in an elementary alternator. Assume that the armature is rotating in a clockwise direction. In position A, the coil is moving parallel to the lines of force and no voltage is produced. As the coil continues to rotate, it begins to cut across the flux, and a voltage is produced. When the coil is moving directly across the lines of force (position B), the maximum voltage is generated. Further rotation of the coil causes the angle to decrease. The induced voltage also decreases. When the coil is again moving parallel with the lines of force (position C), the voltage is zero. As the coil rotates farther, side X is moving up through the field and Y is moving down. This causes the induced voltage to reverse direction. When the coil is again moving at right angles to the field, the maximum voltage is produced. Notice that the polarity of the voltage has reversed. Position E shows the coil back in the original position (one cycle of alternating voltage has been completed).

If the maximum voltage of the alternator is known, the generated voltage can be plotted to form a curve. Draw a circle with the

Figure 13-3
Generation of an alternating voltage. As the loop rotates through the magnetic field, the amount and polarity of the voltage change with the angle and direction of motion.

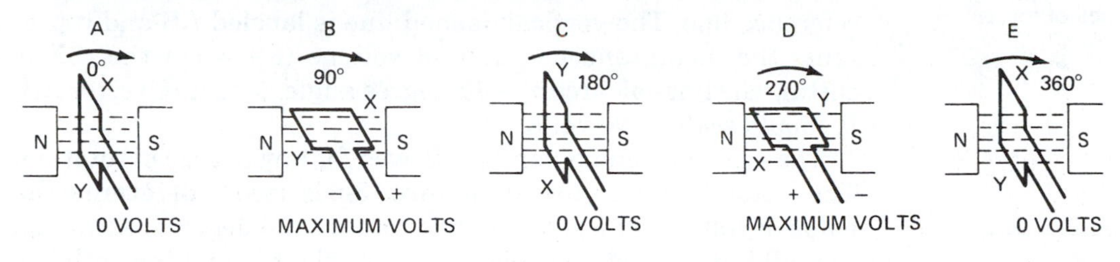

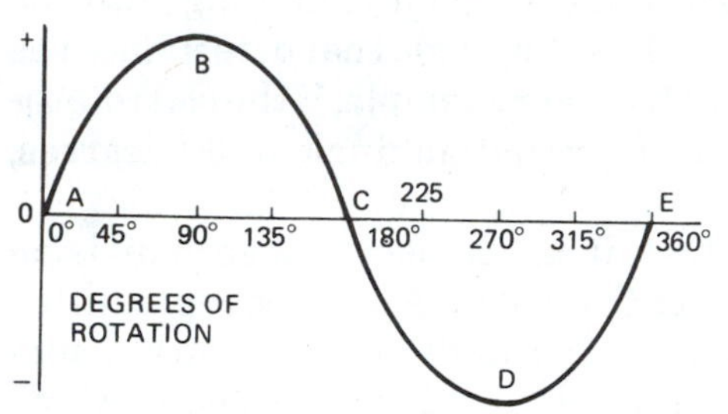

radius representing the maximum value of voltage. Any convenient scale may be used. Divide the circle into equal parts, Figure 13-4. Draw a horizontal line to scale, along which one voltage cycle will be plotted. Divide the line into the same number of equal parts as the circle. Draw horizontal and vertical lines, as illustrated by the dashed lines in Figure 13-4. The intersection of the lines represents the value of voltage at that instant. For example, a horizontal and vertical line intersect at point X. Using the same scale as used for the radius of the circle, one can measure the value of voltage. This value is the emf produced when the coil is cutting the lines of force at a 30-degree angle.

Use of Vector Diagrams

The change that occurs in the value of an alternating voltage and/or current during a cycle can also be shown by using vector diagrams. A *vector* is a line segment that has a definite length and direction. A *vector diagram* is two or more vectors joined together to convey information. Vector diagrams drawn to scale can be used to determine instantaneous values of current and/or voltage.

Figure 13-4 can be analyzed by means of vector diagrams, according to the following procedure. Draw a horizontal line as a reference line, Figure 13-5. Starting at point O, 30 degrees from

Figure 13-4
Plotting a curve of alternating voltage

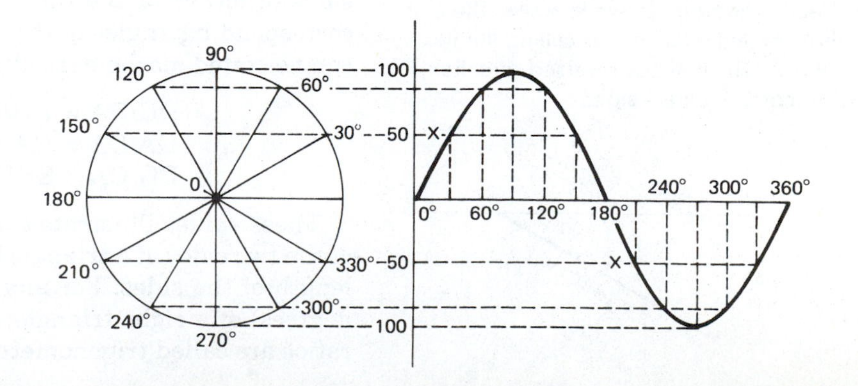

Figure 13-5
Vector diagram drawn to scale to determine the voltage when the coil is moving at a 30° angle to the lines of force

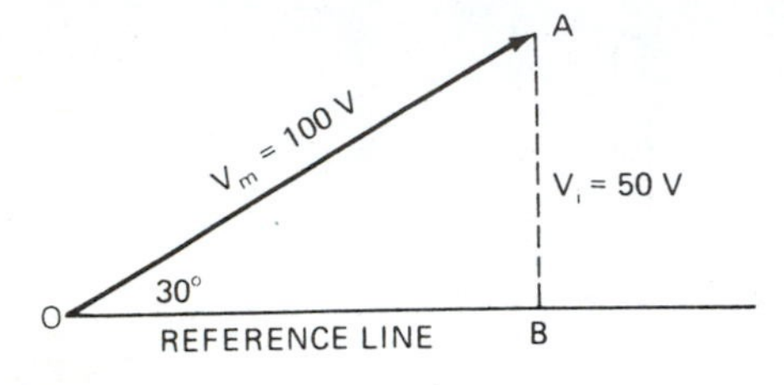

the reference line, draw OA to scale to represent a maximum voltage (V_m) of 100 volts. From the end of vector OA, draw a vertical dashed line. This line should form a 90-degree angle with the reference line. The vertical dashed line is labeled AB and represents the instantaneous value of voltage (E_i) when the coil is cutting the lines of force at a 30-degree angle. Measure vector AB. It should scale to 50 volts.

The same procedure can be followed for any degree of rotation. The vector diagram shown in Figure 13-6 is used to determine the value of voltage when the coil has rotated 120 degrees. Although the coil has rotated 120 degrees, the angle it is making with the lines of force is only 60 degrees. It is this angle that determines the value of the instantaneous voltage. For example, if the coil rotates 210 degrees, it cuts the lines of force at an angle of 30 degrees, Figure 13-7.

Referring back to Figure 13-4, it can be seen that each division of the circle can represent vector OA. Vector AB can be represented by points along the voltage curve. The angle between the horizontal diameter of the circle and the radius V_m is the angle at which the coil is cutting the flux. Although vector diagrams are seldom used alone, they are a simple way of presenting a visual illustration of a problem. Vector diagrams are usually used with trigonometric functions.

Figure 13-6
Vector diagram drawn to scale. The coil has rotated 120° and is cutting the flux at 60°.

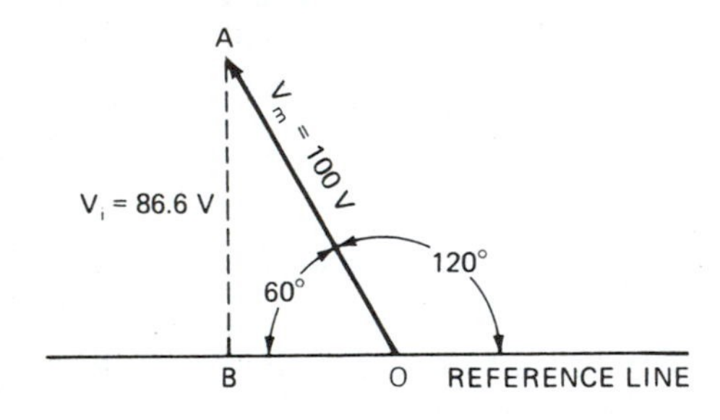

Figure 13-7
Vector diagram drawn to scale. The coil has rotated 210° and is cutting the flux at 30°. The emf has reversed direction, indicated by the – sign.

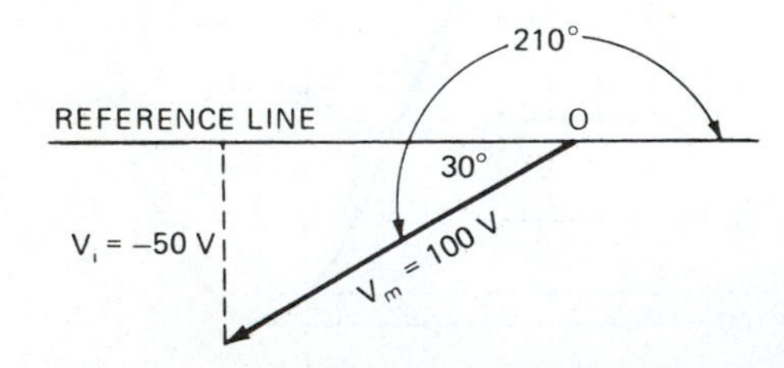

Trigonometry

Many electrical problems are solved through the use of trigonometry. The vector diagrams used with *trigonometric functions* are generally in the form of triangles and/or parallelograms. Table C in the Appendix lists these functions. To understand the table and its use, consider Figure 13-8. Two vectors are drawn to form a vector diagram. Vector AB joins vector AC, forming a 30-degree angle. If three lines are drawn perpendicular to AC, three right triangles are formed. If these triangles are drawn to scale, the ratio of two sides of one triangle is equal to the ratio of the corresponding sides of either of the other two triangles. This is because the corresponding angles of the three triangles are equal. This ratio can be stated mathematically as follows:

$$DG/DA = EH/EA = FI/FA = 0.5$$
$$GA/DA = HA/EA = IA/FA = 0.866$$
$$DG/GA = EH/HA = FI/IA = 0.577$$

These values illustrate that for a given angle A ($\angle A$), the ratio of the two sides of a triangle has the same value regardless of the length of the sides. For any acute angle (an angle less than 90 degrees) of a right triangle, there are six possible ratios. These ratios are called trigonometric functions of the angle.

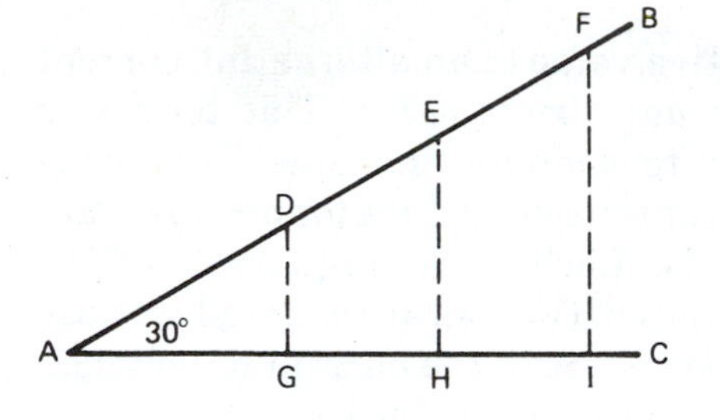

Figure 13-8
Vector diagram forming three 30°-60°-90° triangles

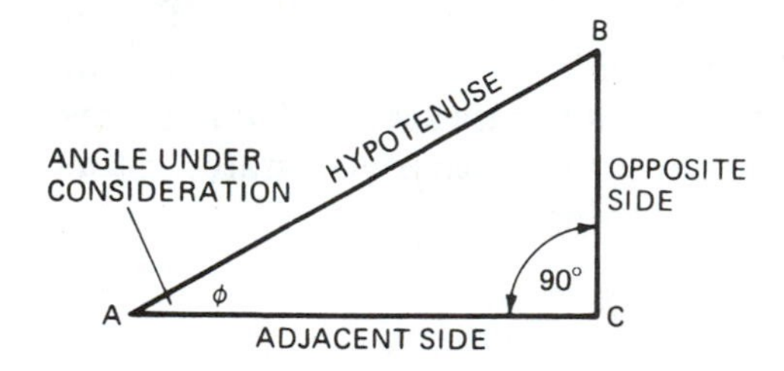

Figure 13-9
Right triangle

In the right triangle ABC in Figure 13-9, AB is the hypotenuse. The *hypotenuse* is always the side opposite to the 90-degree angle (the right angle) and is the longest side of the triangle. BC is opposite angle A (the angle under consideration), and it is called the *opposite side*. AC is touching angle A; therefore, AC is called the *adjacent side*. The ratios are stated as follows:

$$\frac{BC}{AB} = \frac{\text{side opposite } \angle A}{\text{hypotenuse}} = \text{sine of } \angle A\text{, abbreviated sin A}$$

$$\frac{AC}{AB} = \frac{\text{side adjacent } \angle A}{\text{hypotenuse}} = \text{cosine of } \angle A\text{, abbreviated cos A}$$

$$\frac{BC}{AC} = \frac{\text{side opposite } \angle A}{\text{side adjacent } \angle A} = \text{tangent of } \angle A\text{, abbreviated tan A}$$

$$\frac{AC}{BC} = \frac{\text{side adjacent } \angle A}{\text{side opposite } \angle A} = \text{cotangent of } \angle A\text{, abbreviated cot A}$$

$$\frac{AB}{AC} = \frac{\text{hypotenuse}}{\text{side adjacent } \angle A} = \text{secant of } \angle A\text{, abbreviated sec A}$$

$$\frac{AB}{BC} = \frac{\text{hypotenuse}}{\text{side opposite } \angle A} = \text{cosecant of } \angle A\text{, abbreviated csc A}$$

These ratios can be used for either of the acute angles of a right triangle. The ratios are generally written in formula form as follows:

$$\sin \phi = \frac{opp}{h} \qquad \cos \phi = \frac{adj}{h} \qquad \tan \phi = \frac{opp}{adj}$$
$$\cot \phi = \frac{adj}{opp} \qquad \sec \phi = \frac{h}{adj} \qquad \csc \phi = \frac{h}{opp} \qquad \text{(Eq. 13.1)}$$

Because the ratio of the sides of a right triangle is a constant for any given angle, the values for any angle from 0 degrees through 90 degrees can be listed in a table. These values are listed in Appendix C.

ALTERNATING CURRENT AND VOLTAGE VALUES

The voltage and current in ac circuits are always changing in value. This poses the questions of how to measure these values and what is the overall effect of these changing values on the circuit.

Effective Values

The *effective value* of an alternating current is that value that will produce the same heating effect as a specific value of a steady direct current. In other words, an alternating current has an effective value of 1 ampere if it produces heat at the same rate as the heat produced by 1 ampere of direct current, both flowing in the same value of resistance.

Another name for the effective value of an alternating current or voltage is the *root-mean-square* (rms) *value*. This term was derived from one method used to compute the value. The rms is calculated as follows: The instantaneous values for one cycle are selected for equal periods of time. Each value is squared, and the average of the squares is calculated. (Values are squared because the heating effect varies as the square of the current or voltage.) The square root of this answer is the rms value.

A more accurate method requires the use of calculus. Either method shows that the effective or rms value is 0.707 times the maximum value. A simple equation for calculating the effective value is shown here:

$$\text{for voltage, } E = 0.707E_m \qquad \text{(Eq. 13.2)}$$
$$\text{for current, } I = 0.707I_m$$

where subscript m refers to the maximum value

When an alternating current or voltage is specified, it is always the effective value that is meant, unless otherwise stated. Standard ac meters indicate effective values.

Average Values

It is sometimes useful to know the *average value* for one half cycle. If the current changed at the same rate over the entire half cycle, the average value would be one half of the maximum value. However, because current does not change at the same rate, another method must be used. This value can also be found by using calculus. However, it has been determined that the average value is equal to 0.637 times the maximum value. The equations are as follows:

$$\text{for voltage, } E_{av} = 0.637E_m \qquad \text{(Eq. 13.3)}$$
$$\text{for current, } I_{av} = 0.637I_m$$

where subscript av refers to the average value and subscript m refers to the maximum value

Instantaneous Values

The *instantaneous values* of an alternating current or voltage can be determined by drawing a vector diagram to scale and measuring the resultant vector. However, because most alternators produce a voltage that, when plotted on a graph, forms a curve that coincides with the sine table, the following formulas may be used:

$$\text{for voltage, } e = E_m \sin \phi \qquad \text{(Eq. 13.4)}$$
$$\text{for current, } i = I_m \sin \phi$$

where e = instantaneous value of voltage
i = instantaneous value of current
E_m = maximum value of voltage
I_m = maximum value of current
$\sin \phi$ = trigonometric function for the angle at which the flux is being cut

In a plotting of instantaneous values, it should be noted that the distance along the horizontal line represents not only the angle that the coil makes with the lines of force but also the passage of time. Therefore, the values along the horizontal line are called *electrical time degrees.* One cycle represents 360 electrical time degrees.

ADVANTAGES AND DISADVANTAGES OF AC

Alternating current is the primary source of electrical energy today. It is less expensive to generate and transmit than direct current. AC equipment is generally more economical to maintain and requires less space per unit of power than dc equipment.

Alternating current can be generated at higher voltages than dc, with fewer problems of heating and arcing. Some standard values of voltage are 2300, 4600, 6900, 13,800, and 33,100 volts. These values are frequently increased to 100,000, 200,000, and 800,000 volts for transmission over long distances. At the load area, the voltage is decreased to working values of 120, 208, 240, 277, 440, 480, and 550 volts. Figure 13-10 shows a transmission line.

The ease with which the voltage can be raised and lowered makes ac ideal for transmission purposes. Large amounts of power can be transmitted at high voltages and low currents with minimum line loss. Because $P = I^2 R$ (Equation 4.4), the lower the value of transmission current, the less will be the line loss. Because the current is low, smaller transmission wires can be used to reduce the installation and maintenance costs.

Direct current generators, because of their construction, limit their output voltage to 2000 volts or less. The voltage cannot be raised or lowered through the use of transformers. Long-distance transmission requires heavier cables and generally results in greater power loss.

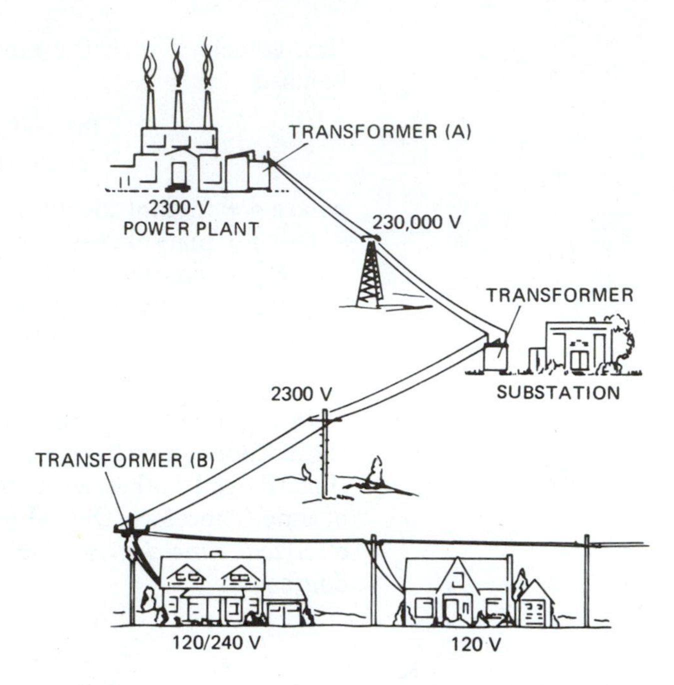

Figure 13-10
Transmission lines from the generating station to residences

Alternating current generators can be driven at high speeds and constructed in large sizes. Because they generally have rotating fields and stationary armatures, their rotor windings are small and light in weight, thus reducing the centrifugal force. Modern alternators are built with capacities of up to 500,000 kilowatts.

Because of the need for commutators, dc generators are limited in capacity. The maximum power available from any one unit is generally 10,000 kilowatts. If dc were the main source of supply, many more generating stations would be needed. Each station would require a source of power to drive the generators. With the supply of fossil fuels dwindling, this would be a gross waste of energy. AC, on the other hand, can be produced in large central stations and distributed over greater distances with maximum efficiency.

An advantage of ac is that it produces a varying magnetic field. This changing field is used in the distribution transformer for raising and lowering the voltage. Lighting units that use transformers produce better and more efficient light. Induction motors utilize the transformer principle for their operation. These motors are less expensive to build, install, and maintain than are dc motors. They also require less space than that required by dc motors of the same horsepower.

DC motors, however, have one distinct advantage over ac motors: they have better speed control. In general, dc motors are used when wide variations of speed and accurate speed adjustments are required. The ac is converted to dc through motor/generator sets or electronic rectifiers.

BASIC TRANSFORMER

Figure 13-11
Elementary transformer

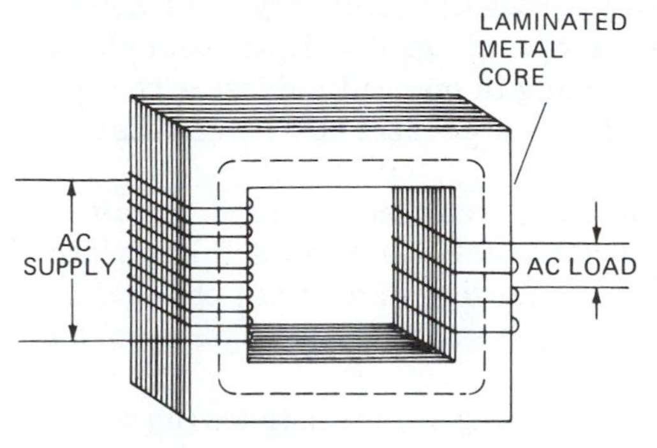

Figure 13-12
Typical power transformer (Courtesy of General Electric Co., Medium Transformer Products Department)

The *transformer* is a device used to increase or decrease alternating voltage. Because transformers have no moving parts, they can perform this operation with very little power loss. An elementary transformer is shown in Figure 13-11. Figure 13-12 shows a picture of a typical power transformer.

Transformers located near the generating station are generally used to increase the voltage. Such transformers are called *step-up transformers*. If they are located near the point of distribution to lower the voltage, they are called *step-down transformers*. (Referring back to Figure 13-10, transformer A is a step-up transformer, and B is a step-down transformer.)

Transformers operate on the principle of electromagnetic induction. Alternating current flowing through the *primary coil* (the coil receiving the power) produces a magnetic field. As the field expands and contracts, it moves across the *secondary coil* (the coil delivering the power). The moving magnetic field induces a voltage into the secondary coil, Figure 13-11.

The ratio of the number of turns of wire on the primary coil to those on the secondary coil determines the voltage output of the transformer. If there are fewer turns on the secondary coil than there are on the primary coil, it is a step-down transformer, Figure 13-13. If the reverse is true, it is a step-up transformer.

If a transformer increases the voltage, the current will decrease in the same proportion. Theoretically, the same amount of power that is supplied to the transformer is delivered to the load. In actuality, there is a slight loss. Most power transformers are between 98 percent and 99.8 percent efficient.

Figure 13-13
Step-down transformer

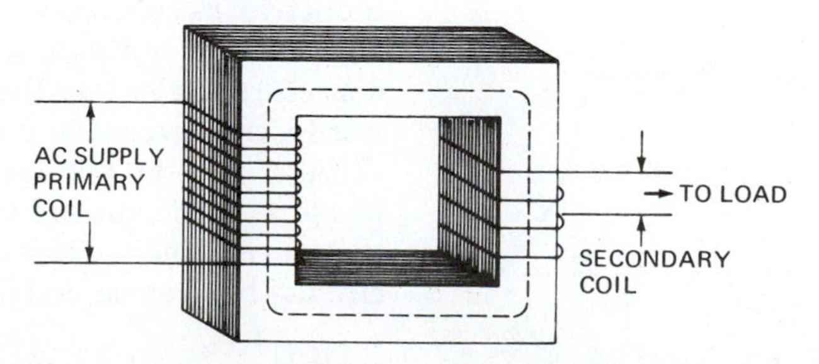

ELECTROMAGNETIC INDUCTION

Electromagnetic induction is the process by which a voltage is produced in a coil as the result of a moving magnetic field's passing across the coil, or the coil's moving through the magnetic field. In a transformer, the ac flowing in the primary produces an alternating magnetic field that moves across the secondary. This action induces an emf in the secondary winding. This is an example of *mutual induction.* Current flowing in one coil induces a voltage into another coil. There is no electrical connection between the two coils.

Another type of induction that takes place in an ac circuit is called *self-induction,* or *self-inductance.* Induction takes place whenever the amount of current flowing in a circuit changes in value. The amount of inductance is generally negligible unless there is a coil in the circuit.

If dc is applied to a circuit containing pure resistance (no coils), the current rises rapidly and then levels off at a steady value, Figure 13-14. If the circuit contains a coil, it takes a longer period of time for the current to reach its maximum value. As the current begins to increase in the coil, the magnetic field begins to expand, moving across the turns of wire. The moving magnetic field induces another voltage into the coil. This voltage of self-induction is in the direction opposite to the applied voltage, and it tends to retard the increasing current. The final result is that it takes the current a longer period of time to reach the maximum value, Figure 13-15.

When the applied voltage is removed from the circuit, the current begins to decrease. The magnetic field contracts, moving across the turns on the coil in the opposite direction. The induced voltage is therefore in a direction that tends to retard the decreasing current, Figure 13-16.

In an ac circuit, the current value is changing continuously. Therefore, the voltage of self-induction is always present. If it were possible to have a pure inductive circuit, then the voltage/current relationship would be as shown in Figure 13-17.

In this circuit, the current reaches its maximum value 90 electrical time degrees after the applied voltage. Thus, in a pure inductive circuit, the current lags the applied voltage by 90 degrees. The voltage of self-induction lags the current by 90 degrees, and also lags the applied voltage by 180 degrees. In other words, the current is 90 degrees out of phase with the applied voltage, and the voltage of self-induction is 180 degrees out of phase with the applied voltage.

A pure inductive circuit cannot be obtained, however, since all circuits have some resistance. Therefore, the phase relationship

Figure 13-14
Graph of dc applied to a pure resistive circuit

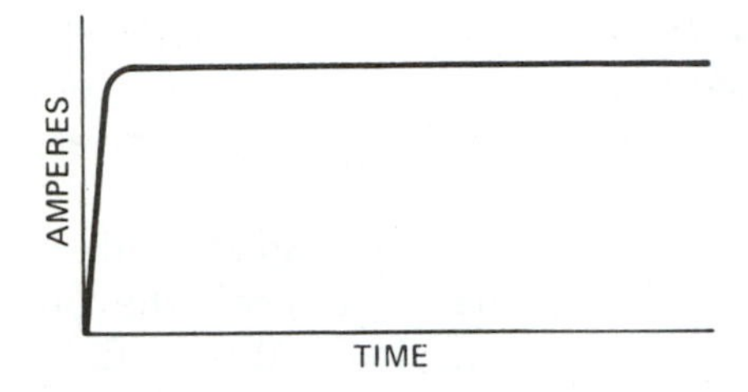

Figure 13-15
Graph of dc applied to an inductive circuit

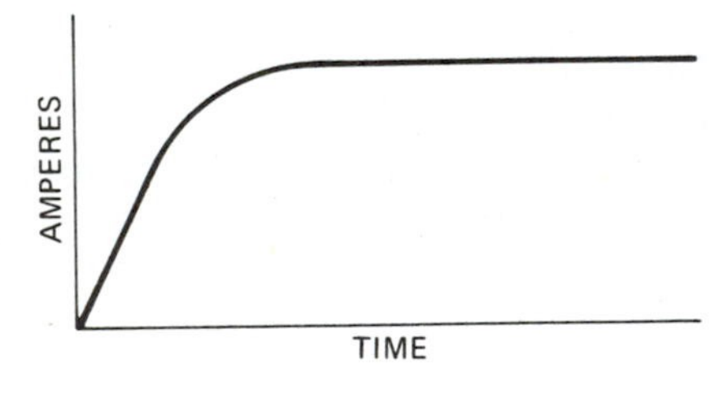

Figure 13-16
Graph of dc when the applied voltage is removed from an inductive circuit

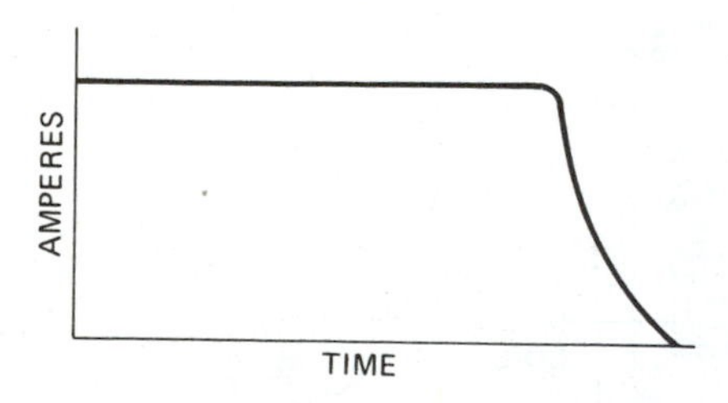

Figure 13-17
Graph of current and voltage in a pure inductive circuit. The current lags the voltage by 90 electrical time degrees. E represents the voltage of self-induction.

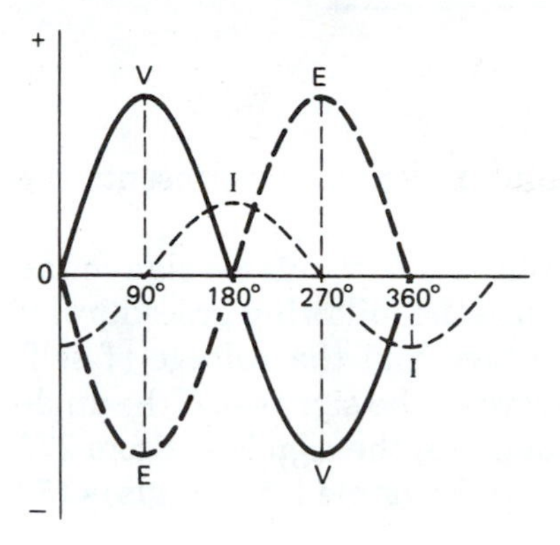

Figure 13-18
Graph of current and voltage in a circuit containing resistance and inductance. The current lags the voltage by 60 electrical time degrees. E represents the voltage of self-induction.

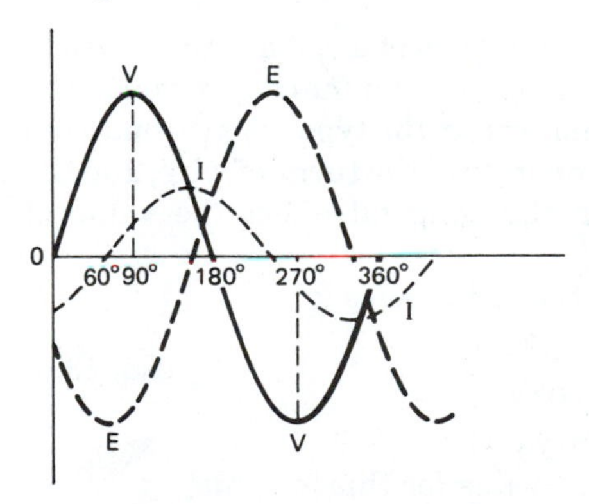

Figure 13-19
Vector diagram of voltage in a circuit containing resistance and inductance

will differ, depending on the value resistance. A more typical circuit is illustrated in Figure 13-18. In this circuit, the current lags the applied voltage by 60 degrees, and the voltage of self-induction lags the current by 90 degrees. The voltage of self-induction lags the applied voltage by 60 degrees + 90 degrees = 150 degrees.

A vector diagram of the two voltages is shown in Figure 13-19. Vector c is the resultant of *a* (applied voltage) and *b* (induced voltage). The resultant voltage can also be determined by the following formula:

$$c = \sqrt{a^2 + b^2 \pm 2ab \cos \phi} \qquad \text{(Eq. 13.5)}$$

where c = resultant value of voltage
a = applied voltage
b = voltage of self-induction
$\cos \phi$ = 0.866 (from the cosine table)

Example 1

If the applied voltage equals 10 V in the circuit in Figure 13-19, what is the resultant voltage if the voltage of self-induction lags the applied voltage by 150 degrees? The voltage of self-induction is equal to the applied voltage (10 V).

$$c = \sqrt{a^2 + b^2 \pm 2ab \cos \phi}$$
$$c = \sqrt{100 + 100 - (2 \times 10 \times 10 \times 0.866)}$$
$$c = \sqrt{200 - 173.2}$$
$$c = \sqrt{26.8}$$
$$c = 5.177 \text{ V}$$

This method of calculation is based on trigonometric functions and the parallelogram method for solving vector diagrams. Figure 13-20 shows how the parallelogram is constructed and the resultant voltage is determined. Lay out vector *a* on the horizontal from point O. Draw vector *b*, beginning at point O, 150 degrees from vector *a*. To form a parallelogram, begin at the end of vector *a*, and draw a dashed line parallel and equal to vector *b*. Beginning at the end of vector *b*, draw a dashed line parallel and equal to vector *a*. The point where the two dashed lines intersect is called point Y.

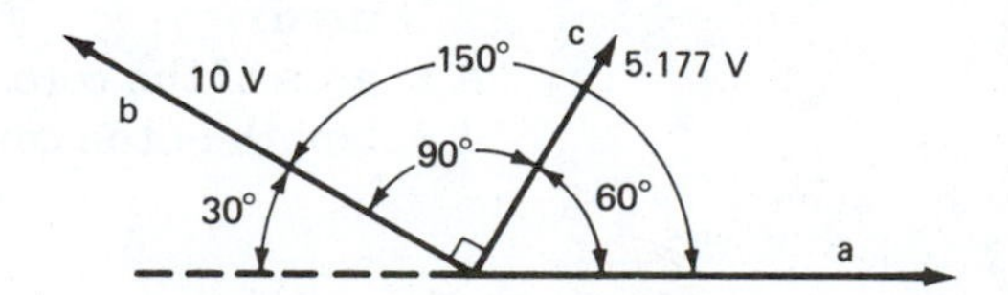

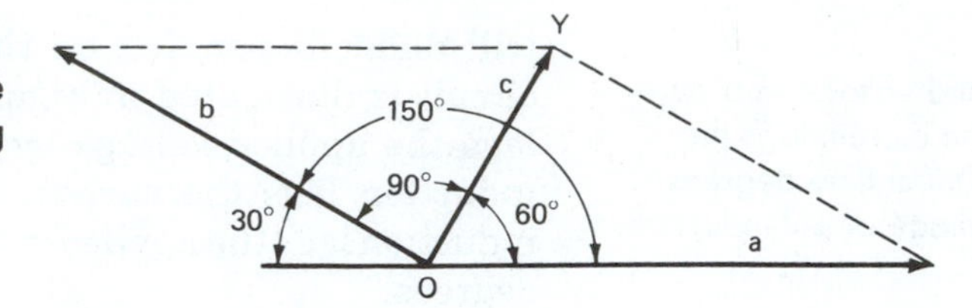

Figure 13-20
Constructing a parallelogram to determine the resultant voltage of a circuit containing resistance and inductance

Draw vector *c* from point O to point Y. Vector *c* represents the resultant voltage.

To determine where to use the plus (+) or minus (–) sign in the formula $c = \sqrt{a^2 + b^2 \pm (2bc \cos \phi)}$, use the following procedure. If the angle between the applied voltage and the voltage of self-induction is from 0 degrees to 90 degrees, the sign is +. If the angle is greater than 90 degrees to 270 degrees, the sign is –. From 270 degrees to 360 degrees, the sign is +. In Example 1, the angle is 150 degrees; therefore, the sign is –.

Inductance

Inductance is a measurement of the induced voltage caused by the changing current. To avoid confusion between the induced voltage and the applied voltage, the unit *henry* (H) has been established. A coil has an inductance of 1 henry if a current changing at the rate of 1 ampere per second produces an induced voltage of 1 volt.

The factors that affect the inductance of a coil are the physical and geometric characteristics of the core and the coil, as well as the frequency of the circuit. In other words, the type, length, and area of the core, the number and proximity of the turns of wire, and the rate at which the current is changing all affect the value of inductance.

A formula for determining the inductance is

$$L = \frac{0.4\pi N^2 \mu A}{10^8 \ell} \qquad \text{(Eq. 13.6)}$$

where L = inductance, in henrys (H)
0.4π = 1.25664 (a constant value for this formula)
N = number of turns of wire on the coil
μ = permeability of the core (determined by the material used)
A = area of the core, in square inches (in.^{2})
ℓ = length of the core, in inches (in.)

Example 2

A coil containing 100 turns of wire is would on a metal core 3 in. long with a surface area of 6.28 in.^{2}. The permeability of the core is 5×10^3. Calculate the inductance of the coil.

$$L = \frac{0.4\pi N^2 \mu A}{10^8 \ell}$$

$$L = \frac{1.25664 \times 100^2 \times 5 \times 10^3 \times 6.28}{10^8 \times 3}$$

$$L = \frac{1.25664 \times 5 \times 6.28}{30}$$

$$L = \frac{39.4585}{30}$$

$$L = 1.315 \text{ H}$$

If the inductance of the coil is known, the voltage of self-inductance can be calculated by using the following formula:

$$E = -L\left(\frac{\Delta I}{\Delta t}\right) \qquad \text{(Eq. 13.7)}$$

where E = voltage of self-inductance
L = inductance, in henrys (H)
ΔI = change in the value of current, in the time Δt
Δt = change, in time

Example 3

Calculate the induced voltage for the coil in Example 2 if the current changes from 5 A to 50 A in 1 s.

$$E = -L\left(\frac{\Delta I}{\Delta t}\right)$$

$$E = -1.3153\left(\frac{45}{1}\right)$$

$$E = -1.3153 \times 45$$

$$E = -59.1885 \text{ V}$$

The negative sign is used to indicate that it is the voltage of self-inductance.

Inductive Time Constant

Inductive coils are often used in dc time delay circuits. The time that it takes a direct current to reach its maximum steady value is determined by the ratio of the inductance to the resistance. One time constant is required for the current to increase from zero to

required for the current to rise from zero to its maximum value. The formula to calculate the inductive time constant is

$$t = \frac{L}{R} \qquad \text{(Eq. 13.8)}$$

where t = inductive time constant, in seconds (s)
L = inductance, in henrys (H)
R = resistance, in ohms (Ω)

Example 4

Calculate the amount of time it will take direct current to reach its maximum value, if the coil in Example 2 has a resistance of 2 Ω.

$$t = \frac{L}{R}$$

$$t = \frac{1.3153}{2}$$

$t = 0.65765$ s = one time constant

$0.65765 \times 5 = 3.28825$ s = the time to reach the maximum value

Many time delay circuits are used in controlling industrial machinery.

Inductive Reactance

Equation 13.5 indicates that the resultant voltage is less than the applied voltage. If equal values of alternating voltage and direct voltage are applied to the same circuit, less current would flow for ac than for dc. With dc, the current is opposed by the induced voltage only when it is rising to its maximum value. Once it has reached a steady value, there is no inductive effect. With alternating voltage, the current is continually changing. Therefore, inductance is present at all times.

This opposition to the flow of ac is called *inductive reactance,* and its unit of measurement is the ohm. The amount of inductive reactance produced by a coil depends upon the frequency of the current and the amount of inductance. The inductive reactance of a circuit or circuit component can be calculated by the following formula:

$$X_L = 2\pi fL \qquad \text{(Eq. 13.9)}$$

where X_L = inductive reactance, in ohms (Ω)
2π = 6.2832 (a constant for this formula)
f = frequency, in hertz (Hz)
L = inductance, in henrys (H)

Figure 13-21
Noninductive coil

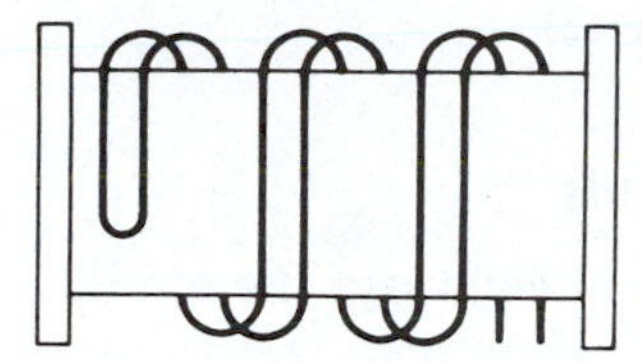

Example 5

Calculate the inductive reactance of the coil in Example 2 if it is connected to a 60-Hz circuit.

$X_L = 2\pi fL$
$X_L = 6.2832 \times 60 \times 1.3153$
$X_L = 496\ \Omega$

Noninductive Coil

It is sometimes necessary to wind a coil so that it will have a high resistance and no inductance. This can be accomplished by winding the coil as shown in Figure 13-21. The current through such a coil is in the opposite direction through adjacent turns. Consequently, the magnetizing action of one half of the turns neutralizes the magnetizing action of the other half. No flux is produced, and the coil is noninductive.

Impedance in Inductive Circuits

Because inductive circuits contain both inductive reactance and resistance, it is apparent that there are two factors that oppose current flow. The combined effect of these two factors determines the amount of current flow. The total opposition to the flow of current in an ac inductive circuit is called *impedance* and is measured in ohms. The Ohm's Law formula for calculating impedance is

$$Z = \frac{E}{I} \qquad \text{(Eq. 13.10)}$$

where Z = total opposition to current flow, in ohms (Ω)
E = applied voltage
I = amount of current flowing in the circuit

Another method for calculating the impedance is used when only the resistance and the inductive reactance are known, Figure 13-22. This method uses vector diagrams and trigonometric functions. Draw vectors at right angles to one another to represent the resistance and the inductive reactance. Scale the resistance vector along the horizontal line and the inductive reactance vector along the vertical line. Connect the ends of the two vectors by a dashed line. The dashed line represents the impedance. Measuring Z will give the value of impedance. A formula for calculating the impedance is

Figure 13-22
Impedance vector diagram

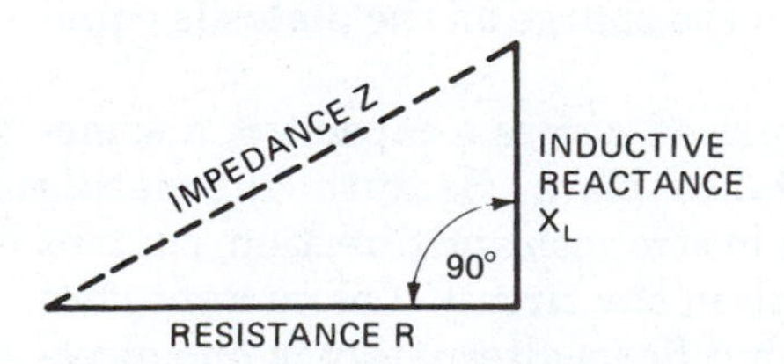

$$Z = \sqrt{R^2 + X_L^2} \qquad \text{(Eq. 13.11)}$$

Equation 13.11 will continue on the following page.

Figure 13-23
Phase relationship of current and voltage in a circuit containing resistance and inductance

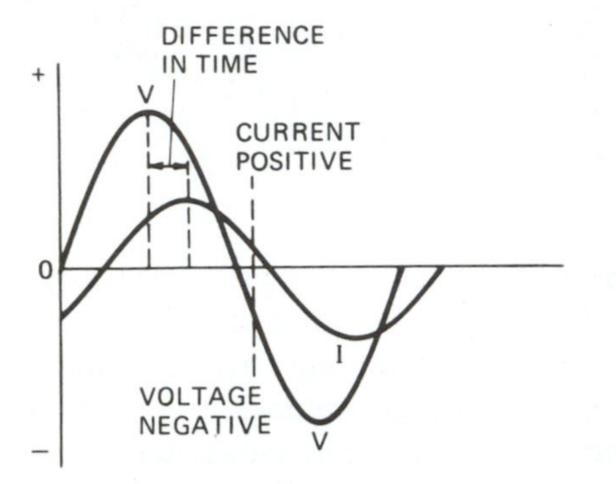

where Z = impedance, in ohms (Ω)
R = resistance, in ohms (Ω)
X_L = inductive reactance, in ohms (Ω)

Phase Relationship in Inductive Circuits

In an ac inductive circuit, the current always lags the applied voltage. That is, the current reaches its maximum and zero values some time after the voltage does. Figure 13-23 illustrates this phenomenon.

Summary of Inductance

Inductance in ac circuits causes the current to lag the voltage. The amount of lag depends upon the ratio of the inductance to the resistance. Inductance also reduces the amount of ac that will flow in the circuit. The greater the value of inductance, the less will be the current. Inductance affects dc only when it is changing in value. The amount of inductance in a coil depends upon the physical and geometric characteristics of the core and the coil, and the frequency of the circuit.

CAPACITANCE

A *capacitor,* also called a *condenser,* consists of two conductors separated by an insulating material. The conductors are called the *plates,* and the insulating material is called the *dielectric.* When a direct voltage is placed across a capacitor, electrons are forced from one plate and caused to accumulate on the other plate. This flow continues until the change built up across the plate is equal to that of the dc source. If the dc is removed, the capacitor remains charged until a conductor is placed across the plates. Placing a conductor across the plates allows the electrons to return to their original plate, and both plates become neutral. Figure 13-24 illustrates a capacitor connected to a battery.

Figure 13-24
Capacitor connected to a battery

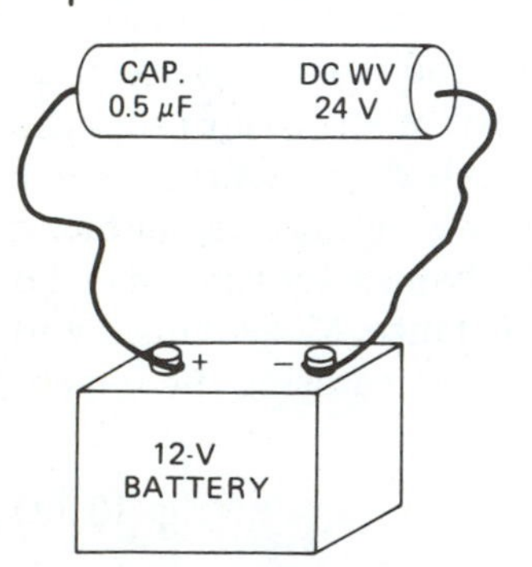

Note that current does not flow through the dielectric. The electrons simply leave one plate and build up on the other plate. The flow of electrons stops when the charge on the plates is equal to the emf of the battery.

If an alternating voltage is placed across a capacitor, a somewhat different phenomenon takes place. Because alternating voltage is continually changing in strength and direction, current continually flows back and forth in the circuit. The current does not flow through the capacitor but flows alternately in one direc-

Figure 13-25A
Device containing a rubber diaphragm

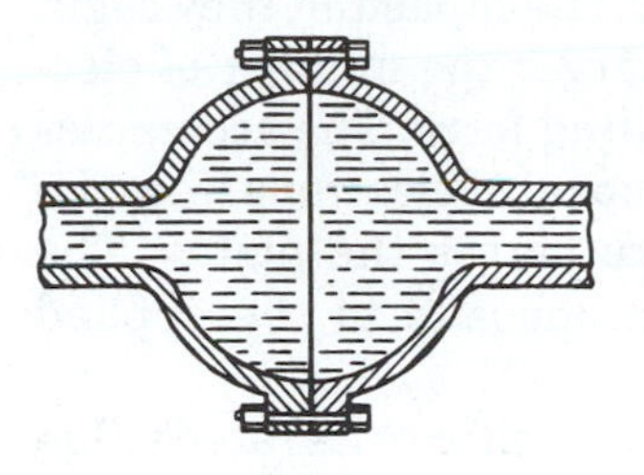

tion and then in the other direction as each plate charges and discharges.

The flow of electrons in a circuit containing a capacitor can be compared to a hydraulic system. The capacitor reaction is similar to a device installed in a closed-loop system that contains a rubber diaphragm stretched across its center. Figure 13-25A illustrates the device, and Figure 13-25B shows the closed-loop system. A centrifugal pump is installed, and the entire system is filled with water.

When the pump is started, there is a flow of water in the pipe while the diaphragm is being stretched. When the pressure of the stretched diaphragm is equal to that of the pump, the flow stops. If the valve is closed while the diaphragm is stretched, the pump may be stopped and no water will flow. Opening the valve allows the water to flow back through the pump until the strain is relieved from the diaphragm.

Figure 13-25B
Closed-loop system containing a rubber diaphragm

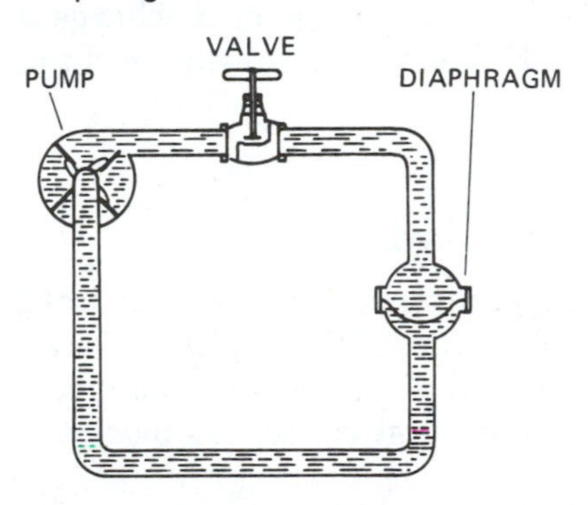

If the valve is left open and the pump is arranged to reverse direction periodically, the water flows first in one direction and then reverses and flows in the opposite direction. Figure 13-26 shows an alternator connected to a capacitor. When the voltage is in one direction, the electrons are pulled off plate A and forced to build up on plate B. When the voltage reverses direction, the electrons are pulled off plate B and forced to build up on plate A. This reversal of electron flow continues as long as the alternator is in operation. Note that the current does not flow through the capacitor.

Figure 13-26
Alternator connected to a capacitor

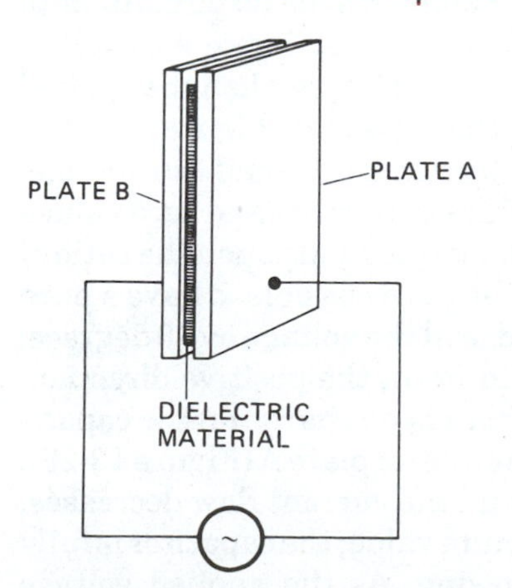

Capacity

The size of the capacitor is determined by the size of the plates, the type of dielectric, and the distance between the plates. A capacitor is said to have a capacity of 1 *farad* (F) when a change of 1 volt per second across its plates produces an average current of 1 ampere. A farad is an extremely large unit of measurement. Therefore, for practical purposes, the *microfarad* (μF) is generally used. One microfarad is equal to one millionth of a farad.

It can also be said that a capacitor has a capacity of 1 farad when 1 volt dc applied to its plates causes it to charge to 1 coulomb. The formula for calculating capacitance is

$$C = \frac{Q}{E} \qquad \text{(Eq. 13.12)}$$

where C = capacitance, in farads (F)
Q = charge on one plate, in coulombs (C)
E = voltage applied to the plates

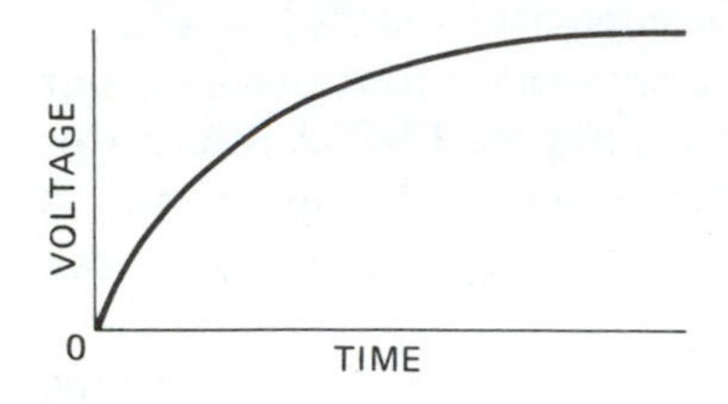

Figure 13-27A
Graph of voltage building up in a dc circuit containing capacitance

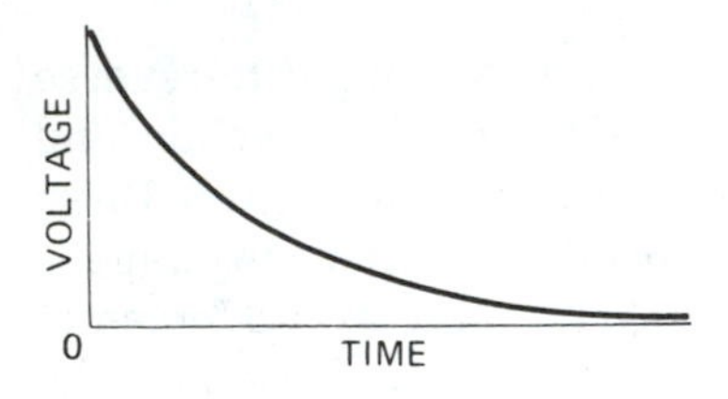

Figure 13-27B
Graph of voltage discharge in a dc circuit containing capacitance

Capacitive Time Constant

As the electrons build up on one plate of the capacitor, they begin to repel any additional electrons. The larger the number of electrons on the plate, the greater the repelling force. The accumulation of electrons on one plate and the removal of electrons from the other plate cause a potential difference across the plates. The potential difference is in the direction opposite to the applied voltage.

The length of time it takes to charge a capacitor to its rated value is a factor of the resistance of the plate circuit and the capacity of the capacitor. The formula for calculating the actual charging rate is

$$t = RC \qquad \text{(Eq. 13.13)}$$

where t = one time constant, in microseconds (μs)
R = resistance of the plate circuit, in ohms (Ω)
C = capacitance, in microfarads (μF)

Five time constants are required to fully charge or discharge a capacitor. Figures 13-27A and 13-27B show typical curves of the voltage buildup and discharge.

Effect of Capacitance on Current Flow

In a dc circuit, a capacitor allows the current to flow only until the charge on the capacitor is equal to that of the applied voltage, Figure 13-28. Once the capacitor has met this requirement, current flow ceases. It can be said, then, that capacitors block the flow of dc.

When both ac and dc are applied to a circuit, but certain parts of the circuit are restricted to ac only, capacitors can be used to block out the dc. Another use for capacitors in dc circuits is to smooth out the ripple. If a pulsating dc is applied to a circuit, a capacitor connected across the circuit discharges when the applied voltage drops below the charge on the capacitor, Figure 13-29.

Figure 13-28
The charge on the capacitor equals the charge on the battery. These forces are equal and opposite, and no current flows.

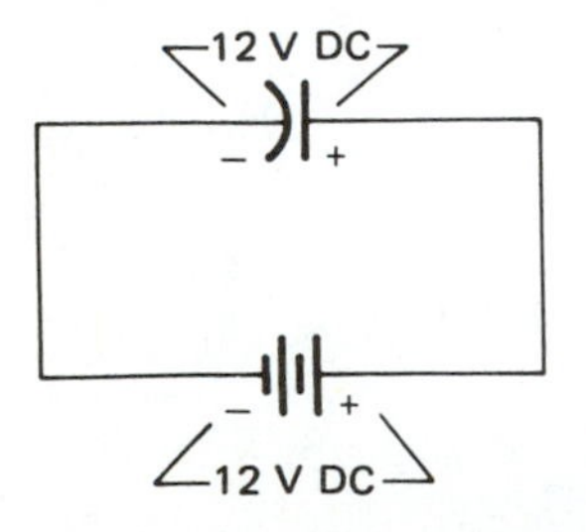

In ac circuits, capacitors cause the current to lead the voltage. In other words, the current reaches its maximum and zero values ahead of the voltage. The amount of lead depends upon the ratio of the resistance to the capacitance. If it were possible to have a pure capacitive circuit, the current world lead the voltage by 90 degrees.

When the voltage begins to build up in the positive direction, maximum current flows because there is no charge on the capacitor. As the electrons begin to accumulate on plate A (Figure 13-26), they oppose any further buildup, and the current flow decreases. When the voltage reaches its maximum value, the capacitor is fully charged and the current stops flowing. As the applied voltage decreases, the charge on the capacitor is greater than the applied voltage, and the current begins to flow from plate A to plate B. Therefore, while the applied voltage is still in the positive direction but decreasing, the current is flowing in the negative direction, Figure 13-30.

When the applied voltage decreases to zero, the charge on the capacitor causes the current to increase to the maximum value. At this point, the applied voltage reverses direction and tries to keep the current flowing in the negative direction (point *c*, Figure 13-30). However, the electrons have now built up on plate B and oppose any further buildup, causing the current flow to decrease (point *d*, Figure 13-30). When the applied voltage reaches the maximum value in the negative direction, the current has dropped to zero. This cycle of current and voltage changes continues as long as ac is applied to the circuit.

Figure 13-29
Graph of a pulsating voltage with a capacitor connected to smooth out the ripple

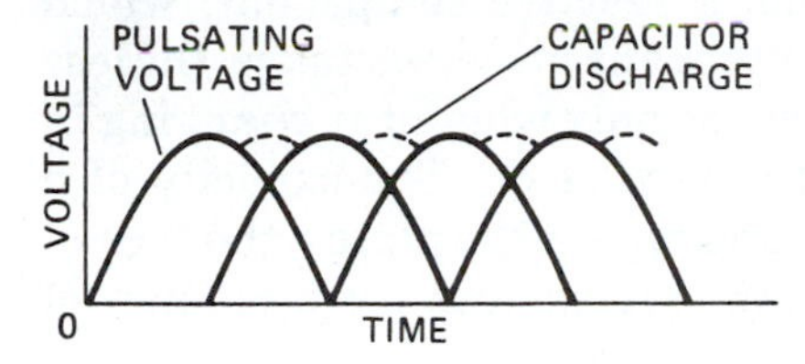

Capacitive Reactance

The opposition to the flow of an alternating current, offered by a capacitor, is called *capacitive reactance*. Its unit of measurement is the ohm, because it opposes current flow. The amount of opposition to current flow produced by a capacitor is determined by its capacitance and the frequency of the circuit. The formula for capacitive reactance is

$$X_C = \frac{10^6}{2\pi fC} \qquad \text{(Eq. 13.14)}$$

where X_C = capacitive reactance, in ohms (Ω)
2π = 6.2832 (a constant for this formula)
f = frequency of the applied voltage, in hertz (Hz)
C = capacitance, in microfarads (μF)

Figure 13-30
Graph showing current/voltage relationship in a pure capacitive circuit

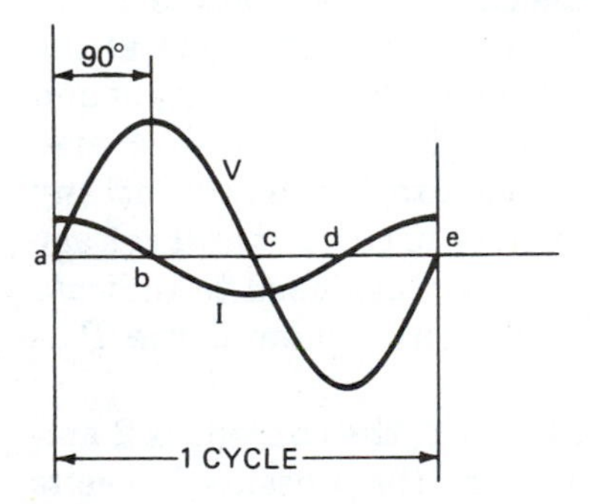

Impedance in Capacitive Circuits

Because capacitive circuits contain both capacitive reactance and resistance, it is apparent that there are two factors that oppose current flow. The combined effect of these factors determines the amount of current flow. Figure 13-31 shows capacitive reactance and resistance combined vectorially. The resultant of the two factors is the impedance of the circuit.

The formula for vector diagrams that form right triangles may be used to solve for impedance. The formula is

$$Z = \sqrt{R^2 + X_C^2} \qquad \text{(Eq. 13.15)}$$

where Z = impedance, in ohms (Ω)
R = resistance, in ohms (Ω)
X_C = capacitive reactance, in ohms (Ω)

Figure 13-31
Vector diagram of resistance, capacitive reactance, and impedance

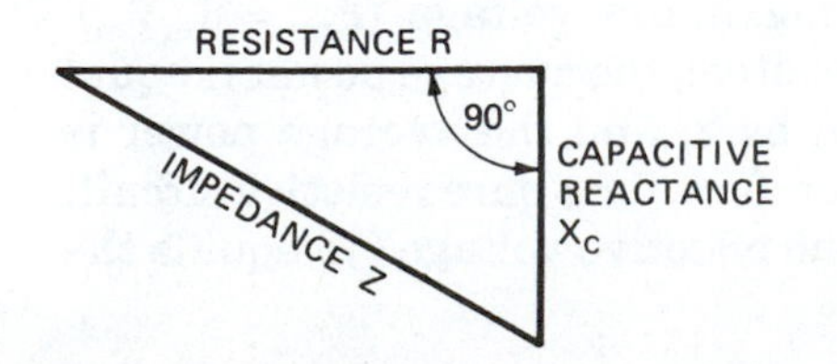

Summary of Capacitance

Capacitance in ac circuits causes the current to lead the voltage. The amount of lead depends upon the ratio of the capacitance to the resistance. Capacitance also has a reactive component, which opposes current flow. The greater the capacitive reactance, the less current flows. Capacitance affects dc only when it is changing in value. A capacitor blocks out the flow of dc. The capacity of a capacitor depends upon the size and type of the plates, the type of dielectric, the distance between the plates, and the frequency of the circuit.

POWER IN AC CIRCUITS

The power expended by a dc circuit is equal to the product of the current flowing in the circuit and the voltage impressed across the circuit. In an ac circuit, both the current and voltage are changing in value. Therefore, it can be said that the power at any instant is equal to the product of the current and voltage at that instant. Figure 13-32 shows a current and voltage wave for a pure resistive circuit. In a pure resistive circuit, the current and voltage are in phase. *In phase* means that the current and voltage start at the same instant, reach their maximum values at the same instant and in the same direction, and then drop back to zero and reverse direction at the same instant. To calculate the power at any instant, multiply the instantaneous values of current and voltage together (Equation 4.2). (Lowercase letters are used to indicate instantaneous values.) With these products a new curve P is plotted. Curve P is the power curve.

Assume that at instant *a* in Figure 13-33, the current is 2 amperes (indicated by *ab*). At the same instant, the pressure is 3 volts (indicated by *ac*). The power at that instant is found by multiplying the instantaneous current by the instantaneous voltage (Equation 4.2), or $p = iv$. Therefore, 2 amperes × 3 volts = 6 watts. Other points are found in a similar manner, and the curve is plotted. The power curve is positive during both alternations (half cycles), because when multiplying like signs together, the product is always positive. Notice that the power curve forms a sine wave having twice the frequency of the current or voltage.

Figure 13-32
Current and voltage curves in a pure resistive circuit

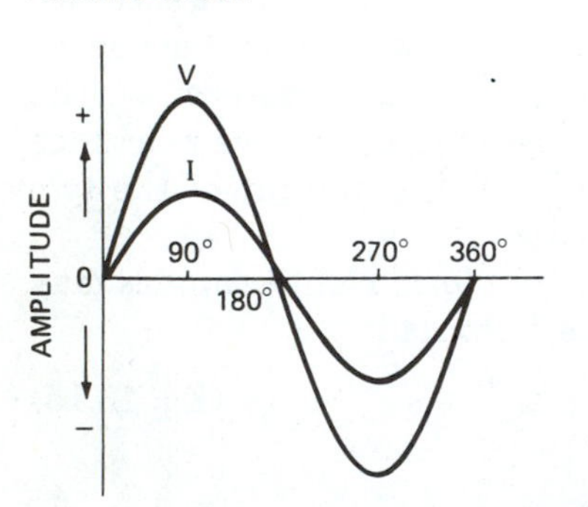

The maximum power in a pure resistive circuit is equal to the maximum current times the maximum voltage ($P_m = I_m V_m$). Because the power curve is all positive, the average power is equal to the maximum power divided by 2, and the average power is equal to the effective power. Therefore, for a pure resistive circuit, the effective current (I) times the effective voltage (V) equals the effective power (P), or $P = IV$.

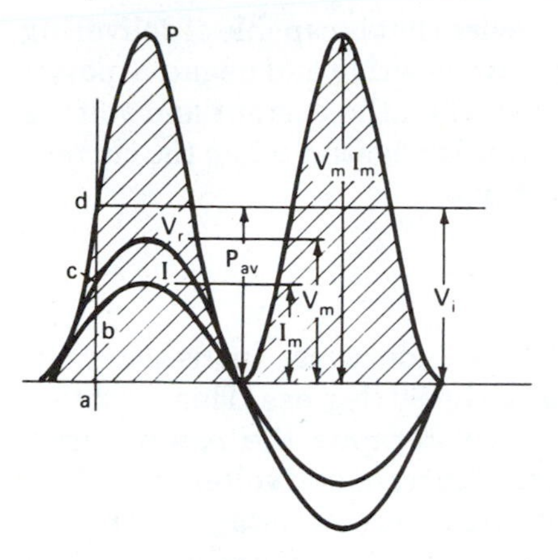

Figure 13-33
Graph of current, voltage, and power in a pure resistive circuit

Positive and Negative Power

In Figure 13-34A, a battery is supplying power to a resistance (R). The emf of the battery, which is 6 volts, forces a current through the circuit from the negative terminal, through the resistance, and back to the positive terminal. In this circuit, both the voltage and the current are in the same direction. If signed numbers are applied to the values of current and voltage, it can be said that both have positive values and that the power delivered to the resistance is also positive.

Figure 13-34B shows a 6-volt battery with a generator connected across its terminals. The positive terminal of the generator is connected to the positive terminal of the battery. The negative terminal of the generator is connected to the negative terminal of the battery. If the emf of the generator is equal to that of the battery, no current will flow.

If the generator emf is lower than that of the battery, current will flow from the negative terminal of the battery to the negative terminal of the generator. In this case, as in the resistive circuit, the current is in the same direction as the battery voltage, and the power is positive. The battery is delivering power to the generator and tends to drive it like a motor. Relative to the current, voltage, and power of the battery, the circuit conditions are the same as in Figure 13-34A.

If the voltage of the generator is adjusted to exceed that of the battery, the current flow will be reversed. The flow will now be from the negative terminal of the generator to the negative terminal of the battery, Figure 13-34C. The emf of the battery remains in the same direction, but the current is reversed. If the battery emf is assigned a positive value, then the current must be negative. The battery is being charged and is receiving power from the circuit. Because the battery is considered to be an energy source, under these conditions it is delivering negative power to

Figure 13-34A
Battery supplying power to a resistance. The arrows indicate the direction of current flow.

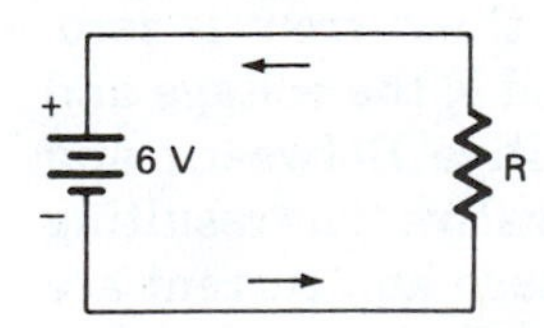

Figure 13-34B
Generator connected across a 6-V battery. The arrows indicate the direction of current flow.

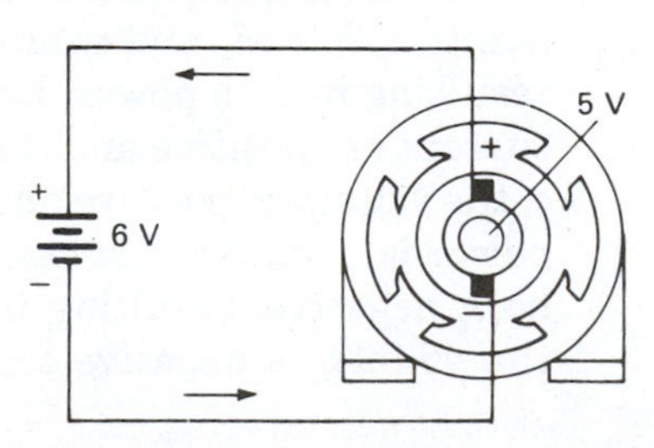

Figure 13-34C
Generator charging a battery. The current flow is in the direction opposite to the battery voltage.

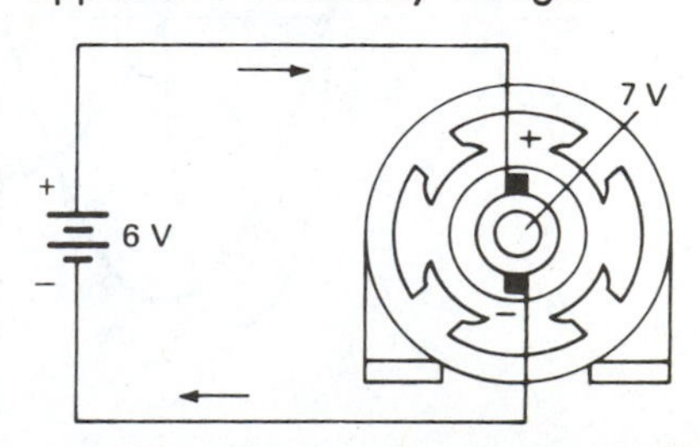

Figure 13-35
Graph of voltage/current relationships in a pure inductive circuit. The current lags the voltage by 90 electrical time degrees.

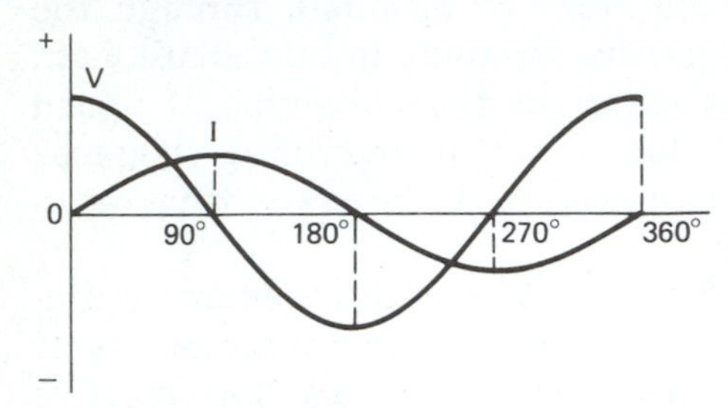

the circuit. Viewing it another way, the current and voltage of the battery are in opposite directions. Therefore, the power is negative.

This example shows that any device that is capable of delivering power and receiving power may have positive and negative power components. The power is positive when the current and voltage are in the same direction. The power is negative when the current and voltage are in opposite directions.

Power in an Inductive Circuit

In a theoretical ac circuit of pure inductance (no resistance), the current lags the applied voltage by 90 degrees. This condition is illustrated in Figure 13-35. To determine the power curve, multiply the instantaneous values of current and voltage together. In Figure 13-36, when either the current or the voltage is zero, the power at those instances is zero (points *a, b, c, d,* and *e*). Between points *a* and *b,* the voltage is positive and the current is negative; thus, the power is negative. This means that the magnetic energy stored in the coil is delivering power to the source, just as the generator is doing in Figure 13-34C. Between points *b* and *c,* the current and voltage are both positive, because they are acting in the same direction. Thus the power during this period of time is positive. The source is supplying power to the inductor. Between *c* and *d,* the voltage is negative and the current is positive. As a result, the power is negative again. Between *d* and *e,* the voltage and current are both negative. Because they are both in the same direction, the power is positive. For one cycle of current, the positive area of the power curve equals the negative area. Therefore, the power received by the inductor from the source is equal to the power returned to the source. The average power input to the inductor is zero, and a wattmeter connected in the circuit indicates zero.

Figure 13-36
Current, voltage, and power curves for a pure inductive circuit

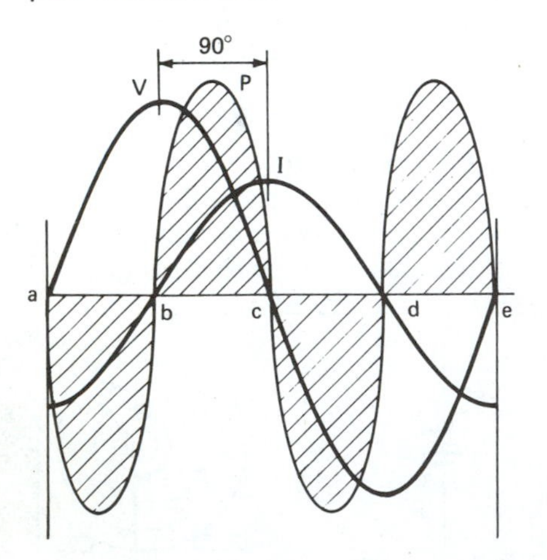

Power in a Capacitive Circuit

Figure 13-37 shows a theoretical circuit of pure capacitance in which the current leads the applied voltage by 90 degrees. At points *a, b, c, d,* and *e,* either the voltage or the current is zero, resulting in zero power. Between points *a* and *b,* the voltage and current are positive and the power is also positive. Between *b* and *c,* the voltage is positive and the current is negative. The resulting power is negative. Between *c* and *d,* the voltage and current are both negative, resulting in a positive power. Between *d* and *e,* the voltage is negative and the current is positive; therefore, a

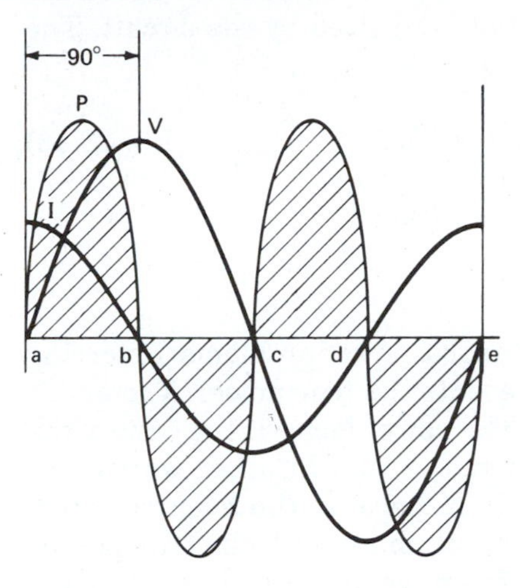

Figure 13-37
Current, voltage, and power curves for a pure capacitive circuit

negative power exists. For one cycle of current, the positive area of power is equal to the negative area of power, and the result is zero power.

Power in Circuits Containing Resistance and Inductance

In ac circuits containing pure resistance, the current and voltage are in phase, and all the power delivered to the circuit by the source is utilized by the circuit.

In an ac circuit of pure inductance, the current lags the voltage by 90 degrees, and all the power delivered to the circuit is returned to the source. The utilized power is zero.

Because all circuits contain some resistance, a circuit containing a coil is a resistive/inductive circuit (RL circuit). In an RL circuit, the current lags the voltage by a certain number of electrical time degrees. The number of degrees is between zero and 90. The angle of lag depends upon the ratio of the resistance (R) to the inductance (L).

In the RL circuit in Figure 13-38, it can be seen that most of the power is positive and is utilized by the circuit. The small amount of negative power is returned to the source by the magnetic energy from the coil. The negative power does not register on a wattmeter. The wattmeter indicates the *true power,* the power utilized by the circuit.

When an ammeter and a voltmeter are connected in this circuit, they indicate the effective values of current and voltage. The product of the effective values of current and voltage in an RL circuit is equal to the apparent power of the circuit. The apparent power is measured in voltamperes and is always equal to or greater than the true power. In an RL circuit, the apparent power is always greater than the true power.

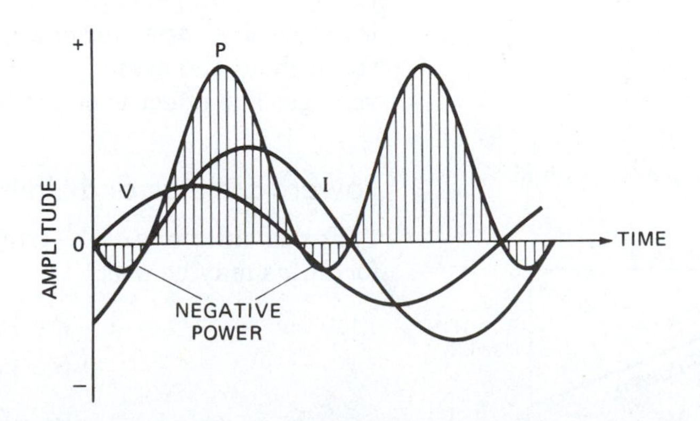

Figure 13-38
Graph of power in an RL ac circuit

The ratio of the true power to the apparent power is called the *power factor* of the circuit. The power factor is usually stated as a decimal value or as a percentage. It is sometimes referred to as the percentage of the total power that is utilized by the circuit. The formula for power factor is

$$Pf = \frac{P}{P_{app}} \times 100 \qquad \text{(Eq. 13.16)}$$

where Pf = power factor, in percent (%)
P = true power
P_{app} = apparent power

Vector diagrams can be used to solve power problems. Referring to Figure 13-39, vector AC represents the true power. Vector AB represents the apparent power. The power factor is the ratio of the true power AC to the apparent power AB. In other words, the power factor is the cosine of angle A. Because the reactive power (negative power) is 90 degrees out of phase with the true power, angle A represents the number of degrees that the current lags the voltage. Vector BC represents the average power that is returned to the source by the magnetic energy stored in the coil.

The more inductive a circuit becomes, the greater is the angle between the true power and the apparent power. As angle A increases, the power factor decreases. This means that the reactive power becomes greater, and the power utilized by the load becomes smaller.

A low power factor is undesirable. With a low power factor, a large amount of power is circulating through the circuit, but only a small amount is transferred to the work load. A power factor lower than 80 percent is considered below industrial standards; steps should be taken to remedy this situation. Many utility companies increase a customer's rates if the power factor drops below a specified value.

Most industrial loads are highly inductive and cause the current to lag the voltage. One way to overcome this lagging power factor is to install a capacitor or a group of capacitors at strategic points throughout the system. Capacitors cause the current to lead the voltage. The effect is opposite to that of an inductor.

Figure 13-39
Vector diagram of power in an RL ac circuit

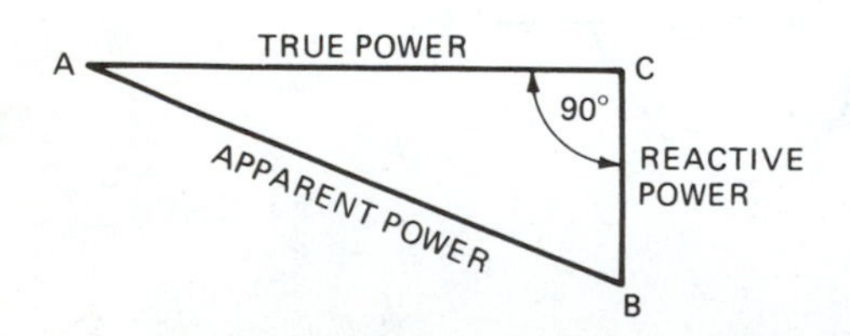

Power Formulas and Units

For calculations of the true power of an ac circuit, the following formulas may be used:

$$P = IE \cos \phi \qquad \text{(Eq. 13.17)}$$

$$P = I^2R \qquad \text{(Eq. 13.18)}$$

$$P = \sqrt{P_{app}^2 - P_{XL}^2} \qquad \text{(Eq. 13.19)}$$

The units of measurement for power in ac systems are as follows:

- True power (P) is measured in watts (W).
- Apparent power (P_{app}) is measured in volt-amperes (VA).
- Reactive power (P_{XL}) is measured in volt-amperes reactive (VAR).

THREE-PHASE SYSTEMS

For most purposes, alternating current is generated and transmitted in the form of three-phase. Three-phase generators produce three separate voltages that are 120 electrical time degrees apart. Figure 13-40 shows the voltage curves for a three-phase system.

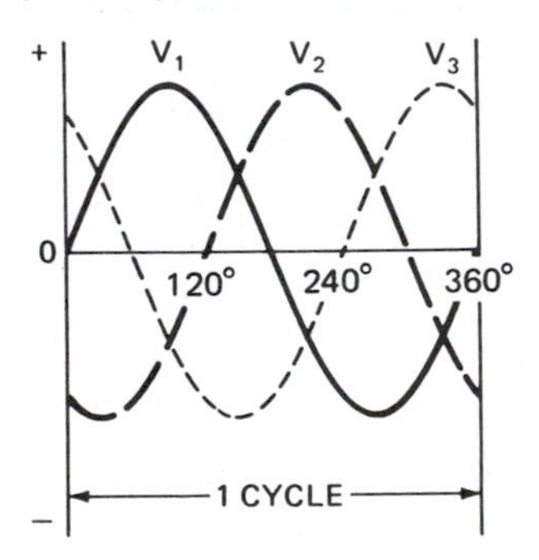

Figure 13-40
Voltage curve for one cycle of a three-phase system

Three-phase systems provide smoother power than single-phase systems. Also, most three-phase equipment requires less space than single-phase equipment of the same rating. Three-phase equipment is more efficient and less expensive. Transmission of three-phase power requires less conductor material than single-phase power, and single-phase circuits can be tapped from three-phase systems.

Figure 13-41 illustrates the coil connections for a three-phase generator. This arrangement is called a *delta connection*. Another frequently used arrangement is the four-wire wye connection, Figure 13-42.

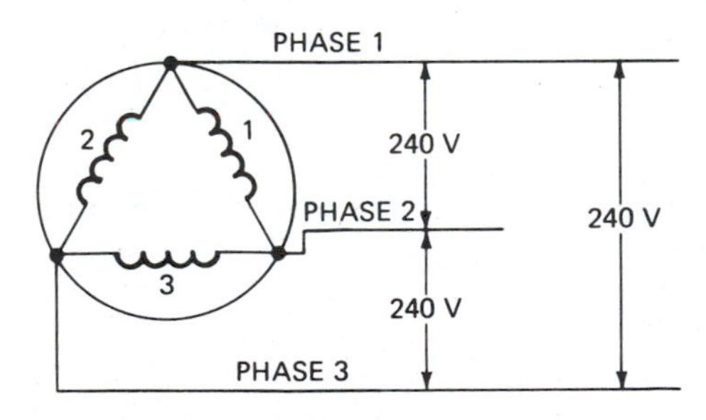

Figure 13-41
Connections for a three-phase generator (delta connection)

With the three-phase delta system, the phase voltage (E_p) is the voltage between any two of the phase conductors. Single-phase equipment can be supplied by using any two of the three conductors. For three-phase equipment, all three conductors are used. The delta system, therefore, supplies the same value of voltage for single-phase and three-phase circuits.

With the three-phase wye system, the phase voltage is the voltage between any one of the phase conductors and the neutral conductor. The voltage between any two phase conductors is called the *line voltage* (E_L). Two values of voltage are available for single-

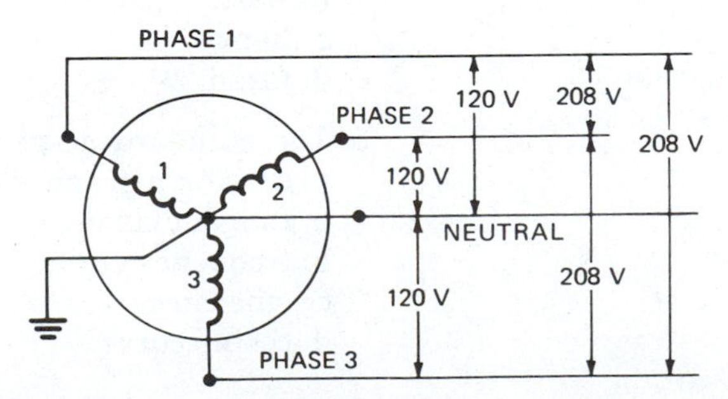

Figure 13-42
Connections for a three-phase generator (wye connection)

conductors. For a wye connection, the line voltage is always 1.73 times as much as the phase voltage. All three-phase conductors are used to supply three-phase circuits.

When a three-phase system is supplying single-phase circuits, the single-phase loads should be balanced between all three phases. On a perfectly balanced system, the current through all three-phase conductors is equal. It is seldom possible to maintain a perfect balance; however, severely unbalanced systems can cause conductors and equipment to overheat and overcurrent devices to operate.

REVIEW

A. Multiple choice.
Select the best answer.

1. An alternating voltage is an emf that is
   a. continually changing in value.
   b. alternately reversing direction.
   c. both a and b.
   d. neither a nor b.
2. The number of cycles of an alternating voltage completed in one second is called the
   a. frequency.
   b. alternation.
   c. hertz.
   d. fluctuation.
3. The most common frequency used in the United States is
   a. 40 cycles per second.
   b. 50 cycles per second.
   c. 60 cycles per second.
   d. 25 cycles per second.
4. The modern unit of measurement for frequency, which means cycles per second, is the
   a. hertz.
   b. watt.
   c. henry.
   d. farad.
5. The standard ac generator produces a voltage that, when plotted on a graph, produces a
   a. tangent curve.
   b. frequency curve.
   c. sine curve.
   d. cosine curve.

REVIEW *(continued)*

A. Multiple choice.
Select the best answer (continued).

6. A vector is
   a. an arc.
   b. a line segment that has a definite length and direction.
   c. curved lines.
   d. a triangle.

7. A vector diagram is
   a. a rectangular diagram.
   b. two or more vectors joined together to convey information.
   c. a triangle.
   d. a circle.

8. The hypotenuse of a right triangle is the
   a. longest side.
   b. shortest side.
   c. sum of all the sides.
   d. side opposite the acute angle.

9. The side of a right triangle that is opposite to the angle under consideration is called the
   a. tangent side.
   b. adjacent side.
   c. opposite side.
   d. hypotenuse.

10. The side of a right triangle that is opposite to the 90-degree angle is called the
    a. tangent side.
    b. adjacent side.
    c. opposite side.
    d. hypotenuse.

11. The effective value of an alternating current is the
    a. value at a particular instant.
    b. value that produces the same heating effect as a specific value of a steady direct current.
    c. average value of one cycle.
    d. maximum value produced.

12. The rms value of an alternating current is the same as the
    a. instantaneous value.
    b. average value.
    c. effective value.
    d. maximum value.

REVIEW *(continued)*

A. Multiple choice.
Select the best answer (continued).

13. Standard ac ammeters and voltmeters indicate
    a. instantaneous values.
    b. average values.
    c. effective values.
    d. maximum values.

14. The maximum value of an alternating voltage is
    a. equal to the effective value.
    b. greater than the effective value.
    c. smaller than the effective value.
    d. none of the above.

15. The primary source of electrical energy today is in the form of
    a. varying current.
    b. alternating current.
    c. pulsating direct current.
    d. steady direct current.

16. AC generators generally produce
    a. higher voltages than dc generators.
    b. lower voltages than dc generators.
    c. voltages equal to those produced by dc generators.
    d. different values of voltage.

17. AC is
    a. less expensive to transmit than dc.
    b. more expensive to transmit than dc.
    c. about the same cost to transmit as dc.
    d. less expensive to transmit than dc but more expensive to maintain.

18. Large alternators have
    a. rotating fields and stationary armatures.
    b. rotating armatures and stationary fields.
    c. rotating fields and rotating armatures.
    d. stationary fields and stationary armatures.

19. Loads requiring variable-speed drives and accurate-speed adjustment generally use
    a. series motors.
    b. ac motors.
    c. dc motors.
    d. universal motors.

REVIEW *(continued)*

A. Multiple choice.
Select the best answer (continued).

20. A transformer is a device that
    a. changes ac to dc.
    b. raises or lower voltage on ac systems.
    c. increase or decreases the power on ac systems.
    d. changes dc to ac.

21. The primary winding of a transformer
    a. is connected to the load.
    b. is connected to the source.
    c. produces the power for the system.
    d. delivers power to the load.

22. In an inductive circuit, the current
    a. lags the voltage.
    b. leads the voltage.
    c. is in phase with the voltage.
    d. is always greater than the voltage.

23. In a capacitive circuit, the current
    a. lags the voltage.
    b. leads the voltage.
    c. is in phase with the voltage.
    d. is always greater than the voltage.

24. The unit of measurement for impedance is the
    a. voltampere.
    b. henry.
    c. farad.
    d. ohm.

25. Power factor is the ratio of the
    a. true power to the reactive power.
    b. true power to the apparent power.
    c. reactive power to the apparent power.
    d. inactive power to reactive power.

26. For most purposes, alternating current is generated and transmitted in the form of
    a. single-phase.
    b. two-phase.
    c. three-phase.
    d. four-phase.

REVIEW *(continued)*

A. Multiple choice.
Select the best answer
(continued).

27. Three-phase systems are generally
    a. more efficient than single-phase systems.
    b. less efficient than single-phase systems.
    c. just as efficient as single-phase systems.
    d. less efficient than dc.

28. Two types of connections for three-phase generators are
    a. delta and wye.
    b. delta and star.
    c. delta and theta.
    d. wye and theta.

29. The three-phase delta system supplies
    a. one value of voltage.
    b. two values of voltage.
    c. three values of voltage.
    d. four values of voltage.

30. On a three-phase system, the voltages are
    a. 60 degrees apart.
    b. 90 degrees apart.
    c. 120 degrees apart.
    d. 180 degrees apart.

B. Give complete answers.

1. Define *alternating current.*
2. Define the term *frequency.*
3. List four factors that determine the amount of voltage induced into a coil that is rotating in a uniform magnetic field.
4. Describe one method used to plot an alternating voltage curve if the maximum value of voltage is known.
5. Define a *vector.*
6. With either acute angle of a right triangle, there are six possible ratios. List these six ratios in formula form.
7. What is meant by the "effective value" of an alternating current?
8. What do the letters *rms* stand for?
9. Write the formula for calculating the effective value of an alternating voltage when the maximum value is known.

REVIEW *(continued)*

B. Give complete answers (continued).

10. Write the formula for calculating the instantaneous value of an alternating current when the maximum value is known.
11. In a plot of a current curve, what unit of measurement is indicated along the horizontal line?
12. List at least three advantages of ac compared to dc.
13. List one advantage of dc compared to ac.
14. What is a transformer?
15. Define the term *step-up* with regard to transformers.
16. Define the term *secondary* with regard to transformers.
17. Describe the basic operating principle of a transformer.
18. What determines the voltage output of a transformer?
19. Define the term *mutual induction.*
20. What is the difference between mutual induction and self-induction?
21. Define *inductance.*
22. What is the unit of measurement of inductance?
23. List three factors that affect the amount of inductance of a coil.
24. What determines the length of time that it takes for a steady direct current to reach a steady value in an inductive circuit?
25. Explain why a specific inductive circuit allows more current to flow when 120 volts dc are applied than when 120 volts ac are applied.
26. Define *inductive reactance.* What is its unit of measurement?
27. How can a coil be wound so that it does not produce an inductive effect?
28. Define *impedance.* What is its unit of measurement?
29. What is the phase relationship between the current and the voltage in an ac inductive circuit?
30. What is a capacitor?
31. Define the term *dielectric.*
32. Describe the effect of connecting a capacitor across a steady dc circuit.

REVIEW *(continued)*

B. Give complete answers (continued).

33. Describe the effect of connecting a capacitor across an ac circuit.
34. Define *capacitance.* What is its unit of measurement?
35. List two factors that determine the length of time necessary to charge a capacitor to its rated value.
36. What is the phase relationship between the current and the voltage in an ac circuit containing capacitance?
37. Under what conditions is it possible to have the applied voltage in one direction while the current is flowing in the opposite direction?
38. Define *capacitive reactance.* What is its unit of measurement?
39. List three types of power present in ac circuits, and state their units of measurement.
40. Define the term *negative power.*
41. What power is indicated on an ac wattmeter?
42. If the current and the voltage of an ac circuit are measured and their values are multiplied together, the result indicates what type of power?
43. Define *power factor.*
44. Draw a power triangle, and label each side.
45. Why is a low power factor undesirable?
46. Are most industrial loads inductive or capacitive?
47. Name one method frequently used to improve the power factor of an industrial load.
48. Why is it sometimes said that the power factor is the cosine of the angle between the current and the voltage?
49. Give one practical use for a capacitor in a dc circuit.
50. List three formulas that may be used to calculate the true power of an ac circuit.
51. What form of alternating current is generally used for the transmission of electrical power?
52. List four advantages of three-phase power compared to single-phase power.

REVIEW *(continued)*

B. Give complete answers (continued).

53. Draw a diagram of a three-phase delta connection.
54. Draw a diagram of a three-phase, four-wire connection.
55. In a three-phase delta system, what is meant by *phase voltage?* What is the symbol for phase voltage?
56. What is the difference between the phase voltage and the line voltage in a three-phase, four-wire wye connection?
57. How many values of voltage are available from a three-phase delta system?
58. How many values of voltage are available from a three-phase, four-wire wye system?
59. What is the symbol for the line voltage in a three-phase system?
60. Why is it necessary to balance a three-phase system?

C. Solve each problem, showing the method used to arrive at the solution.

1. Locate the angle that coincides with the trigonometric functions listed below. (Use Appendix C.)
   a. cosine = 0.8660
   b. sine = 0.5000
   c. tangent = 0.0699
   d. cotangent = 57.29
   e. secant = 1.566
   f. cosecant = 1.701
2. The maximum value of an alternating voltage is 170 V. Calculate the effective value.
3. What is the average value of the voltage in Problem 2?
4. Calculate the instantaneous values of voltage in Problem 2 at the following angles:
   a. 30°
   b. 60°
   c. 90°
   d. 120°
5. A voltage has an effective value of 277 V. What is the maximum value?
6. An ac inductive circuit has an applied voltage of 240 V and the current lags the applied voltage by 45°. Draw a vector diagram showing the applied voltage, the voltage of self-induction, and the resultant voltage. Calculate the resultant voltage.
7. A coil containing 500 turns is wound on a metal core 5 in. long with a cross-sectional area of 10 in.^{2}. The permeability of the core is 6×10^3. Calculate the inductance of the coil.

REVIEW *(continued)*

C. Solve each problem, showing the method used to arrive at the solution (continued).

8. Calculate the induced voltage for the coil in Problem 7 if the current changes from 2A to 80 A in 1 s.
9. Calculate the amount of time it will take for the current in Problem 8 to change from 10 A to 40 A.
10. Calculate the amount of time required for a direct current to reach its maximum steady value if the coil in Problem 7 has a resistance of 5 Ω.
11. What is the inductive reactance of a 20-H coil connected across a 60-Hz supply?
12. What is the inductive reactance of the coil in Problem 11 if it is connected across a 50-Hz supply?
13. If the coil in Problem 11 has a resistance of 10 Ω, what is its impedance?
14. Determine the amount of current that will flow through the coil in Problem 13 if it is connected to a 120-V, 60-Hz supply.
15. Calculate the capacitance of a capacitor that will charge to 0.05 C when 600 V is applied to its terminals.
16. How long will it take to charge a capacitor if the resistance of its plates is 0.05 Ω and it has a capacitance of 100 μF?
17. Calculate the resistance of the plates of a capacitor that can be fully charged to 50 μF in 5 μs.
18. A 100-μF capacitor is connected to a 480-V, 60-Hz supply. What is its capacitive reactance?
19. A capacitor has a resistance of 0.5 Ω and a capacitive reactance of 10 Ω. Calculate the impedance.
20. An ammeter in a circuit indicates 20 A, and the voltmeter indicates 100 V. If a wattmeter indicates 1500 W, what is the reactive power of the circuit?
21. What is the power factor of the circuit in Problem 20?
22. What is the phase angle between the current and the voltage in Problem 20?
23. What is the impedance of the circuit in Problem 20?
24. If the resistance of the circuit in Problem 20 is 0.15 Ω, what is the reactance?
25. What is the power factor of the circuit in Problem 20 if the ammeter indicates 15 A instead of 20 A?

14
AC CIRCUITS

Objectives

After studying this chapter, the student will be able to:

- Discuss the characteristics of various types of alternating current circuits.
- Describe the effects of inductance and capacitance on alternating current circuits.
- Describe the effects of high and low power factors on alternating current circuits.

PURE RESISTIVE CIRCUITS

The current and voltage are in phase in pure resistive circuits. Figure 14-1 illustrates this phase relationship. The capacitive and/or inductive effect is negligible and the characteristics are much the same as for dc. Resistive heating units and incandescent lighting are considered to be pure resistive loads.

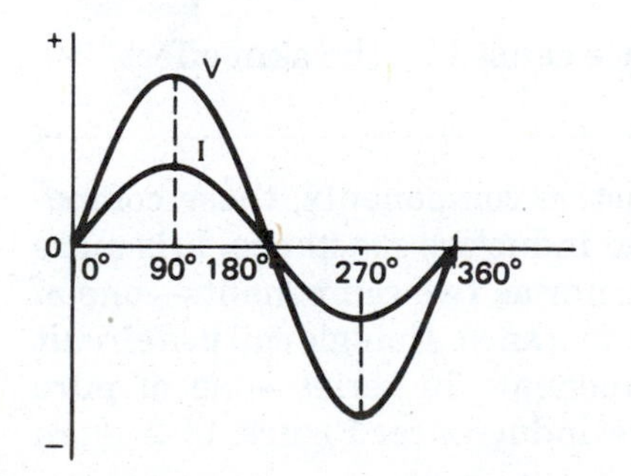

Figure 14-1
Graph of current and voltage in a pure resistive ac circuit

Effective Resistance

In general, pure resistive circuits react much the same for ac or dc. There are, however, some differences that must be considered. These differences vary with the frequency and are generally negligible at low frequencies.

The following five factors affect the amount of current flowing in a pure resistive circuit:

- DC resistance
- Skin effect
- Eddy currents
- Hysteresis effect
- Dielectric stress

The *dc resistance* is the resistance measured with a very accurate ohmmeter. It is the total resistive effect to pure dc.

The fact that alternating current changes in value and direction tends to make it flow along the outer surface of the conductor. This phenomenon, known as *skin effect,* reduces the inner conductive effect of the conducting material and increases the circuit resistance.

Alternating current produces a magnetic flux that changes polarity with each reversal or current flow. The change in polarity causes the molecules in the metal parts near the circuit to be in motion, thus producing heat. The heat either radiates back into the circuit conductors or retards the dissipation of heat produced by current flowing in the conductors. This *hysteresis effect* increases the effective resistance of the circuit.

Eddy currents are caused by voltages induced into the conductors and other surrounding metal parts. They vary with and are directly proportional to the frequency of the supply. Heat produced by these currents tends to increase the effective resistance of the circuit.

As the alternating voltage varies in strength, the stress on the conductor insulation increases and decreases. This variation in *dielectric stress* also produces heat, which increases the circuit resistance.

The effective resistance of an ac circuit is equal to the dc resistance plus the effects of eddy currents, hysteresis, dielectric stress, and skin effect. The effective resistance may be written as follows:

$$R = R_o + R_i + R_m + R_d + R_s \qquad \text{(Eq. 14.1)}$$

where R = effective resistance
R_o = pure dc resistance
R_i = increase in resistance caused by eddy currents
R_m = increase in resistance caused by the hysteresis effect
R_d = increase in resistance caused by dielectric stress variations
R_s = increase in resistance caused by the skin effect

AC SERIES CIRCUITS

When a circuit contains inductive components, these components contain both resistance and inductive reactance. It is more convenient to consider each inductor as two components—one of pure resistance and one of pure inductance. A single coil in a circuit would then appear as two components in series—one of pure resistance and the other of pure inductance. Figure 14-2 is an illustration of this practice.

Figure 14-2
AC circuit containing and inductor

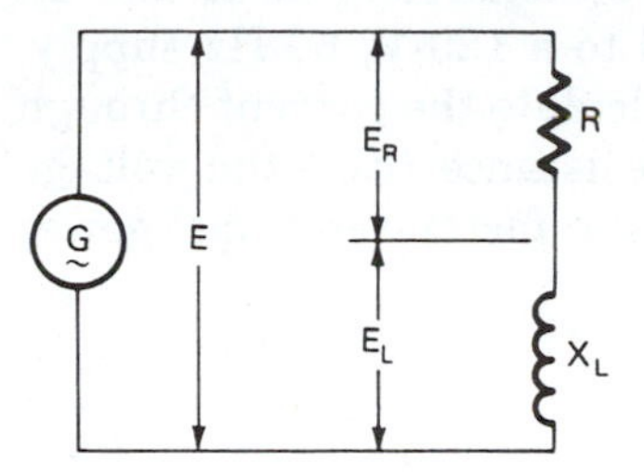

The rules for current, voltage, and resistance in series circuits hold true for ac circuits. However, there are slight variations. The rules for ac series circuits are as follows:

- The instantaneous value of current is the same through all parts of a series circuit.
- The applied voltage is equal to the vectorial (sometimes called phasor) sum of the individual voltages around the circuit.
- The combined resistance of an ac series circuit is equal to the sum of the individual resistances of the circuit.
- The combined reactance of an ac series circuit is equal to the vectorial sum of the individual reactances in the circuit.
- The combined impedance of an ac series circuit is equal to the vectorial sum of the individual impedances of the circuit.

Because the current at any instant has the same value in all parts of a series circuit, the current can be used as a phase reference. Actually, any circuit quantity can be used as a reference, but it is generally more convenient to use the current because it is common to all components of the circuit.

Series RL Circuits

The phase relationship of the applied voltage and the voltage drops for the circuit in Figure 14-2 can be illustrated in a vector diagram, as shown in Figure 14-3.

The phase relationship of the voltages across the parts of the circuit can be expressed with reference to the current vector. Vector E_R is equal to IR and represents the voltage across the resistance R. Because in a pure resistance the current and voltage are in phase, vectors E_R and I_R are laid out along the horizontal. Vector E_L is equal to IX_L and represents the voltage across the inductance L. Because the current lags the voltage by 90 degrees in a pure inductive circuit, vector E_L is laid out 90 degrees out of phase with E_R. Note that E_R is in phase with the current. Vector E_L is drawn vertically upward because vector rotation is always considered to be counterclockwise.

The vector sum of E_R and E_L is equal to the applied voltage E and is calculated by the following formula:

$$E = \sqrt{E_R{}^2 + E_L{}^2} \qquad \text{(Eq. 14.2)}$$

where E = applied voltage
E_R = voltage across the resistance
E_L = voltage across the inductance

Figure 14-3
Vector diagram illustrating the phase relationships of the voltages for the circuit in Figure 14-2

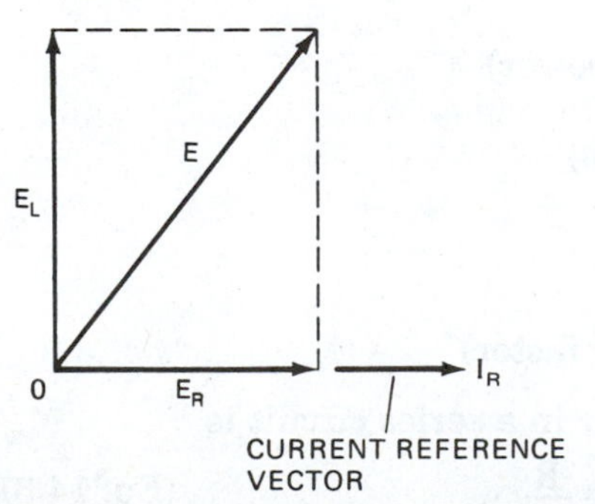

Example 1

The inductor just described has a resistance of 10 Ω and an inductance of 0.04 H. It is connected to a 120-V, 60-Hz supply. What is the impedance of the coil? Calculate the current through the circuit, the voltage across the resistance (E_R), the voltage across the inductance (E_L), the power factor, and the power dissipated.

(a) $X_L = 2\pi fL$ *(Equation 13.9)*
$X_L = 2 \times 3.1416 \times 60 \times 0.04$
$X_L = 15\ \Omega$ (inductive reactance of the coil)

(b) $Z = \sqrt{(R)^2 + (X_L)^2}$ *(Equation 13.11)*
$Z = \sqrt{(10)^2 + (15)^2}$
$Z = \sqrt{325}$
$Z = 18\ \Omega$ (coil impedance)

(c) $I = \frac{E}{Z}$ *(Equation 13.10)*
$I = \frac{120}{18}$
$I = 6.667$ A (circuit current)

(d) $E_R = IR$ *(From Ohm's Law)*
$E_R = 6.667 \times 10$
$E_R = 66.67$ V (voltage across resistance R)

(e) $E_L = IX_L$ *(From Ohm's Law)*
$E_L = 6.667 \times 15$
$E_L = 100$ V (voltage across inductance L)

(f) $P = I^2 R$ *(Equation 13.18)*
$P = (6.667)^2 \times 10$
$P = 44.5 \times 10$
$P = 445$ W (effective power or true power)

(g) $P_{app} = IE$ *(From Equation 4.2)*
$P_{app} = 6.667 \times 120$
$P_{app} = 800$ VA (apparent power)

(h) $Pf = \frac{P}{P_{app}}$ *(Equation 13.16)*
$Pf = \frac{445}{800}$
$Pf = 0.556 = 55.6\%$ (power factor)

Another formula for power factor in a series circuit is

$$Pf = \frac{R}{Z} \qquad \text{(Eq. 14.3)}$$

where Pf = power factor
R = resistance
Z = impedance

The same answer would result if this equation were used in place of Equation 13.16, used in calculation (h) of Example 1.

Example 2

A resistance of 10 Ω is connected in series with a coil of negligible resistance with 0.05-H inductance. If the current through the coil is 10 A, determine the voltage across each component and the supply voltage from a 60-Hz source.

(a) $X_L = 2\pi fL$ *(Equation 13.9)*
$X_L = 2 \times 3.1416 \times 60 \times 0.05$
$X_L = 18.85\ \Omega$ (inductive reactance of coil)

(b) $E_L = IX_L$ *(From Ohm's Law)*
$E_L = 10 \times 18.85$
$E_L = 188.5$ (voltage across coil)

(c) $E_R = IR$ *(From Ohm's Law)*
$E_R = 10 \times 10$
$E_R = 100$ V (voltage across the resistance)

(d) $E = \sqrt{E_R{}^2 + E_L{}^2}$ *(Equation 14.2)*
$E = \sqrt{(100)^2 + (188.5)^2}$
$E = \sqrt{45{,}532}$
$E = 213$ V (applied voltage)

Example 3

Calculate the following for the circuit in Example 2: (a) the impedance, (b) the true power, (c) the apparent power, (d) the reactive power, and (e) the power factor.

(a) $Z = \dfrac{E}{I}$ *(Equation 13.10)*
$Z = \dfrac{213}{10}$
$Z = 21.3\ \Omega$ (impedance)
or
$Z = \sqrt{R^2 + X_L{}^2}$ *(Equation 13.11)*
$Z = \sqrt{(10)^2 + (18.85)^2}$
$Z = 21.3\ \Omega$

(b) $P = I^2 R$ *(Equation 13.18)*
$P = (10)^2 \times 10$
$P = 1000$ W = 1 kW (effective or true power)

(c) $P_{app} = IE$ *(From Equation 4.2)*
$P_{app} = 10 \times 213$
$P_{app} = 2130$ VA = 2.13 kVA (apparent power)

(d) $P_L = IV_L$ *(From Equation 4.2)*
$P_L = 10 \times 188.5$
$P_L = 1885$ VAR (reactive power)

or

$P_L = \sqrt{P_{app}{}^2 - P_{XL}{}^2}$ *(From Equation 13.19)*
$P_L = \sqrt{(2130)^2 - (1000)^2}$
$P_L = 1881$ VAR

The difference of 4 VAR is due to rounding off of numbers and is negligible.

(e) $Pf = \frac{P}{P_{app}}$ *(Equation 13.16)*

$Pf = \frac{1000}{2130}$

$Pf = 0.469 = 46.9\%$ (power factor lagging)

or

$Pf = \frac{R}{Z}$ *(Equation 14.3)*

$Pf = \frac{10}{21.3}$

$Pf = 0.469 = 46.9\%$

When more than one resistance and/or inductance are connected in series, calculate the values for each component and then apply the rules for series circuits.

Series RC Circuits

Figure 14-4 shows a resistance connected in series with a capacitance. The combination is connected to an ac supply. Because this is a series circuit, all the rules for series circuits apply. The voltage across the resistor (E_R) is in phase with the current. The current at any instant is the same value through all the components. Therefore, the vector representing E_R is laid out along the current vector on the horizontal, Figure 14-5. The current leads the voltage in a capacitive circuit. Therefore, vector E_C is laid out to indicate the current leading the voltage by 90 degrees. E_C is drawn vertically downward from E_R. This also indicates that E_C leads E_R by 90 degrees. The applied voltage is the vector sum of these two values:

$$E = \sqrt{(E_R)^2 + (E_C)^2} \quad \text{(Eq. 14.4)}$$

Figure 14-4
Resistance and capacitance connected in series across an ac supply (RC circuit)

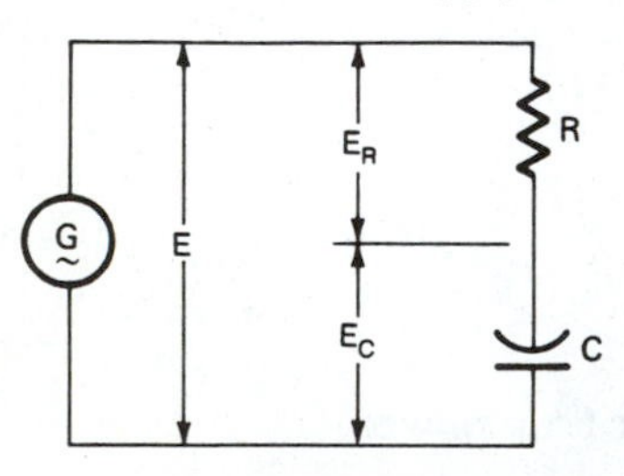

Figure 14-5
Vector diagram illustrating the phase relationships for the voltages in an RC series circuit.
$E = \sqrt{(E_R)^2 + (E_C)^2}$

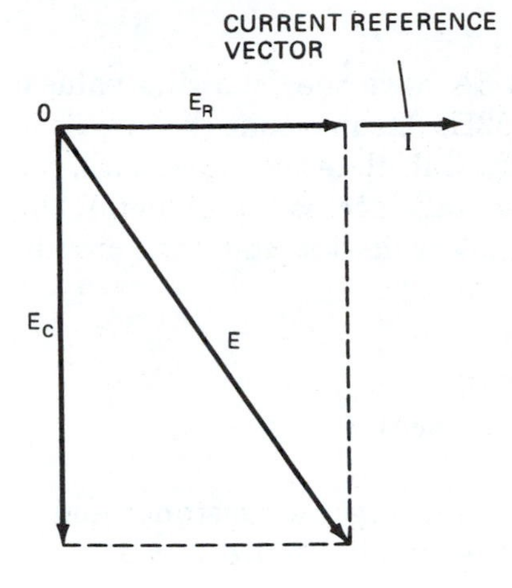

Example 4

The capacitor in Figure 14-4 has a capacity of 900 μF. The resistance R has a value of 5 Ω. Assume the circuit is connected to a 120-V, 60-Hz supply. Calculate the impedance, the current through the circuit, the voltage across the resistance, the voltage across the capacitance, the power dissipated by the circuit, and the power factor.

(a) $X_C = \frac{10^6}{2\pi fC}$ *(Equation 13.14)*

$X_C = \frac{10^6}{2 \times 3.1416 \times 60 \times 900}$

$X_C = \frac{10^6}{339{,}292.8}$

$X_C = 2.947\ \Omega$ (capacitive reactance)

(b) $Z = \sqrt{R^2 + X_C{}^2}$ *(Equation 13.15)*

$Z = \sqrt{(5)^2 + (2.947)^2}$

$Z = \sqrt{25 + 8.6848}$

$Z = \sqrt{33.6848}$

$Z = 5.8\ \Omega$ (impedance)

(c) $I = \frac{E}{Z}$ *(Equation 13.10)*

$I = \frac{120}{5.8}$

$I = 20.7$ A (current)

(d) $E_R = IR$ *(From Ohm's Law)*

$E_R = 20.7 \times 5$

$E_R = 103.5$ V (voltage across the resistance)

(e) $E_C = IX_C$ *(From Ohm's Law)*

$E_C = 20.7 \times 2.947$

$E_C = 61$ V (voltage across the capacitance)

(f) $P = I^2 R$ *(Equation 13.18)*

$P = (20.7)^2 \times 5$

$P = 428.49 \times 5$

$P = 2142$ W (true power)

(g) $P_{app} = IE$ *(From Equation 4.2)*

$P_{app} = 20.7 \times 120$

$P_{app} = 2484$ VA (apparent power)

(h) $Pf = \frac{P}{P_{app}}$ *(Equation 13.16)*

$Pf = \frac{2142}{2484}$

$Pf = 0.862 = 86.2\%$ (power factor)

Figure 14-6A
Resistance, capacitance, and inductance connected in series across an ac supply (RLC circuit)

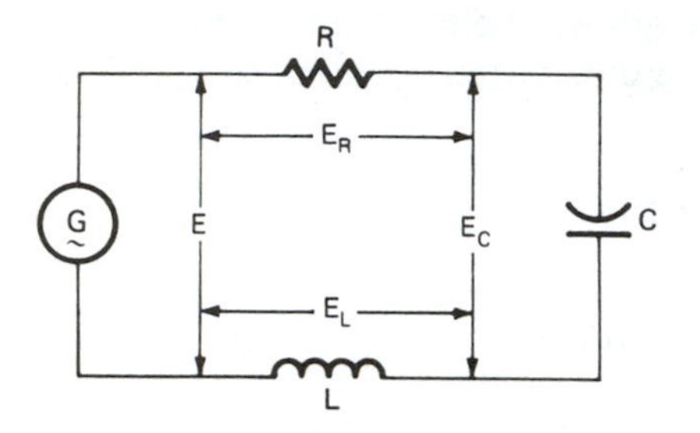

Figure 14-6B
Vector diagram of resistance and inductive reactance

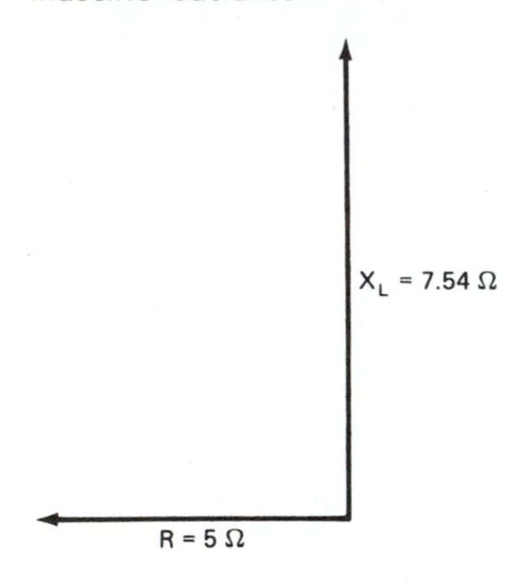

Figure 14-6C
Vector diagram of resistance and capacitive reactance

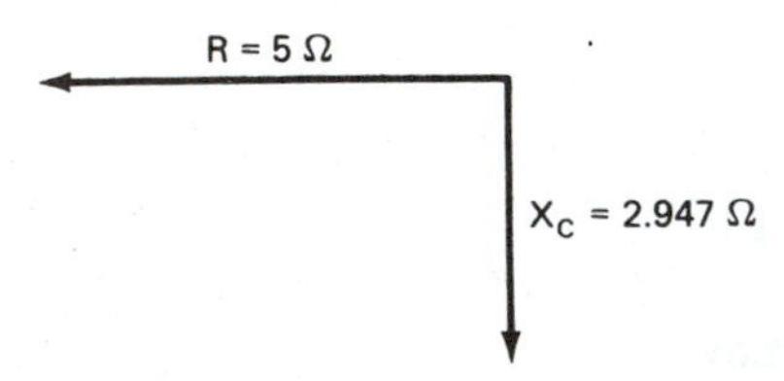

Series RLC Circuits

Figure 14-6A shows a resistance, an inductance, and a capacitance connected in series across an ac supply. By applying the rules for series circuits and vector analysis, we can determine all values.

Example 5

The components in Figure 14-6A have the following values: R = 5 Ω, C = 900 μF, and L = 0.02H. Assume that the circuit is connected to a 120-V, 60-Hz supply. Calculate the impedance, the current through the circuit, the voltage across the inductor, the voltage across the capacitor, the power dissipated by the circuit, and the power factor.

(a) $X_L = 2\pi fL$ *(Equation 13.9)*
$X_L = 2 \times 3.1416 \times 60 \times 0.02$
$X_L = 7.54\ \Omega$ (inductive reactance)

(b) Draw a vector diagram representing the resistance and the inductive reactance of the circuit, Figure 14-6B.

(c) $X_C = \frac{10^6}{2\pi fC}$ *(Equation 13.14)*
$X_C = \frac{10^6}{2 \times 3.1416 \times 60 \times 900}$
$X_C = \frac{10^6}{339{,}292.8}$
$X_C = 2.947\ \Omega$ (capacitive reactance)

(d) Draw a vector diagram representing the resistance and capacitive reactance of the circuit, Figure 14-6C.

(e) Combine the vector diagrams in Steps (b) and (d). See Figure 14-6D.

(f) Calculate the vector sum of X_L and X_C.
$X = X_L + (-X_C)$ *(From Example 3-13)*
$X = 7.54 + (-2.947)$
$X = 4.593\ \Omega$ (reactance)

(g) Draw an impedance vector diagram of the combined reactance and the resistance, Figure 14-6E.

(h) Calculate the impedance.
$Z = \sqrt{R^2 + X^2}$ *(Equation 13.11)*
$Z = \sqrt{25 + 21}$
$Z = \sqrt{46}$
$Z = 6.78\ \Omega$ (impedance)

Figure 14-6D
Vector diagram of resistance, inductive reactance, and capacitive reactance. Because X_L and X_C are drawn in opposite directions, they are given opposite signs (signs of opposition). It is customary to assign a positive sign to X_L and a negative sign to X_C.

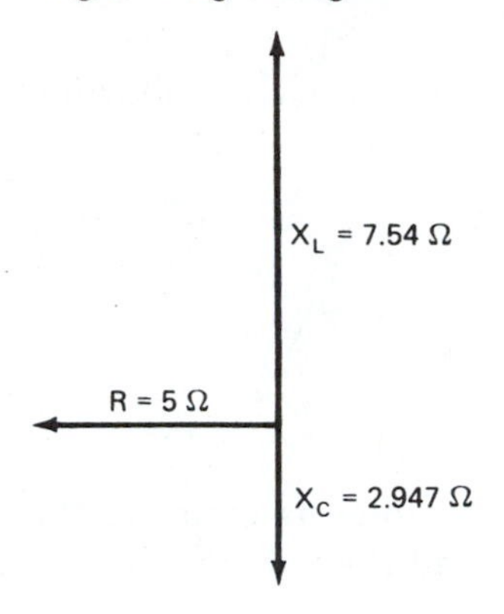

Figure 14-6E
Impedance vector diagram.
The reactance vector is drawn upward from the resistance vector because it is primary inductive.

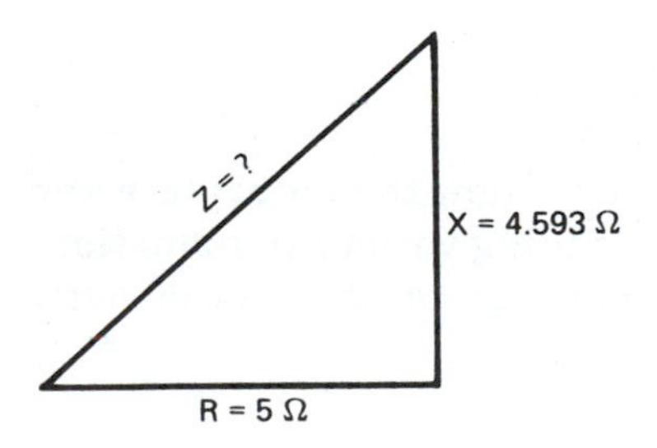

(i) $I = \frac{E}{Z}$ *(Equation 13.10)*

$I = \frac{120}{6.78}$

I = 17.7 A (circuit current)

(j) $E_L = IX_L$ *(From Ohm's Law)*
$E_L = 17.7 \times 7.54$
$E_L = 133.5$ V (voltage across the inductor)

(k) $E_C = IX_C$ *(From Ohm's Law)*
$E_C = 17.7 \times 2.947$
$E_C = 52.2$ V (voltage across the capacitance)

(l) $P = I^2 R$ *(Equation 13.18)*
$P = (17.7)^2 \times 5$
P = 1566 W (true power)

(m) $P_{app} = IE$ *(From Equation 4.2)*
$P_{app} = 17.7 \times 120$
$P_{app} = 2124$ VA (apparent power)

(n) $Pf = \frac{P}{P_{app}}$ *(Equation 13.16)*

$Pf = \frac{1566}{2124}$

Pf = 0.737 = 73.7% (power factor)

In a series ac circuit, the supply voltage is equal to the vector sum of the voltages across the individual components.

Example 6

Perform vector addition of the voltages across the components in Figure 14-6A to determine the supply voltage.

(a) Calculate the voltage drop across R.
$E_R = IR$ *(Ohm's Law)*
$E_R = 17.7 \times 5$
$E_R = 88.5$ V (voltage across R)

(b) Draw a vector diagram of the voltages in Figure 14-6A. See Figure 14-7.

(c) Calculate the vector sum of E_{X_L} and E_{X_C}.
$E_X = E_{X_L} + (-E_{X_C})$ *(From Example 3-11)*
$E_X = 133.5 + (-52.2)$
$E_X = 81.3$ V (reactive voltage)

Figure 14-7
Voltage vector diagram of an RLC circuit

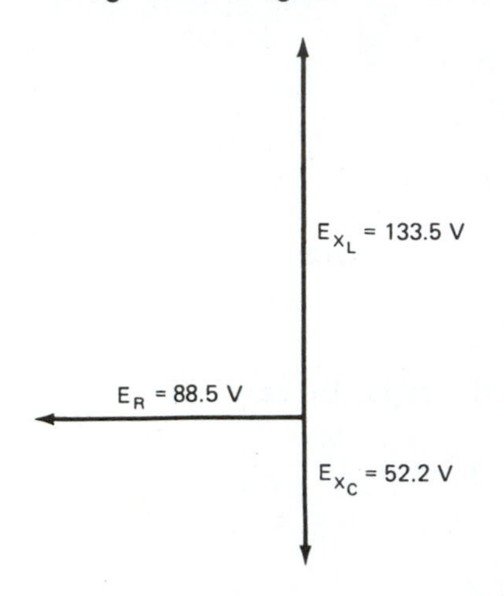

Figure 14-8
Voltage vector diagram

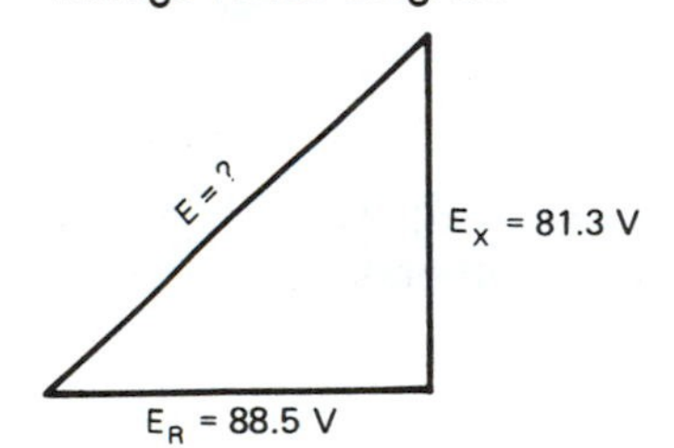

(d) Draw a vector diagram of E_X and E_R, Figure 14-8.

(e) Calculate the supply voltage.

$E = \sqrt{E_R{}^2 + E_X{}^2}$ *(Equation 13.11)*

$E = \sqrt{88.5^2 + 81.3^2}$

$E = \sqrt{14{,}442}$

E = 120 V (supply voltage)

The preceding formulas can be combined into the following equation:

$$E = \sqrt{(E_R)^2 + (E_L - E_C)^2} \quad \text{(Eq. 14.5)}$$

$$E = \sqrt{7832.25 + 6609.69}$$

$$E = \sqrt{14{,}442}$$

$$E = 120 \text{ V}$$

In series ac circuits, the power factor, which is equal to the cosine of the angle between the current and the voltage, may be determined by the following equations:

$$Pf = \frac{E_R}{E} \quad \textit{or} \quad \cos\phi = \frac{E_R}{E} \quad \text{(Eq. 14.6)}$$

$$Pf = \frac{IR}{IZ} \quad \textit{or} \quad \cos\phi = \frac{IR}{IZ}$$

$$Pf = \frac{R}{Z} \quad \textit{or} \quad \cos\phi = \frac{R}{Z}$$

$$Pf = \frac{P}{P_{app}} \quad \textit{or} \quad \cos\phi = \frac{P}{P_{app}}$$

A reexamination of Example 5 shows how the inductance and capacitance affect the power factor. Taking various combinations of the components and connecting them across 120 volts, 60 hertz will bring about varying power factors.

Example 7

(1) Connect the inductor and resistor from Example 5 in series across the 120-V, 60-Hz supply. Compute the power factor of the circuit.

(a) Find the impedance.

$Z = \sqrt{(R)^2 + (X_L)^2}$ *(Equation 13.11)*

$Z = \sqrt{(5)^2 + (7.54)^2}$

$Z = \sqrt{25 + 56.85}$

$Z = \sqrt{81.85}$

Z = 9.05 Ω (impedance)

(b) Calculate the power factor.

$$Pf = \frac{R}{Z} \text{ (Equation 14.3)}$$

$$Pf = \frac{5}{9.05}$$

$$Pf = 0.552 = 55.2\% \text{ (power factor)}$$

(2) Connect the capacitor and the resistor in series across the 120-V, 60-Hz supply.

(a) Calculate the impedance.

$$Z = \sqrt{(R)^2 + (X_C)^2} \text{ (Equation 13.11)}$$

$$Z = \sqrt{(5)^2 + (2.947)^2}$$

$$Z = \sqrt{25 + 8.68}$$

$$Z = \sqrt{33.68}$$

$$Z = 5.8\ \Omega \text{ (impedance)}$$

(b) Determine the power factor.

$$Pf = \frac{R}{Z} \text{ (Equation 14.3)}$$

$$Pf = \frac{5}{5.8}$$

$$Pf = 0.862 = 86.2\% \text{ (power factor)}$$

(3) Connect the capacitor and the inductor in series across the 120-V, 60-Hz supply.

(a) Calculate the impedance.

$$Z = X_L - X_C$$

$$Z = 7.54 - 2.947$$

$$Z = 4.59\ \Omega \text{ (impedance)}$$

Because Z is the result of inductive reactance only, the circuit becomes a pure inductive circuit, and the power factor is zero. This circuit can exist only in theory, because all circuits have some resistance.

It can be seen from the preceding calculations that the RL circuit has a poor power factor (55.2%). Adding the capacitance improves the power factor. The RLC circuit has a power factor of 73.7 percent.

Series Resonance

A resonant circuit contains resistance, inductance, and capacitance. However, the current and voltage of a resonant circuit are in phase. In order to accomplish this, X_L must be equal to X_C.

In other words, a series resonant circuit is an RLC circuit having a power factor of 100 percent.

Example 8

What size capacitor must be connected in series with the resistance and inductance in Example 5 to bring the power factor to unity (100 percent)?

(a) In order to obtain a 100% power factor, X_C must be equal to X_L.

$$X_C = X_L$$

$$\frac{10^6}{2\pi fC} = 2\pi fL \text{ } \textit{(Equations 13.9 and 13.14)}$$

$$C = \frac{10^6}{(2\pi f)^2 L}$$

$$C = \frac{10^6}{(2 \times 3.1416 \times 60)^2 \times 0.02}$$

$$C = \frac{10^6}{2842.46}$$

$$C = 351.8 \ \mu F$$

or

$$C = \frac{10^6}{2\pi fX_C}$$

$$C = \frac{10^6}{2 \times 3.1416 \times 60 \times 7.54}$$

$$C = \frac{10^6}{2842.52}$$

$$C = 351.8 \ \mu F$$

A 351.8-μF capacitor must be connected in series with the resistance and inductance.

Proof:

X_L must equal X_C.

$$X_C = \frac{10^6}{2\pi fC}$$

$$X_C = \frac{10^6}{2 \times 3.1416 \times 60 \times 351.8}$$

$$X_C = 7.54 \ \Omega$$

$$X_L = 7.54 \ \Omega$$

Therefore, $X_C = X_L$.

$$Pf = \frac{R}{Z} \text{ } \textit{(Equation 14.3)}$$

$$Pf = \frac{5}{5}$$

$$Pf = 1 = 100\% = \text{unity power factor}$$

Example 9

What size inductor must be connected in series with the resistance and capacitance in Example 5 to bring the power factor to unity?

$X_L = X_C$
$X_C = 2.947$ *and* $X_L = 2.947$

$L = \frac{X_L}{2\pi f}$ *(From Equation 13.9)*

$L = \frac{2.947}{2 \times 3.1416 \times 60}$

$L = \frac{2.947}{377}$

$L = 0.007817 \text{ H}$

Another way to obtain resonance in a circuit is to adjust the frequency to the proper value. When the frequency is low, a coil has low reactance and a capacitor has high reactance. Under these conditions, the circuit has high impedance and the current is low. When the frequency is high, a coil has high reactance and a capacitor has low reactance. Under these conditions, the circuit again has a high impedance and a low current.

At a certain frequency, the reactance of the coil is equal to the reactance of the capacitor, and the combined reactance is zero. Then the only opposition to current flow is the resistance. When a coil and a capacitor having equal reactance are connected in series, the circuit is in resonance and the power factor is 100 percent.

Example 10

A 5-Ω resistance is connected in series with a 200-μF capacitance and an inductance of 0.02 H. What is the resonant frequency of the circuit?

From the formulas $X_L = 2\pi fL$ and $X_C = \frac{10^6}{2\pi fC}$, and where X_L must equal X_C, as in Example 8, then

$2\pi fL = \frac{10^6}{2\pi fC}$

$f = \sqrt{\frac{10^6}{(2\pi)^2 LC}}$

$f = \sqrt{\frac{10^6}{39.479 \times 0.02 \times 200}}$

$f = \sqrt{\frac{10^6}{157.9}}$

$f = \sqrt{6333}$

$f = 80 \text{ Hz}$ (resonant frequency)

In a series-resonant circuit with low resistance and high reactances, the current will be high and the voltages across the reactances will be equal and high. These voltages may be considerably higher than the supply voltage.

Impedance Vector Diagrams of a Series Circuit

Impedance in a series circuit containing resistance, inductance, and capacitance is found in the following way. Combine, at right angles, the difference between the inductive reactance and the capacitive reactance with the resistance. To draw an impedance vector diagram for Figure 14-6A, first draw a resistance/reactance vector (Figure 14-6D). The resistance is drawn along the horizontal. The inductive reactance is drawn vertically upward from the resistance, and the capacitive reactance is drawn vertically down from the resistance. The impedance diagram is then drawn as shown in Figure 14-6E.

AC PARALLEL CIRCUITS

The following rules apply to parallel ac circuits:

- The line current (total current) is equal to the vector sum of the currents through the individual branches.
- At any instant, the voltage is the same value across all branches of a parallel ac circuit, and is equal to the supply voltage at that instant.
- The combined (equivalent) resistance of a parallel circuit is smaller than the resistance of any of the branches.

Because the voltage at any instant must have the same value across all branches, the voltage is a convenient phase reference. The voltage vector is laid out on the horizontal. The current through each branch may be represented with reference to the common voltage vector. The total current is the vector sum of the individual currents.

The impedance and power factor formulas for parallel circuits are as follows:

The formula for impedance is

$$Z = \frac{RX_L}{\sqrt{(R)^2 + (X_L)^2}} \qquad \text{(Eq. 14.7)}$$

The formula for power factor is

$$Pf = \frac{R}{Z} \qquad \text{(Eq. 14.8)}$$

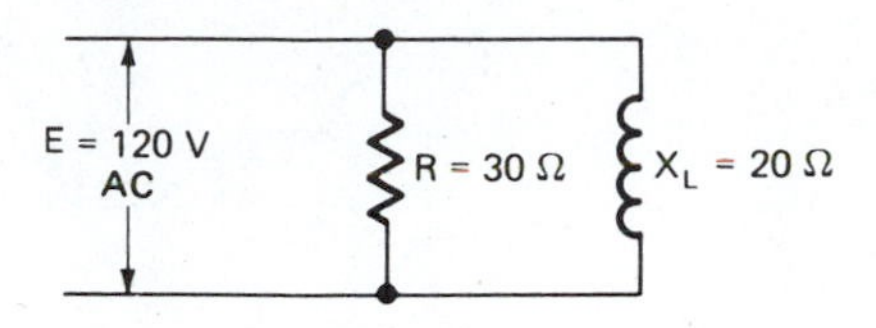

Figure 14-9A
Parallel RL circuit

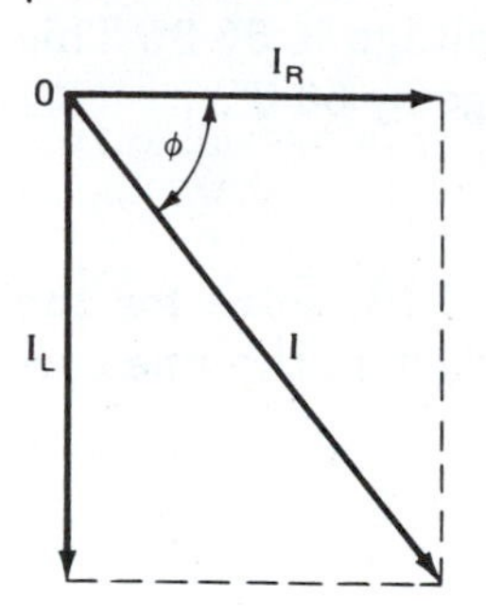

Figure 14-9B
Current vector diagram for an ac parallel RL circuit

RL Parallel Circuits

An RL parallel circuit, Figure 14-9A, is a circuit with branches containing resistance and inductance.

Example 11

In Figure 14-9A, determine the angle of lag between the current and voltage, the impedance and power factor of the circuit, and the value of the line current. The resistance of the inductor is negligible.

The vector diagram for the current is shown in Figure 14-9B.

(a) $I_R = \frac{E}{R}$ *(From Ohm's Law)*

$I_R = \frac{120}{30}$

$I_R = 4$ A (current through R)

(b) $I_L = \frac{E}{X_L}$ *(From Ohm's Law)*

$I_L = \frac{120}{20}$

$I_L = 6$ A (current through X_L)

(c) $Z = \frac{RX_L}{\sqrt{(R)^2 + (X_L)^2}}$ *(Equation 14.7)*

$Z = \frac{30 \times 20}{\sqrt{(30)^2 + (20)^2}}$

$Z = \frac{600}{\sqrt{900 + 400}}$

$Z = \frac{600}{\sqrt{1300}}$

$Z = \frac{600}{36}$

$Z = 16.667\ \Omega$ (circuit impedance)

(d) $I = \frac{E}{Z}$ *(Equation 13.10)*

$I = \frac{120}{16.667}$

$I = 7.2$ A (line current)

(e) $Pf = \frac{Z}{R}$ *(Equation 14.8)*

$Pf = \frac{16.667}{30}$

Pf = 0.555 = 55.5% (power factor of the circuit)

Pf = cosine of the phase angle between the total current and the line voltage

Figure 14-10A
AC parallel RL circuit

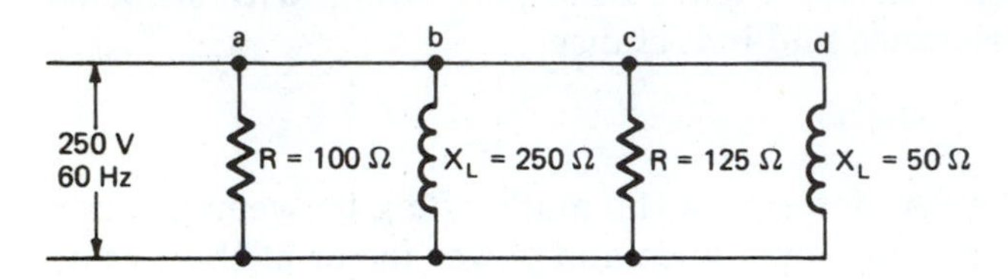

Figure 14-10B
Vector addition of currents in a parallel ac RL circuit

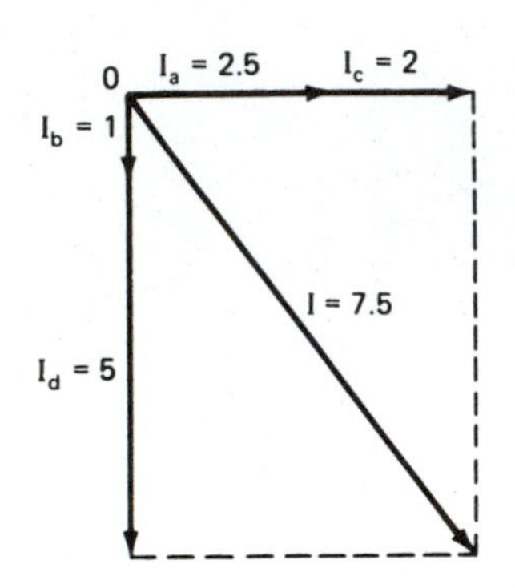

The angle whose cosine is 0.555 falls between 56.0° and 56.5°. The phase angle between the current and the voltage is 56.3°. This indicates that the line current lags the voltage by 56.3°.

Example 12

In the circuit in Figures 14-10A and 14-10B, solve for the following values: the current through each branch, the line current, and the power factor of the circuit.

(a) $I_a = \frac{E}{R}$ *(From Ohm's Law)*

$I_a = \frac{250}{100}$

$I_a = 2.5$ A (current through *a*)

(b) $I_b = \frac{E}{X_L}$

$I_b = \frac{250}{250}$

$I_b = 1$ A (current through *b*)

(c) $I_c = \frac{E}{R}$

$I_c = \frac{250}{125}$

$I_c = 2$ A (current through *c*)

(d) $I_d = \frac{E}{X_L}$

$I_d = \frac{250}{50}$

$I_d = 5$ A (current through *d*)

(e) The currents through *a* and *c* are in phase with the voltage. Therefore, they can be laid out along the horizontal voltage vector end to end. The currents through *b* and *d* are lagging the voltage by 90 degrees, as indicated in Figure 14-10B.

The total current is equal to the square root of the sum of their squares.

$$I = \sqrt{(6)^2 + (4.5)^2}$$
$$I = \sqrt{36 + 20.25}$$
$$I = \sqrt{56.25}$$
$$I = 7.5 \text{ A (line current)}$$

(f) $Z = \frac{E}{I}$ *(From Equation 13.10)*

$$Z = \frac{250}{7.5}$$

$$Z = 33.333\ \Omega \text{ (circuit impedance)}$$

A formula frequently used to calculate the combined resistance of only two resistances connected in parallel is

$$R_t = \frac{R_1 R_2}{R_1 + R_2} \qquad \text{(Eq. 14.9)}$$

where R_t = combined resistance of the two resistances
R_1 = resistance of the first resistance
R_2 = resistance of the second resistance

Using the above Equation 14.9, the combined resistance of resistances in Figure 14-10A can be found as follows:

(g) $R_t = \frac{R_a R_c}{R_a + R_c}$

$$R_t = \frac{12{,}500}{225}$$

$$R_t = 55.56\ \Omega \text{ (resistance of the circuit)}$$

(h) $Pf = \frac{Z}{R}$ *(Equation 14.8)*

$$Pf = \frac{33.33}{55.56}$$

$$Pf = 0.6 = 60\% \text{ (power factor of the circuit)}$$

RC Parallel Circuits

An RC parallel circuit, Figure 14-11, contains capacitance and resistance connected in parallel.

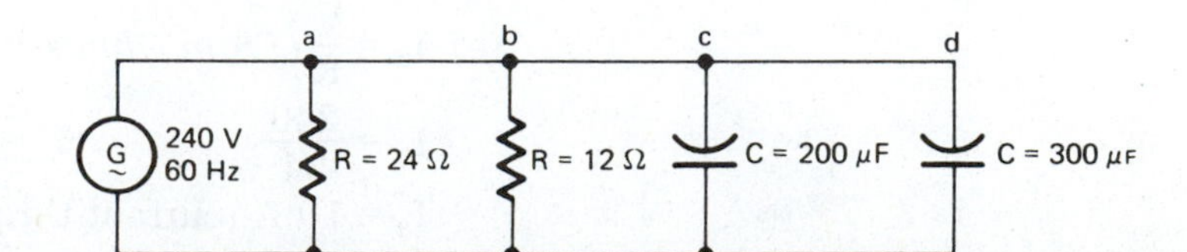

Figure 14-11
RC parallel ac circuit

Example 13

In Figure 14-11, calculate the total resistance, the total capacitance, the capacitive reactance, the current flowing through each branch, the circuit impedance, and the power factor of the circuit.

(a) $R = \frac{R_a R_b}{R_a + R_b}$

$R = \frac{24 \times 12}{24 + 12}$

$R = \frac{288}{36}$

$R = 8\ \Omega$ (circuit resistance)

(b) $C = C_c + C_d$

$C = 200 + 300$

$C = 500\ \mu F$ (circuit capacitance)

(c) $X_C = \frac{10^6}{2\pi fC}$ *(Equation 13.14)*

$X_C = \frac{10^6}{2 \times 3.1416 \times 60 \times 500}$

$X_C = \frac{10^6}{188{,}496}$ *or* $\frac{10^6}{188{,}500}$

$X_C = 5.3\ \Omega$ (circuit reactance)

(d) $X_C = \frac{10^6}{2\pi fC}$ *(Equation 13.14)*

$X_C = \frac{10^6}{2 \times 3.1416 \times 60 \times 200}$

$X_C = \frac{10^6}{75{,}398.4}$ *or* $\frac{10^6}{75{,}400}$

$X_C = 13.3\ \Omega$ (reactance of c)

$X_C = \frac{10^6}{2\pi fC}$

$X_C = \frac{10^6}{2 \times 3.1416 \times 60 \times 300}$

$X_C = \frac{10^6}{113{,}097.6}$ *or* $\frac{10^6}{113{,}100}$

$X_C = 8.84\ \Omega$ (reactance of d)

(e) $I_a = \frac{V}{R}$ *(From Ohm's Law)*

$I_a = \frac{240}{24}$

$I_a = 10$ A (current through a)

$$I_b = \frac{E}{R}$$

$$I_b = \frac{240}{12}$$

I_b = 20 A (current through *b*)

$$I_c = \frac{E}{X_C}$$

$$I_c = \frac{240}{13.3}$$

I_c = 18 A (current through *c*)

$$I_d = \frac{E}{X_C}$$

$$I_d = \frac{240}{8.84}$$

I_d = 27 A (current through *d*)

I_R = 10 + 20 = 30 A through branches *a* and *b*

I_c = 18 + 27 = 45 A through branches *c* and *d*

Because currents I_R and I_c are 90 degrees out of phase with each other, their total is equal to the square root of the sum of their squares.

$$I = \sqrt{(I_R)^2 + (I_c)^2}$$

$$I = \sqrt{900 + 2025}$$

$$I = \sqrt{2925}$$

I = 54 A (line current)

(f) $Z = \frac{E}{I}$ *(Equation 13.10)*

$$Z = \frac{240}{54}$$

Z = 4.4 Ω (circuit impedance)

(g) $Pf = \frac{Z}{R}$ *(Equation 14.8)*

$$Pf = \frac{4.4}{8}$$

Pf = 0.55 = 55% (circuit power factor)

RLC Parallel Circuits

RLC parallel circuits consist of branches containing capacitance, inductance, and resistance. To solve these circuits, the LC portion is usually solved first, and then the RL or RC portion is considered.

Figure 14-12A
RLC parallel ac circuit

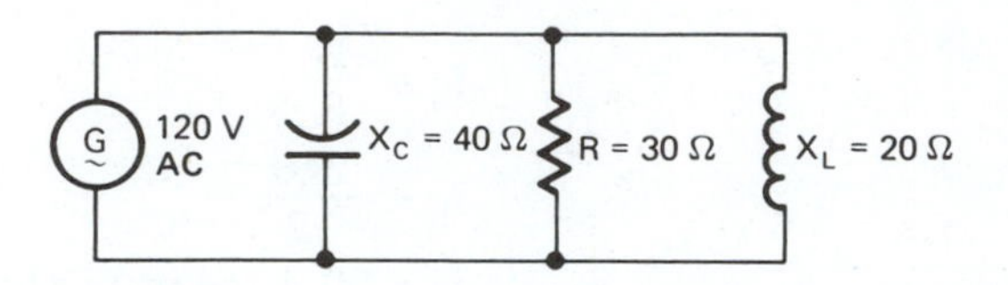

Example 14

Figure 14-12A illustrates an RLC parallel circuit connected to 120 V ac. Calculate the power factor, the line current, and the total power dissipated by the circuit.

(a) $I_R = \frac{E}{R}$ *(From Ohm's Law)*

$I_R = \frac{120}{30}$

$I_R = 4$ A (current through R)

$I_L = \frac{E}{X_L}$

$I_L = \frac{120}{20}$

$I_L = 6$ A (current through X_L)

$I_c = \frac{E}{X_C}$

$I_c = \frac{120}{40}$

$I_c = 3$ A (current through X_C)

The vector diagram for the currents is shown in Figure 14-12B.

$I = \sqrt{(I_R)^2 + (I_L - I_c)^2}$ *(From Equation 14.5)*

$I = \sqrt{16 + (6 - 3)^2}$

$I = \sqrt{16 + 9}$

$I = \sqrt{25}$

$I = 5$ A (line current)

Figure 14-12B
Vector diagram of currents in an RLC circuit

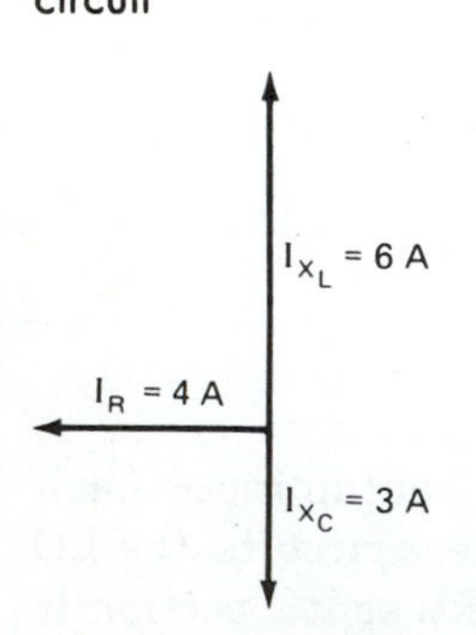

(b) $Z = \frac{E}{I}$ *(From Equation 13.10)*

$Z = \frac{120}{5}$

$Z = 24$ Ω (circuit impedance)

$Pf = \frac{Z}{R}$ *(Equation 14.8)*

$Pf = \frac{24}{30}$

$Pf = 0.8 = 80\%$ (circuit power factor)

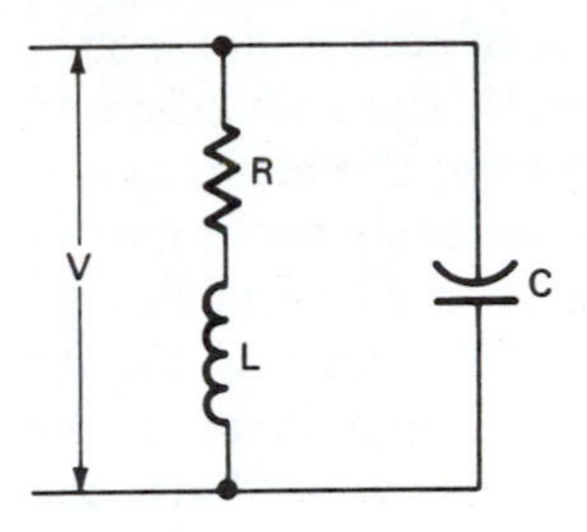

Figure 14-13A
Parallel LC circuit

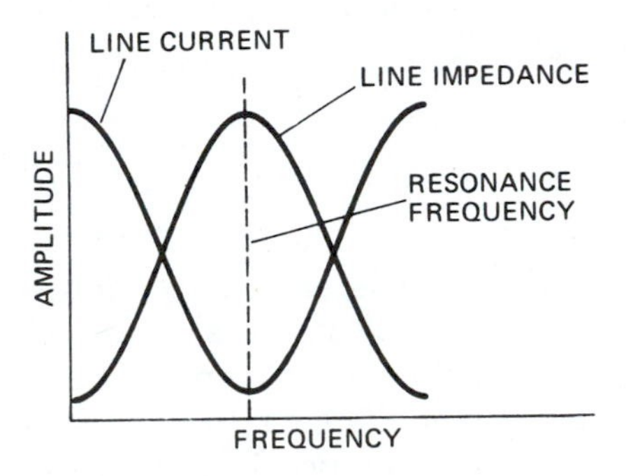

Figure 14-13B
Current/impedance curves at varying frequencies

(c) $P = IV \cos \phi$ *(Equation 13.17)*
$P = 5 \times 120 \times 0.8$
$P = 480$ W (power utilized by the circuit)

Parallel Resonance

Calculations for parallel resonant circuits are similar to those for series resonance. The inductive reactance must be equal to the capacitive reactance ($X_L = X_C$). The following equation can be applied to a parallel circuit as well as to a series circuit:

$$f = \sqrt{\frac{10^6}{(2\pi)^2 LC}} \qquad \text{(Eq. 14.10)}$$

where f = resonant frequency, in hertz (Hz)
π = 3.1416 (a constant)
L = inductance, in henrys (H)
C = capacitance, in microfarads (µF)

Parallel circuits, however, have quite different characteristics from series resonant circuits.

Figure 14-13A shows a circuit containing an inductor and a capacitor connected in parallel. To illustrate that the coil has resistance, a resistance is shown in series with the inductor.

Figure 14-13B shows the current/impedance curves at varying frequencies. When the frequency is low, the inductor takes a high lagging current and the capacitor takes a low leading current. The circuit current is high and lags the voltage; thus, the impedance is low. When the frequency is high, the capacitor takes a high leading current and the inductor takes a low lagging current. The circuit current is again high. In this case, however, the current leads the voltage.

If this circuit has zero resistance at resonant frequency, then the current taken by the inductor will equal the current taken by the capacitor, and the line current will be zero. The capacitor will discharge into the inductor, and the inductor will feed back into the capacitor. There will be a continuous exchange of energy.

To further analyze this condition, assume that the resistance is removed from the circuit in Figure 14-13A and a dc source is momentarily connected across the circuit. The dc supply will charge the capacitor. When the supply is removed, the capacitor will discharge through the inductor. Current flowing through the inductor will set up a magnetic field. As the capacitor discharges, the current will decrease. The decreasing current will produce a moving magnetic field that induces a voltage into the coil. The induced voltage causes the current to continue to flow, charging the capacitor with the opposite polarity. When the field around the

Figure 14-14
Current curve showing the exchange of energy in an LC circuit (tank circuit)

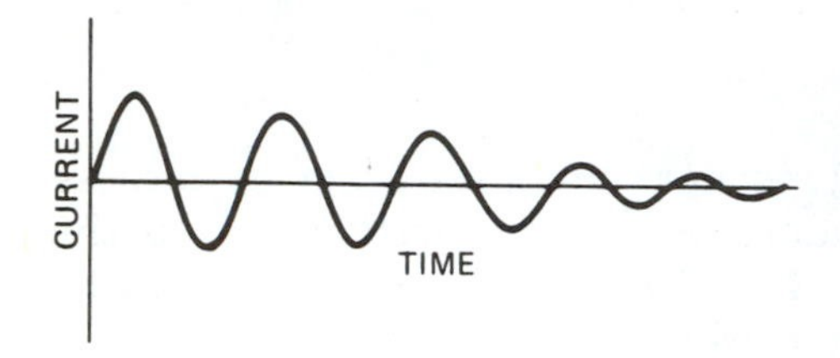

inductor has dissipated, the capacitor will again discharge, but in the direction opposite to the original discharge. This exchange of energy will continue indefinitely in a circuit of zero resistance.

In a practical, parallel LC resonant circuit, the inductor and capacitor have minimum resistance. Therefore, the exchange of energy and the current flow decrease with each exchange of energy. Figure 14-14 shows a current curve for this type of circuit.

If alternating current is applied to the circuit shown in Figure 14-13A, the line current will be just enough to compensate for the initial decrease in current caused by the resistance. Thus, the current between the capacitor and the inductor will be much greater than the line current.

In either series or parallel circuits, resonance may be obtained in any one of three ways: by adjusting the frequency of the supply, by adjusting the inductance, or by adjusting the capacitance. To obtain resonance, X_L must be equal to X_C.

AC COMBINATION CIRCUITS

Most circuits contain more than pure resistance, pure inductance, or pure capacitance; therefore, each path of a parallel circuit must be treated as a separate series circuit. The current for each branch must be determined and a vector diagram drawn. The individual currents are indicated with reference to a common voltage vector.

Example 15

Figure 14-15A shows an inductor and a resistance connected in parallel. The inductor has a reactance of 20 Ω and a resistance of 15 Ω. The pure resistive branch contains 50 Ω of resistance. This resistance and the inductor are connected in parallel across a 130-V ac source. Calculate the current through each branch, the power factor of the circuit, and the total current.

Figure 14-15A
Inductor and resistor connected in parallel across 130 V ac

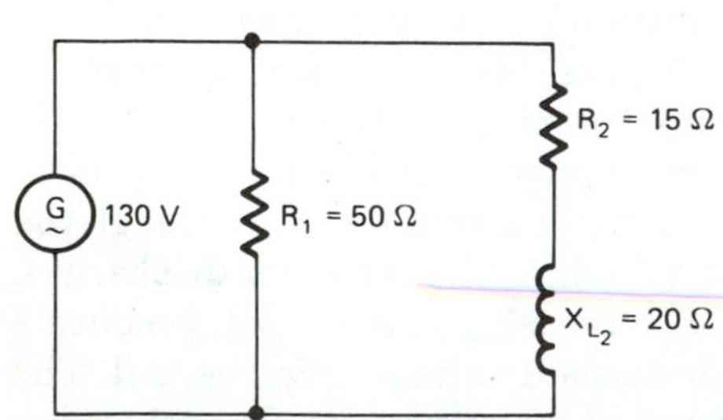

(a) $I_1 = \frac{E}{R_1}$ *(From Ohm's Law)*

$I_1 = \frac{130}{50}$

$I_1 = 2.6$ A (current through branch 1)

(b) $Z_2 = \sqrt{(R_2)^2 + (X_{L_2})^2}$ *(Equation 13.11)*

$Z_2 = \sqrt{(15)^2 + (20)^2}$

$Z_2 = 25\ \Omega$ (impedance of branch 2)

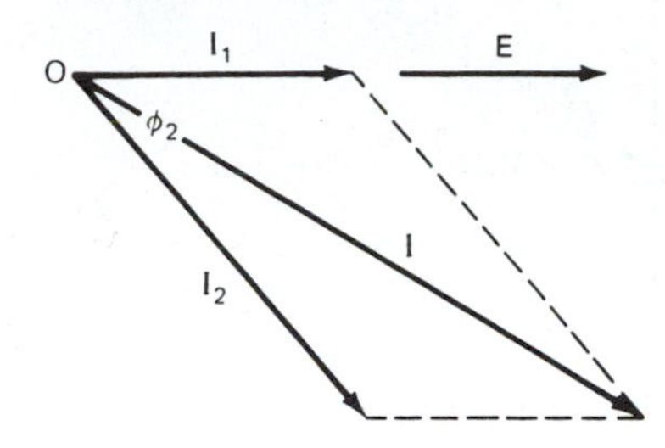

Figure 14-15B
Current vector diagram of inductance and resistance in parallel

(c) $I_2 = \frac{E}{Z_2}$ *(From Ohm's Law)*

$I_2 = \frac{130}{25}$

$I_2 = 5.2$ A (current through branch 2)

Branch 2 contains both resistance and reactance. Therefore, I_2 is neither in phase nor 90° out of phase with E. It is necessary to determine the power factor of branch 2.

(d) $Pf = \frac{R}{Z}$ *(Equation 14.3)*

$Pf = \frac{15}{25}$

$Pf = 0.6$ (power factor for branch 2)

The angle for cosine 0.6 is 53° lagging.

A current vector diagram may be drawn as shown in Figure 14-15B. Using the parallelogram method to solve vector diagrams enables the line current to be calculated.

(e) $I = \sqrt{(I_1)^2 + (I_2)^2 + (2I_1 I_2 \cos \phi)}$ *(Equation 13.5)*

$I = \sqrt{(2.6)^2 + (5.2)^2 + (2 \times 2.6 \times 5.2 \times 0.6)}$

$I = \sqrt{50.024}$

$I = 7.07$ A (line current)

Example 16

Referring to the circuit in Figure 14-16A, calculate the current through each branch, the power factor of the circuit, the line current, the impedance, and the power utilized by the circuit.

(a) $I_1 = \frac{E}{R_1}$ *(From Ohm's Law)*

$I_1 = \frac{240}{10}$

$I_1 = 24$ A (current through branch 1)

(b) $Z_2 = \sqrt{(R_2)^2 + (X_{L_2})^2}$ *(Equation 13.11)*

$Z_2 = \sqrt{(12)^2 + (16)^2}$

$Z_2 = \sqrt{400}$

$Z_2 = 20\ \Omega$ (impedance of branch 2)

$I_2 = \frac{E}{Z_2}$

$I_2 = \frac{240}{20}$

$I_2 = 12$ A (current through branch 2)

Figure 14-16A
RLC parallel ac circuit. Each branch contains resistance.

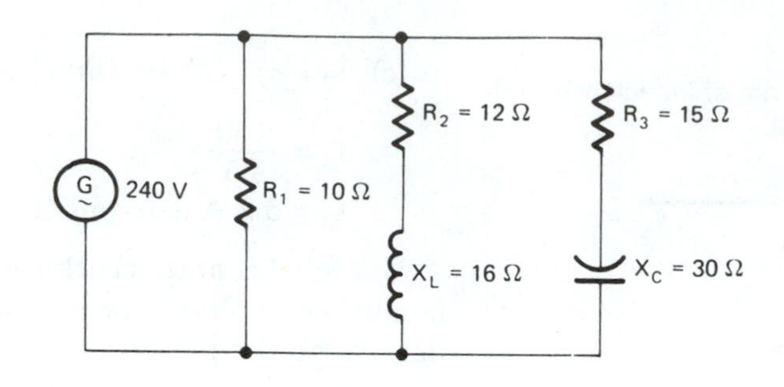

(c) $Z_3 = \sqrt{(R_3)^2 + (X_{C_3})^2}$ *(Equation 13.11)*

$Z_3 = \sqrt{(15)^2 + (30)^2}$

$Z_3 = \sqrt{225 + 900}$

$Z_3 = \sqrt{1125}$

$Z_3 = 33.54\ \Omega$ (impedance of branch 3)

$I_3 = \frac{V}{Z_3}$ *(From Ohm's Law)*

$I_3 = \frac{240}{33.5}$

$I_3 = 7.16$ A (current through branch 3)

(d) $Pf = \frac{R_2}{Z_2}$ *(Equation 14.3)*

$Pf = \frac{12}{20}$

$Pf = 0.6$ (power factor for branch 2)

The angle for cosine 0.6 is 53°.

$Pf = \frac{R_3}{Z_3}$ *(Equation 14.3)*

$Pf = \frac{15}{33.5}$

$Pf = 0.448$ (power factor for branch 3)

The angle for cosine 0.448 is 63.5°.

With complex circuits such as this one, it is simpler to calculate the total in-phase current first. All three branches have resistance. Therefore, all three branches have some in-phase current.

1. Lay out the voltage vector along the horizontal. (See Figure 14-16B).
2. Because branch 1 is pure resistance, I_1 is laid out along the voltage vector.
3. Branch 2 has a resistive component and a reactive component of current. The resistive component is in phase with the voltage and the reactive component is 90° out of phase with the voltage. To construct the vector diagram, begin at

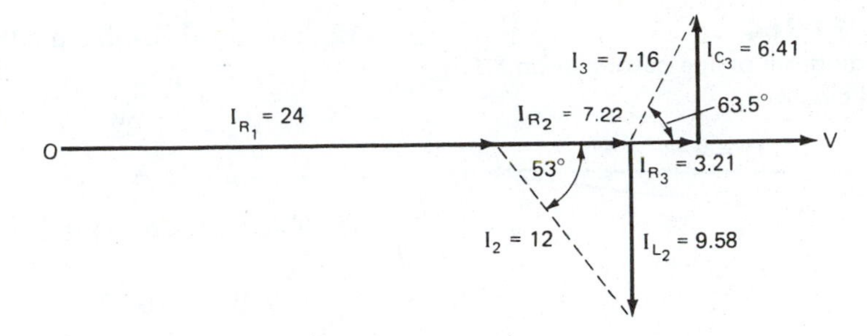

Figure 14-16B
Current vectors for calculating the resistive and reactive components of the total current

the end of I_1, and lay out the in-phase current of I_2 along the voltage vector (I_{R_2}). At the end of I_{R_2} and at a 90° angle to the voltage vector, lay out the reactive current I_{L_2}.

4. Draw vector I_2, representing the total current through branch 2, or 12 A.
5. The power factor for branch 2 is 0.6. The cosine table shows the angle for 0.6 is 53°. Therefore, the angle between I_2 and I_{R_2} is 53°.
6. The values of I_{R_2} and I_{L_2} can be found by using the trigonometric functions.

$$\sin \phi = \frac{\text{opp}}{\text{h}} \qquad \cos \phi = \frac{\text{adj}}{\text{h}}$$

$$\begin{aligned} \text{opp} &= \text{h}(\sin \phi) & \text{adj} &= \text{h}(\cos \phi) \\ I_{L_2} &= I_2(\sin \phi) & I_{R_2} &= I_2(\cos \phi) \\ I_{L_2} &= 12 \times 0.7986 & I_{R_2} &= 12 \times 0.6018 \\ I_{L_2} &= 9.58 \text{ A} & I_{R_2} &= 7.22 \text{ A} \end{aligned}$$

7. Referring to branch 3, lay out the in-phase current vector. Begin at the end of I_{R_2}, and draw I_{R_3} along the voltage vector, as in Figure 14-16B.
8. From the end of I_{R_3} and at a 90° angle to the voltage vector, draw vector I_{C_3}. Because the capacitive component of current is 180° out of phase with the inductive component, vector I_{C_3} must be drawn up.
9. Draw vector I_3, representing the total current through branch 3.
10. The power factor of branch 3 is 0.448. The cosine table shows the angle for 0.448 is 63.5°. Therefore, the angle between I_3 and I_{R_3} is 63.5°.
11. Using the trigonometric formula, solve for I_3 and I_{R_3}.

$$\sin \phi = \frac{\text{opp}}{\text{h}} \qquad \cos \phi = \frac{\text{adj}}{\text{h}}$$

$$\begin{aligned} \text{opp} &= \text{h}(\sin \phi) & \text{adj} &= \text{h}(\cos \phi) \\ I_{C_3} &= 7.16 \times 0.8949 & I_{R_3} &= 7.16 \times 0.448 \\ I_{C_3} &= 6.41 \text{ A} & I_{R_3} &= 3.21 \text{ A} \end{aligned}$$

Figure 14-16C
Vector diagram of line current in an RLC parallel circuit

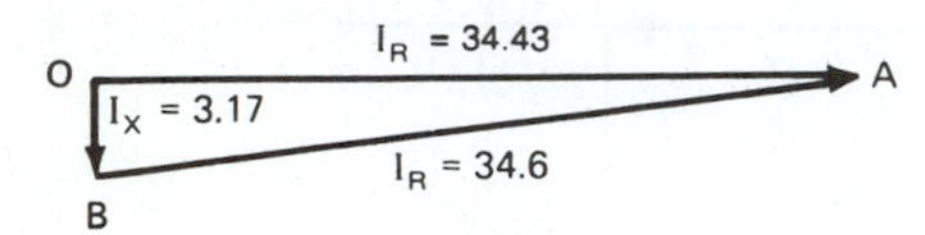

12. The total in-phase current is

$$I_{R_t} = I_{R_1} + I_{R_2} + I_{R_3}$$
$$I_{R_t} = 24 + 7.22 + 3.21$$
$$I_{R_t} = 34.43 \text{ A}$$

13. The total reactive current is

$$I_X = I_L - I_C$$
$$I_X = 9.58 - 6.41$$
$$I_X = 3.17 \text{ A}$$

14. The final current vector diagram is as shown in Figure 14-16C. To solve for the total line current, use the formula for solving right triangles.

$$I = \sqrt{(I_{R_t})^2 + (I_X)^2}$$
$$I = \sqrt{1185.42 + 10}$$
$$I = \sqrt{1195.42}$$
$$I = 34.6 \text{ A (total line current)}$$

15. Figure 14-16C shows that OA represents the in-phase current and is laid out along the voltage vector. AB represents the total line current. Angle A then represents the phase angle between the line current and the supply voltage. The cosine of A is the power factor of the circuit.

$$\cos \phi = \frac{\text{adj}}{\text{h}} \textit{ (Equation 13.1)}$$
$$\cos \phi = \frac{34.43}{34.6}$$
$$\cos \phi = 0.995 = 99.5\% \text{ (power factor of the circuit)}$$

16. $P = IE \cos \phi$ *(Equation 13.17)*

$$P = 34.6 \times 240 \times 0.995$$
$$P = 8262 \text{ W} = 8.262 \text{ kW (power dissipated by the circuit)}$$

17. $Z = \frac{E}{I}$ *(From Equation 13.10)*

$$Z = \frac{240}{34.6}$$
$$Z = 6.94\ \Omega \text{ (circuit impedance)}$$

Figure 14-17
Vector diagram of power in an ac circuit

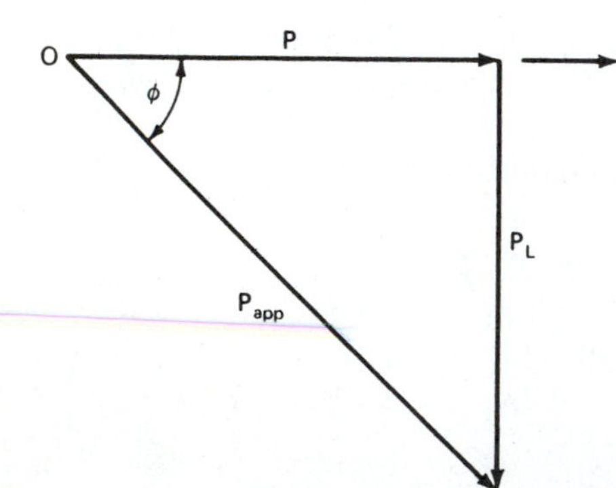

Power Factor Correction

The vector diagram for power is shown in Figure 14-17. Vector P, drawn along the horizontal, designates the true power delivered to the load. The true power is also called effective power because it is the power that has a direct effect on the load. The product of the current and the voltage (IE) is represented by the hypotenuse of the triangle and is ϕ degrees out of phase with the effective power. This power vector represents the apparent power.

The side opposite angle ϕ represents the reactive power. This power circulates between the generator at the power station and the loads. It causes the conductors through which it passes to produce heat, and it performs no useful work. In order to maintain an efficient operation, the reactive power must be kept to a minimum.

Because the power factor of a system varies according to the type of load, certain types of electrical equipment such as alternators and transformers are rated in voltamperes (VA) or kilovolt-amperes (kVA). This type of equipment has a fixed value for the maximum current and voltage ratings. The value depends upon the type of insulation and the size and type of conductors used. The number of watts they can safely deliver depends upon the power factor of the load.

In most industrial establishments, a large percentage of the loads consists of induction motors or other equipment containing coils. Induction motors are simple and rugged and can be manufactured to meet most industrial requirements. These machines, as well as welding machines and induction furnaces, operate with a lagging power factor. In other words, they cause the current to lag the voltage. Although many establishments operate on a power factor as low as 60 percent, anything below 80 percent is considered to be below industrial standards.

A power factor below unity (100 percent) has the following detrimental effects:

1. More current is required in order to deliver a given amount of power to the load. An increase in current causes a greater line loss (I^2R), a greater line drop (IR), and more power loss and voltage drop within the system. If an excessive amount of current is required, it may be necessary to increase the size of the system and circuit conductors.
2. Large out-of-phase currents can cause transformers and alternators to operate with poor voltage regulation.
3. Increased currents caused by a low power factor may cause overcurrent devices to operate.

In an effort to conserve fuel and provide an efficient operation, utility companies strive to sell electrical energy at a high power factor. A consumer whose load has a low power factor may be charged more per kilowatt-hour than if the power factor were high. This induces the consumer to employ various means to improve the power factor.

Customers who have highly inductive loads can install capacitors and/or synchronous motors to improve the power factor. This equipment should be connected on the load side of the supply transformer to keep the transformer losses to a minimum and allow for better voltage regulation. The ideal location for capacitors is as near as practical to the load they are servicing, to help reduce line losses in the feeder and branch circuit conductors. A large industrial establishment may have several capacitors or capacitor banks located throughout the plant.

A capacitor causes the current to lead the voltage, thereby compensating for the lagging current caused by inductive loads. Individual capacitors may be connected directly across the terminals of the inductive load. The action that takes place is the same as that described in the section on parallel resonance in this chapter. The reactive current will circulate between the inductance and capacitance, and the supply current will be kept to a minimum. Frequently, in order to conserve space and reduce installation costs, consumers connect capacitors to the feeder conductors.

Capacitors used for power factor correction are generally rated in vars (VAR) or kilovars (kVAR) instead of farads (F). This is more convenient because when the reactive power of the capacitor is equal to the reactive power of the load, unity power factor has been established.

The reactive power of a synchronous motor depends upon two factors: the dc field excitation and the mechanical load driven by the machine. Maximum leading power factor is developed under maximum field excitation with zero load. Synchronous motors are often more practical for power factor improvement than capacitors are because they can drive a mechanical load as well as develop a leading power factor. When they are serving this dual purpose, it is advisable that they be used to drive a constant load at a constant speed.

Motors and other electrical equipment are provided with information on the nameplate to aid the electrical worker in installing the equipment, analyzing circuits, and solving problems. The nameplate information includes the rating in kilovoltamperes and/or kilowatts, voltage, amperes, and frequency. The total apparent power required by a machine, or the power required from the alternator, may be calculated through vector addition.

Example 17

A load in a factory requires 50 kVA and has a 50% lagging power factor. Another load connected to the same power lines requires 100 kVA and has a lagging power factor of 86.6%. Calculate the effective power, the reactive power, the power factor, and the apparent power of the line.

Figure 14-18A
Vector diagram of the apparent power in two branches of a parallel circuit

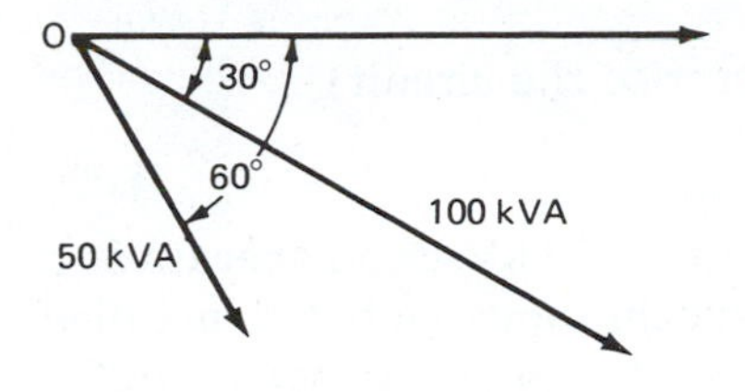

Figure 14-18B
Vector addition of apparent power

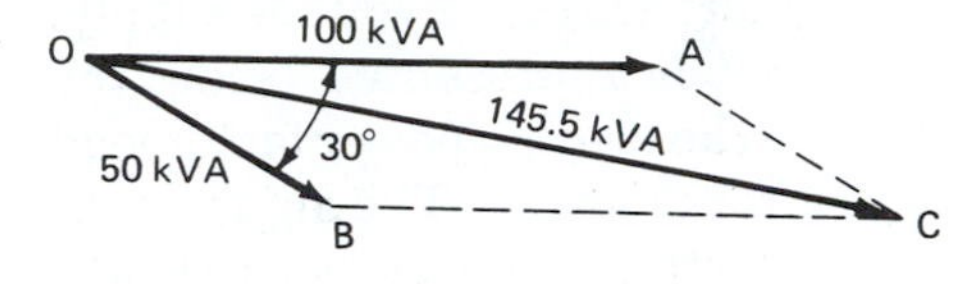

(a) The apparent power taken by each load, expressed with reference to the voltage, is represented in Figure 14-18A. The angle corresponding to a power factor of 50% (cos 0.5000) is 60°. The angle for 86.6% (cos 0.8660) is 30°. To determine the resultant power, draw the vector diagram as shown in Figure 14-18B.

(b) Draw the 100-kVA value along the horizontal and label it OA.

(c) Figure 14-18A shows that the angle between 100 kVA and 50 kVA is 60° –30° = 30°.

(d) Thirty degrees from OA, draw the 50-kVA vector OB (Figure 14-18B).

$$OC = \sqrt{(OA)^2 + (OB)^2 + [2(OA)(OB)(\cos \phi)]}$$
(Equation 13.5)
$$OC = \sqrt{10^4 + (25 \times 100) + (2 \times 100 \times 50 \times 0.866)}$$
$$OC = \sqrt{10^4 + (25 \times 100) + 8660}$$
$$OC = \sqrt{21{,}160}$$
$$OC = 145.5 \text{ kVA (total apparent power)}$$

(e) Calculate the effective power for each load.
$P = IE \cos \phi$ *(Equation 13.17) and* $IE = 50\text{kVA}$
$P = 50 \times 0.5$
$P = 25$ kW

$P = IE \cos \phi$ *and* $IE = 100$ kVA
$P = 100 \times 0.866$
$P = 86.6$ kW

(f) The effective power for each load is in phase with the voltage, and the total effective power is
$25 + 86.6 = 111.6$ kW

Figure 14-18C
Power vector diagram of the total apparent, effective, and reactive power

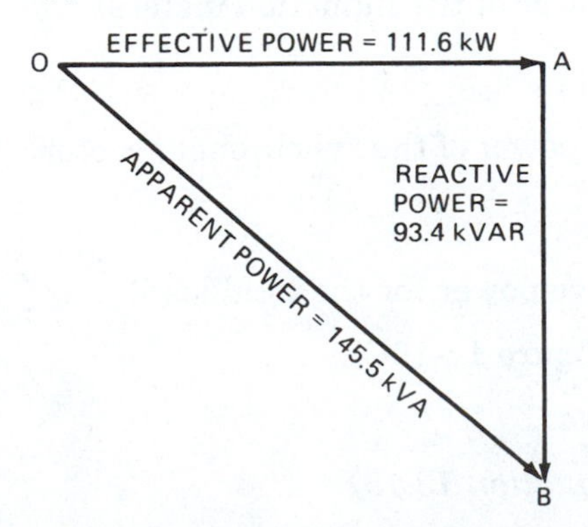

(g) Draw a power vector diagram for the total apparent, effective, and reactive power (Figure 14-18C). Draw the effective power on the horizontal. Draw the reactive power vertically downward from the effective power. The apparent power is the hypotenuse of the right triangle.

(h) Solve for the reactive power.
$$P_L = \sqrt{P_{app}{}^2 - P^2} \quad \textit{(Equation 13.19)}$$
$$P_L = \sqrt{21{,}170 - 12{,}454}$$
$$P_L = \sqrt{8716}$$
$$P_L = 93.4 \text{ kVAR (total reactive power)}$$

Figure 14-19A
Vector diagram of the apparent power with reference to the voltage supply

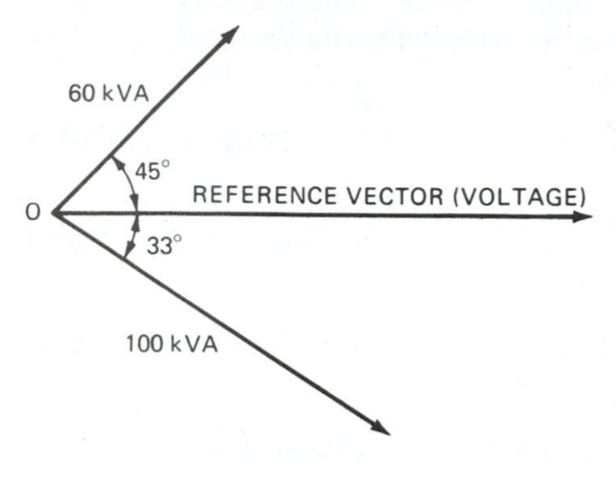

(i) $Pf = \frac{P}{P_{app}}$ *(Equation 13.16)*

$Pf = \frac{111.6}{145.5}$

Pf = 0.767 = 76.7% (power factor of the circuit)

Example 18

A group of induction motors require 100 kVA and operate at a power factor of 84% lagging. If a synchronous motor is installed that requires 60 kVA and operates at a power factor of 70.7% leading, calculate the total effective power, apparent power, reactive power, and power factor of the load.

(a) Figure 14-19A shows a vector diagram of the apparent power of the two loads with reference to the voltage supply. The lagging power factor load is represented below the reference (voltage) vector, and the leading power factor load is represented above the reference vector. The angles indicated correspond to the power factors of the respective loads.

Figure 14-19B
Vector diagram indicating the phase relationship of the apparent power in two branches of a parallel circuit

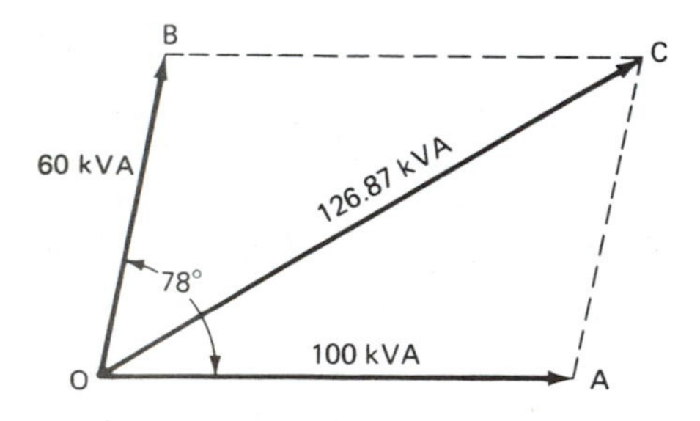

(b) To determine the resultant apparent power, draw a vector diagram indicating the phase relationship with reference to each other, Figure 14-19B. 45° + 33° = 78°.

$C = \sqrt{A^2 + B^2 + 2AB \cos \phi}$ *(Equation 13.5)*

$C = \sqrt{(100)^2 + (60)^2 + (2 \times 100 \times 60 \times 0.2079)}$

$C = \sqrt{10{,}000 + 3600 + 2495}$

$C = \sqrt{16{,}095}$

C = 126.87 kVA (apparent power for the circuit)

(c) Calculate the effective power for each load and the total effective power.

$P = IE \cos \phi$ *(Equation 13.17)*

$P = 100 \times 0.84$

P = 84 kW (effective power of the induction motors)

$P = IE \cos \phi$

$P = 60 \times 0.707$

P = 42.42 kW (effective power of the synchronous motor)

$P_t = P_1 + P_2$

$P_t = 84 + 42.42$

P_t = 126.42 kW (effective power for the total load)

Figure 14-19C
Vector diagram for power in a high power factor circuit

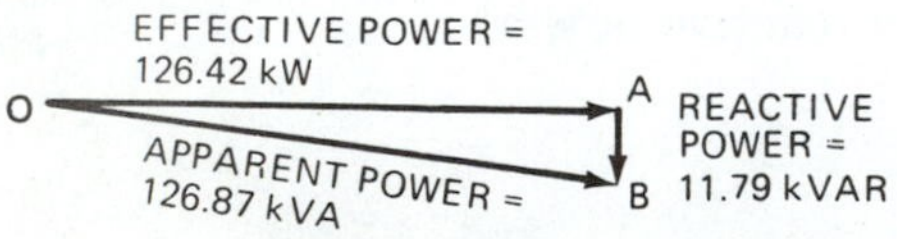

EFFECTIVE POWER = 126.42 kW
O
A
REACTIVE POWER = 11.79 kVAR
APPARENT POWER = 126.87 kVA
B

(d) Draw a power triangle (Figure 14-19C).

(e) Solve for the reactive power.

$P_L = \sqrt{(P_{app})^2 - (P)^2}$ *(Equation 13.19)*

$P_L = \sqrt{(126.87)^2 - (126.42)^2}$

Figure 14-20A
Vector diagram of the apparent power of two inductive loads

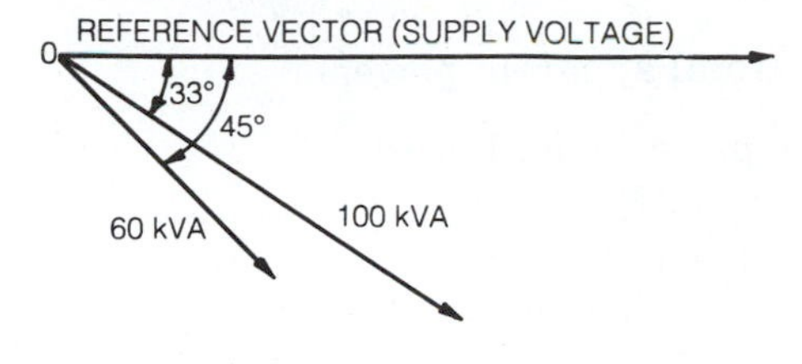

Figure 14-20B
Vector addition of the apparent power of two inductive loads

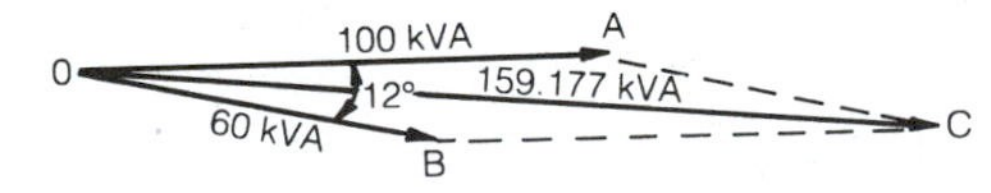

$$P_L = \sqrt{16{,}096 - 15{,}982}$$
$$P_L = \sqrt{114}$$
$$P_L = 10.68 \text{ kVAR (reactive power for the total load)}$$

(f) Solve for the power factor.

$$Pf = \frac{P}{P_{app}} \textit{ (Equation 13.16)}$$

$$Pf = \frac{126.42}{126.87}$$

$$Pf = 0.996 = 99.6\% \text{ (power factor of total load)}$$

An analysis of the circuit in Example 18 shows that the induction motors require 100 kVA and operate at a power factor of 84% *lagging*. The synchronous motor requires 60 kVA and operates at a power factor of 70.7% *leading*.

Because the induction motors have a lagging power factor and the synchronous motor has a leading power factor, the synchronous motor is returning power to the line during most of the cycle when the induction motors are utilizing it. The induction motors return power to the line during most of the cycle when the synchronous motor is utilizing it. The end result is a more efficient operation than if all the motors were of the induction type. Synchronous motors are explained in Chapter 18.

Example 19

Replace the synchronous motor in Example 18 with an induction motor of the same size and operating at the same power factor as the synchronous motor. Calculate the total effective power, apparent power, and reactive power. Determine the power factor of the total load.

(a) The power taken by each load, expressed with reference to the supply voltage, is represented in Figure 14-20A. The angle corresponding to the power factor of 70.7% (cos 0.7071) is 45°. To determine the resultant power, draw a vector diagram as shown in Figure 14-20B. Remember that all induction motors have a lagging power factor.

(b) Draw the 100-kVA value along the horizontal, and label it OA.

(c) Figure 14-20A shows that the angle between the 100-kVA load and the 60-kVA load is 45° – 33° = 12°.

(d) Twelve degrees from OA, draw the 60-kVA vector OB (Figure 14-20B).

(e) Solve for the total apparent power.

$$OC = \sqrt{(OA)^2 + (OB)^2 + 2(OA)(OB)(\cos \phi)}$$

Figure 14-20C
Power vector diagram of the total apparent, effective, and reactive power

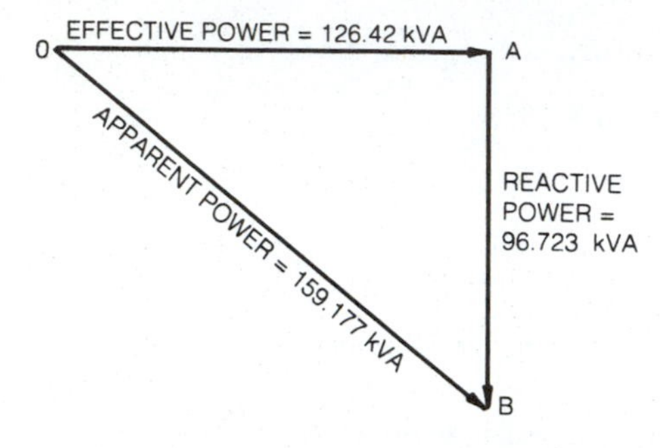

$$OC = \sqrt{(100)^2 + (60)^2 + (2 \times 100 \times 60 \times 0.9781)}$$
$$OC = \sqrt{10{,}000 + 3600 + 11{,}737.2}$$
$$OC = \sqrt{25{,}337.2}$$
$$OC = 159.177 \text{ kVA (total apparent power)}$$

(f) Calculate the effective power of each load.

$P = IE \cos \phi$ $\qquad$ $P = IE \cos \phi$
$P = 60 \times 0.707$ $\qquad$ $P = 100 \times 0.84$
$P = 42.42$ kVA $\qquad$ $P = 84$ kVA

(g) Calculate the total effective power.

$$P_t = P_1 + P_2$$
$$P_t = 42.42 + 84$$
$$P_t = 126.42 \text{ kW (total effective power)}$$

(h) Draw a power vector diagram for the total apparent power, effective power, and reactive power (Figure 14-20C). Draw the effective power on the horizontal. Draw the reactive power vertically downward from the effective power. The apparent power is the hypotenuse of the right triangle.

(i) Solve for the reactive power.

$$P_L = \sqrt{(P_{app})^2 - (P)^2}$$
$$P_L = \sqrt{(159.177)^2 - (126.42)^2}$$
$$P_L = \sqrt{25{,}337.317 - 15{,}982.016}$$
$$P_L = \sqrt{9355.301}$$
$$P_L = 96.732 \text{ kVAR (total reactive power)}$$

(j) Calculate the power factor.

$$Pf = \frac{P}{P_{app}}$$
$$Pf = \frac{126.42}{159.177}$$
$$Pf = 0.794 = 79.4\% \text{ (power factor of the total load)}$$

From the above calculations (Examples 18 and 19) it can be seen that the power factor is improved considerably by using a synchronous motor in place of an induction motor. Maintaining a high power factor provides a more efficient operation.

REVIEW

A. Multiple choice.
Select the best answer.

1. The effective resistance of an ac circuit varies with the
   a. applied voltage.
   b. frequency.
   c. inductive reactance.
   d. impedance.
2. The fact that alternating current tends to flow along the outer surface of the conductor is called
   a. skin effect.
   b. eddy currents.
   c. hysteresis.
   d. inductance.
3. The phase relationship of the voltages across various parts of a series ac circuit can be expressed with reference to the
   a. resistance.
   b. effective power.
   c. current.
   d. apparent power.
4. The voltage across a pure resistive load in an ac series circuit is
   a. out of phase with the current (lagging).
   b. in phase with the current.
   c. out of phase with the current (leading).
   d. none of the above.
5. An RLC circuit is one that contains
   a. resistance and capacitance.
   b. inductance and capacitance.
   c. resistance, inductance, and capacitance.
   d. resistance, reactance, and capacitance.
6. In a resonant circuit, the current and the voltage are
   a. out-of-phase leading.
   b. out-of-phase lagging.
   c. in phase.
   d. out-of-phase with the inductance.
7. A circuit can be brought into resonance by adjusting the
   a. voltage.
   b. current.
   c. frequency.
   d. power.

REVIEW *(continued)*

A. Multiple choice. Select the best answer (continued).

8. In a parallel RLC circuit, when the frequency is low, the inductor takes a
   a. high lagging current.
   b. low lagging current.
   c. high leading current.
   d. low leading current.
9. Another name for true power is
   a. effective power.
   b. reactive power.
   c. apparent power.
   d. inductive power.
10. Most transformers and alternators are rated in
    a. kilowatts.
    b. kilovolt-amperes.
    c. kilovars.
    d. kilograms.

B. Give complete answers.

1. Identify the factors that affect the resistance of an ac circuit but that do not affect a dc circuit.
2. Define *skin effect.*
3. Write the rule for voltage in an ac series circuit.
4. Why is the current vector usually used as a reference vector for ac series circuits?
5. What is the phase relationship of the current and the voltage in a pure resistive ac series circuit?
6. What factors affect the inductive reactance of a coil in an ac circuit?
7. What is the phase relationship of the current and the voltage in a pure inductive ac series circuit?
8. Does the current lead or lag the voltage in a series ac circuit containing a large value of capacitance?
9. List three components that make up impedance.
10. What is a series RC circuit?
11. How does inductance affect the power factor of an ac circuit?
12. Define *resonant circuit.*

REVIEW *(continued)*

B. Give complete answers (continued).

13. List three factors that affect the resonance of a circuit.
14. What is the rule for resistance in a parallel ac circuit?
15. Why is the voltage vector usually used as a reference vector for an ac parallel circuit?
16. What two quantities must be equal in order for a circuit to be in resonance?
17. Describe the exchange of energy between a capacitor and an inductor in an ac resonant parallel circuit.
18. What effect does a low power factor have on a circuit or a system?
19. Do most industrial establishments have a leading or a lagging power factor?
20. Identify two methods frequently used to improve the power factor in an industrial plant.

C. Solve each problem, showing the method used to arrive at the solution.

1. A capacitor with a capacitance of 410 μF is connected to a 120-V, 60-Hz line. Calculate the capacitor current.
2. A resistance of 50 Ω and a capacitance with a reactance of 35 Ω are connected in series across a 120-V, 60-Hz supply. Determine:
   a. The circuit current
   b. The voltage across the resistance
   c. The voltage across the capacitance
   d. The phase angle between the line current and the voltage
   e. The power taken by each component
   f. The effective power of the circuit
   g. The power factor of the circuit
3. A coil with an inductance of 0.1 H and negligible resistance is connected in series with a 500-μF capacitor. The combination is connected to a 230-V, 60-Hz line. Calculate the reactance of the combination and the line current.
4. A 60-μF capacitor and a 0.22-H coil of negligible resistance are connected in series across 120 V, 60 Hz. Determine the reactance of the combination and the line current.

REVIEW *(continued)*

C. Solve each problem, showing the method used to arrive at the solution (continued).

5. A coil with a resistance of 15 Ω and an inductance of 0.1 H is connected in series with a 500-μF capacitor. The combination is connected to a 60-Hz supply. Calculate:
   a. The circuit voltage if 3 A flow in the line
   b. The voltage across each component
   c. The impedance of the circuit
   d. The power factor of the circuit
   e. The power factor of the coil
6. A circuit contains a coil of 0.1-H inductance and 50-Ω resistance. What current will flow through it when it is connected to a 120-V, 60-Hz supply?
7. At what frequency will resonance occur in a series circuit containing an inductor of negligible resistance and 0.1-H inductance and a 2-μF capacitor?
8. What size capacitor must be connected in series with a 5-H coil that has 300-Ω resistance in order to produce resonance at 60 Hz?
9. If the coil in Problem 8 is connected to a 120-V, 60-Hz ac supply, determine:
   a. The line current
   b. The voltage across each component
   c. The power factor of the circuit
   d. The effective power of the circuit
10. A resistance of 10 Ω and an inductive reactance of 20 Ω are connected in parallel across 120 V ac. Calculate:
    a. The power factor of the circuit
    b. The line current
    c. The circuit impedance
    d. The effective power
    e. The apparent power
    f. The reactive power
11. An inductor of negligible resistance and 30-Ω reactance is connected in parallel with a capacitor of 50-Ω reactance. The combination is connected across 150 V ac. What is the line current?

REVIEW *(continued)*

C. Solve each problem, showing the method used to arrive at the solution (continued).

12. A resistance of 20 Ω, an inductive reactance of 15 Ω, and a capacitive reactance of 35 Ω are connected in parallel across a 120-V supply. Determine:
    a. The phase angle between the line current and the voltage
    b. The total current taken by the circuit
    c. The current through each component
    d. The circuit impedance
    e. The effective power of the circuit

13. Determine the impedance of a circuit containing a resistance of 32 Ω and an inductive reactance of 10 Ω connected in parallel.

14. At what frequency will resonance occur in a parallel circuit containing a coil of 20-Ω resistance and 0.5-H inductance that is connected in parallel with a 0.1-μF capacitor?

15. What size capacitor must be connected in parallel with a coil that has 100-Ω resistance and 0.2-H inductance to produce resonance at 180 Hz?

16. A circuit has two paths connected across 110 V. One path has a resistance of 25 Ω; the other path has a resistance of 10 Ω and an inductive reactance of 20 Ω. Calculate:
    a. The circuit power factor
    b. The total current of the circuit
    c. The effective power of the circuit
    d. The impedance of the circuit

17. A 5-hp motor connected to a 120-V, 60-Hz line has an efficiency of 80% at full load and a power factor of 60%.
    a. What size capacitor must be connected across the motor terminals to increase the power factor to unity?
    b. Calculate the line current at unity power factor.
    c. What size capacitor is necessary to increase the power factor to 85% lagging?
    Note: Power factor correcting capacitors are rated in kilovars (kVAR).

18. A circuit has two paths. One path has a coil with a resistance of 8 Ω and an inductive reactance of 20 Ω. The other path has a resistance of 8 Ω in series with a capacitive reactance of 16 Ω. Calculate the impedance of the circuit.

19. An alternator delivers 50 kW and 83.3 kVAR to a load. Determine the apparent power and the power factor of the load.

REVIEW *(continued)*

C. Solve each problem, showing the method used to arrive at the solution (continued).

20. A 40-kW load operating at a power factor of 70% is supplied through 10 miles of No. 6 AWG copper wire. The wire has a resistance of 0.410 Ω per 1000 feet. (The load is 5 miles from the supply.) If the voltage at the load is 13,200 V, calculate:
    a. The reactive power of the load
    b. The apparent power of the load
    c. The line current
    d. The line loss (watts)
    e. The voltage drop
21. Repeat the calculations for Problem 20 using a power factor of 90%.
22. A synchronous motor taking 60 kVA at 10% leading power factor is connected in parallel with an induction motor taking 100 kVA at 70% lagging power factor. Calculate:
    a. The total kW of the circuit
    b. The power factor of the combination
    c. The total kVA
    d. The kVAR of the circuit
23. An induction motor of 30 kW at 75% power factor is in parallel with a synchronous motor taking 20 kVAR at 82% power factor (leading). Determine:
    a. The power factor of the circuit
    b. The total apparent power
    c. The total effective power
24. A load connected to a 60-Hz, 440-V line requires 10 kW. The power factor is 60% (lagging). Calculate the kVAR rating of the capacitor that will increase the power factor to unity.
25. An induction motor is rated at 10 kW. If an ammeter indicates 60 A on a 240-V line, what is the power factor of the motor?

15 AC GENERATORS

Objectives

After studying this chapter, the student will be able to:

- Describe the construction and operating characteristics of various types of alternating current generators.
- Discuss the methods for controlling the output voltage and frequency of alternating current generators.
- Discuss the methods for producing single-phase and multiphase voltages.
- Describe the procedures for connecting alternating current generators in parallel.

AC GENERATORS VERSUS DC GENERATORS

As previously stated, alternating current is the primary source of electrical energy. It is less expensive to produce and transmit than direct current. These advantages, however, do not rule out the use of dc. There are many applications for which direct current is preferred or required. DC generators are often used to provide power for operations such as metal refining, electroplating, and battery charging.

DC generators, however, have certain built-in limitations that restrict their power output. For this reason, and because ac voltage is induced into the armature of all generators, ac generators are generally more practical.

ALTERNATOR CONSTRUCTION

The commutator on a dc generator converts alternating current to direct current. Figure 15-1 shows a single-loop dc generator. If the commutator is replaced with slip rings, Figure 15-2, alternating current will be supplied to the external circuit.

Figure 15-1
Single loop dc generator

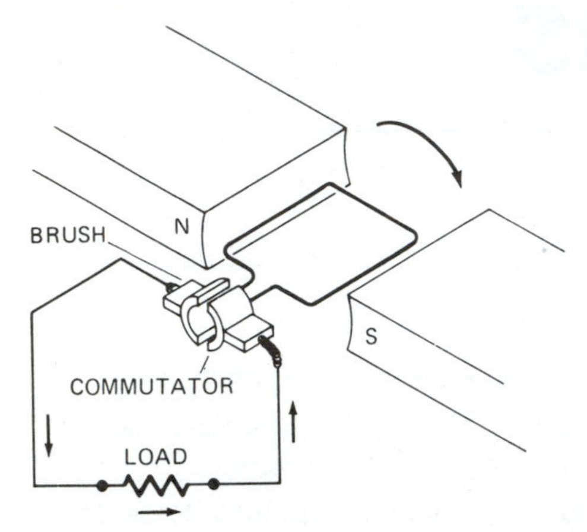

Figure 15-2
Single-loop ac generator

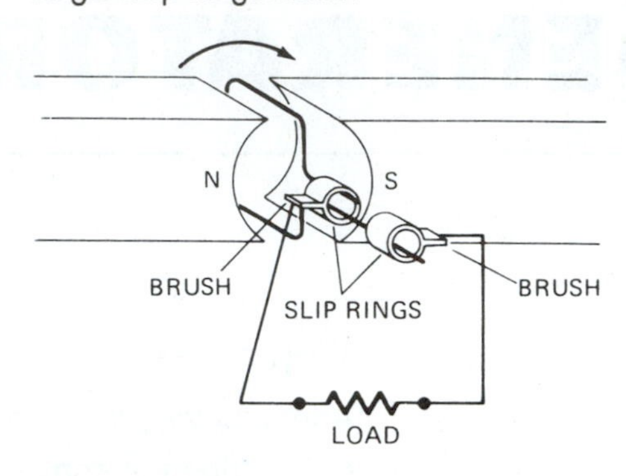

In the two-pole alternator, one cycle is produced for each revolution of the armature. If four poles are used, two cycles are produced for each revolution; six poles produce three cycles; an so on. Figure 15-3 illustrates a four-pole machine.

In order to maintain a uniform magnetic field, it is necessary to provide direct current for the field circuit. A separate dc generator is usually mounted on the same drive shaft as the alternator rotor. This generator, called an *exciter*, provides the dc for the alternator field. Except for the slip rings, the construction of many alternators is similar to that of dc generators. Large alternators, however, have a stationary armature and a rotating field. The field receives its energy from the dc exciter, through slip rings. The exciter voltage is usually 250 volts. At this voltage there is no serious difficulty with insulation breakdown or arcing. Because the armature is stationary, it is called *the stator,* and the revolving field is called the *rotor*. Figure 15-4A shows a typical stator for a slow-speed alternator. Figure 15-4B shows the rotor for the same machine.

Figure 15-3
Four-pole alternator

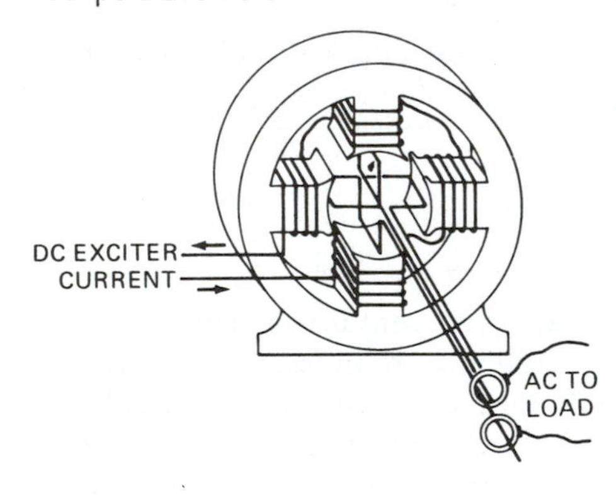

Figure 15-4A
Stator for a slow-speed alternator

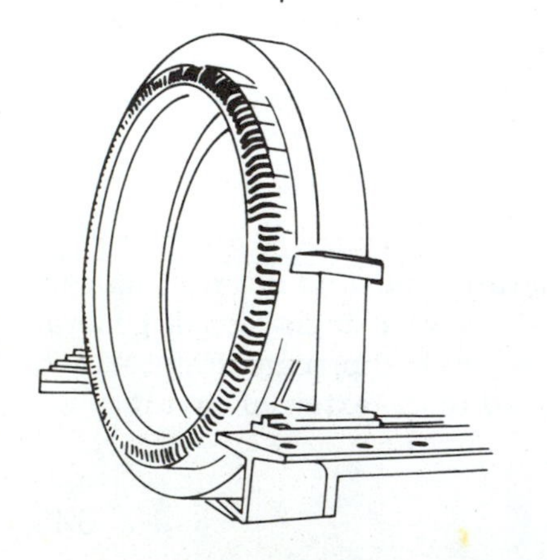

Figure 15-4B
Rotor for a slow-speed alternator

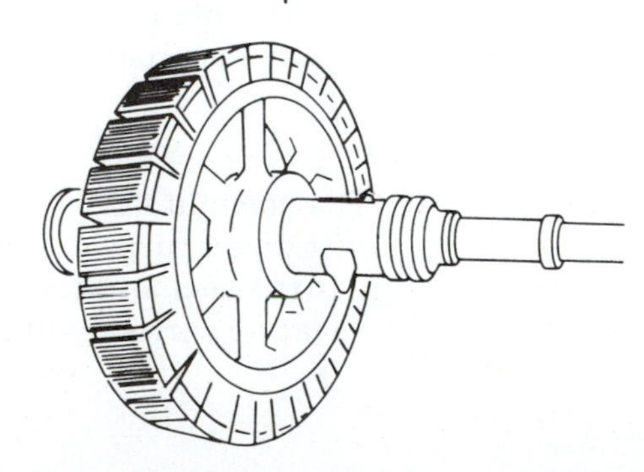

The rotating field of a slow-speed alternator consists of a number of laminated poles dovetailed or bolted to a cast-iron form called a *spider*. The slip rings are attached to and insulated from the spider. The ends of the field windings are connected to the slip rings. Brushes riding on the slip rings connect the dc exciter to the alternator field windings. The field coils for small machines are wound with wire; rectangular copper strips are used for large machines.

When steam turbines are used to drive alternators, the rotors must be constructed for high-speed operation. Alternators of this type generally have only two, four, or six poles. The poles are wound on a cylindrical form. The windings are embedded in the rotor core in order to reduce windage loss. Figure 15-5A shows a cylindrical-type field (rotor) for a steam turbine–driven alternator. The stator for this machine is shown in Figure 15-5B.

Figure 15-5A
Rotor for a steam-driven (high speed) alternator *(Courtesy of Westinghouse Electric Corp., Framingham, MA)*

Figure 15-5B
Stator for a steam-driven (high speed) alternator *(Courtesy of Westinghouse Electric Corp., Framingham, MA)*

These machines produce large quantities of heat because of their speed and design. This heat must be carried away by air currents forced through the passages in the heated parts. A tunnel-like enclosure controls the direction of the air currents and minimizes the noise.

For cooling large turbo-type alternators, hydrogen is often used instead of air. Because hydrogen is lighter than air, the windage losses are reduced and the efficiency of the machine is increased. Other advantages of using hydrogen are reduction in noise and less oxidation in the windings.

ALTERNATOR VOLTAGE OUTPUT

AC generators may be constructed to produce one or more voltages. These voltages and their phase relationship are important factors in the selection and operation of equipment.

Single-Phase Alternator

A *single-phase alternator*, Figure 15-6, is an ac generator that produces only one voltage. The armature coils are connected in "series-additive." In other words, the sum of the emfs induced into each coil produces the total output voltage. Single-phase generators are usually constructed in small sizes only. They are used for standby service in case the main power source is interrupted, for supplying temporary power on construction sites, and for permanent installations in remote locations.

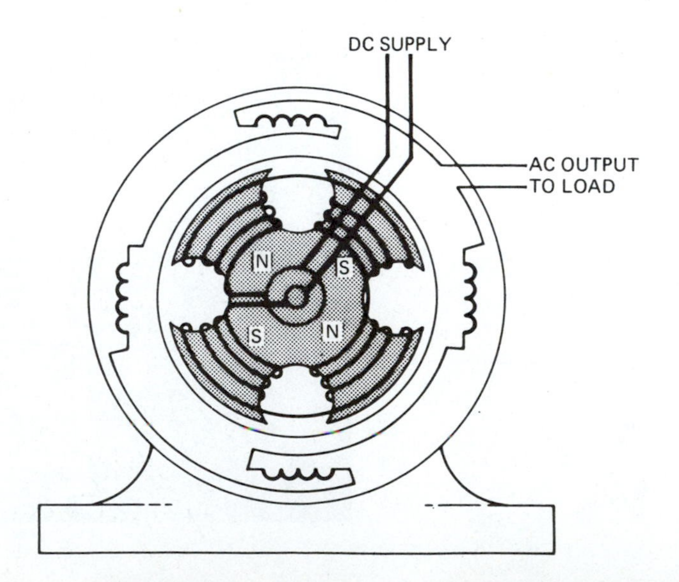

Figure 15-6
Single-phase alternator with a rotating field

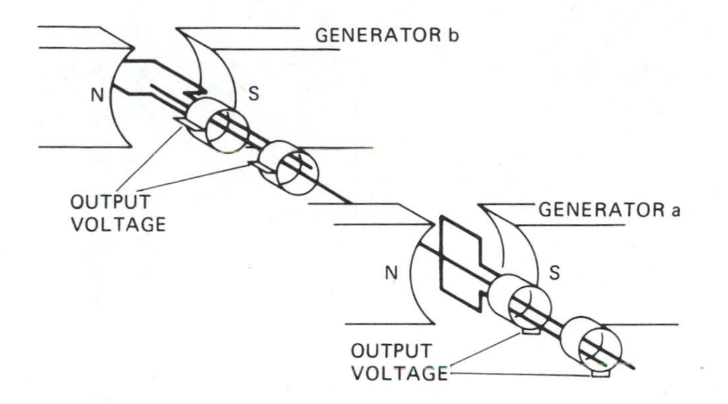

Figure 15-7A
Two single-phase alternators arranged to produce a two-phase configuration

Two-Phase Alternators

The *two-phase alternator* produces two separate voltages 90 degrees out of phase with each other. A two-phase alternator is a combination of two single-phase alternators with their armatures connected so there is a 90-degree phase displacement between the output voltages. Figure 15-7A illustrates this arrangement. The armatures are mounted on the same shaft so they will rotate together. Thus, when generator *a* is producing maximum voltage, generator *b* is producing zero volt. One-quarter turn later, *b* is producing maximum voltage, and *a* is producing zero volt.

Figure 15-7B shows the same two windings installed in one generator with a 90-degree phase displacement. The same effect will take place as before; however, now one generator is producing the two separate voltages. Figure 15-8 illustrates a typical two-phase alternator with a rotating field.

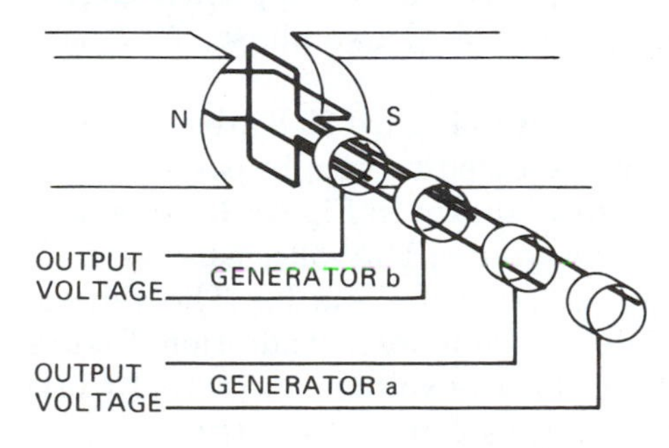

Figure 15-7B
Two-phase alternator with a rotating armature

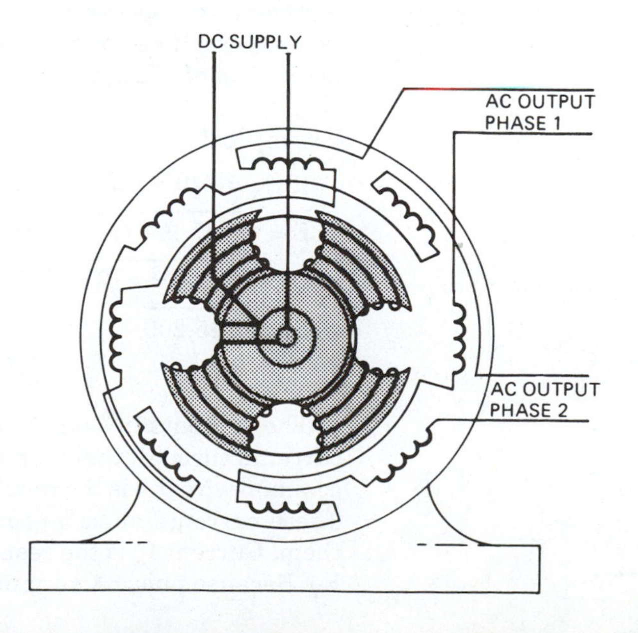

Figure 15-8
Two-phase alternator with a rotating field

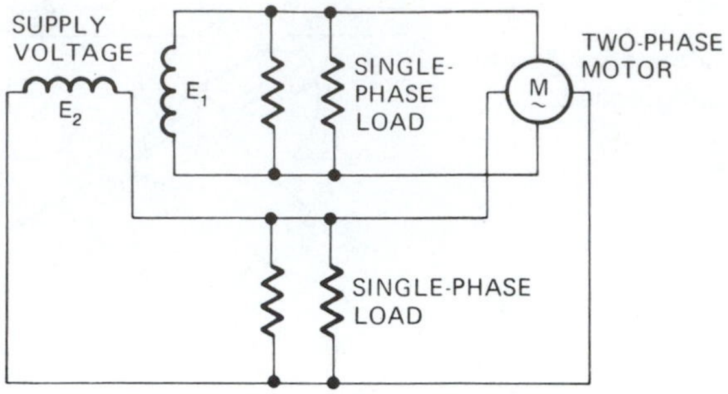

Figure 15-9
Typical two-phase, four-wire system

There are several different methods for connecting the coils of alternators for a two-phase system. One method is to bring the four wires out separately. The coils are not electrically connected together. This arrangement is known as a, *two-phase, four-wire system*. Both single phase and two phase may be supplied from this system. Figure 15-9 illustrates a typical two-phase, four-wire system.

If the two phases of the alternator in Figure 15-8 are connected in series, only three wires are required to supply power to the external load. This arrangement is shown in Figure 15-10 and is known as a two-phase, three-wire system. The voltage between L_X and L_C is equal to the voltage between L_Y and L_C. The voltage between L_X and L_Y may be determined by vector addition. Figure 15-11A shows a vector diagram for the voltages; Figure 15-11B shows the parallelogram method for solving the vectors.

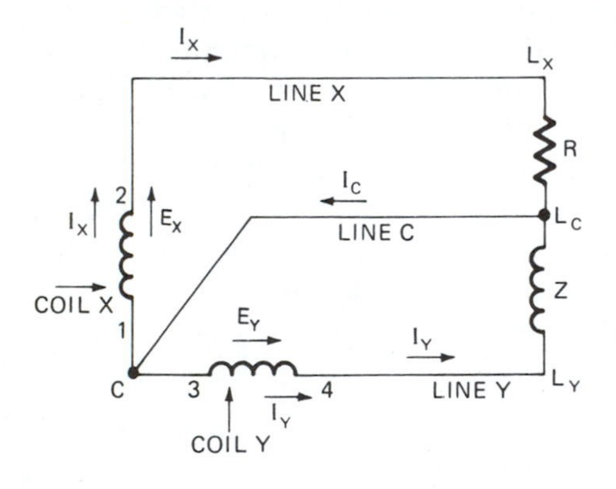

Figure 15-10
Two-phase, three-wire system

By tracing the circuit formed by coils X and Y, it can be seen that the voltage in coil X is in the opposite direction to that of coil Y. To add these voltages vectorially, it is necessary to reverse one of the two vectors. If vector E_X is reversed, Figure 15-11B, then vector V is the sum of $-E_X$ and E_Y.

Example 1

If $E_X = 240$ V and $E_Y = 240$ V, then

$E = \sqrt{(-E_X)^2 + (E_Y)^2}$ *(From Equation 14.4)*

$E = \sqrt{(-240)^2 + (240)^2}$

$E = \sqrt{115{,}200}$

$E = 339$ V

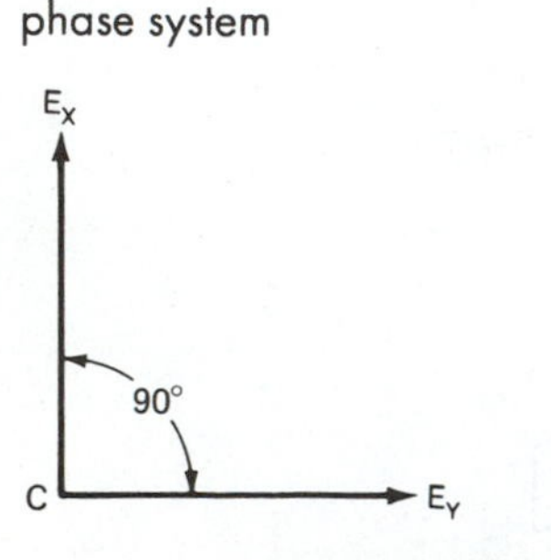

Figure 15-11A
Vector diagram of voltages in a two-phase system

The currents in lines X and Y (Figure 15-10) are equal to the currents through their corresponding phases. The current in the common wire (I_C) is the resultant of the currents in the two phases. These currents must be compared with the voltages that cause them. Current I_X is the result of voltage E_X, and I_Y is the result of E_Y. Because phase X supplies a pure resistive load, I_X is in phase

Figure 15-11B
Parallelogram method for calculating voltage in a two-phase, three-wire system

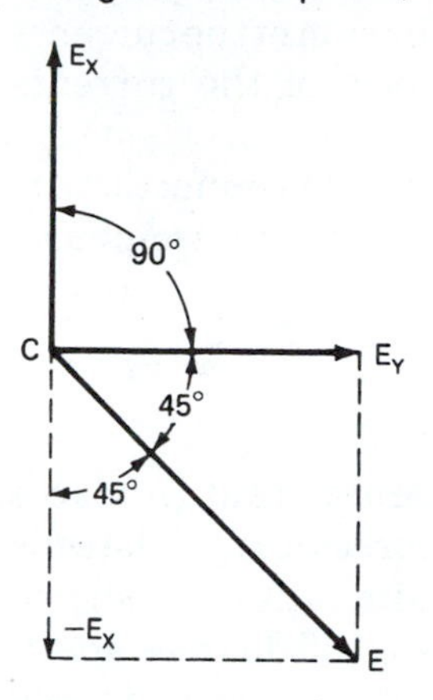

with E_X. Phase Y supplies an inductive load, therefore I_Y lags E_Y by ϕ degrees. The two currents I_X an I_Y are 90 + ϕ degrees out of phase with each other. The current in the common wire may be determined by adding vectorially the currents through the two phases. Because currents I_X and I_Y are both flowing in the same direction through line C, it is not necessary to reverse one of the current vectors.

Example 2

Assume that the current through coil X is 10 A and the current through coil Y is 15 A. If I_Y lags E_Y by 30°, what is the current through line C?

Step 1. Lay out current I_X along the voltage vector E_X.

Step 2. I_Y lags E_Y by 30°. Therefore, I_Y is laid out indicating the 30° lag (Figure 15-12A). The two currents I_X and I_Y are 90° + 30° = 120° out of phase with each other.

Step 3. Form a parallelogram as shown in Figure 15-12B. The resultant vector represents current I_C.

$$I_C = \sqrt{(I_X)^2 + (I_Y)^2 - (2I_XI_Y \cos \phi)} \text{ (From Equation 13.5)}$$

$$I_C = \sqrt{(100) + (225) - (2 \times 10 \times 15 \times 0.5)}$$

$$I_C = \sqrt{325 - 150}$$

$$I_C = \sqrt{175}$$

$$I_C = 13.2 \text{ A}$$

Line C is carrying 13.2 A.

Figure 15-12A
Vector diagram of currents in a two-phase, three-wire system

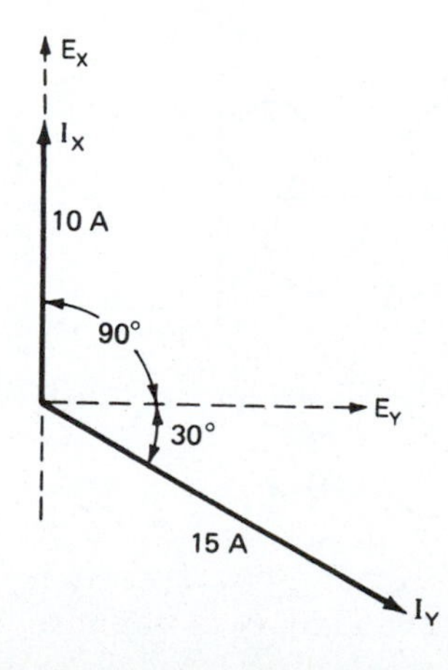

Figure 15-12B
Parallelogram method for calculating the current in the common wire of a two-phase, three-wire system

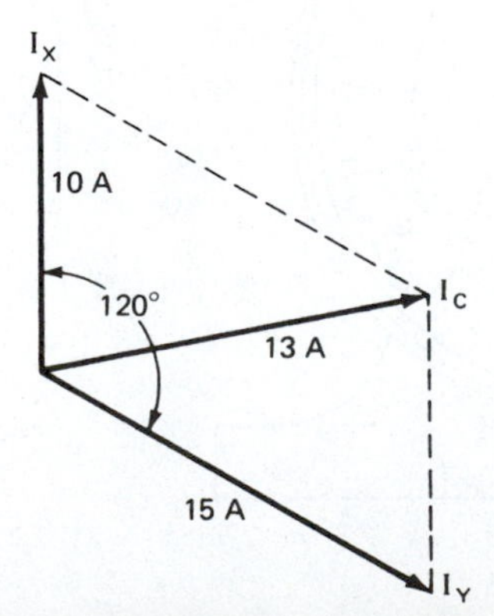

Figure 15-13
Two-phase, five-wire system. Phases A and B supply 240 V, 2 phase. Single phase, 120 V, is supplied between the neutral and any one of the phase conductors. The 170 V is not normally used.

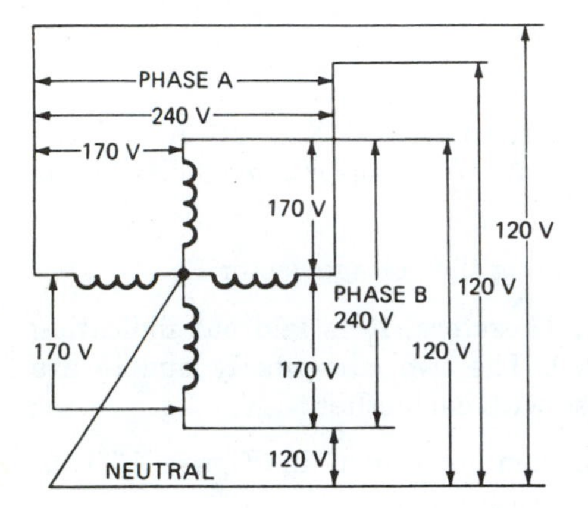

Current I_C is flowing in the common wire and is flowing into point C. The current flowing into point C is equal to the vector sum of the currents flowing through phases X and Y from point C. (Kirchhoff's Current Law states that the vector sum of the currents leaving a junction is equal to the vector sum of the currents entering the junction.)

Another method sometimes used for two-phase connections is shown in Figure 15-13. This arrangement is known as a two-phase, five-wire system.

Three-Phase Alternators

The *three-phase alternator* is an ac generator that produces three separate voltages 120 electrical time degrees apart. The coils are arranged around the stator in a manner similar to that shown in Figure 15-14A. Mechanically they are spaced 60 degrees apart. However, the electrical connections are such that the induced voltages are out of phase by 120 electrical time degrees. A graph of the output voltages is shown in Figure 15-14B.

The coils are usually connected together inside the machine. There are two configurations that are generally used. The *wye connection,* sometimes called the star connection, is illustrated in Figure 15-15A. The *delta connection* is shown in Figure 15-15B.

Figure 15-14A
Three-phase alternator stator with the windings connected into a delta configuration

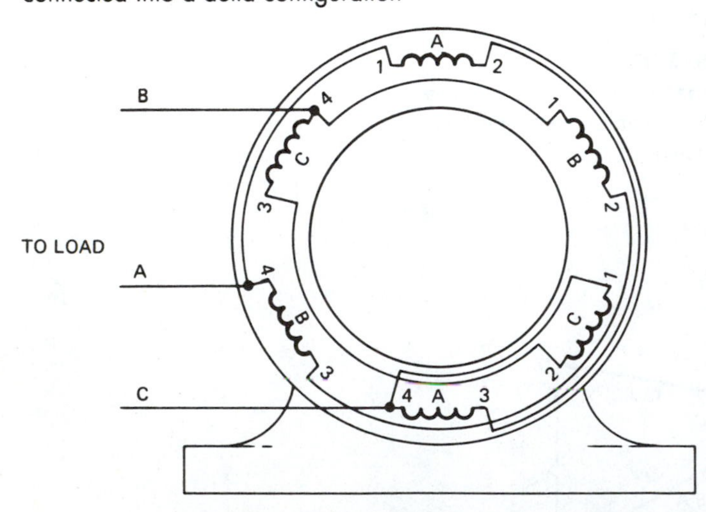

Figure 15-14B
Graph of the output voltages of a three-phase alternator

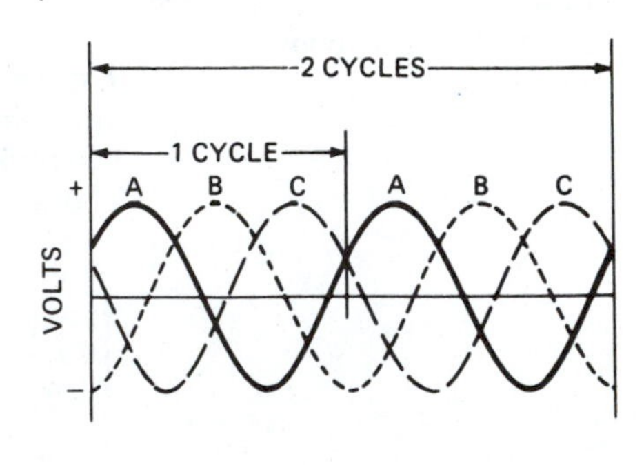

Figure 15-15A
Wye (star) three-phase connection

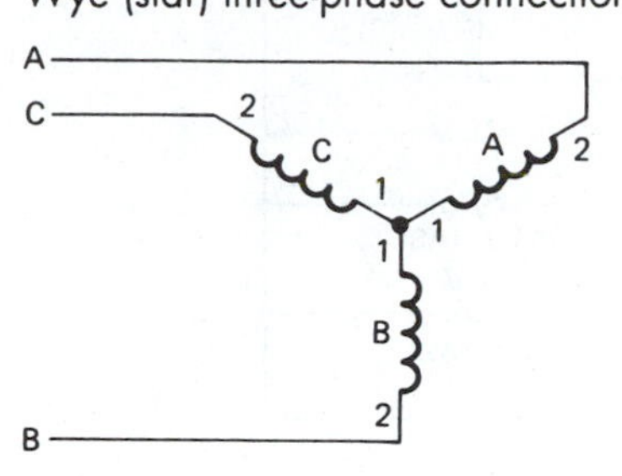

Figure 15-15B
Delta three-phase connection

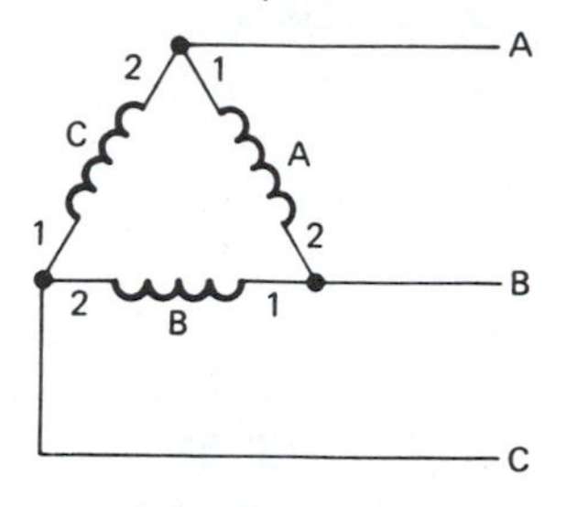

Wye (Star) Connection

With the wye connection, the beginnings of the coils are connected together, and the ends are brought out to the alternator terminals. Another method used to illustrate the wye connection is shown in Figure 15-16. The common voltages for the three-phase, three-wire wye system are 208 volts and 480 volts. This system provides 3 single-phase voltages and 1 three-phase voltage. For example, a 480-volt system can supply 480 volts to a three-phase load and three separate 480-volt circuits for single-phase loads. The single-phase loads are supplied through any two line wires.

Connecting a fourth wire at the junction point of the three phases makes available two values of single-phase voltage. This arrangement is called a three-phase, four-wire wye system. Common voltages obtained from this system are 120 volts and 208 volts, single phase, and 208 volts, three phase. The same type of connection is also used to furnish 277 volts and 480 volts, single phase, and 480 volts, three phase. Figures 15-17A and 15-17B illustrate the load distribution for both the three-wire and four-wire systems.

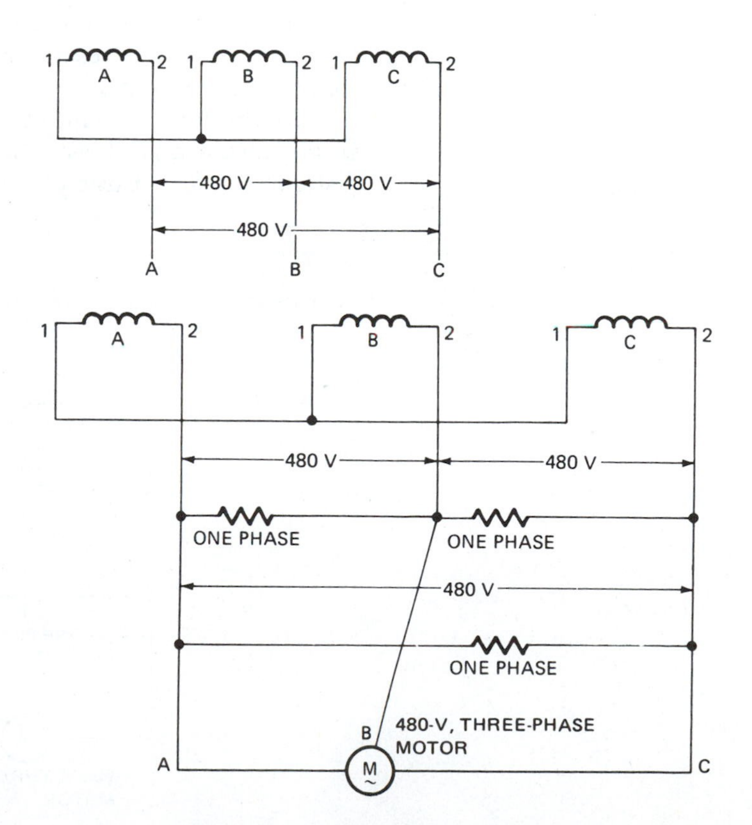

Figure 15-16
Wye (star) three-phase, three-wire connection

Figure 15-17A
Three-phase, three-wire system supplying single-phase and three-phase loads (wye connection)

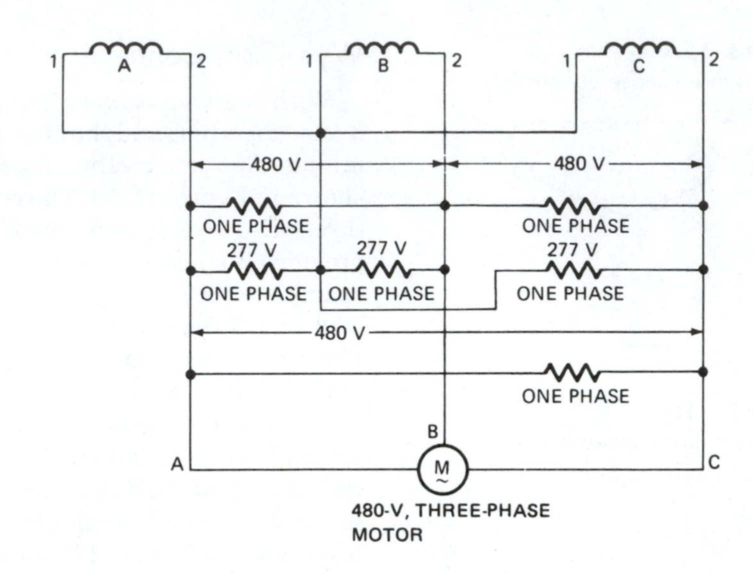

Figure 15-17B
Three-phase, four-wire system supplying single-phase and three-phase loads (wye connection)

Figure 15-18
Three-Phase, three-wire, 440-V delta system

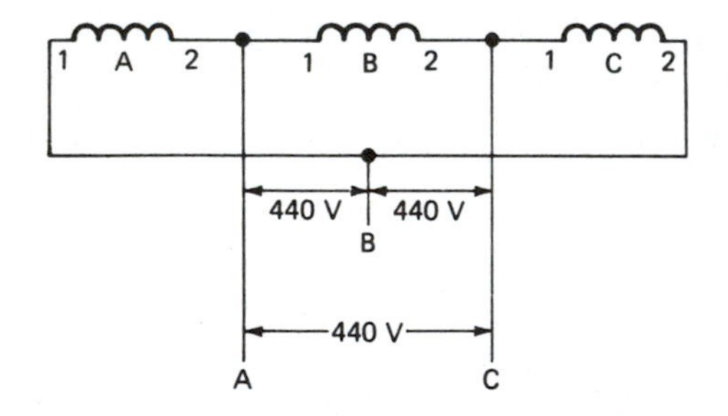

Delta Connection

The delta connection is shown in Figure 15-18. In this configuration, the end of each phase coil is connected to the beginning of the next. The line wires are brought out from the points where the connections are made.

With the delta system, only one value of voltage appears for both single phase and three phase. Figure 15-19 depicts the load distribution for a three-phase, three wire delta system.

Figure 15-19
Three-phase, three-wire delta system supplying single-phase and three-phase loads

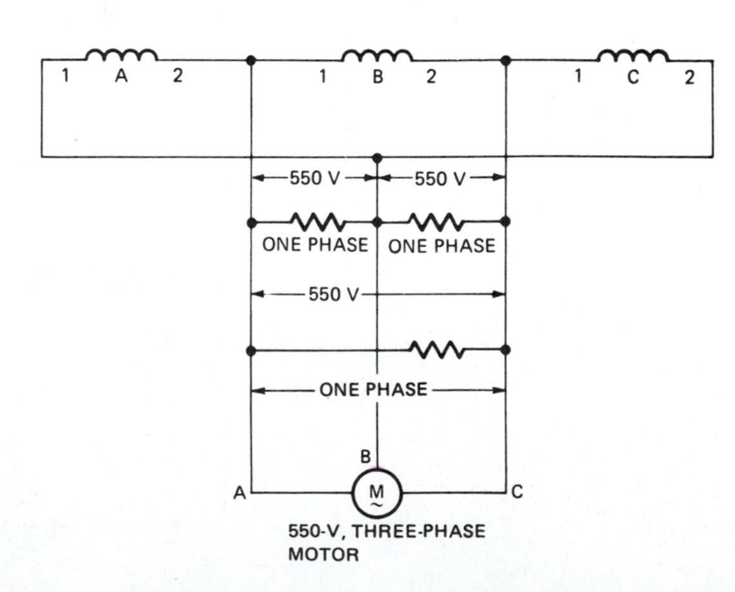

Phase Sequence

The phase sequence is the order in which the voltages follow one another (reach their maximum values). In Figure 15-14B, phase A reaches its maximum value first, followed by B and later by C. The phase sequence is an important factor when connecting polyphase generators in parallel and when connecting polyphase motors.

Voltage and Current in a Delta-Connected Alternator

The voltage induced into the armature coils of a delta-connected alternator is transmitted to the line terminals; therefore the voltage across anyone of the armature coils, Figure 15-15B, is equal to the voltage between any two transmission lines. In other words, in Figure 15-15B, the voltage across coil A is equal to the voltage between lines A and B. The voltage across coil B is equal to the voltage between lines B and C, and the voltage across coil C is equal to the voltage between A and C. The voltage between the lines is called the line voltage (E_L), and the voltage across the coils is called the phase voltage (E_P). For a delta system, the line voltage is equal to the phase voltage ($E_L = E_p$).

The phase currents in Figure 15-15B are the currents through coils A, B, and C. The currents through the line wires are equal to the vector sum of the currents through two of the phases. For example, the current through line A is equal to the vectorial sum of the currents through coils A and C. For a delta system, it has been determined that the vectorial sum of the currents through two of the phases is equal to 1.73 times the current through one phase. Therefore, in a delta-connected alternator, the line voltage is equal to the phase voltage ($E_L = E_p$), and the line current, for balanced loads, is equal to 1.73 times the phase current ($I_L = 1.73 I_p$).

Voltage and Current in a Wye-Connected Alternator

The beginnings of each armature coil of a wye connection are connected together; therefore, the line voltage is equal to the voltage across two of the phase coils. In Figure 15-15A, the voltage between lines A and C is equal to the voltage across coils A and C. If it is assumed that the emf is acting from terminal 1 to terminal 2 in each coil, then the forces are acting in opposition to one another but they are displaced by 120 electrical time degrees. In order to combine the two voltages, they must be subtracted vectorially. It has been determined that the vectorial difference of two voltages in a wye-connected alternator is equal to 1.73 times one phase voltage. Therefore, in a wye-connected alternator, the line voltage is 1.73 times the phase voltage ($E_L = 1.73\ E_p$)

Assume the current is flowing out coil A (Figure 15-15A) from 1 to 2. It then flows through line A to the load. Therefore, the current through line A is equal to the current through coil A. This is true for each coil and each line in a wye system. Thus, for a wye-connected generator, the line current is equal to the phase current ($I_L = I_p$).

Power in a Three-Phase System

The power for a balanced three-phase system can be calculated by the formula $P = 1.73IE \cos \phi$. In a balanced three-phase system, the power factor is the cosine of the angle between the phase current and the phase voltage. This formula may also be used for systems that are only slightly unbalanced. The value of current used must be the average of the three values. With only slightly unbalanced systems, this method will provide a reasonably accurate value of power.

If the load is severely unbalanced, a vectorial analysis of each phase must be done, and then the three phases must be combined. For information on measuring the power of a three-phase system, see Chapter 20.

VOLTAGE AND FREQUENCY CONTROL

The standard frequency for power distribution in the United States is 60 hertz. Europe, Asia, and South America generally operate on a frequency of 50 hertz, and in some cases 25 hertz.

The frequency of the emf produced by an alternator depends upon the number of field poles and the rotor speed. One cycle is generated whenever one pair of poles passes a coil. A simple formula for calculating the frequency is

$$f = \frac{PN}{120} \qquad \text{(Eq. 15.1)}$$

where f = frequency, in hertz (Hz)
P = number of field poles on the alternator
N = speed of the rotor, in revolutions per minute (r/min)

Federal regulations require that utility companies maintain a relatively constant frequency. The maximum variation permitted is 3 percent. Therefore, a speed regulator must be installed on the prime mover in order to maintain 60 hertz under varying load conditions.

The voltage supplied by the utility company varies for different sections of the United States and the world. Some standard voltages supplied by utility companies are 2300 volts, 4800 volts, 6900 volts, and 33,100 volts. Federal regulations also require utility companies to maintain a relatively constant value of voltage. To meet this requirement, utility companies build an automatic voltage regulator into the system. This regulator measures the output voltage of the alternator and adjusts the alternator field current as needed.

ALTERNATOR CHARACTERISTICS

Three factors affect the output voltage of an alternator.

1. The resistance of the stator windings causes an IR drop within the generator. The value of the voltage drop increases with an increase in load.
2. Self-induction takes place within the stator windings, causing an IX_L drop. This voltage drop also varies with the load.
3. The power factor of the load also affects the output voltage.

Figure 15-20 is a graph of the effects of power factor on the output voltage. From this information it can be seen that both the amount of load and the type of load affect the output voltage.

PARALLELING ALTERNATORS

AC power systems generally consist of several alternators connected in parallel to a common distribution bus. Parallel operation of alternators provides for maximum efficiency and allows time for servicing. During the time of day when the load is light, one or more generators can be removed from the line to allow the remaining generators to operate at or near full load. Generators operate at maximum efficiency when fully loaded. Maintenance personnel can take advantage of the time when the generators are not in use to perform preventive maintenance.

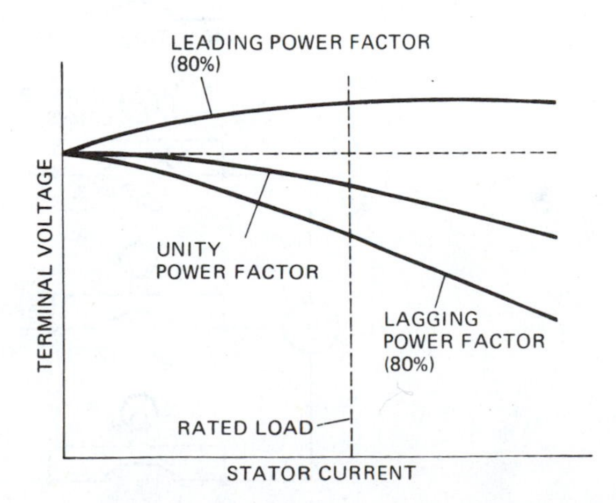

Figure 15-20
Graph of the effects of power factor on the output voltage of an alternator

When alternators are connected in parallel, certain conditions must be fulfilled. These conditions are as follows:

1. The alternator to be placed on the line must produce a voltage wave of approximately the same shape as the voltage wave across the busses to which it is to be connected.
2. The terminal voltage of the alternator must be equal to the voltage across the busses.
3. The frequency of the output voltage must be equal to that of the bus voltage.
4. With reference to the load, the voltage of the alternator must be in phase with the bus voltage.
5. The phase rotation of the alternator and the distribution bus must be the same.

The procedure to follow in order to connect alternators in parallel is called *synchronizing*. Figure 15-21 shows three alternators connected for parallel operation.

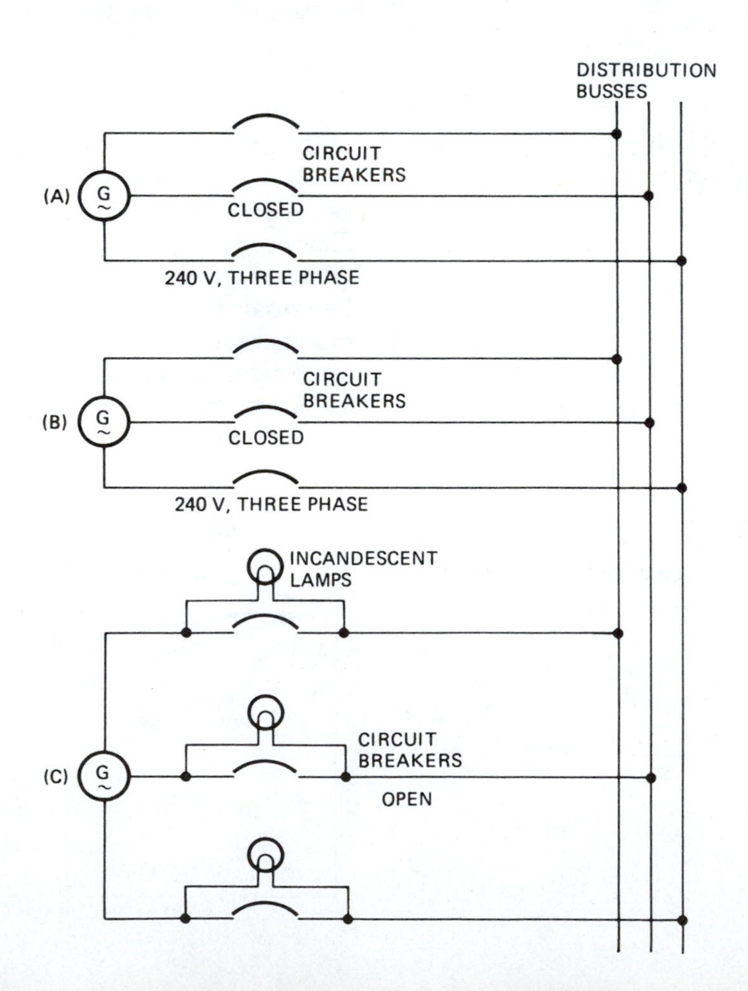

Figure 15-21
Connecting alternators in parallel

Figure 15-22
Synchroscope

Assume that the third alternator (C) is ready to be connected on the line. The following procedure is recommended. Assume that alternators A and B are supplying 240 volts to the distribution busses. Alternator C is started and brought up to speed. The output voltage is adjusted to equal that of the bus voltage. This is accomplished by varying the field rheostat. The frequency of C must be matched to that of the distribution bus. Adjustments in the speed of the rotor accomplish this requirement.

A *synchroscope* is used in order to determine when the frequency of C matches that of the bus. A synchroscope is an instrument that indicates whether it is necessary to increase or decrease the speed of the prime mover, Figure 15-22.

For small alternators, the frequency can be matched using incandescent lamps. *Frequency meters* may also be used.

Once the voltage and frequency are correct, it is necessary to check the phase sequence. This operation can be accomplished by using incandescent lamps or a *phase sequence meter*.

If the lamp method is used, the power rating of the lamps must be equal, and the voltage rating must equal the alternator voltage. One lamp is connected across each pole of the three-pole switch, Figure 15-23. If the phase sequence is correct, the three lamps will become bright and dim together. If the lamps become bright an dim one after another, the phase sequence of the alternator is opposite to that of the distribution bus. Interchanging any two of the three leads from the alternator corrects this error.

If the lamps increase and decrease in brightness simultaneously, the sequence is correct but the frequency is incorrect. If the lamps remain out (dark), the frequency and phase sequence are correct and the alternator may be connected to the distribution bus by closing the switch. For alternators with high voltage outputs, potential transformers can be used to reduce the voltage to that of the lamps.

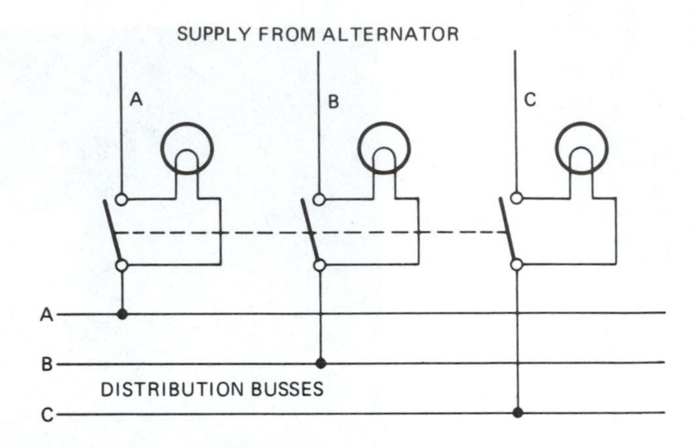

Figure 15-23
Incandescent lamps connected for checking frequency and phase sequence of an oncoming alternator

The "lamp dark" method for synchronizing alternators is generally used only for small operations or for laboratory experiments. Most utility companies have instrument panels equipped with many meters, including synchroscopes and frequency meters. Modern generating stations are equipped with automatic controls that connect or disconnect alternators from the distribution busses with changes in load. All stations, however, must have hand-operated backup equipment in case of automation failure.

Effect of Varying Field Strength

When all the alternators are synchronized and supplying power to the load, they are in phase relative to the load. With reference to one another, they are 180 degrees out of phase. Therefore, no current circulates between the machines.

If the magnetic field of one alternator should increase, the output voltage also increases. This increase in voltage causes a current to circulate between the alternators. Because of the low resistance and high inductance of the alternators, the circulating current is highly reactive and tends to keep the field strength of all the alternators equal. The increase in output voltage is only slight in comparison to the field current, and is generally offset by the impedance drop (IZ) produced by the circulating current.

If two alternators operating in parallel supply equal current to the load and have the same power factor, and the field excitation of one machine is changed, a circulating current is established. This circulating current may increase the current in one machine and decrease the current in the other. However, the power supplied by each alternator is not greatly affected because the power factor changes inversely with the current. (This means that an alternator cannot be made to take a load merely by increasing the field current. But this method can be used to improve the power factor of the machine, resulting in a decrease in armature current.)

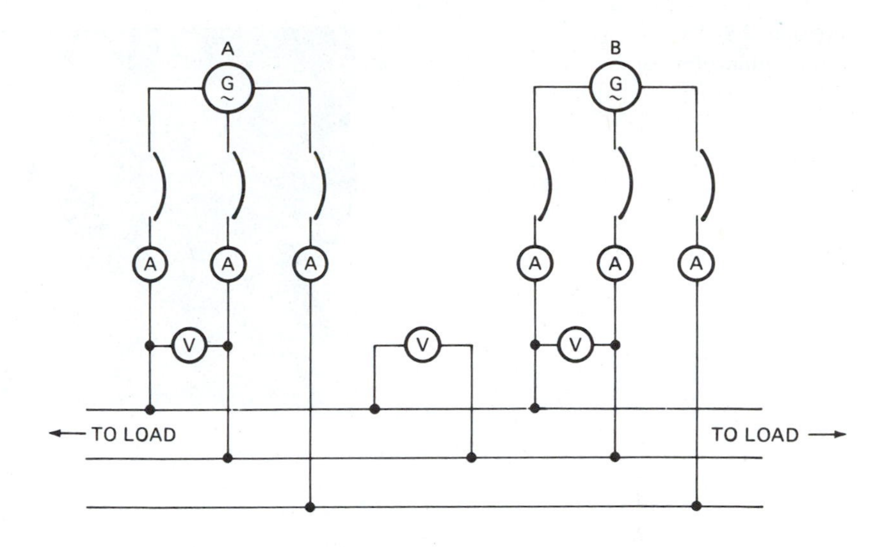

Figure 15-24
Two 3-phase alternators connected in parallel

Effect of Increased Driving Force

Figure 15-24 illustrates two alternators connected in parallel and supplying power to the load. If the torque of alternator B is increased the rotor speed increases for a fraction of a revolution until its output voltage E_B has pulled slightly ahead in phase relationship to E_A. The load on B increases causing a reverse torque, which slows down the rotor to the original speed. Because alternator B has advanced in phase position, the two voltages E_A and E_B no longer neutralize each other, and a resultant voltage exists. The resultant voltage causes a circulating current between the two alternators. Because the resistance of the two alternators is low in comparison to their reactance, the circulating current will lag the resultant voltage by nearly 90 degrees. This current is very close to being in phase with E_B and is nearly 180 degrees out of phase with E_A Because of this phase relationship, alternator B will carry a high current compared to alternator A. (This indicates that if two alternators in parallel supply equal currents to a load, and the driving torque of one machine is increased, the power supplied by this machine is increased and the power supplied by the other machine is decreased. This change occurs without materially affecting the power factor of either machine.)

When one alternator has been synchronized and paralleled with another, it supplies no power to the load. If additional torque is supplied to the oncoming machine, its voltage advances a few degrees, causing it to assume part of the load. This increase in load causes the rotor to slow down to its original speed, thereby maintaining a constant frequency.

Operators in power plants can control the steam or water supplied to the prime movers, thus controlling the torque. This

Figure 15-25
Motor-generator set

permits shifting the load between alternators and adding or removing alternators from the line. These changes are made without interruption of service and with only minor variations of frequency. In modern generating plants, load shifting is accomplished through automated equipment. However, hand-operated backup equipment is installed for emergency purposes.

MOTOR-GENERATOR SETS

A *motor-generator set,* Figure 15-25, is a combination of two machines. An electric motor is coupled to a generator and arranged so that the motor serves as a prime mover. Motor-generator sets serve many different purposes. For example, an ac motor may drive a dc generator to provide dc power for a specific load. A 600-volt dc motor may drive a 115-volt dc generator. A 60-hertz, 120-volt ac motor may drive a 400-hertz ac generator, or a dc motor may drive an ac generator. These are the most common uses for motor-generator sets.

ROTARY CONVERTERS (DYNAMOTORS)

A *rotary converter*, sometimes called a *dynamotor*, is a motor-generator set combined into one housing. The armature for the motor is wound on the same shaft as the generator armature. Rotary converters are used for many of the same purposes as motor-generator sets. They are frequently used to supply dc power in areas where only ac is available. They are also used in communication systems to convert a low-voltage dc to a higher value.

REVIEW

A. Multiple choice.
Select the best answer.

1. Direct current is used for
   a. transformers.
   b. electroplating.
   c. induction motors.
   d. most lighting loads.
2. The voltage induced into the armature of a rotating-type generator is
   a. direct voltage.
   b. alternating voltage.
   c. pulsating voltage.
   d. extremely high voltage.
3. A four-pole alternator will produce
   a. one cycle for each revolution of the rotor.
   b. two cycles for each revolution of the rotor.
   c. three cycles for each revolution of the rotor.
   d. four cycles for each revolution of the rotor.
4. The field circuit of an alternator is supplied with
   a. alternating current.
   b. pulsating current.
   c. direct current.
   d. eddy current.
5. Alternators designed to be driven by steam turbines are built for
   a. high-speed operation.
   b. slow-speed operation.
   c. medium-speed operation.
   d. both a and b.
6. Single-phase alternators are generally used to supply
   a. very large loads.
   b. small loads.
   c. any type of load.
   d. only lighting loads.
7. Two-phase alternators produce two separate voltages that are
   a. 60 electrical time degrees apart.
   b. 90 electrical time degrees apart.
   c. 120 electrical time degrees apart.
   d. 180 electrical time degrees apart.

REVIEW *(continued)*

A. Multiple choice.
Select the best answer (continued).

8. Three-phase alternators produce three separate voltages that are
   a. 60 electrical time degrees apart.
   b. 90 electrical time degrees apart.
   c. 120 electrical time degrees apart.
   d. 180 electrical time degrees apart.
9. Another name for the three-phase wye connection is
   a. delta.
   b. alpha.
   c. star.
   d. beta.
10. The order in which the voltages of a three-phase system follow one another is called
    a. phase order.
    b. phase sequence.
    c. phase pattern.
    d. phase system.
11. The line voltage of a delta-connected system is equal to
    a. 1.73 times the phase voltage.
    b. the phase voltage.
    c. the phase voltage divided by 1.73.
    d. the phase voltage plus 1.73.
12. The line current of a delta system is equal to
    a. 1.73 times the phase current.
    b. the phase current.
    c. the phase current divided by 1.73.
    d. the phase current plus 1.73.
13. The line voltage for a wye-connected system is equal to
    a. 1.73 times the phase voltage.
    b. the phase voltage.
    c. the phase voltage divided by 1.73.
    d. the phase voltage plus 1.73.
14. The line current of a wye-connected system is equal to
    a. 1.73 times the phase current.
    b. the phase current.
    c. the phase current divided by 1.73.
    d. the phase current plus 1.73.

REVIEW *(continued)*

A. Multiple choice. Select the best answer (continued).

15. The most common frequency used in the United States is
    a. 50 hertz.
    b. 60 hertz.
    c. 80 hertz.
    d. 120 hertz.

16. The most common frequency in use in Europe and Asia is
    a. 50 hertz.
    b. 60 hertz.
    c. 80 hertz.
    d. 120 hertz.

17. Federal regulations require that utility companies maintain a relatively constant frequency. The maximum variation permitted is
    a. 1 percent.
    b. 3 percent.
    c. 5 percent.
    d. 7 percent.

18. AC power systems generally consist of
    a. one large alternator.
    b. several alternators connected in series.
    c. several alternators connected in parallel.
    d. several alternators connected in series-parallel.

19. Generators operate at maximum efficiency when carrying
    a. 75 percent of their rated load.
    b. 100 percent of their rated load.
    c. 25 percent of their rated load.
    d. 125 percent of their rated load.

20. A synchroscope is an instrument that indicates differences in
    a. voltage.
    b. phase.
    c. speed.
    d. current.

21. When two alternators are connected in parallel, one alternator will assume more load if
    a. its field strength is increased.
    b. its speed is increased.
    c. both its speed and field strength are increased.
    d. its armature current is increased.

REVIEW *(continued)*

A. Multiple choice.
Select the best answer (continued).

22. A motor-generator set is
    a. a generator supplying power to a motor.
    b. an electric motor supplying power to a generator.
    c. both a and b.
    d. neither a nor b.
23. The cost of producing and transmitting alternating current, compared to direct current, is
    a. less expensive.
    b. more expensive.
    c. about the same.
    d. slightly more expensive.
24. A rotary converter is sometimes called a
    a. dynamotor.
    b. rotating generator.
    c. phase shifter.
    d. rectomotor.
25. Rotary converters are frequently used in
    a. air-conditioning systems.
    b. electric railway systems.
    c. communication systems.
    d. heating systems.

B. Give complete answers.

1. Why is ac the primary source of electrical energy?
2. List five applications in which dc is preferred or required over ac.
3. Why is dc necessary for the alternator field circuit?
4. What name is given to the generator that provides dc for the alternator field?
5. Describe the construction of the rotating field of a slow-speed alternator.
6. Why is hydrogen used for cooling large turbo-type alternators?
7. Explain how the armature coils are connected on a single-phase alternator.
8. List three uses for single-phase alternators.
9. List three methods for connecting the stator coils on a two-phase alternator.

REVIEW *(continued)*

B. Give complete answers (continued).

10. What is a two-phase alternator?
11. What is a three-phase alternator?
12. List two common connections that are used with three-phase alternators.
13. Can single-phase loads be connected to three-phase systems?
14. List the common voltages supplied by a three-phase, four wire wye system.
15. Describe how the coils are connected for a three-phase, three-wire delta system.
16. List some common transmission voltages used in the United States.
17. Why do power companies prefer to use several smaller alternators connected in parallel rather than one very large alternator?
18. List five conditions that must be fulfilled in order to connect alternators in parallel.
19. Describe the procedure for connecting alternators in parallel.
20. What is meant when it is said that the alternators are being synchronized?

C. Solve each problem, showing the method used to arrive at the solution.

1. The voltage across each phase of a two-phase alternator is 440 V. If the coils are connected to form a two-phase, three-wire system, what is the voltage across the combination?
2. If the current through phase X in the alternator in Problem 1 is 25 amperes, and the current through phase Y is 30 amperes, how much current flows in the common wire? Assume that the power factor of both loads is 100%.
3. If the power factor for load Y in Problem 2 is 86.6% lagging, how much current flows in the common wire?
4. On a two-phase, five-wire system, the phase voltage is 440 V. Calculate the voltage between each line wire and the neutral.
5. A four-pole alternator is operating at a speed of 1000 r/min. What is the frequency of the output voltage?

REVIEW *(continued)*

C. Solve each problem, showing the method used to arrive at the solution (continued).

6. If the phase voltage of a wye-connected generator is 277 volts, what is the line voltage?
7. If the line voltage of a wye-connected system is 208 volts, what is the phase voltage?
8. The line current of a delta-connected system is 600 amperes. What is the phase current?
9. The current flowing through one phase of a delta-connected generator is 800 amperes. What is the line current?
10. How much power is taken from a three-phase alternator if the current is 200 amperes per line and the line voltage is 480 volts? The power factor of the load is 90%.

16

TRANSFORMERS

Objectives

After studying this chapter, the student will be able to:

- Describe the construction and operating characteristics of transformers
- Explain the theory of operation of various types of transformers.
- Describe the various methods used to prevent transformers from overheating.
- Illustrate various types of transformer connections and discuss the results of these connections.
- Describe the construction, use, and operating characteristics of special transformers.

TRANSMISSION EFFICIENCY

Alternating current can be transmitted over great distances much more economically than direct current because of the transformer. The *transformer* is a device used to raise or lower voltage. It has no moving parts and is simple, rugged, and durable. The efficiency of a transformer may be as high as 99 percent.

Most electrical equipment is designed to be used on relatively low voltages. Common operating voltages are 120 volts, 208 volts, 240 volts, 277 volts, 440 volts, and 480 volts. Though it may be reasonably safe to do so, it would be very inefficient to transmit power over long distances at these voltages. The large quantities of power needed by cities, towns, and rural areas would require very high currents. The end result would be large power ($I^2 R$) losses in the transmission lines, the use of very large conductors, or both.

Power-station alternators usually generate voltages in the vicinity of 13,000 volts to 22,000 volts. Power is transmitted for moderate distances at this range of voltage. At the load center, transformers are used to lower the voltage to the rated voltage of the equipment. If the power is being transmitted a long distance, the voltage is increased at the generating station and then is decreased at the load center. When power is being transmitted 10 miles or more, the voltage is usually raised 1000 volts for each mile transmitted. In rural areas, transmission voltages can be as high as 1,000,000 volts.

TRANSFORMER PRINCIPLE

The operation of a transformer depends upon electromagnetic induction. Most transformers consist of two or more coils. The coil receiving the energy is called the *primary,* and the coil delivering the power to the load is called the *secondary.* When an alternating voltage is applied to the primary, an alternating current flows. This current sets up an alternating magnetic field, which moves across the secondary coil, Figure 16-1. The moving magnetic lines of force induce a voltage into this coil.

In the transformer shown in Figure 16-1, there is no load connected to the secondary. Thus, no current is flowing in the secondary coil. Because the primary circuit has high inductance and low resistance, the current lags the voltage by nearly 90 degrees. The magnetic field produced by this current establishes a cemf that is nearly 180 degrees out of phase with the applied voltage. Figure 16-2 illustrates this relationship.

The voltage induced into the secondary is a result of the current flowing in the primary and therefore is in phase with the cemf in

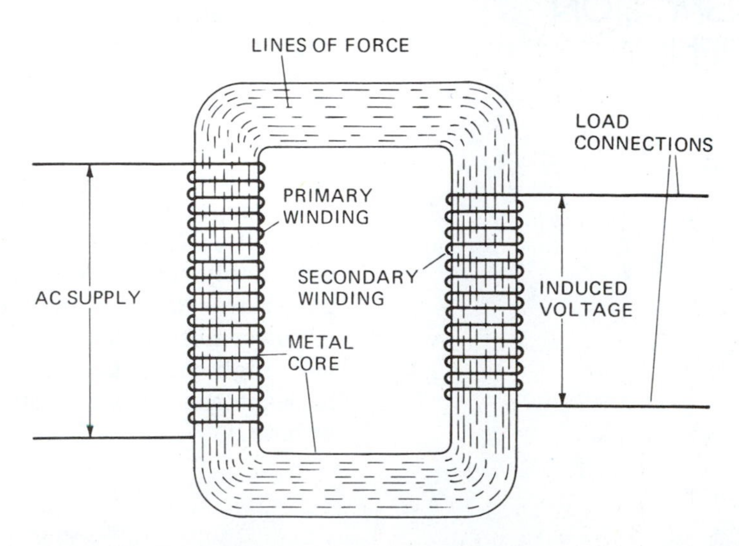

Figure 16-1
Basic transformer: AC flowing in the primary winding establishes an alternating magnetic field, which induces a voltage into the secondary winding. No electrical connection is made between the primary winding and the secondary winding.

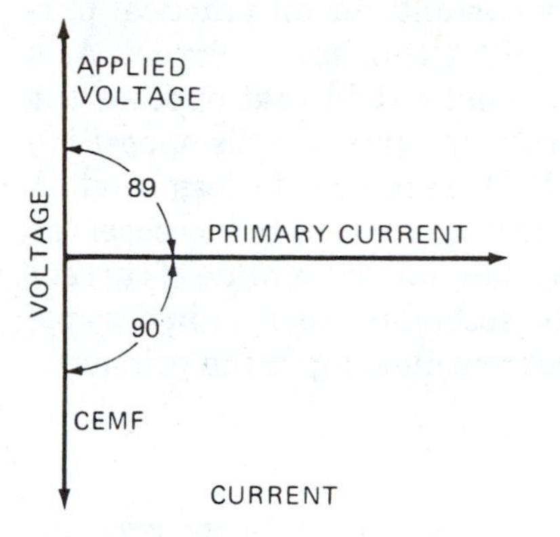

Figure 16-2
Vector diagram of the relationship between the applied voltage, the primary current, and the cemf of a transformer at no load

the primary. In other words, at any instant, the secondary voltage of a transformer is in the direction opposite to the voltage applied to the primary.

The amount of voltage produced by electromagnetic induction depends upon the rate of speed at which the magnetic field moves across the loops and on the number of loops of wire on the coil. The rate of speed depends upon the frequency of the primary circuit (which is a constant value). Therefore, the number of loops on the secondary coil must determine the value of the induced voltage. The more turns of wire on the secondary coil, the greater will be the induced voltage.

Because the secondary voltage is a result of the primary current, their frequencies must be equal. Because the primary current is a result of the primary voltage, the frequency of the secondary voltage is equal to that of the primary voltage. Also, the number of volts per turn on the primary coil is equal to the number of volts per turn on the secondary coil. In other words, a transformer with 100 turns of wire on the primary and 200 turns on the secondary produces a voltage across the secondary that is double the voltage of the primary. If a pressure of 500 volts is applied to the primary, 1000 volts will appear across the secondary. Through basic mathematics it can be seen that there are 5 volts per turn on both the primary coil and the secondary coil.

If a transformer increases the voltage, it is called a *step-up transformer*. If it decreases the voltage, it is called a *step-down transformer*. The same transformer can be used for either stepping up or stepping down the voltage. For example, if the transformer described previously is supplied with 1000 volts to the 200-turn coil, that coil becomes the primary coil. The 100-turn coil then becomes the secondary coil and will deliver 500 volts.

With no load on the secondary of a transformer, only a very small current flows in the primary winding. This is because the cemf is equal and almost opposite to the applied voltage. For example, if the applied voltage is 100 volts and the cemf lags the applied voltage by 179 degrees, then, from Equation 13.5,

$$E = \sqrt{(E_a)^2 + (E_i)^2 - (2E_a\,E_i \cos\phi)}$$

$$E = \sqrt{(100)^2 + (100)^2 - (2 \times 100 \times 100 \times 0.9998)}$$

$$E = \sqrt{20{,}000 - 19{,}996}$$

$$E = \sqrt{4}$$

$$E = 2\text{ V}$$

The value of current flowing in the primary winding is the result of 2 volts.

Current Regulation Under Load

When a load is connected across the secondary of a transformer, current flows in the secondary winding. The amount of current depends upon the impedance of the load. If the secondary load is primarily resistive, the secondary current, for all practical purposes, is 180 degrees out of phase with the primary current. As a result, this current produces a magnetic field that opposes the magnetic field produced by the primary current. This opposition weakens the primary magnetic field, resulting in less cemf. A reduction in the cemf allows the primary current to increase; the amount it increases is directly proportional to the value of current flowing in the secondary. Thus, the secondary current in a transformer regulates the amount of current flowing in the primary.

Voltage Regulation Under Load

The current flowing in the primary winding of a transformer produces a magnetic field that induces a voltage into the secondary winding. If the secondary current causes the primary flux to become weaker, how is the secondary voltage affected? To answer this question, refer to the statement about the primary current: "A reduction in the cemf allows the primary current to increase; . . ." This increase in current strengthens the primary field flux just enough to maintain the necessary voltage across the secondary windings.

Capacitive and Inductive Effect

The power factor of the load determines the transformer's ability to maintain good secondary voltage regulation. As the load becomes more inductive, the secondary current begins to lag the voltage across the load. This changes the phase relationship between the primary current and the secondary current. The more inductive the load becomes, the greater is the change. As a result, highly inductive loads cause poor secondary voltage regulation even if the primary voltage is constant. Loads taking large lagging currents result in a high IX_L drop in the secondary winding of a transformer. Capacitors or synchronous motors must be installed to overcome this problem. If enough capacitance is installed to bring the power factor back to 100 percent, the circuit is in resonance. Under this condition the transformer will maintain a stable secondary voltage.

Turns—Voltage—Current Relationships

The relationship between the voltage and the number of turns on each winding of a transformer is called the *turns ratio.*

When a transformer is supplying a load at or near 100 percent power factor, the voltage ratio between the primary and the secondary depends upon the number of turns of wire on each winding. If there are more turns on the secondary than on the primary, the secondary voltage will be greater than the voltage of the primary. If there are fewer turns on the secondary than on the primary, the secondary voltage will be less than the voltage of the primary. This ratio of voltage and turns can be expressed mathematically as follows:

$$\frac{E_p}{E_s} = \frac{N_p}{N_s} \qquad \text{(Eq. 16.1)}$$

where E_p = primary voltage
E_s = secondary voltage
N_p = number of turns of wire on the primary winding
N_s = number of turns of wire on the secondary winding

Example 1

A transformer is being designed to decrease the voltage from 120 V to 12 V. If the primary requires 400 turns of wire, how many turns are required on the secondary?

$$\frac{E_p}{E_s} = \frac{N_p}{N_s}$$

$$\frac{120}{12} = \frac{400}{N_s}$$

$$N_s = \frac{4800}{120}$$

$$N_s = 40 \text{ turns (secondary turns)}$$

The amount of current flowing in the secondary depends upon the load connected to the transformer. This secondary current regulates the primary current. Therefore, there is a definite mathematical relationship between the two. Transformers are used to transfer power from one circuit to another with minimum loss. For most practical calculations, it can be assumed that transformers are 100 percent efficient. With this in mind, it can be observed that when the voltage increases, the current must decrease. Power is equal to current times voltage ($P = IE$). Thus, to maintain the same value of power, an increase in voltage will require a corresponding decrease in current. Likewise, a decrease in voltage will require an increase in current. This relationship between the voltage and the current in the two windings of a transformer may be expressed mathematically as follows:

$$\frac{E_p}{E_s} = \frac{I_s}{I_p} \qquad \text{(Eq. 16.2)}$$

where E_p = primary voltage
E_s = secondary voltage
I_s = secondary current
I_p = primary current

Example 2

If the transformer in Example 1 supplies a load requiring 5 A, how much current flows in the primary circuit?

$$\frac{E_p}{E_s} = \frac{I_s}{I_p}$$

$$\frac{120}{12} = \frac{5}{I_p}$$

$$I_p = \frac{60}{120}$$

$$I_p = 0.5 \text{ A (primary current)}$$

The combined relationship between voltage, current, and turns can be expressed mathematically as follows:

$$\frac{E_p}{E_s} = \frac{N_p}{N_s} = \frac{I_s}{I_p} \qquad \text{(Eq. 16.3)}$$

When the term *transformer ratio* is used, it generally refers to the turns ratio. It is a common practice in industry to state the primary number first. For example, the transformer in Example 1 has a ratio of 10 to 1 (written 10:1). This indicates that the transformer is being used as a step-down transformer. If the primary winding has 200 turns and the secondary winding has 600 turns, the ratio is 1:3, indicating a step-up transformer.

LOSSES AND EFFICIENCY

The power losses in a transformer are made up of copper losses and magnetic losses. Manufacturers strive to keep losses to a minimum. The losses generally range from 1 percent to 15 percent, depending on the application. Transformers that are used to transfer large amounts of power are designed to be at least 98 percent efficient. Small transformers, such as those used for doorbell and signal circuits, are usually rather inefficient (even less than 85 percent). These transformers transfer small amounts of power for very brief periods of time; therefore, efficiency is not important.

Copper Losses

The windings on transformers consist of copper wire, which has some resistance. Although the wire is sized to keep the resistance to a minimum, it cannot be completely eliminated. Copper losses are proportional to the square of the current ($P = I^2 R$). Therefore, the proper sizing of the wire is very important.

When there is no load on a transformer, zero current flows in the secondary and only a very small current flows in the primary. Under this condition, the $I^2 R$ loss is negligible.

Magnetic Losses

Figure 16-3
Magnetic flux lags the magnetizing force (current) that produces it.

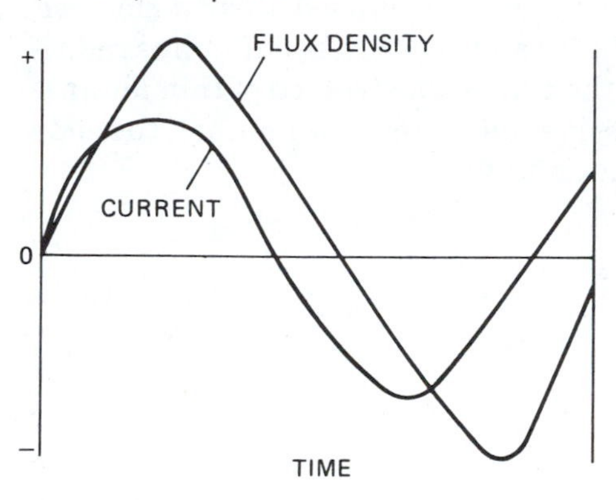

The current flowing in the windings of a transformer is the source of the magnetic energy. The transformer is designed so that the magnetic lines of force will pass through the core. However, some of the magnetic lines leak out into the surrounding air. These serve no useful purpose and thus are wasted.

The core of a transformer is used to provide a low reluctance path for the magnetic lines of force. The type of material from which the core is made affects the efficiency of the transformer. Because an alternating current flows in the windings, the magnetic flux produced by the current is also alternating. This alternating flux tends to lag behind the current that produces it. Figure 16-3 is a graph illustrating the magnetizing current and its time relationship to the flux density. The shape of the flux curve and the amount of lag depend upon the material of the core.

When the magnetizing current begins to flow and increases in value, the flux tends to increase in a direction that is determined by the direction of the magnetizing current. The flux reaches a maximum value shortly after the current does. At this point, the flux tends to remain constant somewhat longer than the current. The amount of time depends upon the retentivity of the core. When the current reaches zero, the core still retains some magnetic flux. As the current reverses its direction of flow, it must produce a magnetic field that is strong enough to demagnetize the core before it can reverse the flux direction. The power required to demagnetize the core is considered a loss and therefore reduces the transformer efficiency.

Because the magnetizing and demagnetizing of the core cause molecular motion, the resulting friction produces heat. This heat loss plus the power loss caused by the magnetic lag is called the hysteresis loss.

Another magnetic loss caused by electromagnetic induction takes place in the core. The alternating magnetic field produced by the primary current induces a voltage into the core of the transformer. The voltage causes small currents to flow within the core. These eddy currents produce heat, which is dissipated into the surrounding air, resulting in a power loss.

A third magnetic loss is caused by a condition called *core saturation*. When the flux lines within the core become so dense

Figure 16-4
The relationship of the magnetic lines of force to the current

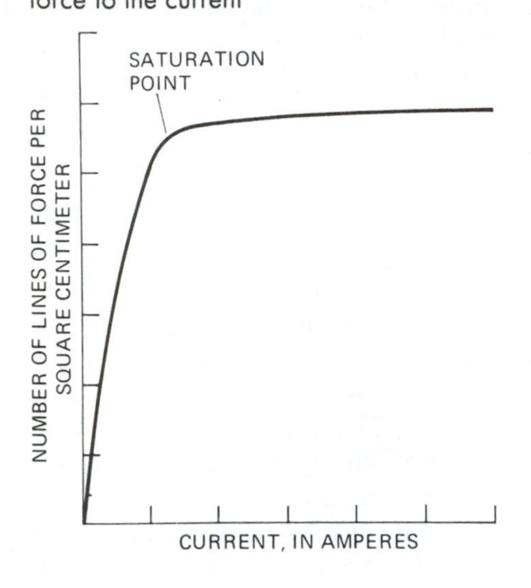

that it is difficult to develop any more lines within it, the core is said to be magnetically saturated. At this point a large increase in current will cause only a slight increase in the magnetic flux. After the saturation point is reached, a further increase in the primary current results in greater I^2 R loss, with very little increase in the magnetic effect. Figure 16-4 is a graph showing the relationship of the current and the lines of force through the saturation point.

Hysteresis and eddy current losses have a direct relationship to the frequency of the supply voltage. Therefore, in the design of transformers, the frequency should be of major concern in determining the type of core. When a metal core is used in a transformer, it should have good permeability and poor retentivity, thus reducing the hysteresis losses. If the core is constructed of thin laminations and the laminations are insulated from one another, the eddy currents can be kept to a minimum.

TRANSFORMER CONSTRUCTION

The windings of commercial transformers are not placed on separate legs (as shown in Figure 16-1). A more efficient method, which reduces flux leakage, is to place the windings on top of one another, Figure 16-5. If a transformer is wound as shown in Figure 16-1, much of the flux produced by the primary current cannot reach the secondary winding. The leakage flux will induce a back voltage in the primary, causing a primary reactance drop. Similarly, much of the secondary flux will not reach the primary and therefore will not neutralize the primary flux but will produce a reactance drop in the secondary. The overall effect is similar to connecting a reactance in series with each winding.

In addition to reactance, each winding has resistance. The reactance and resistance of the windings may be represented as shown in Figure 16-6. In each winding of a transformer there are an IX_L drop and an IR drop. The IX_L drop is kept to a minimum by placing the primary and secondary windings on the same leg of the core. Frequently they are made cylindrical in form and placed one inside the other. Another method is to build up thin, flat sections called *pancake coils*. These sections are sandwiched between layers of insulation. Figure 16-7 shows the cylindrical method; Figure 16-8 shows the pancake method. In the cylindrical method, the low-voltage winding is placed next to the core and the high-voltage winding is placed on the outside. This arrangement requires only one layer of high-voltage insulation placed between the two windings.

Figure 16-5
Transformer with two high-voltage windings and two low-voltage windings. The windings may be connected either in series or in parallel, depending upon the voltage ratings and the desired results.

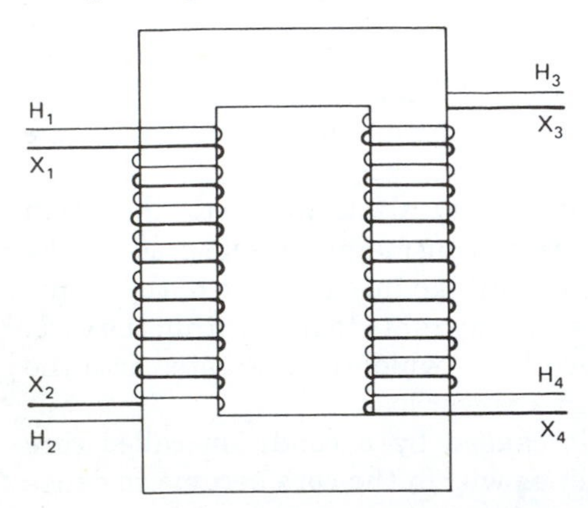

Figure 16-6
Diagram showing the resistance and reactance of both windings of a transformer

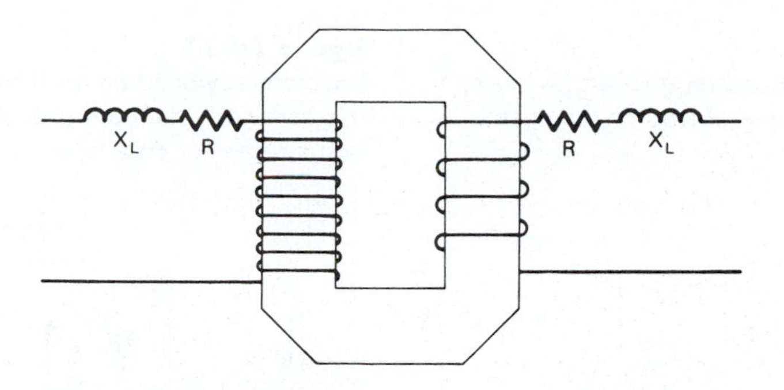

Figure 16-7
Cutaway view of a transformer. The windings are made in the form of a cylinder and placed inside one another.

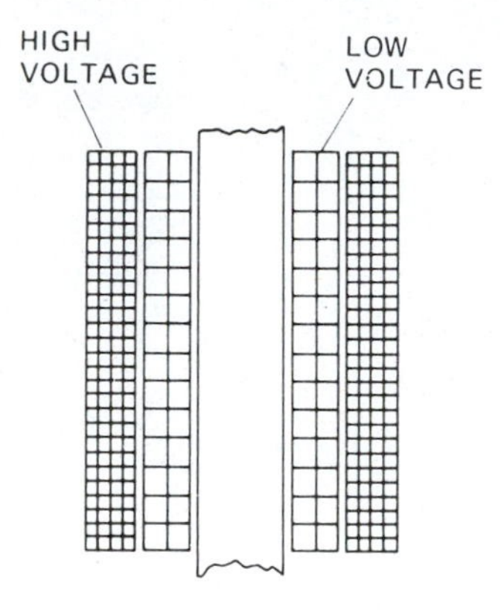

Transformer Core Structure

Transformers used for power distribution are wound on cores made of iron or steel. The cores are built up in thin laminations, which are insulated from one another. Figure 16-9 illustrates one type of core construction. The purpose of the laminated core is to reduce eddy current losses.

Iron/Steel Core Transformers

There are several different types of iron/steel cores. The transformer in Figure 16-9 uses a style known as the *core type*. With this arrangement, one half of each winding is placed on each leg.

Figure 16-8
Cutaway view of a transformer. The windings consist of thin, flat sections sandwiched between layers of insulation.

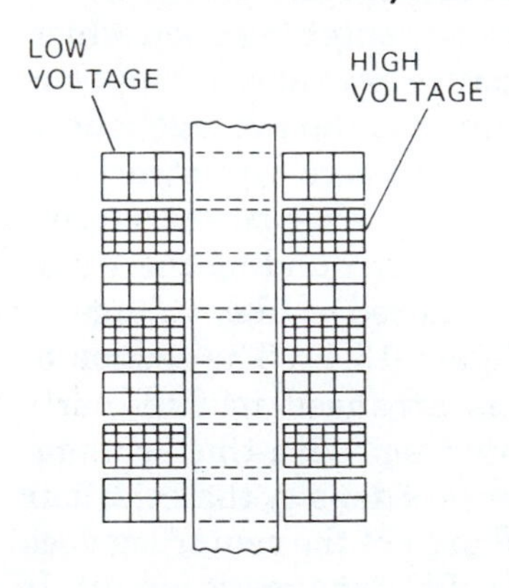

Figure 16-9
The transformer core is made up of several layers of sheet steel, which are bolted together. This arrangement is known as core-type construction.

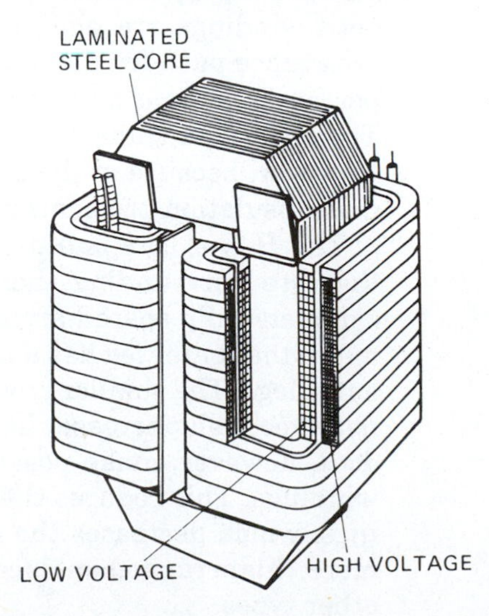

Figure 16-10
Transformer wound on a shell-type core. Both windings are placed on the center leg.

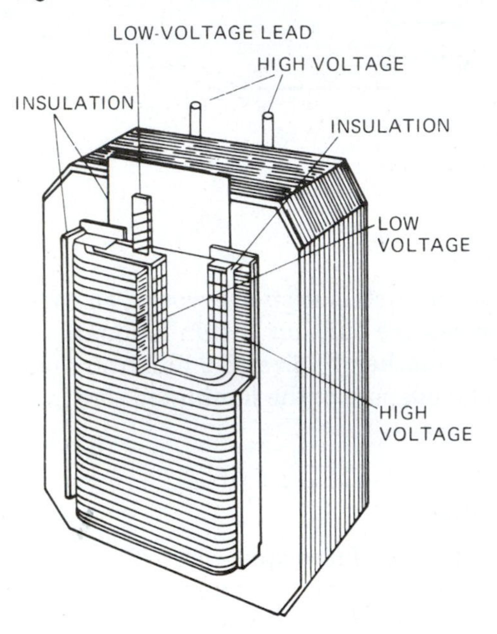

Figure 16-11
Transformer wound on an H-type core. This core can retain a higher flux density than the shell or core types.

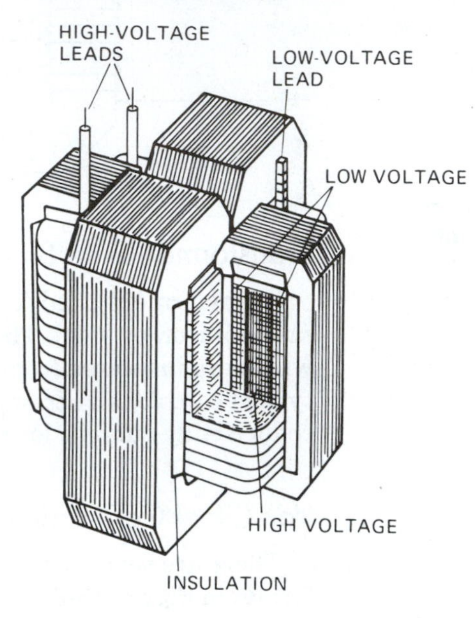

Generally, leads are brought out from each half, permitting either series or parallel connections. The core-type transformer is one of the most economical types to build and is particularly suited to installations requiring small power transformers.

The *shell-type core,* Figure 16-10, has a center leg upon which both windings are placed. This structure provides a very low reluctance path for the magnetic circuit, but the average length per turn of wire is much greater than for the core-type structure. The shell-type transformer is suitable for a large power output. However, because of the additional steel surrounding the windings, insulation problems may be encountered at high voltages.

The *H-type core* is illustrated in Figure 16-11. The section of the core surrounding the windings is arranged in four parts symmetrically spaced around the center leg. With this arrangement, the center leg has a much higher flux density than the four outer legs. The smaller cross-sectional area of the center leg does not effectively increase the reluctance of the magnetic circuit. It does, however, make possible a shorter length per turn on the windings. This reduces the length of the wire used on the windings, which decreases the resistance and the IR drop. Furthermore, this arrangement is compact and more easily cooled than the other types.

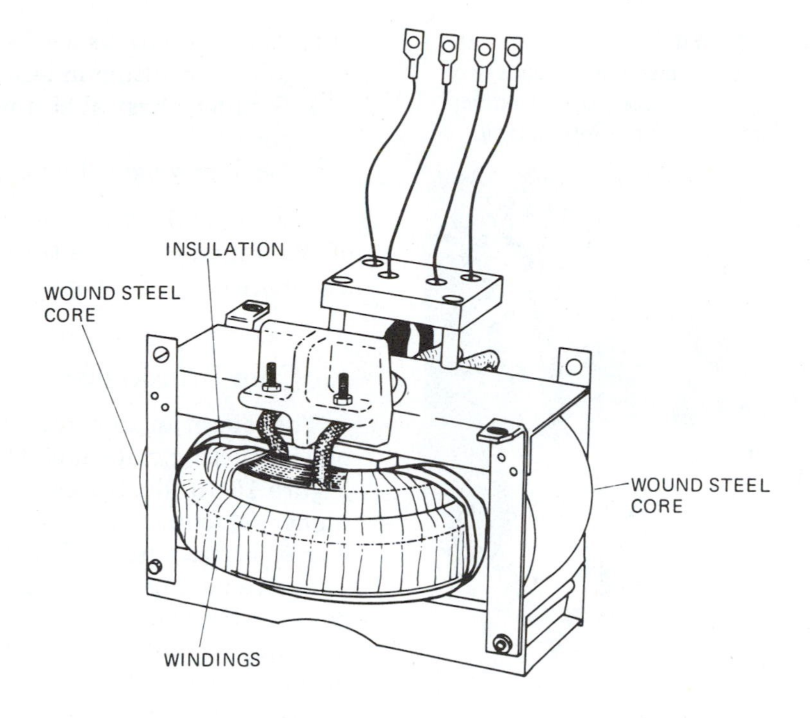

Figure 16-12
Wound core-type transformer.
Many layers of steel are wound to form a thick core.

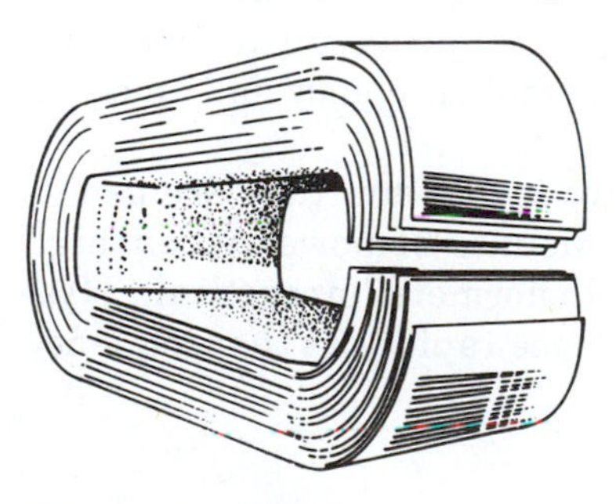

Figure 16-13A
Wound core cut and ready to be placed around the transformer windings

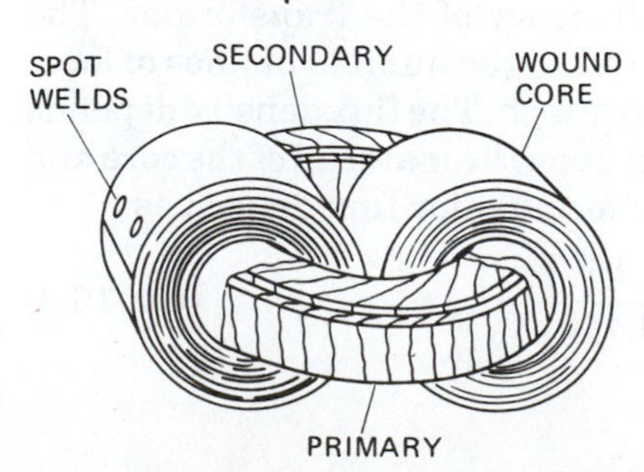

Figure 16-13B
Wound-core transformer after the core has been placed around the windings and secured in place

The *wound core-type* transformer is shown in Figure 16-12. Two methods are used for assembling the core and placing it around the windings. One method uses strips of silicon steel wound tightly together to form a coil. The core is then cut and placed around the transformer windings (Figures 16-13A and 16-13B). This assembly is held together firmly by steel straps, which are spot welded to prevent slipping.

The second method is to wind a continuous strip of silicon steel around the windings. This provides a magnetic path of steel with no joints to increase the reluctance. The coils for the high- and low-voltage windings are wound and insulated separately. The high-voltage winding is then sandwiched between two low-voltage windings. The steel core is wound around the windings, Figure 16-14.

The advantages of the wound-core construction are as follows:

1. Low construction cost
2. Little or no waste
3. Lower reluctance than stacked cores because the flux is not required to pass through the tiny air gaps at the joints
4. Less hysteresis loss than stacked cores because the flux flows in the same direction as the metal
5. Because the reluctance has been decreased, the inductance is increased, resulting in a lower exciting current (no-load current).

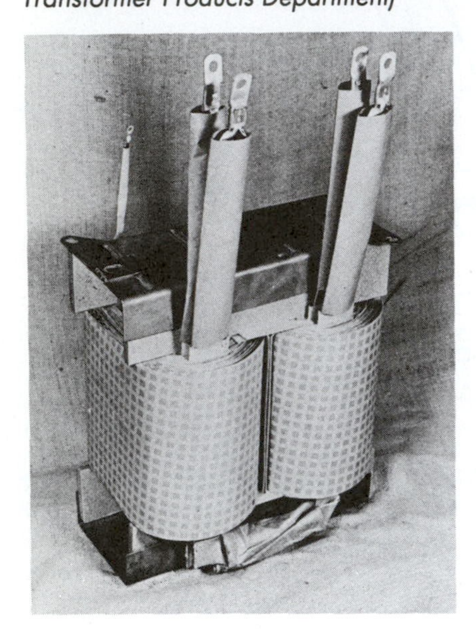

Figure 16-14
Wound-core transformer *(Courtesy of General Electric Company, Medium Transformer Products Department)*

6. The laminations are held tighter together, with fewer or no joints, resulting in less noise (transformer hum).
7. Smaller physical size per kilovolt-ampere than the stacked core
8. Lighter weight than the stacked core

Because of the machines and methods used in the construction of the wound core-type transformer, it is less expensive to build and more efficient to operate.

Air Core Transformer

The coils of an *air core transformer* are wound on forms made of nonconducting materials. The magnetic circuit is through the air. Figure 16-15 illustrates an air core transformer. Air has much higher reluctance than iron or steel; therefore, this type of transformer is much less efficient than the iron/steel core.

A core having properties to form a perfect magnetic coupling between the primary and the secondary would have a coefficient of 1. Good-quality iron or steel cores used in distribution transformers have coefficients ranging from 0.975 to 0.988. The coefficients for most air core transformers range from 0.500 to 0.685.

The efficiency of an air core transformer is quite low. As a result, Equation 16.3, established for iron/steel core transformers ($E_p/E_s = N_p/N_s = I_s/I_p$), does not apply to air core transformers. These equations are based on 100 percent efficiency and an equal exchange of power.

Because air core transformers are inefficient, they are generally used for transferring small amounts of power when high frequencies make iron/steel core transformers impractical. (Hysteresis and eddy current losses increase rapidly as the frequency increases.)

Figure 16-15
Transformer wound on a nonconductive, nonmagnetic hollow tube (air core transformer)

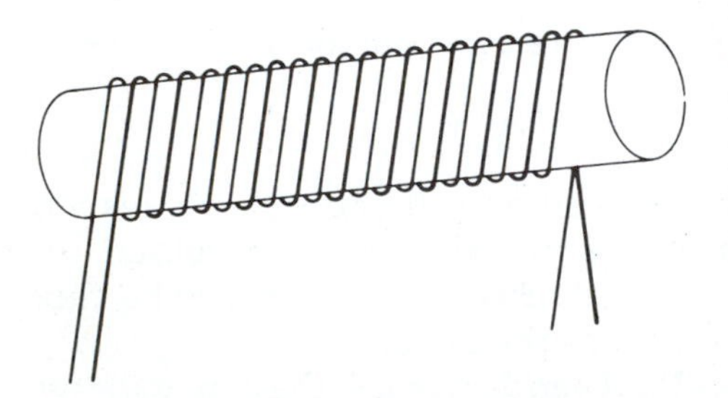

General Transformer Equation

The maximum flux density must be given careful consideration in the design of transformer cores. It has a major effect on the operating characteristics and efficiency of the transformer. The flux density is frequently measured by the number of lines of force per square centimeter or per square inch. The flux density depends upon many factors, but primarily upon the material of the core and the method of construction. The formula for flux density is

$$B_m = \frac{10^8 E_p}{4.44 f A N_p} \qquad \text{(Eq. 16.4)}$$

where B_m = flux density, in lines per square centimeter (lines/cm_2) or lines per square inch (lines/in.^{2})
E_p = voltage applied to the primary
f = frequency, in hertz (Hz)
A = area, in square centimeters (cm^2) or square inches (in.^{2})
N_p = number of turns on the primary

From this equation, it can be seen that the primary voltage, primary turns, area of the core, and type of material affect the flux density. The equation is based on the standard iron/steel core material in general use.

The design engineer works from specifications that list the voltage and frequency for the design. The material is selected from a manufacturer's list of B_m ratings. The area (A) and the number of turns (N_p) must be determined. If the area is made large and the number of turns small, the voltage regulation will be good, but the physical size of the transformer will be rather large for a given kilovolt-ampere (kVA) rating. If the area is made small, a large number of turns will be required. This arrangement provides a smaller transformer, but the voltage regulation is poor. No specific equation can be provided for calculating the best value for A or N_p. Some standard curves have been developed that provide reasonably good reference data. Figure 16-16 shows curves for 25 hertz and 60 hertz at various kilovolt-ampere ratings. Using these curves and Equation 16.4, the design engineer can calculate the area of the core.

Example 3

An engineer is requested to design a shell-type transformer, rated at 10 kVA, 440 V to 110 V, at 60 Hz. The core material has a flux density of 60,000 lines of force per square inch (9300 lines per square centimeter). (a) What will be the number of turns of wire required for the primary and secondary coils? (b) What cross-sectional area is required for the core?

(a) For a transformer of this rating, Figure 16-16 recommends 3.6 V per turn. Therefore,

$$N_p = \frac{440}{3.6}$$

N_p = 122 turns on the primary

$$\frac{N_p}{N_s} = \frac{E_p}{E_s}$$

Equation continued on following page.

Figure 16-16
Recommended volts per turn at various values of kVA and frequency

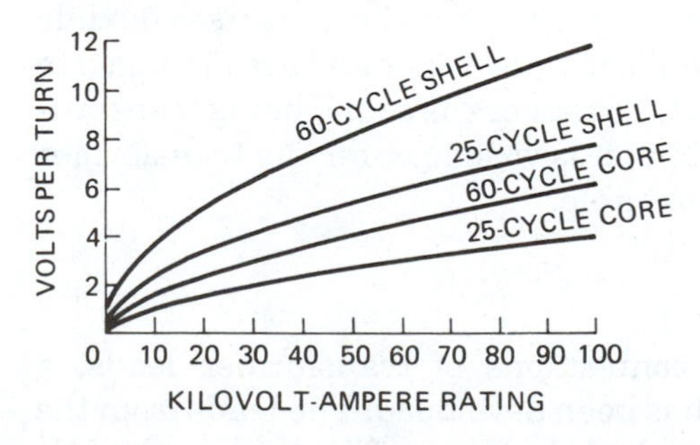

$$\frac{122}{N_s} = \frac{440}{110}$$

$$N_s = \frac{13{,}420}{440}$$

$N_s = 30.5$ turns on the secondary

(b) $$A = \frac{10^8 E_p}{4.44fN_pB_m}$$

$$A = \frac{10^8 \times 440}{4.44 \times 60 \times 122 \times 6 \times 10^4}$$

$$A = \frac{440}{19.5}$$

$A = 22.56$ in.^{2} (cross-sectional area of center leg)

The center leg requires 22.56 square inches of sheet steel. The remainder of the magnetic circuit conducts only one half of the flux and therefore requires only one half of the cross-sectional area, or 11.28 square inches. These calculations do not allow for an IR or IX_L drop in the windings. Therefore, enough additional turns of wire are required on the secondary to compensate for these voltage drops.

Noise Level

Because of their construction and their principle of operation, all transformers produce an audible "hum." The hum is a result of the laminations and windings vibrating as the magnetic field alternates. The amount of sound emitted depends upon the design and method of constructing the transformer. Wound core-type transformers are probably the least noisy. The noise level must be considered when recommending transformers for specific installations. The noise level should always be lower than the ambient noise level for the area in which the transformer is to be installed.

The hum of a transformer will be amplified if it is not mounted properly. The sound may also be transmitted through the conduit system. One method of mounting large transformers uses flexible mounts that are designed to absorb vibrations and are arranged to avoid metal-to-metal contact. To reduce the sound being transmitted through the wiring system, it is wise to enter the transformer through a flexible conduit or cable.

Polarity

To ensure the correct connections of transformer leads, a standard marking system has been developed. The leads from the high-voltage winding are marked H_1, H_2, etc. The lead marked H_1 must always be located on the left when viewing the transformer

Figure 16-17
Connections for polarity test

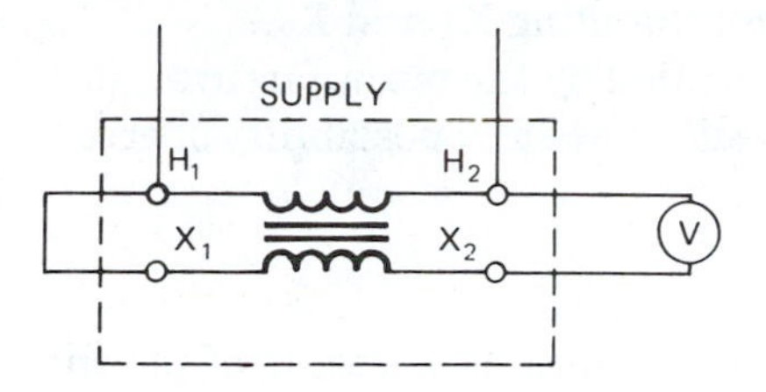

from the side through which the low-voltage leads are brought out. The low-voltage leads are marked X_1, X_2, and so on.

A standard test procedure has been developed to ensure correct labeling. This procedure is as follows:

1. Connect one end of the high-voltage winding to one end of the low-voltage winding, Figure 16-17.
2. Connect a voltmeter between the two open ends.
3. Apply a voltage no greater than the rated voltage of the winding to either the high- or low-voltage winding.

Figure 16-18 illustrates this method.

Figure 16-18
Using the voltmeter method to determine the polarity of a transformer

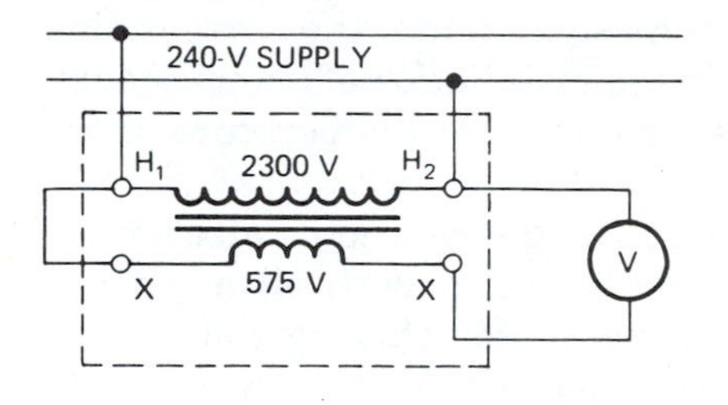

CAUTION: Notice that the high-voltage winding is rated at 2300 volts and the low-voltage winding is rated at 575 volts. Because these voltages are not always available and may present hazardous working conditions, lower voltages are used to perform the test.

It can be determined by basic mathematics that this transformer has a ratio of 4 to 1. Therefore, if 240 volts are impressed across the high-voltage winding, 60 volts will be induced into the low-voltage winding. The voltmeter selected for the test must be able to measure the sum of the two voltages (240 volts + 60 volts = 300 volts).

If the voltmeter indicates the sum of the two voltages, the transformer is additive and should be labeled as shown in Figure 16-19 (H_1 on the left and X_1 on the right). If the voltmeter indicates the difference between the two voltages (240 volts – 60 volts = 180 volts), the transformer is subtractive. Therefore, the leads should be marked as shown in Figure 16-20 (H_1 on the left and X_1 on the left).

The polarities of transformers should be such that at the instant when the current is entering H_1, it is leaving X_1 (see Figures 16-19 and 16-20). The polarity of the transformer can be changed by interchanging the low-voltage leads.

Figure 16-19
Distribution transformer arranged for additive polarity

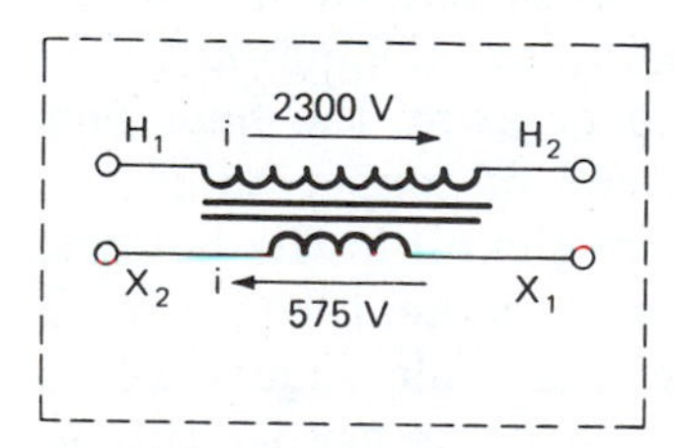

Some specific standards for transformer polarity are as follows:

1. Single-phase transformers with power ratings of up to 200 kilovolt-amperes and voltage ratings of not over 9000 volts shall be additive.
2. Single-phase transformers with power ratings, of up to 200 kilovolt-amperes and voltage ratings greater than 9000 volts, shall be subtractive.
3. Single-phase transformers with power ratings of over 200 kilovolt-amperes, regardless of the voltage rating, shall be subtractive.

Figure 16-20
Distribution transformer arranged for subtractive polarity

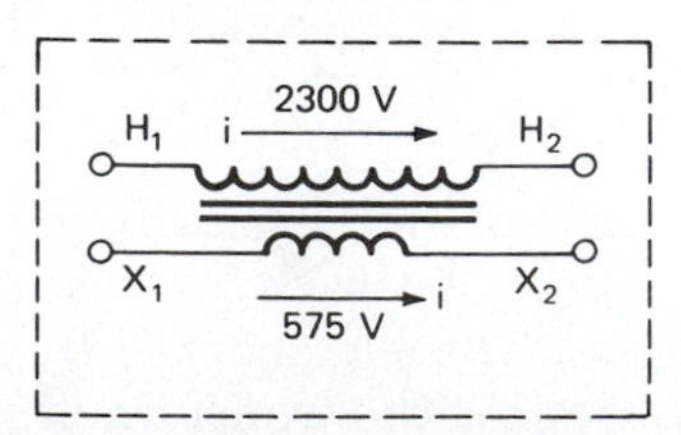

If the transformer shown in Figure 16-19 is rated at 400 kilovolt-amperes, the leads should be arranged for subtractive polarity. This can be accomplished by interchanging X_1 and X_2.

Even though the leads are identified by the manufacturer, it is wise to check them in the field to eliminate any possibility of error.

Methods of Cooling

Heat is generated within transformers as a result of (1) the current flowing in the windings, (2) eddy currents, and (3) hysteresis. The transformer must dissipate this heat as rapidly as it is generated or the transformer will overheat, causing the insulation to break down.

Transformers rated at 5 kilovolt-amperes and less are generally air cooled. The heat produced is dissipated into the surrounding air through natural radiation. In the installation of transformers, it is very important to consider the ambient temperature of the installation area.

Small and medium-sized power and distribution transformers are cooled by being housed in tanks filled with oil or with a synthetic, nonflammable insulating liquid. This type of coolant serves two purposes. First, it carries the heat from the windings to the surface of the tank, where it is dissipated into the air. Second, it serves as an insulation between the windings.

A moderately sized oil-cooled transformer is shown in Figure 16-21A. Figure 16-21B shows a larger size transformer, which requires the use of tubes to increase the radiating surface. Some transformers require external radiators as shown in Figure 16-21C. With very large capacity transformers (20,000 kilovolt-amperes and higher), the oil cannot carry away the heat fast enough. To overcome this problem, a coil of copper tubing is installed near the top of the tank, where the oil is the hottest. Water circulating through the coil absorbs the heat from the oil and dissipates it into the air outside of the tank. Figure 16-22 is a cutaway drawing of a transformer with the copper tubing installed.

Figure 16-21A
Oil-cooled distribution transformer
(Courtesy of General Electric Company, Medium Transformer Product Department)

Transformers are sometimes cooled with air forced up from the bottom. The air circulates around the core and windings and exhausts the excess heat out the top. These types of transformers are frequently used when cost and space are major factors. They are generally lighter than water-cooled transformers, take up less space, and cost less. Their main disadvantage is the necessity for clean, dry air. Using forced air increases a dry transformer's capacity by about 33 percent over the self-cooled (natural circulation) type. A *forced-air-cooled transformer,* sometimes called an *air-blast transformer,* is shown in Figure 16-23.

Figure 16-21B
Oil-cooled distribution transformer with tubes added to increase the ability to radiate heat

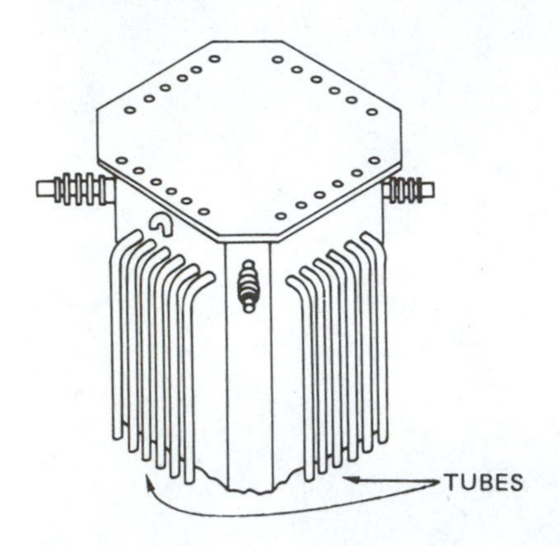

Figure 16-21C
Oil-cooled transformer with radiators added to achieve a further increase of the ability to radiate heat

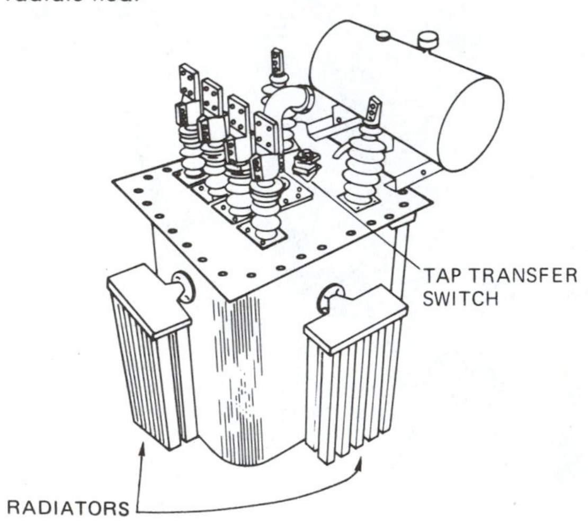

Figure 16-22
Cutaway view of a water-cooled transformer with copper tubing installed in the top portion of the tank

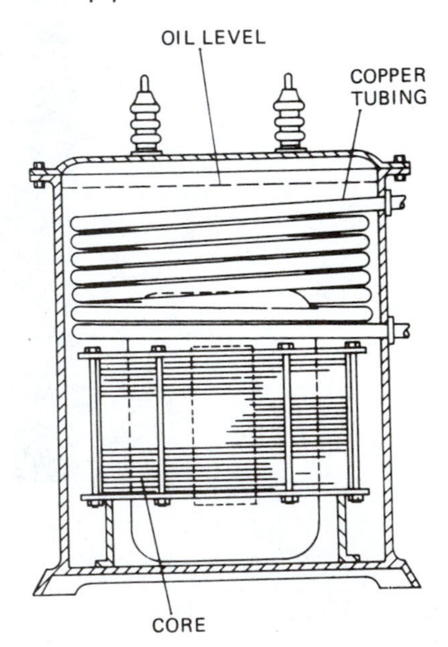

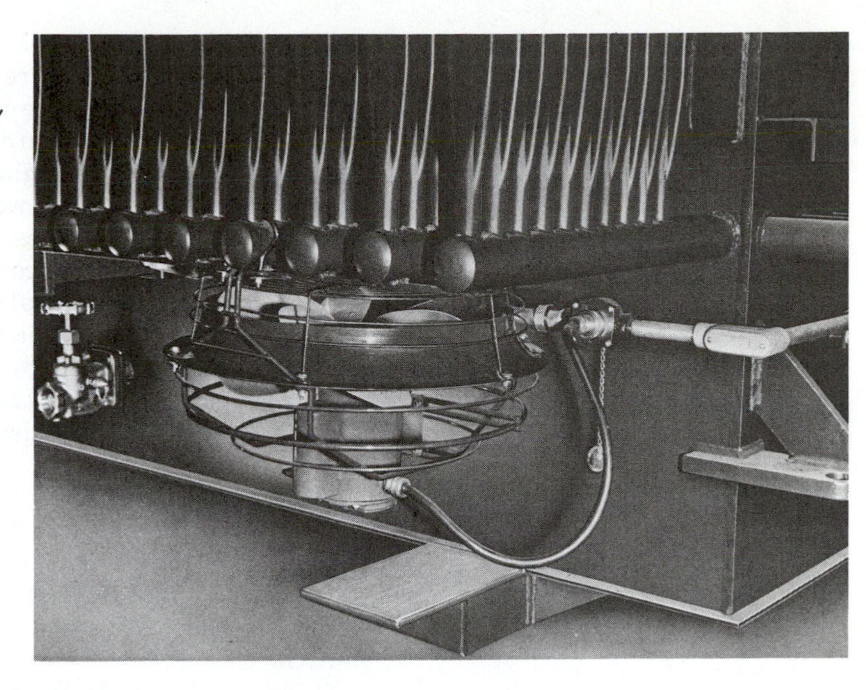

Figure 16-23
Forced-air-cooled transformer (air-blast transformer) *(Courtesy of McGraw-Edison, Power Systems Division)*

SPECIAL TRANSFORMERS

Many installations require special types of transformers. For example, transformers installed a long distance from the supply may require a method to adjust the secondary voltage to compensate for varying line drops as the load increases or decreases. Transformers installed where occasional heavy loads are used may require infrequent voltage adjustments. Some installations may require a constant current; other may require a constant voltage. Special apparatuses may require only a slightly different voltage from the supply voltage. Instruments, relays, lighting fixtures, and electronic equipment require various types of transformers.

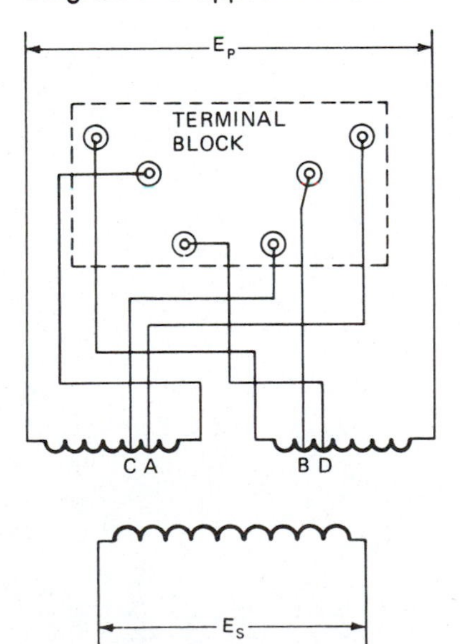

Figure 16-24
Diagram of a tapped transformer

Tapped Transformers

A *tapped transformer* is used for installations where heavy loads are used infrequently or where it is apparent that future demand may require additional loading. For such installations, tap changes are required infrequently and are made by hand. Figure 16-24 is a schematic diagram of a tapped transformer. To change the taps, the transformer must be disconnected from the line, the change made, and then the transformer placed back in service. A switch installed in the primary lines permits this operation to be performed with ease and safety.

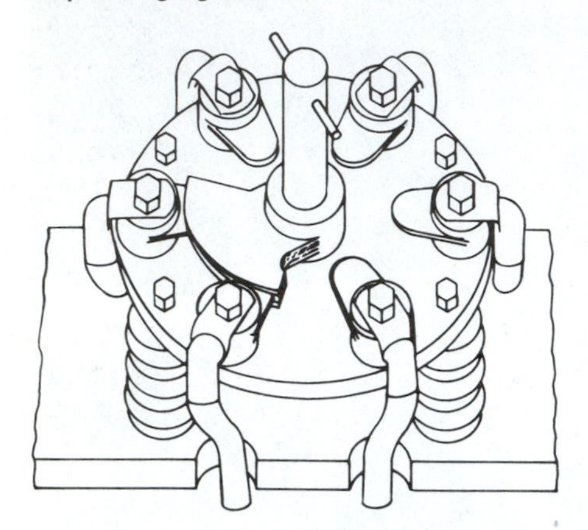

Figure 16-25A
Tap-changing switch

The taps from the secondary are connected to studs. On oil-filled transformers, these studs are inside the tank. To make the change, it is necessary to remove the cover and make the changes while the studs are immersed in oil. In addition to the inconvenience, certain hazards exist. Incorrect connections may result, the oil is exposed to contamination, and the worker may drop parts into the tank. Despite such disadvantages, however, this method is sometimes used because of the lower costs of initial installation.

A safer and more practical method for changing the connections is to use a tap-changing switch. With this method, the taps may be connected to either the high-voltage or low-voltage winding. It is not necessary to open the tank to operate the switch. The transformer, however, must be disconnected from the supply before any changes are made.

The switch is mounted on a terminal block, usually located above the transformer core. It is operated from outside the tank by a shaft, which protrudes through the cover. A knob and indicating plate are arranged to identify the positions in order to obtain the correct voltage. Figure 16-25A shows one type of tap-changing switch.

A diagram showing the connections to the switch contacts is shown in Figure 16-25B. When the movable contact is in the position shown, all the primary winding is connected into the circuit. In this position, the transformer is delivering the lowest

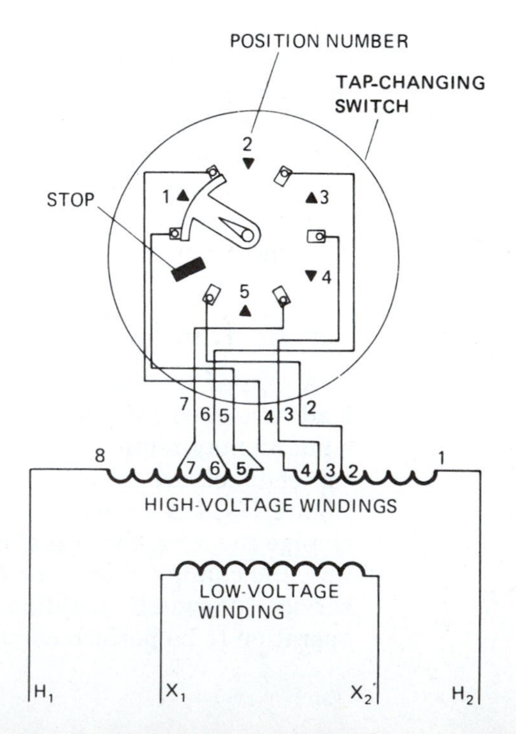

Figure 16-25B
Transformer connections to a tap-changing switch

Figure 16-26
Tap-changing (rotary) switch and tapped transformer with reactor winding. The switch is arranged for various connections of the high-voltage and reactor windings.

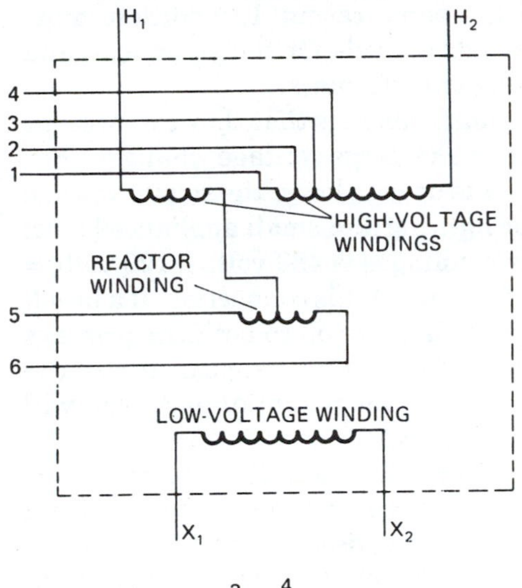

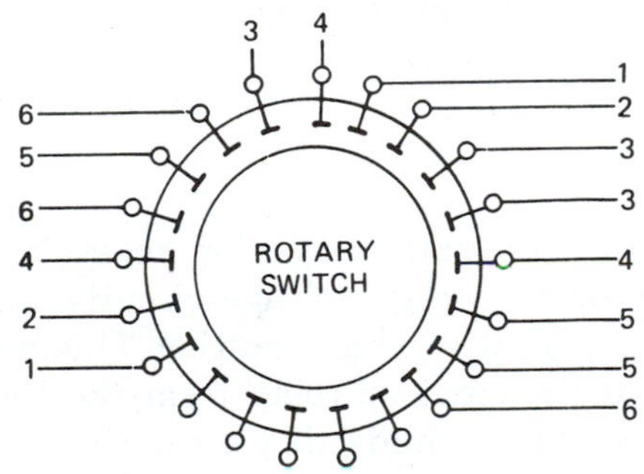

voltage. When the contactor is moved to position 2, section 5 is removed from the circuit, causing the secondary voltage to increase. Further movement of the contactor to positions 3, 4, and 5 eliminates more of the primary turns, thereby increasing the secondary voltage. This increase can be verified by using the turns ratio equation ($E_p/E_s = N_p/N_s$).

Many installations require that adjustments in voltage and, therefore, tap changing, take place while the transformer is supplying a load. For this operation, the transformer usually has at least two high-voltage windings plus a reactor winding. The reactor shunts the switch during the tap-changing process. Figure 16-26 is a diagram of this operation. There is no open position on the switch. This means that at least one set of contacts is made at all times, eliminating accidental opening of the high-voltage winding. Various combinations of switching arrangements increase or decrease the voltage across X_1 and X_2.

Autotransformers

The *autotransformer* differs from the standard transformer in that it has only one winding. Figures 16-27A and 16-27B show how the windings are connected. One portion of the winding serves for both the primary and the secondary. If the transformer is being used to lower the voltage, the turns between H_1 and H_2 constitute the primary winding and those between L_1 and L_2 constitute the secondary winding. The ratio of the voltage, as in a two-winding transformer, is equal to the ratio of the primary turns to the secondary turns.

If the exciting current is neglected, the primary and secondary ampere-turns in an autotransformer are equal. Therefore, the turns ratio equation ($E_p/E_s = N_p/N_s = I_s/I_p$) also applies to autotransformers.

At any instant, the currents in the primary and secondary of a transformer flow in opposite directions. In an autotransformer, the current in the portion of the winding that is common to both the

Figure 16-27A
Autotransformer

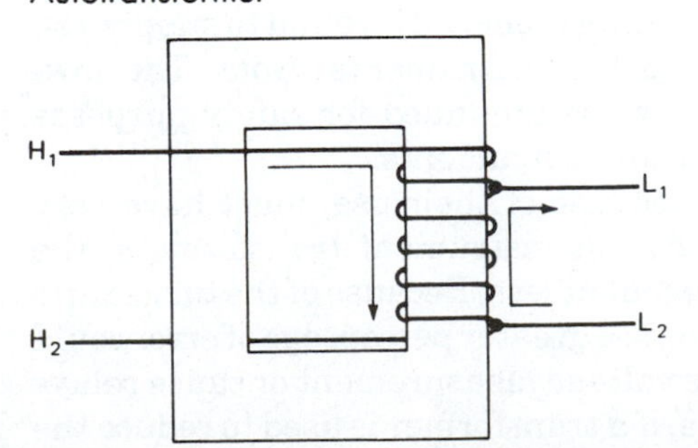

Figure 16-27B
Autotransformer connected to a load

H1
L1
240-V SUPPLY
230-V LOAD
H2
L2

Figure 16-28A
Autotransformer used as a step-down transformer

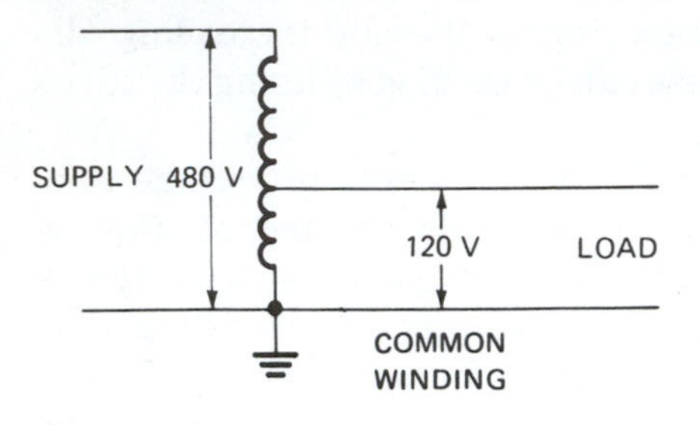

Figure 16-28B
Diagram of autotransformer illustrating possible voltages to ground

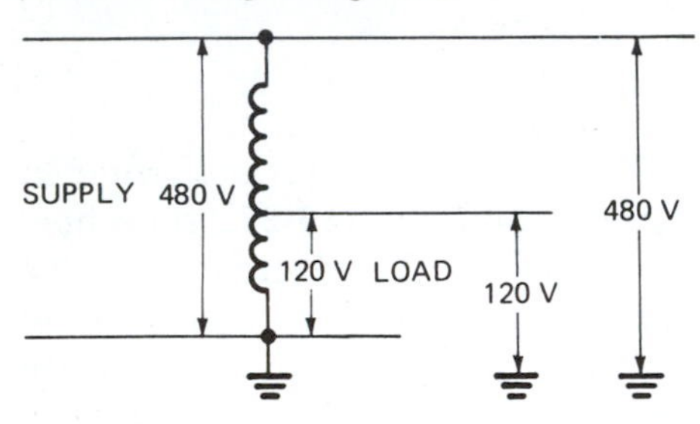

Figure 16-28C
Autotransformer used to increase the voltage from 208 V to 240 V for special equipment

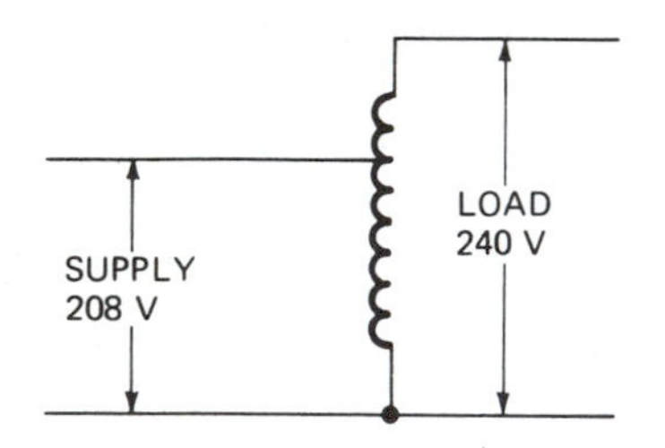

primary and secondary is equal to the difference between the source current and the load current.

Autotransformers require less copper; thus, they are less expensive to manufacture than conventional transformers. They operate at higher efficiency and are smaller in physical size than a two-winding transformer of the same rating. In addition, most autotransformers have lower noise levels for the same load and frequency than do two-winding transformers.

One disadvantage of autotransformers is that they can present hazards if they are used to make large voltage changes. For example, if an autotransformer is used to lower the source voltage from 480 volts to 120 volts for lighting and small appliance loads, the difference between the two voltages is 480 volts – 120 volts = 360 volts. Figure 16-28A is a diagram of this connection. If a break should occur in the winding that is common to both the primary and secondary, approximately 480 volts will appear across the load. Under ground-fault conditions, the potential to ground will be 480 volts, not 120 volts, Figure 16-28B.

Autotransformers are used to reduce the voltage to ac motors during the starting period and to increase the voltage in special situations. They are also used to compensate for voltage drops in transmission lines and to make minor changes in voltage to meet certain load requirements, Figure 16-28C.

Instrument Transformers

Certain measuring instruments and relays for ac systems should not be connected directly to high-voltage circuits. The *instrument transformer* is used to satisfy this condition. This type of transformer isolates the instrument or relay from the high voltage. It also makes it possible to standardize most electrical instruments for operation at lower values of current and voltage, such as 120 volts and 5 amperes.

Voltage transformers, also called *potential transformers,* are similar to the standard two-winding transformer. They are built in small sizes, usually not larger than 500 voltamperes. They have a high-voltage winding, which is connected directly across the supply, and a low-voltage winding (generally wound to supply 120 volts), which is connected to the instruments. *Note:* The low-voltage winding should always be grounded for safety purposes and to eliminate static from the instruments.

Potential transformers, because of their use, must have very accurate voltage ratios. For most instrument transformers, the percentage of error is 0.5 percent or less. Because of the turns ratio on most potential transformers, a greater percentage of error could produce a major error in the voltage measurement or cause relays to malfunction. For example, if a transformer is used to reduce the

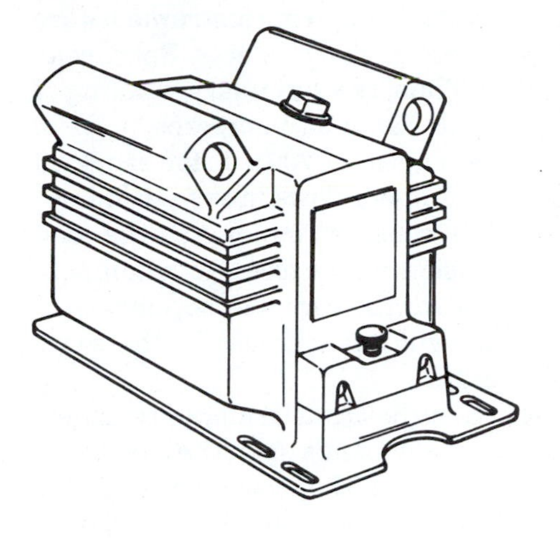

Figure 16-29
Potential transformer

voltage from 2400 volts to 120 volts, the ratio is 20 to 1. A voltmeter connected across the 120-volt winding should indicate 120 volts. This value, when multiplied by 20, is equal to 2400 volts. If there is a –4 percent error in the voltage ratio, the voltmeter will indicate 115.2 volts, and 20 × 115.2 volts = 2304 volts, resulting in an error of 96 volts. Figure 16-29 shows a potential transformer.

To measure large direct currents, ammeters with separate shunts are used. Most of the current flows through the shunt; only a small percentage of the current flows through the meter. The use of shunts for ac is generally unsatisfactory because of the inductive effect. A *current transformer* is a much better device for this application.

The primary winding of a current transformer consists of one or more turns of heavy wire wound on an iron/steel core and connected in series with the supply line. Sometimes a cable or bus bar, passing through the center of the core, serves as the primary winding. Figure 16-30A shows a current transformer; Figure 16-30B illustrates the connections.

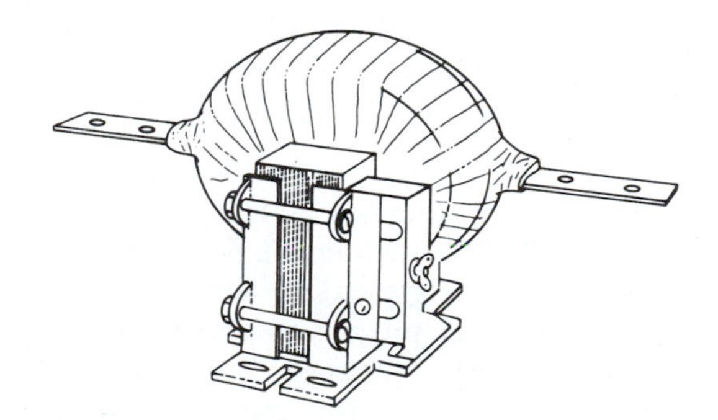

Figure 16-30A
Current transformer

Figure 16-30B
Current transformers connected to service conductors *(Courtesy of Massachusetts Electric Company)*

The secondary winding has many turns of fine wire. The current coils of instruments and/or relays are connected to this winding and act as the load on the transformer. Measuring instruments are usually designed for full-scale deflection at 5 amperes. Relay coils are generally rated at this value. Thus, the secondary winding of the current transformer is generally rated at 5 amperes.

If the transformer has a current ratio of 20 to 1, for every 20 amperes flowing in the primary, 1 ampere flows in the secondary. This current ratio remains nearly constant as long as the load being served does not cause the transformer to exceed its voltampere (VA) rating. If the rating is exceeded, the core will become saturated and the secondary current will not increase in the same proportion as the primary current.

Current transformers differ from the transformers previously described in that the primary current depends upon the load being served and not upon the secondary load. If the secondary is opened while the primary is carrying a heavy load, the demagnetizing effect of the secondary current will no longer exist and the flux in the core will increase. This increase in flux induces a high voltage into the secondary winding, which may damage the insulation or cause a severe shock to anyone coming in contact with it.

CAUTION: The secondary circuit of a current transformer should never be opened while the primary is energized.

Most current transformers have impedance that is high enough to limit the current to a safe value even when the secondary is short-circuited. Figure 16-31 shows a typical ac service using instrument transformers for metering purposes. The terminal block above the meters is arranged so that the secondary windings of the current transformers can be short-circuited in order to remove the meters without disconnecting the supply.

Figure 16-32 is a diagram showing the connections for a voltmeter, ammeter, and wattmeter to a 13,200-volt line. For permanent installations, the instruments are calibrated for direct reading.

In order to avoid confusion regarding transformer polarity, manufacturers have adopted a standard that requires that all instrument transformers have subtractive polarity.

Buck-Boost Transformers

A *buck-boost transformer* is a two-winding transformer that is generally connected to operate as an autotransformer. Figure 16-33 is a schematic diagram of one of the most common connections used.

Figure 16-31
AC service using instrument transformers for metering purposes *(Courtesy of Massachusetts Electric Company)*

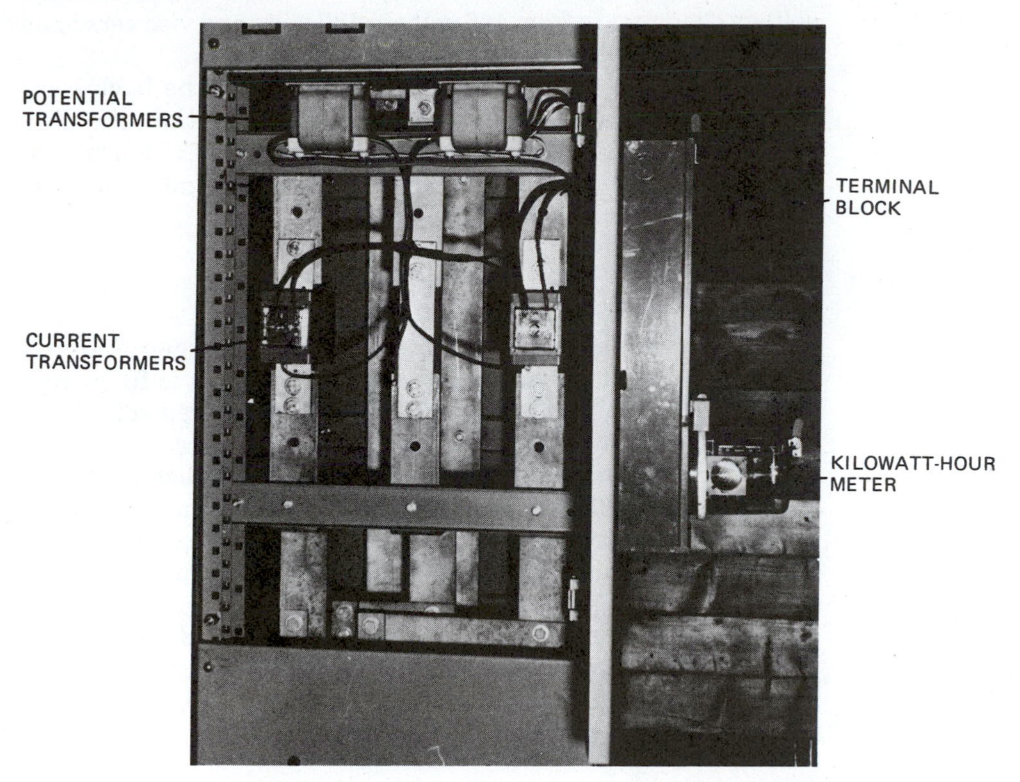

Figure 16-32
Instrument transformers connected to a 13,200-V supply, reducing the voltage and current for metering purposes

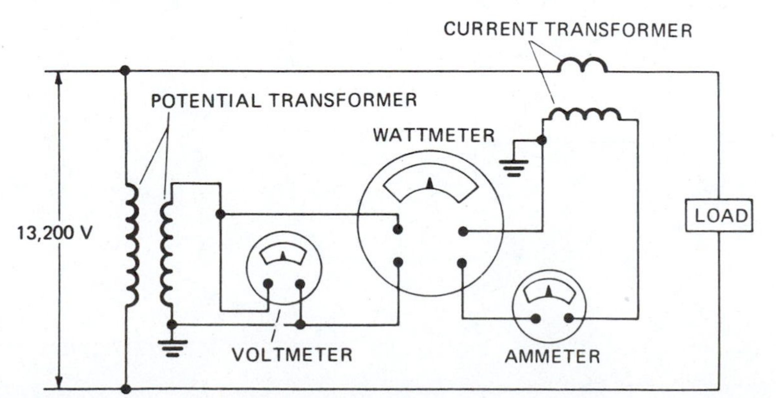

Figure 16-33
Connections for a buck-boost transformer

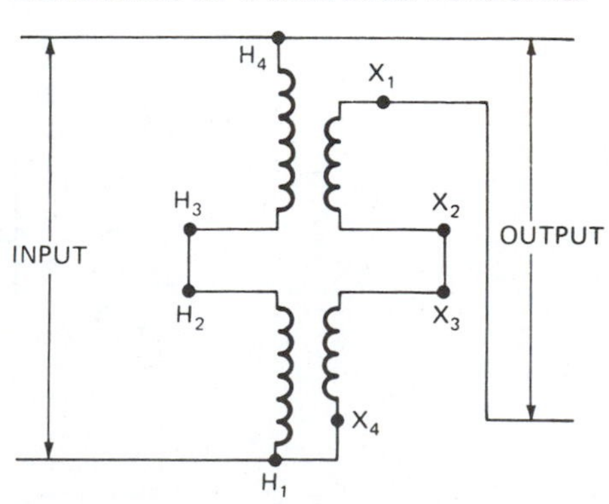

Figure 16-34
Control transformer with primary taps for connections to various voltages

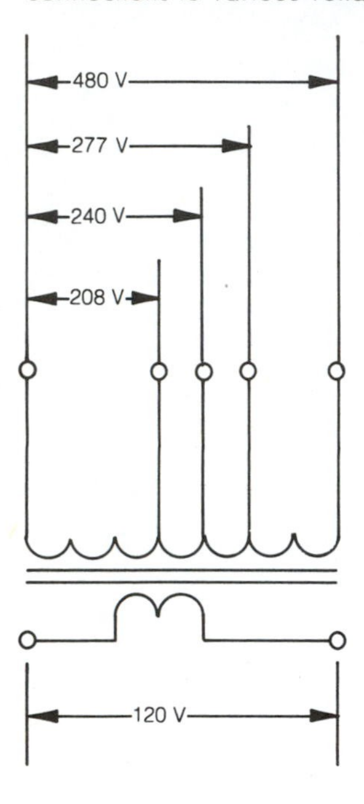

The buck-boost transformer is used to increase or decrease the supply voltage by a small amount. For example, it may be used to increase the voltage from 208 volts to 240 volts. Other common voltage changes are from 440 volts to 480 volts and vice versa and from 240 volts to 277 volts.

Because small voltage changes are involved, the buck-boost transformer is smaller in physical size and rating than the standard two-winding transformer for the same load. The buck-boost transformer is more efficient and produces less noise, and its purchase price is lower.

Control Transformer

A control transformer is a two-winding transformer that provides a lower voltage supply for control circuits. It is frequently used in industrial establishments to provide reduced voltage for the control circuits of motor starters and controllers.

Common voltages supplied to industrial establishments are 208, 240, 277, 480, and 575 volts. The more common industrial controllers are manufactured with control circuits of 120 volts or less. These controls are less expensive to manufacture than those with higher voltage ratings.

The control transformer may have a primary winding with taps for various voltages, Figure 16-34, or the windings may be arranged for only one or two voltages. Figure 16-35 illustrates a transformer with two coils on the primary.

Figure 16-35
Control transformer with two 240-V coils connected in series across 480 V

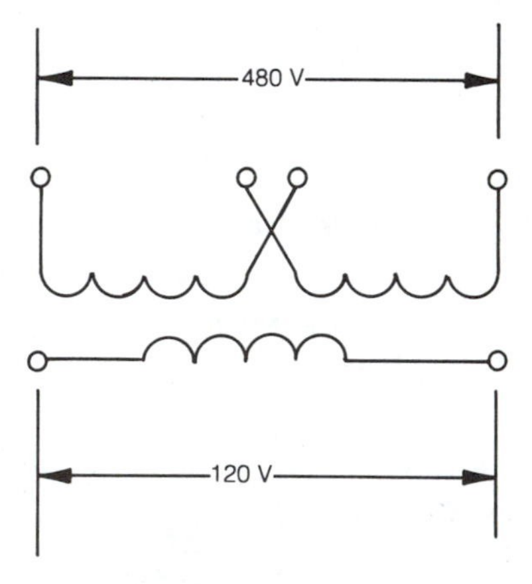

TRANSFORMER CONNECTIONS

Transformers, like other electrical devices, may be connected into series, parallel, two-phase, or three-phase arrangements. When they are grouped together in any of these arrangements, the group is called a *transformer bank*.

In order to group transformers, it is necessary to comply with the following requirements:

1. Their voltage ratings must be equal.
2. Their impedance ratios (percent impedance) must be equal.
3. Their polarities must be determined and connections made accordingly.
4. Transformers are seldom connected in series. When they are, however, their current ratings must be large enough to carry the maximum current of the load. For the most efficient operation, their current ratings must be equal.

Figure 16-36
A transformer with two low-voltage windings

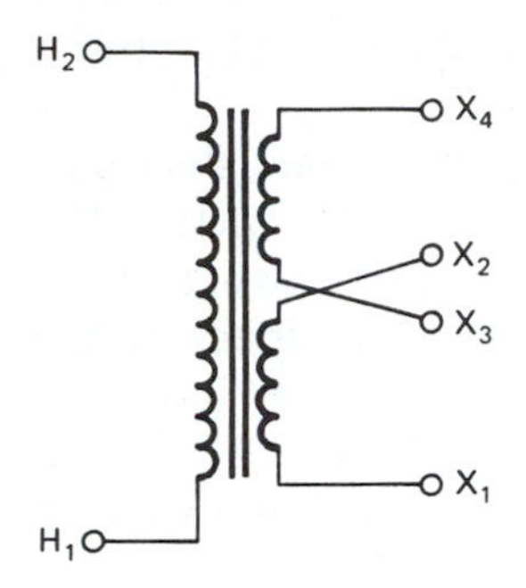

Series Connections

The main purpose of connecting transformers in series is to obtain higher voltage ratings. For example, if the supply voltage is 480 volts and the load requires two values of voltage (120 and 240), a transformer with two low-voltage windings is required, Figure 16-36. It is possible, however, to obtain the same result with two transformers connected in series. To perform this operation, the following procedure is used:

1. Check the voltage ratings of both transformers. They must be equal.
2. Check the percentage impedances of both transformers. They must be equal.
3. Check the current ratings of both transformers. They must be high enough to carry the maximum load.
4. Determine the polarity of each transformer.
5. Connect the high-voltage windings as shown in Figure 16-37A.
6. Connect the high-voltage windings to the supply voltage, Figure 16-37B.
7. Measure the voltage across the secondary of each transformer. The voltage should be the same for each transformer.
8. Connect the secondaries as shown in Figure 16-37C. Be sure the polarities are correct.
9. Measure the voltage across the secondary of each transformer and across the two transformers, Figure 16-37D.

Figure 16-37A
Two transformers with the high-voltage windings connected in series

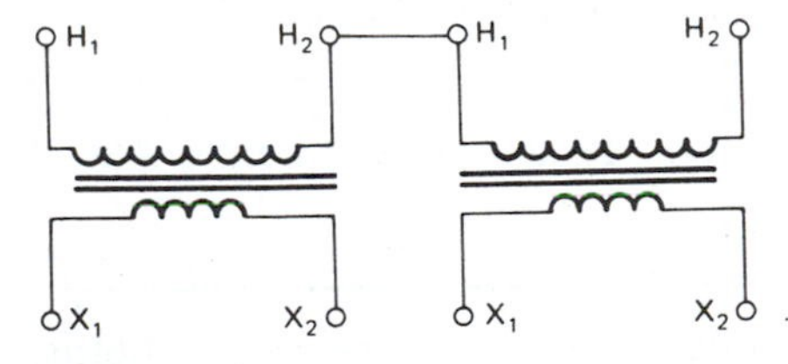

Figure 16-37B
Two transformers with the high-voltage windings connected in series across a 480-V supply

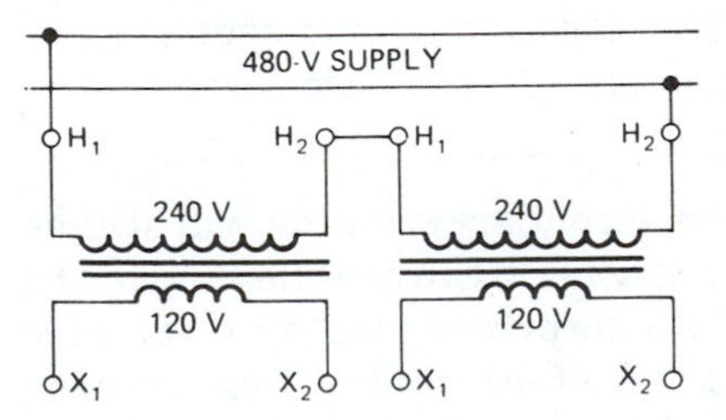

Figure 16-37C
Two transformers with series connections on both the primary and secondary windings

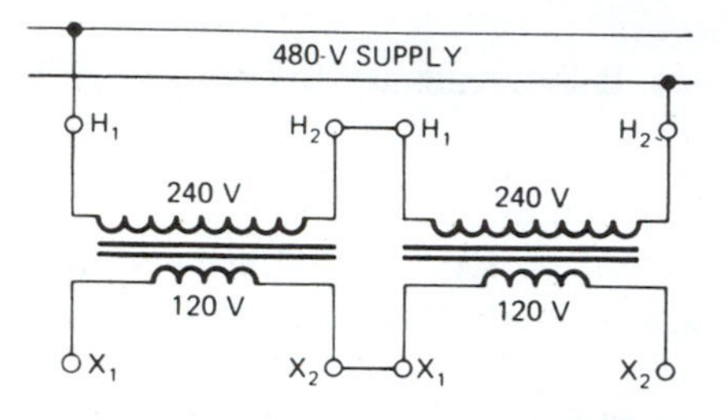

Figure 16-37D
Two transformers connected to form a three-wire system

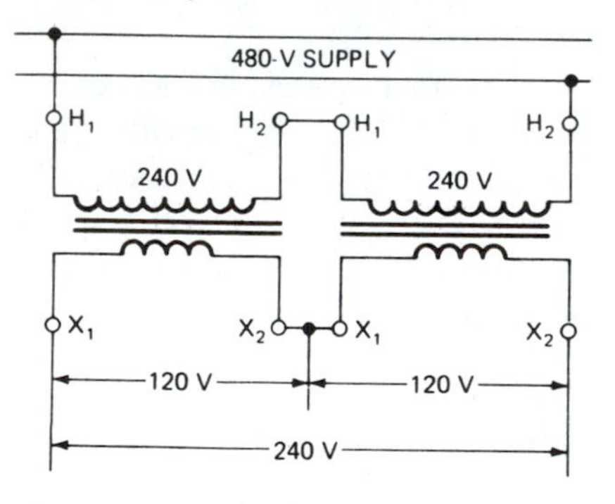

Parallel Connections

Two transformers of equal ratings connected in parallel carry twice the kilovolt-ampere rating of either one. In other words, the kilovolt-ampere rating of transformers in parallel is equal to the sum of the individual ratings.

When an additional load is being installed in a plant, it is sometimes more practical to add another transformer rather than to change the existing one. In order to divide the load according to the rating of the transformers and to avoid circulating currents, it is necessary to meet the requirements previously stated for grouping transformers.

Before the final connections are made, a polarity test should be performed. This can be accomplished without disconnecting the original transformer. The second transformer is connected as shown in Figure 16-38A. Note that the secondary leads of the second transformer are not marked, but the voltmeter indicates 480 volts. The sum of the two secondary voltages is 480 volts. Therefore, the secondaries are connected in series rather than in parallel.

From the results of this test, it can be assumed that transformer A is subtractive and transformer B is additive. The secondary connections are changed as shown in Figure 16-38B. The voltmeter now should indicate zero volt. Remove the voltmeter and connect the transformers as shown in Figure 16-38C.

CAUTION: Never connect together leads that have a potential difference between them.

Parallel operation of transformers offers the advantage of being able to remove one transformer from service for replacement or maintenance without interrupting the entire operation.

CAUTION: When it is necessary to remove the transformer from the line, always be sure that both the primary and secondary switches are open before disconnecting the transformers.

If only the primary is open, the secondary winding will still be energized through the other transformers. Current flowing in the secondary will induce a voltage into the primary winding. It is also a safe practice to disconnect the low-voltage winding first, thereby ensuring that when the high-voltage winding is disconnected, there will be no induced emf across it.

Figure 16-38A
Procedure for connecting transformers in parallel

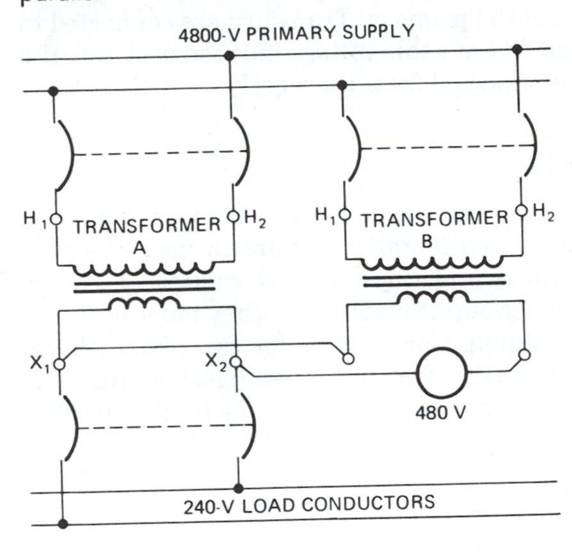

Figure 16-38B
Connecting transformers in parallel

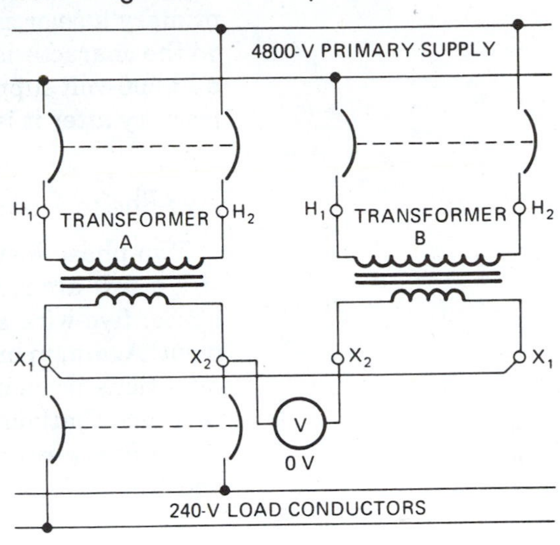

Figure 16-38C
Final connections for two transformers in parallel

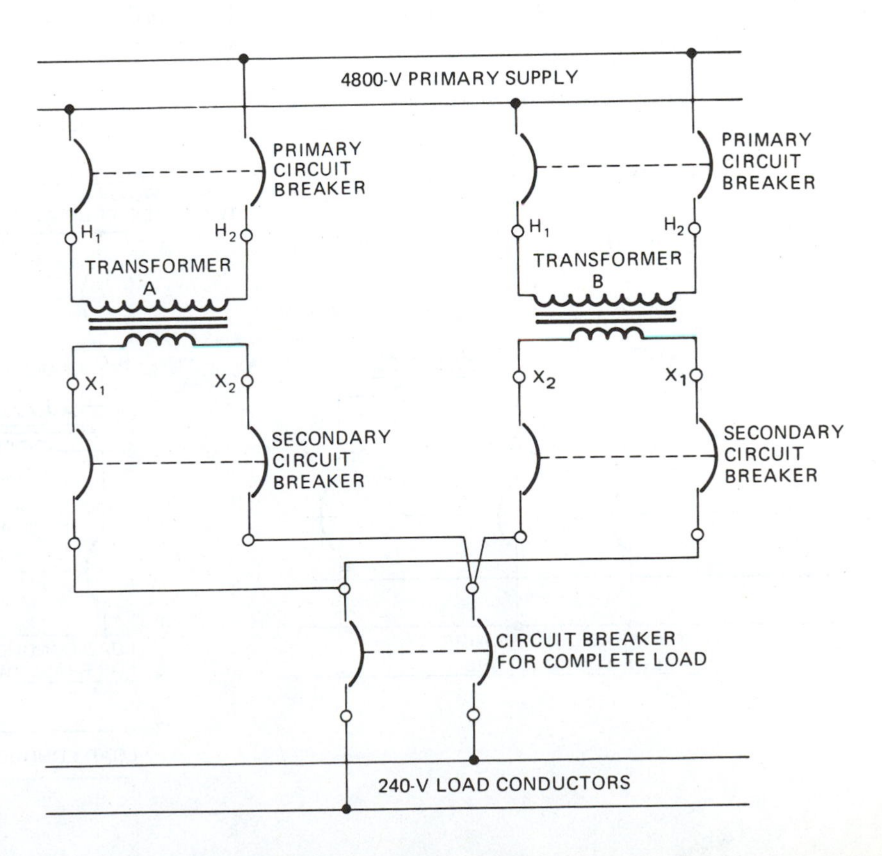

On step-down transformers in parallel, a very hazardous condition can exist because of the possibility of this back feed. If the primary fuse or circuit breaker should open, the secondary takes on the characteristics of the primary. Transformers connected to a 13,200-volt supply could have this voltage induced back into the primary after it is disconnected from the supply.

Two-Phase Connections

Two-phase installations are not common. However, there are still a few areas where such installations are in use. The two-phase, five-wire system is probably the most common arrangement. Again, in order to group transformers, they must meet the conditions previously stated. Connections for two-phase, three-wire installations; two-phase, four-wire installations; and two-phase, five-wire installations are shown in Figures 16-39A, 16-39B, and 16-39C.

Figure 16-39A
Two transformers connected to a two-phase, three-wire system

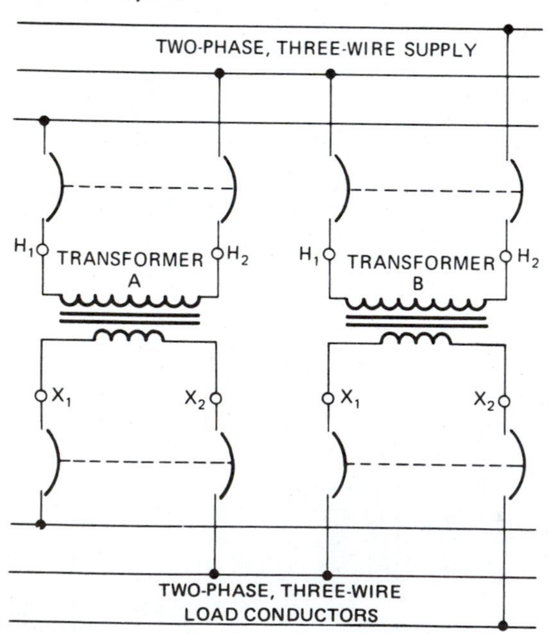

Figure 16-39B
Two transformers connected to a two-phase, four-wire system

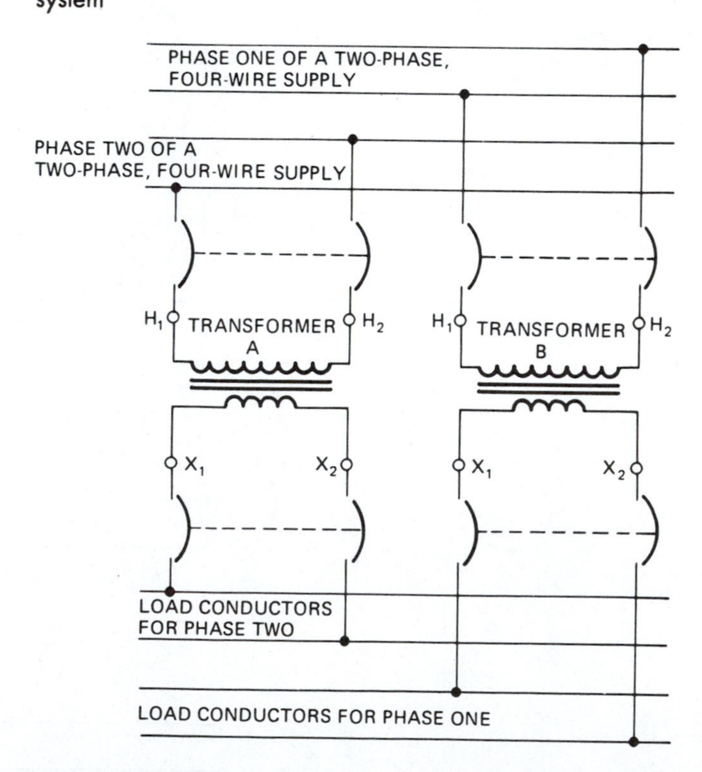

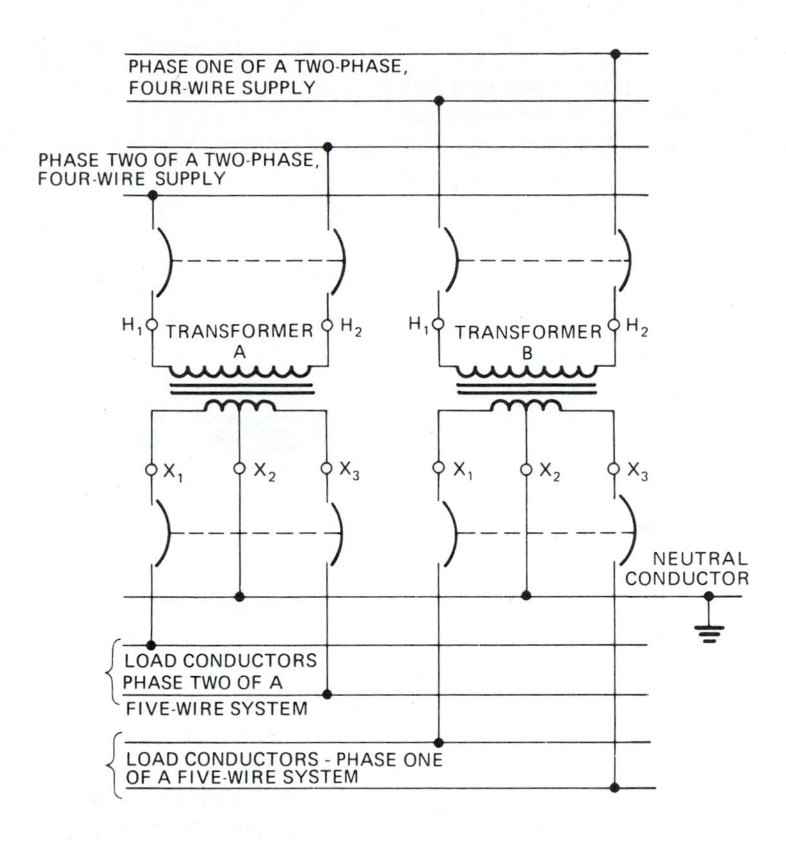

Figure 16-39C
Two transformers used to lower the voltage and convert from a two-phase, four-wire system to a two-phase, five-wire system

Three-Phase Connections

It is common practice to connect single-phase transformers into a three-phase bank. There are many different methods that may be utilized. One very common method is the *delta/delta connection,* Figures 16-40 and 16-41.

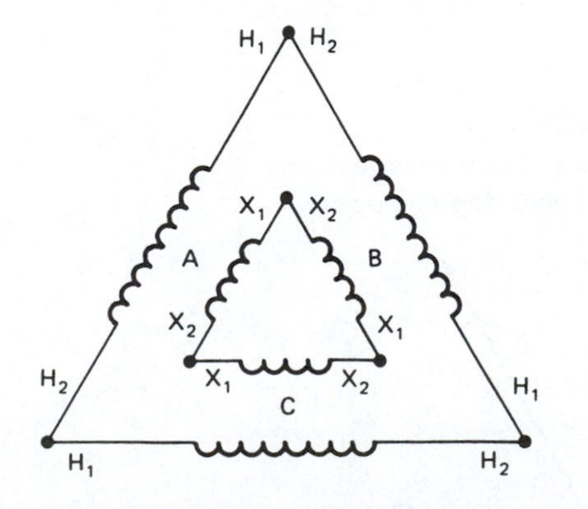

Figure 16-40
Transformer windings connected to form a three-phase, delta/delta configuration

Three-Phase Delta/Delta Connections. For connecting single-phase transformers into a three-phase bank, the following procedure is used. Assume that the transformers are rated for 4800 volts and 240 volts. They are to be installed as step-down transformers for a three-phase delta system.

1. Connect the high-voltage windings as shown in Figures 16-40 and 16-41.
2. Measure the voltage across each secondary winding. The meter should indicate 240 volts.
3. Connect X_2 of transformer A to X_1 of transformer B. Measure the voltage across the open ends (Figure 16-42A). The meter should indicate the vector sum of the voltage induced into each winding, which is 240 volts.

Figure 16-41
Three single-phase transformers with their high-voltage windings connected to form a three-phase, delta configuration

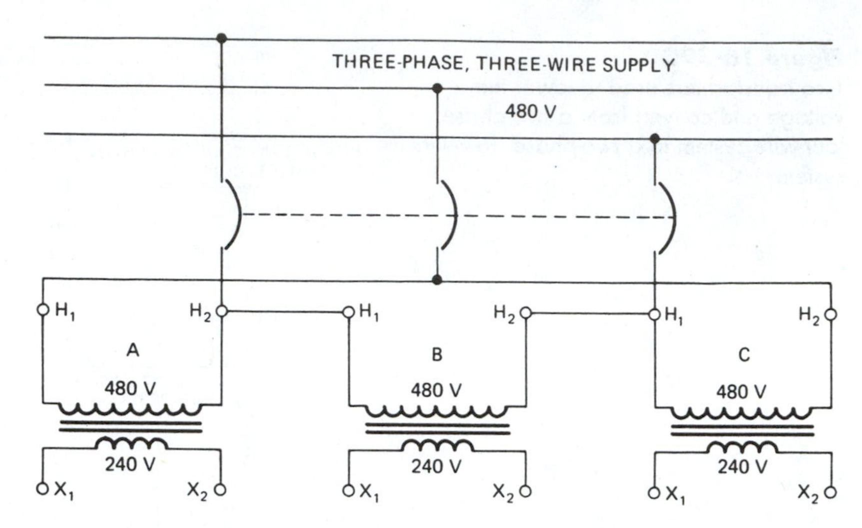

Figure 16-42A shows the correct connection, and the arrows indicate the direction of the instantaneous voltage. Both voltages are in the same direction; therefore, vector addition is performed. Figure 16-42B is a vector diagram of the two voltages, which are 120 degrees out of phase. Figure 16-43A illustrates incorrect connections. Note that the instantaneous voltages are in opposite directions. Therefore, in order to perform vector addition, one vector must be reversed. Figure 16-43B shows the vector diagram for this condition. The mathematical solutions for Figures 16-42B and 16-43B are as follows:

For Figure 16-42B:

$$E = \sqrt{(E_1)^2 + (E_2)^2 - (2E_1\,E_2 \cos \phi)}$$

(From Equation 13.5)

$$E = \sqrt{(240)^2 + (240)^2 - (2 \times 240 \times 240 \times 0.5)}$$

$$E = \sqrt{57{,}600}$$

$$E = 240\ \text{V}$$

Figure 16-42A
Secondary windings of two transformers, illustrating the beginning of a three-phase, closed-delta connection

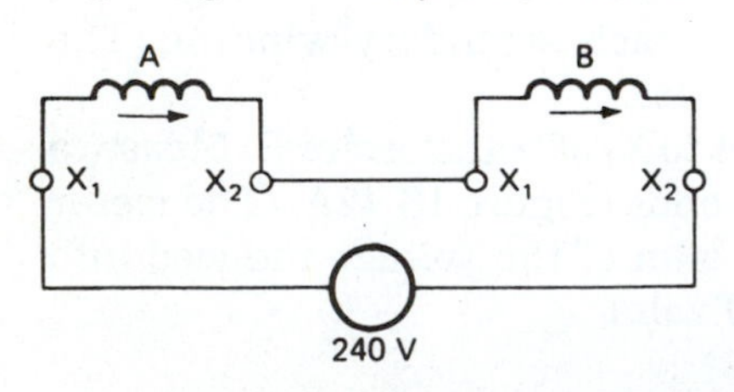

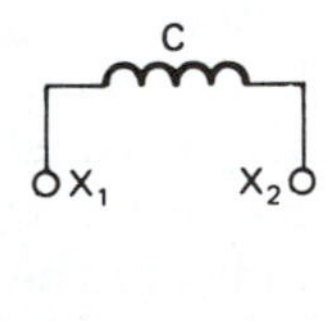

Figure 16-42B
Vector diagram of two voltages 120 electrical time degrees apart

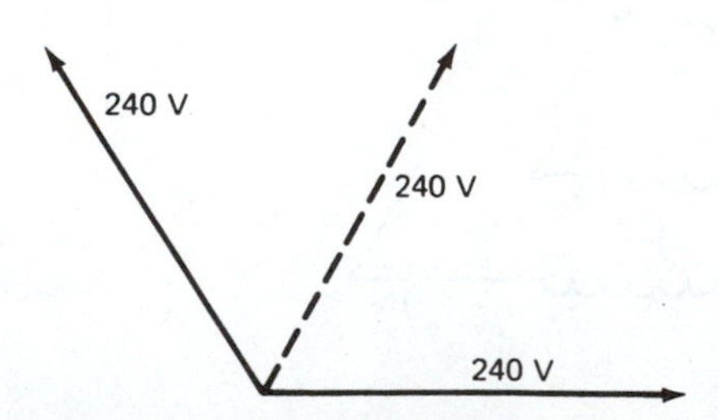

Figure 16-43A
Incorrect connection for two transformers being connected into a closed-delta configuration

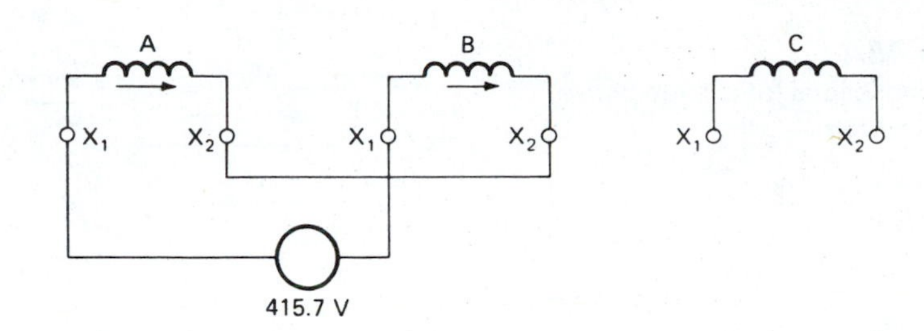

Figure 16-43B
Vector diagram illustrating the resulting value of two voltages of opposite polarity 120 electrical time degrees apart

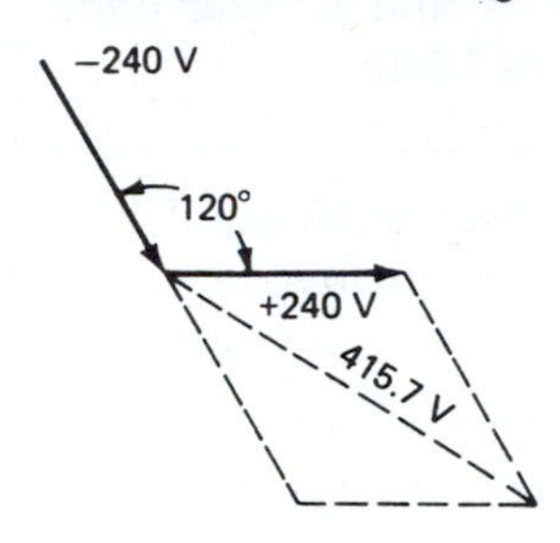

For Figure 16-43B:

$$E = \sqrt{(E_1)^2 + (E_2)^2 + (2E_1\ E_2 \cos \phi)}$$

(From Equation 13.5)

$$E = \sqrt{(240)^2 + (240)^2 + (2 \times 240 \times 240 \times 0.5)}$$

$$E = \sqrt{172{,}800}$$

$$E = 415.7\ V$$

4. With transformers A and B connected as shown in Figure 16-42A, connect X_2 of transformer B to X_1 of transformer C (Figure 16-44A). Measure the voltage between X_1 of transformer A and X_2 of transformer C. The meter should indicate zero volt. Figure 16-44B shows the vector diagram for determining their sum.

Figure 16-44A
Connections for a three-phase, closed-delta system. The voltmeter indicates 0 V.

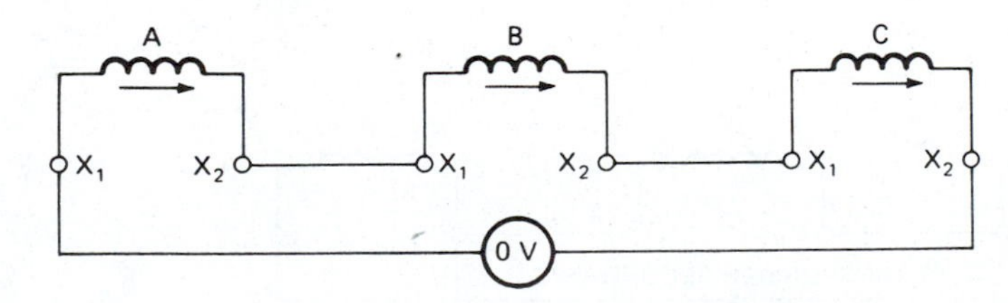

Figure 16-44B
Vector diagram of three equal voltages 120 electrical time degrees apart. The resultant voltage is zero.

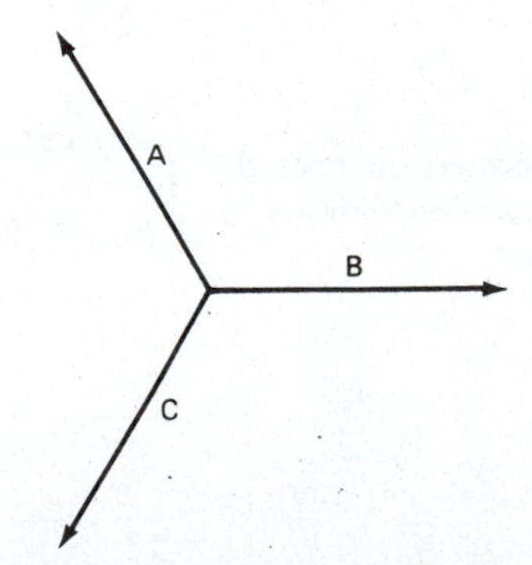

Figure 16-45A
Incorrect connections for a three-phase, closed-delta system

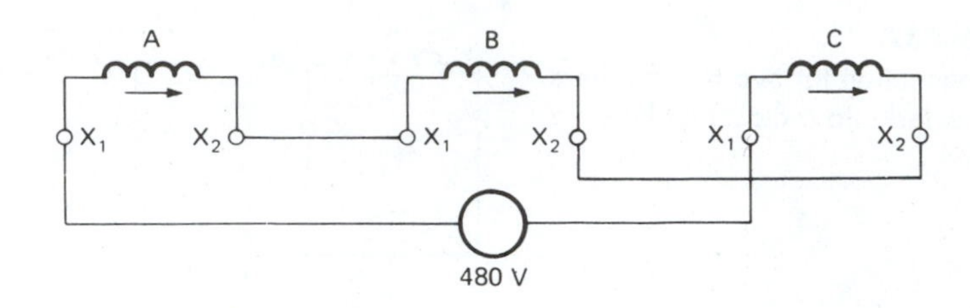

Figure 16-45B
Vector diagram of the voltages in Figure 16-45A

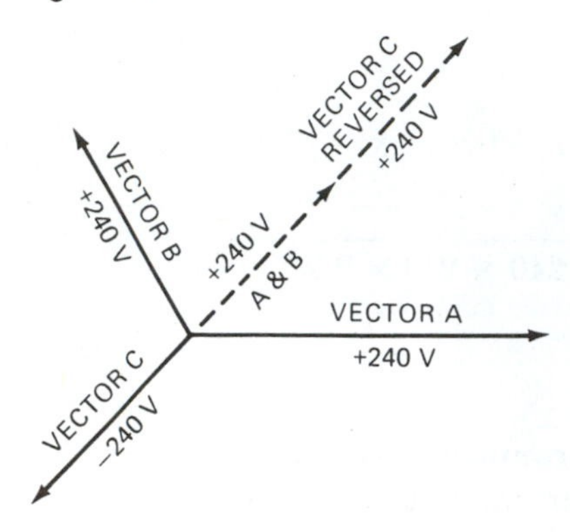

Figure 16-45A illustrates incorrect connections. Note that the instantaneous voltage of transformer C is opposite to that of transformer A and B. To perform vector addition, vector C must be reversed. This operation puts vector C in phase with the resultant voltage of transformers A and B, and adds arithmetically to equal 480 volts, Figure 16-45B.

> **CAUTION:** Never connect together two terminals if a potential difference exists between the two points.

5. With the transformer connected as shown in Figure 16-44A, zero potential difference between X_2 of transformer C and X_1 of transformer A, connect X_2 and X_1 together.
6. Connect the load conductors to the junction points of X_1 and X_2 as shown in Figures 16-46A and 16-46B.

The preceding connections are for a delta/delta, three-phase, three-wire system. *Delta/delta* means that both the primary and secondary windings are connected in a delta configuration.

Figure 16-46A
Three single-phase transformers connected to form a three-phase, closed-delta bank

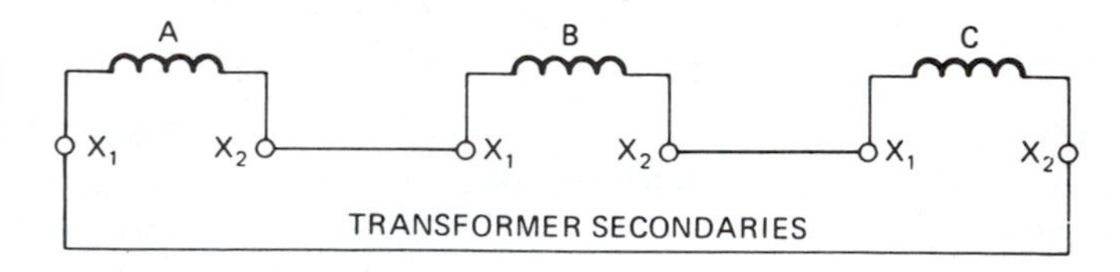

Figure 16-46B
Three single-phase transformers connected into a closed-delta configuration with the load conductors connected

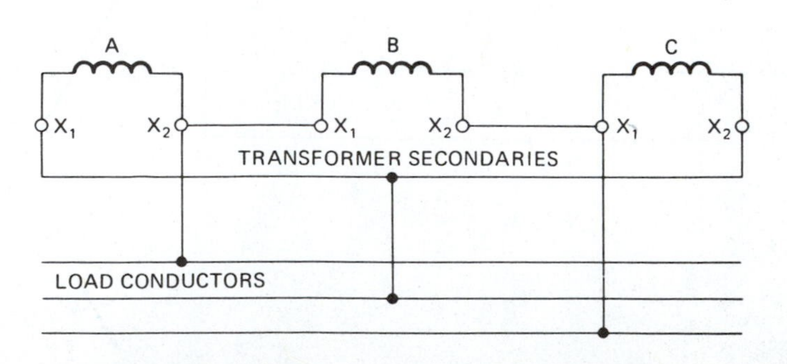

Figure 16-47A
High-voltage windings of three single-phase transformers connected into a wye configuration and connected to a three-phase, three-wire supply

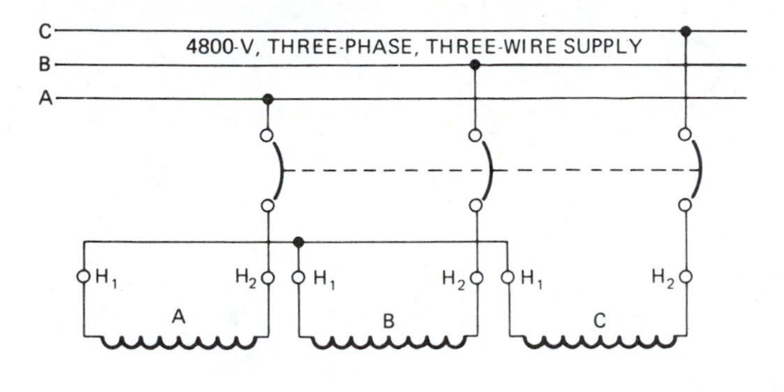

Figure 16-47B
Primary windings of a three-phase transformer bank connected into a wye configuration

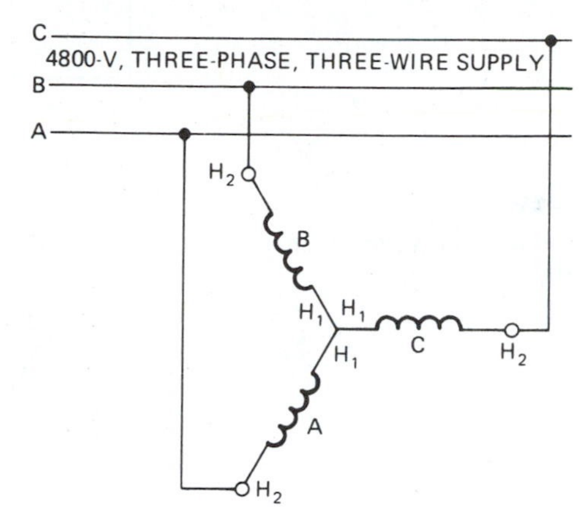

Figure 16-48A
Connections for two of three single-phase transformers that are being connected into a three-phase wye configuration

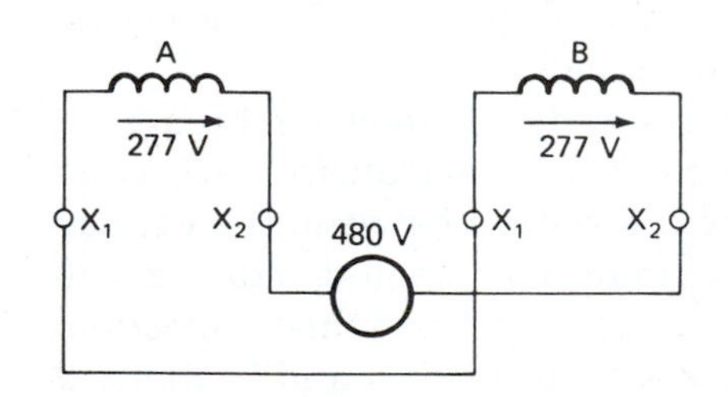

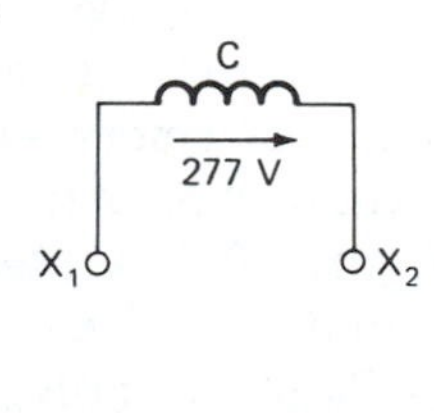

Figure 16-48B
Vector diagram of the voltages in Figure 16-48A

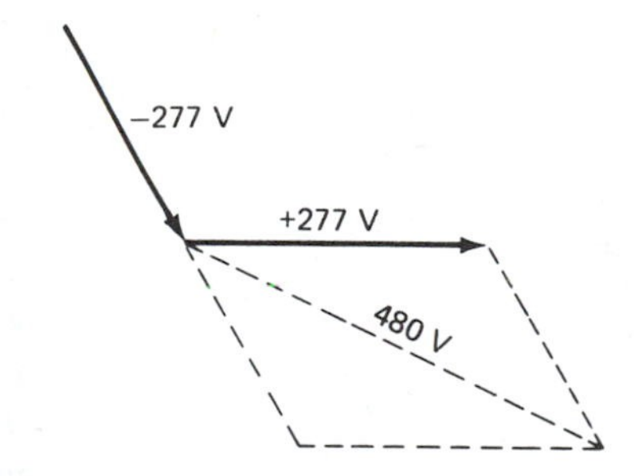

Three-Phase Wye/Wye Connections. Another method for connecting transformers is *wye / wye*. The following procedure is used. Assume that the transformers are rated at 4800 volts and 277 volts.

1. Connect H_1 of all three transformers together, Figures 16-47A and 16-47B.
2. Connect H_2 of each transformer to a supply conductor.
3. Measure the voltage across the secondary of each transformer. The meter should indicate 277 volts.
4. Connect X_1 of transformer A to X_1 of transformer B, Figure 16-48A. Measure the voltage between X_2 of transformer A and X_2 of transformer B. The meter should indicate 480 volts. Figure 16-48B shows the vector diagram for this connection. Note that the instantaneous voltages are in the opposite

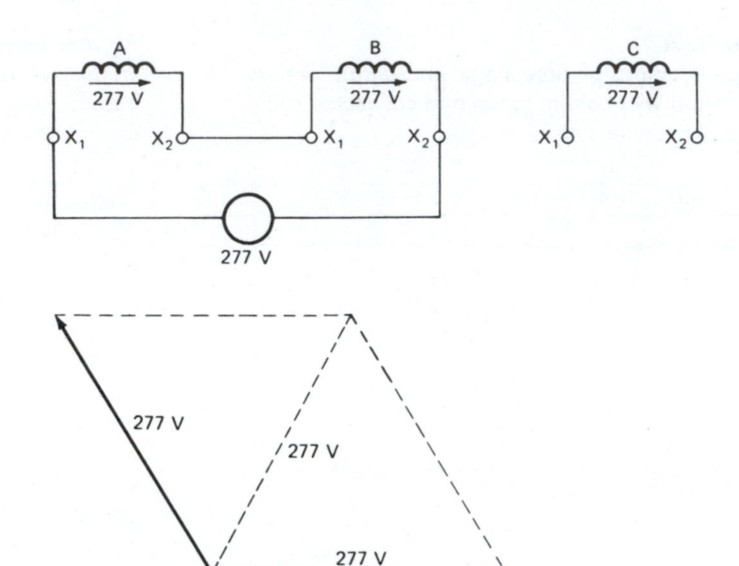

Figure 16-49A
Incorrect connections for two of three single-phase transformers being connected into a three-phase wye configuration

Figure 16-49B
Vector diagram of the voltages in Figure 16-49A

directions. Therefore, to perform the vector addition, one vector must be reversed.

Figure 16-49A shows incorrect connections, which would result in a voltage of 277 volts. Figure 16-49B shows the vector diagram for this connection.

5. With transformers A and B connected as shown in Figure 16-48A, connect X_1 of transformer C to X_1 of transformers A and B, Figure 16-50A. Measure the voltage between the various combinations of X_2. All combinations should indicate 480 volts. Figure 16-50B is the vector diagram of this connection. Phase 1 is the vector sum of transformers A and B. Phase 2 is the vector sum of transformers B and C. Phase 3 is the vector sum of transformers C and A.
6. Connect the leads X_2 to the load conductors. A conductor connected to the midpoint (X_1) will provide a second voltage. Between conductor N (Figure 16-51) and any of the load conductors, the voltage will be 277 volts.

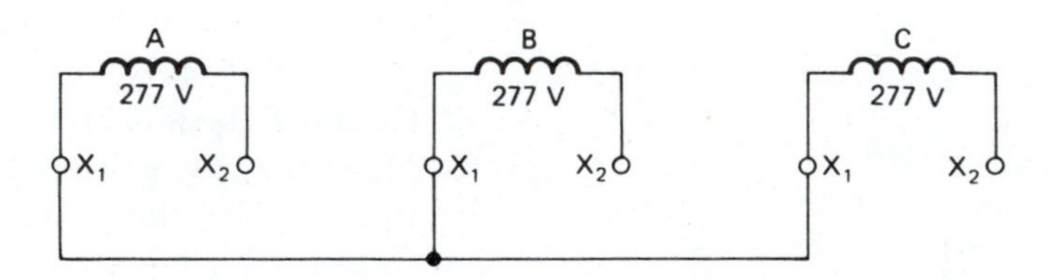

Figure 16-50A
Low-voltage windings of three single-phase transformers connected to form a three-phase wye configuration

Figure 16-50B
Vector diagram of three separate voltages obtained from a three-phase, three-wire wye connection

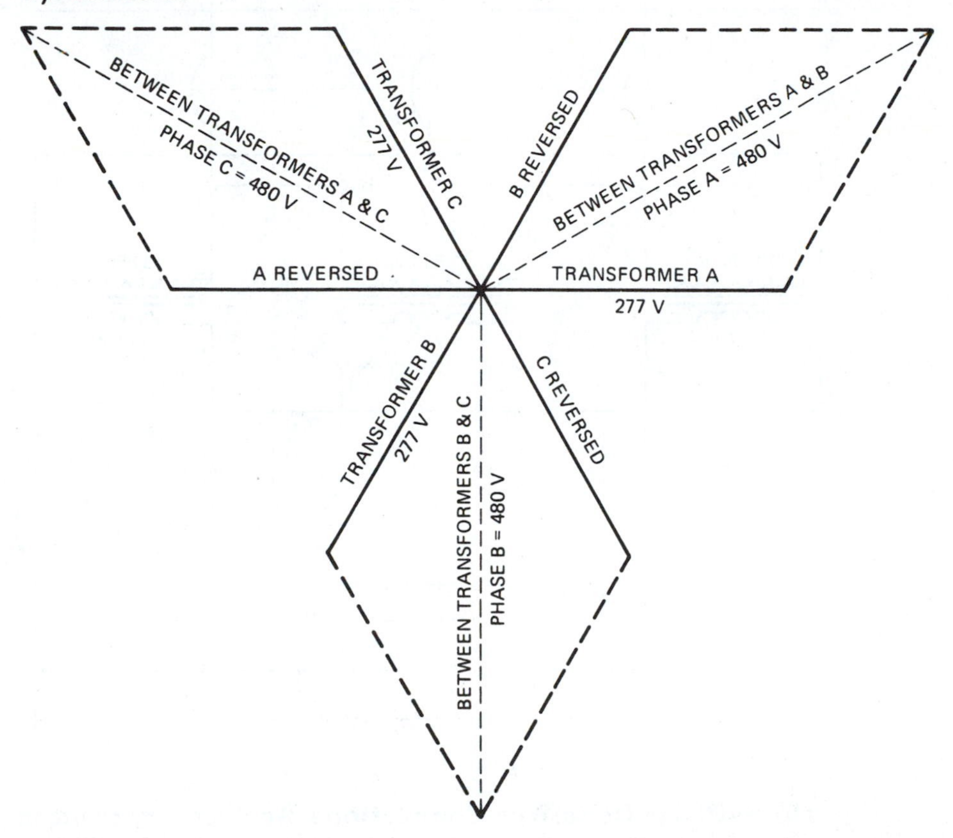

Figure 16-51
Connections for the low-voltage windings of three single-phase transformers arranged to form a three-phase, four-wire wye system

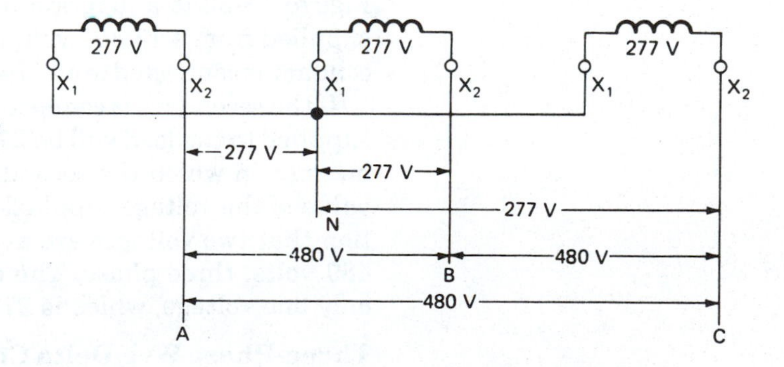

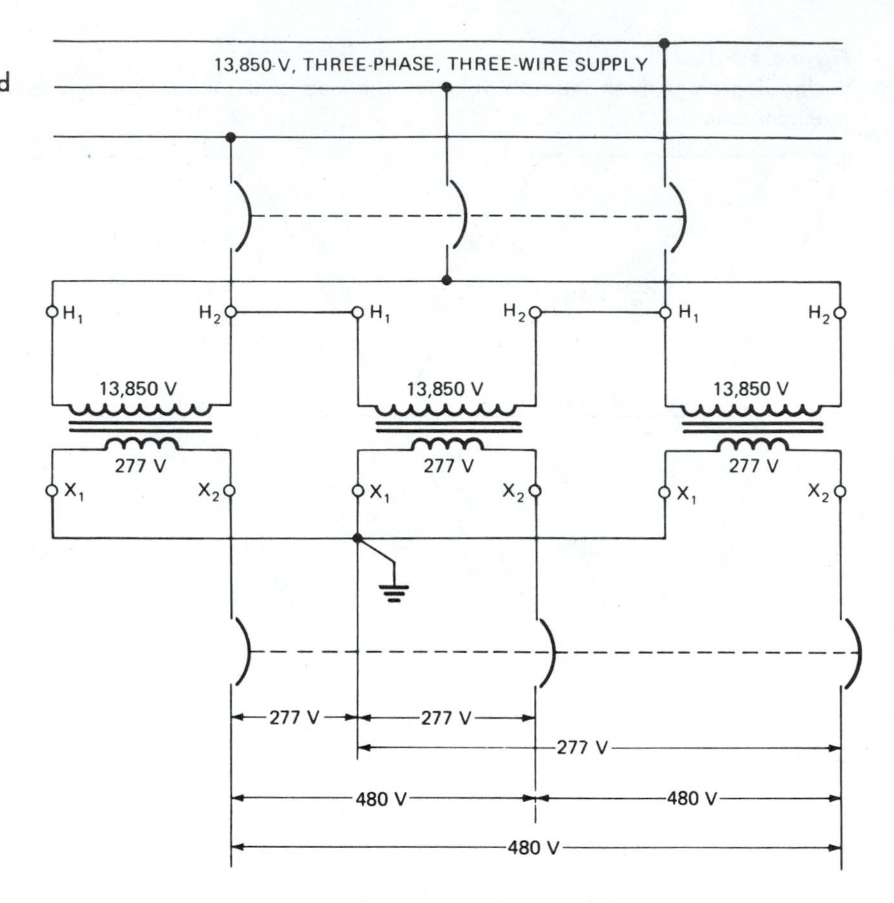

Figure 16-52
Three single-phase transformers arranged to form a three-phase delta/wye bank

Three-Phase Delta/Wye Connections. Another very common connection is the *delta / wye.* The primary is connected to form a delta configuration, and the secondary is connected to form a wye. Figure 16-52 is a diagram of this arrangement. The primary is supplied from a 13,850-volt, three-phase, three-wire line. The secondary is connected to a 277/480-volt, three-phase, four-wire load.

If the secondary is connected in delta instead of wye, the voltage supplied to the load will be 277 volts, three phase, three wire. The manner in which the secondary is connected will determine the value of the voltage supplied to the load. Note in the wye connection that two voltages are available: 277 volts, single phase, and 480 volts, three phase. The delta connection, however, supplies only one voltage, which is 277 volts, three phase.

Three-Phase Wye/Delta Connections. Transformers may also be connected in *wye / delta.* Figure 16-53 illustrates this connection. Note that the secondary of each transformer has two separate windings. This is a common practice and permits a midpoint tap from one transformer.

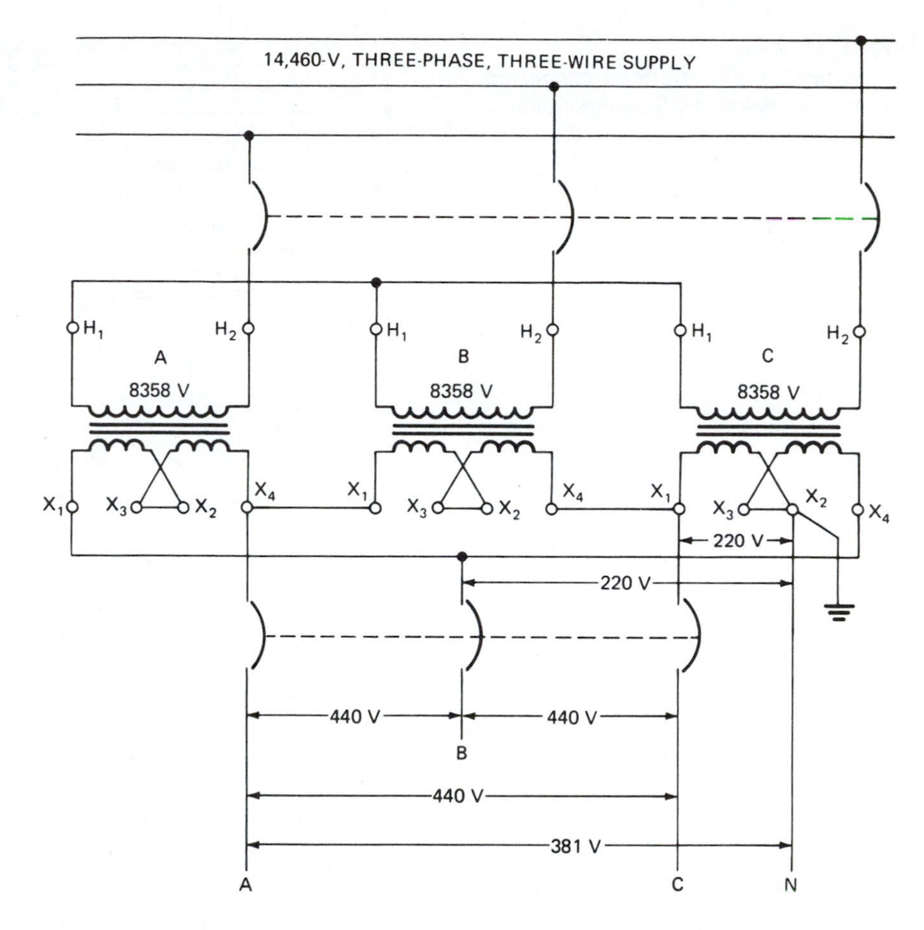

Figure 16-53
Three single-phase transformers arranged into a three-phase wye/delta bank

The voltage supplied to the primary is 14,460 volts, three phase. The voltage across each primary winding is 14,460/1.73 = 8358 volts. If the ratio is 19 to 1, then the voltage across each secondary winding is 8358/19 = 440 volts. The secondary supplies 440 volts, three phase to the load. The voltage between the midpoint of transformer C and phase wires B and C is 220 volts, single phase. The voltage between the midpoint of transformer C and phase wire A is the vector sum of 440 and 220, which is 381 volts.

Three-Phase, Open-Delta Connections. Another method sometimes used for three-phase transformer connections is called the *open delta,* or *V,* and requires only two transformers. In areas where additional loading is anticipated, two transformers connected in open delta can be installed for the initial load. When the additional load is installed, a third transformer can be added to form a closed-delta system. The open-delta arrangement will carry 58 percent of the load that can be applied to a closed delta when using the same size of transformers.

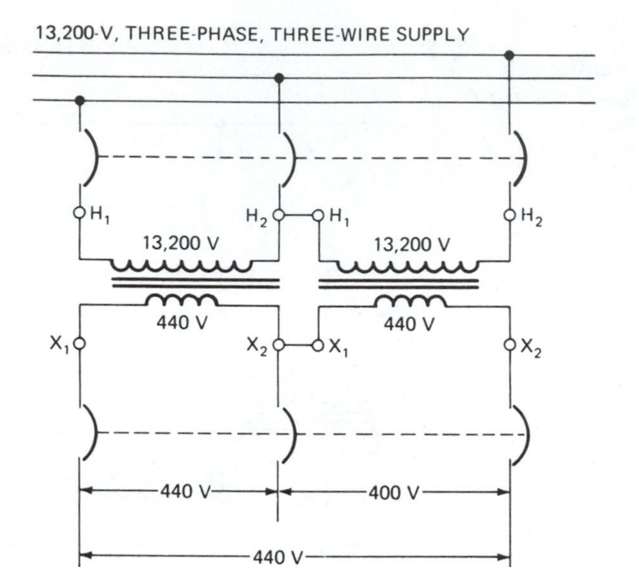

Figure 16-54
Two single-phase transformers connected to form an open-delta (V) configuration

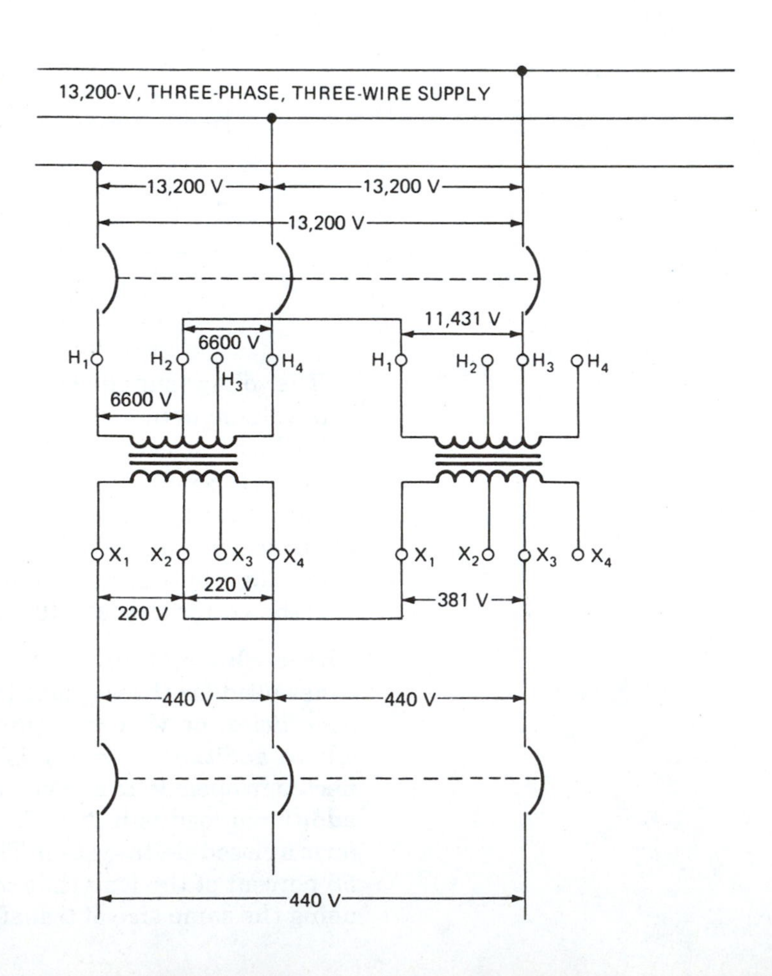

Figure 16-55
Schematic diagram of a three-phase T connection

The open-delta arrangement is also useful for maintenance purposes. If one transformer of a closed-delta system becomes defective, the transformer can be removed and the system reconnected in open delta. Under these conditions, the system can maintain 58 percent of its total load. Figure 16-54 shows an open-delta system.

Three-Phase T Connection. In this arrangement, two transformers can be used with a three-phase system. The transformers must have a midpoint tap and an 86.6 percent tap. Figure 16-55 is a schematic diagram of the T connection.

Scott Connection. Most electrical equipment is constructed to operate on either single phase or three phase. In areas where the supply is two phase, transformers can be used to convert it to three phase. Figure 16-56 illustrates this arrangement, called a *Scott connection.*

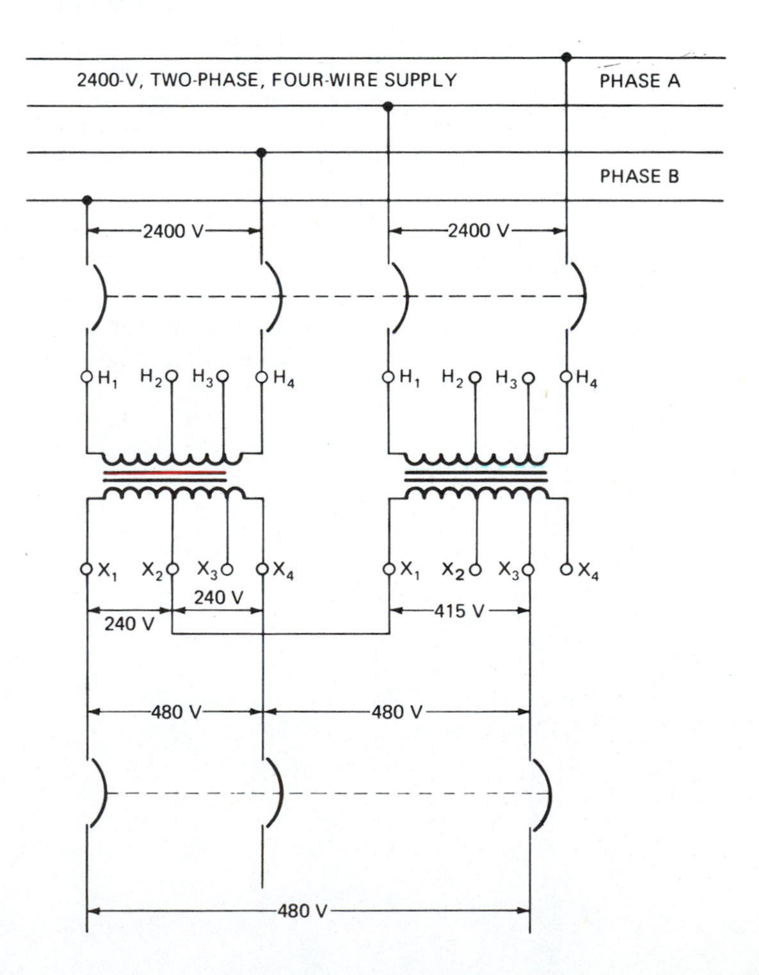

Figure 16-56
Scott connection using two single-phase transformers to convert two phase, four wire, to three phase, three wire

Figure 16-57A
Three-phase wye system. The line current is equal to the phase current. The line voltage is equal to 1.73 times the phase voltage.

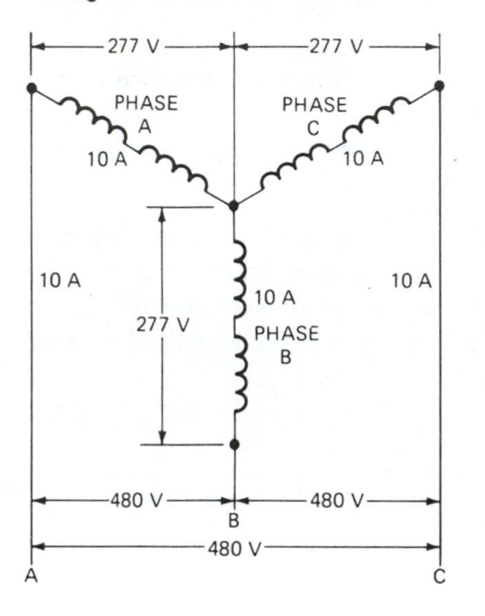

Figure 16-57B
Three-phase delta system. The line current is equal to 1.73 times the phase current. The line voltage is equal to the phase voltage.

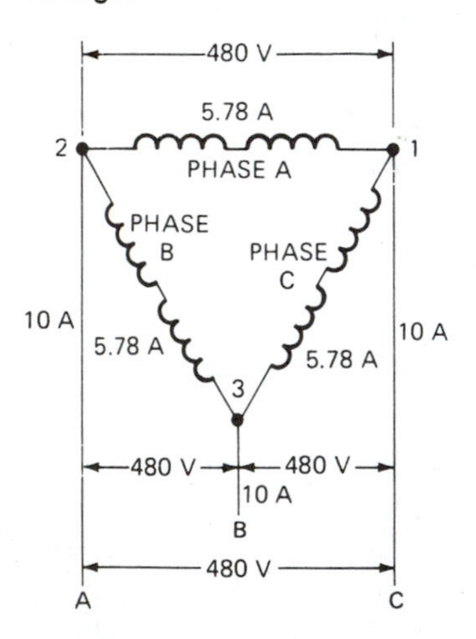

Isolation Transformers. Another common use of transformers is to isolate a system or part of a system from the supply. Frequently this requirement is fulfilled without a change in voltage. Transformers with a ratio of 1 to 1 are used. When used for this purpose, they are called *isolation transformers.*

Three-Phase Voltage and Current Characteristics

There are certain voltage and current characteristics associated with three-phase connections. For example, when transformers are connected wye, the line voltage is always 1.73 times as large as the phase voltage. However, the phase current and the line current are equal. Figure 16-57A illustrates this condition. Notice that the voltage between lines B and C is equal to the vector sum of the phase voltages B and C. The line current is 10 amperes and is equal to the phase current.

The delta system illustrated in Figure 16-57B shows that the line voltages and the phase voltages are equal, but the line current is 1.73 times as great as the phase current. In Figure 16-57B, the current through line A is 10 amperes. At point 2 the current divides, with some flowing through phase A and some flowing through phase B ($5.78 \times 1.73 = 10$ amperes). At some other instant, 10 amperes are flowing in line B. At point 3, the current divides, and 5.78 amperes flow through each phase B and C. This condition continues to change between the lines as each current varies from zero to maximum. The change takes place so rapidly that the ordinary ammeter cannot detect it. Thus, an ammeter placed in any line will always indicate a value 1.73 times the phase current.

REVIEW

A. Multiple choice.
Select the best answer.

1. Alternating current can be transmitted over great distances much more economically than can direct current because of a device called a
   a. transformer.
   b. transducer.
   c. motor-generator set.
   d. transmitter.

2. The operation of a transformer depends upon
   a. electrodynamics.
   b. electromagnetic induction.
   c. electromagnetic attraction.
   d. electromechanical induction.

3. The primary winding of a transformer is the coil that
   a. transfers the energy
   b. receives the energy from the supply.
   c. changes magnetic energy into electrical energy.
   d. changes electrical energy into mechanical energy.

4. A transformer that is used to lower voltage is called a
   a. step-down transformer.
   b. step-up transformer.
   c. voltage-reducing transformer.
   d. voltage-lowering transformer.

5. The number of volts per turn on the primary winding of a transformer is
   a. equal to the number of volts per turn on the secondary winding.
   b. less than the number of volts per turn on the secondary winding.
   c. greater than the number of volts per turn on the secondary winding.
   d. equal to or greater than the number of volts per turn on the secondary winding.

6. When the secondary winding of a transformer is disconnected from the load,
   a. zero current flows in the primary.
   b. very little current flows in the primary.
   c. the primary current remains the same as under load.
   d. the primary current becomes very high.

REVIEW *(continued)*

A. Multiple choice.
Select the best answer (continued).

7. The phase relationship between the primary voltage and the secondary voltage of a transformer is determined by the
   a. amount of load.
   b. demand factor of the load.
   c. power factor of the load.
   d. none of the above.

8. The iron/steel core of a transformer is
   a. solid.
   b. laminated.
   c. perforated.
   d. hollow.

9. Air core transformers are used when the frequency is
   a. high.
   b. low.
   c. medium.
   d. steady.

10. Air core transformers are generally used
    a. for distribution transformers.
    b. for signal transformers.
    c. in electronic equipment.
    d. in lighting transformers.

11. Transformer hum is a result of
    a. laminations and windings vibrating.
    b. the magnetic field alternating.
    c. high current.
    d. both a and b.

12. Heat generated by a transformer is a result of
    a. eddy currents and hysteresis.
    b. high voltages.
    c. high resistance.
    d. low frequency.

13. Air-cooled transformers are generally rated at
    a. 5 kilovolt-amperes or less.
    b. 5 kilovolt-amperes or more.
    c. 15 kilovolt-amperes or less.
    d. 15 kilovolt-amperes or more.

REVIEW *(continued)*

A. Multiple choice.
Select the best answer (continued).

14. The purpose of oil in transformers is to
    a. lubricate.
    b. insulate and cool.
    c. lubricate and cool.
    d. insulate and lubricate.

15. A tapped transformer is one that has
    a. a tap for draining the oil.
    b. a T tap on the primary.
    c. several terminal posts to allow for voltage adjustments.
    d. taps for connecting the primary winding to the secondary winding.

16. Autotransformers are used
    a. to make small changes in voltage.
    b. in automobiles.
    c. to provide automatic voltage changes.
    d. to make large voltage changes.

17. Potential transformers are used
    a. with fluorescent lighting.
    b. with instruments and relays.
    c. to make small changes in voltage.
    d. to transfer large amounts of power.

18. Current transformers are used to
    a. increase the current to the load.
    b. decrease the current in special situations.
    c. maintain a constant current in the system.
    d. decrease large amounts of voltage.

19. A buck-boost transformer is a
    a. single-winding transformer.
    b. two-winding transformer.
    c. type of instrument transformer.
    d. heavy-duty transformer.

20. A control transformer is used to
    a. control the voltage to fluorescent lights.
    b. reduce the voltage for motor circuits.
    c. control the current to motors.
    d. reduce the voltage for motor control circuits.

REVIEW *(continued)*

B. Give complete answers.

1. What is a transformer?
2. Describe the basic construction of a distribution transformer.
3. Explain the basic principle of transformer operation.
4. What determines the amount of voltage produced in the secondary winding of a transformer?
5. Define the terms *primary winding* and *secondary winding.*
6. What determines the amount of current that will flow in the primary winding of a distribution transformer?
7. What determines the value of current that will flow in the secondary winding of a distribution transformer?
8. Define the term *current regulation* with regard to a distribution transformer.
9. How does the power factor of the load affect the operation of the transformer?
10. Define *turns ratio.*
11. What is the relationship of the volts per turn on the primary and secondary of standard transformers?
12. What causes power loss in a transformer?
13. Discuss hysteresis loss with regard to transformers.
14. Explain eddy currents and what causes them with regard to transformers.
15. How does core saturation affect the transformer operation?
16. List three types of core construction.
17. Describe a wound-core-type transformer.
18. List five advantages of the wound-core construction over the standard method of construction.
19. When is an air core transformer used?
20. What causes a transformer to hum?
21. Define the term *polarity* with regard to transformers.
22. Define the terms *additive* and *subtractive* with regard to transformers.

REVIEW *(continued)*

B. Give complete answers (continued).

23. Transformers should have certain polarities according to their ratings. List the polarities and ratings.
24. List four methods used to cool transformers.
25. What is a tap-changing switch?
26. Where are tapped transformers generally used?
27. Describe an automatic tap-changing switch.
28. What is an autotransformer?
29. List two advantages of autotransformers.
30. What is the main disadvantage of an autotransformer?
31. For what purposes are autotransformers generally used?
32. What is the purpose of an instrument transformer?
33. Describe a potential transformer.
34. According to industrial standards, what is the maximum percentage of error for instrument transformers?
35. Describe a current transformer.
36. What determines the value of current flowing in the primary winding of a current transformer?
37. What determines the value of current flowing in the secondary winding of a current transformer?
38. List one important safety rule that must be followed when working with current transformers.
39. What is a buck-boost transformer?
40. What is a control transformer?
41. List four requirements necessary for connecting transformers into a transformer bank.
42. Describe the procedure to be followed when connecting two transformers in parallel.
43. Draw a diagram of two single-phase transformers connected to provide a two-phase, five-wire system.
44. Describe the procedure for connecting three single-phase transformers into a three-phase delta/delta bank.

REVIEW *(continued)*

B. Give complete answers (continued).

45. Draw a diagram of three single-phase transformers connected into a delta/wye, three-phase, four-wire system.
46. Describe a three-phase, open-delta system.
47. Draw a diagram of two single-phase transformers used to convert two phase to three phase.
48. What is an isolation transformer?
49. Describe the T connection of two single-phase transformers used in a three-phase system.

C. Solve each problem, showing the method used to arrive at the solution.

1. A transformer is being designed to increase the voltage from 120 V to 480 V. If the primary winding requires 200 turns of wire, how many turns are required on the secondary?
2. If the load on the transformer in Problem 1 is 60 A, what is the primary current?
3. An engineer has been requested to design a core-type transformer. The core material has a flux density of 10,000 lines of force per square centimeter. The transformer is to be rated at 15 kVA, 480 V to 120 V, at 60 Hz. What will be the cross-sectional area of the core?
4. Calculate the number of turns required for the primary and secondary coils in Problem 3.
5. If the transformer in Problem 3 is supplying its rated load at 100% power factor, how much current is flowing in the primary and secondary coils?

17

ELECTRICAL DISTRIBUTION

Objectives

After studying this chapter, the student will be able to:

- Describe primary distribution systems.
- Describe residential, commercial, and industrial distribution systems.
- Explain the methods and purpose of grounding electrical systems.
- Explain the methods and purpose of grounding electrical equipment.
- Describe ground-fault protection.
- Describe the operating characteristics of unbalanced three-phase systems.
- Discuss the effect of harmonics on multiwire systems.

PRIMARY DISTRIBUTION SYSTEMS

The wiring between the generating station and the final distribution point is called the *primary distribution system.* There are several methods used for transmitting the power between these two pints. The two most common methods are the radial system and the loop system.

The Radial System

The term *radial* comes from the word *radiate,* which means to send out or emit from one central point. A *radial system* is an

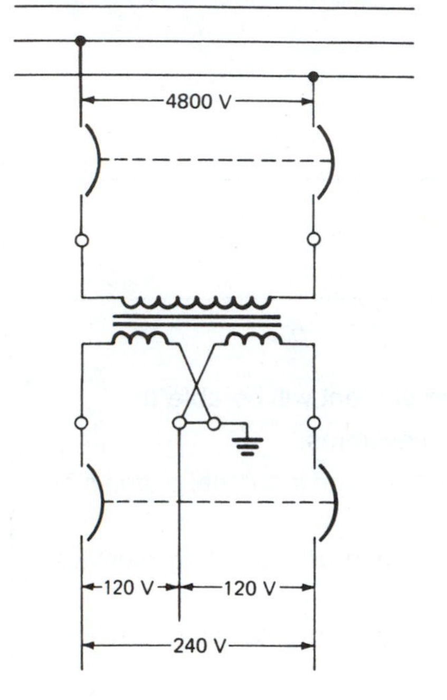

Figure 17-1
Single-phase, three-wire, 120/240-V system

electrical transmission system that begins at a central station and supplies power to various substations.

In its simplest form, a radial system consists of a generating station that produces the electrical energy. This energy is transmitted from the generator(s) to the central station, which is generally part of, or adjacent to, the generating station. At the central station, the voltage is stepped up to a higher value for long-distance transmission.

From the central station, several lines carry the power to various substations. At the substations, the voltage is usually lowered to a value more suitable for distribution in populated areas. From the substations, lines carry the power to distribution transformers. These transformers lower the voltage to the value required by the consumer.

The Loop System

The *loop system* starts from the central station or a substation and makes a complete loop through the area to be served and back to the starting point. This results in the area's being supplied from both ends, allowing sections to be isolated in case of a breakdown. An expanded version of the loop system consists of several central stations joined together to form a very large loop.

CONSUMER DISTRIBUTION SYSTEMS

The type of distribution system that the consumer uses to transmit power within the premises depends upon the requirements of the particular installation. Residential occupancies generally use the simplest type. Commercial and industrial systems vary widely with load requirements.

Single-Phase Systems

Most single-phase systems are supplied from a three-phase primary. The primary of a single-phase transformer is connected to one phase of the three-phase system. The secondary contains two coils connected in series with a midpoint tap to provide a single-phase, three-wire system. This arrangement is generally used to supply power to residential occupancies and some commercial establishments. A schematic diagram is shown in Figure 17-1.

For residential occupancies, the service conductors are installed either overhead or underground. Single-family and small multifamily dwellings have kilowatt-hour meters that are installed on the outside of the building. From the kilowatt-hour meter, the conductors are connected to the main disconnect. Figures 17-2A and 17-2B show this arrangement.

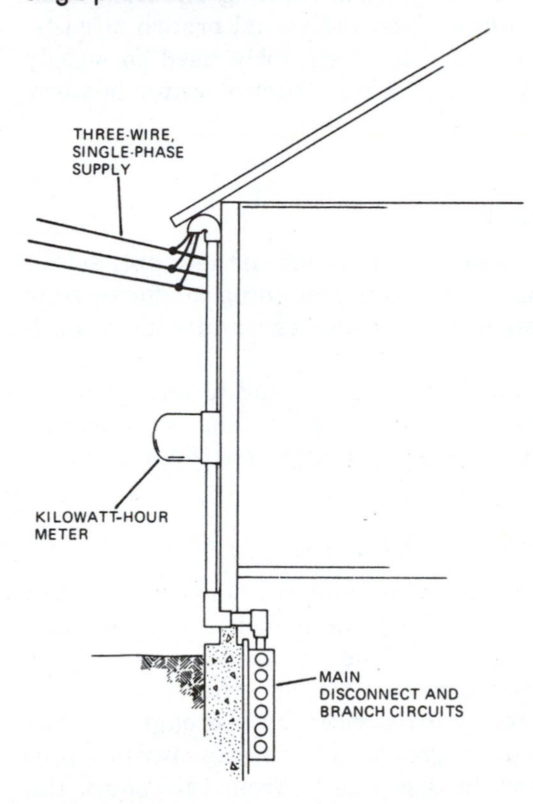

Figure 17-2A
Single-family residence with a three-wire, single-phase service

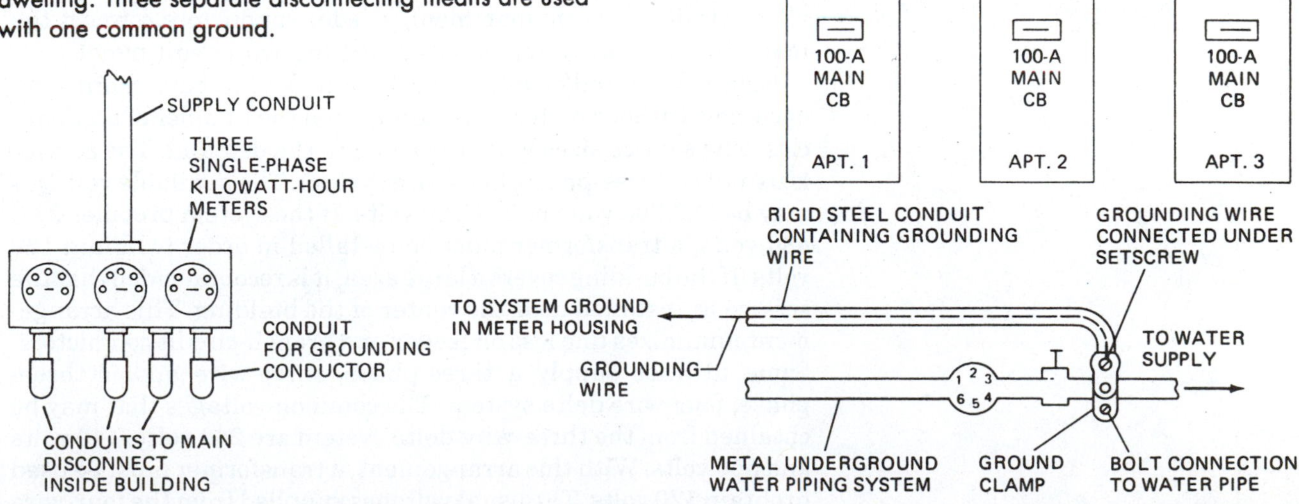

Figure 17-2B
Three-wire, single-phase service for a multifamily dwelling. Three separate disconnecting means are used with one common ground.

From the main disconnect, the conductors supply power to the branch circuit panels. For dwelling occupancies, there are three basic types of branch circuits: general lighting circuits, small appliance and laundry circuits, and individual branch circuits. The individual branch circuits are frequently used to supply central heating and/or air-conditioning systems, water heaters, and other special loads.

Grounding Requirements

All ac services are required to be grounded on the supply side of the service-disconnecting means. This grounding conductor runs from the combination system and equipment ground to the grounding electrode.

For multifamily occupancies it is permitted to use up to six service-disconnecting means. A single grounding conductor of adequate size should be used for the system ground (Figure 17-2B).

Commercial and Industrial Installations

Commercial and industrial installations are more complex than small residential installations. Large apartment complexes and condominiums, although classified as residential occupancies, often use commercial-style services. A single-phase, three-wire service or a three-phase, four-wire service may be brought into the building, generally from underground. The service-entrance conductors terminate in a main disconnect. From this point, the conductors are connected to the individual kilowatt-hour meters for each apartment and then to smaller disconnecting means and overcurrent protective devices. Branch-circuit panels are generally installed in each apartment. Feeder conductors connect the individual disconnecting means to the branch-circuit panels.

Commercial and/or industrial buildings may have more than one kilowatt-hour meter, depending upon the number of occupancies. The service sizes vary according to the demand. The service is usually a three-phase, four-wire system. The available voltages may be 120/208 volts or 277/480 volts. If the system provides 277/480 volts, a transformer must be installed in order to obtain 120 volts. If the building covers a large area, it is recommended that the service be installed near the center of the building. This arrangement minimizes line loss on feeder and branch-circuit conductors. Some utilities supply a three-phase, three-wire and/or three-phase, four-wire delta system. The common voltages that may be obtained from the three-wire delta system are 240 volts, 440 volts and 550 volts. With this arrangement, a transformer must be used to obtain 120 volts. The usual voltages supplied from the four-wire delta system are 240 volts, three-phase, and 120 volts, single phase.

Many large consumers purchase electrical energy at the primary voltage, and transformers are installed on their premises. Three-phase voltages up to 15 kilovolts are often used.

The service for this type of installation generally consists of metal cubicles called a *substation unit*. The transformers are installed either within the cubicle or adjacent to it. Isolation switches of the drawer type are installed within the cubicle. These switches are used to isolate the main switch or circuit breaker from the supply during maintenance or repair.

Consumer Loop Systems

Although the radial system of distribution is probably the most commonly used system of transmitting power on the consumer's property, the loop system is also employed. A block diagram of both systems is illustrated in Figures 17-3A and 17-3B. There are several variations of these systems in use in the industry, but the systems illustrated here show the basic structure.

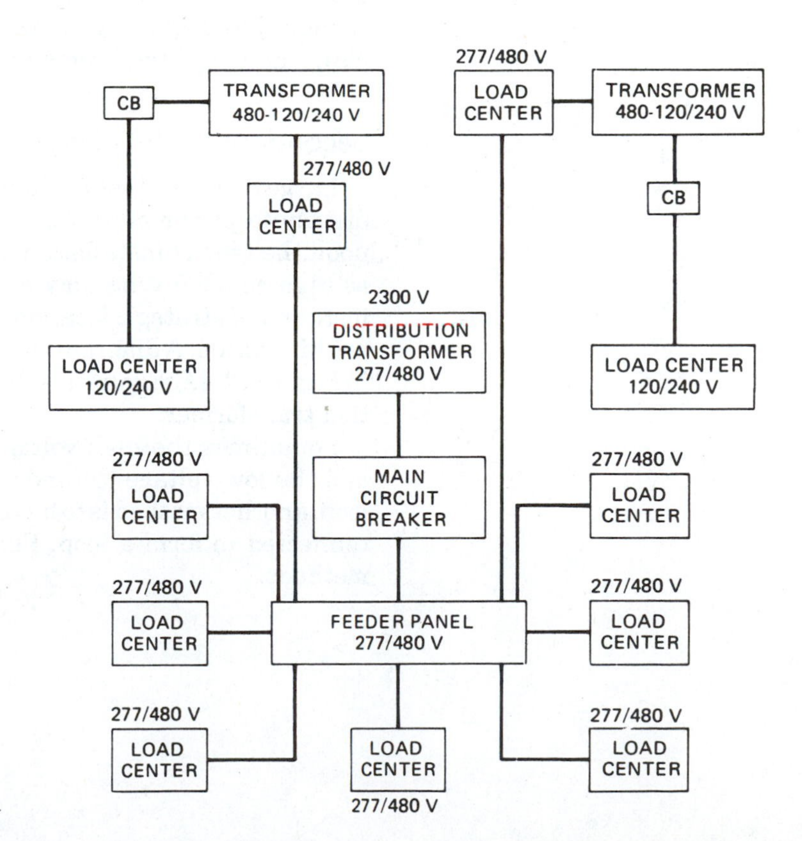

Figure 17-3A
Consumer radial distribution system

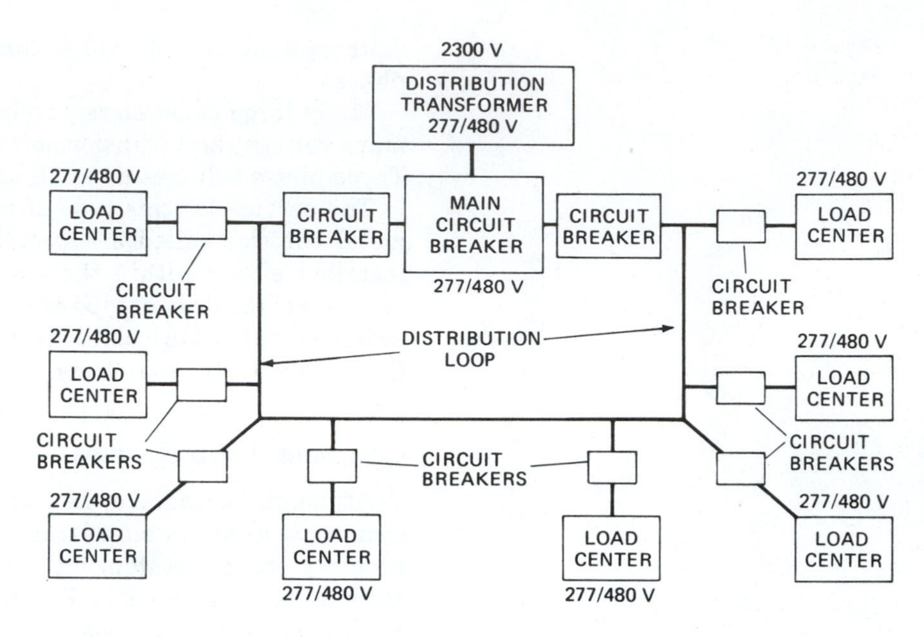

Figure 17-3B
Consumer loop distribution system. Disconnecting means may be installed anywhere in the distribution loop to provide for isolating sections.

In the installation of any system, overcurrent protection and grounding must be given primary consideration. Electrical personnel who design and install these systems must comply with the *NEC®* and local requirements.

Secondary High-Voltage Distribution

Large industrial establishments may find it more economical to distribute power at voltages higher than 600 volts. Depending upon the type of installation and the load requirements, voltages as high as 2300 volts may be used. Step-down transformers are installed in strategic locations to reduce the voltage to a practical working value. A diagram of a high-voltage radial system is shown in Figure 17-4A. Figure 17-4B illustrates a pad-mounted distribution transformer.

Sometimes the high-voltage (primary) system may be radial, and the low-voltage (secondary) system may be connected into a loop. another method is to have both the primaries and secondaries connected to form a loop. Figures 17-5A and 17-5B show these methods.

Figure 17-4A
Secondary high-voltage radial distribution system

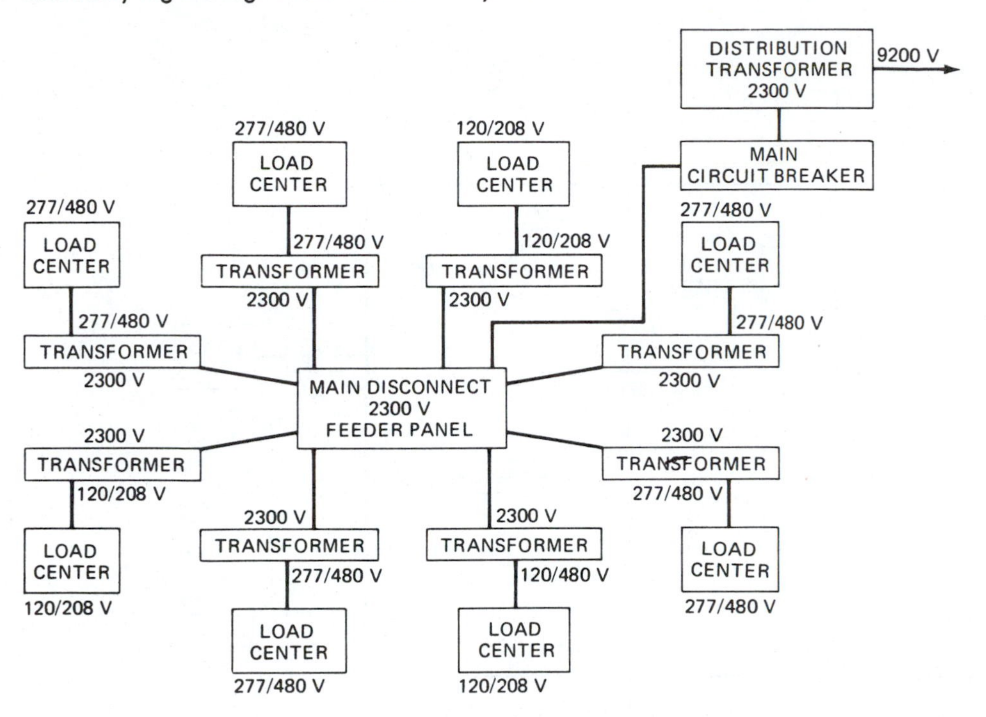

Figure 17-4B
Pad-Mounted distribution transformer
(Courtesy of Niagara Mohawk)

Figure 17-5A
Secondary high-voltage distribution system; high-voltage radial, low-voltage loop

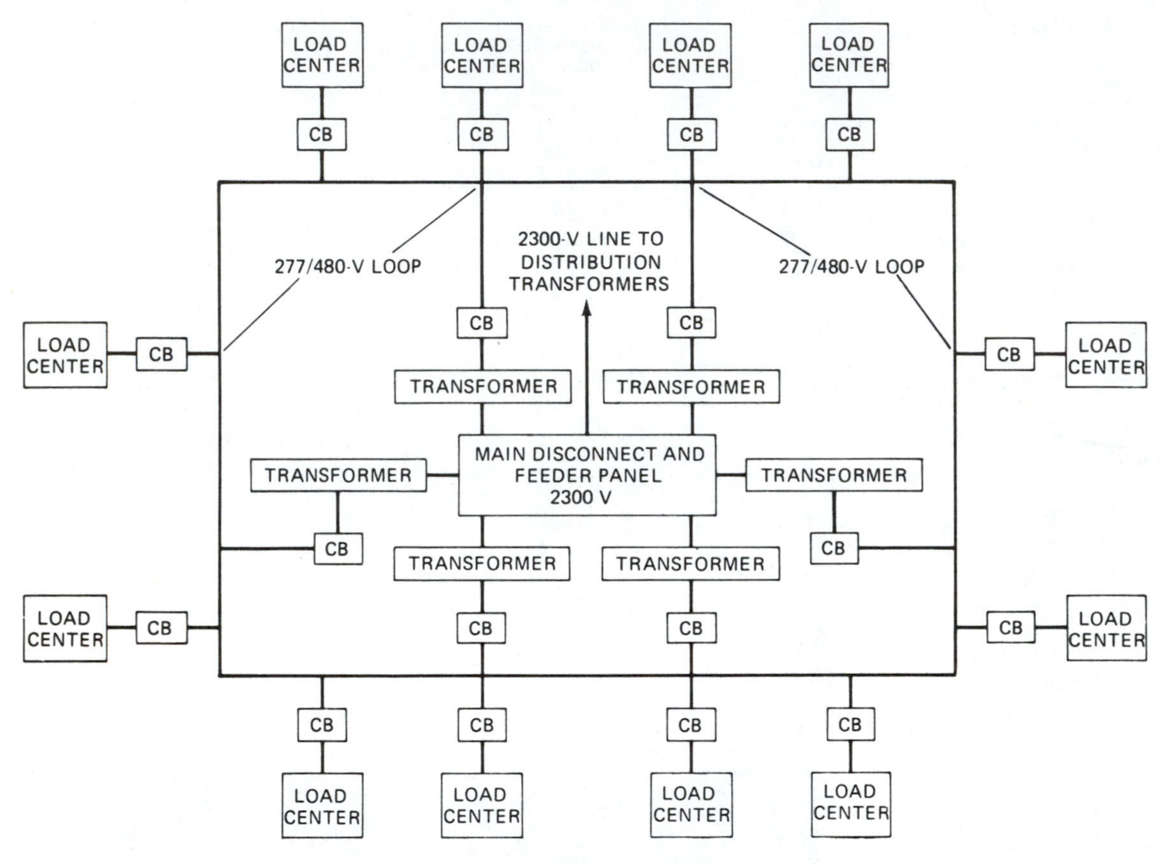

Secondary Ties Loop System

It is frequently convenient to connect loads to the secondary conductors at points between transformers. These conductors are called *secondary ties*. *Article 450* of the *NEC*® gives specific requirements regarding the conductor sizes and overcurrent protection.

Figure 17-5B
Consumer distribution system with high-voltage and low-voltage loops

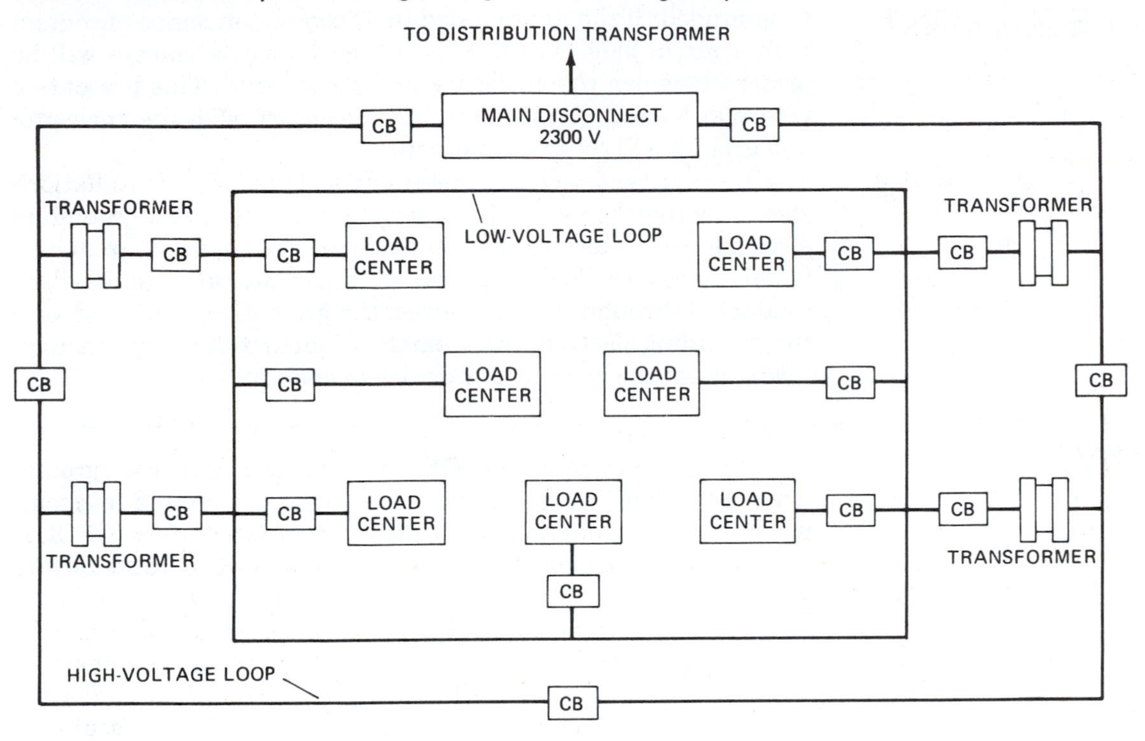

GROUNDING OF ELECTRICAL SYSTEMS

In general, most electrical systems must be grounded. The purpose of grounding is to limit the magnitude of voltage caused by lightning, momentary surges and accidental contact with higher voltages. System grounds must be arranged to provide a path of minimum impedance in order to ensure the operation of overcurernt devices when a ground fault occurs. Current should not flow through the grounding conductor during normal operation.

Direct current systems generally have the grounding conductor connected to the system at the supply station and not at the individual service. Alternating current systems, on the other hand, must be grounded on the supply side of the main disconnect at each individual service. For specific information on the location and method of grounding, refer to *NEC® Article 250.*

GROUNDING OF ELECTRICAL EQUIPMENT

Metal conduit and cases that enclose electrical conductors must be grounded. If the ungrounded (hot) conductor comes in contact with a metal enclosure that is not grounded, a voltage will be present between the enclosure and the ground. This presents a potential hazard. Persons coming in contact with the enclosure and ground will complete a circuit.

All noncurrent-carrying metal parts of electrical installations should be tightly bonded together and connected to a grounding electrode. Good electrical continuity should be ensured through all metal enclosures. The current caused by accidental grounds will be conducted through the enclosures, the grounding conductor, and the grounding electrode to the earth. If the current is large enough, it will cause the overcurrent device to open.

GROUND-FAULT PROTECTION

A *ground-fault protector* (GFP) is a device that senses ground faults and opens the circuit when the current to ground reaches a predetermined value. A *ground-fault circuit interrupter* (GFCI) is a device that opens the circuit when very small currents flow to ground.

There is no way to determine in advance the impedance of an accidental ground. Most circuits are protected by 15-ampere or larger overcurrent devices. If the impedance of a ground fault is low enough, such devices will open the circuit. What about currents of less than 15 amperes? It has been proved that currents as small as 50 milliamperes through the heart, lungs, or brain can be fatal.

Electrical equipment exposed to moisture or vibration may develop high-impedance grounds. Arcing between a conductor and the frame of equipment may cause a fire, yet the current may be less than 1 ampere. Leakage current caused by dirt and/or moisture may take place between the conductor and the frame. Portable tools are frequently not properly grounded, and the only path to ground is through the body of the operator.

The GFCI was developed to provide protection against ground-fault currents of less than 15 amperes. The GFCI is designed to operate on two-wire circuits in which one of the two wires is grounded. The standard circuit voltages are 120 volts and 277 volts. The time it takes to operate depends upon the value of the ground-fault current. Small currents of 10 milliamperes or less may flow for up to 5 seconds before the circuit is opened. A current of 20 milliamperes will cause the GFCI to operate in less than 0.04 seconds. This time/current element provides a sufficient margin of safety without nuisance tripping.

The GFCI operates on the principle that an equal amount of current is flowing through the two wires. When a ground fault occurs, some of the current flowing through the ungrounded (hot) wire does not flow through the grounded wire; it completes the circuit through the accidental ground. The GFCI senses the difference in the value of current between the two wires and opens the circuit. GFCI's may be incorporated into circuit breakers, installed in the line, or incorporated into a receptacle outlet or equipment.

GFPs are generally designed for use with commercial and/or industrial installations. They provide protection against ground-fault currents from 2 amperes (special types go as low as 50 milliamperes) up to 2000 amperes. GFPs are generally installed on the main, submain, and/or feeder conductors. GFCIs are installed in the branch circuits. GFPs are generally used for three-wire, single-phase, and for three-phase installations; GFCIs are used for two-wire, single-phase circuits.

A GFP installed on supply conductors must enclose all the circuit conductors, including the neutral, if present. When operating under normal conditions, all the current to and from the load flows through the circuit conductors. The algebraic sum of the flux produced by these currents is zero. When a phase-to-ground fault occurs, the fault current returns through the grounding conductor. Under this condition, an alternating flux is produced within the sensing device. When the fault current reaches a predetermined value, the magnetic flux causes a relay to actuate a circuit breaker.

Sometimes the GFP is installed on the grounding conductor of the system. Under this condition, the unit senses the amount of phase-to-ground current flowing in the grounding conductor. When the current exceeds the setting of the GFP, it will cause the circuit breaker to open.

The GFP is actually a specially designed current transformer connected to a solid-state relay.

THREE-PHASE SYSTEMS

The various three-phase systems in normal use are described in Chapter 16. Under ideal conditions, these systems operate in perfect balance, and if a neutral conductor is present, it carries zero current. In actual practice, perfectly balanced system are seldom seen. The electrical worker, therefore, must be able to calculate values of current and voltage in unbalanced systems. Single-phase loads are frequently supplied from three-phase systems. The single-phase load requirements vary considerably, making it virtually impossible to maintain a perfect balance.

Figure 17-6
Three-phase, unbalanced delta connection

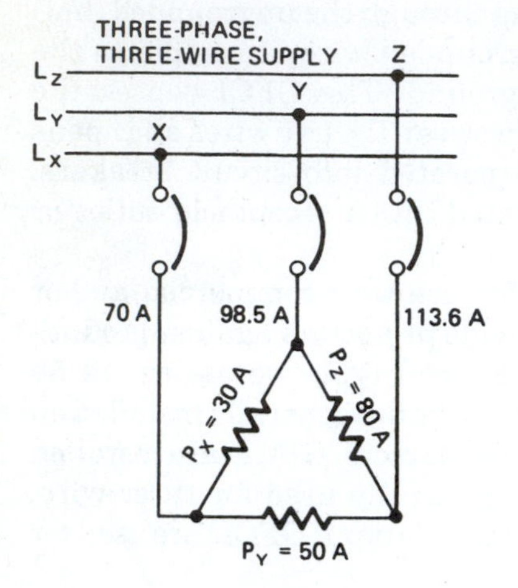

In a balanced three-phase system, the currents in the three lines are equal. The currents in the three phases are also equal. In other words, $I_{LX} = I_{LY} = I_{LZ}$ and $I_{pX} = I_{pX} = I_{pZ}$. If, however, $I_{LX} \neq I_{LY} \neq I_{LZ}$, then $I_{pX} \neq I_{pY} \neq I_{pZ}$ and the system is unbalanced. See Figure 17-6.

To calculate the line currents in an unbalanced three-phase system, the method in the following example may be used.

Example 1

Three pure resistance, single-phase loads are connected in a delta configuration across a three-phase supply, as illustrated in Figure 17-6. Load X requires 30 amperes, load Y requires 50 amperes, and load Z requires 80 amperes. Calculate the current through each line wire.

1. Line $X = \sqrt{X^2 + Y^2 + (2XY \cos \phi)}$ *(From Equation 13.5)*
   $X = \sqrt{30^2 + 50^2 + (2 \times 30 \times 50 \times 0.5)}$
   $X = \sqrt{900 + 2500 + 1500}$
   $X = \sqrt{4900}$
   $X = 70$ amperes
2. Line $Y = \sqrt{X^2 + Z^2 + (2XZ \cos \phi)}$
   $Y = \sqrt{30^2 + 80^2 + (2 \times 30 \times 80 \times 0.5)}$
   $Y = \sqrt{900 + 6400 + 2400}$
   $Y = \sqrt{9700}$
   $Y = 98.5$ amperes
3. Line $Z = \sqrt{Z^2 + Y^2 + (2ZY \cos \phi)}$
   $Z = \sqrt{80^2 + 50^2 + (2 \times 80 \times 50 \times 0.5)}$
   $Z = \sqrt{6400 + 2500 + 4000}$
   $Z = \sqrt{12{,}900}$
   $Z = 113.6$ amperes

Example 1 applies to loads of 100 percent power factor connected in delta. With loads of different power factors, the phase angle will vary from 120 degrees. For a wye connection, the line current is equal to the phase current.

Some connections may be a combination of single-phase and three-phase loads. Under these conditions, the phase angle between the three-phase load and the single-phase load must be considered.

HARMONICS

Most distribution systems in the United States and Canada operate on a frequency of 60 hertz. Certain types of electronic equipment, such as computer systems and some fluorescent lighting systems, produce secondary frequencies, which are multiples of the supply frequency. These secondary frequencies are called *harmonics*. For example, the second harmonic of 60 hertz is 120 hertz, the third harmonic is 180 hertz, and so on.

The alternating flux developed by some transformers that are used in fluorescent lighting ballasts produces a voltage with a frequency of 180 hertz. This problem, however, has been reduced to a minimum in high-quality ballasts.

Equipment such as computers and programmable controls also produces voltages with frequencies that are harmonics of the supply frequency. These harmonics cause additional current to flow in the supply conductors. The additional current in the phase conductors is usually only a small percentage of the supply current. This harmonic current adds to the supply current, causing a greater heating effect in the conductors. The increase in heating effect is usually rather small, possibly in the vicinity of 3 percent to 5 percent.

The effect on the neutral conductor is quite different. The harmonic currents from the phase conductors add together, causing a large increase in the neutral current. The heating effect may be as much as 90 percent greater than if there were no harmonic effect.

CAUTION: In the installation of supply, feeder, and branch circuit conductors for heavy fluorescent loads and computer loads, the size of the neutral conductor may have to be increased to allow for harmonic currents. See *Section 210-4* of the *NEC*®.

REVIEW

A. Multiple choice.
Select the best answer.

1. A primary distribution system consists of the wiring between the
   a. generator and the substation.
   b. transformer and the final branch circuit.
   c. generating station and the final distribution point of the utility company.
   d. generating station and the transformer.
2. A radial distribution system is an arrangement of transmission lines that
   a. extend in many directions from the central station to various substations; at the substation the lines divide again to carry the power to the distribution transformers.
   b. leave the central station, supplying many substations in succession, and return to the central station.
   c. distribute electrical power from one central station to another.
   d. extend in only one direction from the central station.
3. A loop-type distribution system is an arrangement of transmission lines that
   a. make a complete loop through each substation.
   b. start from the central station or a substation, make a complete loop through the area to be served, and return to the starting point.
   c. start from the central station, feed a substation, and loop back to the central station.
   d. make a complete loop to each transformer in the central system.
4. Most single-phase consumer distribution systems are supplied
   a. from a two-phase primary.
   b. directly from the generating station.
   c. from a three-phase primary.
   d. from a single-phase, three-wire primary.
5. Most single-family and small multifamily dwellings have the kilowatt-hour meters installed
   a. in the utility room.
   b. on the outside of the dwelling.
   c. in the basement.
   d. in the attic.

REVIEW *(continued)*

A. Multiple choice.
Select the best answer (continued).

6. Central air-conditioning systems are generally connected to
   a. lighting circuits.
   b. small appliance circuits.
   c. individual branch circuits.
   d. any of the above.
7. All ac services are required to be grounded
   a. on the load side of the service disconnect.
   b. on the supply side of the service disconnect.
   c. within the housing for the kilowatt-hour meter.
   d. anywhere on the system.
8. Multifamily occupancies are permitted to use
   a. up to six service-disconnecting means.
   b. up to ten service-disconnecting means.
   c. only one main disconnecting means.
   d. not more than three service-disconnecting means.
9. Large apartment complexes and condominiums are classified as
   a. combination occupancies.
   b. residential occupancies.
   c. commercial occupancies.
   d. industrial occupancies.
10. The supply conductors for condominiums are generally installed
    a. overhead
    b. in wireways.
    c. in busways.
    d. underground.
11. The most common type of service for commercial and industrial installations is the
    a. three-phase, four-wire wye system.
    b. three-phase, four-wire delta system.
    c. three-phase, three-wire, open-delta system.
    d. three-phase, three-wire delta system.
12. The service for a large industrial plant is made up of metal cubicles. When the primary voltage is brought to transformers located in or near these cubicles, the arrangement is called
    a. a distribution center
    b. a substation.
    c. a load center.
    d. either a or b.

REVIEW

A. Multiple choice.
Select the best answer
(continued).

13. Consumer distribution systems are arranged to form
    a. a radial system.
    b. a loop system.
    c. a commercial system.
    d. either a or b.

14. Secondary high-voltage distribution systems are consumer distribution systems that use voltages as great as
    a. 300 volts.
    b. 600 volts.
    c. 1000 volts.
    d. 2300 volts.

15. Secondary ties are the conductors
    a. of a loop system that connect several transformer secondaries together.
    b. of a radial system that connect the transformer secondaries to the loads.
    c. between the kilowatt-hour meter and the main disconnect.
    d. that connect the primary of several transformers together.

16. Systems are grounded
    a. because the ground attracts electricity.
    b. to limit the magnitude of voltage caused by lightning.
    c. to eliminate inductance.
    d. to reduce the resistance of the supply conductors.

17. Alternating current systems are grounded
    a. at each individual service.
    b. at the supply station.
    c. only at the substation.
    d. only at the alternator.

18. Electrical conduits and other noncurrent-carrying metal equipment are grounded to
    a. eliminate possible hazards that may be caused by accidental grounds.
    b. conduct eddy currents to ground.
    c. reduce inductance.
    d. eliminate harmonics.

REVIEW *(continued)*

A. Multiple choice.
Select the best answer (continued).

19. A ground-fault protector is a device that
    a. senses ground faults and opens the circuit when the current to ground reaches a predetermined value.
    b. prevents ground faults from occurring.
    c. senses the location of the ground fault and eliminates it.
    d. opens the grounding conductor when a fault occurs.

20. A ground-fault circuit interrupter is sometimes called a
    a. GFCI.
    b. GCI.
    c. GFC.
    d. ICFG.

21. When a ground-fault protector is being installed on the supply conductors, it must
    a. be installed on each conductor separately.
    b. enclose all the circuit conductors.
    c. enclose only the neutral conductor.
    d. be installed in series with the neutral conductor.

22. The ground-fault protector is a type of
    a. buck-boost transformer.
    b. potential transformer.
    c. current transformer.
    d. control transformer.

23. Harmonics produced by electronic equipment
    a. decrease the current in the neutral conductor of four-wire, three-phase systems.
    b. decrease the current in the supply conductors of four-wire, three-phase systems.
    c. cause current in the grounding conductor of four-wire, three-phase systems.
    d. increase the current in the neutral conductor of four-wire, three-phase systems.

B. Give complete answers.

1. What is the difference between a primary distribution system and a secondary distribution system?
2. Where does one obtain information regarding the type of electrical service and systems available?
3. What is the function of an electric utility company?

REVIEW *(continued)*

B. Give complete answers (continued).

4. Describe a radial transmission system.
5. Describe a loop transmission system.
6. What advantage does the loop system have over the radial system?
7. What type of service is usually supplied to a single-family residence?
8. What two points on an ac wiring system are joined by the grounding conductor?
9. What type of service is usually supplied to an industrial or commercial installation?
10. Why is it recommended to install the main service near the center of a large building?
11. Why would a consumer wish to distribute electrical power at a voltage greater than 600 volts?
12. What is a secondary tie?
13. Why are electrical systems grounded?
14. Why are metal parts of electrical equipment grounded?
15. What is a ground-fault protector?
16. What is a ground-fault circuit interrupter?
17. List two common causes of accidental grounds.
18. What type of hazards can occur because of accidental grounds.
19. Describe the operation of a GFCI.
20. Explain the principle operation of a GFP.
21. What causes unbalances in a three-phase system?
22. What are harmonics?
23. Explain the effect of harmonics in ac systems.
24. What type of equipment causes harmonic problems?
25. What does *Section 210-4* of the *NEC®* state regarding harmonics?

18

AC Motors

Objectives

After studying this chapter, the student will be able to:

- Describe the construction of various types of ac motors.
- Explain the principle of operation of various types of three-phase motors and single-phase motors.
- Discuss the reasons for the difference in the values of starting and running currents in ac motors.

AC MOTOR CONSTRUCTION

The *induction motor* is the most common type of ac motor. Its simple, rugged construction makes it relatively inexpensive to manufacture, and it meets most industrial requirements. Its two main components are the stator and the rotor. The stator consists of electromagnets secured to the frame, spaced equal distances apart. The rotor is made of steel lamination in the shape of a cylinder. Windings are placed in slots on the rotor surface. The stator of an ac motor is shown in Figure 18-1A; Figure 18-1B shows the rotor.

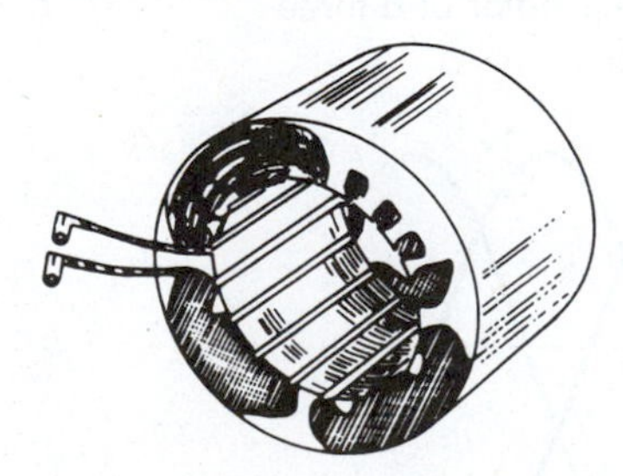

Figure 18-1A
AC motor stator

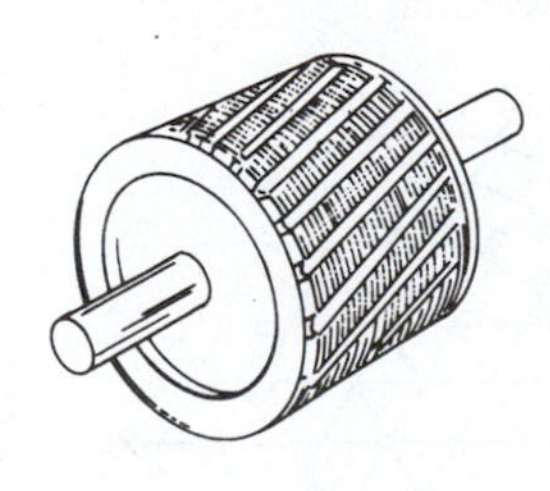

Figure 18-1B
AC motor rotor

Squirrel-Cage Rotor Winding

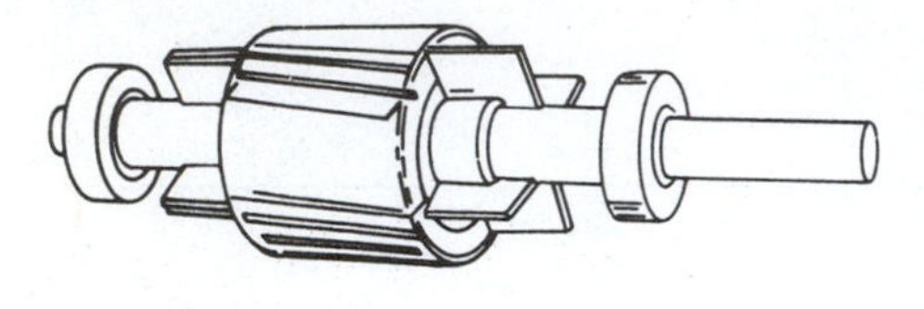

Figure 18-2
Squirrel-cage rotor

A *squirrel-cage rotor winding,* Figure 18-2, consists of heavy copper bars connected together at each end by copper or brass end rings. The bars are welded to the end rings.

Some manufactures use a casting process to construct the rotor. The entire rotor, including the bars, is placed in a mold, and the bar ends are cast to the copper end rings. For small squirrel-cage rotors, the bars, end rings, and fan blades are cast in one piece. Generally, aluminum is used for this process.

Wound-Rotor Motor

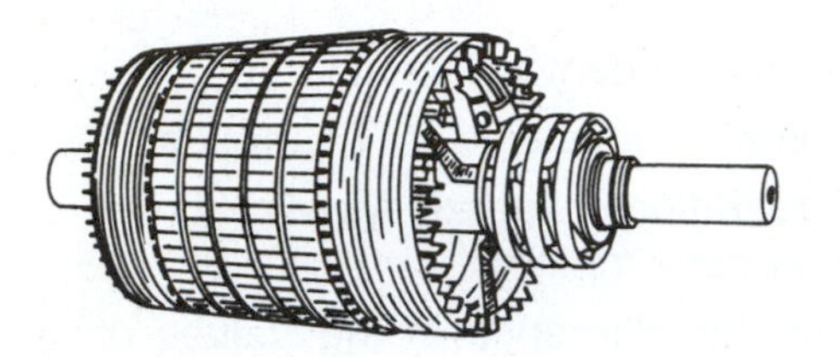

Figure 18-3
Rotor of a wound-rotor motor

Some industrial applications require a wire-wound rotor. Copper wire is wound into the slots of the rotor as shown in Figure 18-3. For the three-phase motor, the windings are generally connected in wye, and the open ends are connected to slip rings mounted on the shaft.

THREE-PHASE MOTOR THEORY

The stator of an induction motor has no projecting poles. The windings are embedded in slots (Figure 18-1A). On the three-phase motor, the windings are arranged to produce a rotating magnetic field when connected to a three-phase source.

Figure 18-4A shows a two-pole, three-phase winding. When this winding is energized from a three-phase source, the three-phase currents vary as shown in Figure 18-4B. The currents are 120 electrical time degrees apart and are continuously increasing and decreasing in value and changing direction. The effect of this variation in strength and change of direction produces a rotating field. Figure 18-4C illustrates one complete revolution.

Figure 18-4A
Two-pole, three-phase stator winding

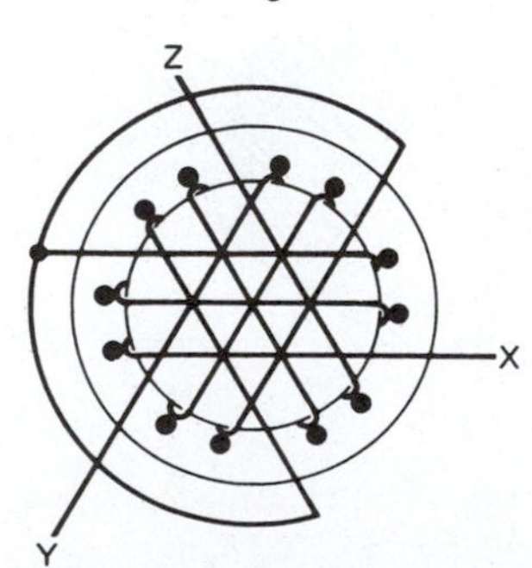

Figure 18-4B
Graph of three-phase currents flowing in the stator of a three-phase motor

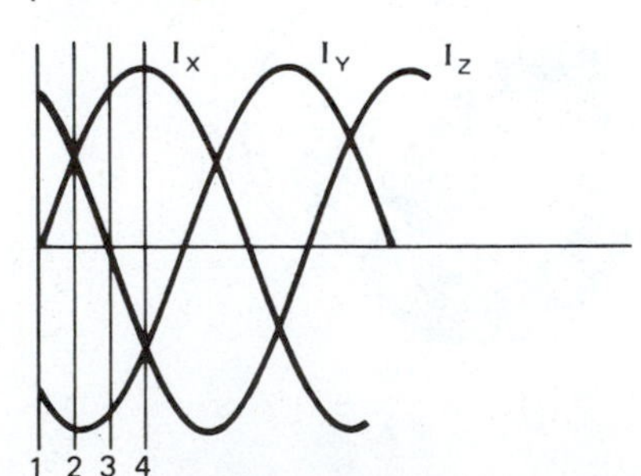

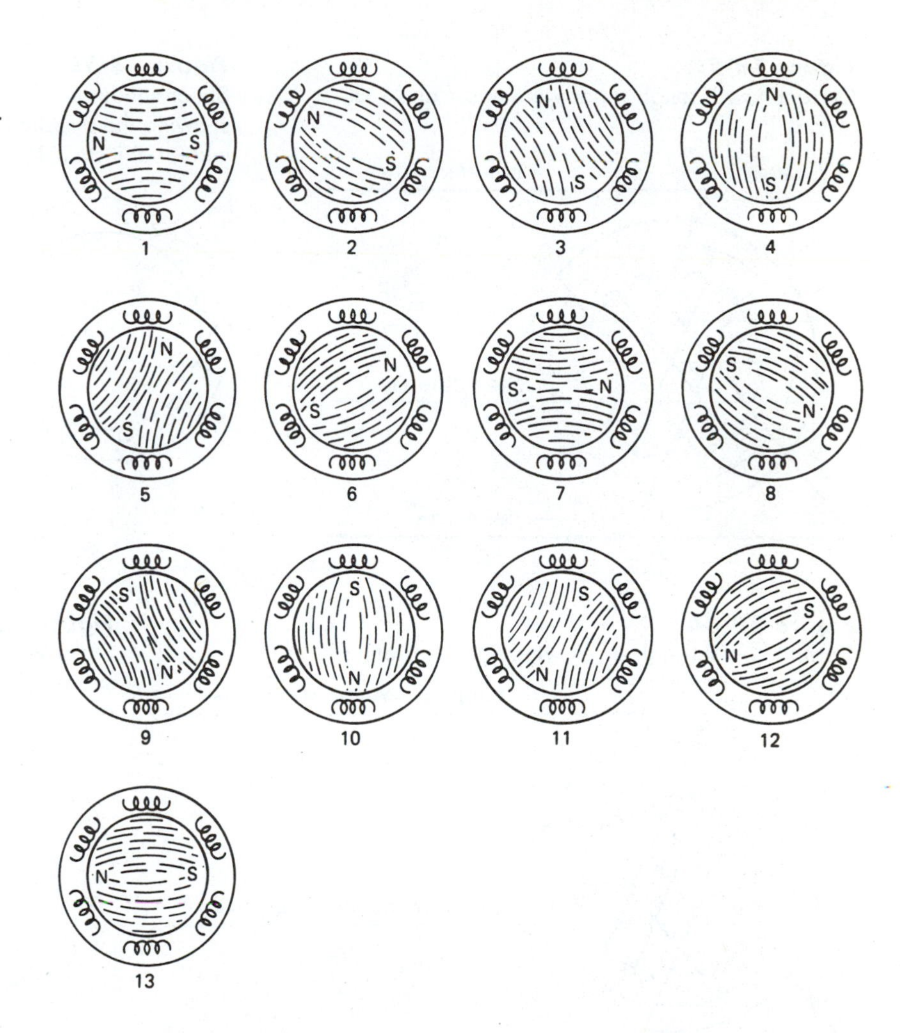

Figure 18-4C
Rotating magnetic field produced by three-phase currents flowing in the stator of a three-phase motor

At instant 1 in Figures 18-4B and 18-4C, the current in phase X is zero, and the currents in Y and Z are equal and opposite. In otherwords, the current is flowing into winding Y and out winding Z. Figures 18-5A and 18-5B illustrate the current flow at this instant. The magnetic field established by these currents is shown in Figure 18-5C. At instant 2 in Figure 18-4B, the current in phase Y is a negative maximum value, and the currents in X and Z are 50 percent of the positive maximum value. This change in current value per phase causes the flux to shift 30 degrees in a clockwise direction. Figures 18-6A and 18-6B illustrate the current flow at instant 2, and Figure 18-6C shows the magnetic field. Instant 3 indicates zero current in phase Z, and phases X and Y are equal and opposite. The currents for instant 3 are shown in Figures 18-7A and 18-7B. Under these conditions, the magnetic field has shifted another 30 degrees clockwise, Figure 18-7C. At instant 4, the

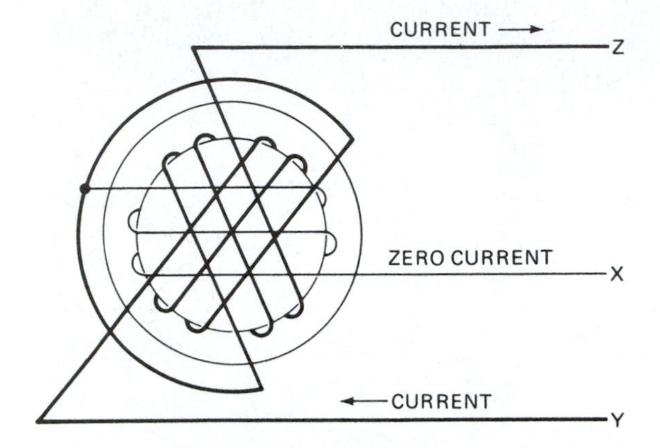

Figure 18-5A
Current flow through stator windings Y and Z at instant 1

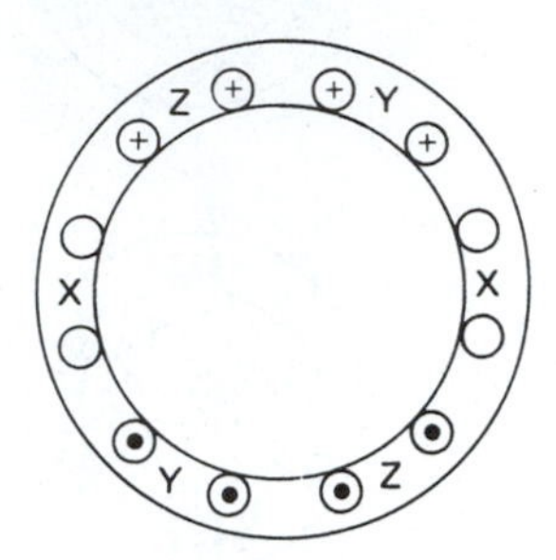

Figure 18-5B
Stator windings showing the direction of current at instant 1

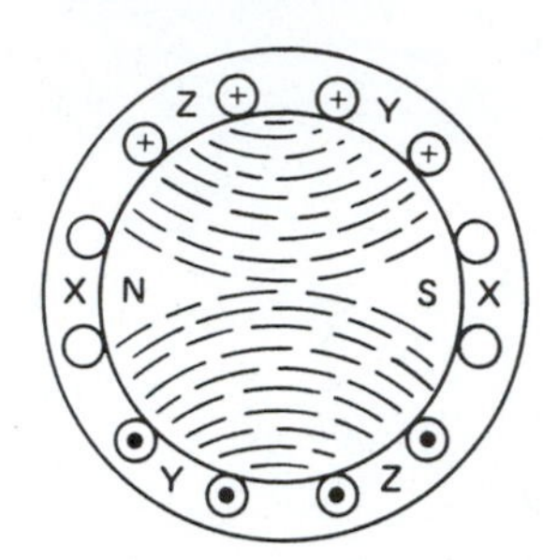

Figure 18-5C
Magnetic field established by three-phase currents at instant 1

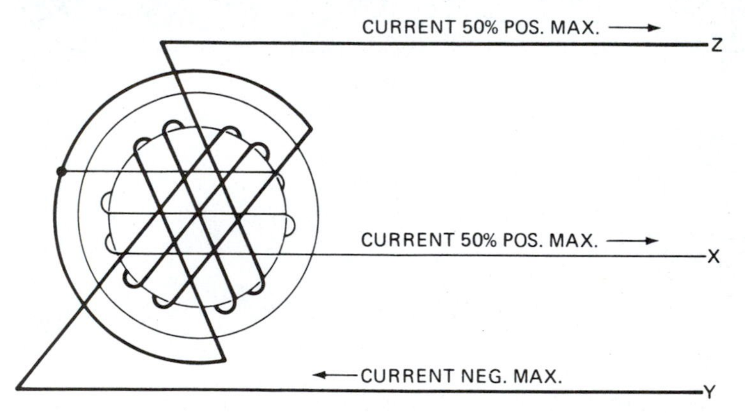

Figure 18-6A
Current flow through stator windings at instant 2

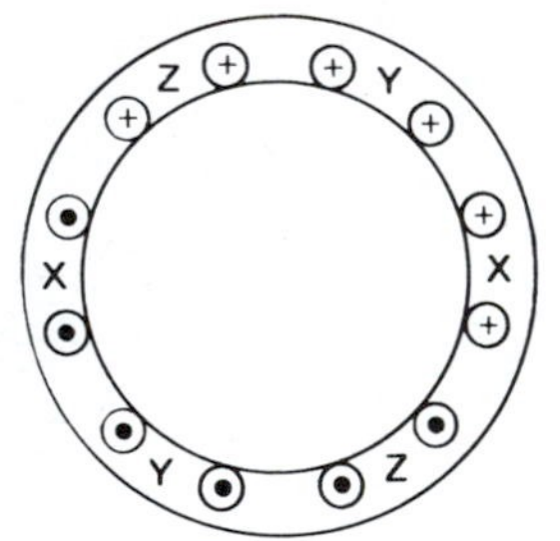

Figure 18-6B
Stator windings showing the direction of current at instant 2

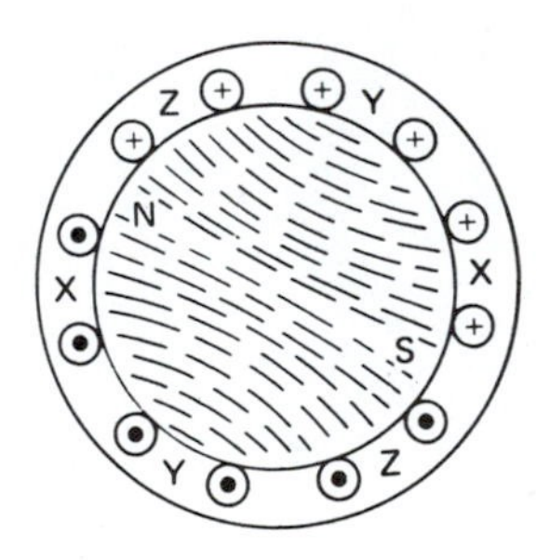

Figure 18-6C
Magnetic field established by three-phase currents at instant 2

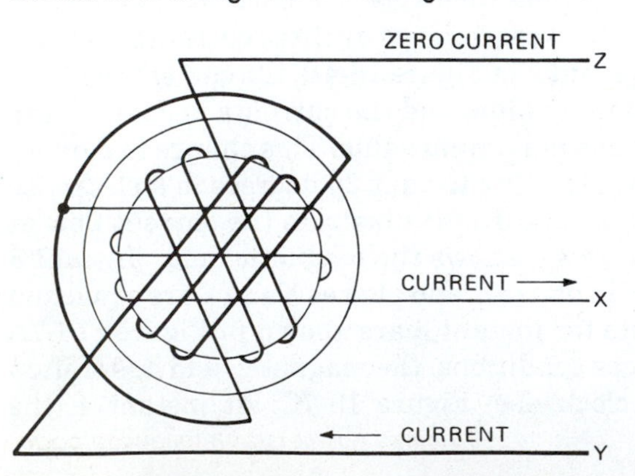

Figure 18-7A
Current flow through stator windings at instant 3

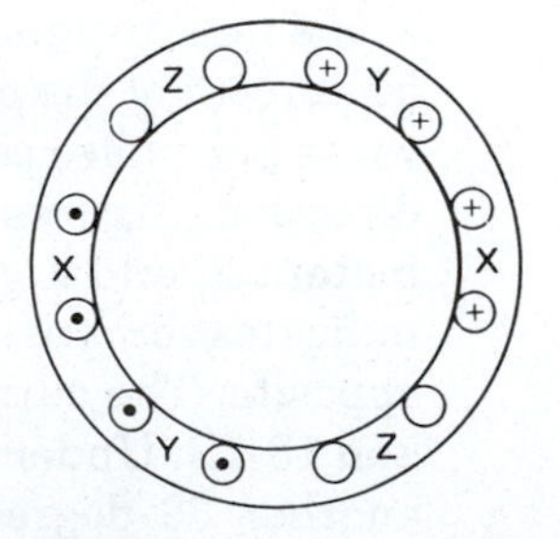

Figure 18-7B
Stator windings showing the direction of currents at instant 3

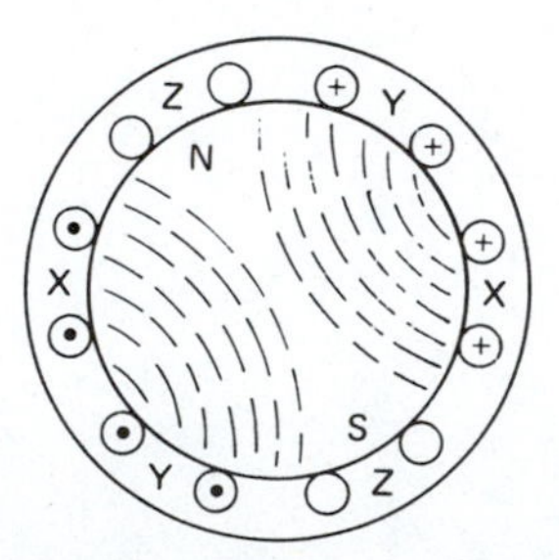

Figure 18-7C
Magnetic field established by three-phase currents at instant 3

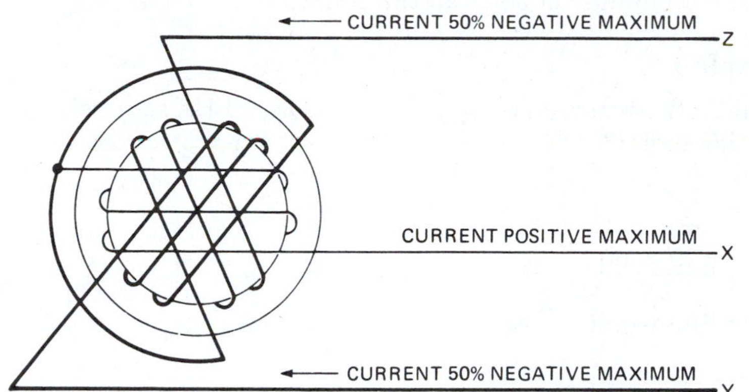

Figure 18-8A
Current flow through stator windings at instant 4

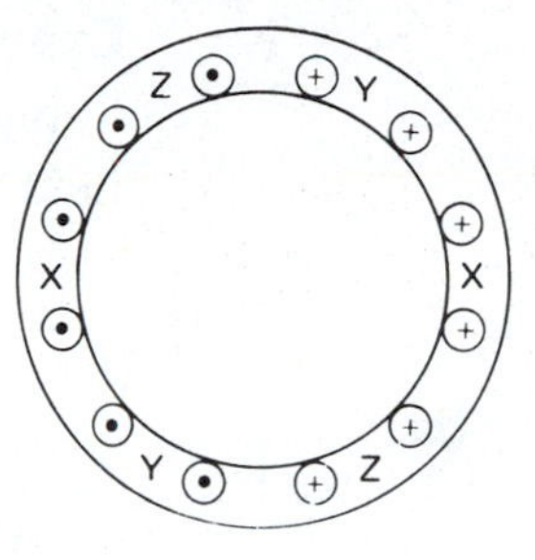

Figure 18-8B
Stator windings showing the direction of currents at instant 4

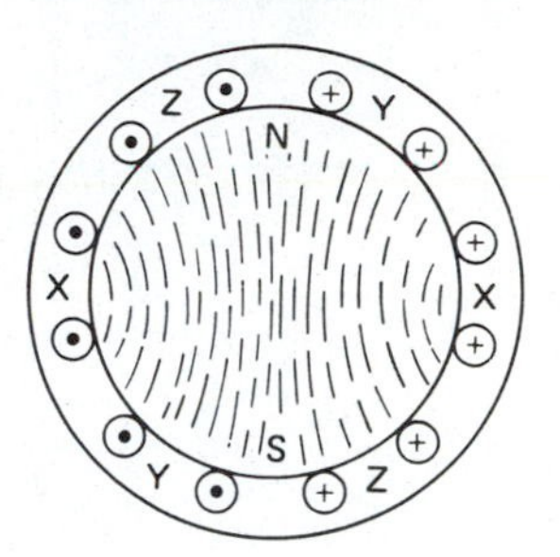

Figure 18-8C
Magnetic field established by three-phase currents at instant 4

current in phase X is the maximum positive value, and the currents in Y and Z are 50 percent of the maximum negative value. Note that the current in phase Z has reversed direction. Figures 18-8A and 18-8B illustrate this condition, and Figure 18-8C shows the magnetic field, which has again shifted 30 degrees clockwise. Following the procedure for one complete cycle shows that the magnetic field rotates 360 degrees, or one complete revolution, in a clockwise direction. At this point, the cycle of current begins again and the field continues to rotate in a clockwise direction.

As the magnetic field rotates around the stator, the lines of force move across the conductors of the rotor. This action induces a voltage into the rotor conductors. The induced voltage causes a current to flow in the rotor conductors, and the current establishes a magnetic field in the rotor. The poles of the rotor are attracted by the poles of the stator, and a rotation is produced.

Speed of the Rotating Magnetic Field

The magnetic field of a two-pole induction motor completes 60 revolutions per second (r/s), or 3600 revolutions per minute (r/min), when supplied with 60 hertz. This speed is referred to as the *synchronous speed* of the motor. To obtain lower speeds, it is necessary to increase the number of poles on the stator. The synchronous speed of a four-pole motor is 1800 revolutions per minute; a six-pole motor produces 1200 revolutions per minute. To determine the synchronous speed of a motor, the following formula may be used:

$$n_1 = \frac{120f}{P} \qquad \text{(Eq. 18.1)}$$

Equation continued on following page.

where n_1 = synchronous speed, in revolutions per minute (r/min)
f = frequency of the supply current, in hertz (Hz)
P = number of poles on the stator

Example 1

What is the synchronous speed of a 12-pole, 60-Hz, squirrel-cage induction motor?

$$n_1 = \frac{120f}{P}$$

$$n_1 = \frac{120 \times 60}{12}$$

$$n_1 = 600 \text{ r/min}$$

Rotor Speed

The speed of the rotor of an induction motor depends upon the synchronous speed and the load it must drive. The rotor does not revolve at synchronous speed but tends to slip behind. The amount of slip at full load may be as great as 5 percent. At no load, the amount of slip may be as low as 1 percent. For example, the synchronous speed of a two-pole, 60-hertz induction motor is 3600 revolutions per minute. At no load, the rotor revolves at approximately 3564 revolutions per minute, and at full load with 5 percent slip, it revolves at 3420 revolutions per minute.

Direction of Rotor Rotation

The direction of rotation of the stator field of a three-phase induction motor depends upon the phase sequence. The rotor field is attracted by the field of the stator and therefore revolves in the same direction as the stator field. Interchanging any two of the three-phase leads that supply current to the stator reverses the phase sequence in the motor and the direction of the stator magnetic field. Reversing the direction of the stator field causes the rotor to reverse direction.

Torque

The torque produced by an induction motor varies with the strength of the stator and rotor fields. The phase relationship between the two fields also affects the torque.

When a three-phase current is first supplied to the stator, the rotor is at a standstill. The three-phase current produces a rotating field in the stator. At this instant, there is 100 percent slip and the stator field is sweeping across the rotor conductors at maximum speed. The voltage induced into the rotor is maximum, and the frequency is the same as that of the stator. Under this condition, the inductive reactance of the rotor is high compared to its resistance. Thus the rotor current lags the voltage by a large amount. There is only a slight lag between the stator voltage and current, which results in the rotor's current lagging the stator current by a considerable amount. The magnetic fields produced by these currents are also considerably out of phase, resulting in a low starting torque.

As the rotor begins to revolve, the slip decreases, and the rate at which the stator field sweeps across the rotor conductors decreases. This causes a decrease in the rotor voltage and frequency. Further increases in the rotor speed cause further reductions in the frequency and voltage of the rotor. If the rotor speed could increase until it revolved at exactly the same speed as the rotating magnetic field, there would be no induced voltage, no rotor current, and, therefore, no torque.

As the rotor frequency decreases, the inductive reactance of the rotor also decreases, resulting in a decrease in the phase angle between the stator and rotor currents. The two magnetic fields pull closer together. The attractive force between he magnetic fields increases, producing a greater torque.

When a load is applied to the motor, the rotor speed decreases until the induced emf in the rotor reaches a value that will develop enough torque to drive the load at a constant speed.

The equations for determining the torque of an induction motor is

$$T = K_T \phi_S I_R \cos \phi_R \qquad \text{(Eq. 18.2)}$$

where T = torque, in pound-feet (lb·ft)
K_T = torque constant
ϕ_S = stator flux
I_R = rotor current
$\text{Cos } \phi_R$ = rotor power factor

Slip

The difference between the synchronous speed and the rotor speed is called the *slip* of the rotor and may be stated in revolutions per minute or as a percentage. The percentage of slip may be calculated as follows:

$$\%S = \frac{n_1 - n_2}{n_1} \times 100 \qquad \text{(Eq. 18.3)}$$

where $\%S$ = percentage of slip
n_1 = synchronous speed, in revolutions per minute (r/min)
n_2 = rotor speed, in revolutions per minute (r/min)

Slip in revolutions per minute is calculated as follows:

$$S = n_1 - n_2 \qquad \text{(Eq. 18.4)}$$

where S = slip, in revolutions per minute (r/min)
n_1 = synchronous speed, in revolutions per minute (r/min)
n_2 = rotor speed, in revolutions per minute (r/min)

The rotor frequency is directly proportional to the slip. Therefore,

$$f_r = Sf_s \qquad \text{(Eq. 18.5)}$$

where fr = rotor frequency
S = slip percentage
fs = slip frequency

Example 2

Calculate the percentage of slip and the rotor frequency of a 60-Hz, 8-pole motor operating at 840 r/min.

(1) $n_1 = \dfrac{120f}{P}$

$n_1 = \dfrac{120 \times 60}{8}$

$n_1 = 900 \text{ r/min}$

(2) $\%S = \dfrac{900 - 840}{900} \times 100$

$\%S = 0.067 \times 100$

$\%S = 6.7\%$

(3) $f_r = Sf_s$

$f_r = 0.067 \times 60$

$f_r = 4.02 \text{ Hz}$

THREE-PHASE MOTOR STARTING AND RUNNING CURRENT

The induction motor is basically a transformer in which the stator is the primary and the rotor is a short-circuited secondary. When a three-phase voltage is applied to the stator (primary), the rotating magnetic field induces a voltage into the rotor (secondary). The rotor current develops a flux that opposes and therefore weakens the stator flux. This allows more current to flow in the stator windings, just as an increase in the current in the secondary of a transformer results in a corresponding increase in the primary

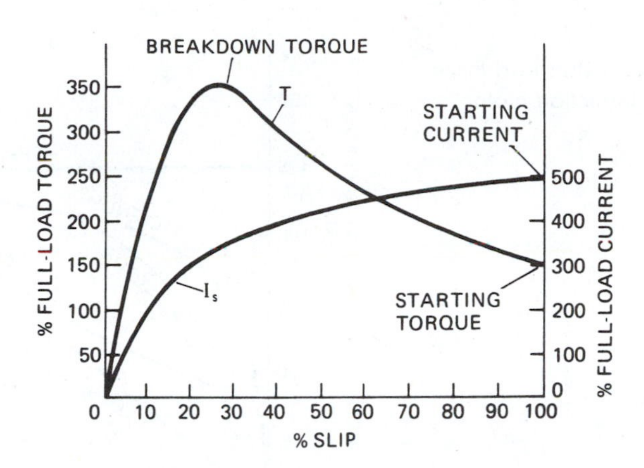

Figure 18-9
Torque/current curves of a standard squirrel-cage induction motor

current. Because of the air gap between the stator and the rotor of an induction motor, sufficient flux leakage occurs to limit the starting current to a value approximately 4 to 6 times the full-load current.

Loading a Squirrel-Cage Motor

Because of its rugged construction, the three-phase squirrel-cage motor is capable of handling the starting current without damage to itself. Very large motors, however, may require a value of starting current that will cause line drops, which may affect other equipment operating from the same system. For such installations, reduced voltage starters are used. The reduced voltage limits the starting current to a lower value.

Starting induction motors under reduced voltage will reduce the starting torque considerably. It is sometimes necessary to start the motor without the load. In this case, once the motor has reached its rated speed, the load is applied. A 50 percent reduction in voltage will reduce the torque to only 25 percent of its normal value.

A torque/current curve of a standard squirrel-cage motor is depicted in Figure 18-9. The maximum torque is reached at about 25 percent slip. Beyond this value, the power factor of the rotor decreases faster than the current increases, causing the torque to decrease. If the motor is loaded beyond maximum torque (breakdown torque), it will quickly slow down and stop. In Figure 18-9, the value of T at starting (at 100 percent slip) is about 150 percent of the full-load torque. The starting current is about 5 times the full-load current. This motor is essentially a constant-speed machine with speed characteristics similar to those of a dc motor. The

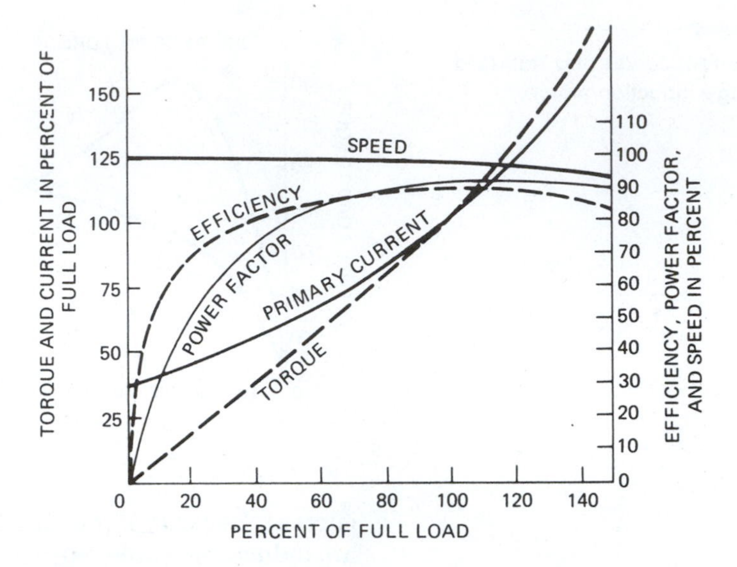

Figure 18-10
Performance curves of a standard three-phase, squirrel-cage induction motor

performance curves for a three-phase, squirrel-cage motor are shown in Figure 18-10. This motor is generally used where average starting torque and relatively constant speed are required.

Double Squirrel-Cage Rotor

One type of induction motor with excellent operating characteristics has a rotor with two squirrel-cage windings. Figure 18-11 illustrates this type of winding. The bars of the inner winding are made of a low-resistance metal, usually copper surrounded by iron/steel (except for the space between the two windings). This construction results in a winding with low resistance and high inductance. The bars of the outer winding are made of small copper or aluminum strips having a high resistance when compared with the inner winding. Only the sides of the outer winding of the rotor core are made of iron/steel. This results in a winding with high resistance and low inductance.

When the rotating magnetic field of the stator sweeps across the two windings of the rotor, an equal amount of emf is induced into each winding. At the instant of starting, the rotor frequency is the same as the line frequency. As a result, the reactance of the inner winding is much greater than that of the outer winding. Thus, there is a large current and a high power factor in the outer winding. This combination produces a high starting torque. As the rotor approaches synchronous speed, its frequency decreases rapidly, and the current division between the two windings is governed by their resistance. This results in most of the current's flowing through the inner winding, and the machine operates as a standard squirrel-cage motor.

Figure 18-11
Section of a double squirrel-cage winding

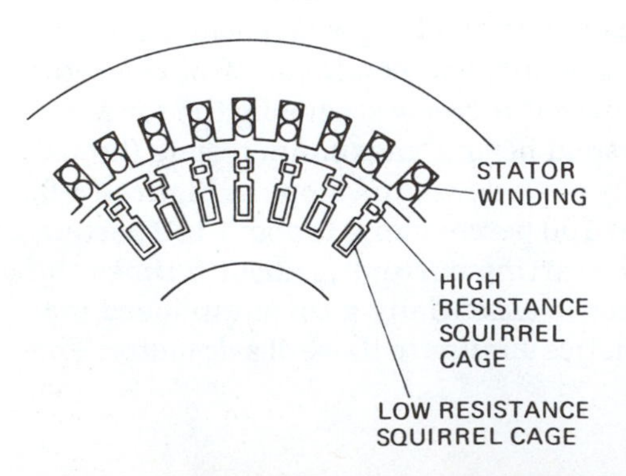

TYPES OF THREE-PHASE MOTORS

Special types of three-phase motors have been designed to provide for variances in speed, frequency, and power factor. The type of motor selected depends upon the service requirements.

Multispeed Squirrel-Cage Motors

The speed of the standard squirrel-cage induction motor is inherently constant. However, special squirrel-cage motors are manufactured with stator windings in which the number of poles may be changed by changing the external connections. For example, a stator may be wound so that one connection will provide four poles and another connection will provide eight poles. The end of these windings may be connected to a switching device and arranged so that one position will provide the slow speed and the other position will provide the fast speed. If the motor is supplied with 60-hertz power, the synchronous speeds will be 900 revolutions per minute and 1800 revolutions per minute. If more variations of speed are required, more windings may be placed on the stator. One set of windings may provide connections for 4 or 8 poles; the other set may provide connections for 6 or 12 poles. This arrangement will provide synchronous speeds of 600, 900, 1200, and 1800 revolutions per minute.

Wound-Rotor Induction Motors

A *wound-rotor induction motor*, sometimes called a *slip-ring motor*, is an induction motor with a wire-wound rotor. The windings are usually connected in wye, and the open ends are connected to slip rings. Figure 18-12A illustrates this connection. Connecting variable resistances between the slip rings inserts resistance into the rotor circuit during the starting period and removes it from the

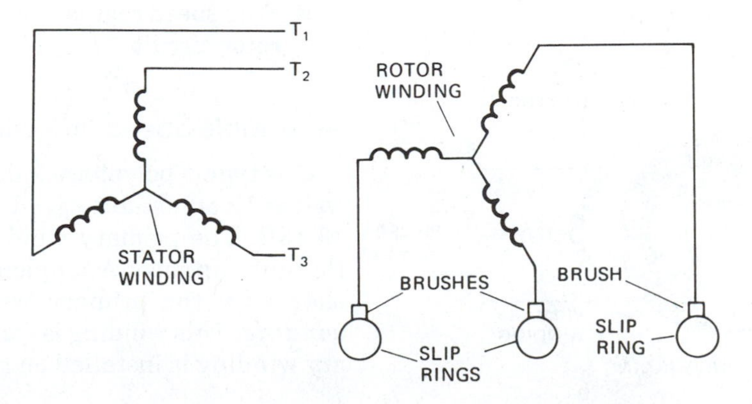

Figure 18-12A
Connections for a wound-rotor, three-phase induction motor

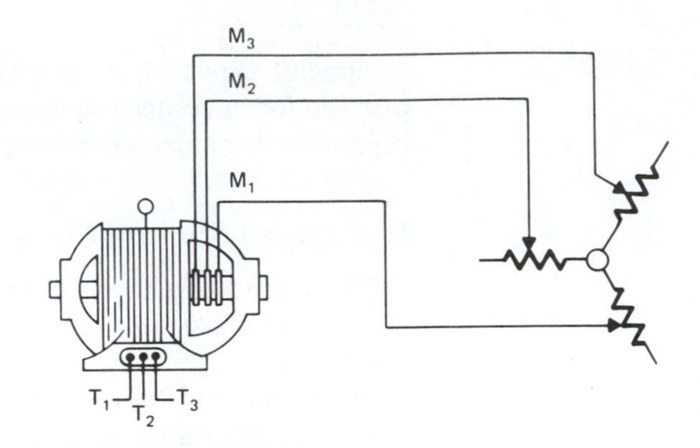

Figure 18-12B
Wound-rotor motor connected to a three-phase rheostat

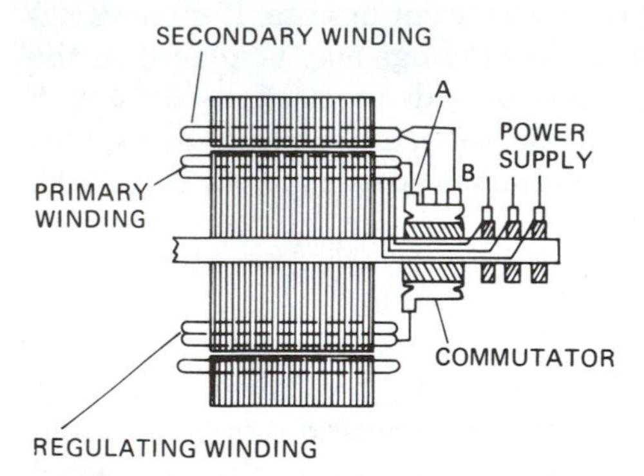

Figure 18-13A
Cutaway view of an adjustable-speed induction motor (brush-shifting type)

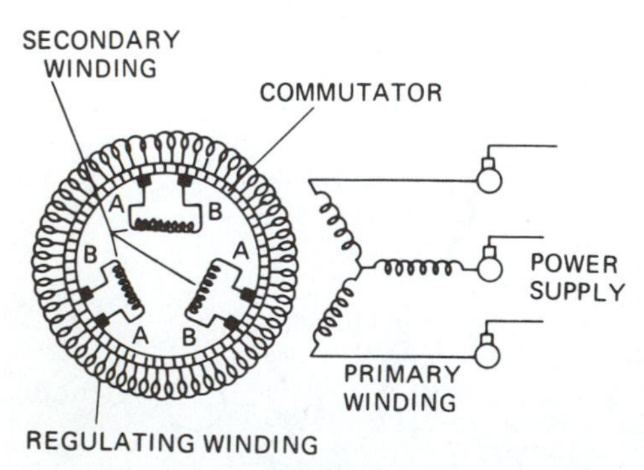

Figure 18-13B
Schematic diagram of the windings on an adjustable-speed induction motor (brush-shifting type)

circuit as the rotor accelerates to rated speed. This method of starting produces a high starting torque at a low current value. When the resistance is removed from the circuit, the slip rings are short-circuited, and the motor operates like a squirrel-cage motor.

When the motor is operating under a constant load, the speed can be varied by varying the resistance of the rotor circuit. Figure 18-12B shows a wound-rotor connected to a three-phase rheostat.

The wound-rotor induction motor has the following advantages compared to the squirrel-cage motor:

- High starting torque with low starting current
- No abnormal heating during the starting period
- Smooth acceleration under heavy loads
- Good speed adjustment when operating under a constant load

The disadvantages of the wound-rotor induction motor are as follows:

- Greater initial cost and maintenance costs than those of the squirrel-cage motor
- Poor speed regulation when operating with resistance in the rotor circuit

Adjustable-Speed Induction Motor (Brush-Shifting Motor)

One type of polyphase induction motor that operates reasonably well as an adjustable-speed motor is shown in Figures 18-13A and 18-13B. The primary winding is on the rotor and is energized through slip rings. A supplementary winding is wound in the same slots with the primary winding and is called the *regulating winding*. This winding is connected to a commutator. The secondary winding is installed on the stator, and the ends of each phase

Figure 18-13C
Adjustable-speed induction motor with the brushes set for below-normal speed

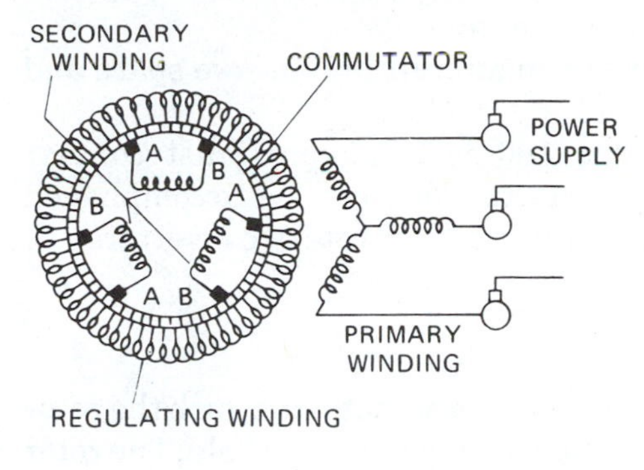

Figure 18-13D
Adjustable-speed induction motor with the brushes set for above-normal speed

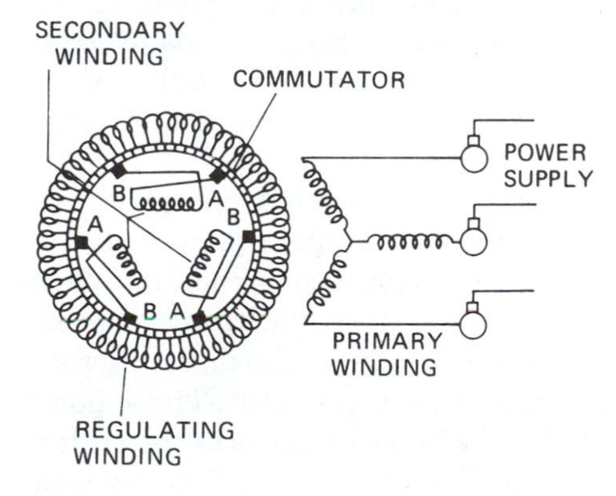

winding are connected to brushes that slide on the commutator. Figure 18-13B illustrates this *brush-shifting motor*. The brush holders are secured to a yoke and arranged so that they can be shifted around the commutator. When the brushes marked A are shifted in one direction, the brushes marked B are moved the same distance in the opposite direction. When the primary winding is energized, it develops a rotating magnetic field. This field sweeps across the secondary winding and induces a voltage into the secondary coils. A voltage is also induced into the regulating winding. However, with the brushes positioned as shown in Figure 18-13B the voltage has very little effect on the motor operation. Current flowing in the secondary winding establishes a rotation flux, which reacts with the primary flux, causing the rotor to turn.

When brushes A and B of each phase are moved apart, the voltage induced into the regulating winding is impressed across the secondary winding. The greater the distance is between A and B, the greater the emf is across the brushes and across the secondary windings. The voltage of the regulating winding is of the same frequency as that of the secondary. Depending upon the brush arrangement, this voltage may be in phase or 180 degrees out of phase with the secondary emf. With the brushes located as shown in Figure 18-13C, it can be assumed that the voltages are of the opposite polarity. Under this condition, the secondary current will decrease, causing a decrease in the magnetic flux of the secondary, an increase in the slip, and reduced speed.

If brushes A and B in each phase are shifted to the position shown in Figure 18-13D, the emf across the brushes will be in phase with the secondary emf, and the secondary current will increase. The result is a stronger secondary magnetic field and an increase in the rotor speed.

The adjustable-speed induction motor can provide a wide range of speeds, depending upon the position of the brushes.

High-Frequency Motors

Certain installations require motors that will operate at speeds greater than those obtainable from a 60-hertz supply. One method used to meet this requirement is to construct motors to operate on higher frequencies. The next higher standard frequency is 180 hertz.

The synchronous speed of the standard 60-hertz, two-pole motor is 3600 revolutions per minute. The same motor constructed to operate on 180 hertz has a synchronous speed of 10,800 revolutions per minute. Lower speeds are obtainable by increasing the number of poles.

Motors operating at frequencies above 60 hertz are generally used where it is necessary to reduce the size or avoid the use of transmission gears. As the frequency is increased, the physical size and weight of the motor decrease.

Motors of 400 hertz are used on aircraft to conserve space and weight.

Industrial plants that use 180-hertz motors must convert their 60-hertz supply to 180 hertz. This can be accomplished through the use of motor-generator sets or rotary converters.

Synchronous Motors

A *polyphase synchronous motor* is an ac motor in which the rotor revolves at the same speed as the rotating magnetic field. The rotor generally has two windings: an ac winding, which may be of the squirrel-cage or the wound-rotor type, and a dc winding. The stator windings are similar to those of the polyphase, squirrel-cage, or wound-rotor motors.

A synchronous motor cannot be started with the field energized. Under this condition, an alternating torque is produced in the rotor. As the stator field sweeps across the rotor, it tends to cause the rotor to try to turn—first in the direction opposite to that of the rotating field and then in the same direction. This action takes place so rapidly that the rotor remains stationary.

To start the synchronous motor, the rotor is left deenergized and the motor is started in the same manner as the squirrel-cage or wound-rotor motor, depending upon the rotor construction. When the rotor reaches approximately 95 percent of synchronous speed, direct current is applied to the exciter winding. The direct current produces definite north and south poles in the rotor. These poles are attracted to, and lock in with, the opposite poles of the stator. Because the rotor is turning at synchronous speed, the magnetic field of the stator is no longer sweeping across the conductors of the rotor. Thus, the only current flowing in the rotor is the dc exciter current. If the rotor momentarily slips below synchronous speed, an induced current will flow and the rotor current will increase. This brings the rotor back to synchronous speed.

During the starting period, the rotating field induces a voltage into both the ac winding and the exciter (dc) winding. Because of the speed ratio and the large number of turns on the exciter winding, this voltage can be very high. The voltage may be so high that an insulation breakdown may occur. A high voltage is also induced when the dc is removed from the exciter winding. For this reason, a low resistance is connected across the exciter field whenever the dc supply switch is open. This resistance is called a *field discharge resistor*. Figure 18-14 is a schematic diagram of this arrangement.

Figure 18-14
Schematic diagram of the dc field of a synchronous motor. A field discharge resistor (FDR) is connected across the dc field when the supply switch is open.

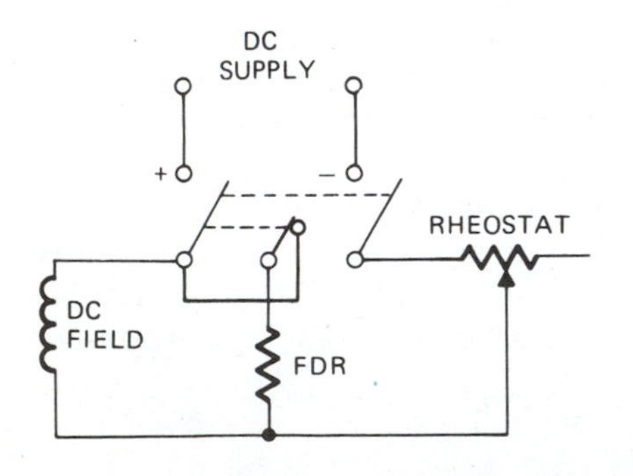

The synchronous motor has two very important features:

- It operates at a constant speed from no load to full load.
- The power factor of the motor can be controlled by varying the amount of current in the exciter winding.

In plants where the load consists chiefly of induction motors, the power factor is usually very low. Synchronous motors are frequently used to correct this problem. If the motors are operated without a load, they can be adjusted to have a leading power factor, which may be in the vicinity of 10 percent. Motors operated in this manner are referred to as *synchronous capacitors*. When operated without a load, the synchronous motor requires very little effective power (watts) but develops a high leading reactive power. This leading reactive power compensates for the lagging reactive power of the induction motors. Synchronous motors operating under load can also develop a leading power factor, but not to the extent that they can when operating without a load.

It is not advisable to use synchronous motors when the application requires sudden and heavy loads. Such operation can cause the rotor speed to decrease and remain out of step with the rotating stator field. Synchronous motors are generally used for driving loads requiring constant speeds and infrequent starting and stopping. Some common types of loads are dc generators, blowers, and compressors.

SINGLE-PHASE MOTORS

Figure 18-15A shows a *single-phase induction motor* with a squirrel-cage type rotor and a single winding on the stator. When this motor is energized by a single-phase, 60-hertz source, alternating current will flow. At the instant that the current is flowing

Figure 18-15A
Single-phase, squirrel-cage induction motor

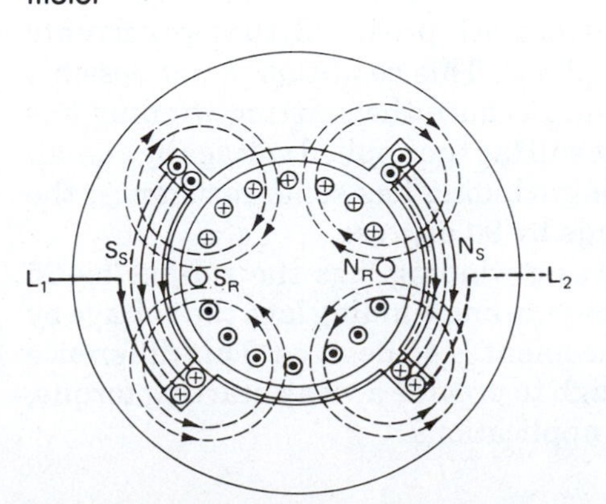

Figure 18-15B
Change in rotor flux caused by rotor rotation

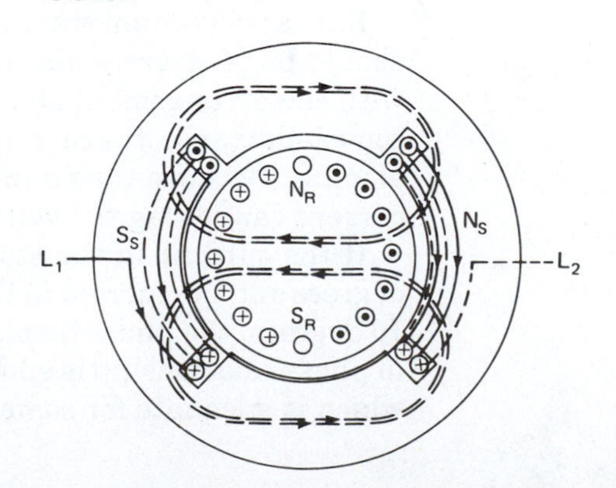

from L_1 to L_2 and increasing, a flux is established, which induces an emf into the rotor winding and causes a rotor current as indicated. The current will develop poles in the rotor, as illustrated by N_R and S_R. The poles are lined up with the stator poles N_S and S_S. No torque is developed because the force is in a straight line. This condition is true for any instant of the ac cycle. It is apparent that this type of single-phase motor is not self-starting.

If a means is provided for turning the rotor by hand, such as by a rope and pulley arrangement, a different condition will exist. Because of the rotating action, the rotor conductors will now cut the stator flux. This results in an emf's being induced into the rotor conductors, causing a current to flow, Figure 18-15B. Because the rotor is highly inductive, the rotor current lags the rotor voltage by approximately 90 degrees. Therefore, the rotor flux is nearly 90 degrees out of phase with the stator flux (Figure 18-15B). Because the two fields are nearly 90 degrees out of phase, a rotating force is developed.

After a single-phase motor reaches its rated speed, its performance is nearly the same as that of a polyphase motor. The single-phase motor, however, does not develop as smooth a torque as the polyphase motor. Other advantages of the polyphase motor are that it is generally smaller in physical size, is more efficient, and has a higher operating power factor than a single-phase motor of the same horsepower.

STARTING SINGLE-PHASE MOTORS

One method of making a single-phase motor self-starting is to apply the *phase-splitting principle*. Figure 18-16A shows an arrangement in which the stator has two windings. The main winding, which is connected across the line in the usual manner, has low resistance and high inductance. The auxiliary winding (starting winding) has high resistance and low inductance. The result is that the currents in the two windings are out of phase with each other.

So that maximum starting torque is produced, the two currents should be 90 degrees out of phase. This condition is not possible with the arrangement shown. Because the starting winding has some inductance, the current will lag the applied voltage by a small amount. Because the running winding has some resistance, the current cannot lag the voltage by 90 degrees.

If the current in the starting winding lags the voltage by 25 degrees and the current in the running winding lags the voltage by 70 degrees, the phase displacement is 45 degrees. This difference in phase relationship is enough to provide a weak starting torque, which is adequate for some applications.

Figure 18-16A
Single-phase motor with two stator windings

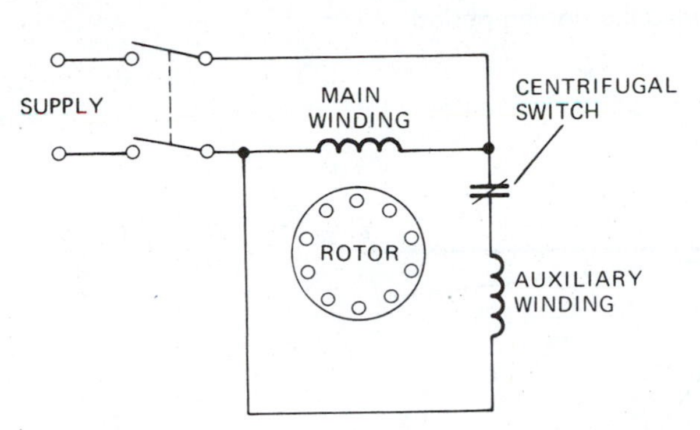

Figure 18-16B
Capacitor-start, split-phase induction motor with the centrifugal switch in the run position

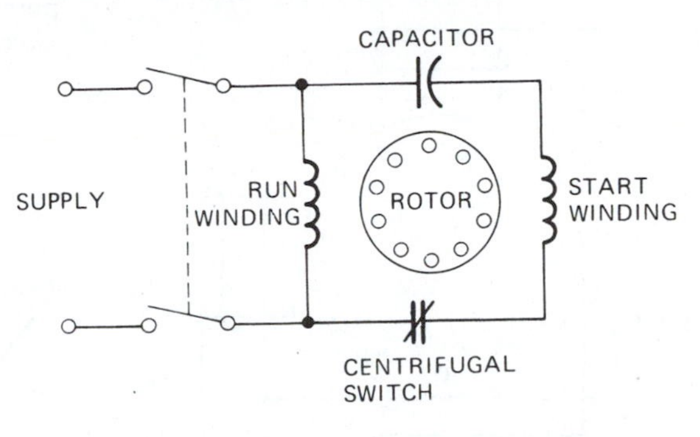

The auxiliary winding is generally designed to remain in the circuit for only a short time. If energized too long, overheating will result and will damage the insulation. A centrifugal switch connected in series with this winding (Figure 18-16A) is designed to open when the rotor reaches approximately 75 percent of synchronous speed. The motor then operates as a single-phase induction motor.

Resistance Split-Phase Motor

In order to obtain better starting torque, *resistance split-phase motors* are designed with a resistance connected in series with the auxiliary winding. This increases the resistance of the starting circuit and produces a greater phase difference between the stator and rotor currents. The results are a greater displacement between the two magnetic fields and better starting torque.

Capacitor Split-Phase Motors

Some single-phase motor installations require higher starting torque than is available from the resistance split-phase motor. For this purpose, a capacitor-start, induction-run motor, called a *capacitor split-phase motor*, is frequently used.

In the standard split-phase motor, the current in the running winding may lag the supply voltage by 70 degrees. A properly sized capacitor connected in series with the starting winding can cause a leading current of 20 degrees. With this arrangement, a 90-degree phase displacement is obtained and maximum starting torque is produced. Once the motor has reached approximately 75 percent of synchronous speed, the centrifugal switch opens and the motor operates as a single-phase induction motor, Figure 18-16B.

Figure 18-17A
Permanent capacitor, split-phase motor

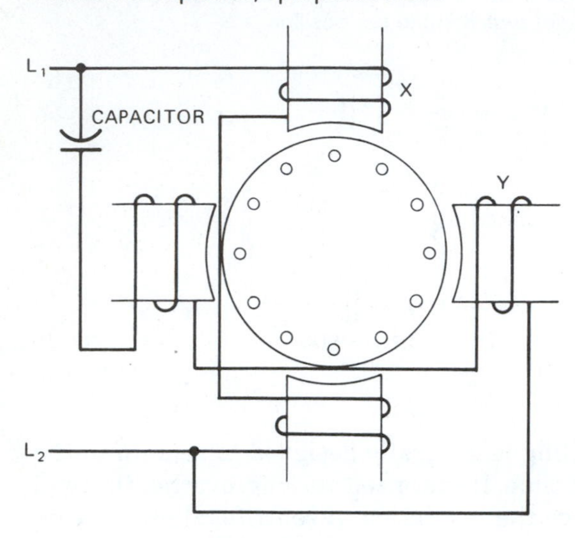

Figure 18-17B
Vector diagram illustrating the phase relationship of the currents in a permanent capacitor, split-phase motor during the starting period

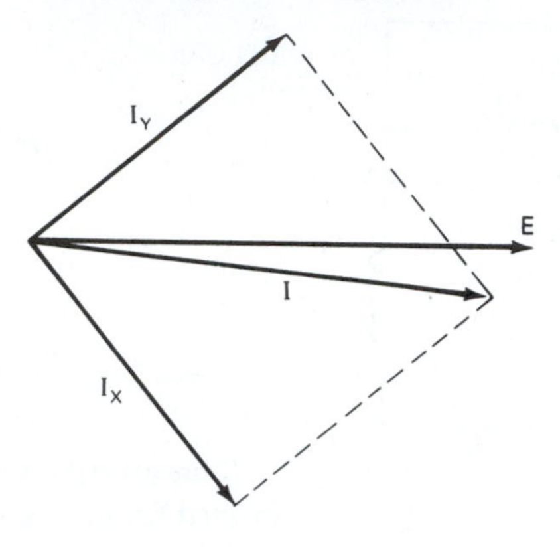

Figure 18-18A
Permanent capacitor, split-phase motor with two capacitors

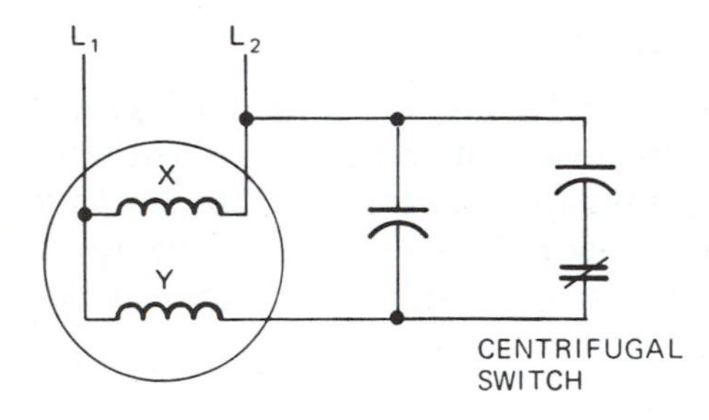

Figure 18-18B
Permanent capacitor, autotransformer split-phase motor

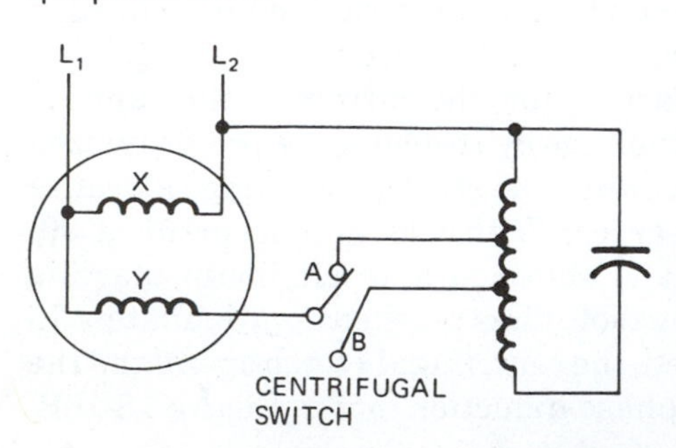

Figure 18-17A illustrates one type of permanent capacitor motor. This type of motor has both windings connected directly across the line, but the auxiliary winding has a capacitor connected in series with it.

A vector diagram for the starting conditions is shown in Figure 18-17B. In coil X, the current lags the voltage by approximately the same angle as in the main line. The current in coil Y leads the current in coil X by 90 degrees. The total current I is the vector sum of currents I_x and I_y and is nearly in phase with the supply voltage. This results in a power factor of almost unity (1.0) and a high starting torque.

Capacitor motors, as described, develop high starting torque but require much larger capacitors than are necessary during normal operation. To solve this problem, a way must be provided to reduce the capacitance once the rotor attains 75 percent of full speed. There are two methods frequently used to accomplish this. One method is shown in Figure 18-18A. Two capacitors are connected in parallel and the pair is connected in series with the auxiliary winding. A centrifugal switch is connected in series with one capacitor. When the motor reaches approximately 75 percent of its rated speed, the centrifugal switch opens, disconnecting the larger of the two capacitors. The motor now operates with only the smaller capacitor in the circuit. This arrangement provides excellent starting torque and a high power factor under normal operation.

The arrangement shown in Figure 18-18B is another method that provide high starting torque. During the starting period, the centrifugal switch is closed and the auxiliary winding is connected to point A on the autotransformer. The result is a voltage across the capacitor that is from 2.5 times to 5 times the supply voltage. This high transformer ratio provides a current through the auxiliary winding that is about 20 times as great as the current that would flow if the capacitor were connected directly in series with the auxiliary winding. The high current produces flux strong enough to develop a high starting torque.

When the rotor reaches approximately full speed, the centrifugal switch connects the auxiliary winding to point B of the autotransformer. The transformer ratio then becomes about 1 to 2. The current through the auxiliary winding becomes about twice the value that would flow if the capacitor were connected directly in series with the winding. This type of motor has operating characteristics very similar to a three-phase, double squirrel-cage motor.

Capacitor motors are generally divided into the following three classes:

1. Low starting torque
2. Capacitor start, induction run (medium to high starting torque)
3. Capacitor start, capacitor run (high starting torque, high power factor)

Reversing Split-Phase Motors

Some applications require that the direction of rotation of the rotor be reversed. The direction of rotation depends upon the instantaneous polarities of the main field flux and the flux produced by the auxiliary winding. Therefore, reversing the polarity of one of the fields reverses the torque.

The standard direction of rotation of the rotor on single-phase motors when viewed from the shaft end is counterclockwise. To reverse the direction, interchange the connections to one of the windings.

Figure 18-19
Shaded-pole motor

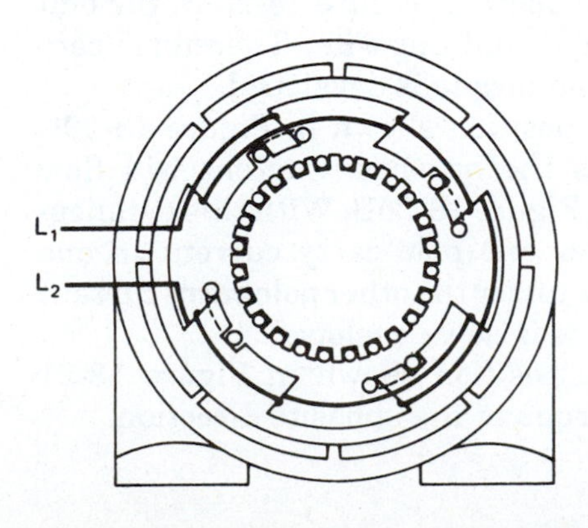

Shaded-Pole Motors

Single-phase *shaded-pole motors* are started by means of a low-resistance, short-circuited coil placed around one tip of each pole, Figure 18-19. When the current and therefore the field flux are increasing, a portion of the flux sweeps across the shading coil. This induces a voltage into the coil, causing a current to flow. The current in the shading coil establishes a flux that opposes the main field flux. During this instant, lines of force are flowing only in the unshaded portion of the pole pieces.

Figure 18-20A
Repulsion motor — the induced emf is additive.

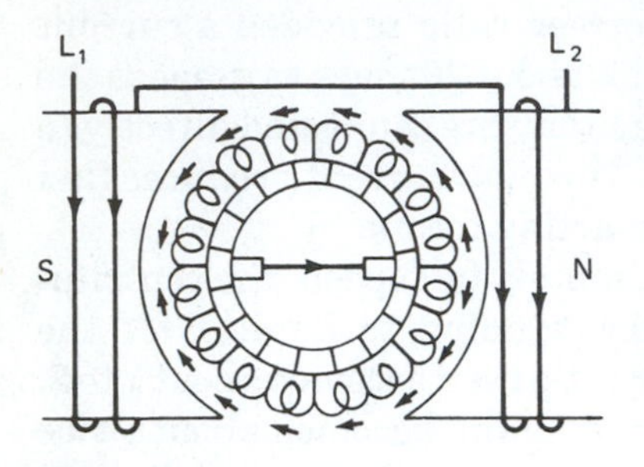

Figure 18-20B
Repulsion motor — the induced emf is equal and opposite. No torque is developed.

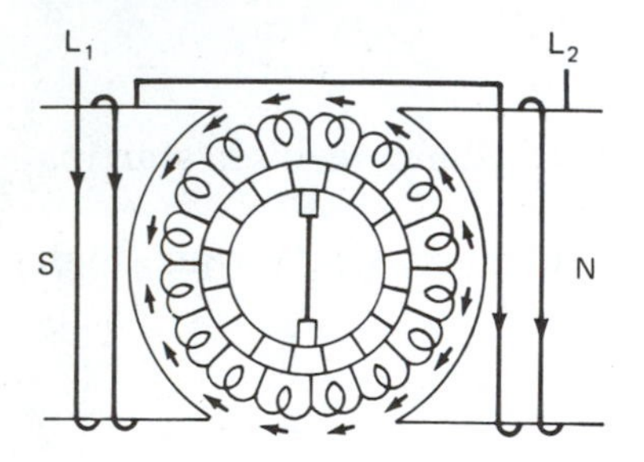

Figure 18-20C
Repulsion motor — an emf is developed across the brushes.

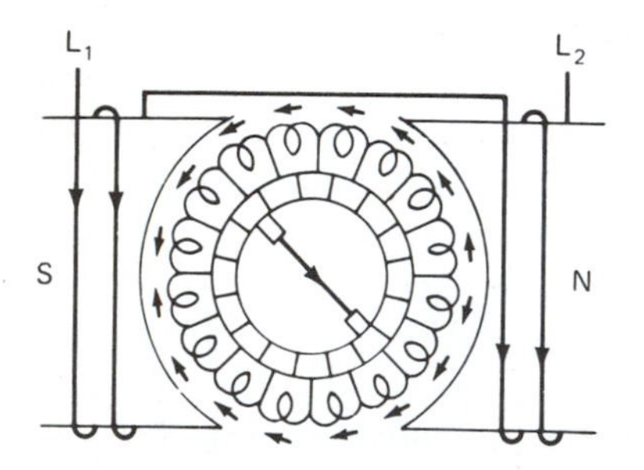

When the main field flux reaches maximum value, the magnetic field is stationary, and zero volt is induced into the shading coil. With zero current flowing in the shading coil, there is no opposing flux, and the lines of force from the main field also flow through the shaded pole.

At the instant when the main field flux is decreasing, an emf is again induced into the shading coil but this time in the opposite direction. The current now flowing establishes a flux that is in the same direction as the main field flux, causing a higher flux in the shaded pole.

The overall effect is that the flux in the shaded pole is always out of phase with the main field flux, resulting in a weak rotating force. This type of motor has poor starting torque and is manufactured only in very small sizes.

Repulsion Motors

The *repulsion-start, induction-run motor* is a single-phase motor that starts based on the principle that like poles repel. The rotating member contains windings similar to those of a dc motor. The windings are connected to commutator segments that are secured to, and insulated from, the rotor shaft. The brushes riding on the commutator do not connect to the line but are connected to one another to form a complete circuit through one set of rotor coils.

The method for starting this motor is shown in Figures 18-20A through 18-20D. Although the stator current is ac, it can be assumed that at the instant shown, it is rising from zero to a maximum in the positive direction. The flux produced will induce an emf into the rotor conductors, as indicated in Figure 18-20A. The voltages are additive on each side of the brushes. Therefore, a high current is forced through the armature and the short-circuited brushes. No torque will be developed, however, because one half of the conductors under each pole carry current in one direction and one half carry current in the opposite direction.

If the brushes are shifted 90 degrees, Figure 18-20 B, the emf induced into each path is equal and opposite. Therefore, zero current flows in the rotor and no torque is developed.

Shifting the brushes to the position shown in Figure 18-20C causes a resultant emf across the brushes. Current will flow through the rotor and brushes, Figure 18-20D. With this arrangement, all the conductors under one pole carry current in one direction, and all the conductors under the other pole carry current in the opposite direction. Torque is now developed.

Shifting the brushes to the position shown in Figure 18-21 causes the motor to develop torque in the opposite direction.

Figure 18-20D
Repulsion motor — current flowing in the rotor coils develops a torque in a counterclockwise direction.

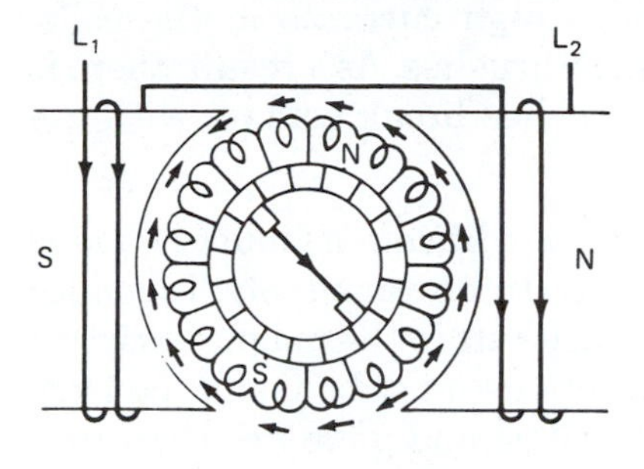

Figure 18-21
Repulsion motor — the current flowing through the rotor coils develops a torque in a clockwise direction.

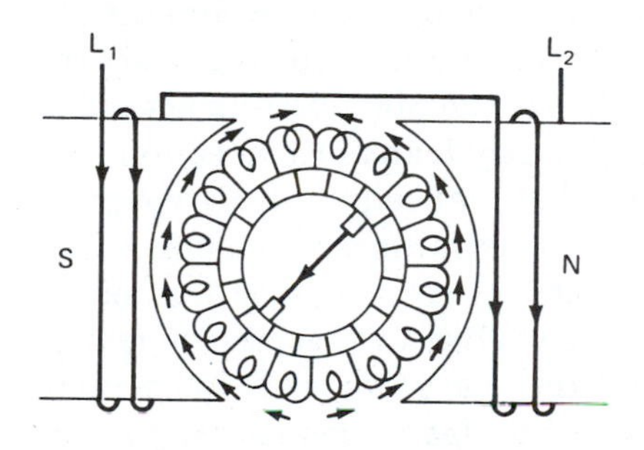

The machine starts as a repulsion motor, which develops a very high starting torque. As soon as the rotor reaches approximately 75 percent of the rated speed, a centrifugal device causes a short-circuiting ring to connect all the segments of the commutator. This converts the machine to an induction motor. The centrifugal device also lifts the brushes off the commutator, thus decreasing the brush wear.

The *repulsion-induction motor* differs from the repulsion-start, induction-run motor in that it has two windings on the rotor. One winding is either a squirrel-cage or a wound-rotor-type winding. The other winding is the repulsion type and is connected to the commutator and short-circuited brushes. This motor starts as a repulsion motor, but there is no short-circuiting ring to short out the commutator segments. Thus, this machine operates as a combination repulsion and induction motor. It is considered to be a constant-speed motor, which develops very good starting and running torque.

The repulsion motor is expensive and requires considerable maintenance; thus it is rarely used.

Series AC Motors

If the direction of current supplying a dc motor is reversed, it will not affect the direction of rotor rotation. In order to reverse the direction of rotor rotation of a dc motor, it is necessary to change the direction of the current through the field or the armature, but not both. Even if a way were developed to reverse the dc supply very rapidly, the torque would continue to be developed in only one direction. Thus, if a dc motor is supplied with ac, a unidirectional torque will be developed.

With an *ac shunt motor*, only a very low current will flow because of the high resistance of the field winding coupled with a high inductive reactance caused by alternating current. As a result, a very weak magnetic field is produced. In addition,the inductive effect causes the field flux to be considerably out of phase with the armature flux, resulting in very little torque. These undesirable effects can be reduced to some extent in small motors. Therefore, ac shunt motors are built only in small sizes.

In a *series motor*, the armature current and field current are in phase, resulting in a machine that can develop very high torque. The ordinary dc series motor, however, does not function satisfactorily on ac for the following reasons:

1. The alternating flux produces large eddy currents in the unlaminated parts of the machine, resulting in excessive heating.

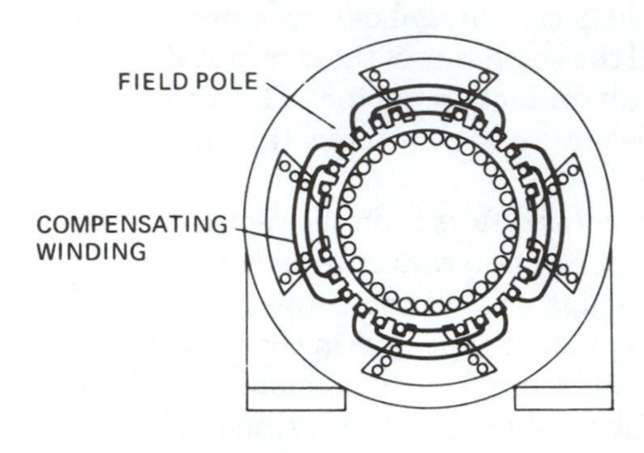

Figure 18-22
Series ac motor

2. The high field reactance establishes a large voltage drop across the field winding. This reduces the input current and power factor to such an extent that it makes the motor impractical.
3. The alternating flux develops high currents in the coils, which are short-circuited by the brushes. As a result, there is excessive arcing when the brushes break contact with the commutator segments.

It is obvious from these facts that modifications must be made in order for the series motor to operate from an ac supply. To reduce eddy currents, all metal parts that are within the magnetic circuit must be laminated, and the laminations must be insulated from one another. The laminated field poles and yoke are then supported in a cast-steel housing.

A common method of reducing armature reaction is to make use of a compensating winding. The winding is embedded in the field pole faces, Figure 18-22. It is arranged to supply a magnetizing action that is equal and opposite to that of the armature coils, regardless of the load. This is accomplished either by connecting the compensating winding in series with the armature or by short-circuiting it on itself. In the latter case, the magnetizing action of the compensating winding is obtained by transformer action.

The excessive arcing at the brushes, caused by flux sweeping across the short-circuited coils, can be eliminated by using high-resistance leads to connect the coils to the commutator segments. Each coil short-circuited by the brushes has two resistance leads, but for the main armature current there are two leads in parallel at each brush. Only those resistance leads connected to the commutator segments in contact with the brushes carry current at any instant. Thus, the resistance leads do not affect the resistance of the armature as a whole.

Because of these modifications, ac series motors are more complex in structure and are heavier per horsepower. They are also more expensive than dc motors of the same ratings. The operating characteristics of ac series motors are very similar to those of dc motors.

UNIVERSAL MOTOR

A *universal motor* is a series motor that will operate on both ac and dc. This motor combines the features of both the ac and dc series motors. The uncompensating winding must be connected in series with the series field. Because of modification problems, it is generally manufactured only in small sizes, usually only fractional horsepower ratings.

Universal motors are frequently used for vacuum cleaners, fans, portable electrical tools, and other small household and office appliances.

SELECTION OF SINGLE-PHASE MOTORS

Small fans, phonographs, and measuring instruments require very little starting torque. The shaded-pole motor is satisfactory for these machines. Larger machines that start without a load, such as small lathes, mills, grinders, and drills, can be started and operated most economically by split-phase motors.

Machines that are required to be started under load must use motors that develop high torque. Two common types of high starting-torque motors are the capacitor motor and the repulsion motor.

TORQUE MOTORS

Motors used to open and close valves, dampers, doors, gates, windows, drive tool returns, and chuck devices must be designed to stall at a fixed torque. These motors, called *torque motors*, must have their greatest torque output when stalled. Large torque motors are made to operate on three phase, and small ones are usually versions of the universal motor. These motors are not intended for continuous duty. The nameplate of such a motor usually lists the running time, full-load speed, stalling torque, and power supply. Most torque motors can remain stalled for brief periods of time without overheating. Specially designed motors can remain stalled for long periods of time without overheating.

DUAL-VOLTAGE WINDINGS

It is common practice to manufacture motors that can be operated on either of the two voltages. This can be accomplished, using either a wye or delta connection, by dividing each phase into two sections.

Figure 18-23A illustrates windings that may be connected either in series or in parallel wye. If the series connection is suitable for 440 volts, then the parallel connection will be suitable for 220 volts.

The series connection is formed by joining T_4 to T_7, T_5 to T_8, and T_6 to T_9. T_1, T_2 and T_3 are connected to the three-phase supply. The parallel connection is obtained by connecting T_1 to T_7, T_2 to T_8, and T_3 to T_9. T_4, T_5, and T_6 are connected together to form a separate junction. The three-phase supply conductors are connected to the junctions of T_1 and T_7, T_2 and T_8, and T_3 and T_9.

Figure 18-23B illustrates the connections for a two-voltage delta. The terminal block diagrams shown below the schematic diagrams illustrate the method of identifying the connections on the nameplate of the motor.

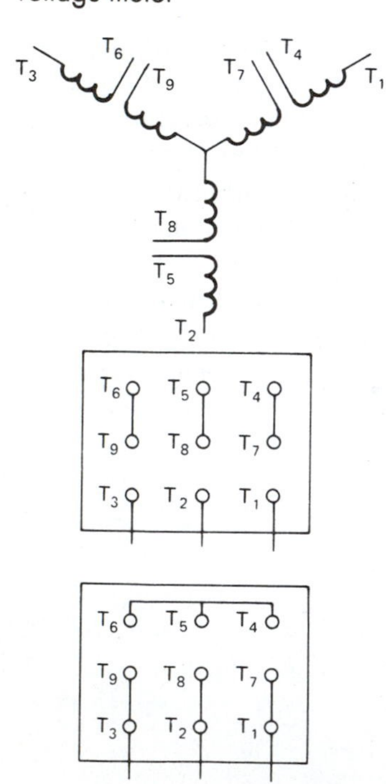

Figure 18-23A
Wye connections for a dual-voltage motor

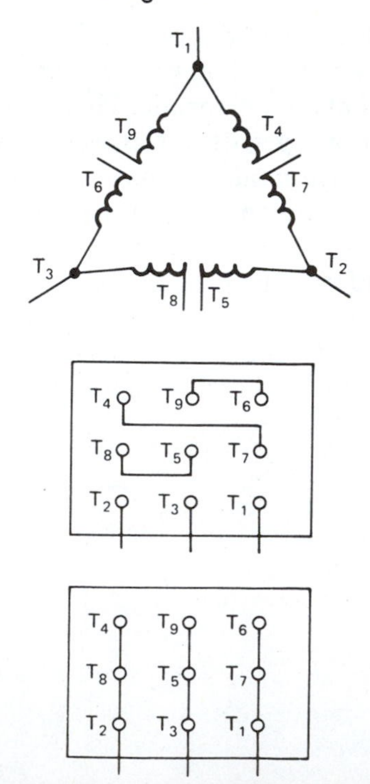

Figure 18-23B
Delta connections for a dual-voltage motor

MULTISPEED INDUCTION MOTORS

The synchronous speed of an induction motor depends upon the supply frequency and the number of poles. Changing the speed by varying the frequency requires the use of frequency converters and a means of adjusting the motor current to meet the change in inductive reactance.

This can be accomplished with a solid-state controller. The alternating voltage supply is changed to direct voltage and then converted back to alternating voltage at a variable frequency. Varying the frequency provides a smooth speed variation similar to that obtained from a dc motor.

Changing the number of poles provides definite speeds that correspond to the number of poles selected.

Squirrel-cage motors, with windings that may be connected for different numbers of poles, offer an economical and simple means for obtaining definite speeds with minimal additional equipment. These motors are generally manufactured for two, three, and four speeds.

Figure 18-24A
Two-speed motor connected for high speed

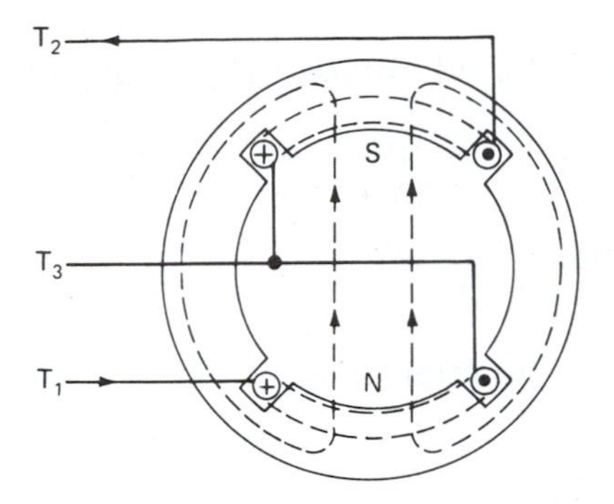

Figure 18-24B
Two-speed motor connected for low speed

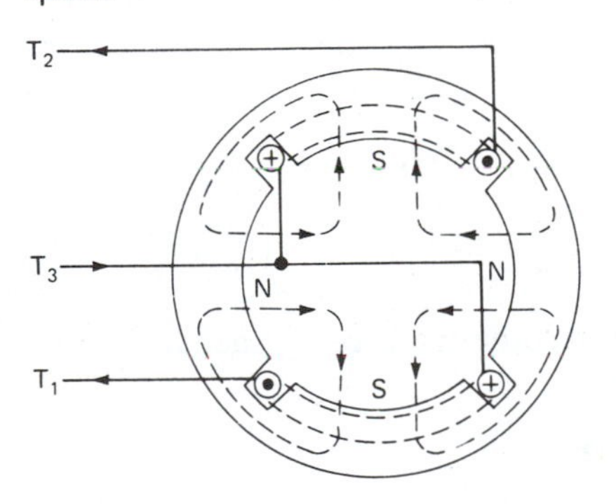

Two-speed motors with single windings are in general use because they require only a few leads and a very simple control. The control serves to change the connections of the stator windings. The two speeds are obtained by producing twice as many poles in the stator for the low-speed operation as are needed for the high-speed operation. To understand how this is accomplished, consider a single-phase motor with only two stator coils, Figures 18-24A and 18-24B. In Figure 18-24A, terminal 3 is left open and the current enters T_1 and flows through the coils to T_2 to produce one north pole and one south pole. With this connection, the flux passes from the north pole, through the rotor, and through the south pole, then returns to the north pole through the stator core. If the motor is designed to be connected to a 60-hertz, single-phase supply, the two poles will produce a synchronous speed of 3600 revolutions per minute.

If T_1 and T_2 are connected together and the current enters T_3 and flows back through T_1 and T_2, a parallel connection results. The current flows through the lower coil in the direction opposite to that shown in Figure 18-24A. This change in current direction in the lower coil results in two north poles' being established in the stator, Figure 18-24B. Under this condition, the flux cannot follow the same path as before, but must form a new magnetic path, Figure 18-24B. The overall result is to establish two north poles and two south poles in the stator, or a total of four poles. The synchronous speed for 60-hertz now becomes 1800 revolutions per minute.

Variations in the design of a multispeed motor result in different operating characteristics. Some provide the same maximum horsepower at all speeds; others will produce the same maximum torque at all speeds.

REVIEW

A. Multiple choice.
Select the best answer.

1. A squirrel-cage rotor has windings that consist of
   a. copper wire wound on a core similar to a dc shunt motor.
   b. copper bars placed in slots around the surface of a cylinder.
   c. projecting poles similar to an ac alternator.
   d. either a or b.
2. On a three-phase motor, the stator windings are arranged to produce
   a. an oscillating magnetic field.
   b. a rotating magnetic field.
   c. a fluctuating magnetic field.
   d. a stationary magnetic field.
3. The synchronous speed of an induction motor refers to the speed of the
   a. rotor.
   b. armature.
   c. rotating magnetic field.
   d. rotor flux.
4. A three-phase, two-pole induction motor operating from a 60-hertz supply will produce a synchronous speed, in revolutions per minute, of
   a. 3600.
   b. 1800.
   c. 2400.
   d. 1200.
5. The direction of rotation of the rotor of a three-phase induction motor is determined by the
   a. frequency.
   b. phase sequence.
   c. number of stator poles.
   d. number of phases.
6. When a load is applied to a three-phase, squirrel-cage induction motor, the rotor speed will
   a. increase.
   b. decrease slightly.
   c. remain the same.
   d. fluctuate.

REVIEW *(continued)*

A. Multiple choice.
Select the best answer (continued).

7. The starting current of an induction motor is
   a. equal to the full-load current.
   b. less than the full-load current.
   c. greater than the full-load current.
   d. one half of the full-load current.

8. Starting an induction motor under reduced voltage will
   a. reduce the starting torque.
   b. increase the starting torque.
   c. have no effect on the starting torque.
   d. increase the horsepower.

9. The speed of a squirrel-cage motor can be varied by varying the supply
   a. voltage.
   b. frequency.
   c. current.
   d. wattage.

10. High-frequency motors are used to obtain
    a. greater speeds than are available from 60 hertz.
    b. lower speeds than are available from 60 hertz.
    c. more torque per horsepower.
    d. greater horsepower.

11. A polyphase synchronous motor requires
    a. both ac and dc.
    b. two values of voltage.
    c. only ac.
    d. only dc.

12. The field discharge resistor for a synchronous motor is connected into the circuit during the
    a. starting period.
    b. running period.
    c. entire operation.
    d. shutdown period.

13. A synchronous motor is frequently used
    a. to improve the power factor.
    b. to drive variable speed loads.
    c. to maintain a steady frequency.
    d. none of the above.

REVIEW *(continued)*

A. Multiple choice.
Select the best answer (continued).

14. In order for a single-phase induction motor to develop maximum starting torque, the stator field should have phase displacement of
    a. 30 degrees.
    b. 45 degrees.
    c. 60 degrees.
    d. 90 degrees.

15. A resistance, split-phase induction motor has resistance
    a. added to the rotor circuit.
    b. connected in series with the running winding.
    c. connected in series with the starting winding.
    d. connected in series with the motor circuit.

16. Capacitors are used on split-phase motors
    a.to increase the starting torque.
    b. to improve power factor.
    c. both a and b.
    d. neither a nor b.

17. The direction of rotation on a split-phase motor can be reversed by interchanging the connections to
    a. the main winding.
    b. the auxiliary winding.
    c. either a or b.
    d. both a and b.

18. Shaded-pole motors are used to drive
    a. compressors.
    b. printing presses.
    c. small fans.
    d. pumps.

19. A repulsion motor is a
    a. three-phase motor.
    b. single-phase motor.
    c. dc motor.
    d. universal motor.

20. A universal motor is designed to operate from
    a. an ac single-phase supply.
    b. a dc supply.
    c. both a and b.
    d. a three-phase supply.

REVIEW *(continued)*

B. Give complete answers.

1. Why is the squirrel-cage induction motor the most common type of motor used in industry?
2. Describe the rotor construction of a squirrel-cage induction motor.
3. Explain how a rotating magnetic field is developed in the stator of a three-phase induction motor.
4. What causes current to flow in the rotor of a three-phase induction motor?
5. What factors affect the speed of the rotating field that is produced in the stator of a three-phase induction motor?
6. Define the term *synchronous speed.*
7. How does the full-load rotor speed compare with the synchronous speed of a three-phase induction motor?
8. What determines the direction of rotation of the stator field of a three-phase induction motor?
9. Describe the operating principle of a three-phase, squirrel-cage induction motor.
10. How does one reverse the direction of rotation of the rotor of a three-phase, squirrel-cage induction motor?
11. What determines the amount of torque developed by a three-phase, squirrel-cage induction motor?
12. Define the term *slip* as it pertains to an induction motor.
13. Describe the operating characteristics of a three-phase, squirrel-cage induction motor.
14. Define *starting current* and *full-load current.*
15. Describe the causes and effects on the supply current for a three-phase, squirrel-cage induction motor from the initial start to the rated speed at full load.
16. Why is it sometimes necessary to start a three-phase induction motor with reduced voltage?
17. How does starting an induction motor with reduced voltage affect the torque?
18. Define the term *breakdown torque.*
19. Describe a double squirrel-cage rotor.

REVIEW *(continued)*

B. Give complete answers (continued).

20. What is the purpose of a double squirrel-cage rotor?
21. List two methods used to vary the speed of a three-phase, squirrel-cage induction motor.
22. Which of the methods listed in the answer to Problem 21 provides a smooth variation of speed?
23. Describe a wound-rotor induction motor.
24. List the advantages of a three-phase, wound-rotor induction motor compared to the three-phase, squirrel-cage induction motor.
25. List two disadvantages of the three-phase, wound-rotor induction motor.
26. Describe the three-phase, brush-shifting induction motor.
27. What is the purpose of the brush-shifting arrangement on a brush-shifting induction motor?
28. What advantages are gained by using induction motors that operate on frequencies greater than 60 hertz?
29. Identify two uses for induction motors that operate on frequencies greater than 60 hertz.
30. Describe a three-phase, ac synchronous motor.
31. Explain the operation of a three-phase synchronous motor.
32. List two advantages of a three-phase synchronous motor compared to a three-phase, squirrel-cage induction motor.
33. Name two disadvantages of the three-phase synchronous motor.
34. Why would a three-phase synchronous motor be operated without a load?
35. Describe the procedure to follow when starting a three-phase synchronous motor.
36. What is the purpose of a field discharge resistor when used in conjunction with a three-phase synchronous motor?
37. Define *synchronous capacitor*.
38. Explain why a single-phase induction motor with a squirrel-cage-type rotor and a single winding on the stator does not develop a starting torque.

REVIEW *(continued)*

B. Give complete answers (continued).

39. Describe the construction and operation of a standard split-phase motor.
40. Describe the construction and operation of a capacitor-start, induction-run, split-phase motor.
41. What is the advantage of the capacitor-start, induction-run, split-phase motor as compared to the standard split-phase motor?
42. Describe the construction and operation of a capacitor-start, capacitor-run induction motor.
43. What are the advantages of the capacitor-start, capacitor-run induction motor compared to the capacitor-start, induction-run motor?
44. Name two types of capacitor-start, capacitor-run induction motors, and describe the difference between the two.
45. Describe the capacitor, autotransformer, split-phase induction motor, and explain its operation.
46. Define the *centrifugal switch*, and describe its operation.
47. Explain how to reverse the direction of the rotor rotation on a split-phase induction motor.
48. Describe the construction and operation of a shaded-pole motor.
49. Describe the construction and operation of a repulsion induction motor.
50. Describe the construction and operation of a universal motor.

C. Extended Study

1. Describe the difference between the repulsion-start, induction-run motor and the repulsion induction motor.
2. Explain how to reverse the direction of rotor rotation of a repulsion-start, induction-run motor.
3. Describe the difference between a series ac motor and universal motor.
4. Define the term *torque motor*.
5. List five common uses for a torque motor.
6. Describe a three-phase, double-voltage, squirrel-cage induction motor.

19

AC Motor Controllers

Objectives

After studying this chapter, the student will be able to:

- Explain the purpose of motor controllers.
- Describe the construction and operation of various types of motor controllers and motor protective devices.
- List various maintenance procedures for motor controllers.

ACROSS-THE-LINE STARTERS

If an induction motor is started by connecting it directly across the line, it will develop a higher starting torque than if it is started with a reduced voltage. This provides rapid acceleration to rated speed and permits starting under load. The starting current, however, will range from 4 times to 6 times the full-load current. Because of their rugged construction and method of operation, squirrel-cage induction motors will not be damaged by high starting currents. With very large motors, however, the high starting current may cause too much voltage fluctuation in the power lines or may impose too great a stress on the driven machinery. Under these conditions, the voltage must be reduced during the starting period.

Many factors must be considered when selecting starting equipment. These factors include starting current, voltage (line) drop, type of load, motor protection, and operator safety.

Motors that are to be started under full voltage generally use a control called an *across-the-line starter,* Figure 19-1. This mechanism usually consists of a magnetic relay, an overload

Figure 19-1
Three-phase, across-the-line starter *(Courtesy of Square D Company, Milwaukee, WI)*

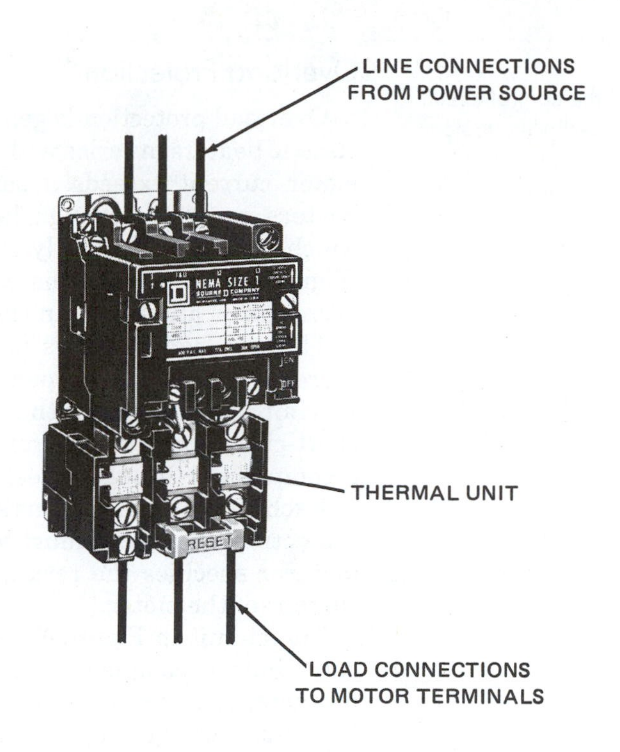

Figure 19-2A
Momentary push-button station *(Courtesy of Square D Company, Milwaukee, WI)*

Figure 19-2B
Schematic symbol for a momentary push-button station (stop/start)

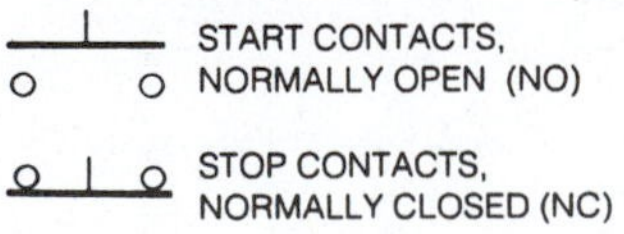

START CONTACTS, NORMALLY OPEN (NO)

STOP CONTACTS, NORMALLY CLOSED (NC)

protective device, and an arrangement that can provide low-voltage protection.

A *momentary push-button station* is often used in conjunction with the across-the-line starter. Figure 19-2A shows one type of momentary push-button station, and Figure 19-2B shows the schematic symbol. This type of installation consists of two circuits—the main circuit and the control circuit. The *main circuit* contains the conductors and devices from the supply to the motor. The *control circuit* contains the magnetic coil, the overload contacts, the auxiliary contacts, the push buttons, and the conductors that connect them together. Figure 19-3A is a schematic diagram of the control circuit. Figure 19-3B is a schematic diagram of the main circuit. Figure 19-3C shows the two circuits combined.

For many installations, control transformers are used to provide a lower voltage for the control circuits. If a control transformer is used, it may be contained within the same enclosure as the across-the-line starter, or it may be located elsewhere, supplying the power for the control circuit of more than one starter.

Figure 19-3A
Schematic diagram of the control circuit for a three-phase, across-the-line motor starter

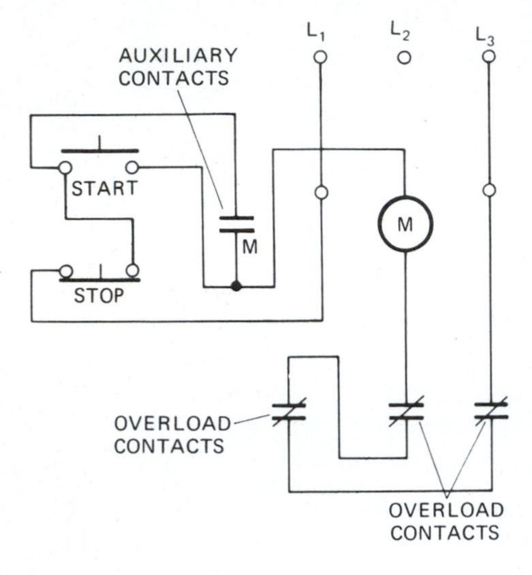

Overload Protection

Overload protection is generally achieved by connecting small electric heaters in series with the motor (Figure 19-3C). When the motor current exceeds a safe value, the current through the heaters generates enough heat to activate the thermal device, which opens the normally closed contacts in the control circuit (Figure 19-3A). When the control circuit is opened, the coil is deenergized and the main contacts open (Figure 19-3C).

The *overload device* is sensitive to current and time. Small currents take longer to open the contacts than do large currents. It is important to realize that the overload device does not provide short-circuit protection. Protection against short circuits must be accomplished through fuses, circuit breakers, or other devices.

Each heating element has a specific rating. To provide adequate protection, the heater must be matched to the motor. The manufacturer specifies the rating according to the full-load running current of the motor.

The circuit in Figure 19-3C operates as follows. Pushing the start button completes the circuit from L_1 through the stop contacts, start contacts, coil M, and the overload contacts to L_3. Energizing coil M causes all M contacts to close. Auxiliary contacts M are in parallel with the start contacts. Therefore, the control

Figure 19-3B
Main circuit for a three-phase, across-the-line motor starter

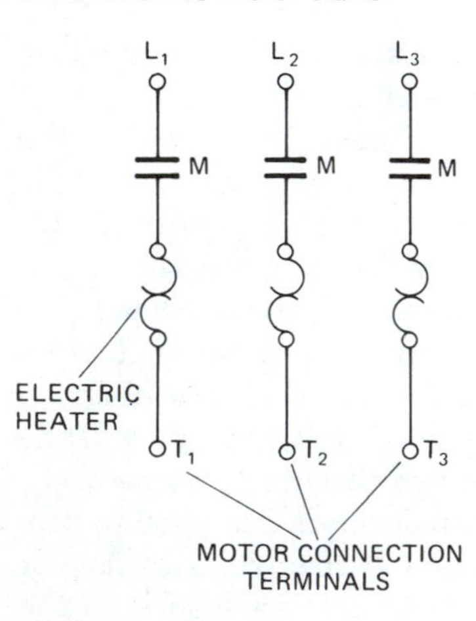

Figure 19-3C
Stop/start station connected to an across-the-line motor starter

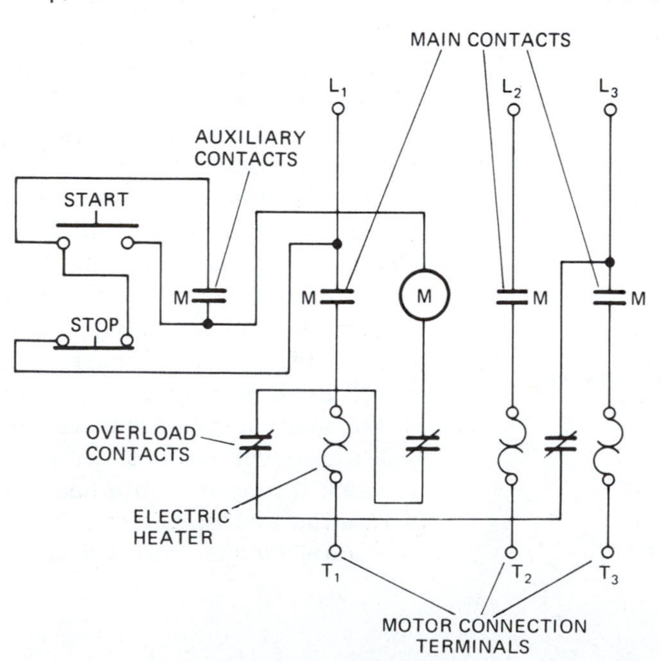

circuit is completed through the auxiliary contacts, permitting the release of the start button. The main contacts M are also closed, completing the circuit to the motor. Pressing the stop button opens the control circuit and deenergizes the coil, causing contacts M to open.

If the contacts are in the operating position and the motor becomes overloaded, the heating elements will produce enough heat to activate the overload contacts in the control circuit. When these contacts open, the coil is deenergized, allowing contacts M to open.

LOW-VOLTAGE PROTECTION

> **CAUTION:** Voltage fluctuations can be hazardous to motors and equipment. Low voltages can cause motors to overheat, melting the insulation on the windings. Motors that restart suddenly after a loss of voltage can also be hazardous.

Two types of low-voltage protection are in common use in industry. For the safety of personnel and equipment, serious consideration must be given to the type selected.

If it is preferable that the motor restart automatically upon restoration of normal voltage, an *undervoltage release device* is used. *Note:* This device should never be used if there is the slightest chance that personnel could be injured. It is frequently convenient, however, when no hazards exist, to allow the motor to restart automatically.

If starting automatically presents a hazard, then an *undervoltage protection device* should be used. This device requires manual starting after the normal operating voltage is restored.

Referring again to Figure 19-3C, the auxiliary contacts serve to maintain the coil circuit after the start button is released. With this arrangement, the M contacts open with voltage failure. Upon restoration of normal voltage, it is necessary to press the start button in order to restart the motor.

Relay Chatter

The current through the electromagnet, Figure 19-3C, varies from zero to maximum twice for every cycle. Since the pull is proportional to the current, the armature starts to drop out every time the current nears zero. Just enough movement takes place to cause a "chatter." To prevent chattering, a shading coil is installed around part of the pole face, Figure 19-4. The voltage induced into the shading coil causes a current that produces a flux out of phase

Figure 19-4
Core and armature of the electromagnet in an across-the-line starter *(Courtesy of Square D Company, Milwaukee, WI)*

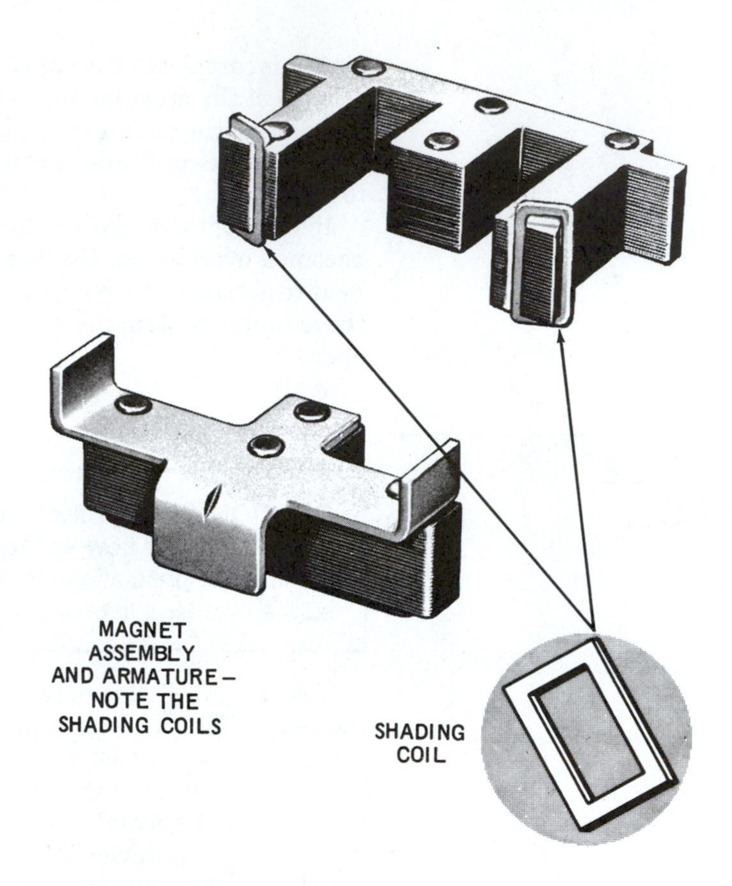

with the main flux. As a result, a flux exists in the shaded portion of the magnet after the main flux has dissipated. The magnetic field produced in this manner is just strong enough to prevent chattering.

Construction and Operation of Overload Relays

The overload devices protect the motor against excessive momentary surges, normal overloads existing for long periods, and high currents caused by an open phase. These relays have characteristics similar to those shown in Figure 19-5. The curve indicates that before the relay opens the circuit, 500 percent of the motor full-load current can flow for about 10 seconds, or 150 percent of the full-load current can flow for approximately 4 minutes.

One type of relay consists of mica-insulated nichrome heating elements placed adjacent to a cylinder containing a low melting alloy and the shaft of a ratchet wheel, Figure 19-6A. The relay is set by placing its spring under tension and is held by a latch engaged in the teeth of the wheel, Figure 19-6B. When an overload

Figure 19-5
Graph of motor heating curve and overload relay trip curve *(Courtesy of Square D Company, Milwaukee, WI)*

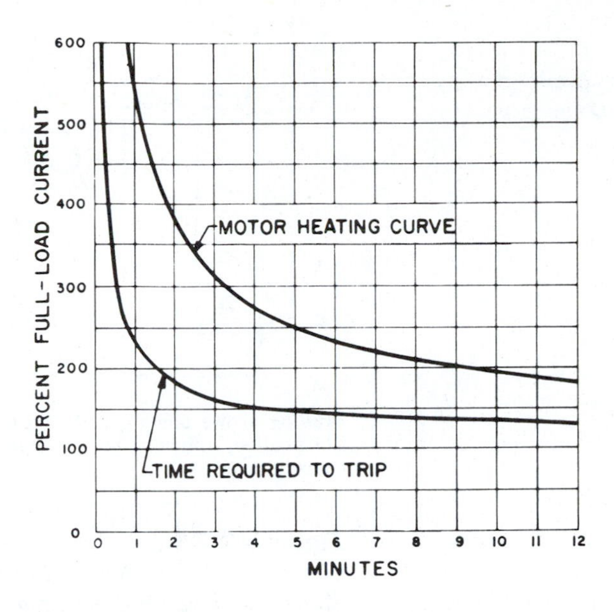

Figure 19-6A
Thermal overload relay with heating elements enlarged (left) to illustrate construction *(Courtesy of Square D Company, Milwaukee, WI)*

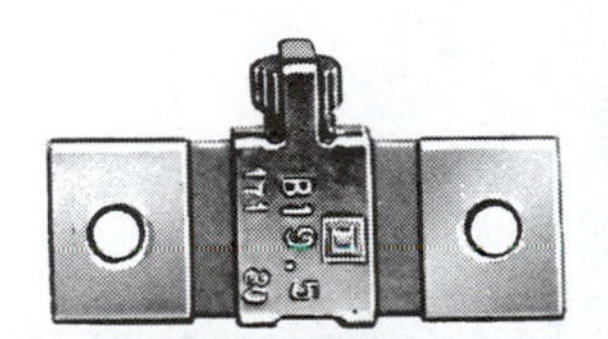

One Piece Thermal Unit

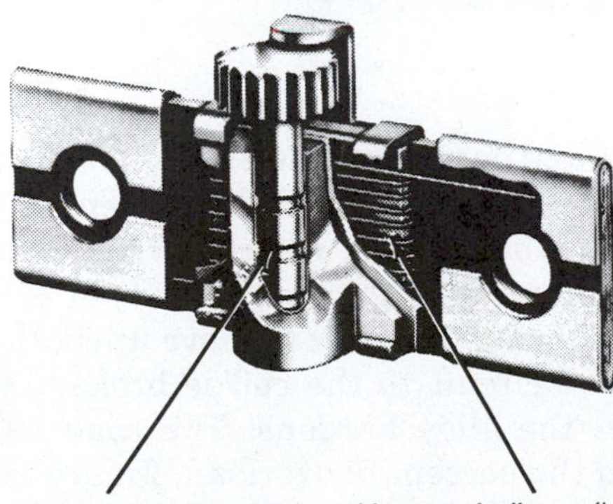

Solder Pot (heat-sensitive element) is an integral part of the thermal unit. It provides accurate response to overload current, yet prevents nuisance tripping.

Heater winding (heat-producing element) is permanently joined to the solder pot, so proper heat transfer is always ensured. No chance of misalignment in the field.

Three-Pole Melting Alloy Thermal Overload Relay

Figure 19-6B
Drawing of a thermal overload relay *(Courtesy of Square D Company, Milwaukee, WI)*

Thermal Relay Unit
to Motor
to Magnet Coil

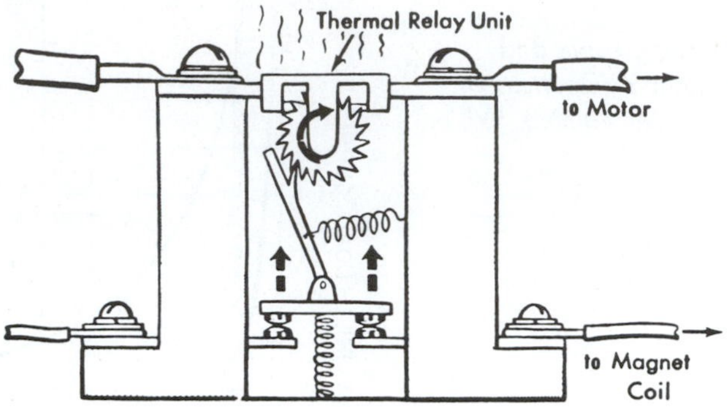

Drawing shows operation of melting alloy overload relay. As heat melts alloy, ratchet wheel is free to turn – spring then pushes contacts open.

Figure 19-6C
Graph of hardening characteristics of metal alloy used in an overload relay

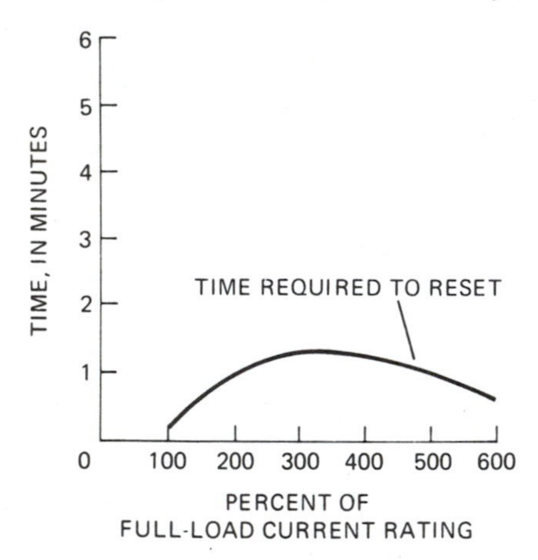

Figure 19-7
Bimetallic overload relay with side cover removed *(Courtesy of Square D Company, Milwaukee, WI)*

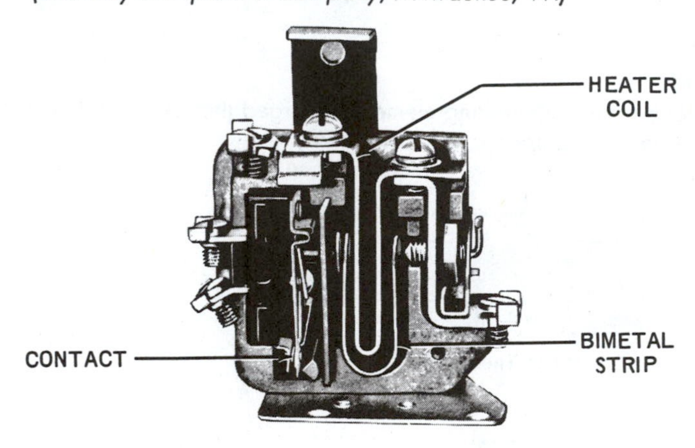

occurs, the heat generated by the heating element melts the alloy. The time it takes to melt depends upon the percent of overload. When the alloy melts, the ratchet wheel is free to turn. The compression on the bottom spring forces the wheel to turn in a clockwise direction, allowing the lever to move upward and open the contacts. Thus, the circuit to the coil is broken. The relay cannot be reset until the alloy hardens. The time required to harden depends upon the percent of overload, Figure 19-6C.

Another common type of thermal relay, Figure 19-7, has an element that heats a bimetal strip. When an overload occurs, the heat generated by the element causes the bimetal strip to bend, thus opening the overload contacts.

Heating elements for thermal relays are manufactured in many different current ratings for each size of starter. It is possible to use the same starter for motors of different sizes merely by changing the size (current rating) of the heating element to match that of the motor. The current rating of the heater should not exceed 125 percent of the motor full-load current rating.

Manual Across-the-Line Starters

A *manual starter,* as referred to in this text, is a starter that requires hand operation at the location. It is operated by a push button or lever that operates the contacts directly. Manual across-the-line starters are generally used only on small motors.

The manual starter usually provides overloaded protection and undervoltage release. It does not provide undervoltage protection. The main advantage of the manual controller is its low initial cost.

Contact Maintenance

Atmospheric conditions can have adverse effects on electrical contacts. Dust and corrosion produce high contact resistance. The resistance increases with time until eventually it reduces the current to a value that is too low for proper operation. To prevent dust accumulation, the starters should be cleaned with a high-power vacuum cleaner. Compressed air is sometimes used to blow away any dust. However, compressed air contains moisture, which may cause other problems.

After the dust is removed from the starter, the contacts should be cleaned with a nontoxic cleaning agent. When using cleaning agents, always follow the directions supplied by the manufacturer.

CAUTION: Do not use carbon tetrachloride. It is highly toxic and may cause permanent physical damage.

If, after a thorough cleaning, the contact resistance is not reduced sufficiently, the contacts should be replaced.

Most contacts are made of silver or are silverplated. Arcing can produce a coating of silver oxide, which gives the contacts a slightly yellow color. This coating should not be removed because it is a fairly good conductor and is self-reducing by wear and heat. Pitting of silver contacts can be removed with very fine sandpaper. No more material should be removed than is absolutely necessary. The contacts should be restored to their original shape and then cleaned to remove all dust particles. Extreme care should be taken to not remove all the plating from silverplated contacts. If too much plating is removed, the contacts must be replaced.

If the contact arms require adjustment, be sure that the contacts close squarely once the adjustment is completed. Multiple contact relays should have the arms adjusted so that all contacts close and/or open simultaneously.

A common contact fault is the transfer of metal from one contact to the other. The extent to which metal transfer takes place depends upon the type of metal and the amount and kind of current (ac or dc) being interrupted. Some metal transfer will take place whether or not an arc is formed.

When an arc persists for more than one-half cycle on ac, or for about 0.01 second on dc, contact life is shortened considerably.

The arc on ac usually extinguishes as the current passes through the zero point on the cycle. A magnetic blowout coil or arc chute is frequently used with dc.

When low-voltage dc circuits containing inductance are broken, a high voltage developed across the inductance can cause arcing. Two methods are used to prevent this arcing. Connecting a resistance across the inductance at the instant the circuit is broken makes the emf appear as a voltage drop across the resistance. Connecting a capacitor across the contacts at the instant the contacts open makes the current that produces the arc to flow into the capacitor and charge it.

A double-break contact, which has about 3 times the interrupting capacity of a single-break contact, is often used to prolong contact life. For high current and low voltage, the contacts are connected in parallel. For low current and high voltage, the contacts are connected in series.

JOGGING CONTROL

Some industrial applications require that machine components stop at an exact position. This operation is difficult to accomplish with only stop and start buttons. If, however, a *jog button* is added to the station, the operation becomes relatively simple.

Figure 19-8A is a diagram of this circuit. When the jog button is pressed, it opens the circuit to the auxiliary contacts and completes the circuit through coil M. Energizing coil M causes contacts M to close, supplying power to the motor. Releasing the jog button breaks the circuit. Before the circuit can be completed through the auxiliary contacts, coil M is deenergized, opening contacts M. With the M contacts now open, the circuit is again completed to the auxiliary contacts and is ready to be operated by either the start button or the jog button.

Pressing the start button energizes coil M, closing contacts M. The auxiliary contacts close to complete the circuit through the normally closed jog contacts and bypass the start contacts. This allows the release of the start button, and the closed auxiliary contacts maintain the coil circuit.

When the start button is pressed, the motor will start and continue to run until the stop button is pressed. Pressing the jog

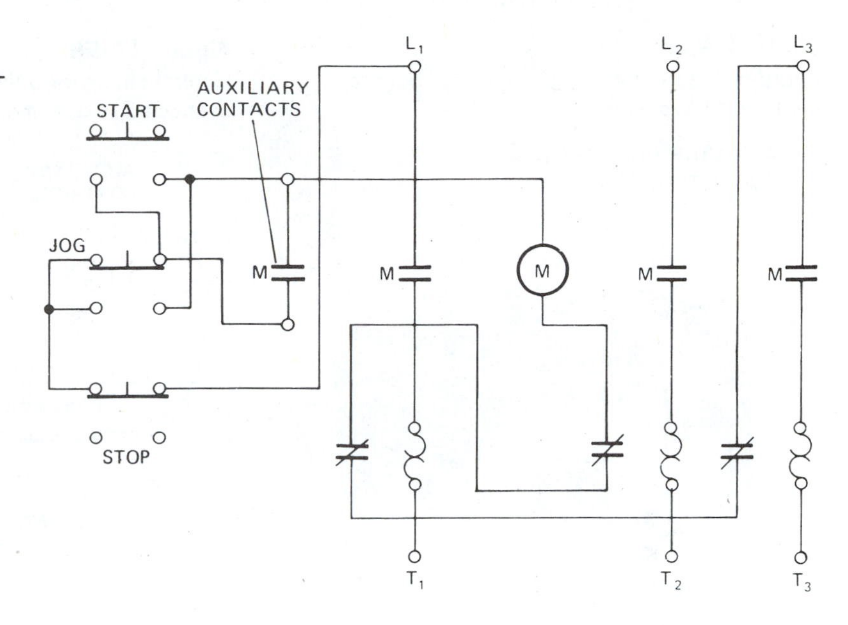

Figure 19-8A
Stop-start-jog station connected to a three-phase, across-the-line starter

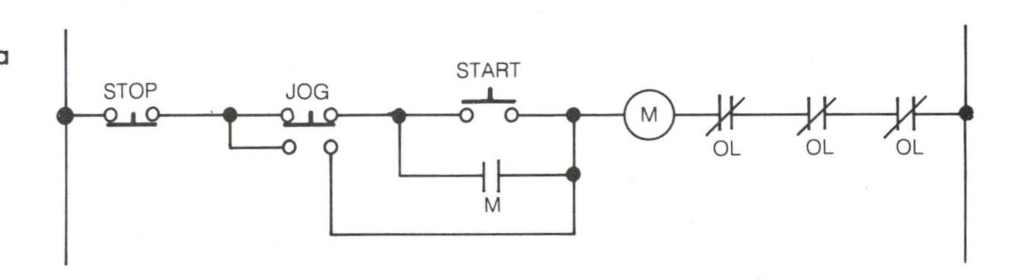

Figure 19-8B
Ladder diagram of the control circuit of a stop-start-jog station

button, however, energizes the motor only as long as the button is being pressed. Releasing the button breaks the circuit and de-energizes coil M, causing contacts M to open. When contacts M open, the motor is disconnected from the supply. Figure 19-8B shows a ladder diagram of the control circuit used in Figure 19-8A.

REVERSING CONTROLLERS

Some applications require repeated reversals of the direction of rotation of the rotor. This can be accomplished by using a *reversing controller,* which interchanges the connections to two of the three supply lines. This type of controller is shown in Figure 19-9A. Figure 19-9B is a diagram for the control circuit, and Figure 19-9C shows both the control circuit and the main circuit. One button is provided for each direction of rotation, and only one stop button is necessary.

Pressing the forward button, Figure 19-9C, completes the circuit from L_1 through the forward coil to L_3, closing the F

Figure 19-9A
Forward-reverse controller *(Courtesy of Square D Company, Milwaukee, WI)*

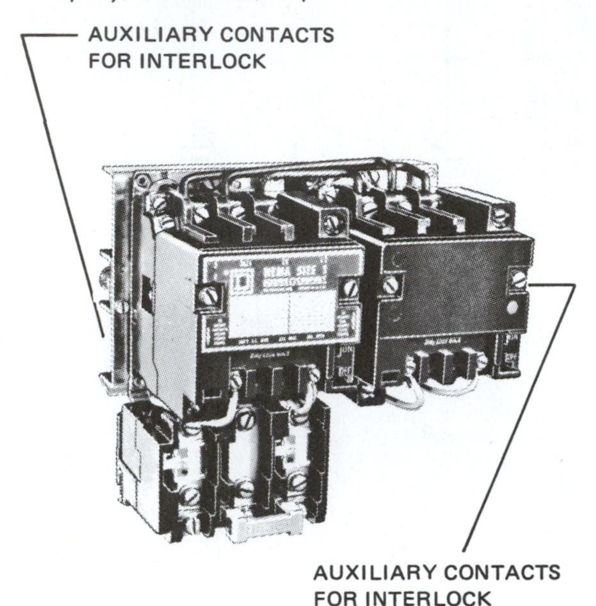

Figure 19-9B
Control circuit for a three-phase, forward-reverse controller connected to a momentary push-button station

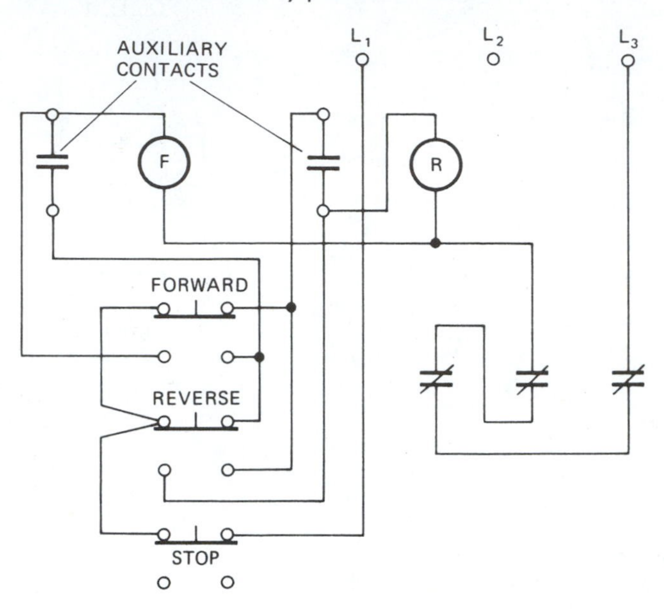

contacts. When contacts F close, the motor is connected to the line leads for forward operation. Auxiliary contacts F bypass the forward push-button contacts to maintain the control circuit after the forward button is released.

If the reverse button is pressed, the circuit to coil F is opened and the circuit to coil R is completed. Energizing coil R closes contacts R, interchanging the motor connections to line wires L_1 and L_3. Interchanging the connections to L_1 and L_3 reverses the phase sequence to the motor and causes the rotor to reverse its direction of rotation. The control circuit for coil R is maintained through auxiliary contacts R, and the reverse button can be released.

Interlocks

Interlocks are used to prevent faulty operation and to ensure the proper operating sequence. Two methods of interlocking are mechanical and electrical.

A *mechanical interlock* on a forward/reverse starter is arranged to prevent the reverse contacts from closing when the motor is in forward operation. It will also prevent the forward contacts from closing when the motor is in reverse operation.

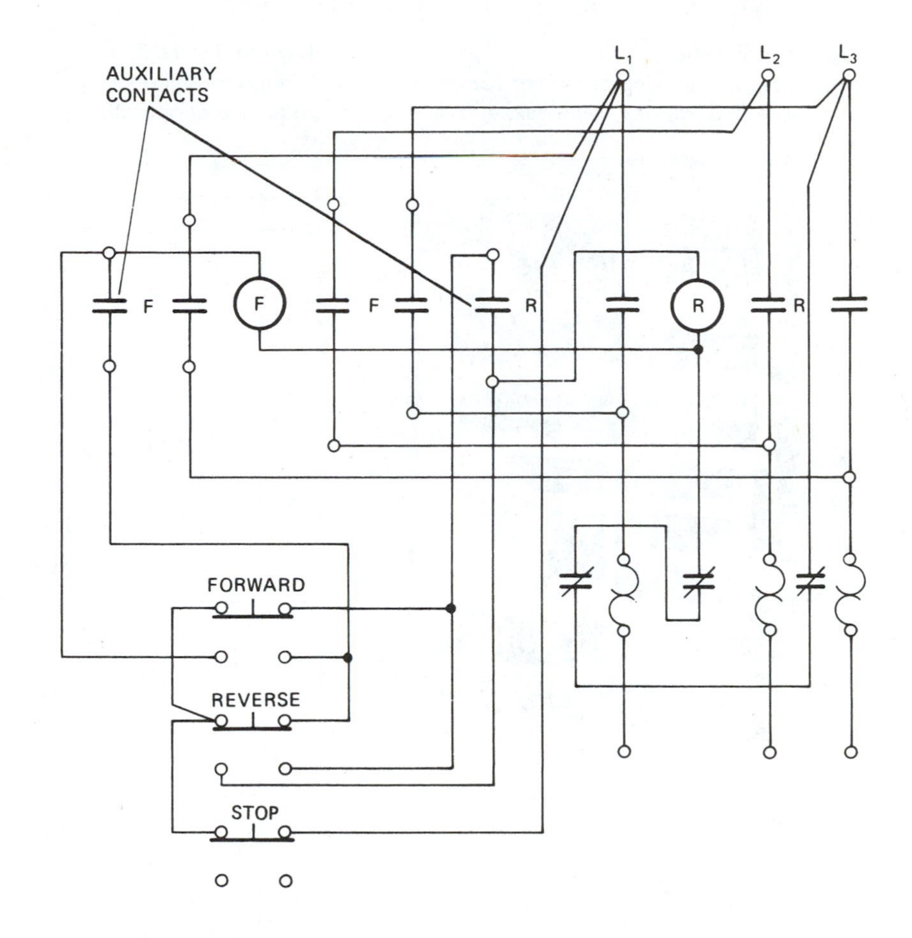

Figure 19-9C
Control circuit and main circuit for a forward-reverse, across-the-line starter with one forward-reverse-stop station

Electrical interlocks are auxiliary contacts mechanically actuated by the main contactor. Figure 19-9A shows a forward/reverse controller with electrical interlocks. When coil F is energized, coil R cannot be energized without deenergizing coil F. When coil R is energized, coil F cannot be energized without deenergizing coil R.

INTERLOCKING MOTORS

Machines are often equipped with several motors and/or other electrically operated equipment. Such applications require the use of many interlocking circuits. A controller for this purpose is shown in Figure 19-10A. Examples of this type of installation are illustrated in Figures 19-10B, 19-10C, and 19-10D. The circuit in Figure 19-10B is used when the second motor must not be started unless the first motor is stopped and vice versa. The circuit in Figure 19-10C is used when the second motor must not be started unless the first motor is operating. The circuit in Figure 19-10D is used if one motor must not be operating without the other.

Figure 19-10A
Sequence, interlocking controller *(Courtesy of Square D Company, Milwaukee, WI)*

Figure 19-10B, C, and D
Circuits for interlocking motors. The controller ensures proper sequence of operation.

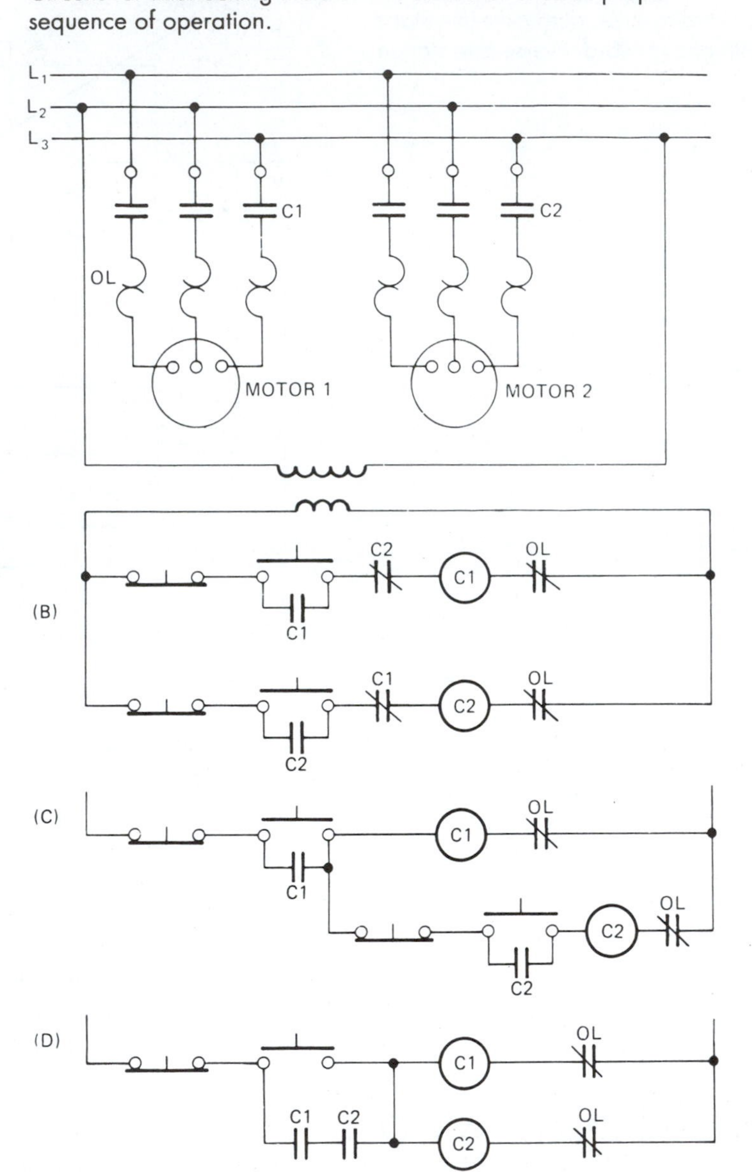

DYNAMIC BRAKING

Dynamic braking for dc motors is discussed in Chapter 12. Dynamic braking for induction motors is accomplished in a much different manner. Figure 19-11 shows a method used with ac motors. When the stop button is pressed, the motor is disconnected from the power line and the stator is energized with dc. The dc in the stator produces a stationary magnetic field. As the rotor revolves through this field, a high voltage is induced across the rotor circuit. This voltage causes a high current to produce a magnetic field. The magnetic field locks in with the stator field and stops the rotor. All of these actions take place very quickly.

Figure 19-11 also shows a limit switch that is connected in order to limit the distance the moving component can travel. The switch is spring loaded and normally closed. As the moving mechanism approaches the maximum distance of travel, a cam arrangement

Figure 19-11
Connections for dynamic braking of a three-phase, squirrel-cage induction motor

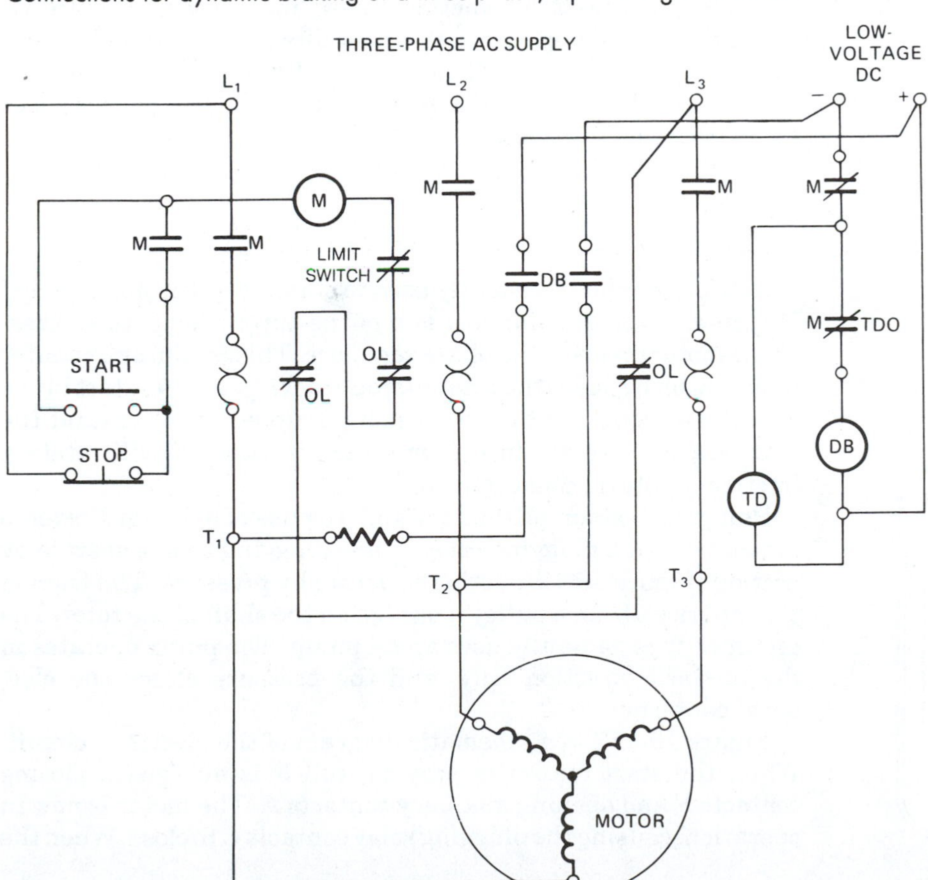

momentarily opens the control circuit. This action deenergizes coil M, disconnecting the motor from the power source, and energizes the stator with dc. Dynamic braking thus takes place, stopping the motor quickly.

FRICTION BRAKE

Friction brakes are frequently used to stop the rotor. Although they do not act as quickly as dynamic braking, they are satisfactory for many operations. The braking mechanism is a drum-type apparatus, which locks around a pulley attached to the shaft of the rotor. An electromagnet is connected between two of the motor leads. When the motor is energized, the magnet acts on an armature, causing the brake shoes to spread, which allows the rotor to revolve. When the power to the motor is interrupted, the electromagnet is deenergized. The brake spring then forces the brake shoes against the pulley to stop the rotor.

Friction brakes are sometimes used in conjunction with dynamic braking. For example, if a hoist lifts a load several feet and dynamic braking is applied, the motor will stop, but the weight will cause the load to drop. The incorporation of a friction brake causes the brake to hold the load.

PLUGGING

Figure 19-12A
Friction-type plugging relay

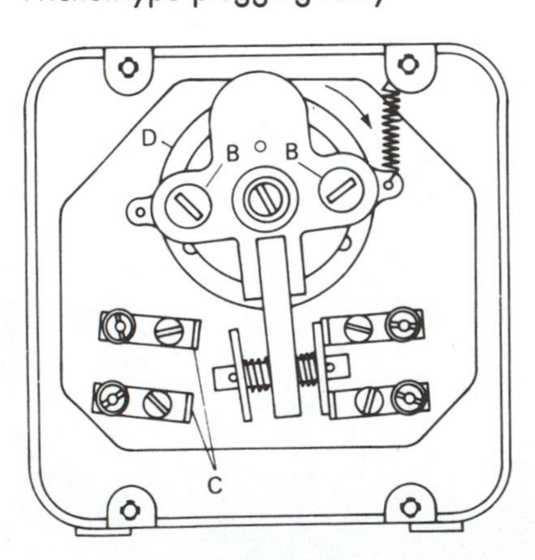

Another method frequently used to obtain a quick stop is called *plugging*. Interchanging any two of the supply leads to a three-phase motor reverses the phase sequence. This produces a counter torque, which causes the rotor to stop quickly and then start in the opposite direction. If the power is interrupted at the instant the rotor starts to rotate in the opposite direction, it will rotate a fraction of a turn, and then stop.

Plugging is accomplished through the use of a forward/reverse controller and a plugging relay. Some plugging relays operate by friction (Figure 19-12A); others operate by pressure. The friction type operates from a pulley mounted on the shaft of the rotor. The pressure type generally uses an oil pump. The pump operates in the forward direction only, and the pressure closes the electrical contacts.

Figure 19-12B is a schematic diagram of the electrical circuit. When the start button is pressed, coil F is energized, closing contacts F and opening auxiliary contacts A. The motor is now in operation, causing the plugging relay contacts C to close. When the

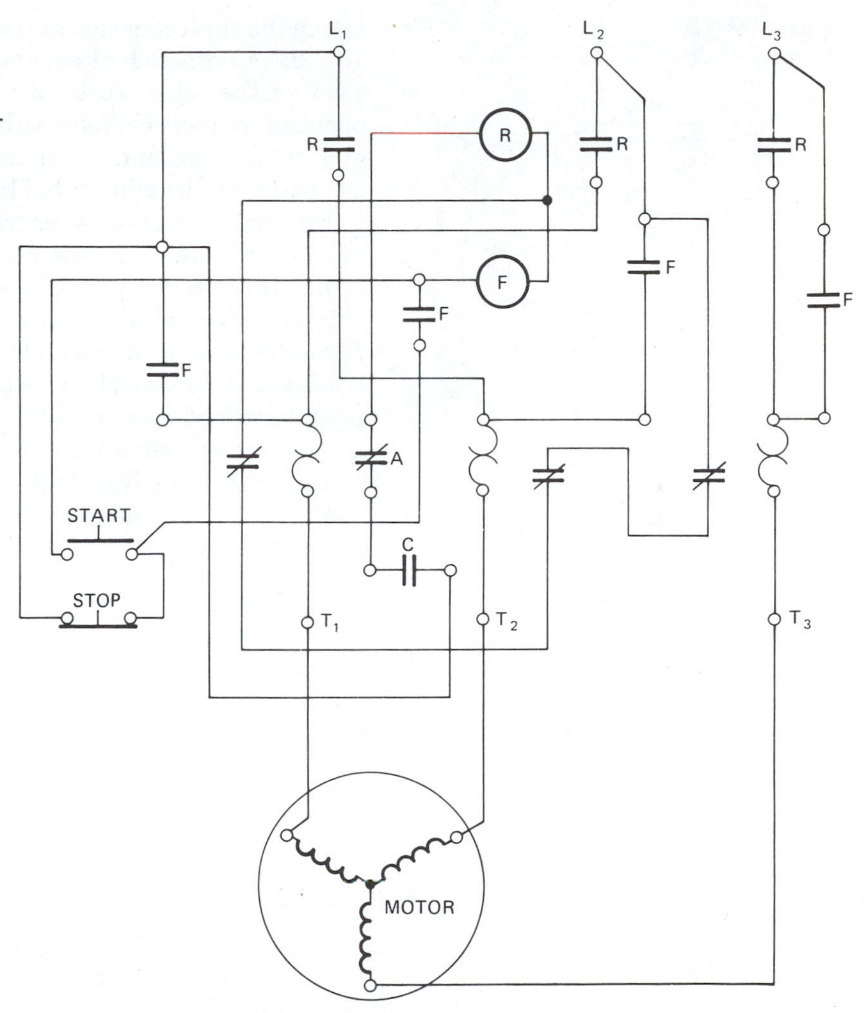

Figure 19-12B
Connections for a three-phase, squirrel-cage induction motor with a momentary push-button, stop/start station, across-the-line starter, and a plugging relay

start button is released, the circuit to coil F is maintained through the auxiliary contacts F. When the stop button is pressed, coil F is deenergized, opening contacts F and closing auxiliary contacts A. Because the plugging relay contacts are still closed, the circuit to coil R is complete, and contacts R close. This interchanges two of the motor terminal connections to the supply, and a counter torque is developed. The counter torque causes the rotor to stop quickly and then begin to rotate in the opposite direction. At the instant the rotor reverses, contacts C open, coil R is deenergized, and main line contacts R open, disconnecting the motor from the supply.

In the servicing or retooling of a machine, the motor shaft is sometimes turned by hand. If the motor is wired for plugging, ro-

Figure 19-13
Centrifugal relay

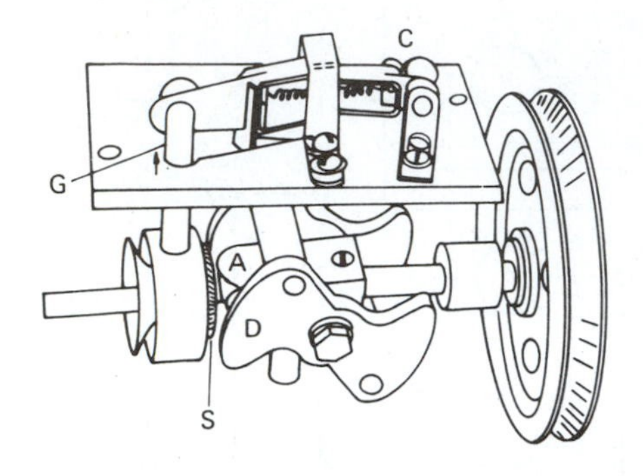

tating the shaft causes contacts C to close, energizing coil R (Figure 19-12B). Contacts R close, and power is supplied to the motor. As soon as the rotor starts to rotate, the plugging relay operates, opening contacts C. *Note:* Although the rotor turns only a fraction of a turn, it presents a hazard to personnel and equipment. The condition can be eliminated by using a centrifugal relay.

Figure 19-13 shows a centrifugal relay. When the relay shaft is revolving above a predetermined slow speed, the centrifugal device D forces component A against spring S. This establishes a friction drive, which moves finger G in the direction indicated. Finger G closes contacts C of the control relay (Figure 19-12B) and holds them closed as long as the rotor is revolving faster than the predetermined slow speed. When the stop button is pushed, the motor decelerates rapidly. When the rotor speed drops below the setting of the centrifugal relay, contacts C open. Contacts C remain open until the rotor speed is equal to the setting of the centrifugal relay.

Figure 19-14
Synchronous motor connected for dynamic braking

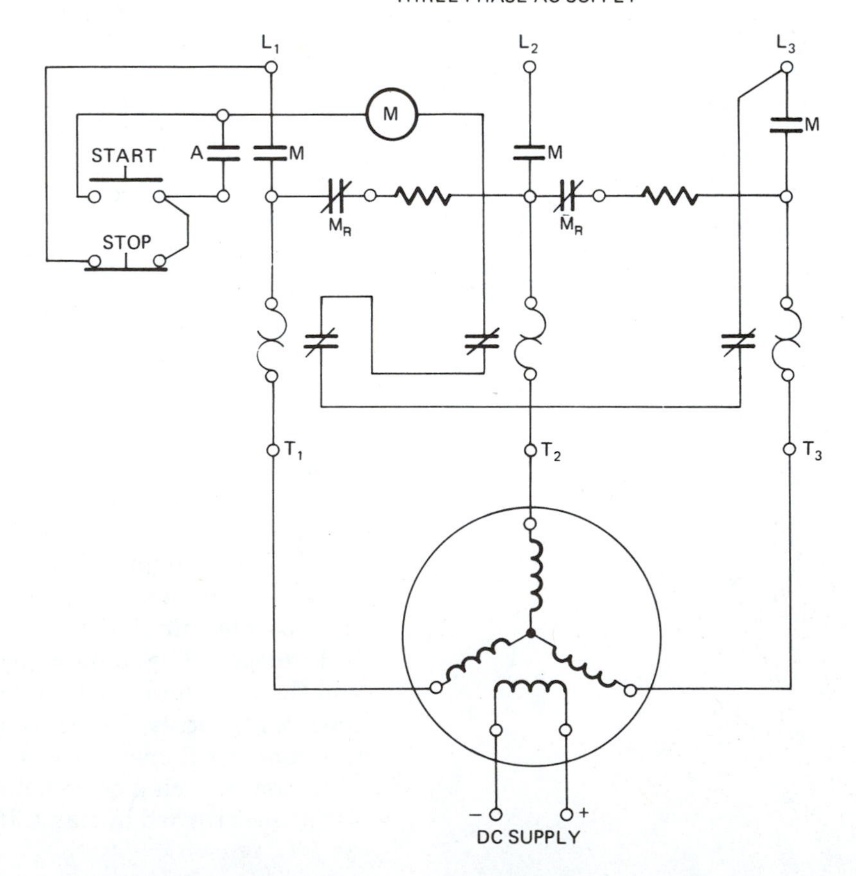

DYNAMIC BRAKING OF SYNCHRONOUS MOTORS

Figure 19-14 illustrates a method used to apply dynamic braking to three-phase synchronous motors. When the stop button is pressed, coil M is deenergized, opening the main line contacts M and closing the resistance contacts M_R. This results in the stator's becoming disconnected from the three-phase supply and the resistances' becoming connected across the stator windings. The dc field remains energized. This action is similar to dynamic braking of dc motors.

WYE-DELTA CONTROLLER

The *wye-delta controller* is a device that changes the stator connections from wye to delta. The controller provides a reduced voltage for starting when in the wye position and a normal operating voltage when in the delta position, Figure 19-15.

Pressing the start button completes the circuit to the *timing relay coil* (TR). Energizing coil TR starts the timing operation and closes the instantaneous contacts TR, completing the circuit to coil S. Coil S opens the normally closed (NC) control contacts S and closes the normally open (NO) contacts S. At the same time, the main line contacts S close, connecting the stator windings into a wye configuration. When the control contacts S close, the circuit is completed to coil 1M. Coil 1M closes contacts 1M, connecting the motor to the supply and bypassing control contacts S. The motor is now operating in the start position with about 58 percent of the operating voltage supplied to each phase winding.

Figure 19-15
Elementary diagram of a wye-delta starter with open transition starting *(Courtesy of Square D Company, Milwaukee, WI)*

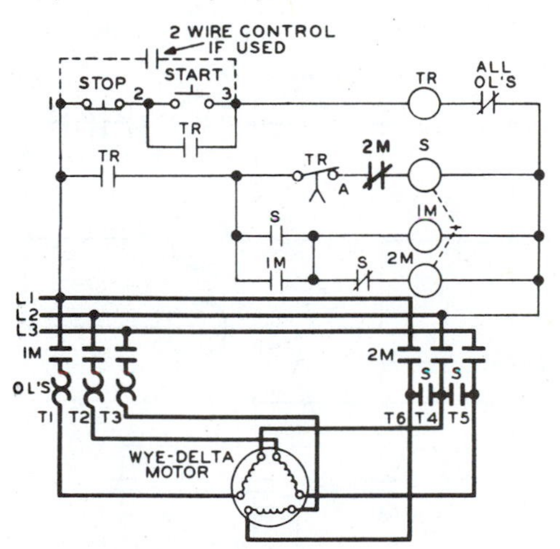

When the time-delay relay TR-A completes its timing, it opens the circuit to coil S, causing contacts S to open. The circuit to coil 1M is maintained through the auxiliary contacts 1M. The NC contacts S close, completing the circuit to coil 2M and thus closing contacts 2M. The motor is now connected to the supply, with the stator windings connected in delta. This arrangement supplies full voltage to the stator windings, permitting the motor to operate at full torque. It should be noted that when coil S is deenergized, contacts S and 2M operate simultaneously.

Pressing the stop button opens the contacts to coil TR, causing contacts TR to open. This deenergizes coil 2M, and all contacts return to their normal position. The motor is now disconnected from the supply.

When connecting a wye-delta controller, it is important to ensure that the same phase sequence be maintained for both wye and delta connections. If the sequence is reversed, the rotor will stop and attempt to start in the opposite direction when the connections are changed from wye to delta. This results in a large current flow, which causes the overload relays to operate, disconnecting the motor from the supply.

RESISTANCE STARTER

The *resistance starter* is a controller that reduces the motor starting current by inserting an equal amount of resistance into each supply conductor. The resistance is simultaneously removed when the rotor reaches a predetermined constant speed. Figure 19-16 is a schematic diagram of the wiring for a resistance starter.

When the start button is pressed, coil M is energized, closing contacts M and completing the circuit to the motor through the resistors. The circuit is also completed to coil T, starting a time-delay relay to operate. After a specific time, contacts B close, completing the circuit to coil N. Coil N closes contacts N, shunting out the resistance and supplying full voltage to the motor.

When the stop button is pressed, the circuits to coils M and T are opened. Contacts M open, interrupting the power supplied to the motor and allowing the time-delay to return to the start position.

The starting current flowing through the resistance causes a high voltage drop across the resistance. Therefore, a rather low

Figure 19-16
Three-phase, primary resistor-type starter

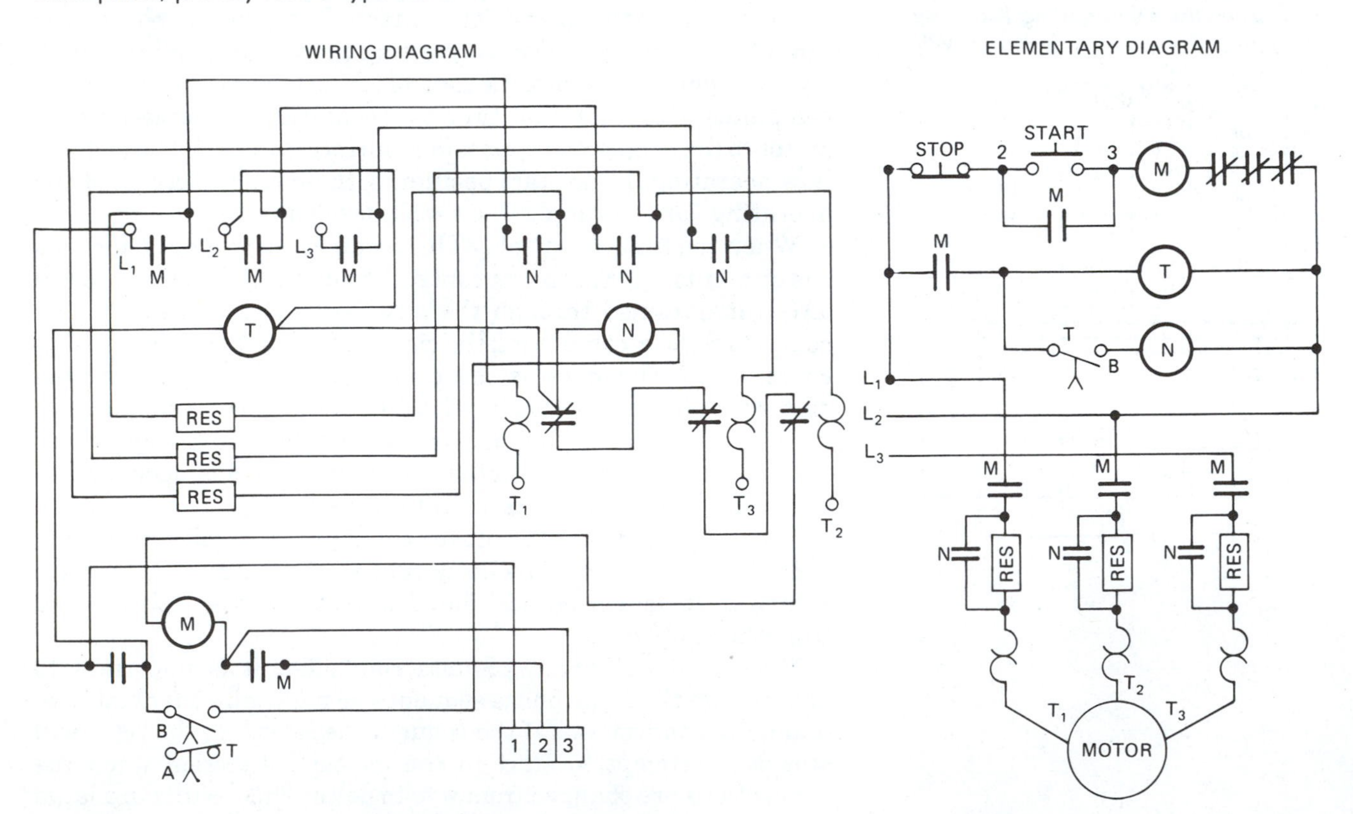

voltage is available at the motor terminals. As the rotor speed increases, the current and, therefore, the voltage drop across the resistances decrease. This results in a gradual increase in the terminal voltage, producing a smooth acceleration.

The advantages of this starter are simplicity of construction, low initial cost, and minimum maintenance. The disadvantages are low starting torque, particularly under heavy loads, and low starting economy.

FACEPLATE CONTROLLER

The *faceplate controller* is a type of resistance controller used with wound-rotor induction motors. This controller provides a means of limiting the starting current as well as varying the speed. It consists of three resistances connected in wye and installed in the rotor circuit, Figure 19-17. Because the controller serves only to vary the resistance of the rotor circuit, an across-the-line starter must be installed to connect the stator to the supply.

At the start, all the resistance is inserted into the rotor circuit and the stator is energized. As the rotor speed increases, the resistance should be reduced gradually. However, to operate the motor below normal speed, some resistance is left in the rotor circuit.

When the rotor resistance is used only for starting purposes, the magnetic control, Figure 19-18, can be used. With this arrangement, the rotor resistance is automatically shunted out in steps. Multiple contacts are arranged to operate in sequence at definite time intervals.

Pressing the start button, Figure 19-18, completes the circuit to coil P, causing contacts P to close. In addition, the timing operation of time-delay relay P is started. The motor is now connected to the supply, and full resistance is in the rotor circuit. After a definite time, time-delay relay P completes the circuit to coil S_1, causing contacts S_1 to close as well as starting the timing operation of time-delay relay S_1. When the main line contacts S_1 close, the first set of resistances is shunted out, permitting the rotor speed to in-

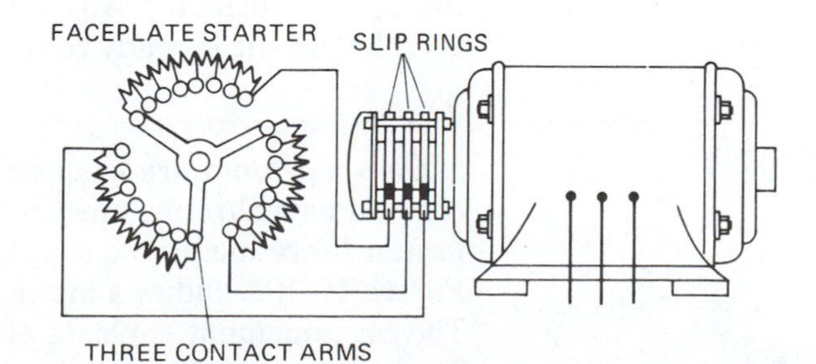

Figure 19-17
Wound-rotor motor connected to a faceplate controller

Figure 19-18
Wound-rotor motor starter with three points of acceleration

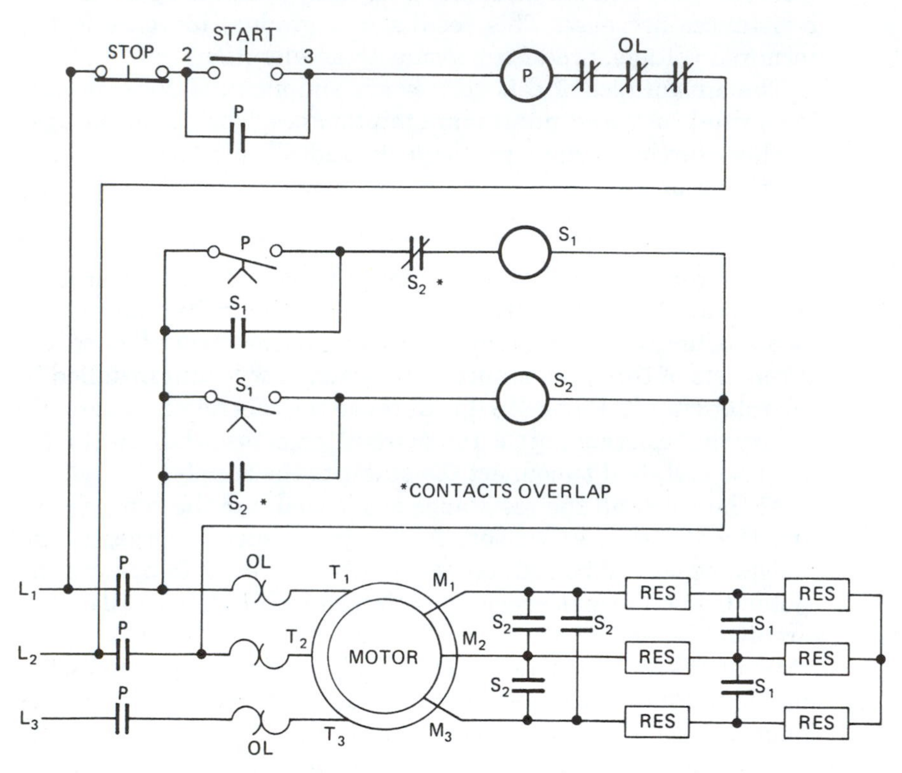

crease. When time-delay relay S_1 has completed its timing, it completes the circuit to S_2. Energizing S_2 causes all the normally open S_2 contacts to close, removing all the resistance from the rotor circuit and allowing the rotor to accelerate to full speed. An instant after the normally open S_2 contacts close, the normally closed S_2 contacts open, deenergizing coil S_1 and opening the S_1 contacts.

When the stop button is pressed, the circuit to coil P is opened, allowing contacts P to open. Opening contacts P deenergizes coil S_2 and the motor. All contacts return to their normal position, and the circuit is ready to repeat the starting operation.

COMPENSATORS

Compensators are frequently used for starting large squirrel-cage and synchronous motors. This controller uses an autotransformer for reducing the supply voltage during the starting period. Figure 19-19A shows a manually operated starting compensator. The arrangement consists of a wye-connected autotransformer. The schematic diagram is shown in Figure 19-19B. When the

Figure 19-19A
Manually operated starting compensator

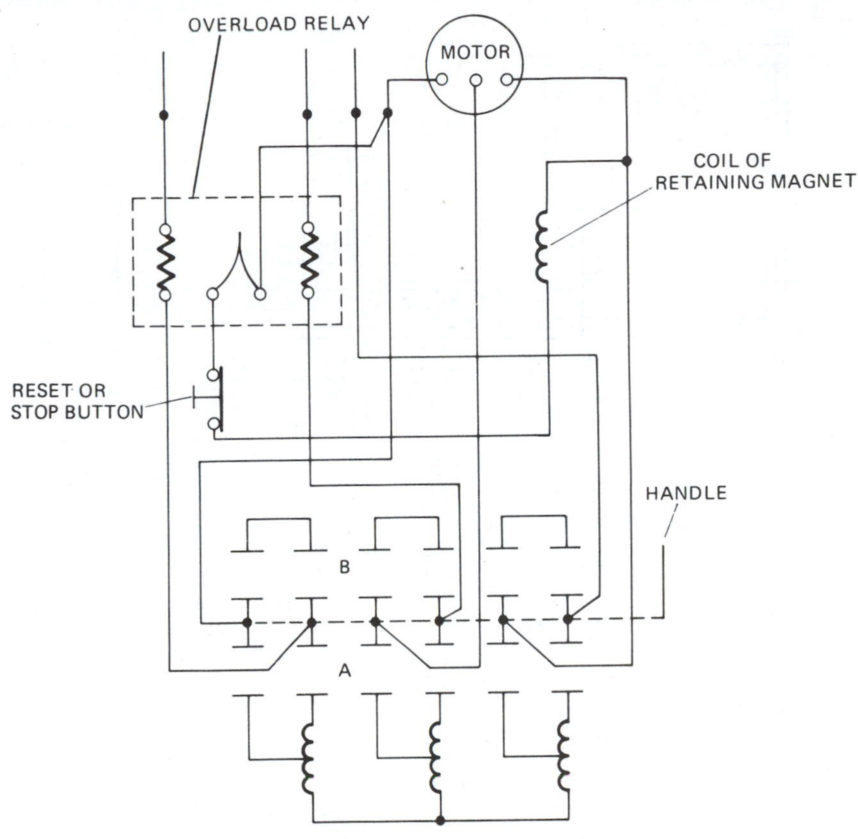

Figure 19-19B
Schematic diagram of an induction motor connected to a three-phase starting compensator

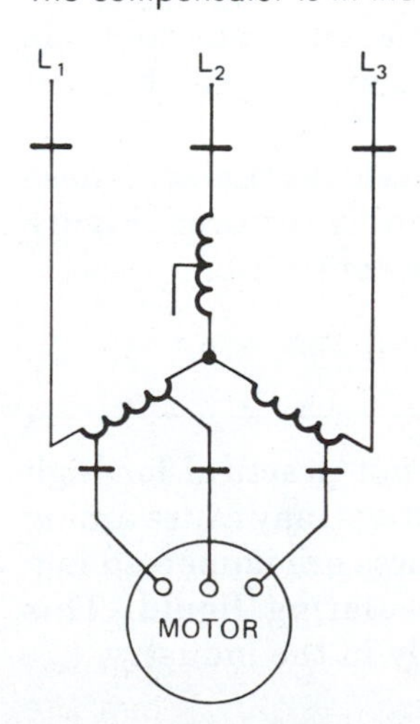

Figure 19-19C
Induction motor connected to a three-phase starting compensator. The compensator is in the start position.

handle is moved to the start position, contacts A close to form the circuit illustrated in Figure 19-19C. The autotransformer is connected across the line and supplies a reduced voltage to the motor. When the rotor is accelerated to a constant speed, the handle is moved to the run position. This opens contacts A and closes contacts B, Figure 19-19B, connecting the motor directly across the line. The thermal overload relays are now in the circuit.

Undervoltage protection is provided by a retaining magnet, which holds contacts B closed. The coil of this electromagnet is connected across the line, through the overload contacts and the stop contacts. The stop button serves to stop the motor and to reset the overload contacts.

A mechanism is provided that permits the handle to be operated only if it is first moved to the start position. Then, when the rotor has accelerated to the proper speed, the handle is quickly

Figure 19-20
Reduced-voltage autotransformer-type starter, closed transition starting

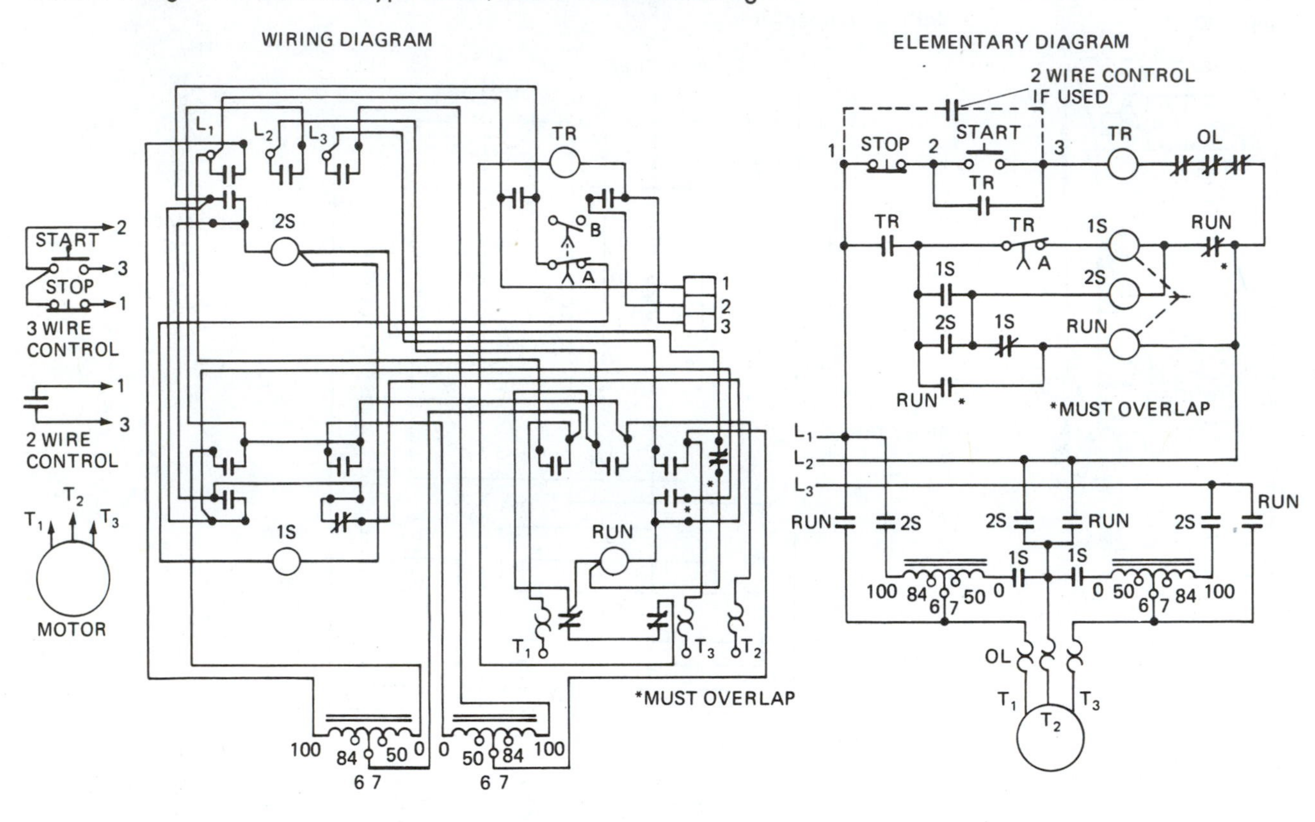

moved to the run position. The autotransformer generally has several taps so that various starting voltages may be obtained according to the specific installation.

Automatic starting compensators are also available. These compensators are used where remote control is required. Figure 19-20 is a schematic diagram of this type of controller.

OIL SWITCHES AND CONTACTORS

Ordinary contacts operating in air are not practical for high voltages (more than 600 volts). The high voltage may cause arcing across the contacts. To avoid this, the contacts are immersed in a special oil or another nonflammable insulating liquid. This method of arc quenching is used extensively in the industry.

MULTISPEED CONTROL

Multispeed control of squirrel-cage motors can be obtained through push-button stations and magnetic relays. The controller is designed to change the external connections of the motor to obtain the various speeds. Figure 19-21 shows the connections for a two-speed controller, which may be used with constant-horsepower motors.

A mechanical interlock is provided between the two sets of contacts so that the high-speed and low-speed coils cannot be energized at the same time. If this should happen, a short circuit would be established between the three supply conductors.

With certain installations, excessive stress is imposed on the driven equipment when starting the motor at high speed. Also, the starting current may be large and long enough to operate the

Figure 19-21
Connections for a two-speed controller controlling a constant-horsepower, squirrel-cage induction motor

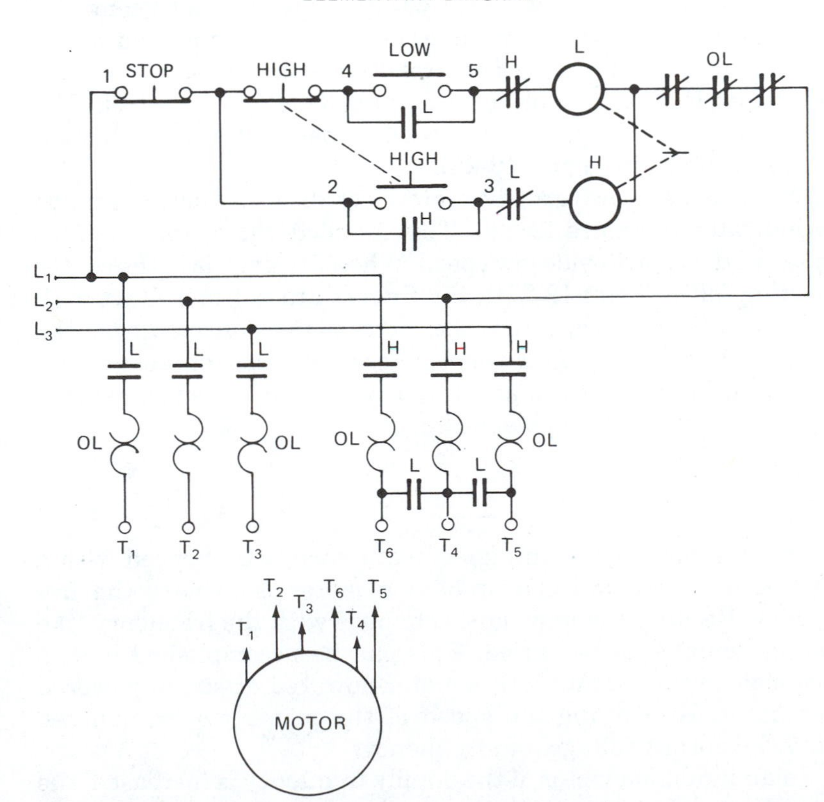

Figure 19-22A
Small drum controller capable of providing connections for two-speed control

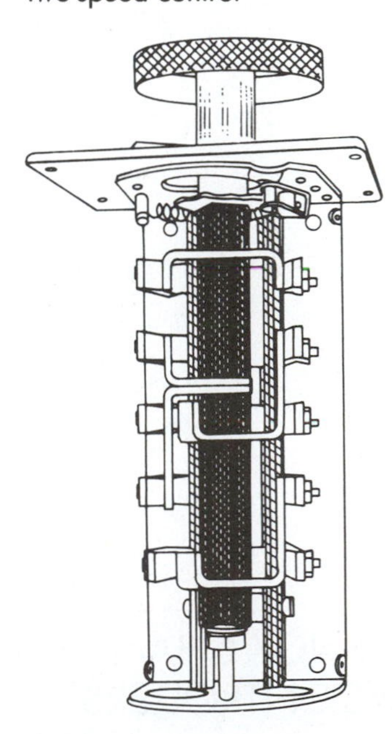

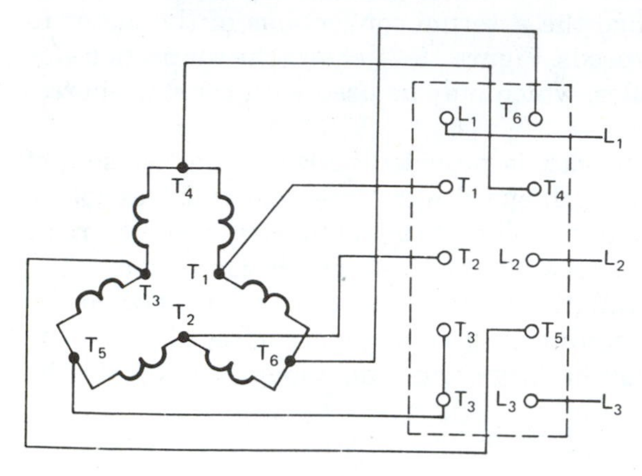

Figure 19-22B
Connections for motor and controller of a two-speed, constant-torque induction motor

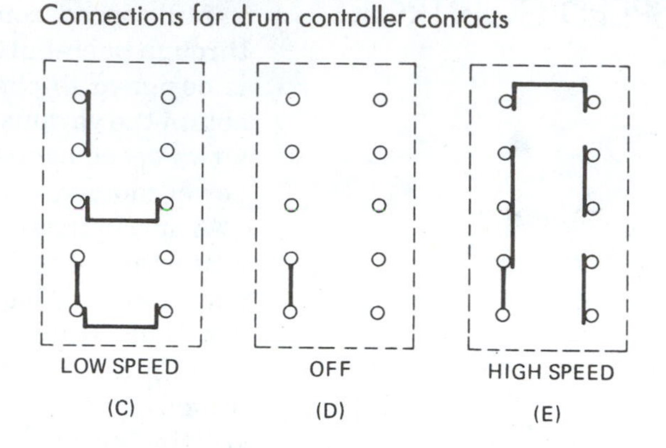

Figure 19-22C, D and E
Connections for drum controller contacts

overload relay. This condition can be avoided by starting the motor at low speed. To ensure that the motor is started at low speed, an auxiliary relay is incorporated into the speed control.

Drum controllers are sometimes used with medium- and small-sized motors to obtain speed control. Figure 19-22A shows a drum controller that is capable of providing two-speed control. The schematic diagram for a two-speed, constant-torque induction motor is shown in Figure 19-22B.

When the knob is turned counterclockwise, contacts are made as indicated in Figure 19-22C. This connects the motor windings in series delta to provide low speed. When the knob is in the center position (off), Figure 19-22D, the fingers are not in contact with the segments. Consequently, the motor is disconnected from the supply. When the knob is turned clockwise, the contacts are as shown in Figure 19-22E, and the motor is connected in parallel wye, which provides high speed.

FREQUENCY CHANGE SPEED CONTROLLERS

One method for obtaining a smooth variation of speed with a three-phase, squirrel-cage induction motor is to vary the frequency. Because the impedance changes with the frequency, the voltage must also be varied. This can be accomplished with a motor-generator set or with a water-powered or steam-powered generator. By varying the speed of the prime mover, one can vary the output voltage and frequency.

In an induction motor, if the supply frequency is increased, the impedance of the motor will also increase. When the supply frequency is decreased, the impedance of the motor decreases. This condition lends itself well to the *variable speed alternator*. When the speed of the alternator is increased, both the frequency and the voltage increase and vice versa. If the voltage/frequency curve does not match the impedance/frequency curve of the motor, supplementary voltage control can be used.

Another method used for varying the frequency is the *electronic control system*. This system takes up much less space and is less expensive to operate than the variable speed alternator. It usually consists of a silicon-controlled rectifier and an inverter. The 60-hertz ac is converted to dc through the silicon-controlled rectifier. The dc is then changed into a variable frequency ac by the inverter.

To vary the speed, the input frequency to the motor is varied. To maintain the proper value of current with changes in frequency, it is necessary to change the voltage with each change in frequency. The silicon-controlled rectifier speed controller will provide the necessary frequency-voltage variations. This will be discussed further in Chapters 23 and 24.

REVIEW

A. Multiple choice.
Select the best answer.

1. An induction motor that is started directly across the line will develop
   a. a higher starting torque than if started at a reduced voltage.
   b. a lower starting torque than if started at a reduced voltage.
   c. the same starting torque as if started at a reduced voltage.
   d. a pulsating starting torque.
2. Motors that are started at full voltage generally employ a control known as
   a. an across-the-line starter.
   b. a full-voltage starter.
   c. a reduced-voltage starter.
   d. a resistance starter.
3. Overload protection for motors is generally achieved by connecting
   a. small electric heaters in series with the motor.
   b. a reset button in series with the motor.
   c. indicating lights in series with the motor.
   d. auxiliary contacts in series with the motor.

REVIEW *(continued)*

A. Multiple choice. Select the best answer (continued).

4. The start button on a momentary push-button station completes the circuit to the main control coil
   a. only while the button is being pressed.
   b. continually once the button has been pressed.
   c. only after the button has been released.
   d. none of the above.

5. The stop button on a momentary push-button station completes the circuit to the main control coil
   a. only while the button is being pressed.
   b. continually once the button has been pressed.
   c. only when the button is not being pressed.
   d. none of the above.

6. Low-voltage protection refers to a means of disconnecting the load from the supply when the
   a. voltage drops below a specific value.
   b. circuit breakers trip.
   c. power for the entire system is lost.
   d. all of the above.

7. Relay chatter refers to the noise made when the
   a. contacts of the electromagnetic relay close.
   b. contacts of the electromagnetic relay open.
   c. armature of an electromagnetic relay vibrates.
   d. coil of an electromagnetic relay vibrates.

8. Overload relays protect the motor against
   a. short circuits.
   b. excessive momentary surges.
   c. grounds.
   d. high voltage.

9. A manual starter is operated by hand
   a. from a remote location.
   b. at the starter location.
   c. at a central location.
   d. from many locations.

10. The contacts in motor starting equipment should be cleaned periodically with
    a. carbon tetrachloride.
    b. acid.
    c. a nontoxic cleaning agent.
    d. soap and water.

REVIEW *(continued)*

A. Multiple choice.
Select the best answer (continued).

11. Silver oxide is a
    a. film that forms on electrical contacts.
    b. cleaning agent for electric contacts.
    c. material used in the manufacture of electrical contacts.
    d. paint put on electrical contacts.
12. A double-break contact has
    a. twice the interrupting capacity of a single-break contact.
    b. one half of the interrupting capacity of a single-break contact.
    c. three times the interrupting capacity of a single-break contact.
    d. the same interrupting capacity as a single-break contact.
13. Jogging refers to
    a. a motor that is not capable of developing a smooth torque.
    b. a method used to stop a motor for exact positioning of components.
    c. a motor that starts and stops periodically.
    d. none of the above.
14. An interlocking arrangement on a forward/reverse controller
    a. prevents short circuits.
    b. ensures proper operating sequence.
    c. provides overload protection.
    d. prevents the motor from overheating.
15. Dynamic braking on three-phase induction motors is accomplished
    a. in the same manner as on dc motors.
    b. by energizing the rotor with dc.
    c. by energizing the stator with dc.
    d. by reversing the phase sequence.
16. A friction brake has
    a. a hand-operated release.
    b. a hydraulic release.
    c. an electromagnetic release.
    d. a pneumatic release.
17. Plugging on a three-phase induction motor is acco
    a. energizing the stator with dc.
    b. reversing the phase sequence to the stator.
    c. reversing the phase sequence to the rotor.
    d. energizing the rotor with dc.

REVIEW *(continued)*

A. Multiple choice.
Select the best answer (continued).

18. A plugging relay is operated from
    a. an electromagnet.
    b. a push-button station.
    c. the rotation of the rotor.
    d. the current in the stator.

19. Dynamic braking on a three-phase synchronous motor is accomplished
    a. in a manner similar to that of a dc motor.
    b. by energizing the rotor with dc.
    c. by energizing the stator with dc.
    d. by changing the phase sequence.

20. A wye-delta controller is used for starting three-phase induction motors. When in the start position, the motor windings are connected in
    a. wye.
    b. delta.
    c. open delta.
    d. open wye.

21. A primary resistance starter provides
    a. sudden spurts of torque as the motor accelerates.
    b. smooth acceleration to the rated speed.
    c. sudden spurts of speed as the motor accelerates.
    d. rough acceleration but good speed control.

22. The faceplate controller is used in conjunction with a
    a. three-phase, wound-rotor motor.
    b. single-phase series motor.
    c. three-phase, squirrel-cage induction motor.
    d. three-phase synchronous motor.

23. A starting compensator uses
    a. resistance to reduce the starting voltage.
    b. a voltage divider to reduce the starting voltage.
    c. an autotransformer to reduce the starting voltage.
    d. a standard two-winding transformer to reduce the starting voltage.

24. Oil switches and contactors are frequently used when the voltage exceeds
    a. 300 volts.
    b. 600 volts.
    c. 1000 volts.
    d. 1500 volts.

REVIEW *(continued)*

A. Multiple choice.
Select the best answer (continued).

25. A smooth variation of speed can be obtained with a three-phase, squirrel-cage induction motor by changing the
    a. number of poles on the stator
    b. frequency of the supply voltage.
    c. value of the supply voltage.
    d. value of the rotor voltage.

B. Give complete answers.

1. What is a motor controller?
2. Describe an across-the-line starter.
3. What is the difference between a manual across-the-line starter and a remotely operated across-the-line starter?
4. List five factors that must be considered when selecting motor starting equipment.
5. Describe a momentary push-button station.
6. What safety precaution is provided by a momentary push-button station that cannot be provided with a single-pole switch?
7. What two circuits make up a typical installation of a momentary push-button station, an across-the-line starter, and an ac motor?
8. Describe a method that is commonly used to achieve overload protection when using an across-the-line starter.
9. Describe two types of thermal overload relays used in conjunction with motor starting equipment.
10. What is the purpose of low-voltage protection in motor circuits?
11. List two types of low-voltage protection, and describe the difference between the two.
12. What causes relay chatter in a motor starter, and how can it be overcome?
13. Describe the characteristics of overload relays used in conjunction with motor starting equipment.
14. Describe the effects of dust and corrosion on electrical contacts.
15. Describe how control equipment and contacts are cleaned.

REVIEW *(continued)*

B. Give complete answers (continued).

16. What type of cleaning agent should not be used to clean contacts?
17. Describe how to care for pitted electrical contacts.
18. Describe how double-break contacts are connected for the following:
    a. high-current and low-voltage circuits
    b. low-current and high-voltage circuits
19. What is the purpose of a jogging control? How is jogging accomplished?
20. Describe the construction and operation of a reversing controller.
21. What is the purpose of interlocks on a forward/reverse controller?
22. How is dynamic braking accomplished with a three-phase induction motor?
23. Describe the operation of a friction brake.
24. Define the term *plugging,* and explain how it is accomplished with a three-phase induction motor.
25. Explain how dynamic braking is accomplished with a three-phase synchronous motor.

REVIEW *(continued)*

C. Complete the diagrams.

1. Across-the-line starter with a single-pole switch controlling a three-phase induction motor

L1
L2
L3
M
SP SWITCH
MOTOR

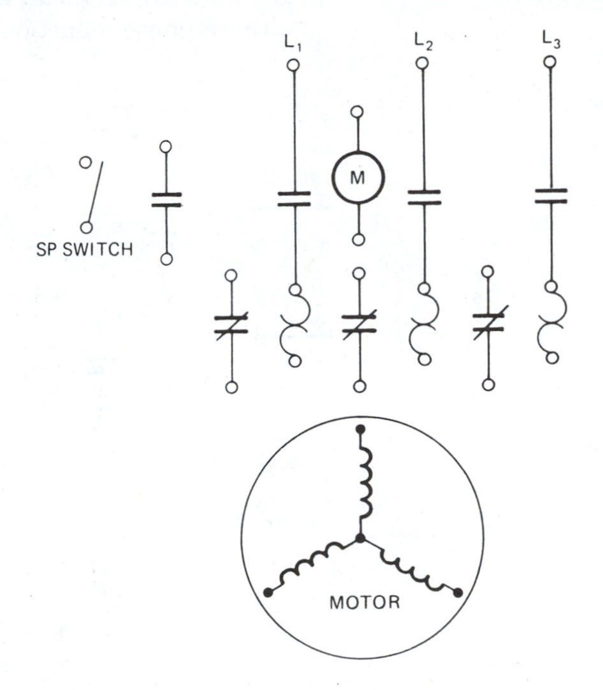

2. Across-the-line starter with a stop-start station controlling a three-phase induction motor

L1
L2
L3
AUXILIARY CONTACTS
START
STOP
M
MOTOR

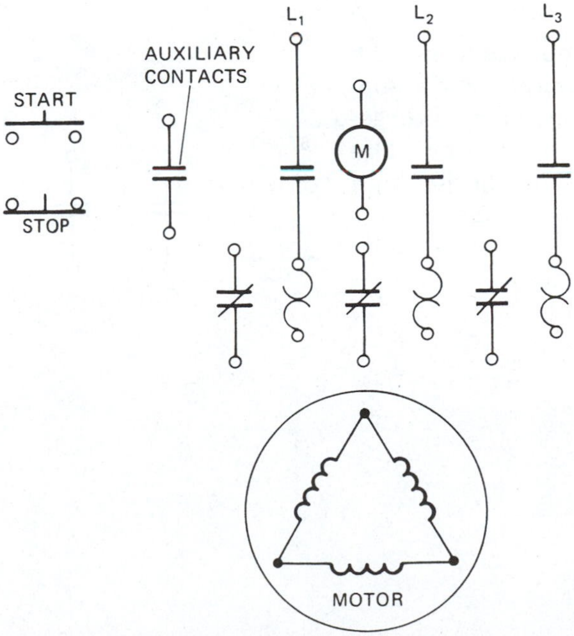

REVIEW *(continued)*

C. Complete the diagrams (continued).

3. Across-the-line starter with a stop-start-jog station controlling a three-phase induction motor

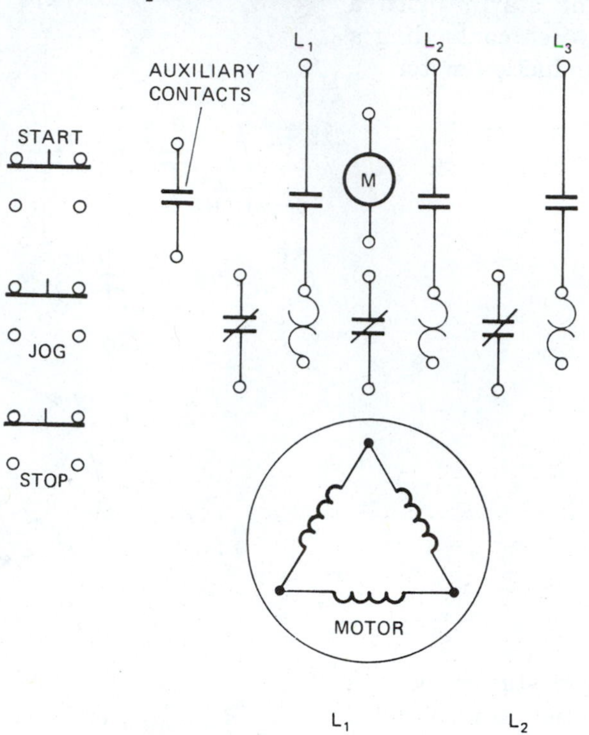

4. Across-the-line starter with double-break contacts and a stop-start station controlling a three-phase induction motor. The connections should be for high current and low voltage.

L_1 L_2 L_3

START

M

STOP

T_1 T_2 T_3

REVIEW *(continued)*

C. Complete the diagrams (continued).

5. Across-the-line starter with double-break contacts and a stop-start station controlling a three-phase induction motor. The connections should be for low current and high voltage.

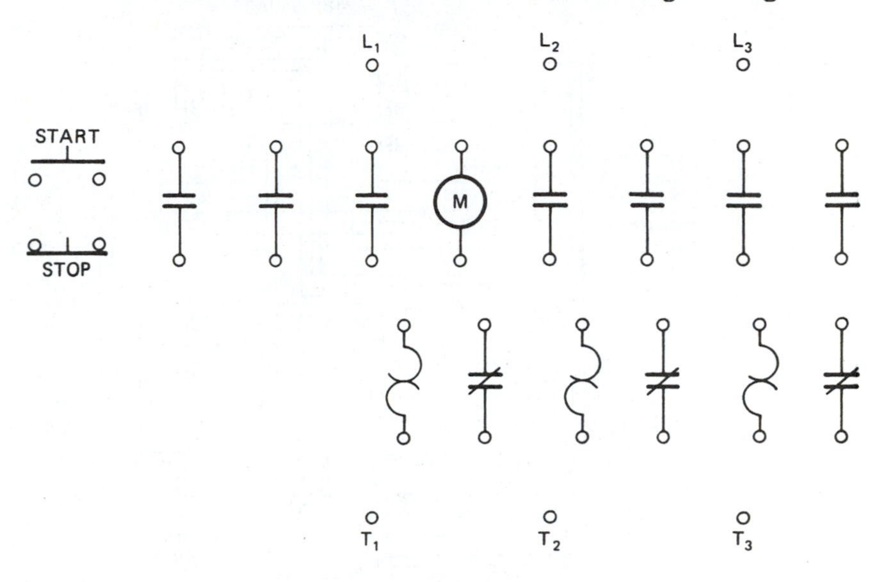

6. Reversing controller with forward-reverse-jog station controlling a three-phase induction motor

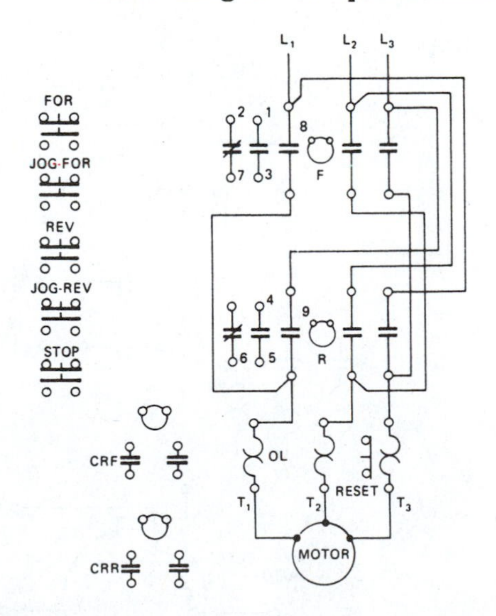

REVIEW *(continued)*

C. Complete the diagrams (continued).

7. Reduced-voltage autotransformer-type starter

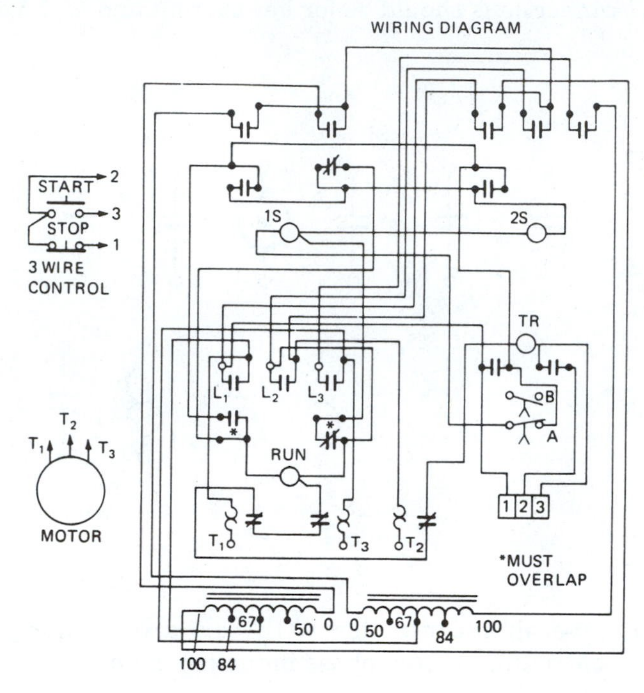

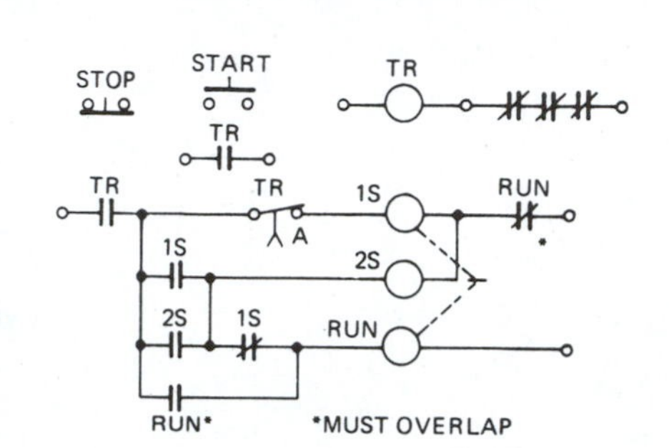

RUN 2S 2S RUN 2S RUN

1S 1S

100 84 67 50 0 0 84 67 50 100

OL OL

T_1 T_2 T_3

MOTOR

8. Wye-delta controller with a stop-start station controlling a three-phase induction motor

L_1 L_2 L_3

TR OL

TR TR TR A

S

1M

S

1M S 2M

S

1M

OL

2M

T_2 T_4

WYE-DELTA MOTOR

T_1 T_5 T_3 T_6

REVIEW *(continued)*

C. Complete the diagrams (continued).

9. Primary resistance starter for starting a three-phase induction motor

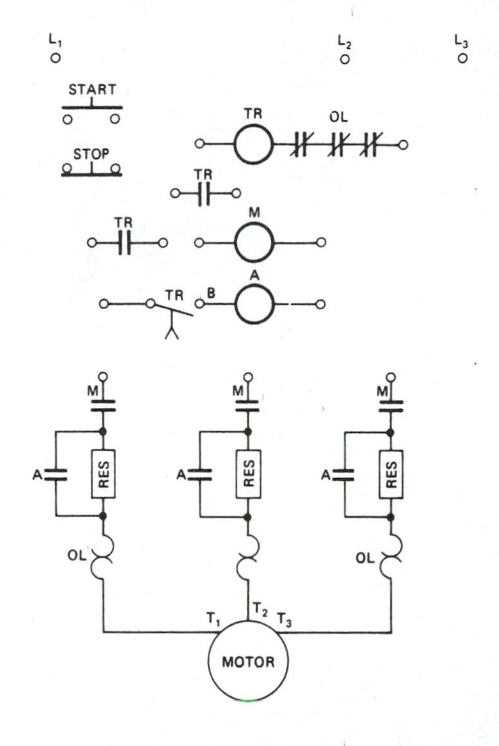

10. Wound-rotor motor starter with three-point acceleration

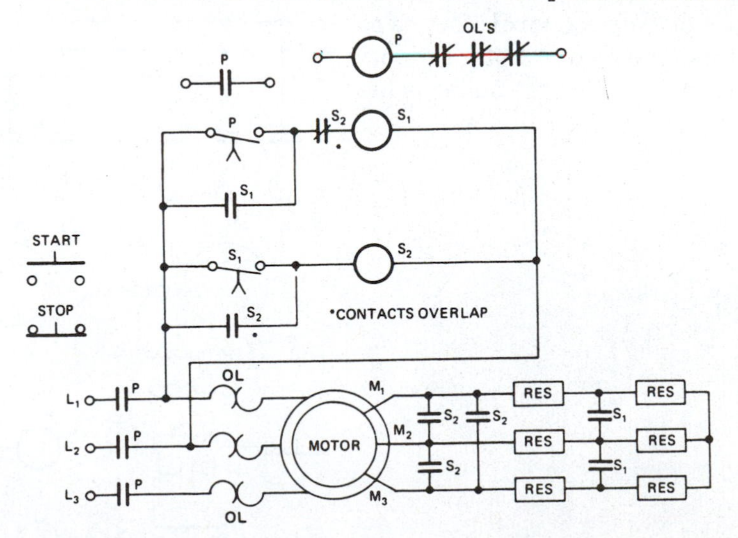

REVIEW *(continued)*

C. Complete the diagrams (continued).

11. Faceplate controller for a wound-rotor induction motor

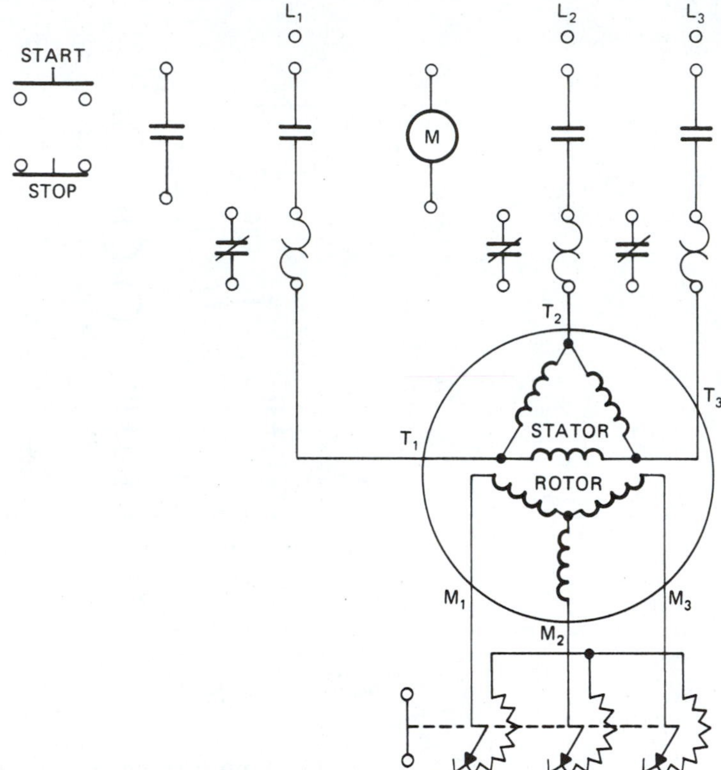

12. Interlocking control for two motors. The second motor cannot start unless the first motor is in operation.

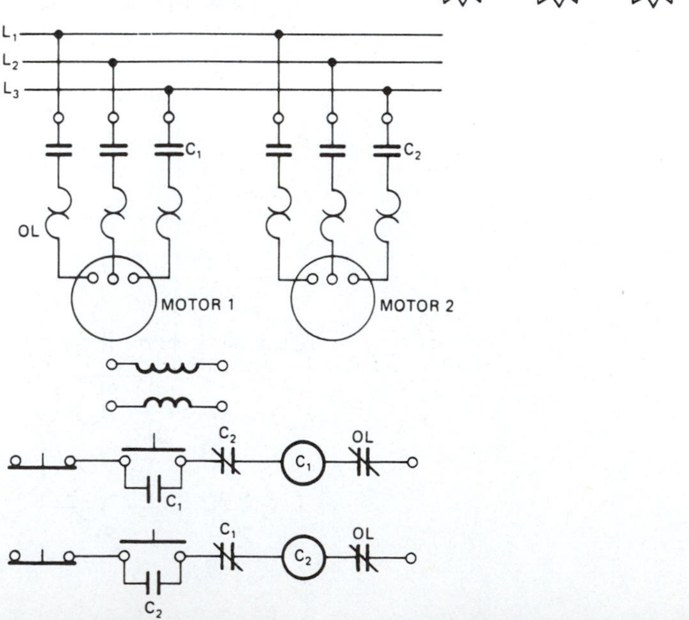

REVIEW *(continued)*

C. Complete the diagrams (continued).

13. Manually operated starting compensator

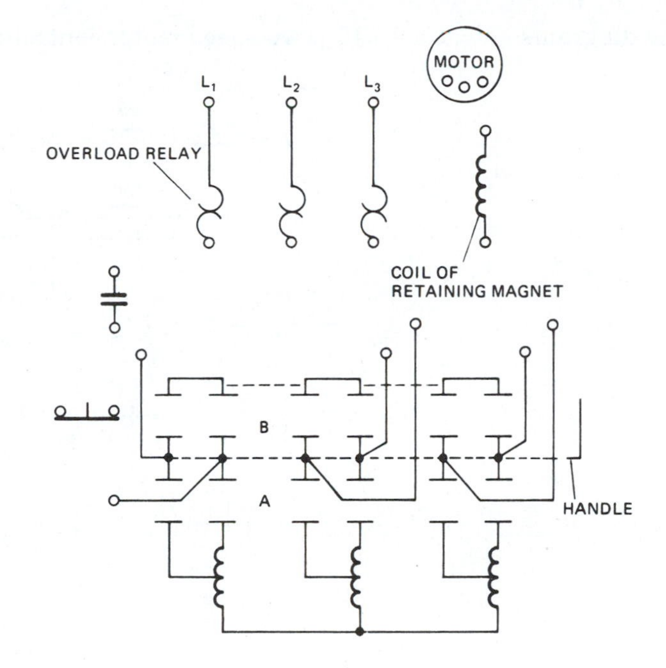

14. Starting compensator with start-stop station connected to a three-phase induction motor

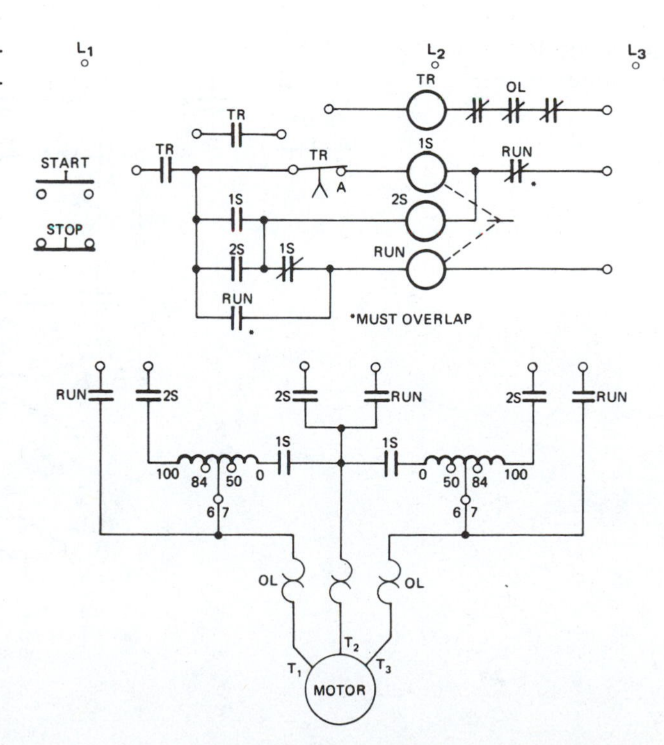

REVIEW *(continued)*

C. Complete the diagrams (continued).

15. Two-speed motor controller

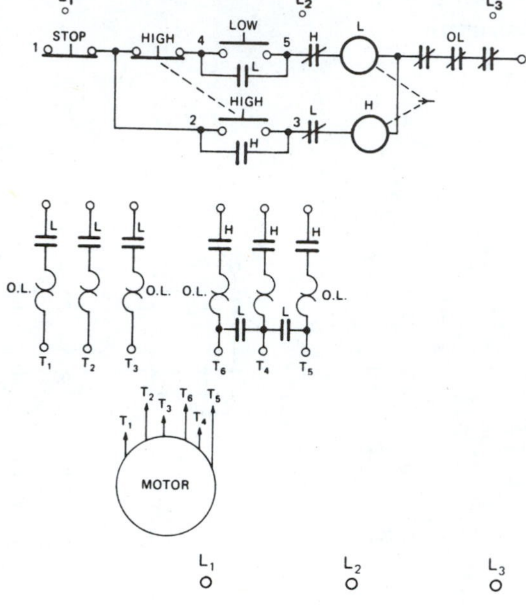

16. Dynamic braking for a three-phase synchronous motor

L1
L2
L3
START
STOP
A
M
M
M
M_R
M_R
T1
T2
T3
DC SUPPLY

REVIEW *(continued)*

C. Complete the diagrams (continued).

17. Plugging controller with a centrifugal relay

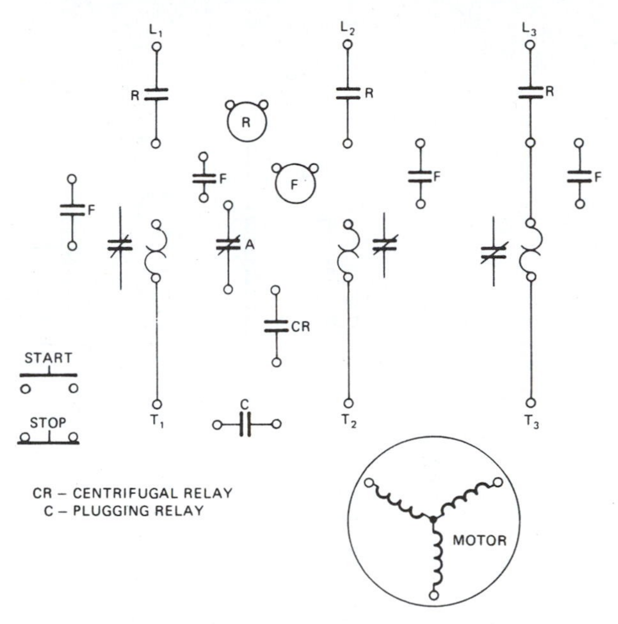

18. Dynamic braking for a three-phase, squirrel-cage induction motor

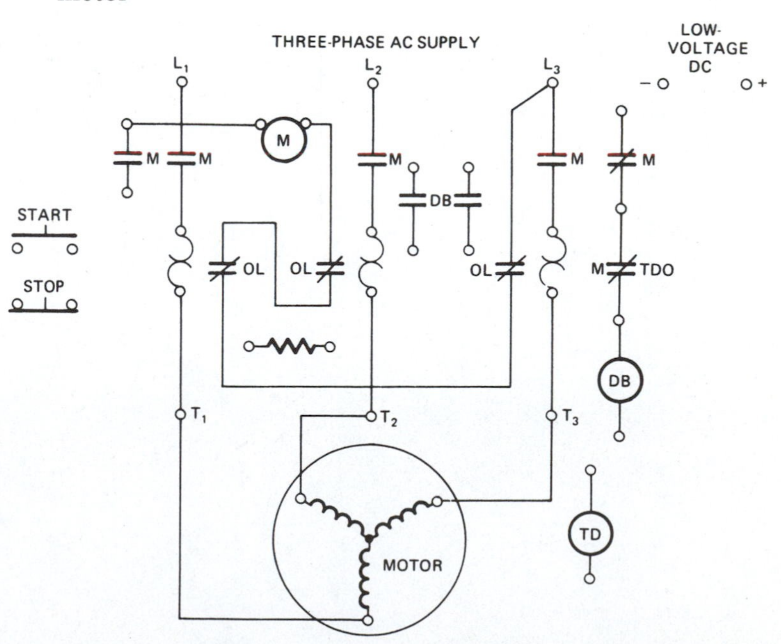

REVIEW *(continued)*

C. Complete the diagrams (continued).

19. Reversing controller with electrical interlocking

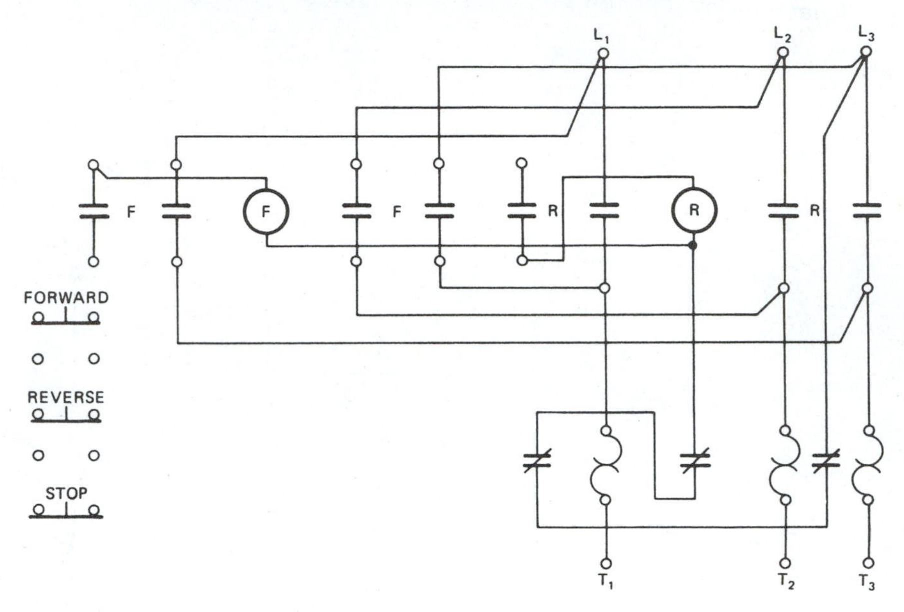

20. Ladder diagram of the control circuit for an across-the-line starter connected to a stop-start station

L_1 L_2

START

STOP

M

M

OL OL OL

20
LIGHTING

Objectives

After studying this chapter, the student will be able to:

- Describe the different units of measurement for light.
- Determine the amount of light required for various areas and types of work.
- Lay out and select the correct lighting fixtures for various areas.

LIGHTING MEASUREMENTS

When electric lighting was first developed, it was necessary to express in a specific unit the amount of light produced. Because the candle was in common use prior to electric lighting, it was selected as a standard reference. Candles differ in the amount of light they emit, and therefore a candle made to specific standards was used.

The unit for light intensity is based on the amount of light emitted from a standard candle in a horizontal direction. This unit is called the *candlepower*. The candlepower of a lamp does not indicate the total light emitted but indicates the intensity in a given direction. The *mean spherical candlepower* is the average of the candlepowers in all directions from the lamp.

The unit for the total light emitted from a lighting source is the *lumen*. This unit can be better understood by using the spherical candlepower as a reference. A light source producing 1 candlepower in all directions is placed at the center of a sphere that has a radius of 1 foot. To prevent light from reflecting inside the sphere, the sphere is blackened on the inside. An opening of 1 square foot is cut into the sphere. The amount of light passing through the opening is equal to 1 lumen (1 lm.), Figure 20-1.

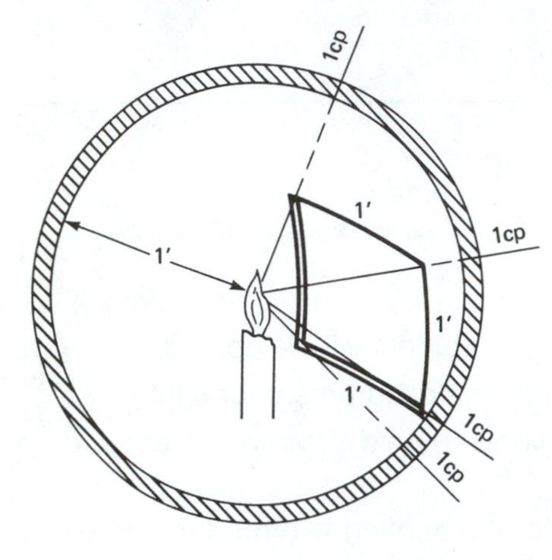

Figure 20-1
Sphere containing a light source of 1 candlepower

One lumen is the amount of light passing through an opening 1 square foot in area and located 1 foot from a light source that emits 1 candlepower in all directions.

The area of a sphere that has a radius of 1 foot is 12.57 (rounded off from 12.5664) square feet. A light source that emits 1 candlepower in all directions, placed in the center of the sphere, emits 1 lumen for each square foot of sphere surface or a total of 12.57 lumens. This relationship between candlepower and lumens can be expressed mathematically as follows:

$$L = 12.57C_m \quad \text{(Eq. 20.1)}$$

where L = lumens
C_m = means spherical candlepower

Candlepower and lumens are units of measurement of light emitted form a source. In the design of lighting systems, the amount of illumination on a given surface is of great importance. This unit of measurement is the *foot-candle*. One foot-candle is the intensity of illumination on a plane 1 foot away from a lighting source of 1 candlepower and at a right angle to the light

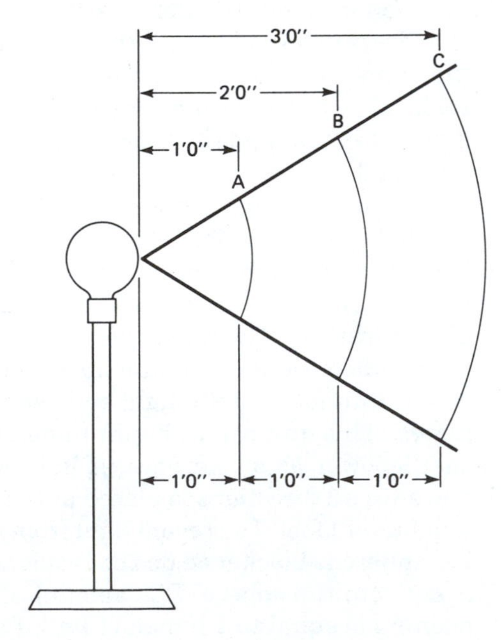

Figure 20-2
Light source of 1 candlepower

rays from the source. Figure 20-2 illustrates a light source of 1 candlepower illuminating every point on the surface A to an intensity of 1 foot-candle. In this case, 1 lumen of light is striking surface A. *One lumen of light covering an area of 1 square foot provides illumination at an intensity of 1 foot-candle.*

INVERSE SQUARE LAW

In Figure 20-2, if plane A is removed, the light source now illuminates plane B, which is 2 feet away from the source. The same amount of light now covers four times the area, which is 4 square feet. The illumination per square foot on B is now 1/4 the amount of A, or 1/4 foot-candle. Removing B permits plane C to be illuminated. Plane C is 3 feet away from the light source, and thus the same amount of light will spread over an area nine times as large as A and illuminate plane C to 1/9 foot-candle. This illustration shows that *the intensity of illumination from a light source varies inversely with the square of the distance from the source.* Known as the *inverse square law*, it may be expressed mathematically as

$$f_c = \frac{C}{d^2} \qquad \text{(Eq. 20.2)}$$

where f_c = foot-candles
C = candlepower of the source
d = the distance in feet from the light source to a point on the plane that the light strikes.

Example 1

A ceiling fixture mounted 8 feet above a workbench emits light at an intensity of 100 candlepower downward. Calculate the illumination produced on the workbench at a point directly below the fixture.

$$f_c = \frac{C}{d^2}$$

$$f_c = \frac{100}{8^2}$$

f_c = 1.56 foot-candles

Equation 20.2 applies only when the light strikes the surface at a right angle, as in Figure 20-3.

Figure 20-3
Lighting illuminating a work surface

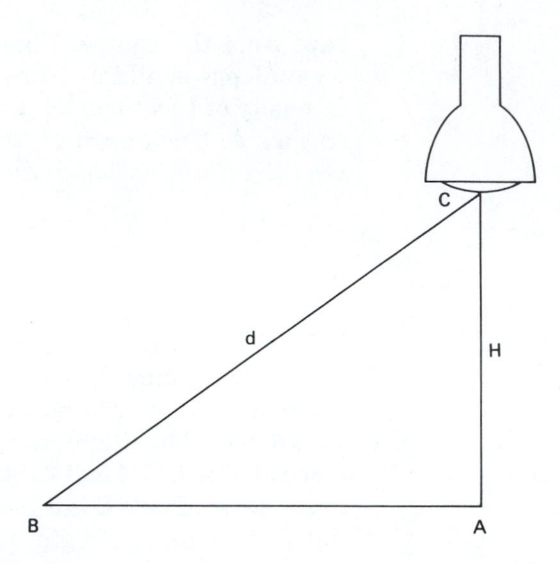

For many installations, the light source does not strike the surface at a right angle. For any other angle, Equation 20.2 must be modified as in Equation 20.3:

$$f_c = \frac{CH}{d^3} \qquad \text{(Eq. 20.3)}$$

where f_c = foot-candles
C = candlepower of the source
H = the distance in feet from the surface being illuminated to the light source.
d = the distance in feet from the point being considered to the light source.

Example 2

What is the intensity of light at a point on a work surface 12 feet from a point directly under a lighting fixture? The fixture is 9 feet above the surface and emits a uniform illumination of 200 candlepower. (See Figure 20-3.)

Distance BC is equal to the square root of the sum of the squares of AB and AC.

$$d = \sqrt{12^2 + 9^2}$$
$$d = \sqrt{225}$$
$$d = 15 \text{ feet}$$

$$f_c = \frac{CH}{d^3}$$

$$f_c = \frac{200 \times 9}{15^3}$$

$f_c = 0.53$ foot-candles

When the light is from long tubes like fluorescent lamps placed end to end, the rays strike at right angles to the surface, producing illumination that varies inversely with the distance from the source.

$$f_c = \frac{C}{d} \qquad \text{(Eq. 20.4)}$$

When the light rays from florescent lamps placed end to end strike the surface at any angle other than 90 degrees, the equation becomes

$$f_c = \frac{CH}{d^2} \qquad \text{(Eq. 20.5)}$$

To determine the relationship between foot-candles and lumens, refer back to Figure 20-2. Surface A, 1 square foot in area, is illuminated to 1 foot-candle. After removing plane A, surface B, which is 4 square feet in area, is illuminated to 1/4 foot-candle. Removing A nd B illuminates surface C, which is 9 square feet in area, to an intensity of 1/9 foot-candle. Each of these surfaces was illuminated by 1 lumen of light. The formula that expresses this relationship is

$$L = f_cA \qquad \text{(Eq. 20.6)}$$

where L = total light in lumens
f_c = foot-candles of light intensity
A = the area of the surface in square feet

Example 3

An area of 50 square feet must be illuminated at an average intensity of 32 foot-candles. What size floodlight is required?

$L = f_cA$
$L = 32 \times 50$
$L = 1600$ lumens

To determine the wattage rating of a 1600-lumen lamp, refer to the manufacturer's chart. Table 20-1 illustrates a typical chart.

TABLE 20-1: Chart indicating approximate lumens *(Courtesy of GTE Sylvania)*

Watts	Bulb	Base	VENDOR ID NO. DCI 046135 UPC 46135 Syl. Item No.	Ordering Abbreviation Except Volts	Volts	Pkg. Qty.	Description	Class and Fil.	Avg. Rated Hours Life	Approx. Lumens & Beam Spread	L.C.L.	M.O.L.
75	R-30	Med.	15095	**75R30/SP/6**	125	6	•Reflector Spot, I.F.Ⓟ 19,32	C, CC-6	2000	1600⟨50°†CP		5⅜
			15041	**75R30/FL**	120	24	•Reflector Flood, I.F.Ⓟ 19,32	C, CC-6	2000	470⟨130°†CP		5⅜
							esp ***WE SUGGEST*** **50ER30 120 Volts, An Energy Saving Product**					
	Reflector		15068	**75R30/FL/RP**	120	6	•Reflector Flood, I.F.Ⓟ 19,32	C, CC-6	2000	470⟨130°†CP		5⅜
			15079	**75R30/FL/6**	120	6	•Reflector Flood, I.F.Ⓟ 19,32	C, CC-6	2000	470⟨130°†CP		5⅜
			15080	**75R30/FL/RP**	125	6	•Reflector Flood, I.F.Ⓟ 19,32	C, CC-6	2000	470⟨130°†CP		5⅜
			15043	**75R30/FL**	125-130	24	•Reflector Flood, I.F.Ⓟ 19,32	C, CC-6	2000	470⟨130°†CP		5⅜
							esp ***WE SUGGEST*** **50ER30 130 Volts, An Energy Saving Product**					
	Reflector		15045	**75R30/FL/SL**	125-130	24	•Rfl. Flood, I.F., Safeline® CoatedⓅ 19,32	C, CC-6	2000			5⅜
			15087	**75R30/A/RP**	115-125	6	•Rfl. Light I.F. AmberⓅ 12,32	C, CC-6	2000			5⅜
			15083	**75R30/B/RP**	115-125	6	•Rfl. Light I.F. BlueⓅ 19,32	C, CC-6	2000			5⅜
			15082	**75R30/G/RP**	115-125	6	•Rfl. Light I.F. GreenⓅ 19,32	C, CC-6	2000			5⅜
			15084	**75R30/PK/RP**	115-125	6	•Rfl. Light I.F. PinkⓅ 19,32	C, CC-6	2000			5⅜
							esp ***WE SUGGEST*** **50ER30/PK 120 Volts, An Energy Saving Product**					

			15085	**75R30/R/RP**	115-125	6	•Rfl. Light I.F. Red (P)[19,32]	C, CC-6	2000			5⅜
			15086	**75R30/Y/RP**	115-125	6	•Rfl. Light I.F. Yellow (P)[19,32]	C, CC-6	2000			5⅜
			15104	**75R30/GRO**	120	6	•Reflector Spot-GRO	C, CC-6	2000			5⅜
	R-40	Med.	15063	**75R/FL**	120	24	•I.F., Reflector Flood[19]	C, CC-6	2000	460⟨120 †CP		6½
			15064	**75R/FL/RP**	120	6	I.F., Reflector Flood[19]	C, CC-16	2000	460⟨120 †CP		6½
			15062	**75R/FL**	125-130	24	•I.F., Reflector Flood[19]	C, CC-6	2000	460⟨120 †CP		6½
76	A-23	Med.	12997	**76A23/47**	125	120	Clear, St. Ltg., Gr. Replace.	C, C-9	3000	920	4⅜	6¹⁄₁₆
80	PAR-38	Med. Skt.	13899	(S) **100/80PAR/FL/SS**	120	15	★Flood, SuperSaver® (I/O)[116]	C, CC-6	2000	816		5⁵⁄₁₆
(esp)			13898	(S) **100/80PAR/SP/SS**	120	15	★Spot, SuperSaver® (I/O)[116]	C, CC-6	2000	816		5⁵⁄₁₆
			13962	(S) **100/80PAR/FL/SS**	130	15	★Flood, SuperSaver® (I/O)[116]	C, CC-6	2000	816		5⁵⁄₁₆
			13961	(S) **100/80PAR/SP/SS**	130	15	★Spot, SuperSaver® (I/O)[116]	C, CC-6	2000	816		5⁵⁄₁₆
85	A-23	Med.	12989	**85A23/48**	120	120	Clear, Street Lighting[21]	C, C-9	1500	1143	4⅜	6¹⁄₁₆
90 (esp)	A-17	Med.	17753	**100A/90/SS**	120	120	•Inside Frost, SuperSaver®	C, CC-8	750	1620	3⅛	4⁵⁄₁₆

(S) Exclusive Sylvania Design.
• Indicates Aluminum Base.
† Field Angle: All Other Beam Angle.

★ Heat Resistant Glass Weather Durable (Hard Glass).
(P) Burn only in porcelain sockets.
(esp) *WE SUGGEST* An Energy Saving Product.

(I/O) PAR lamps are suitable for indoor and outdoor use.
CP CandlePower
Numbered Footnotes are listed on page 88.

LIGHT DISTRIBUTION

Light intensity in various directions around a lamp or fixture is shown by a graph. For the fixture in Figure 20-4, the variations are indicated by the intensity curve. The length of each radial line within the curve indicates the intensity of light in that direction.

The direction of the rays is expressed by means of angles measured from a vertical line down from the source. In Figure 20-4, the intensity of light at 0 degrees is 1450 candlepower. The intensity of light at 15-degree increments is 1400, 1300, 1050, 800, and 600 candlepower.

For a given height, it is possible to determine the foot-candle intensity for various distances from the lighting source.

Example 4

In Figure 20-5, determine the foot-candle intensity of the surface at point A. The distribution curve indicates that the light intensity at point A is 1240 candlepower. The height to the light source is 8 feet. The horizontal distance from the source is 6 feet. See Figure 20-6.

$$d = \sqrt{8^2 + 6^2}$$

$$d = \sqrt{64 + 36}$$

$$d = 10 \text{ feet}$$

$$f_c = \frac{1240 \times 8}{10^3}$$

$$f_c = \frac{9920}{10^3}$$

$$f_c = 9.92 \text{ foot-candles}$$

The light intensity at each of various points along the surface is found in the same manner.

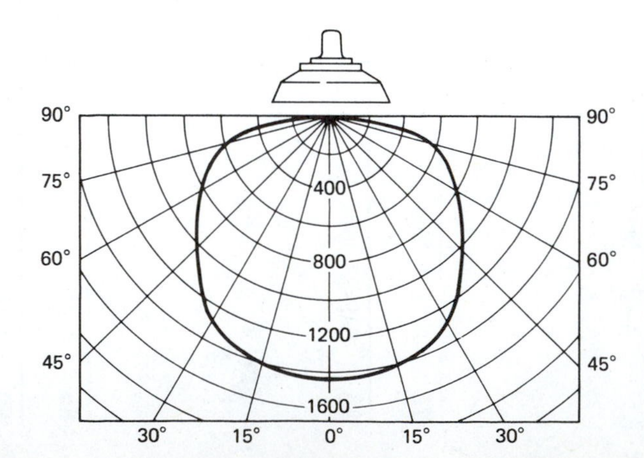

ANGLE	CANDLE POWER
0°	1450
15°	1400
30°	1300
45°	1050
60°	800
75°	600
90°	0

Figure 20-4
Intensity curve showing variation of light from a lighting source

Figure 20-5
Candlepower distribution curve

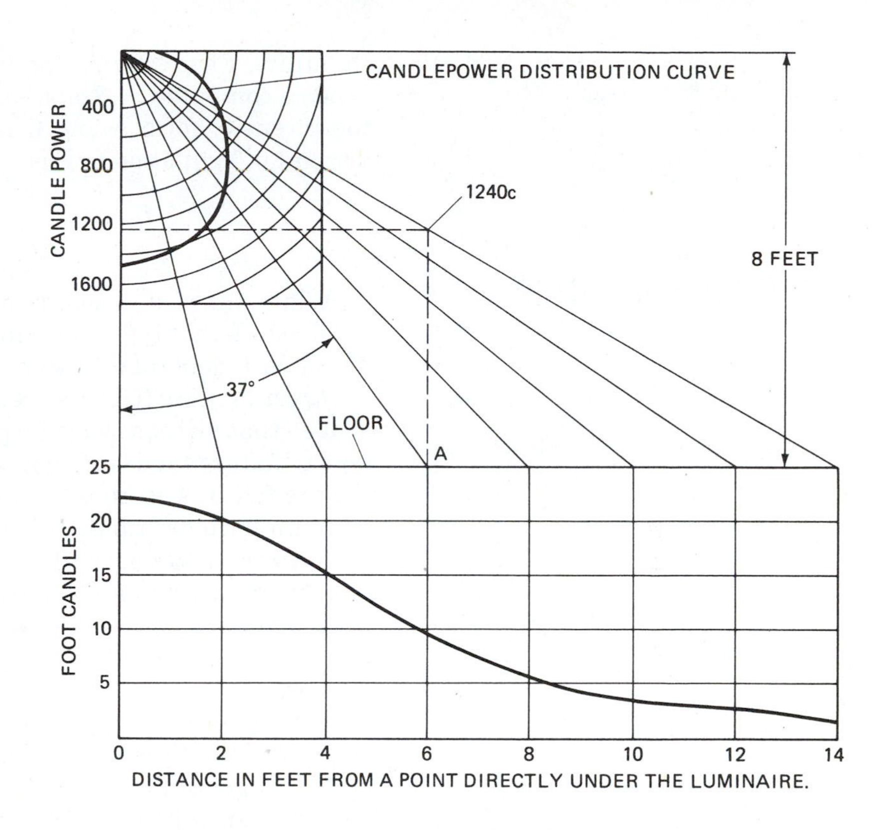

When designing a lighting system, one must consider many lighting sources at specific distances. Figure 20-7 is an example of more than one source illuminating a surface. The amount of illumination at point P is the sum of the light rays coming from all

Figure 20-6
Light source illuminating a surface

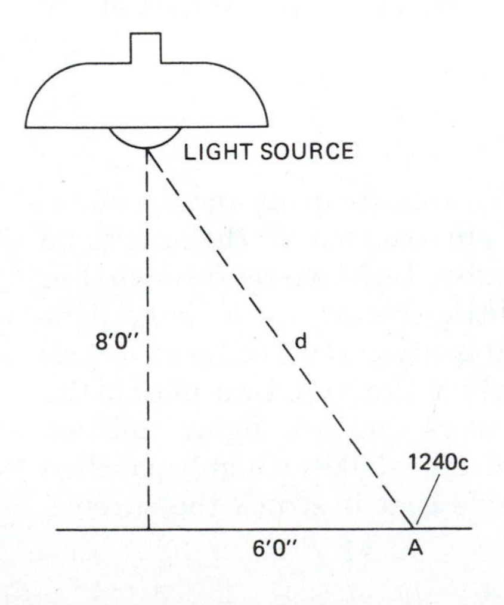

Figure 20-7
More than one lighting source illuminating a surface

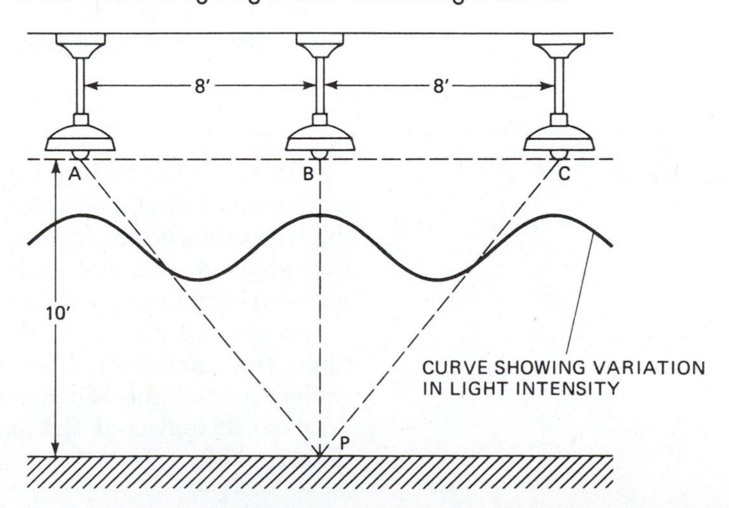

the units, Figure 20-7. The intensity from fixture B is 10 foot-candles computed by Equation 20.2. Equation 20.3 indicates that the intensity form A and C is 4.76 foot-candles each. The total amount of light at point P is 10 + 4.76 + 4.76 = 19.52 foot-candles.

TYPES OF LUMINAIRES

A *luminaire* is a source of electric light. In general, there are five different types of luminaires. They are classified as direct, semidirect, general diffusing, semi-indirect, and indirect.

Direct lighting fixtures are usually equipped with reflectors or shades that direct at least 90 percent of the light downward. This type of fixture tends to produce glare and spotty lighting. Installing the fixtures as high as possible from the surface results in a more uniform distribution of light and less glare.

Fixtures that transmit 60 percent to 90 percent of the light downward are classified as *semidirect*. They usually consist of a lamp or lamps enclosed in a prismatic or opalescent glass. This arrangement conceals the light source, diffuses the light, and reduces glare.

Fixtures classified as *general diffusing* direct 40 percent to 60 percent of the light downward and the rest upward. They often consist of glass diffusing globes or louvered units. Nearly all the light is transmitted up or down and practically none sideways. This arrangement is often used in office areas.

Semi-indirect lighting fixtures transmit 60 percent to 90 percent of the light upward. This arrangement produces very little glare and is more efficient than indirect types because more light is directed downward.

Indirect lighting fixtures transmit 90 percent or more of the light upward. The ceiling serves as a secondary light source by reflecting the light downward. The result is a diffused light of low intensity over a large area.

REFLECTION

Materials from which lighting fixtures are made and the material, smoothness, and color of the surfaces within the area to be lighted all determine the lighting effect. Light passes through thin transparent material with very little scatter. Little or no light passes through opaque materials; it is either absorbed or reflected.

An opaque, highly polished, bright material reflects most of the light that strikes it. Even dark colors that are highly polished reflect much light. When a beam of light strikes a highly polished surface, it leaves at the same angle that it struck the surface, Figure 20-8.

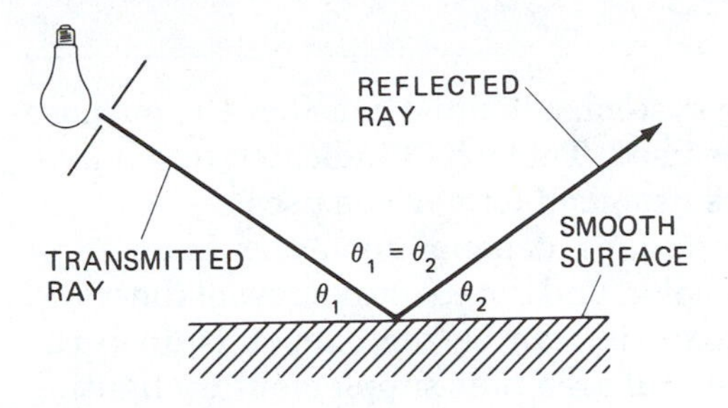

Figure 20-8
Light ray reflected from a smooth, polished surface

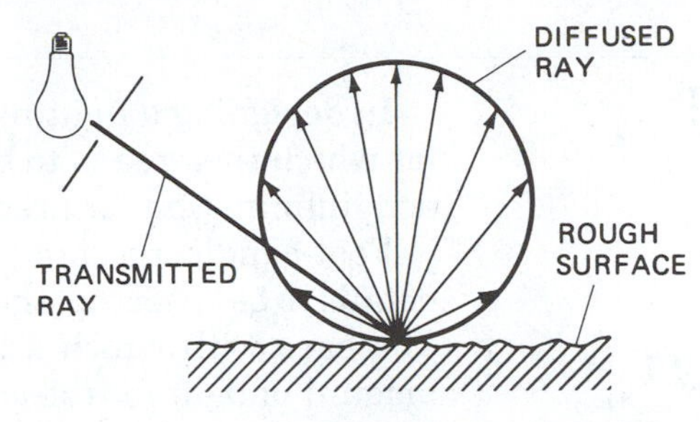

Figure 20-9
Light ray reflected from a rough, opaque surface

A person positioned so that the eyes are in the path of the reflected beam will be subject to glare, causing eye fatigue. Light striking a rough, opaque surface reflects in all directions, Figure 20-9. The same intensity of brightness will be observed from the entire surface, thus reducing glare.

For uniformity of light and reduced glare, ceilings and walls should have a rough surface. The approximate percentages of light reflected by surfaces of different colors are given in Table 20-2.

TABLE 20-2:
Percent of light reflected

Surface	% of Light Reflected	Color	% of Light Reflected	Color	% of Light Reflected
Polished silver	92	Flat white	80	Dark gray	22
Mirrored glass	85	Bright yellow	65	Olive green	14
Polished chromium	65	Light green	62	Black	3

GLARE

When lighting is so bright that it causes eye discomfort, it is said to produce glare. There are many causes of glare, including a light source directly in line of vision, light reflected from highly polished surfaces, and high contrast between the light source and the background.

Brightness depends upon the intensity of the light from the source and the ability of a surface to reflect the light. Too much brightness causes discomfort, and too little causes eye strain.

A clear incandescent lamp emits a large amount of light from a small central point. It appears very bright and produces glare. Most incandescent lamps are frosted on the inside, thus scattering the rays and causing the light to appear distributed throughout the bulb.

Color also has an effect on brightness. If two sheets of paper, one white and one blue, are placed under a light, the white sheet appears brighter because it reflects more light.

LIGHTING LAYOUT

In designing a lighting system, one must consider the purpose for which the area is to be used. Table 20-3 indicates recommended illumination for areas designed for various uses.

Foot-candle recommendations depend upon the fineness of details to be observed, the color, and the color contrast of the work. The ease with which a material can be seen depends upon the amount of light for the general area plus supplementary lighting in specific areas.

Other factors to consider are the coefficient of utilization, the maintenance factor, and color requirements. The lighting fixtures, ceilings, walls, floors, and other objects all absorb light. Thus the light striking the working plane is always less than the amount produced at the source.

The ratio between the light that reaches the working plane and the light produced by the luminaire provides a constant known as the *coefficient of utilization*. As a general rule, the coefficient of utilization for large rooms (20 feet by 20 feet or larger) may be considered 40 percent for indirect lighting, 50 percent for semi-indirect, 60 percent for semidirect, and 70 percent for direct lighting.

Consideration must also be given to the decrease in illumination caused by depreciation of the lamp. In addition, as the reflecting surfaces get dirty, more light is absorbed and less

TABLE 20-3:
Recommended illumination in foot-candles

Area	Foot-candles	Area	Foot-candles
Blue Print Reading	30	Locker Room	5
Cafeterias	15	Offices:	
Corridors, Stairways and Aisles	5	Casual Desk Work	20
Factories:		Close Work	40
Aisle	2	Coarse Work	10
Color Inspection	200	Desk Work	50
Drafting Room	50	Reception Room	10
Extra Fine Work	200	Schools:	
Fine Work	150	Auditorium	10
Medium Work	30	Class Room	20
Rough Work	10	Library	20
Wash Room	10	Sight Saving	50
Hotels:		Stores:	
Bed Rooms	10	General Merchandising	50
Dining Room	5	Show Windows	100
Lobby	10	Stock Rooms	15
Laundry	20	Wall Cases	100

reflected. This absorption of light by dirty reflecting surfaces must be considered. It is accomplished by applying the maintenance factor. *The maintenance factor* is the percent of initial light to be expected with the usual cleaning and/or painting of the luminaires and reflecting surfaces.

Because some luminaires collect dust more readily than others, and some locations are dirtier than others, maintenance factors vary. In general, a maintenance factor of 70 percent can be used for normal conditions; for clean areas, a factor of 80 percent can be used; and for dirty areas, 60 percent should be adequate. For more accurate values of both the coefficient of utilization and the maintenance factor, refer to the manufacturer's reference manual.

To determine the total lumens required for a certain area, one must consider both the coefficient of utilization and the maintenance factor. Equation 20.7 gives the total lumens without consideration to the coefficient of utilization or the maintenance factor. This equation can be revised as

$$L = \frac{f_c A}{K_u M} \qquad \text{(Eq. 20.7)}$$

where L = total lumens
f_c = foot-candles
A = area in square feet
K_u = coefficient of utilization
M = maintenance factor

Example 5

A machine shop that is 1296 square feet in area has a fixture height of 15 feet and is used for medium work. Calculate the number of lumens needed for adequate lighting.

$$L = \frac{f_c A}{K_u M}$$

$$L = \frac{30 \times 1296}{0.70 \times 0.70}$$

$$L = \frac{38880}{0.49}$$

$$L = 79{,}349 \text{ lumens}$$

In the selection of lighting fixtures, certain general factors must be considered. They are:

1. Type of work area to be lighted
2. Required number of lumens
3. Height of fixture from work surface
4. Spacing between fixtures and spacing between fixtures and wall
5. Size of lamps required

COLOR

Color is a very important consideration in the design of lighting systems. Merchandise displayed in a salesroom may appear one color under artificial light and another color under natural light. In most areas, it is important to install a lighting system that will produce the same effect as natural daylight. Special installations may warrant colors that produce unnatural effects.

TYPES OF LIGHTING FIXTURES

In general, there are two categories of lighting fixtures: *incandescent* and *electric discharge.* These two categories have many types.

Incandescent lamps are manufactured in many different styles and colors. These lamps produce light form a filament that is heated to incandescent. The filament is made of a thin tungsten wire that has high resistance. Current flowing through the filament produces heat at a temperature high enough to cause the emission of white light. The white light from the lamp consists of the following colors: red, orange, yellow, green, blue, and violet. It also includes two invisible forms of light, known as *infrared* and *ultraviolet.* Ultraviolet light will not pass through ordinary glass.

White light striking a material is partly reflected, partly absorbed, and partly passed through the material. Colored light is usually produced by passing light through glass that absorbs light rays of all colors except one. If a light of a certain color, say blue, is desired, a source of white light is enclosed in a blue-colored glass that allows only blue rays to pass through. Some lamps are manufactured with a tint of blue in the bulb. As the light is transmitted through the bulb, some of the reds and yellows are absorbed, causing the appearance of a whiter light.

Incandescent lamps are manufactured in many different sizes and shapes. The four general types of screw-base lamps are shown in Figure 20-10. Figure 20-11 illustrates lamps of different shapes and bases. The letters and numbers beneath the lamps refer to the manufacturers' code for indicating the type of lamp.

Figure 20-10
Screw bases for incandescent lamps

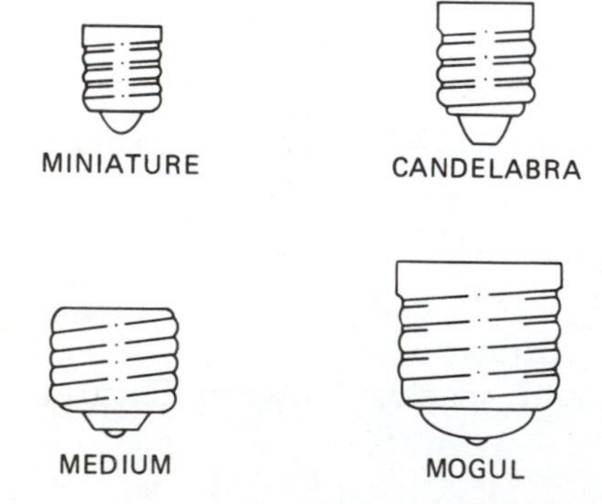

In general, smaller-base lamps have a lower power rating than those with larger bases. The medium-base lamps are manufactured in ratings of 5 watts through 300 watts. Those with lower ratings are made with miniature or candelabra bases. The mogul base is used for lamps of greater than 300 watts rating. Lamps of very high power ratings, 1,500 watts or more, are designed with bases to meet their requirements.

A very special type of incandescent lamp is the quartz lamp. Because the filament enclosure is made of quartz and filled with an iodine vapor, the filament does not deteriorate as rapidly as the one in the ordinary incandescent lamp.

Figure 20-11
Incandescent lamps of different shapes and sizes

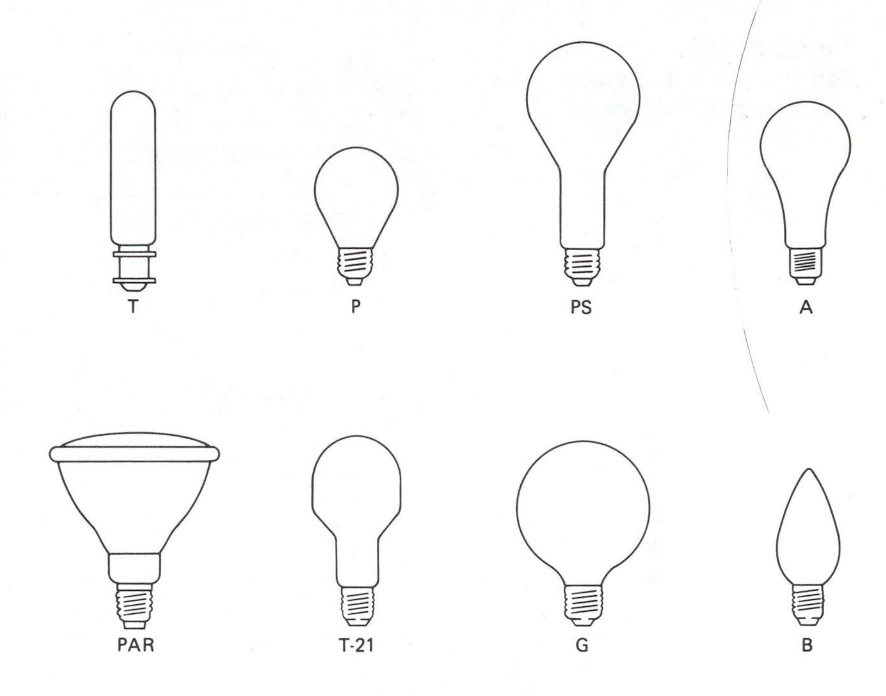

The most common type of electric discharge lamp is the *fluorescent*. This type of lighting is used extensively in commercial and industrial establishments. Its popularity is the result of its efficiency and diversity of colors. The average fluorescent lamp delivers about seven times the number of lumens per watt as the average incandescent.

Fluorescent lamps are manufactured in tubular form. Figure 20-12 illustrates some of the more common styles. The letters and numbers refer to the manufacturers' code. The inside of the glass tube is coated with phosphors. A small amount of argon or krypton gas and a bit of mercury are sealed within the tube. Each end contains an electrode, and under proper conditions, current flows from one electrode to the other through the ionized gas. Coating the inner surface of the glass with different phosphors makes it possible to obtain lamps that emit light of most any desired color.

Figure 20-13 illustrates the basic circuit for a fluorescent lamp. When the on switch is pressed, current flows through the circuit and heats the filaments, causing them to emit electrons into the space around them. When the switch is released, the magnetic field around the induction coil collapses rapidly, inducing a high voltage across the filaments A and B. This emf causes the emitted electrons to flow through the tube, ionizing the gas. The ionized gas becomes a good conductor, and current flow is established. The lamp may be turned off by pressing the off button, which breaks the circuit from the power source.

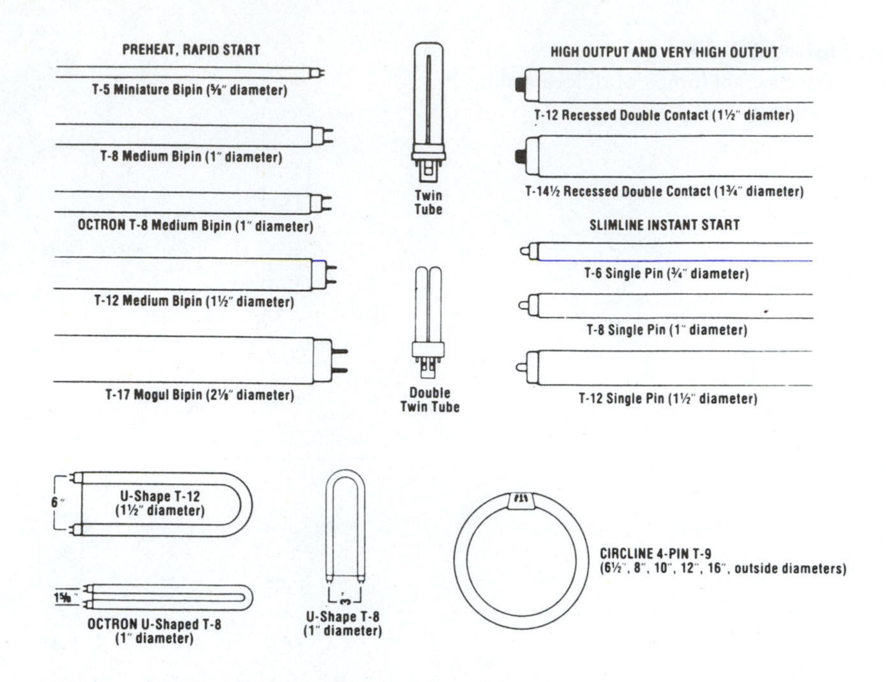

Figure 20-12
Various styles of fluorescent lamps *(Courtesy of GTE Sylvania)*

The operation of a fluorescent fixture depends upon the design. One type of construction is the *preheat*, Figure 20-14. This fixture has a glow switch. This switch, commonly called a starter, is similar to a neon light with a bimetal electrode, Figures 20-14 and 20-15. When a voltage is applied to the circuit, the starter is energized and glows, heating the electrodes. The bimetal electrode bends and makes contact with the other electrode. The circuit is now complete from L_1 through the induction coil, filament A, the starter, and filament B to L_2.

Current flowing through the circuit heats the filaments and causes them to emit electrons. Because the starter contacts are closed, heat is no longer generated by the neon light. The bimetal contacts cool and break the circuit. The magnetic field around the induction coil collapses rapidly, inducing a high voltage across filaments A and B. This high emf establishes current flow through the lamp.

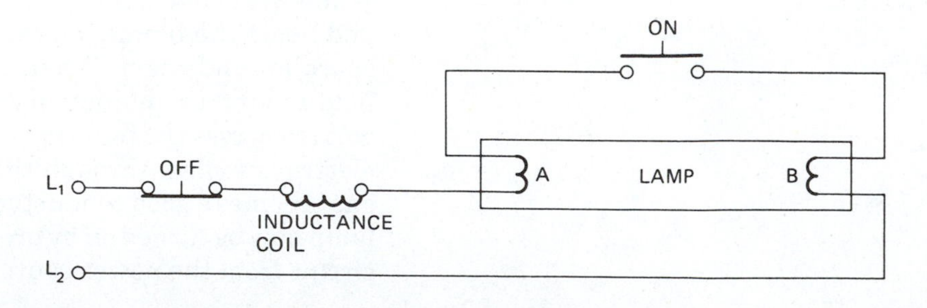

Figure 20-13
Basic circuit for a fluorescent lamp

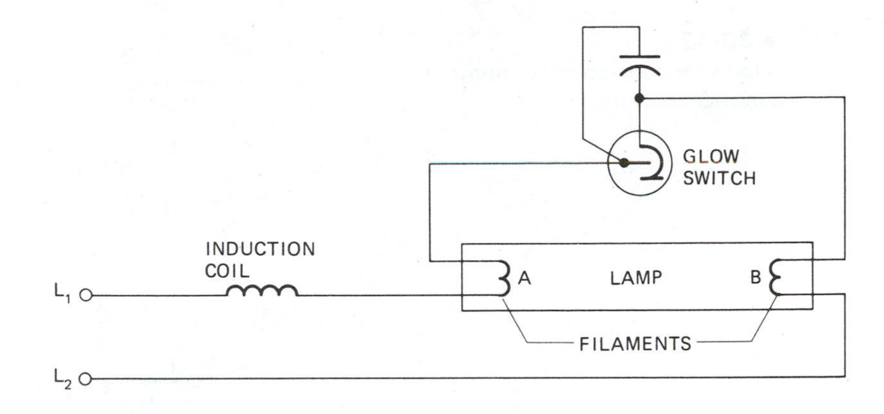

Figure 20-14
Schematic diagram for a preheat fluorescent fixture

Besides producing a high voltage across filaments A and B, the induction coil serves as a current-limiting device. The heat produced by the filaments causes the mercury in the tube to turn to a gas, changing the electrical characteristics within the tube. As current flows from A to B, the resistance within the tube begins to decrease, allowing more current to flow. The greater the current, the greater the decrease in resistance. If no means is provided to limit the value of current, this action will continue until the circuit breaker opens or the tube explodes.

Figure 20-15
Glow switch (starter) for a preheat fluorescent fixture

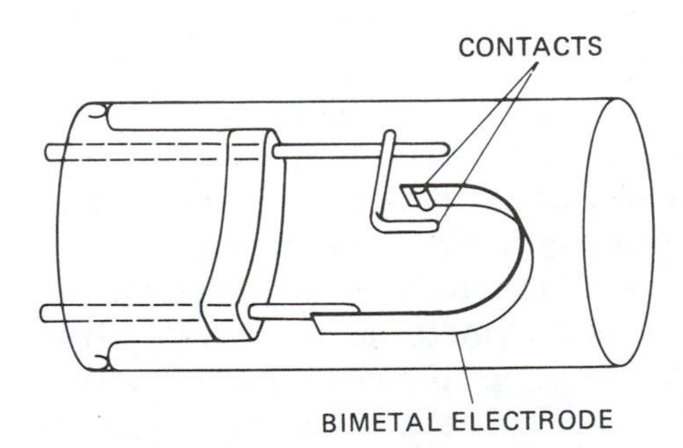

The inductance coil, connected in series with the lamp, provides the impedance necessary to limit the current to a safe value. The major disadvantage of this type of fixture is the time required for preheating. The primary advantage is the initial cost of the fixture and the life span of the lamps compared to other types of fluorescent fixtures.

Two types of fixtures that require less starting time than the preheat are the instant-start and the rapid-start. The *hot-cathode, instant-start, bipin lamp* has two electrodes, Figure 20-16. These electrodes contain filaments that are short-circuited. The lamp is started without preheating the filaments, thus eliminating the need for a starter. After the lamp has started, the voltage supplied to it must be reduced to a value just large enough to maintain a filament temperature sufficient to emit electrons. This reduction in voltage is accomplished by a specially designed transformer. Because of the high voltage available from the transformer, the fixture is wired so that the primary circuit is open when the lamp is removed. Figure 20-17 illustrates the connections for the hot-cathode, instant-start fixture.

Figure 20-16
Electrodes connected to a short-circuited filament in a hot-cathode, instant-start, bipin fluorescent lamp

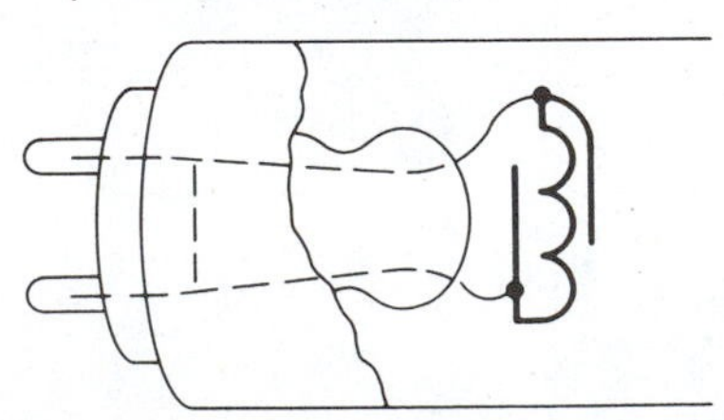

Because of the short-circuited filaments, these lamps cannot be used with ballasts for preheat starting. Conversely, a preheat lamp cannot be used with a ballast intended for an instant-start lamp. If this is attempted, the high-voltage surge would cause the

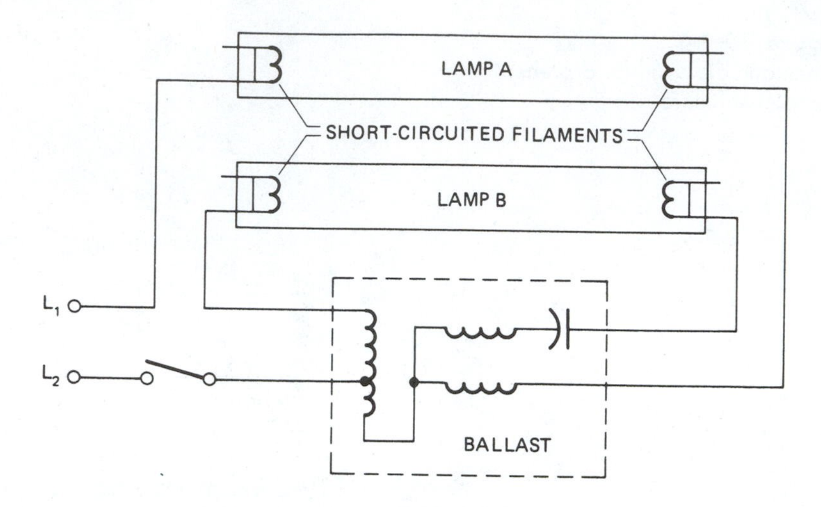

Figure 20-17
Connections for a hot-cathode, instant-start, fluorescent fixture

filaments of the preheat lamps to reach a temperature high enough to cause rapid deterioration of the filaments.

Another type of hot-cathode, instant-start lamp has a single pin and is called *slimline*. It consists of a long, slender tube with only one prong at each end, Figure 20-18. This lamp is manufactured in specific lengths from 42 inches to 96 inches and 3/4 inch or 1 inch in diameter. The slimline lamp operates at a higher voltage and lower current than the preheat lamp. In addition, this fixture is wired so that the primary circuit is open when the lamp is removed.

The *cold-cathode lamp* is also a type of instant-start. In place of the filament, it has large, thimble-type iron cathodes. These cathodes are coated with an active material to provide large emitting areas, Figure 20-19. This design provides longer lamp life than the hot-cathode type and still has all the advantages of instant start. Other advantages are simplicity of design and less flicker. The circuitry for this fixture is the same as for the slimline.

The *rapid-start, hot-cathode lamp* is one of the most popular types of fluorescent lamps in use today. It has longer life than the

Figure 20-18
Tube for a slimline, fluorescent fixture

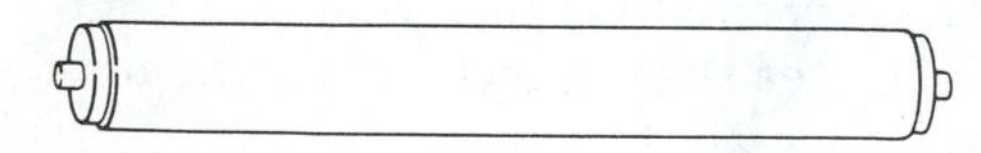

Figure 20-19
Thimble-type iron cathodes used in a cold-cathode fluorescent lamp

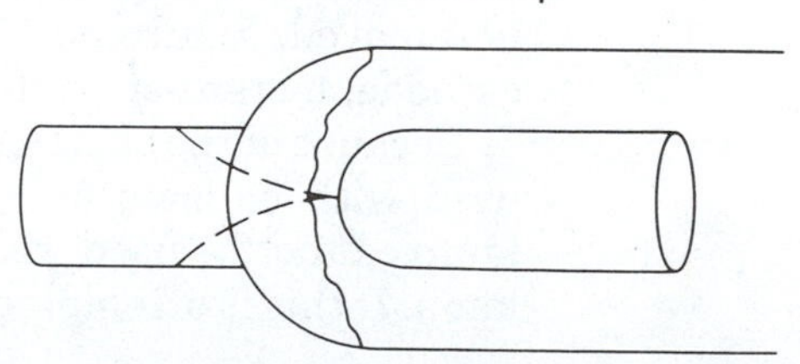

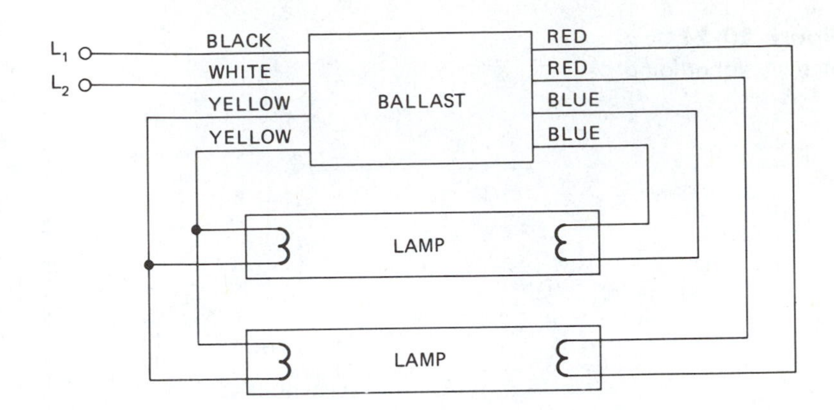

Figure 20-20
Connections for a rapid-start fluorescent fixture

instant start, yet requires minimal starting time. This lamp requires a special ballast and a metal starting aid, which is at ground potential. One popular arrangement is to use the fixture reflector for this purpose.

Rapid-start ballasts are designed to provide smooth, quick starting. The lamp reaches full brightness in about two seconds. The ballast of this fixture must be grounded. This is usually accomplished by contact with the metal of the fixture, which is grounded through the equipment ground.

Proper polarity is very important when connecting a rapid-start fixture. The black wire from the ballast must be connected to the ungrounded (hot) wire from the supply. Figure 20-20 illustrates the connections for a rapid-start fixture.

Another type ballast know as *trigger-start* is designed to operate with regular fluorescent lamps and requires no starter. It contains preheat windings that bring the filaments to temperature within one second. To provide quick starting, the trigger-start ballast must provide a higher open-circuit voltage than the preheat or rapid-start type.

The average fluorescent fixture operates best in ambient temperatures between 65°F and 80°F (18°C and 27°C). At other temperatures, the efficiency and lighting output are reduced.

For outdoor installations, the fixtures must be weatherproof, and the ballast must be designed for operation at ambient temperatures for the area. If temperatures drop below 50°F (10°C) special ballasts must be used. Some fixtures are designed to confine the air around the lamps. When the lamp is in operation, it heats the air. Continuous operation can heat the air to a point of maximum light output.

Fluorescent fixtures are designed to operate on various source voltages. One must always check the input voltage rating to ensure

Figure 20-21
Mercury vapor lamp

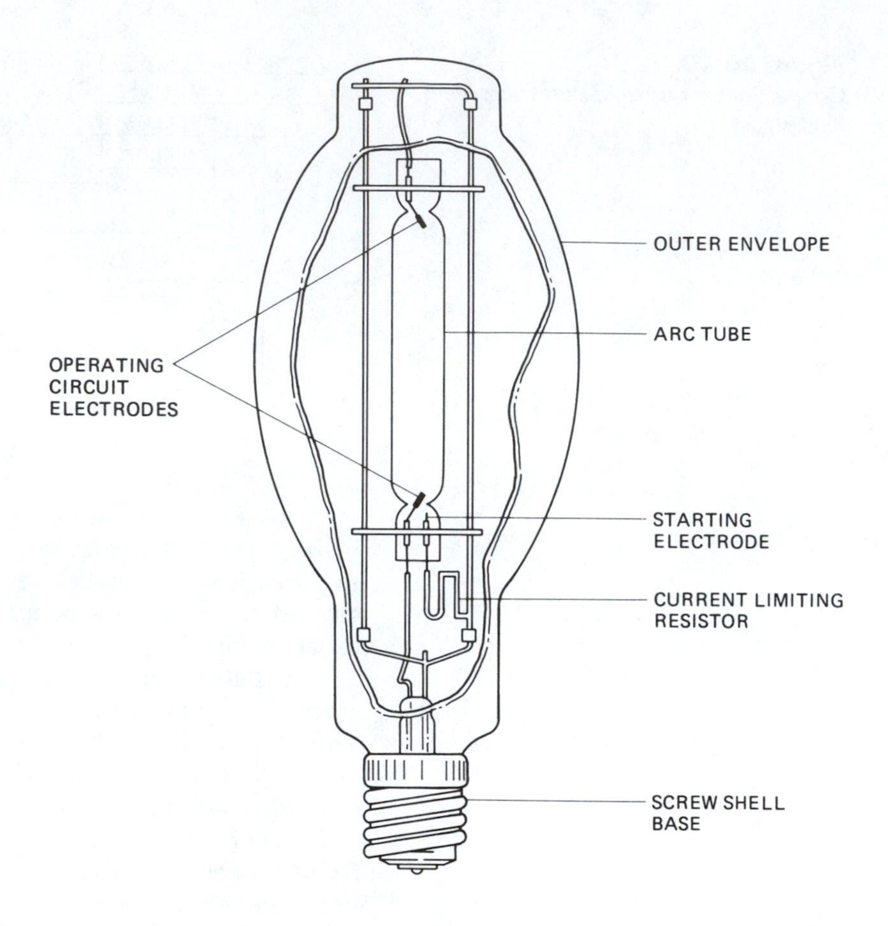

proper operation. Some common voltage ratings are 120 volts and 277 volts.

Another type of electric discharge lighting uses the *mercury vapor lamp,* Figure 20-21. These lamps are available with screw bases in either medium or mogul sizes. They require a special ballast designed to meet the requirements of the lamp.

Mercury vapor lamps are more efficient than incandescent and fluorescent lamps; however, the initial cost is greater. They require longer starting times and close voltage limitations.

LIGHTING MAINTENANCE

As previously stated, the lumen output of a lamp decreases with usage. The amount of depreciation varies with the type of lamp and the conditions of use. The average life of an incandescent lamp is 1,000 hours of usage. The average life of a fluorescent lamp is 7,500 hours of usage. As a lamp approaches its maximum life span, the lumen output decreases rapidly.

Another factor that affects the amount of illumination is the reflecting quality of the fixture, ceilings, walls, and other surfaces. As these surfaces become dirty, their ability to reflect the light rays decreases. The amount of dirt collecting on lamps and globes also reduces the amount of light passing through to the area to be illuminated. All these factors must be considered in a maintenance program.

The illumination should be checked periodically to ensure adequate lighting for the type of use. This can be accomplished by placing a foot-candle meter on the surface under consideration. If the meter indicates a light measurement less than that recommended for the type of use, the condition must be evaluated.

Many manufacturers provide, free of charge, computerized analysis of a lighting system. A computer analysis will not only determine the necessary amount of illumination for a specific work area but will also recommend ways to improve the lighting efficiency.

Computerized instruments measure the amount of illumination, determine if the light is evenly distributed, and make recommendations for improvement. Corrective measures may be to clean the fixtures and the lamps. If the lighting is still inadequate after performing the above, it may be necessary to clean the ceilings, walls, and other surfaces.

A record should be maintained showing when new lamps have been installed and indicating the number of hours the fixtures are in use each day. As the lamps near the end of their life expectancy, they should be replaced before they deteriorate to a point of inadequate lighting.

Other than cleaning and replacing lamps, incandescent fixtures require very little maintenance. Socket contacts sometimes become loose, causing overheating. Replacement of the socket may be necessary if damage has occurred because of overheating.

After lamps have been replaced many times, the center contact of the socket may lose its spring tension. This causes it to lie flat in the bottom of the socket, providing a poor connection or no connection. If the lamp does not complete the connection to the center contact, an open circuit exists and the lamp will not glow. If the lamp makes a poor connection, current flowing through the lamp will cause the socket to overheat.

In the cleaning of fixtures, the sockets should be checked for loose connections. The center contact should be checked and the socket inspected for signs of overheating.

Fluorescent fixtures also require periodic cleaning. Although the sockets on fluorescent fixtures are not of the screw shell type, they also must be checked for loose connections and dirty, loose, or pitted contacts.

Other components of fluorescent fixtures that must be considered are the ballast and the starter. If only the ends of the lamp glow, and the lamp does not blink, the problem is likely to be the starter. Replace it with a new one. A lamp that blinks while the ends glow generally indicates lamp failure. If, however, the lamp is known to be good, then the trouble may be a defective ballast. It is always good practice to ensure that the lamp and starter are good before replacing the ballast.

An arrangement in the maintenance shop whereby lamps, starters, and ballasts can be tested will eliminate a lot of work in the area being illuminated.

Electric discharge lamps, especially the fluorescent type, can be hazardous if mishandled. They are under a partial vacuum and, if dropped or damaged, may explode. Also, the material contained in some lamps is poisonous if inhaled, if ingested, or if it enters the bloodstream through a cut.

Methods of disposing of lamps vary. One method is to place them into the original cartons, to be disposed of with similar hazardous waste.

REVIEW

A. Multiple choice.
Select the best answer.

1. The unit for the total light emitted from a lighting source is the
   a. candlepower.
   b. lumen.
   c. foot-candle.
   d. watt.

2. The amount of illumination on a given surface is measured in units called
   a. candlepower.
   b. lumen.
   c. lunars.
   d. foot-candle.

3. The intensity of illumination from a lighting source
   a. varies inversely with the distance from the source.
   b. varies directly with the square of the distance from the source.
   c. varies inversely with the square of the distance from the source.
   d. varies directly with the distance from the source.

4. A direct lighting fixture directs at least
   a. 90% of the light downward.
   b. 85% of the light downward.
   c. 85% of the light upward.
   d. 80% of the light downward.

REVIEW *(continued)*

A. Multiple choice.
Select the best answer (continued).

5. Fixtures classified as general diffusing direct
   a. 70% to 80% of the light downward and the rest upward.
   b. 60% to 70% of the light downward and the rest upward.
   c. 40% to 60% of the light downward and the rest upward.
   d. 30% to 40% of the light downward and the rest upward.

6. Thin, transparent material allows light to pass through with
   a. very little scatter.
   b. a great amount of scatter.
   c. no scatter.
   d. a moderate amount of scatter.

7. Opaque, highly polished, bright material
   a. absorbs most of the light that strikes it.
   b. passes most of the light that strikes it.
   c. reflects most of the light that strikes it.
   d. defuses most of the light that strikes it.

8. Glare is caused by
   a. poor placement of the light source.
   b. insufficient illumination.
   c. rough surfaces.
   d. dull surfaces.

9. The intensity of illumination needed for a specific area depends upon
   a. the type of work to be performed in the area to be illuminated.
   b. the number of people working in the area.
   c. the type of fixtures to be used.
   d. the height of the lighting fixture.

10. The coefficient of utilization refers to
    a. the ratio of the power input to the power output.
    b. the ratio of the power input to the lighting output.
    c. the ratio between the light reaching the working plane and the light produced by the luminaires.
    d. the ratio of the life expectancy of the lamp to the actual life of the lamp.

REVIEW *(continued)*

A. Multiple choice.
Select the best answer (continued).

11. In a determination of the amount of illumination provided by lighting fixtures, a maintenance factor must be considered because
    a. the lighting output deteriorates with the life of the lamp.
    b. the reflecting surfaces and the lighting fixtures become dirty over a period of time.
    c. the lighting fixtures must be disconnected from the supply when performing maintenance.
    d. the lighting output increases with time.

12. In general, there are two categories of lighting fixtures. They are
    a. resistance an fluorescent.
    b. electric discharge and resistance.
    c. incandescent and electric discharge.
    d. resistance and incandescent.

13. A lamp that produces light from a filament that is heated to a very high temperature is called
    a. fluorescent.
    b. incandescent.
    c. thermal.
    d. resistance.

14. Lamps that are rated greater than 300 watts have a
    a. candelabra base.
    b. mogul base.
    c. medium base.
    d. miniature base.

15. The most common electric discharge lamp is the
    a. incandescent.
    b. thermal.
    c. fluorescent.
    d. resistance.

16. The glow switch is required with
    a. rapid-start fluorescent fixtures.
    b. preheat fluorescent fixtures.
    c. cold-cathode fluorescent fixtures.
    d. all of the above.

17. The cold-cathode fluorescent lamp is a type of
    a. preheat lamp.
    b. rapid-start lamp.
    c. instant-start lamp.
    d. resistance-start lamp.

REVIEW *(continued)*

A. Multiple choice.
Select the best answer (continued).

18. The average fluorescent lamp operates best in ambient temperatures between
    a. 32°F and 65°F.
    b. 45°F and 75°F.
    c. 65°F and 80°F.
    d. 65°C and 80°C.

19. The average life of an incandescent lamp is
    a. 800 hours of usage.
    b. 1000 hours of usage.
    c. 5000 hours of usage.
    d. 9000 hours of usage.

20. The average life of a fluorescent lamp is
    a. 7500 hours of usage.
    b. 8000 hours of usage.
    c. 9500 hours of usage.
    d. 15,000 hours of usage.

21. It is good maintenance practice to check the illumination of work areas on a regularly scheduled basis. This can be accomplished with a
    a. lumen meter.
    b. foot-candle meter.
    c. candlepower meter.
    d. lunar meter.

22. A record should be maintained showing when new lamps have been installed. The purpose of this record is to determine
    a. if the manufacturer's ratings are correct.
    b. when the lamps should be replaced.
    c. how long the lamps last.
    d. the cost of lighting equipment.

23. When performing regular maintenance on lighting fixtures, one should always inspect the
    a. reflectors.
    b. circuit connections.
    c. lamp sockets.
    d. circuit breakers.

24. A fluorescent lamp that blinks while the ends glow usually indicates
    a. a defective starter.
    b. a lamp failure.
    c. a defective ballast.
    d. all of the above.

REVIEW *(continued)*

B. Give complete answers to the following questions.

1. What is the unit of measurement for the total light emitted from a lighting source?
2. Define the term *foot-candle*.
3. Write the inverse square law for light.
4. What does a light intensity curve indicate?
5. List five types of luminaires.
6. Describe a general diffusing fixture.
7. List four factors that affect light distribution.
8. Under what conditions do opaque materials reflect rather than absorb light?
9. Define glare.
10. What factors affect the brightness of illumination?
11. List three factors that must be considered when determining the amount of illumination required for a specific area.
12. Define the coefficient of utilization.
13. What is meant by the term *maintenance factor?*
14. What maintenance factor is usually applied under normal conditions of use?
15. List five factors that should be considered when selecting lighting fixtures.
16. List two general categories of lighting fixtures.
17. Describe how light is produced with an incandescent lamp.
18. List four general types of screw shell bases.
19. What is the advantage of a quartz lamp over an ordinary incandescent lamp?
20. Describe the construction of a fluorescent lamp.
21. How is it possible to obtain different colors of light from fluorescent lamps?
22. Draw a diagram illustrating the basic circuit for a fluorescent fixture.
23. List two functions of the induction coil in a fluorescent fixture.
24. What is the main difference between a preheat fluorescent fixture and a rapid-start fluorescent fixture?

REVIEW *(continued)*

B. Give complete answers to the following questions (continued).

25. List two types of fluorescent fixtures that require less starting time than the preheat type.
26. Can fluorescent lamps designed for use with preheat ballast be used in fixtures designed for instant start?
27. Between what ambient temperatures do fluorescent fixtures provide their best performance?
28. Under what conditions can fluorescent fixtures be installed exposed to the weather?
29. List two general types of electric discharge lamps.
30. What is the average life of an incandescent lamp? A fluorescent lamp?
31. List four factors that affect the amount of illumination in an area.
32. Describe a good maintenance program for lighting systems.
33. List four common causes of improper operation of fluorescent fixtures.

C. Solve each problem, showing the method used to arrive at the solution.

1. A fixture is to be positioned 6 feet directly above a secretary's desk. According to the illumination table (Table 20-2), what is the amount of illumination in foot-candles required on the desk surface? What must be the minimum candlepower of the fixture?
2. A fixture is secured 10 feet above a reading table in a library. If it emits light at an intensity of 1000 candlepower downward, does it provide adequate illumination on the table surface?
3. What is the intensity of light on a workbench if a fixture is 8 feet above the work surface and 6 feet horizontally from the work area and emits a uniform illumination of 1000 candlepower?
4. A room 20 feet long and 15 feet wide must be illuminated to an average intensity of 20 foot-candles. Calculate the number of lumens required.
5. Using Figure 20-7, calculate the light at point P if the candlepower of each fixture is 3000.

21

ELECTRIC HEAT

Objectives

After studying this chapter, the student will be able to:

- Describe three methods used for heat transfer.
- Discuss the factors that must be considered when calculating heat losses.
- Describe various types of electric heating equipment.
- Explain the operation and function of controls used in conjunction with electric heating systems.

SPACE HEATING

Space heating refers to heating of specific areas of commercial and industrial buildings. Heat energy is generally transferred from one place or object to another by one of three methods: conduction, convection, or radiation. In most instances all three methods are used in varying degrees. The system is named according to the primary method used.

The *conduction method* transfers heat from one object to another. One might say the object conducts heat. An example of this method is cooking. A pan is placed on the range in contact with the heating element. Heat is transferred from the element to the pan and from the pan to the ingredients in the pan.

The *convection method* transfers heat through a fluid or air. Hot water, steam, and warm-air heating systems use the convection method.

Heat radiation is accomplished when the heat rays are transmitted through space. The rays are absorbed by the objects they contact, thereby warming the objects. They do not warm the air but warm the objects into which they are absorbed.

One common source of radiant heat is the sun. All objects radiate some heat; therefore as heat rays strike objects, some are absorbed and some are radiated back into space.

All three methods of space heating can be employed when using electric energy as a heat source.

CALCULATIONS

The unit of measurement for heat is the *British thermal unit (Btu).* A Btu is the amount of heat required to raise the temperature of 1 pound of water 1°F.

To provide a comfortable area, it is recommended to maintain a temperature of 72°F (22°C) with a relative humidity of 35 percent.

The greater the difference of temperature between two areas, the quicker heat is transferred from the warm area to the cold area. Other factors to consider when designing a heating system are the amount and type of materials separating the warm area from the cold area. The type, density, thickness, and color of the material all affect heat transfer. To determine the amount of heat required to maintain body comfort, all the aforementioned factors must be considered.

Because temperatures vary considerably in different parts of the world, it is necessary to refer to engineer's recommendations for the specific area. For example, in the northeastern United States, it is generally recommended to maintain a temperature of 72°F (22°C) when the outside temperature is –10°F (–23°C) with a wind of 10 miles per hour (16 kilometers per hour).

The heat that is transferred from the area being heated to the colder surrounding area is considered a heat loss. To calculate this loss, it is necessary to determine the square foot area of the walls, ceilings, and floors that are exposed to the colder area. Window and door areas must also be considered, in addition to the type and amount of insulation to be used.

In many areas, the local utility company provides charts and formulas for calculating heat loss. Some utility companies do the calculations as an incentive to install electric heat. And they often provide estimates of the annual heating cost.

Electric heat cost is based on the number of kilowatt hours of energy dissipated. Many utility companies give special rates for heating. In many areas, particularly in the colder regions, the price of electric heat may not be competitive with other fuels. It has, however, the advantage of providing quick, clean heat at a very even temperature. Electric heat requires little or no space for the heating plant and needs no area for storage of fuel. Another advantage is that electric heating systems are far more efficient than fossil fuel systems.

HEATING EQUIPMENT

One very efficient and flexible system is *resistance baseboard heating*. The units consist of resistance heaters placed along the wall near the floor line. They are installed in areas of greatest heat loss (i.e., outside walls, under windows). Air circulates through the units, warming the walls and the surrounding air. This method of heating represents examples of both conduction and natural convection. A small amount of radiation also takes place. It is referred to as a convection system because the major portion of the heat is transmitted by convection. The heat is produced by current flowing through a resistance. The amount of heat produced is directly proportional to the resistance and the square of the current.

Perimeter heating provides a uniform heat throughout an area. It is extremely efficient because all of the heat produced is dissipated into the area to be heated. With this method of heating, each room can have its own thermostat.

Electric heating cables, often installed in ceilings, are used only when cost is not a factor. The heat from this system is transmitted primarily by radiation. Another type of radiant heating consists of plasterboard panels containing a conductive film that produces heat when energized. Both of these systems are also very efficient, producing the heat within the area being heated and providing uniform heat throughout the area. An individual thermostat is usually provided in each room.

An *electric furnace* is a central unit in which resistance heating elements are installed. A fan is placed near the elements. When the thermostat calls for heat, the circuit is completed to the heaters. When the air around the heaters reaches a predetermined temperature, the fan is energized. The fan pulls the cold air from the area to be heated and forces the warm air into the area. The cooler air entering the furnace is then heated by the elements. This cycle of operation continues until the thermostat is satisfied.

Duct heaters are similar to the electric furnace. Instead of having all the elements in one location, they are installed in the ducts for each specific area. This type of installation provides the flexibility of zone control.

The *heat pump* is a machine that can heat an area during cold weather and can cool the area during warm weather. It operates on a principle similar to a refrigerator. The pump circulates a liquid (refrigerant) through two sets of coils. A compressor is

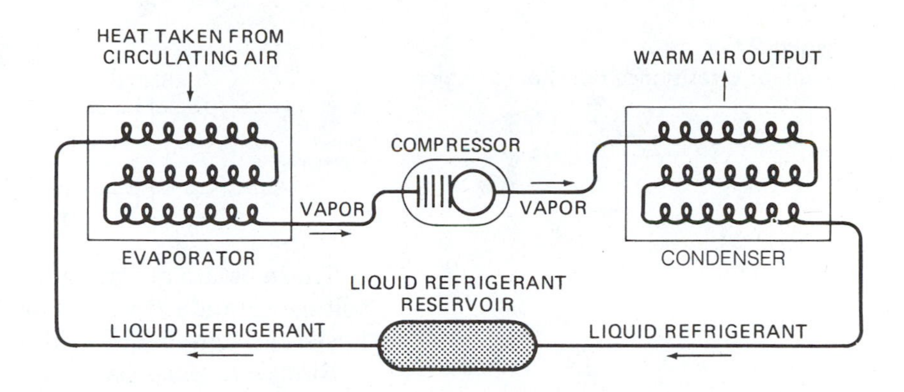

Figure 21-1
Simplified sketch of a warm-air heating system using a heat pump

connected between the coils, Figure 21-1. The liquid is the type that boils at a very low temperature. When it boils, it vaporizes, just as water turns to steam when it boils. This cold vapor absorbs heat from the surrounding air. By compressing the vapor, the temperature is raised further. Before leaving the compressor, the vapor reaches a temperature of 100°F (37.8°C) or higher. The heated vapor enters the other coil, called a condenser. The air surrounding the condenser is cooler than the vapor and absorbs the heat. As the vapor cools, it returns to liquid form, and the process is repeated. For air conditioning, the process is reversed. The method used to transmit the heated air to a specific area is much the same as described for an electric furnace.

The heat pump is probably the most efficient type of electric heat. Depending on the type of installation, the operating cost can be competitive with, or even less than, systems using fossil fuels.

SYSTEM CONTROLS

All automatic heating systems must have controls for their operation. There are two basic types: operating controls and safety controls. These two types can be further divided according to their particular function.

One of the more simple systems is used in conjunction with baseboard resistance heaters. Here only one type of operating control and one type of safety control are required.

The room thermostat functions as the operating control. When the temperature of the room drops below a predetermined value, the thermostat contacts close, completing the circuit to the heater(s). When the room temperature reaches the thermostat setting, the contacts open, disconnecting the heater(s) from the supply.

The accuracy of the thermostat depends upon its design and quality. The sensitivity of the heat-sensing device and the ability of the contacts to make and break quickly play important roles in maintaining a constant temperature.

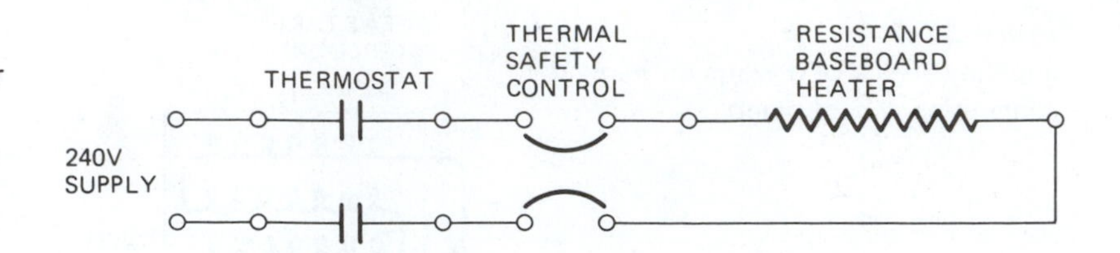

Figure 21-2
Circuit for a resistance baseboard heater

Thermostats are manufactured for use with various values of voltage. Standard values are 24 volts, 120 volts, and 240 volts. Resistance baseboard heaters generally require 240 volts.

Built into each baseboard heater is a safety control. It is a thermal device that opens the circuit if excessive heat is produced. The purpose of this control is twofold: It prevents damage to the unit caused by overheating, and it reduces the possibility of fire.

Baseboard resistance heating units, ceiling panels, and heating cables usually require only a thermostat and a thermal device. Figure 21-2 shows the circuit for a 240-volt baseboard heater.

The electric furnace requires additional controls. Beside the thermostat, an operating control is installed near the heaters in the plenum chamber. The purpose of this control is to stop and start the fan, Figure 21-3.

When the thermostat calls for heat, the circuit is completed to the heaters. When the air temperature in the plenum reaches 85°F (30°C), the fan control completes the circuit to the fan motor. The air is circulated through the area to be heated and through the system. If the temperature in the plenum drops below 82°F (28°C), the fan control opens the circuit and stops the fan. With adequate design, the air temperature in the plenum will remain above 82°F (28°C) until the thermostat is satisfied and for a short time thereafter.

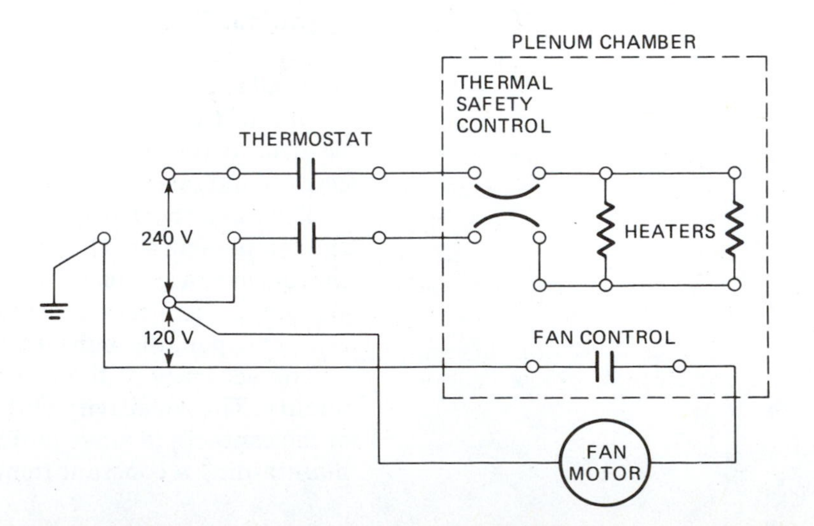

Figure 21-3
Basic circuit for an electric furnace

When the thermostat is satisfied, the circuit to the heaters is opened, disconnecting them from the supply. The air in the plenum quickly cools, causing the fan control to open the circuit to the fan motor.

Another control that is placed in the plenum chamber is for safety. This device opens the circuit to the heaters if the temperature in the plenum exceeds a safe value.

Each installation is unique in its design. In addition, the type and number of controls vary according to the requirements. Manufacturers of various systems provide instructions and diagrams to meet their requirements. It is important to study this information carefully.

HUMIDITY CONTROL

Humidity is a very important consideration in the design of space heating. Many heating systems tend to remove humidity from the air. The result is dry, unhealthy air and the need for higher temperature to attain body comfort. To compensate for the moisture loss, humidifiers must be installed. Humidifiers replace lost moisture.

Other heating systems tend to cause excessive moisture. When this occurs, exhaust fans must be installed. Such fans pull the moist air to the outside. The moist air is then replaced by drier air.

Another method of removing moisture is to install a dehumidifier. This machine removes the moisture from the air and collects it in a reservoir or disposes of it into a drain.

For health and comfort, a relative humidity of 30 to 40 percent is desirable. A humidistat is a switch that makes and breaks contacts as the humidity changes. Humidistats should be installed to control the humidifier, dehumidifier, or exhaust fans, whichever is needed. The design of a building and the type of heating system are the determining factors as to what is needed. The amount of insulation and the type of vapor barriers are also factors. Some designs may include both exhaust fans and humidifiers. The controls are arranged to operate whichever is needed.

If exhaust fans are installed, they should have good dampers to prevent heat loss when the fans are not in operation. Non-metallic ducts that are insulated are also recommended. Such an arrangement reduces heat loss and prevents condensation from forming inside the duct and returning to the heated air.

DIELECTRIC HEATING

Dielectric heating is based on the principle of capacitance. An insulating material is placed between two conducting plates. The plates are connected to a high-frequency ac supply. The alternating polarity of the plates causes the electrons in the insulating material to strain first in one direction, then in the other. The continuous and rapid strain reversal produces heat.

This method of heating has many applications, such as cooking, softening of materials such as plastics, and rapid drying of materials. It is also used for medical purposes. Intense pain caused by strain or tense muscles can often be relieved by warming the affected part of the body. This treatment is called *diathermy*. The portion of the body to be treated is placed between the plates. A warm sensation is felt throughout the area being treated. The electron strain and the heat cause the muscles to relax, thus reducing pain.

The advantage of dielectric heating is its ability to instantly produce heat uniformly within a material. It is also easy to control by varying the frequency of the applied voltage.

INDUCTION HEATING

Induction heating is used for metal treatment and in some electric furnaces. Induction heating, as the name implies, uses electromagnetic induction to produce heat. A conducting material is placed within a coil that is attached to a high-frequency ac supply. The increase and decrease of the magnetic field around the coil cause relative motion between the field and the conductor. The result is to produce eddy currents on the surface of the conducting material. The eddy currents produce heat.

The amount of heat produced depends upon the frequency of the supply voltage and the amount of current flowing in the coil. The depth to which the heat penetrates depends upon the frequency of the supply voltage. The higher the frequency, the less the depth. With lower frequencies, a longer heating time is required, allowing the heat to penetrate to a greater depth.

Other factors that affect the amount of heat and depth penetration are the resistivity and permeability of the material being heated.

RESISTANCE HEATING

Resistance heating utilizes the fact that current flowing through a material produces heat. The amount of heat depends upon the value of current and the size and type of material through which the current is flowing. Some of the most common materials are aluminum, nickel, chromium, copper, iron, and cobalt in various proportions. Resistance heating elements are commonly used in cooking appliances, electric space heating, clothes dryers, and irons.

Some types of electric welders also utilize the theory of resistance heating. Two types of resistance welders are the spot welder and the seam welder.

With spot welding, thin sheets of metal are joined together in spots. The metal sheets are placed between two electrodes. The electrodes are moved together, clamping the sheets between them. A high current passes through the sheets from one electrode to the other. The current produces heat, causing the metal to weld. As soon as the spot reaches welding temperature, the circuit is opened. When the material cools, the electrodes are released.

Seam welding is similar to spot welding except the weld is continuous. The metal is joined together and placed between two electrodes in the form of rollers. As the metal is moved between the electrodes, a high current passes from one roller to the other. A welding heat is produced and forms one continuous seam.

ARC HEATING

An electric arc is produced by ionizing the air between two electrodes. The heat from the arc is used for welding and melting metal.

With arc welding, the material to be welded forms one electrode, and the welding rod the other. To form the arc, the welding rod is touched to the material to be welded. The heat produced by the current flowing through the point of contact causes the metal to vaporize, causing the surrounding air to ionize. When the rod is pulled away from the material, the ionized air conducts the current, forming an arc. Once the arc is produced, the heat generated will cause the material and the rod to melt. The metal from the rod flows with that from the material to form a continuous seam.

INFRARED HEAT

Infrared heating is a type of radiant heating. Electromagnetic waves transmitted through space are absorbed by the materials they strike.

These rays are usually produced by electric lamps containing a filament. Current flowing through the filament produces electromagnetic rays whose wavelengths are greater than those that produce visible light but shorter than microwaves. They have a frequency slightly below the visible light range. The lamps that are used to contain the filament are generally of the quartz type, in tubular shape or shaped like the lamp in Figure 20-11 (PAR).

Infrared heating is used for drying operations, for food warming, and for medical purposes.

REVIEW

A. Multiple choice.
Select the best answer.

1. Convection is a method of transferring heat
   a. by contact.
   b. through a fluid or air.
   c. by rays traveling through space.
   d. through a vacuum.

2. The unit of measurement for heat is
   a. Fahrenheit.
   b. Celsius.
   c. American thermal unit.
   d. British thermal unit.

3. To maintain a comfortable body temperature, it is recommended to maintain
   a. a temperature of 75°F (24°C) with a relative humidity of 32%.
   b. a temperature of 72°F (22°C) with a relative humidity of 32%.
   c. a temperature of 72°F (22°C) with a relative humidity of 35%.
   d. a temperature of 70°F (21°C) with a relative humidity of 35%.

4. In the northeastern United States, it is recommended to install a heating system capable of maintaining a temperature of
   a. 75°F (24°C) when the outside temperature is –10°F (–23.3°C) with a 10-mile-per-hour (32-kilometer-per-hour) wind.
   b. 72°F (22°C) when the outside temperature is –10°F (–23.3°C) with a 10-mile-per-hour (16-kilometer-per-hour) wind.
   c. 70°F (21°C) when the outside temperature is –10°F (–23.3°C) with a 10-mile-per-hour (16-kilometer-per-hour) wind.
   d. 70°F (21°C) when the outside temperature is –10°F (–23.3°C) with no wind.

5. A room becomes cooler because
   a. cold seeps into the heated area.
   b. cold air is radiated through the outside surfaces.
   c. the heat is absorbed by the surrounding cold surfaces.
   d. the heat loses its strength over a period of time.

REVIEW *(continued)*

A. Multiple choice.
Select the best answer
(continued.

6. Charts for calculating heat loss may be obtained from
   a. the local utility company.
   b. manufacturers of the heating equipment.
   c. the National Fire Protection Association.
   d. both a and b.
7. A major advantage of electric heat is that
   a. it is quick and clean.
   b. it is very inexpensive.
   c. it never fails.
   d. it is available in quantity.
8. The best location for baseboard heaters is
   a. along the inside walls.
   b. anywhere on an outside wall.
   c. under windows.
   d. beside the door.
9. Electric heating cables installed in the ceiling transfer heat by
   a. conduction.
   b. convection.
   c. radiation.
   d. all of the above.
10. An electric furnace transfers heat by
    a. conduction.
    b. induction.
    c. convection.
    d. radiation.
11. The liquid refrigerant in a heat pump boils at a
    a. very high temperature.
    b. very low temperature.
    c. temperature of 212°F (100°C).
    d. temperature of 200°F (93.3°C).
12. Compression of the vapor within a heat pump causes the temperature of the vapor to
    a. increase.
    b. decrease.
    c. remain the same.
    d. decrease slightly.

REVIEW *(continued)*

A. Multiple choice.
Select the best answer (continued).

13. Two general classifications for heating controls are
    a. operating controls and sensing controls.
    b. safety controls and sensing controls.
    c. operating controls and safety controls.
    d. operating controls and monitoring controls.

14. The operating control that is installed in the plenum chamber of an electric furnace serves
    a. to operate the heaters.
    b. to stop and start the fan motor.
    c. to disconnect the heaters from the supply if the temperature exceeds a predetermined value.
    d. all of the above.

15. Common voltage ratings for thermostats are
    a. 24 V, 120 V, 240 V.
    b. 120 V, 208 V, 480 V.
    c. 32 V, 120 V, 277 V.
    d. 16 V, 150 V, 280 V.

16. The room thermostat for an electric furnace controls the current to the
    a. fan.
    b. heaters.
    c. fan and heaters.
    d. entire heating system.

17. Duct heaters are similar to
    a. baseboard heating units.
    b. the electric furnace.
    c. ceiling panels.
    d. electric cables.

18. A heat pump
    a. heats the area to which it is connected.
    b. cools the area to which it is connected.
    c. both a and b.
    d. neither a nor b.

19. A humidity control is used to regulate
    a. the moisture content in the air.
    b. the temperature of the air.
    c. both moisture and temperature.
    d. the temperature of the moisture.

REVIEW *(continued)*

A. Multiple choice. Select the best answer (continued).

20. For health and comfort, the relative humidity should be between
    a. 25% and 35%.
    b. 30% and 40%.
    c. 35% and 45%.
    d. 40% and 50%.
21. A machine used to add moisture to the air is called a
    a. dehumidifier.
    b. humidifier.
    c. exhaust fan.
    d. humidistat.
22. Dielectric heating is based on the principle of
    a. capacitance.
    b. resistance.
    c. inductance.
    d. conductance.
23. Induction heating is based on the principle of
    a. capacitance.
    b. inductance.
    c. resistance.
    d. conductance.
24. Arc heating occurs when the air between two electrodes of opposite polarity becomes
    a. moistened.
    b. dry.
    c. ionized.
    d. polarized.

B. Give complete answers to the following.

1. List three methods used to transmit heat.
2. Describe one method to transmit heat.
3. Define the term *British thermal unit.*
4. What temperature is generally recommended to maintain body comfort?
5. What percent humidity is recommended for good health and body comfort?

REVIEW *(continued)*

B. Give complete answers to the following (continued).

6. What temperature and wind factor should be used when designing a heating system for a house in the northeastern United States?
7. Whom might one contact for assistance in calculating heat loss?
8. List three advantages of electric heating systems.
9. What locations are recommended for electric baseboard units?
10. List three types of electric heating systems that have the advantage of individual control for each room.
11. What is an electric furnace?
12. Describe the operation of an electric furnace.
13. What are duct heaters?
14. What is a heat pump?
15. What is the purpose of the compressor used in conjunction with a heat pump?
16. List two types of controls designed for use with electric heating systems.
17. What is the purpose of the control that is built into every baseboard heating unit?
18. List the controls that must be used in conjunction with an electric furnace.
19. What is the purpose of a humidifier?
20. List two methods used to removed moisture from the air.
21. What is a humidistat?
22. Describe dielectric heating.
23. What is inductive heating?
24. What method of heating is used for spot welding?
25. List two uses for electric arc heating.

22
Electrical Instruments and Equipment

Objectives

After studying this chapter, the student will be able to:

- Describe various types of electrical measuring instruments.
- Discuss industrial controls.
- Explain the construction and operation of a self-synchronous system.
- Describe various industrial relays.

ELECTRICAL MEASURING INSTRUMENTS

Electrical measuring instruments are commonly referred to as *meters* and are designed in many types, shapes, and sizes. The standard meter has a scale calibrated in the unit to be measured and an indicating needle. Other styles are the digital type, showing the exact numerical value, and the graph type.

Permanent-Magnet Instruments

The permanent-magnet meter, Figure 22-1, is used almost exclusively for dc measurements. The movable coil is generally wound on an aluminum bobbin, which is pivoted to turn between the poles of a permanent magnet. The bobbin, in addition to supporting the coil, dampens (retards) its movement. The coil is connected to the terminals of the meter through two spiral springs. The springs offer an opposing force proportional to the amount of deflection. When current flows in the coil, it rotates until the electromagnetic force is equal to the spring tension.

Figure 22-1
Permanent-magnet meter

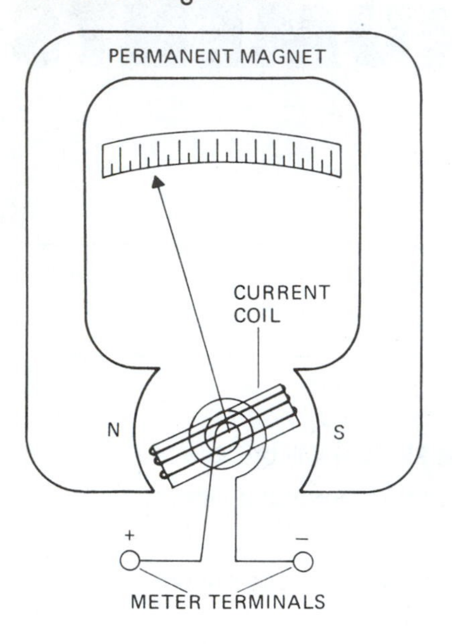

Figure 22-2
Permanent-magnet meter with an external shunt

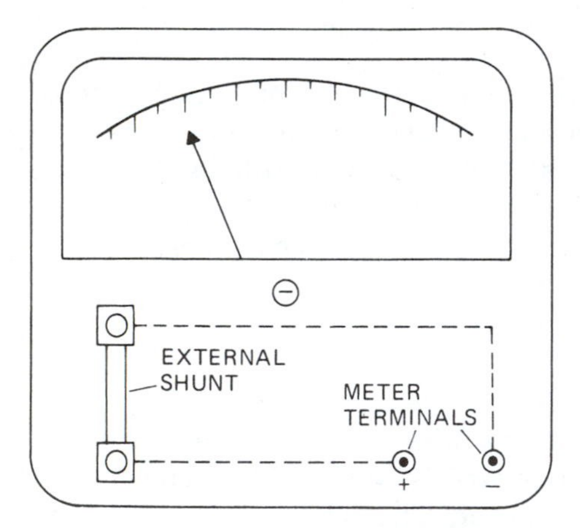

Permanent-magnet meters give full-scale deflection with coil currents ranging from 10 microamperes (0.000010 ampere) to 20 milliamperes (0.020 ampere). To measure currents higher than 20 milliamperes, a shunt is connected in parallel with the coil, Figure 22-2. The shunt is made of materials that are good conductors and produce minimum heat at high currents.

For measuring low currents, the shunt is connected internally. If the meter is designed for several ranges, more than one shunt is installed. A selector switch is provided and arranged to select the proper shunt or combination of shunts to give the correct full-scale deflection. Meters for measuring large currents may have external shunts. They are usually rated for 50 millivolts (mV) for full-scale currents.

These shunts may be in the form of a bar or a wire and are made especially for the specific instrument. If they are equipped with connecting leads, the leads must not be altered. Any change in the length or size will result in incorrect measurements. It is possible to use the same instrument for current measurements of any range, provided the proper shunt is used.

Keeping the resistance of the shunt to a minimum value keeps most of the current flowing through the shunt, and only a very small amount flows through the coil. For example, if the coil is rated at 20 milliamperes and the meter is designed to provide a full-scale deflection when the circuit current is 20 amperes, it is necessary to install a shunt that will carry 19.98 amperes when 20 milliamperes flow through the coil.

Example 1

A meter with a resistance of 2.5 Ω gives full-scale deflection when 20 mA flow in the coil. In order to provide full-scale deflection at 10 A, what is the resistance of the shunt?

Determine the voltage across the coil, from Ohm's Law:

$$E = IR$$
$$E = 0.02 \times 2.5$$
$$E = 0.05 \text{ V}$$

The shunt is to be connected in parallel with the coil. Therefore, the voltage across the shunt is 0.05 V.

The current through the shunt is 10 A – 0.02 A = 9.98 A. The resistance of the shunt can be found by Ohm's Law.

$$R = \frac{E}{I}$$
$$R = \frac{0.05}{9.98}$$
$$R = 0.005\ \Omega$$

Example 2

A meter is designed to use a coil rated for a maximum current of 10 mA. The instrument is intended to provide a full-scale deflection at 50 A. What is the resistance of the shunt if the meter has a resistance of 100 Ω?

$$E = IR$$
$$E = 0.01 \times 100$$
$$E = 1\ V$$

$$I_s = I - I_c$$
$$I_s = 50 - 0.01$$
$$I_s = 49.99\ A$$

$$R = \frac{E}{I}$$
$$R = \frac{1}{49.99}$$
$$R = 0.02\ \Omega$$

This same instrument can be used to measure voltage. Connecting the proper resistance in series with the coil limits the current to the rated maximum value of the coil. Because a voltmeter is connected in parallel with the apparatus under consideration, the current divides according to the resistance.

Example 3

What resistance must be connected in series with the coil of the meter in Example 1 in order to obtain full-scale deflection at 150 V?

The total resistance is

$$R = \frac{E}{I}$$
$$R = \frac{150}{0.02}$$
$$R = 7500\ \Omega$$

The coil has a resistance of 2.5 Ω, and 7500 Ω – 2.5 Ω = 7497.5 Ω.

For direct readings, the instrument should be equipped with a 150-volt-range scale. Voltage measurements for other ranges may be obtained by providing the proper size of series resistors. These resistors are frequently called *multipliers*. A multiple range meter can be constructed by providing various sizes of multipliers and a selector switch.

A well-designed meter measures voltage with minimum current flowing through the meter. So that consumers may compare various voltmeters, voltmeters are rated in the number of ohms per volt (Ω/V) at full-scale deflection. In other words, a meter rated at 20,000 ohms per volt will cause less effective change in the circuit characteristics than will a meter rated at 10,000 ohms per volt. The ohms-per-volt rating of a meter is referred to as the *sensitivity* of the meter.

Sensitivity of Instruments

A voltmeter with high sensitivity has a high resistance per volt of scale. Instruments with high sensitivity are very desirable for measuring voltage drops in circuits containing high resistance or reactance.

Figure 22-3
Voltage drop in a series circuit

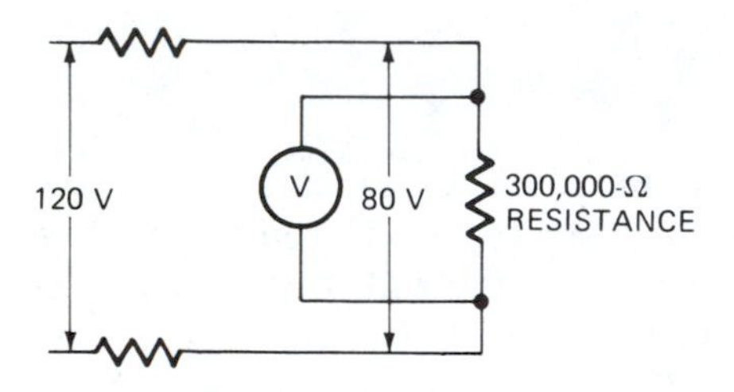

As an example, suppose it is necessary to measure the voltage drop across a 300,000-ohm resistance in a series circuit. If the voltage is measured with a meter having low sensitivity (1000 ohms per volt) and a 300-volt range, the voltage indicated by the meter will be quite different from the actual voltage drop across the resistor. Figure 22-3 illustrates this circuit. The meter has a resistance of 1000 ohms × 300 volts = 300,000 ohms. If it is connected across a 300,000-ohm resistance, the current will divide equally between the meter and the resistance. Also, the combined resistance of the resistor and the meter will be half of 300,000, or 150,000 ohms. This allows much more current to flow in the circuit, changing the circuit characteristics and indicating an incorrect voltage drop.

If the voltage is measured with a meter having 20,000 ohms per volt and a 300-volt range, the resistance of the meter will be 20,000 ohms × 300 volts = 6,000,000 ohms. With this meter, very little current flows through the meter in comparison to the current flowing through the resistance. The circuit characteristics are only slightly changed and the meter indicates very close to 80 volts.

With reference to ammeters, the sensitivity is the amount of current required for full-scale deflection. If a meter requires 1 milliampere for full-scale deflection, it is referred to as a *one-mil meter*. Some meters are manufactured to have a sensitivity as low as 10 microamperes.

The sensitivity of ammeters is also expressed in the amount of voltage drop across the coil at full-scale deflection. For example, if the meter indicates full-scale deflection with a 50-millivolt drop across the coil, the meter has a sensitivity of 50 millivolts.

Figure 22-4A
Simplified diagram of an electrodynamometer

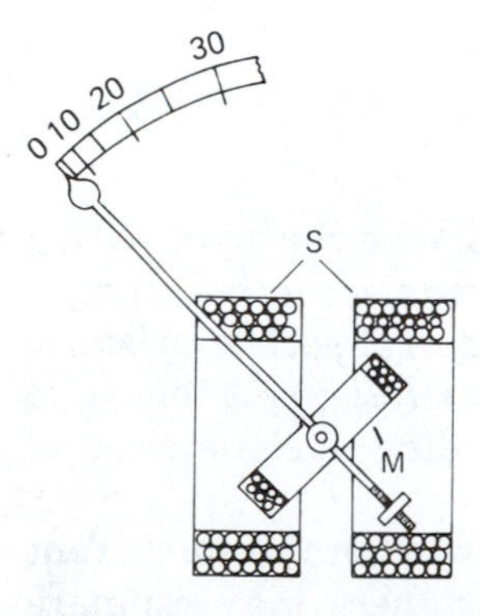

Electrodynamometer

The electrodynamometer, Figure 22-4A, consists of two stationary coils and a movable coil. The indicating needle is attached to

Figure 22-4B
Electrodynamometer indicating needle and vane attached to the movable shaft

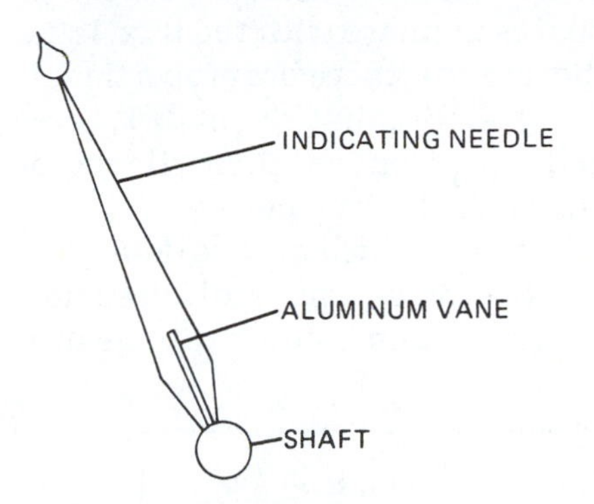

the movable coil. The three coils are connected in series through two spiral springs. When current is flowing through these coils, the reaction between the magnetic fields causes the movable coil to turn in a clockwise direction.

If the direction of the current through the meter is reversed, the torque will remain in a clockwise direction. Thus, an instrument of this type can be used to measure alternating current as well as direct current.

The torque produced within the meter varies as the square of the current. For this reason, the scale is not divided uniformly. The divisions near the lower end of the scale are small and difficult to read, and those near the upper end are large and can be read very accurately.

Most ac meters are dampened. This is generally accomplished by an aluminum vane placed in an enclosure and attached to the same shaft as the needle, Figure 22-4B. When the coil in Figure 22-4A, and therefore the vane, moves, air is forced from one side of the vane to the other, producing a dampening effect.

Wattmeter

The wattmeter is generally a dynamometer-type instrument. It is constructed like the electrodynamometer in Figure 22-4A, with the coils connected as shown in Figure 22-5A. The stationary coils (current coils) are connected in series with the load, and the movable coil (voltage coil) is connected across the line through a high resistance. The circuit through the voltage coil has negligible reactance and high resistance. Therefore, its current can be considered to be in phase with the line voltage.

At unity power factor, the flux of the current coil is in phase with that of the voltage coil, and maximum wattage is indicated. When the current lags the voltage, as with an inductive load, less

Figure 22-5A
Connections for an electrodynamometer-type wattmeter

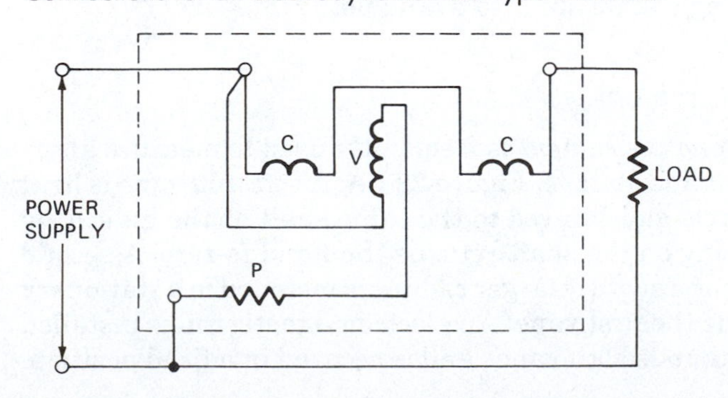

Figure 22-5B
Flux vector diagram of an inductive circuit. Φ_c lags Φ_v by ϕ degree ($P = IV \cos \phi$).

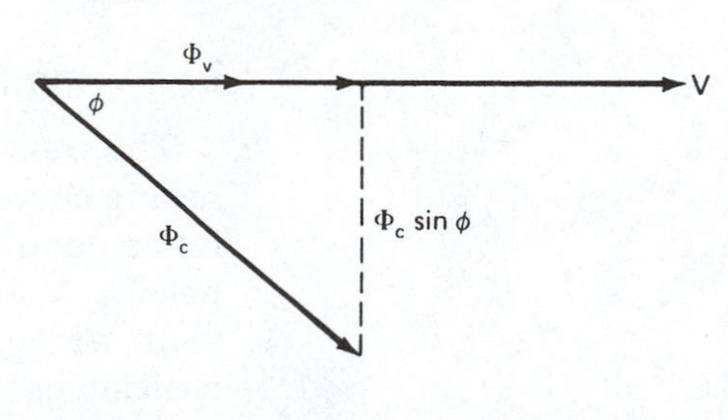

torque is developed and lower wattage is indicated. Figure 22-5B is a vector diagram of an out-of-phase condition. The flux from the current coil is indicated by Φ_c and from the voltage coil by Φ_v. Only the component of flux Φ_c that is in phase with the flux Φ_v is effective in developing torque. The torque is therefore proportional to the product of Φ_v, Φ_c, and cos ϕ. Because Φ_v and Φ_c are respectively proportional to E and I, the torque is proportional to EI cos ϕ, which, from equation 13.17, is the true power.

A wattmeter has two circuits, either of which may be damaged by too much current, even though the indicating needle does not move off scale. For this reason, current and voltage ratings are shown on the meter.

CAUTION: These ratings must not be exceeded.

Two-phase power can be measured by using two single-phase wattmeters with one meter connected in each phase. The total power is the arithmetic sum of the power indicated on each meter.

Two single-phase wattmeters can also be used to measure the power of a three-phase circuit. For circuits having a power factor greater than 0.50, the three-phase power is equal to the arithmetic sum of the power indicated on each meter. If the power factor is less than 0.50, the three-phase power is equal to the difference between the two values. If the power factor is 0.50, the three-phase power is the value indicated on one of the meters. (Both meters will indicate the same value.)

Two single-phase meters may be combined into one three-phase meter. In this case, the two voltage coils are mounted on the same shaft and rotate with their respective current coils. The three-phase power is indicated on a single scale.

For three-phase, four-wire systems, it is necessary to use three meters. Connect one meter between each phase and the neutral. The three-phase power is the arithmetic sum of the three values. Three-phase, four-wire meters that indicate the three-phase power on a single scale are also available.

Iron-Vane Instrument

The *iron-vane instrument* is frequently used to measure alternating current and voltage, Figure 22-6A. A soft iron vane is bent into a semicircle and secured to the same shaft as the indicating needle. A spring on the shaft returns the hand to zero. A second vane, which has a slightly larger radius, is fastened in a stationary position beside the first vane. An electromagnetic coil is installed so that it surrounds both vanes and is secured in a fixed position.

Figure 22-6A
Iron-vane meter

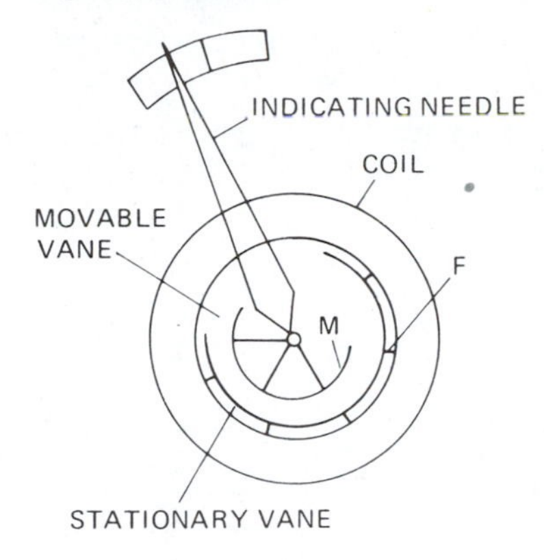

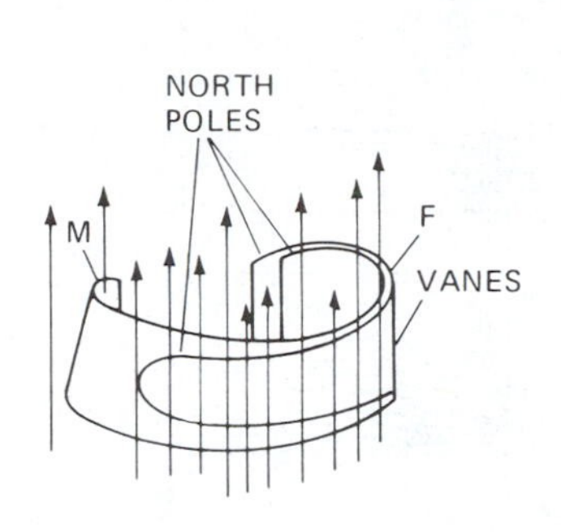

Figure 22-6B
Iron vanes

STATIONARY VANE

MOVABLE VANE

Figure 22-7A
Inclined-coil meter

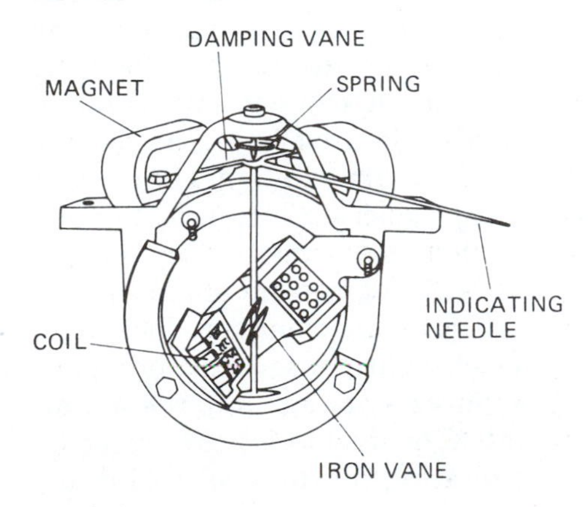

Figure 22-7B
Arrangement of coils, vanes, spring, and needle for an inclined-coil meter

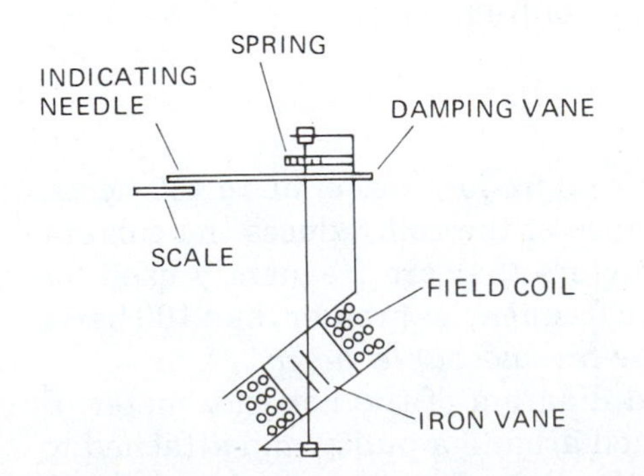

If the instrument is to be used as an ammeter, it will have only a few turns of heavy wire. If it is to be used as a voltmeter, it will have many turns of fine wire connected in series with a resistor.

When the meter is connected into a circuit, current flowing through the coil produces a field that magnetizes the two vanes. The magnetic polarity of the two vanes is such that a repelling action takes place, causing the movable vane to turn in a clockwise direction. Because the vane is secured to the shaft, both the shaft and the indicating needle will move. The degree of rotation depends upon the strength of the repelling force. This force, in turn, depends upon the amount of current flowing through the coil. Figure 22-6B shows both vanes before they are bent into a semicircle. An instrument of this type can be designed for use on either dc or ac.

Inclined-Coil Instrument

The *inclined-coil instrument* has a coil at an oblique angle to the shaft. Secured to the shaft are the indicating needle and two soft iron vanes arranged at oblique angles to both the shaft and the coil, Figure 22-7A. A spring attached to the shaft holds the indicating needle at zero when there is no current flowing through the coil. When current flows, the iron vane attempts to line up with the magnetic field. This causes the shaft to rotate against the spring action, and the indicating needle moves up scale. The amount of current and the spring tension determine the degree of deflection, Figure 22-7B.

Both types of iron-vane meters can be used for either ac or dc measurements. Because of their simple construction and low cost, they are used extensively in the electrical industry.

Figure 22-8A
Induction-type meter

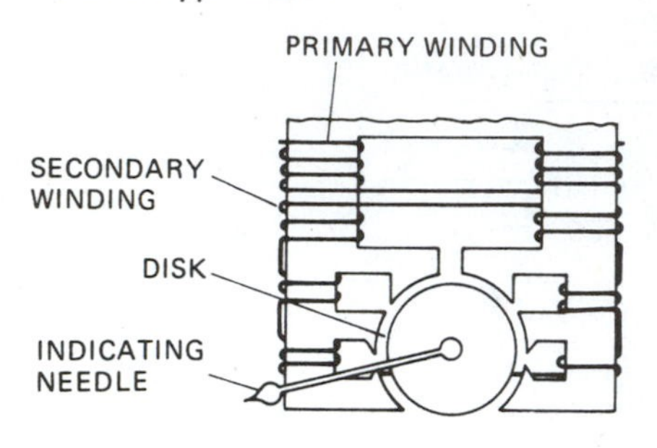

Figure 22-8B
Flux path established by the primary current in an induction-type meter

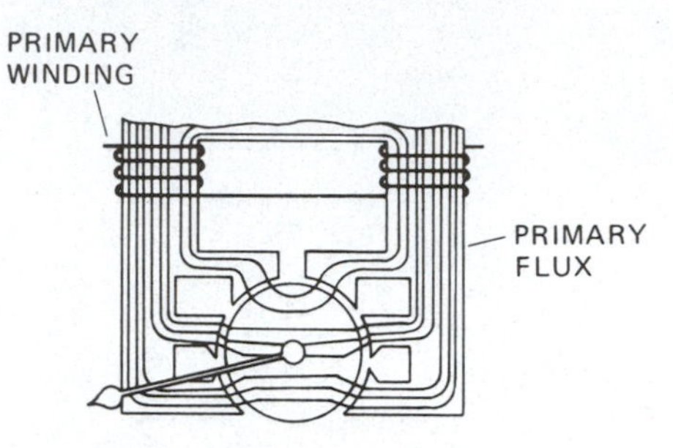

Figure 22-8C
Flux path established by the secondary current in an induction-type meter

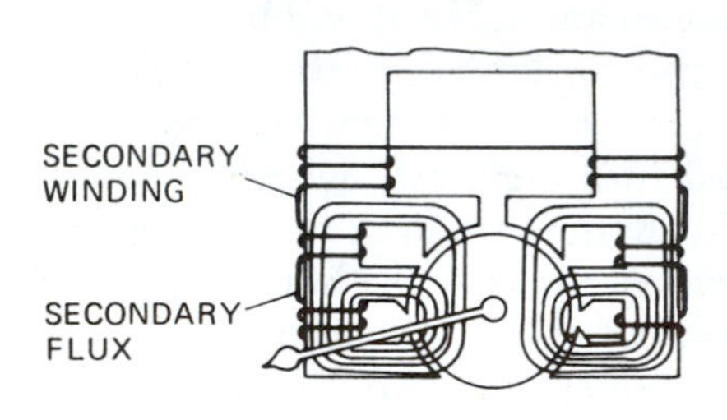

Induction-Type Instrument

The general construction of an induction-type instrument is shown in Figure 22-8A. The principle of operation is similar to that of a split-phase motor. The indicating needle is attached to an aluminum disk, which serves as the rotor. A spring holds the needle at zero when there is no current flowing in the stator coils. When the meter is connected into a circuit, current flows through the primary winding, Figure 22-8B. The flux produced by this current induces a voltage into the secondary winding. The current flowing in the secondary winding produces a magnetic flux, as shown in Figure 22-8C. The instrument is designed so that the flux produced by the primary current is 90 degrees out of phase with the flux of the secondary winding. The arrangement produces a rotating magnetic field similar to that of a split-phase capacitor motor.

The rotating magnetic field induces a voltage into the aluminum disk, producing eddy currents. The magnetic field caused by the eddy currents reacts with the rotating field to develop a torque. This torque, which is proportional to the amount of current flowing in the primary winding, causes the disk to rotate until the spring tension is equal to the rotating force.

This type of instrument is designed for ac measurements and is very accurate over a wide range of frequencies.

Figure 22-9
Simplified diagram of a hot-wire meter

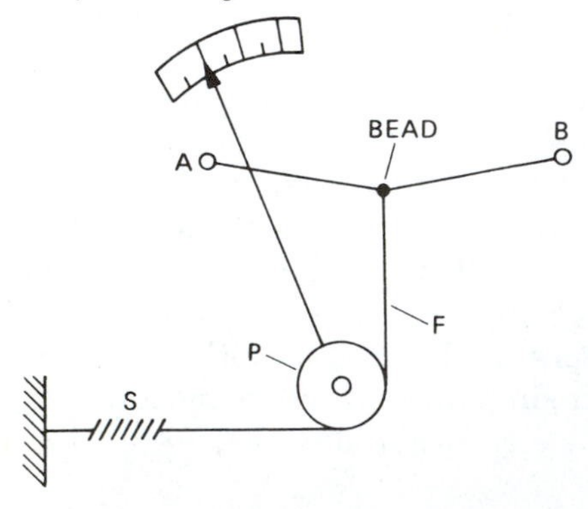

High-Frequency Meters

Most ac meters are accurate on frequencies of up to 100 hertz. Beyond this point, the reactance of the coil reduces the current considerably. Two types of meters that are frequently used for measurements on circuits with frequencies greater than 100 hertz are the *hot-wire meter* and the *thermocouple meter*.

Figure 22-9 is a simplified diagram of the hot-wire meter. F represents a silk thread formed around a pulley and attached to a spring. The upper end of the thread is connected to a bead, which is threaded onto a wire connected to terminals A and B. The wire is made of a platinum alloy, which stretches when heated. When the meter is connected into a circuit, the current flowing through the platinum wire produces heat. The spring tension on the thread causes the wire to stretch, forcing the pulley to rotate. The indicating needle, which is secured to the pulley, moves up scale as the wire is stretched. The amount of stretch is proportional to the amount of current flowing in the wire.

This type of meter is slow in reacting, and the tension on the wire changes with a change in the ambient temperature. It is necessary, therefore, to ensure that the indicating needle is on zero before using the instrument.

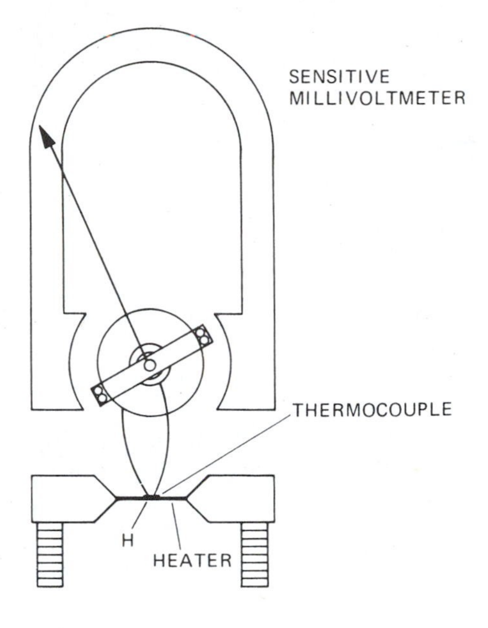

Figure 22-10
Basic construction of a thermocouple meter

The basic construction of a thermocouple meter is shown in Figure 22-10. Because the instrument depends upon the amount of heat produced at the thermojunction, it operates on either ac or dc and functions satisfactorily over a wide range of frequencies well over 100 hertz.

The thermocouple meter is connected into the circuit so that the current will flow through the heater. The heat is transferred by conduction to the thermojunction, developing an emf across the ends of the thermocouple. This voltage is proportional to the difference in temperature between the junction and the open ends. A sensitive milliammeter is connected across the open ends of the thermocouple. Generally, the milliammeter is calibrated to indicate the value of current through the heater. For very low values of current, the heater and thermocouple are enclosed in an evacuated glass bulb to eliminate oxidation.

Another instrument that is suitable for high frequencies uses the principle of rectification. A *rectifier* is a device that changes ac to dc. Four rectifiers connected to form a circuit are installed in the meter, Figure 22-11A. Figure 22-11B illustrates the meter

Figure 22-11A
Rectifier-type, high-frequency meter

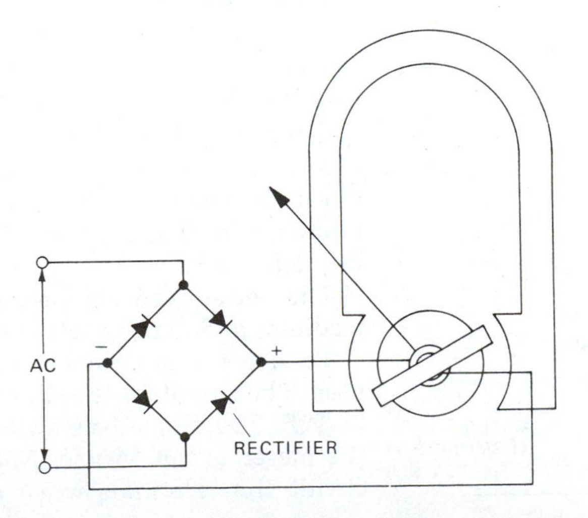

Figure 22-11B
Dry-type rectifier connections for a high-frequency measuring instrument

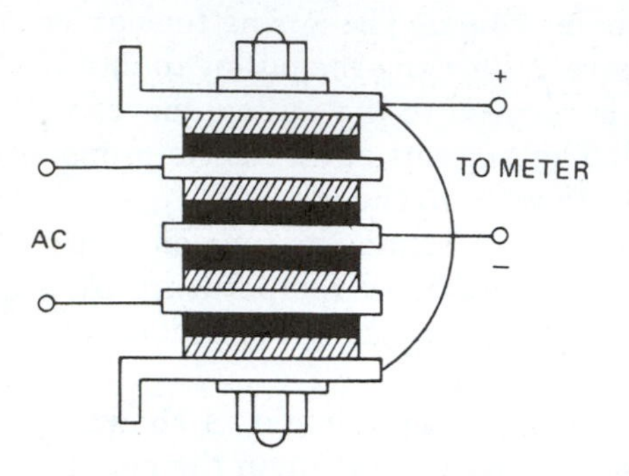

Figure 22-11C
Current flow through a meter circuit during the positive half cycle of ac

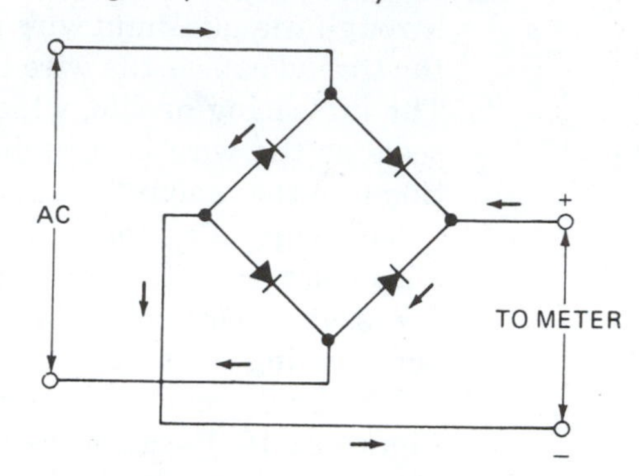

Figure 22-11D
Current flow through a meter circuit during the negative half cycle of ac

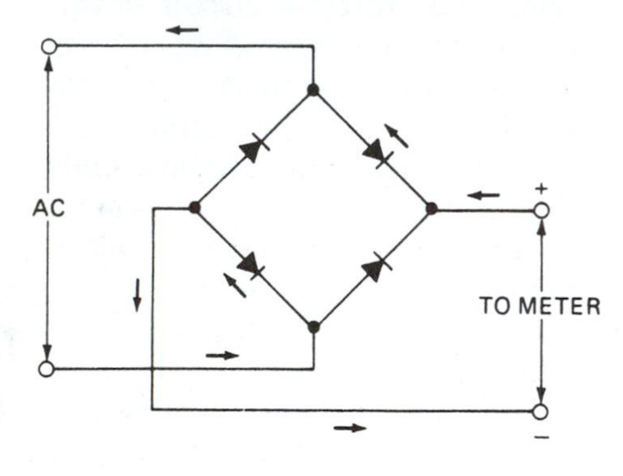

connections. Figure 22-11C indicates the flow of current during the positive half cycle, and Figure 22-11D shows the flow during the negative half cycle. Note that the flow through the meter is always in the same direction.

This meter has a high sensitivity and is very accurate over a wide range of frequencies.

Ohmmeters

There are two basic types of ohmmeters: the *series type* and the *shunt type*. The circuit used in the series type is shown in Figure 22-12A. The rheostat is used to zero the meter before making measurements. When the meter leads are connected together, maximum current flows, giving full-scale deflection and indicating zero resistance. If the leads are connected across an unknown resistance, the current in the circuit will decrease by an amount, depending upon the value of the unknown resistance. The greater the resistance, the less the indicating needle will be deflected. The meter scale is calibrated to indicate the value of resistance, in ohms. Note that the meter scale is in a counterclockwise direction, even though the needle moves in a clockwise direction. The reason for this arrangement is that, as the resistance increases, the current through the circuit decreases, causing less deflection.

The series-type ohmmeter is generally used for measuring medium- and high-resistance values.

To measure low values of resistance, the shunt-type meter is used. The circuit for the shunt-type ohmmeter is shown in Figure 22-12B. The single-pole switch is used to open the circuit when the meter is not in use. *Note:* With this meter, do not short-circuit the terminals when adjusting the needle for full-scale

Figure 22-12A
Series-type ohmmeter

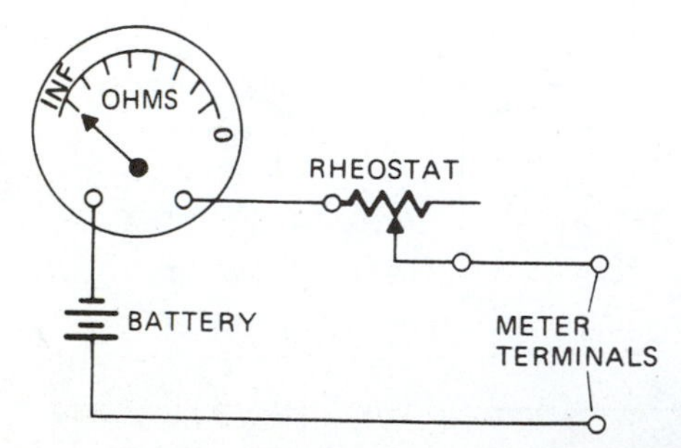

Figure 22-12B
Shunt-type ohmmeter

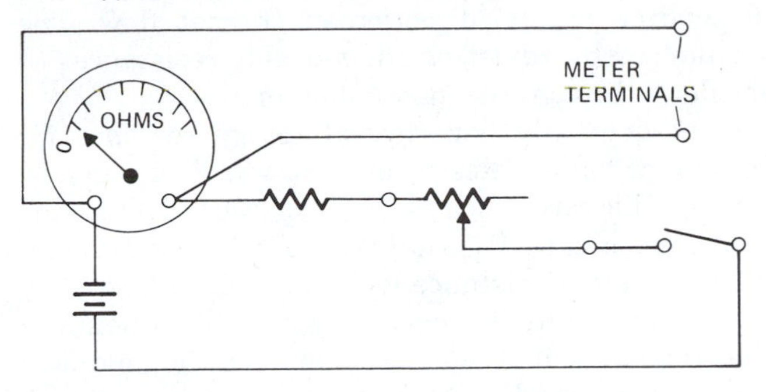

Figure 22-13A
Two series circuits as used in a Wheatstone bridge

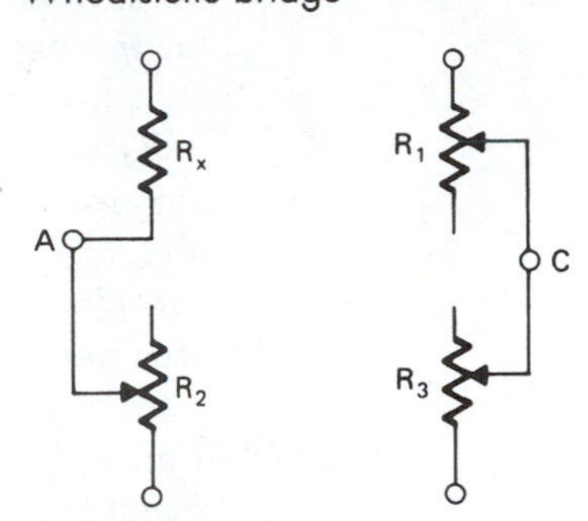

Figure 22-13B
Circuit with galvanometer used to measure resistance

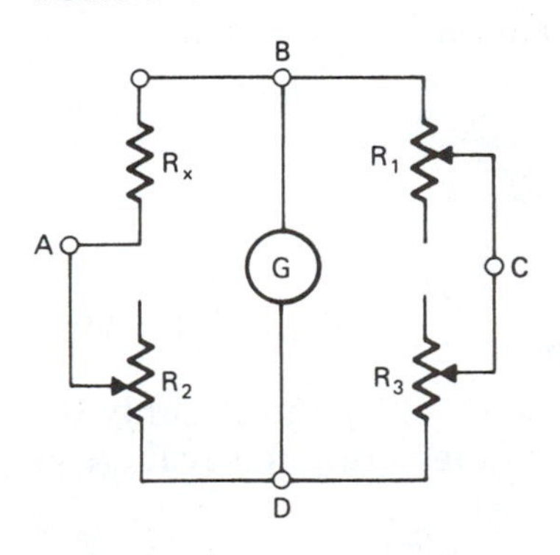

Figure 22-13C
Bridge circuit for measuring resistance

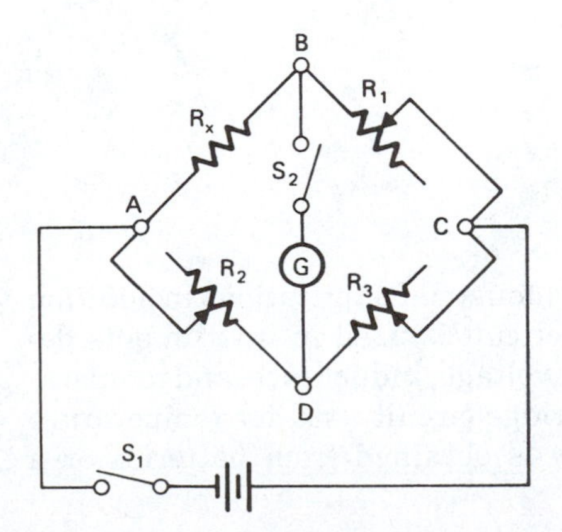

deflection. Closing the switch completes the circuit through the rheostat, series resistor, and meter movement. The rheostat can then be adjusted to cause the needle to indicate full-scale deflection. When the meter leads are placed across an unknown resistance, part of the current will be shunted from the meter through the resistance. The torque on the meter movement will decrease, causing less deflection. For very low values of resistance, most of the current will flow through the resistance and little current will flow through the meter movement. This condition produces less torque, and the needle will move down scale. With this type of meter, the scale reads in the normal direction.

Resistance Bridge

An instrument used to obtain very accurate resistance measurements is the *resistance bridge.* The two most common types are the *Wheatstone bridge* and the *slide-wire bridge.*

The metering circuit for the Wheatstone bridge consists of three adjustable resistances whose values are indicated for each setting. (See Figures 22-13A, 22-13B, and 22-13C.) The resistance to be measured is connected across two terminals, A and B. The circuit is supplied from a battery. A galvanometer (an instrument for measuring very low currents) is also connected into the circuit. The resistances are arranged to form two series circuits, Figure 22-13A. The two series circuits and the galvanometer are connected in parallel, and a dc supply is connected to points A and C, as indicated in Figure 22-13C. This arrangement is known as a bridge circuit. Switches S_1 an S_2 are added to control current flow.

When S_1 is closed, current flows to point A, where it then divides, with some current flowing through R_x and R_1, and some current flowing through R_2 and R_3. At point C, the current paths join and the total current returns to the battery. Closing

S_2 forms another path through the galvanometer. When current flows through the galvanometer, the needle will deflect to the right or the left, depending on the direction of current flow. The bridge can be balanced by adjusting the variable resistances so that no current flows through the galvanometer.

This instrument generally has several rotary switches for varying the resistance values. Resistance, R_1 and R_3 are generally adjusted to a specific ratio, such as 10 to 1 or 100 to 1. Once the ratio is established, resistance R_2 is adjusted until the galvanometer needle indicates zero. Resistance R_2 is generally adjusted by rotating several rotary switches, each making adjustments of 1 ohm, 10 ohms, 100 ohms, and so on. The value of the unknown resistance R_x is a direct reading of units, tens, and hundreds, as indicated.

When the bridge is balanced, the potential difference form B to D is zero. In order to obtain this balance, the voltage drop from A to B must be equal to the voltage drop from A to D., and the voltage drop from B to C must be equal to the voltage drop from D to C.

Expressed algebraically, $I_1R_x = I_2R_2$ and $I_1R_1 = I_2R_3$. These two equations can be combined:

$$\frac{I_1R_x}{I_1R_1} = \frac{I_2R_2}{I_2R_3}$$

Therefore,

$$R_x = \frac{R_1R_2}{R_3} \qquad \text{(Eq. 22.1)}$$

It is not necessary to know the values of R_2 and R_3; only the value of their ratio is necessary. This value multiplied by R_1 gives the unknown resistance.

Example 4

A bridge is balanced when $R_2 = 1000\ \Omega$, $R_3 = 10\ \Omega$, and $R_1 = 34\ \Omega$. What is the value of R_x?

$$R_x = \frac{R_1R_2}{R_3}$$

$$R_x = \frac{34 \times 1000}{10}$$

$$R_x = 34 \times 100$$

$$R_x = 3400\ \Omega$$

The bridge circuit has many industrial applications in addition to measuring resistance. The circuit is used in instruments designed to measure capacitance, voltage, inductance, and temperatures. Figure 22-14 shows a bridge circuit used for temperature measurements. The dc supply is obtained from batteries or a

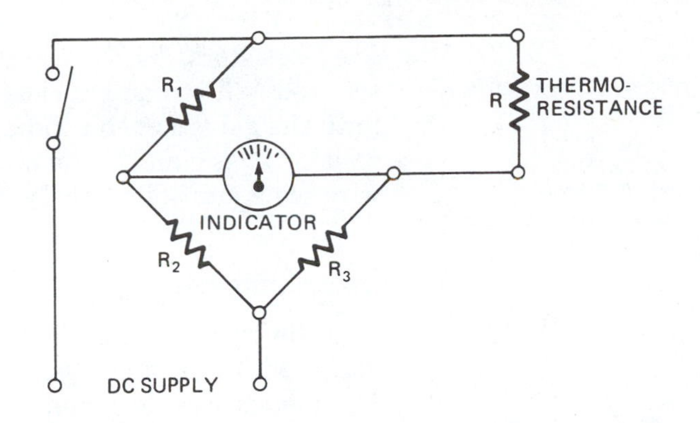

Figure 22-14
Bridge-type temperature-measuring instrument

rectifier. The indicator is a milliammeter calibrated in degrees Fahrenheit or Celsius. Resistor R is placed at the point of measurement. This resistor is made of heat-sensitive materials such as platinum, nickel, and copper. The materials are said to have a high temperature coefficient (a numerical value that indicates the relationship between temperature change and time). Resistors R_1, R_2, and R_3 are made of manganese (which has an extremely low temperature coefficient). The resistance of this material varies proportionally with changes in temperature. Therefore, it is very adaptable to temperature measurements. The resistors are designed and arranged so that at normal temperature, the bridge is balanced and the indicating needle is at the midpoint of the scale. When the temperature is above or below normal, the needle will deflect either to the right or to the left, indicating the number of degrees above or below normal. Variations in the supply voltage will not affect the meter.

The circuit for the slide-wire bridge is shown in Figure 22-15. It is similar to the Wheatstone bridge, except R_2 and R_3 are replaced with a high-resistance wire that is stretched tightly and placed parallel with a scale. The resistance to be measured is connected between terminals A and B. R_1 consists of several

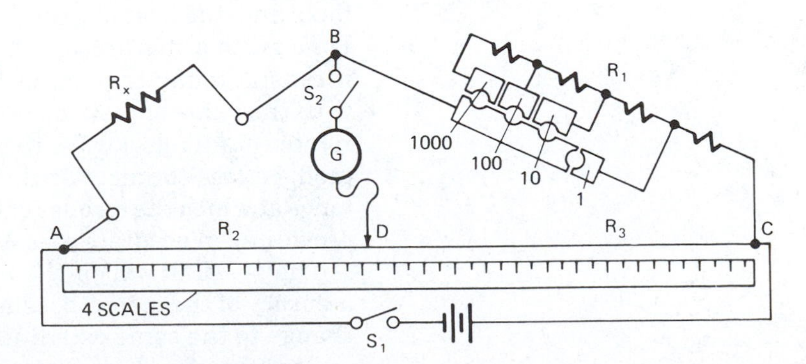

Figure 22-15
Slide-wire bridge for measuring resistance

Figure 22-16
Basic construction and circuitry of a megohmmeter

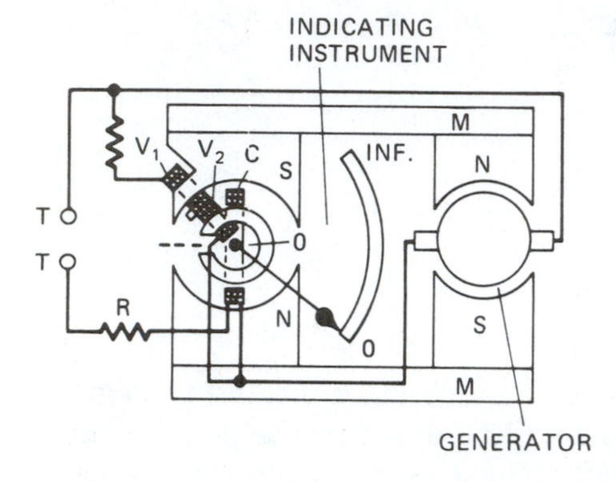

fixed resistors, used as multipliers, any one of which may be selected. With S_1 and S_2 closed, contact D is moved along the wire until the galvanometer indicates zero. The value of the unknown resistor R_x is equal to the numerical value indicated on the slide-wire scale multiplied by the value of the selected multiplier.

Megohmmeter

Ordinary ohmmeters and resistance bridges are not generally designed to measure extremely high values of resistance. The instrument designed for this purpose is the *megohmmeter*, Figure 22-16. It consists of a small generator arranged to drive a meter calibrated to measure high resistances.

The generator, commonly called a magneto, is often designed to produce various voltages, depending upon the value of the resistance to be measured. The output may be as low as 500 volts or as high as 1 megavolt. The current is generally limited to 50 milliamperes, regardless of the value of the resistance being measured. The meter scale is calibrated in kilohms (kΩ) and megohms (MΩ).

The permanent magnets supply the flux for both the generator and the metering device. The voltage coils are connected in series across the generator terminals. The current coil is arranged so that it will be in series with the resistance to be measured. The unknown resistance is connected between the terminals T. When the armature of the magneto is rotated, an emf is produced. This causes current to flow through the current coil and the resistance being measured. The amount of current is determined by the value of the resistance and the output voltage of the generator. The torque exerted on the meter movement is proportional to the value of current flowing through the current coil.

The current through the current coil, which is under the influence of the permanent magnet, develops a clockwise torque. The flux produced by the voltage coils reacts with the main field flux, and the voltage coils develop a counterclockwise torque. For a given armature speed, the current through the voltage coils is constant, and the strength of the current coil varies inversely with the value of resistance being measured. As the voltage coils rotate counterclockwise, they move away from the iron core and produce less torque. A point is reached for each value of resistance at which the torques of the current and voltage coils balance, providing an accurate measurement of resistance.

The speed at which the armature rotates does not affect the accuracy of the meter, because the currents through both circuits change to the same extent for a given change in voltage.

Because megohmmeters are designed to measure very high value of resistance, they are frequently used for insulation tests.

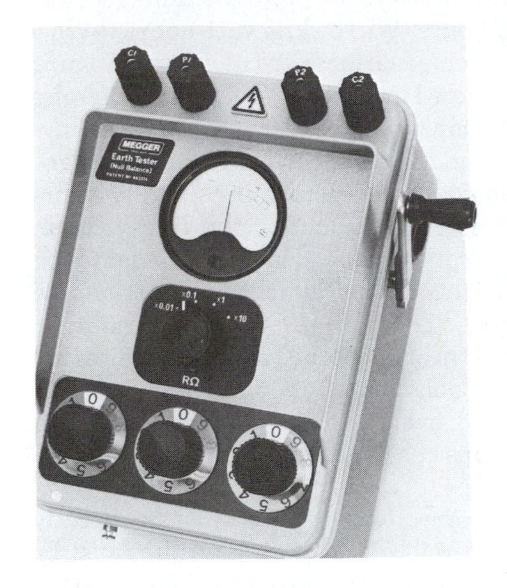

Figure 22-17
Null balance earth tester
(Courtesy of James G. Biddle Co., Plymouth Meeting, PA)

In tests of insulation, the instrument is generally connected between the conductor and the ground. It follows, then, that a good ground is a very important part of the testing procedure. The grounding system should be checked with the megohmmeter as well as with a low-range ohmmeter.

Ground Tester

The *NEC*® requires that the impedance of a system ground be sufficiently low to limit the voltage to ground and to facilitate the operation of the circuit protective device. To ensure good continuity to ground, an earth resistance test should be performed at the time of installation and periodically thereafter. Various manufacturers of resistance-measuring instruments will supply information on ground testing. One such instrument, the *null balance earth tester*, is illustrated in Figure 22-17.

Digital Meters

Digital meters have many useful applications in industry. They are fast acting and provide a numerical readout, thus eliminating the possibility of error caused by improper sighting of a pointer. They are constructed of solid-state circuitry and require less maintenance than the meters previously mentioned. They are capable of indicating millisecond power surges, making them very useful for troubleshooting. Figure 22-18 illustrates some types of digital multimeters.

Figure 22-18
Digital multimeters *(Courtesy of Simpson Electric Co.)*

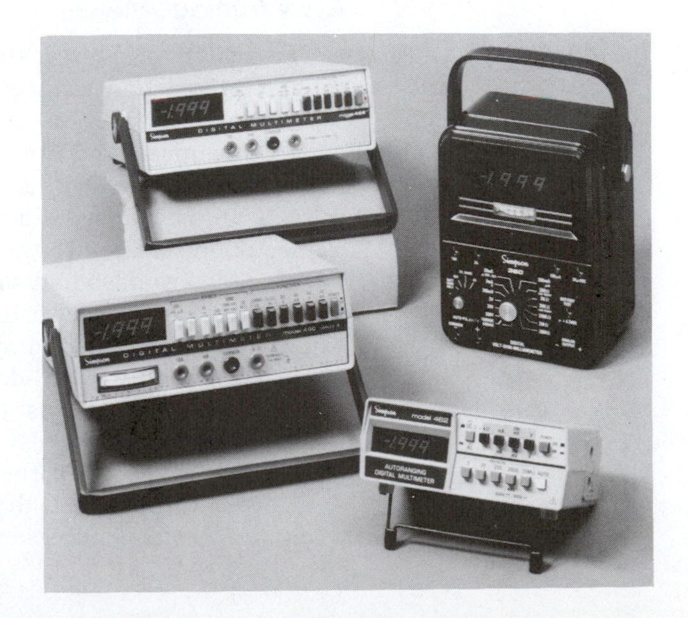

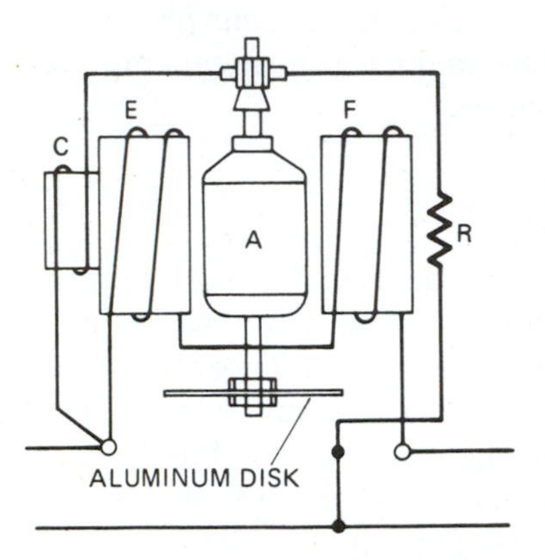

Figure 22-19
DC watthour meter

DC Watthour Meter

Electrical energy is the measurement of the amount of power used times the period of time it is used. Electrical energy, therefore, is the product of the power in watts (W) and the time in hours (h), and is expressed in watthours (Wh) or kilowatt-hours (kWh). The *watthour meter* is used to measure electrical energy, Figure 22-19. The field coils are connected in series with the load, thereby producing a magnetic field that is proportional to the load current. A smaller coil of fine wire, the armature, and a resistor are arranged in series and then connected across the line. The current through the armature is proportional to the line voltage. With this arrangement, the torque produced is proportional to the main field flux and the armature current (which is proportional to the power supplied to the load). This torque causes the armature to rotate and drive indicators through a train of gears, indicating the number of kilowatt-hours delivered.

Coil C produces a weak field, even when no current is supplied to the load. This field develops just enough torque to overcome friction.

For the meter to indicate correctly, its speed must be directly proportional to the power. Therefore, a resisting torque must be provided that is directly proportional to the speed. The resisting torque is obtained by mounting an aluminum disk on the shaft and arranging it to rotate between the poles of two permanent magnets. Eddy currents produced in the disk develop a counter torque, which is directly proportional to the speed.

AC Watthour Meter

The watthour meter shown in Figure 22-19 will operate satisfactorily when connected in an ac system. The induction-type meter, however, is generally used because it is simpler, more economical, and more reliable, and requires less maintenance. Figure 22-20 illustrates this type of instrument. The voltage coil is connected across the line, and the current coil is connected in series with the load.

The coils are designed so that the current in the voltage coil lags that of the current coil by 90 electrical time degrees. Thus, when connected into the system, a rotating magnetic field is produced. To ensure a 90-degree phase displacement, an adjustable resistance (R) is installed on the core of the voltage coil. Current flowing in this resistance tends to delay the buildup and decay of the flux in the core to obtain the 90-degree displacement.

In order to provide a drive that is proportional to the load, a counter torque is necessary. This is accomplished by installing

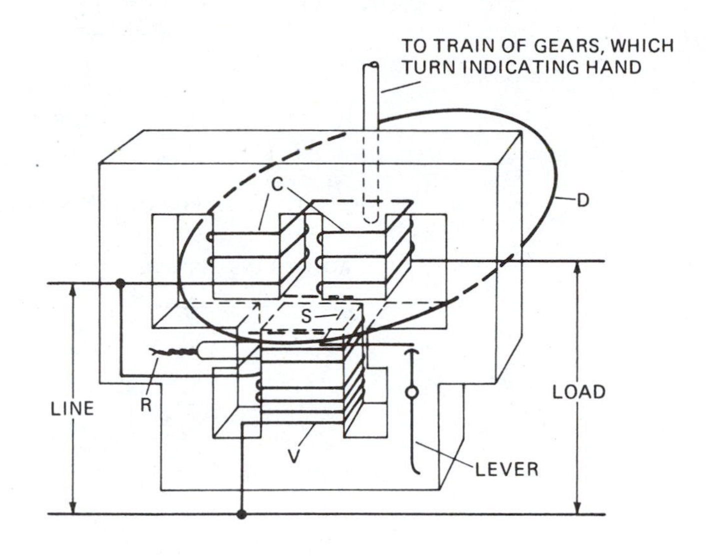

Figure 22-20
AC watthour meter. D indicates the aluminum disk, C indicates the current coils, V indicates the voltage coil, and R indicates a variable resistance

a permanent magnet arranged so that the aluminum disk rotates within its field. Voltages induced into the disk produce eddy currents proportional to the speed of the disk. The field produced by these currents develops a counter torque in proportion to the load current. As a result, the speed of the disk is directly proportional to the load being measured.

The rotating disk drives a gear train, which in turn drives the indicating hands of the meter. To compensate for the friction caused by the gear train, a shading coil is installed in the air gap. This coil produces just enough positive torque to compensate for the friction.

In a three-phase system, energy, like power, may be measured by means of two single-phase instruments. A three-phase meter, Figure 22-21, is simply a combination of two single-phase meters, with their moving elements mounted on the same shaft. Under this arrangement, the total driving torque is the sum of the torques exerted on each element. Thus, only one registering mechanism is required.

Figure 22-21
Three-phase watthour meter

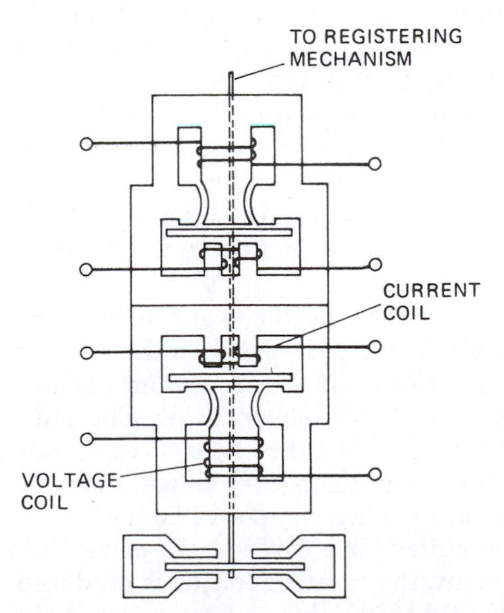

Reactive Power Measurements

Reactive power is the power that circulates in the system but does not produce useful work. This power is referred to as the reactive volt-amperes, or vars.

Reactive power can be measured with a standard ac wattmeter and phase-shifting transformers. The arrangement consists of two autotransformers connected in open delta, Figure 22-22. The transformers shift the voltage applied to the potential coils by

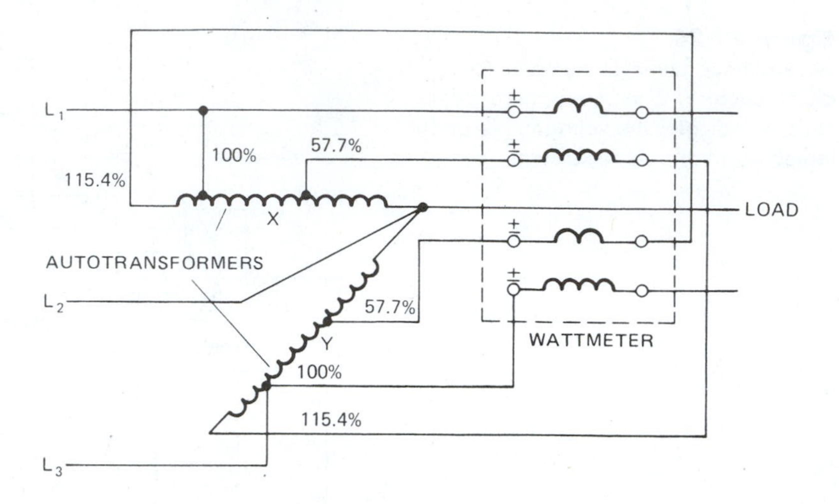

Figure 22-22
Circuit for measuring reactive power

90 electrical time degrees. The value of the voltage remains the same. The result is that the torque developed by the meter is proportional to VI sin ϕ, which is the reactive power.

Power Factor Meter

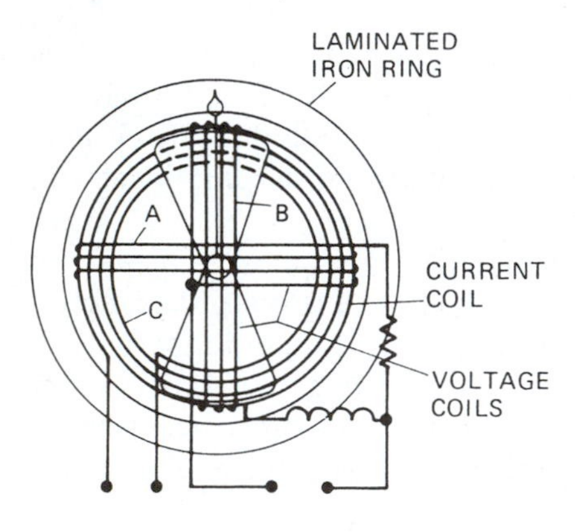

Figure 22-23A
Connections for a power factor meter

Figure 22-23A illustrates a *power factor meter*. Two coils, A and B, are identical and are arranged at right angles to each other. One coil is connected in series with a high inductance. The currents through the two coils are nearly 90 degrees out of phase with each other, thereby producing a rotating magnetic field. Coil C is wound at right angles to both A and B, and is connected in series with the load. A soft iron vane, which is constructed as shown in Figure 22-23B, is mounted so that its axis coincides with the axis of coil C. The vane is magnetized by the current through C. A laminated iron ring provides a return path for the flux. The attraction between the iron vane and the rotating magnetic field causes the vane to take a position parallel to the rotating magnetic field at the instant that the magnetization of the vane is at a maximum.

At unity power factor, the indicating needle is at the midpoint of the scale, indicating a power factor of 1.00. When the current leads the voltage, the magnetization of the iron vane reaches maximum sooner than it does at unity power factor. The indicating needle moves to the left. The distance it moves depends upon the amount of lead. If the current lags the voltage, the vane will deflect to the right, indicating a lagging power factor.

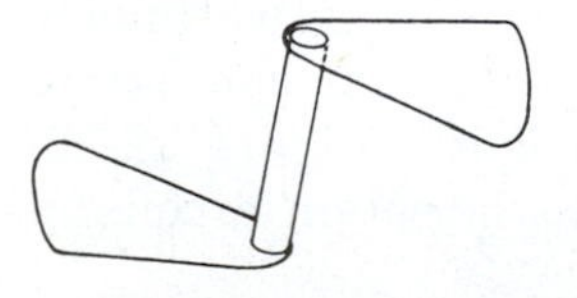

Figure 22-23B
Soft iron vane

Figure 22-24 shows the circuitry for a three-phase power factor meter. With this instrument, the rotating torque is produced by the three-phase magnetic field as in a three-phase motor. As in

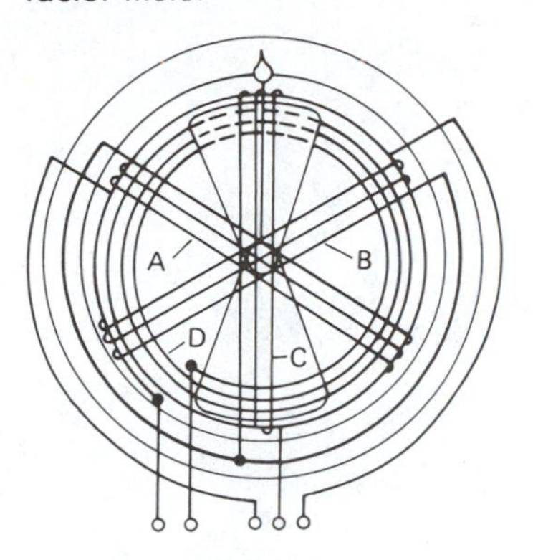

Figure 22-24
Circuitry for a three-phase power factor meter

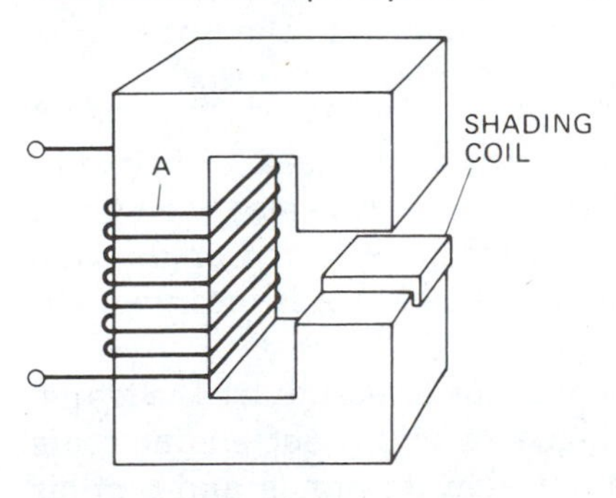

Figure 22-25A
Coil used in a frequency meter

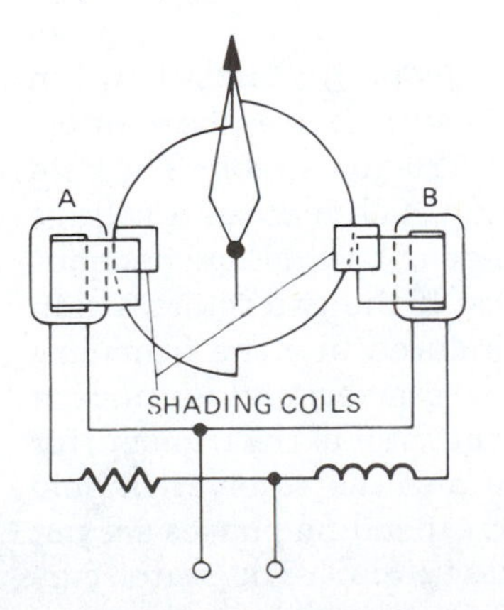

Figure 22-25B
Basic construction and circuitry for a frequency meter

the case of the single-phase instrument, coil D produces an alternating field in the iron vane. This causes it to take a position parallel to the rotating magnetic field when the magnetizing action of the iron vane is maximum, thereby showing the power factor.

Frequency Meter

A common type of *frequency meter* is shown in Figures 22-25A and 22-25B. This instrument requires two electromagnets, each with a shaded pole. When the meter is connected into the circuit, one coil develops a torque in one direction and the other develops a torque in the opposite direction. The indicating needle is secured to the disk and indicates the normal frequency at the midpoint on the scale. The two electromagnets are connected in parallel, but one is in series with a high resistance and the other is in series with a high inductance. At a given voltage, the effective current through coil A is constant and the current through coil B varies with the frequency.

When the frequency increases, the current through B decreases because of the increased inductive reactance: $X_L = 6.28fL$ (Equation 13.9). The decrease in current is accompanied by a decrease in the torque produced by coil B. The disk turns clockwise until the torque of A is equal to that of B.

Because each half of the disk is off center from the other and from the electromagnet, the amount of torque varies according to the area within the magnetic field. Rotating the disk causes more or less of the metal to be in the magnetic field. This arrangement provides equal torque at some point on the meter scale for any frequency within the range of the meter.

When the frequency decreases, the current through B increases, thus developing a stronger torque and causing the disk to turn counterclockwise until the two torques are equal.

The Synchroscope

The *synchroscope* is an instrument used to indicate the phase relationship between two voltages. It is used when paralleling two alternators or when paralleling an alternator with the power line. Its construction is similar to a power factor meter. If the current coil in the power factor meter is replaced by a voltage coil and the leads are brought out, as shown in Figure 22-26, the instrument will indicate the phase displacement between two voltages.

Coils A and B are connected through a transformer to the alternator, and coil C is connected to the line bus. If the frequency of the alternator is higher than the line frequency, the indicating needle will move in one direction. If the frequency is lower,

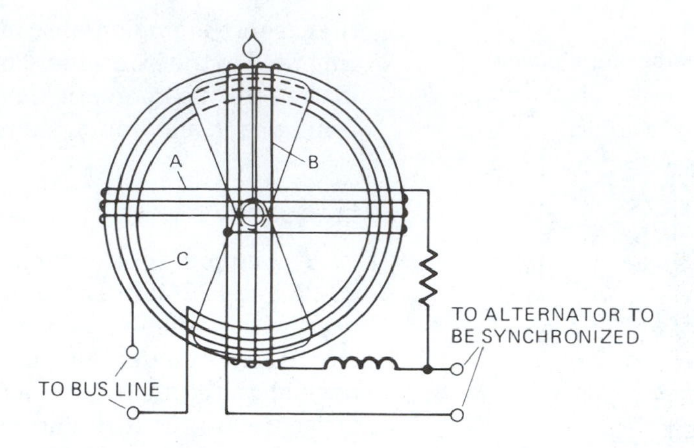

Figure 22-26
Basic circuit for a synchroscope

the needle will move in the opposite direction. If the frequencies are equal but the voltages are out of phase, the needle will fluctuate. When the frequencies are equal and in phase, the needle will remain at the midpoint. When the synchroscope remains at the midpoint, the alternator may be connected to the line.

SELF-SYNCHRONOUS SYSTEMS

A *self-synchronous transmission system* is used to transmit information or signals to some distant point. It is also used when it is necessary to rotate devices through preselected angles and when these devices are located in areas where it is inconvenient to operate them mechanically.

Some uses of synchro (self-synchronous) systems are the transmission of messages from the bridge of a ship to the engine room and the rotation of radar and television antennas and similar equipment. Synchro systems are used to indicate wind direction and velocity as well as to synchronize the speed between motors.

AC Synchronous Systems

A wiring diagram of an *ac self-synchronous system* is shown in Figure 22-27. The rotors are energized with single-phase alternating current, usually from a 60-hertz, 120-volt supply. The field current establishes an alternating flux, which induces a voltage into the stator windings. The magnitude of the voltage depends upon the relative position of the rotors. If the two rotors are in corresponding positions, the voltage induced into the transmitter is equal and opposite to that of the receiver, and no current flows in the stator circuit. If, however, the rotor of the transmitter is rotated a given number of degrees and the receiver is held stationary, the induced emfs in the corresponding phases are not equal and current will flow in the stator circuit. The stator cur-

Figure 22-27
AC self-synchronous system

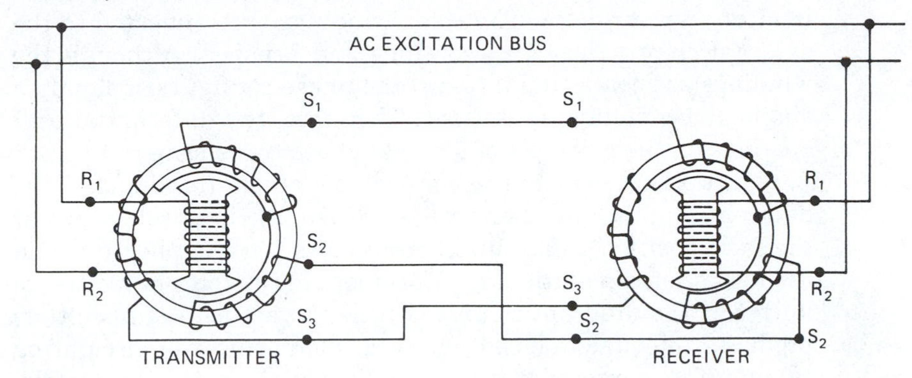

rent establishes poles in the receiver that are in the same position, and of the same polarity, as that of the transmitter. This develops a torque in the receiver that causes the rotor to rotate to a position corresponding to that of the transmitter. When the rotors are again in like positions, the stator voltages are again equal and opposite.

When the rotor of the transmitter is moved, the rotor of the receiver will move to the same position. In other words, this system uses electrical transmission to obtain mechanical motion.

Differential Self-synchronous Systems

A *differential self-synchronous system*, Figure 22-28, consists of three or more units. One unit is a receiver, one is a differential unit, and the rest are transmitters.

Figure 22-28
Differential self-synchronous system

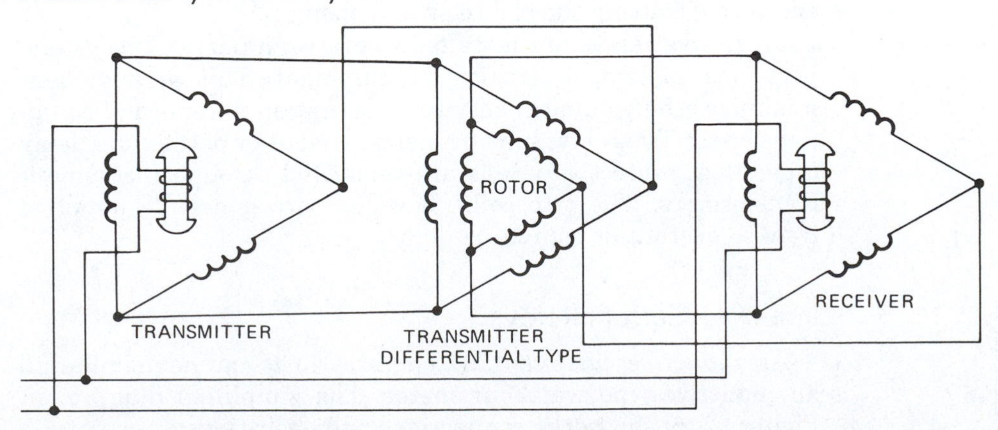

The differential unit can be compared to a wound-wire motor. The stator and rotor have three windings, which may be connected in either wye or delta. The rotor windings are connected to the external circuit through slip rings and brushes. Although the windings are connected into a three-phase configuration, only a single-phase voltage is present. Therefore, the differential unit operates on the principle of a single-phase transformer.

Figure 22-28 depicts the connections of the transmitter, the differential unit, and the receiver. When all three rotors are at the same position, minimum current flows between the units and the rotors remain stationary. Movement of the rotor of the transmitter or the differential unit (which can act as a transmitter) produces unequal induced voltages, increasing the circulating current. The increased current establishes poles in the stator of the receiver in the same position as that of the transmitter. This causes the rotor to move to a position that is the same as that of the transmitter. If both the transmitter and the differential unit are rotated simultaneously,the receiver rotation will be equal to the sum of their movements. Also, the receiver can be used as a transmitter, causing the other two units to rotate through various angles. The differential unit will indicate the difference between the two angles.

PROTECTIVE RELAYS

A *protective relay* is an device designed to protect circuits and equipment from abnormal conditions of current, voltage, or frequency. For high currents and voltages, protective relays are safer and more practical than the ordinary circuit breaker. Oil switches (switches with their contacts immersed in a nonflammable insulating liquid) are frequently used in conjunction with protective relays. The relay operates the oil switch to open the main circuit.

Under normal operating conditions, the solenoid of the relay is deenergized. When an abnormal condition occurs, the solenoid is energized, causing the coil to switch open.

Protective relays are installed to protect a particular system, part of a system, a circuit, or equipment. This arrangement minimizes the volume of service interruption if abnormal conditions occur. The relays may be instantaneous or of the time-delay type. Most protective relays are connected through instrument transformers. The trip coils, however, are generally supplied from a constant dc source.

Figure 22-29
Circuit for an inductive current relay

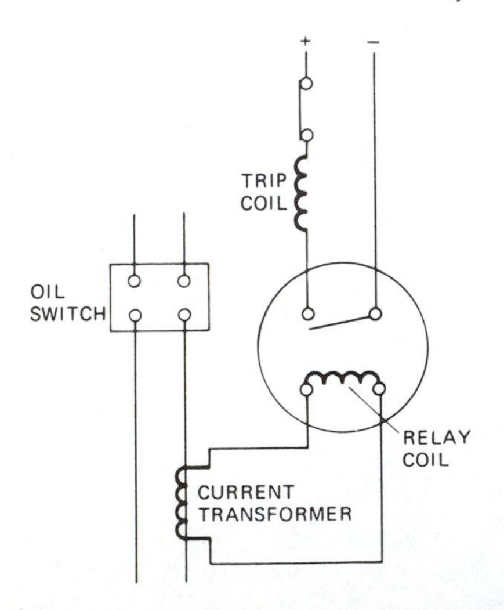

Inductive Current Relays

An *inductive current relay* operates in a manner similar to an inductive-type watthour meter. The simplified diagram in Figure 22-29 shows the connections and operation.

When an abnormal current flows in the main circuit, a high voltage is induced into the secondary of the current transformer. This causes an increase in the current in the relay coil. The increase in current produces a greater flux, causing the disk to rotate enough to close the relay switch and complete the circuit through the relay tripping coil. Current flowing in the relay tripping coil establishes a magnetic field that opens the main line switch.

To make sure that the disk remains stationary and in the starting position until an overload occurs, holes are punched in the section of the disk that lies between the poles of the relay coil. Because of the holes, the torque is low and a light spring tension causes it to remain stationary. When the current through the relay coil becomes high enough to cause the disk to move only slightly, the holes move out from between the poles and the solid part of the disk produces a high torque. The disk rotates to a point where it closes the contacts. Movement ceases at this point, because the movable contact attached to the disk has met the stationary contact. The torque now causes the contacts to remain solidly closed until the line switch opens.

This type of relay usually has the inverse time-limit characteristics. It is generally adjusted to trip at the end of 4 seconds on a 300 percent overload, and at the end of 2 seconds on a 400 percent overload. For loads above 400 percent, the action is almost instantaneous. Most relay coils are provided with a means of adjusting the value of trip current.

Differential Relay

The *differential relay* is very similar to the inductive current relay. The major difference is that the differential relay has an additional electromagnet, which develops a counter torque when the circuit or equipment is operating under normal conditions. Differential relays compare the supply current to the return current. These currents are either equal to or in direct proportion to each other. When a fault occurs, an unbalanced condition exists and the difference between the two currents flows through the control relay, closing the control contacts. The tripping relay establishes a magnetic field that opens the circuit breaker, protecting the equipment.

Voltage Relay

An overvoltage causes an excessive current, producing enough heat to damage the insulation on conductors and equipment and cause lamps to burn out. Undervoltages result in less current and reduced lighting. When applied to motors, the reduced current will not develop enough torque to accelerate the motors to their rated speed. Because the motor is operating at a reduced speed, a lower counter electromotive force (cemf) is produced and an excessive current flows. The relays that are used to guard against these conditions are called *voltage relays*. They are designed as either undervoltage relays or overvoltage relays. Both types are connected in the same manner as the inductive-type current relay, except that a potential transformer is used in place of a current transformer.

The overvoltage relay has a spring that holds the contacts open, up to and including the normal operating voltage. If the voltage increases beyond the setting of the relay, sufficient torque is developed to overcome the spring tension, allowing the contacts to close. The trip coil then operates the line switch.

The undervoltage relay is designed so that the spring holds the control contacts closed when the potential transformer is deenergized. When the voltage across the transformer is normal, the contacts are held open by the torque on the disk. If the voltage drops below the setting of the relay, the torque decreases and the spring closes the control contacts. Then the trip coil becomes energized and operates the line switch.

Power Relays

A power relay operates in the same manner as an inductive-type watthour meter, with the exception that the shaft attached to the revolving disk operates a contact. The power relay senses the direction of the supply power and is used to protect against the reversal of power. It is used in powerhouses or substations. The torque developed holds the control contacts open when the power is from the correct direction; it closes the contacts when the power is from the wrong direction. This relay can also be used to protect against excessive power from one direction. When used for this purpose, the connections to either the current coil or voltage coil must be reversed.

Power-Directional Relays

Power-directional relays are generally used by electric utility companies to protect transmission lines. Figure 22-30 shows one method of installing power-directional relays. If a short circuit occurs at point S, the power will be supplied not only through feeders A but also through feeders B, through the substation buses, and through switch 2 to point S. If the only protection for

Figure 22-30
Power-directional relays used to protect power distribution lines

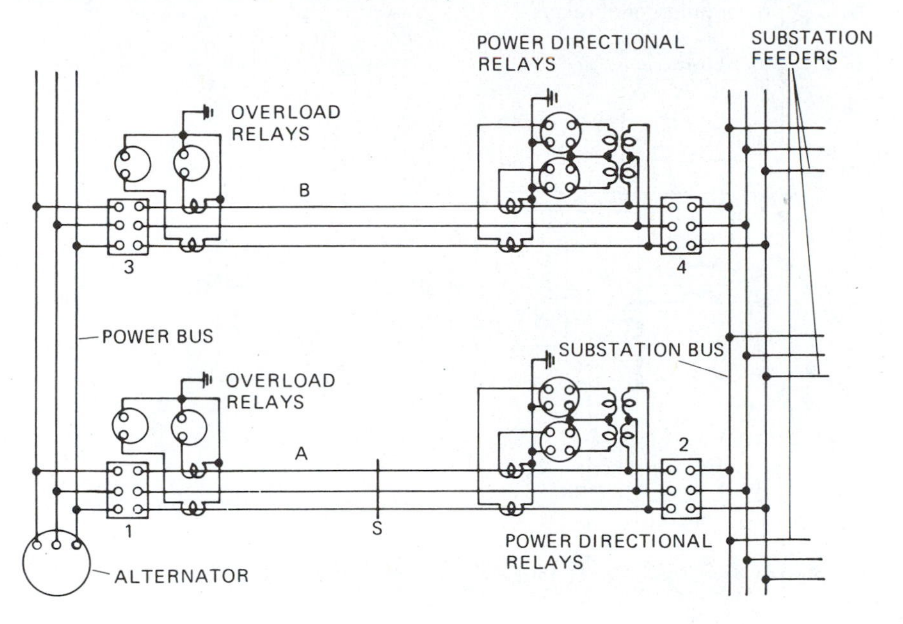

this system consists of the overload relays located at switches 1 and 3, these relays will operate, opening the circuits through both A and B. This action will result in the loss of power to the entire system.

If, however, power relays are located at switches 2 and 4, the power through switch 2 will be in the wrong direction and the power relay will operate, opening switch 2. The power through switch 4 is in the right direction. Therefore, the power relay contacts will remain open. The relay coil is not energized, permitting the line contacts to remain closed. Power will continue to be supplied through the B feeders.

This arrangement provides adequate protection for the feeders between the overload relays and the power relays. If, however, a short circuit occurs near the power relays of switch 2, the emf impressed across the voltage coils of the relays will drop to a very low value. This value will be so low that the relay will not operate.

Another disadvantage of this arrangement is that when it is used with a three-phase system and a short circuit occurs between two lines only, the currents through the voltage and current-actuating relays are frequently nearly 90 degrees out of phase. Under this condition, not enough torque is developed to cause the relays to operate.

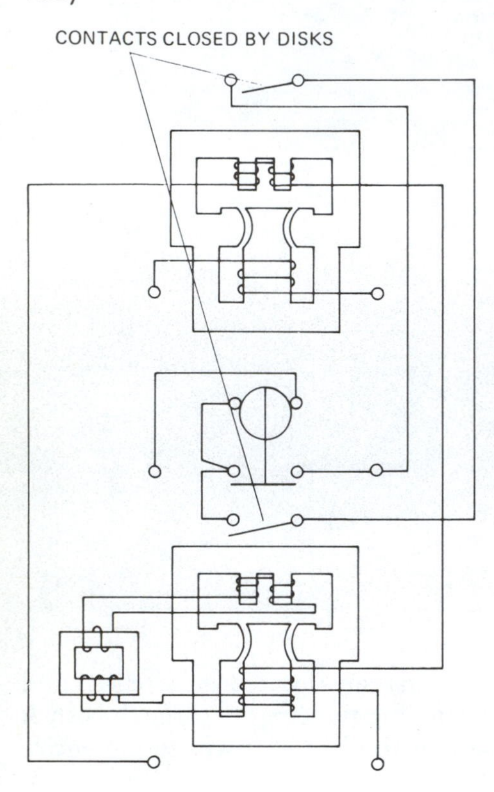

Figure 22-31A
Specially designed power-differential relay

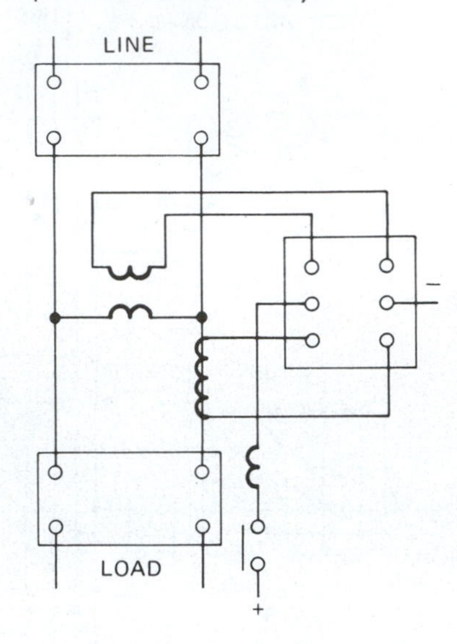

Figure 22-31B
Single-phase connections for power-differential relays

Specially designed power-differential relays are manufactured to eliminate this problem. Wiring diagrams of this type of relay are shown in Figures 22-31A and 22-31B. For single phase it is connected as shown in Figure 22-31B. The lower unit of the relay is the same as an inductive-type overcurrent relay. The upper unit is similar to the power relay discussed previously, except that it requires only a very small amount of power in the opposite direction to cause the control contacts to close. This relay has two contacts connected in series; thus, both must be closed to energize the trip coil.

If a short circuit occurs, the direction of the supply power reverses and the power unit closes its contacts instantly (Figure 22-31A). The lower unit, however, being a current relay, closes only after a predetermined amount of time. When the current relay contacts close, the circuit is complete to the trip coil and the main line switch opens.

Phase-Balance Current Relay

The *phase-balance current relay* is used to protect a motor against an open phase, a reversal of phase, and a short circuit in any phase.

The internal connections of the relay are shown in Figure 22-32A. The lower part is similar to the inductive-type over-

Figure 22-32A
Internal connections for a phase-balance current relay

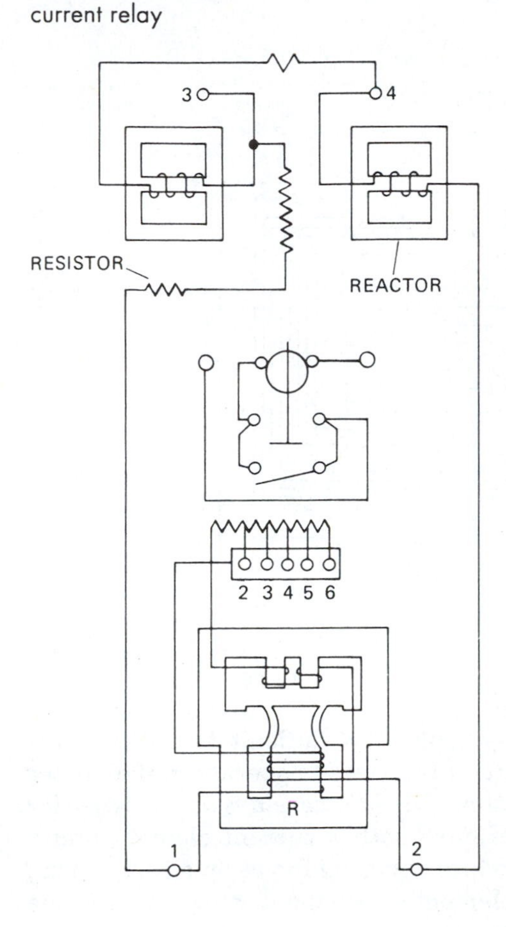

Figure 22-23B
External connections for a phase-balance current relay

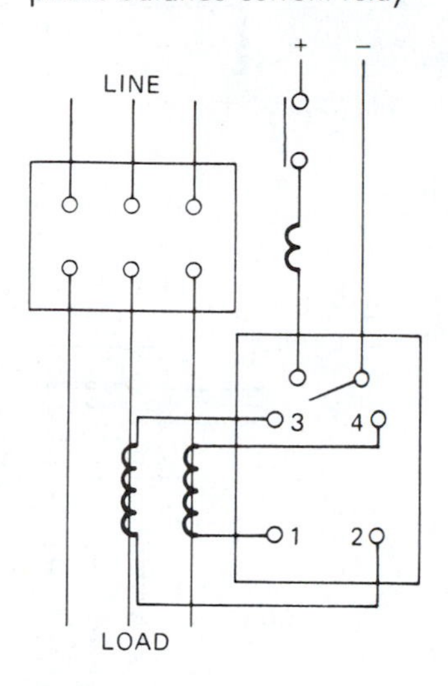

Figure 22-32C
Simplified diagram of internal and external connections for a phase-balance current relay

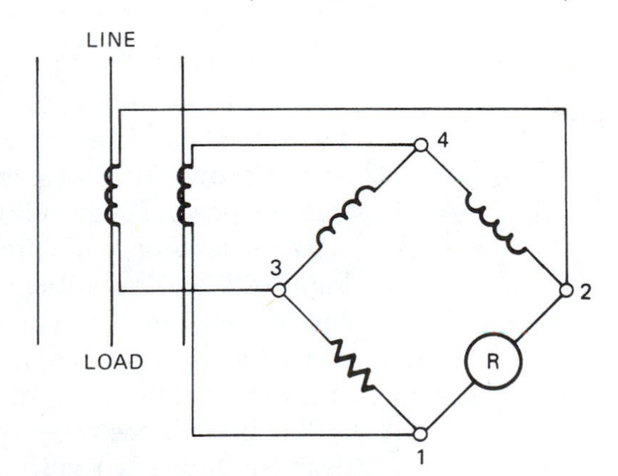

current relay; the upper part consists of stationary resistors and reactors. It is connected through current transformers, Figure 22-32B and 22-32C. The resistors and reactors are designed so that no current flows through R when all conditions are normal. If an abnormal condition occurs, such as an open phase, current will flow through coil R and the relay will operate, opening the main line switch to the motor.

This relay is designed to operate when the currents in the three phases become unequal even by a small amount.

Figure 22-33
Impedance relay

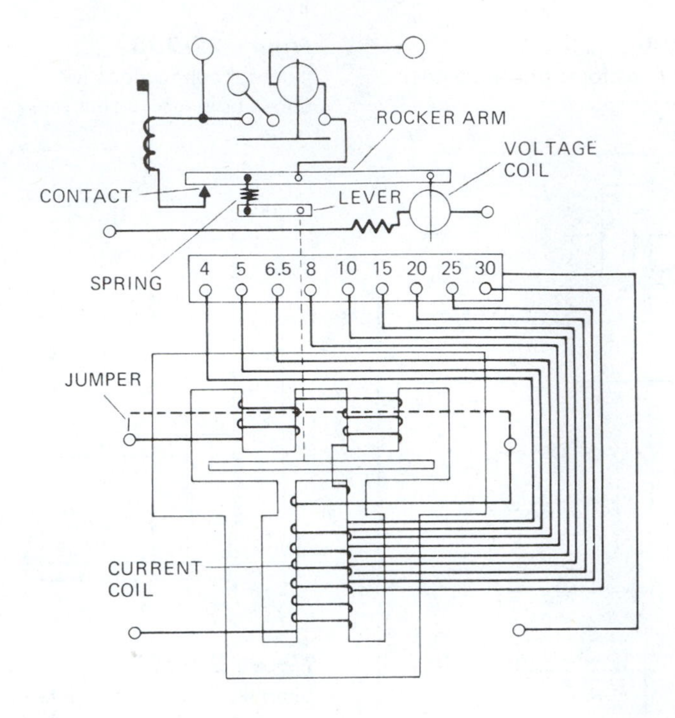

Impedance Relays

With some industrial systems it is difficult to adjust overcurrent relays, fuses, or circuit breakers to operate in the proper sequence and within a safe time. *Impedance relays* are designed for this purpose. These relays have both a current element and a voltage element, which produce opposing forces on a rocker arm, Figure 22-33. The voltage element is a solenoid, which pulls on one end of the arm. The current element is an overcurrent relay in which the aluminum disk drives a gear train that produces a force on an adjustable spring attached to the other end of the rocker arm.

The impedance relay requires very little time to operate if it is near the fault. It requires a much longer time if it is some distance from the fault. This feature causes the two relays nearest the fault to operate first, thus isolating the fault.

CURRENT-LIMITING REACTORS

Current-limiting reactors are frequently used to limit the maximum current flow to a safe value, even under short-circuit conditions. They are used to protect conductors and equipment from damage caused by mechanical and magnetic stresses and/or high temperatures that can occur during short-circuit conditions.

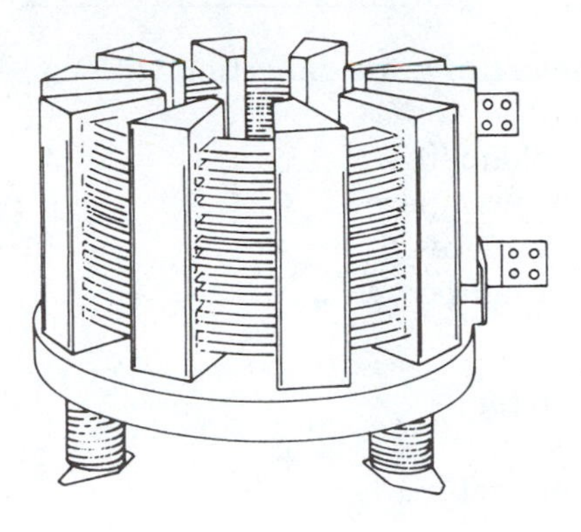

Figure 22-34
Dry-type, current-limiting reactor

Types of reactors

A *current-limiting reactor* is an induction coil designed to produce a specific amount of reactance at a specific frequency. The two standard types of reactors are the *dry type* and the *oil-immersed type*. Neither contains an iron core. An iron core is unnecessary and undesirable because at low currents the reactance should be a minimum, and at high currents the core becomes saturated, having little effect in retarding the current.

The dry-type reactor consists of a cable wound in several layers on a concrete cylinder that has concrete disks on each end. The coil is held in place by a nonconducting, nonflammable material. The reactor must have enough mechanical strength to withstand the powerful forces produced by the magnetic field established when the high currents flow. Figure 22-34 shows a dry-type, current-limiting reactor.

The windings on an oil-immersed reactor are enclosed in a laminated magnetic shield. This shield forms the magnetic circuit for the flux that extends beyond the coil. If the shield were omitted, the field would expand into the case and induce eddy currents into it, producing heat. This leakage of flux into the case would also produce a large power loss.

The reactor is immersed in oil to provide cooling and additional insulation between the turns of cable. Figure 22-35 shows an oil-immersed reactor.

By limiting the maximum amount of current under short-circuiting conditions, reactors make it possible to use overcurrent devices of lower interrupting capacity. This results in lower material and maintenance costs, as well as conservation of space.

Figure 22-35
Oil-immersed, current-limiting reactor

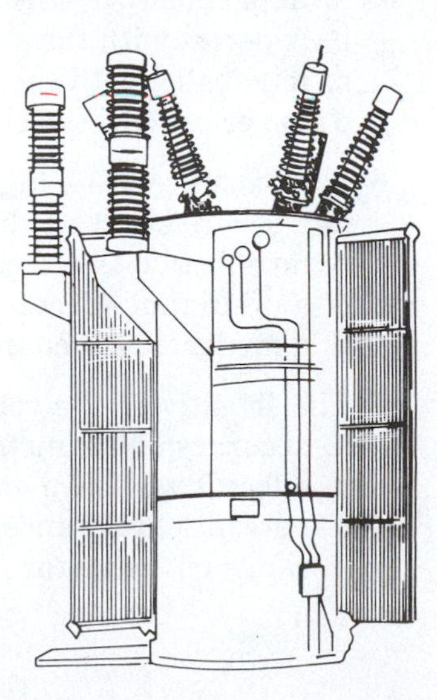

REVIEW

A. Multiple choice.
Select the best answer.

1. Permanent-magnet meters give full-scale deflection with coil currents ranging from
   a. 10 microamperes to 20 milliamperes.
   b. 100 microamperes to 200 milliamperes.
   c. 10 milliamperes to 20 megamperes.
   d. 10 microamperes to 200 milliamperes.
2. Meter shunts are used to
   a. increase the range of the meter.
   b. decrease the range of the meter.
   c. increase the accuracy of the meter.
   d. bypass the meter.
3. External shunts for meters are usually rated for 50
   a. microamperes.
   b. microvolts.
   c. millivolts.
   d. milliamperes.
4. A permanent-magnet meter can be used for measuring
   a. current.
   b. voltage.
   c. wattage.
   d. either a or b.
5. When a permanent-magnet meter is designed for measuring voltage, a resistance is connected
   a. in parallel with the coil.
   b. in series with the coil.
   c. in parallel with the circuit.
   d. in series with the circuit.
6. A voltmeter with a high resistance per volt of scale will have
   a. a severe effect on the circuit characteristics.
   b. no effect on the circuit characteristics.
   c. a minimum effect on the circuit characteristics.
   d. a moderate effect on the circuit characteristics.
7. The sensitivity of a voltmeter indicates the
   a. accuracy of the instrument.
   b. effect it will have on the characteristics of the circuit.
   c. percent of tolerance.
   d. range of the meter.

REVIEW *(continued)*

A. Multiple choice.
Select the best answer (continued).

8. The sensitivity of an ammeter indicates
   a. the amount of current necessary for full-scale deflection.
   b. the voltage drop across the coil at full-scale deflection.
   c. either a or b.
   d. neither a nor b.
9. An electrodynamometer is an instrument designed to measure
   a. direct current.
   b. alternating current.
   c. pulsating current.
   d. all of the above.
10. A wattmeter is generally
    a. an induction-type instrument.
    b. a dynamometer-type instrument.
    c. a capacitive-type instrument.
    d. a reactance-type instrument.
11. A wattmeter has
    a. one circuit.
    b. two circuits.
    c. three circuits.
    d. four circuits.
12. Two-phase power is measured with
    a. a two-capacitor meter.
    b. a two-phase wattmeter.
    c. two single-phase wattmeters.
    d. either b or c.
13. Three-phase power is measured with
    a. a three-phase wattmeter.
    b. two single-phase wattmeters.
    c. either a or b.
    d. a three-capacitor meter.
14. An iron-vane meter is used to measure
    a. alternating current.
    b. direct current.
    c. either a or b.
    d. none of the above.
15. An induction-type instrument is designed to measure
    a. alternating current.
    b. direct current.
    c. either a or b.
    d. only induced currents.

REVIEW *(continued)*

A. Multiple choice.
Select the best answer (continued).

16. Most ac meters are accurate on frequencies of up to
    a. 1000 hertz.
    b. 500 hertz.
    c. 100 hertz.
    d. 60 hertz.

17. One type of meter used for current measurements on circuits having supply frequencies greater than 100 hertz is the
    a. induction-type meter.
    b. hot-wire meter.
    c. aluminum-vane meter.
    d. copper-vane meter.

18. A rectifier is a device that
    a. changes ac to dc.
    b. changes dc to ac.
    c. both a and b.
    d. none of the above.

19. Two types of ohmmeters are the
    a. shunt and series.
    b. wye and delta.
    c. inductive and capacitive.
    d. inductive and resistive.

20. A resistance bridge is
    a. a bus used to shunt resistance.
    b. an instrument used to measure resistance.
    c. a type of rheostat.
    d. an inductive meter.

21. A megohmmeter is an instrument used to measure
    a. very low values of resistance.
    b. very high values of resistance.
    c. ac resistance.
    d. ac reactance.

22. Reactive power is measured with a
    a. reactance meter.
    b. wattmeter and transformers.
    c. voltmeter and capacitors.
    d. wattmeter and capacitors.

REVIEW *(continued)*

A. Multiple choice. Select the best answer (continued).

23. The synchroscope is an instrument used to indicate the
    a. speed of a motor.
    b. phase relationship between the current and the voltage.
    c. phase relationship between two voltages.
    d. phase relationship between the true power and the reactive power.

24. The impedance relay requires very little time to operate if it is
    a. a long distance from the fault.
    b. near the fault.
    c. connected to a circuit breaker.
    d. connected to a GFCI.

25. The primary purpose of a current-limiting reactor is to limit the current during
    a. overloads.
    b. short circuits.
    c. power surges.
    d. overvoltages.

B. Give complete answers.

1. Describe a permanent magnet-type meter.
2. What is the purpose of a shunt used in conjunction with a meter?
3. What is the purpose of a multiplier used in conjunction with a voltmeter?
4. Define the term *sensitivity* as used in relation to a voltmeter.
5. Describe the construction of an electrodynamometer.
6. Define the term *dampened* with reference to an ac meter.
7. Describe the construction of a wattmeter.
8. Explain how to measure the power of a two-phase system using two single-phase wattmeters.
9. Describe how to measure the power of a three-phase system using two single-phase wattmeters.
10. For a three-phase, four-wire system, how can the power be measured using single-phase meters?
11. Describe the construction and operation of an iron-vane meter.

REVIEW *(continued)*

B. Give complete answers (continued).

12. Describe the construction and operation of an inclined-coil meter.
13. Explain the principle of operation of an induction-type meter.
14. Why is a thermocouple meter not sensitive to frequency?
15. Why is a rectifier-type meter not sensitive to frequency?
16. What type of ohmmeter is used for measuring low values of resistance?
17. Describe the operation of the series-type ohmmeter.
18. What is a resistance bridge?
19. Describe the Wheatstone bridge.
20. How does a slide-wire bridge differ from a Wheatstone bridge?
21. What type of instrument is used to measure extremely high resistances?
22. What is a magneto?
23. Describe the dc watthour meter.
24. Describe an inductive-type watthour meter.
25. How can the reactive power of a circuit be measured?
26. Describe the construction of a single-phase power factor meter.
27. Explain the operation of a single-phase power factor meter.
28. How is the torque developed in a three-phase power factor meter?
29. How many electromagnets are required in a frequency meter?
30. In a frequency meter, why is an electromagnet connected in series with a high inductance?
31. How is the aluminum disk in a frequency meter arranged to produce a varying torque?
32. What is a synchroscope?
33. Where is a synchroscope generally used?
34. What is the purpose of a self-synchronous transmission system?
35. List three uses of a self-synchronous transmission system.

REVIEW *(continued)*

B. Give complete answers (continued).

36. Draw a diagram of an ac self-synchronous system.
37. Describe the operation of a self-synchronous system.
38. How many individual units are required in a differential self-synchronous system?
39. Describe the differential unit of a differential self-synchronous system.
40. Draw a diagram of a differential self-synchronous system.
41. Describe the operation of a differential self-synchronous system.
42. What is a protective relay?
43. Describe an inductive current relay.
44. Name two types of voltage relays.
45. What is the purpose of a power relay?
46. What is the purpose of a power-differential relay?
47. What is the purpose of a phase-balance current relay?
48. When are current-limiting reactors required?
49. List two types of current-limiting reactors.
50. Describe the operation of a current-limiting reactor.

23

BASIC INDUSTRIAL ELECTRONICS

Objectives

After studying this chapter, the student will be able to:

- Describe a transistor and explain its operation.
- Explain the use of transistors and other electronic devices in industry.
- Explain the operation of electronic motor controls.
- Describe the cathode ray tube.

Electronics is the branch of electrical science that deals with the control of electrons in circuits containing transistors, tubes, and amplifiers. The primary functions of electronic equipment are to amplify, rectify, and relay signals. Industrial electronics refers to the use of electronic equipment to control equipment and machine operations.

Electronic controls have many functions similar to electromagnetic controls. They are also usually more accurate, take up less space, and require less maintenance. Electronic controls are rapidly taking over many of the functions of electromagnetic controls.

SEMICONDUCTORS

The previous chapters address the flow of electrons through conductors and resistors. Electronic circuits include materials known as semiconductors. A *semiconductor* is a material that is neither a good conductor nor a good insulator.

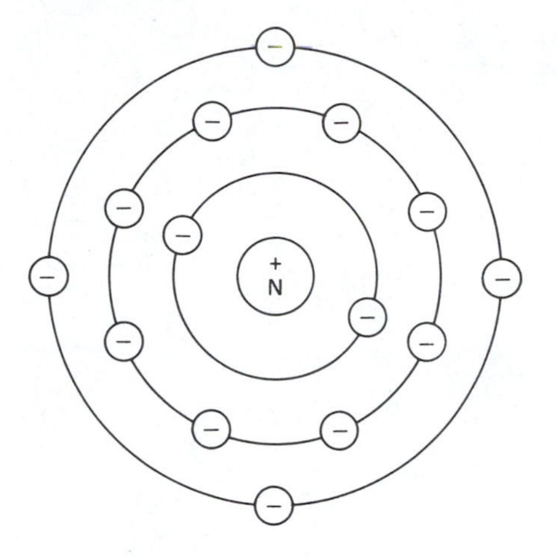

Figure 23-1
Silicon atom

Two of the most common materials used to make semiconductors are germanium and silicon. These materials, in their pure state, are not of much use in the electronic industry; however, by adding an impurity such as arsenic or indium, they take on very different characteristics. This process is called *doping.*

If arsenic is added to pure silicon, the atomic structure is altered. A silicon atom contains 14 electrons, 4 of which are in the outer shell (valance), Figure 23-1.

As stated in Chapter 2, atoms with 1 to 4 electrons in the outer shell are generally good conductors of electricity. Atoms with 5, 6, or 7 electrons in the outer shell are classified as poor conductors. Those with 8 electrons in the outer shell are insulators. The structure of silicon, however, presents a different phenomenon.

Figure 23-2 illustrates a small section of silicon. To simplify the drawing, only the valance electrons are shown for each atom. As the electrons orbit around the nucleus of their respective atoms, their valances overlap. This arrangement allows a sharing of valance electrons. A close examination shows that each atom has 4 of its own electrons in the valance and 4 electrons from the atoms surrounding it. This makes a total of 8 valance electrons. Pure silicon might then be classified as an insulator. This arrangement is called *covalent bonding.*

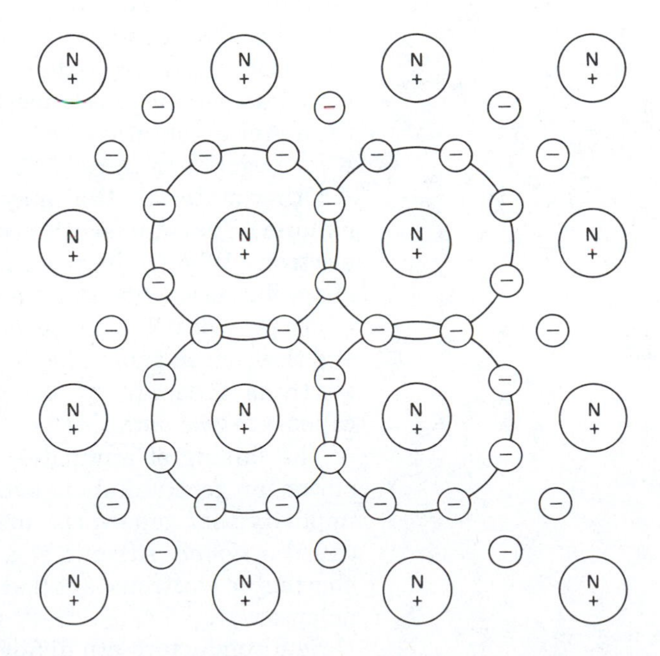

Figure 23-2
Small section of silicon showing covalent bonding.

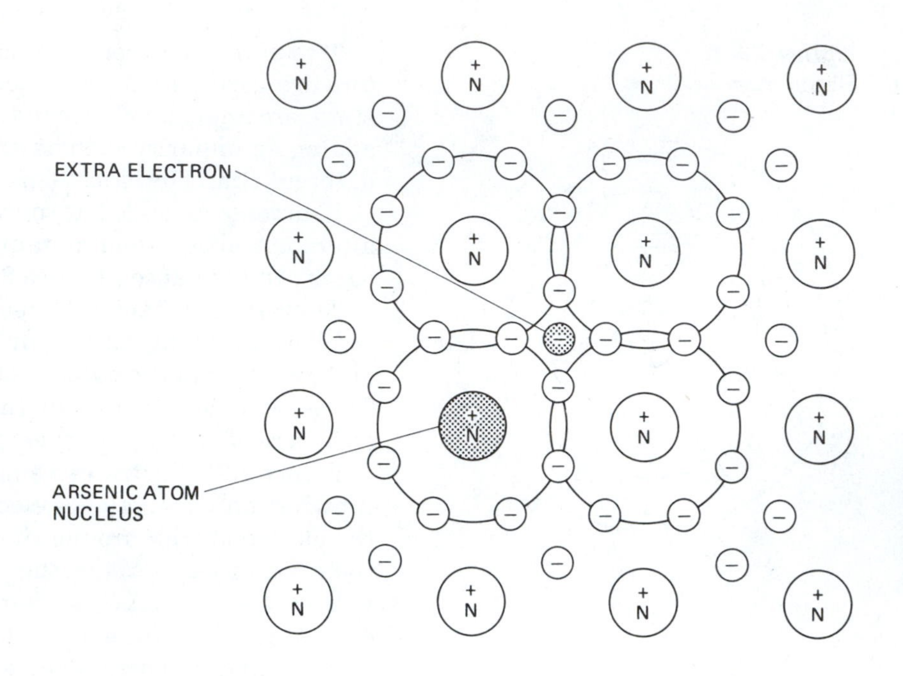

Figure 23-3
Silicon doped with arsenic to form an N-type material

An arsenic atom has 5 electrons in its valance. When arsenic is combined with silicon, 4 of the 5 valance electrons share valances with the adjacent silicon atoms. This leaves 1 electron free to drift. In this semiconductor there are many silicon and arsenic atoms that share valance electrons, leaving many free electrons. Because the material contains free electrons, it is called an *N-type material,* N for negative, Figure 23-3.

Other materials that may be used for doping are indium and gallium. The atoms of these impurities have only 3 valance electrons. When indium is added to silicon, the 3 valance electrons share valances with the 4 adjacent silicon atoms. This arrangement provides 7 valance electrons, leaving a hole where the eighth electron should be, Figure 23-4. Because of the absence of electrons, the material now has a positive charge. It is therefore called a *P-type material.*

The mixing of impurities (arsenic, indium, or gallium) with silicon or germanium is a chemical process called *doping.* An impurity that causes the material to contain extra electrons is called a *donor.* Arsenic is a donor. An impurity that causes a shortage of electrons is called an *acceptor.* Indium and gallium are acceptors.

Semiconductors are divided into three groups: unipolar, bipolar, and diode. They are also classified according to the arrangement of the P and N materials, such as PN, PNP, and NPN.

Figure 23-4
Silicon doped with indium to form a P-type material

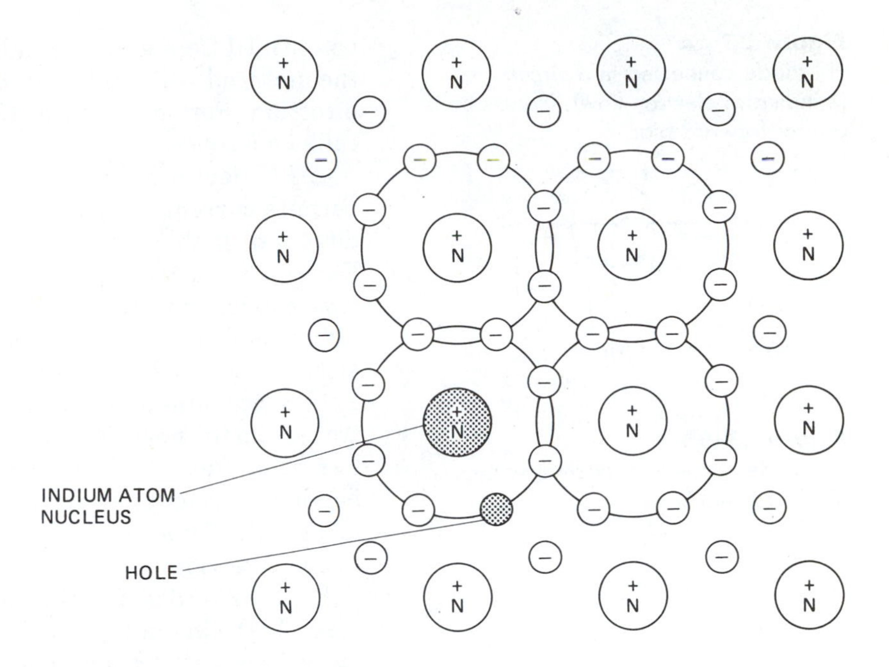

Figure 23-5A
PN diode and its schematic symbol

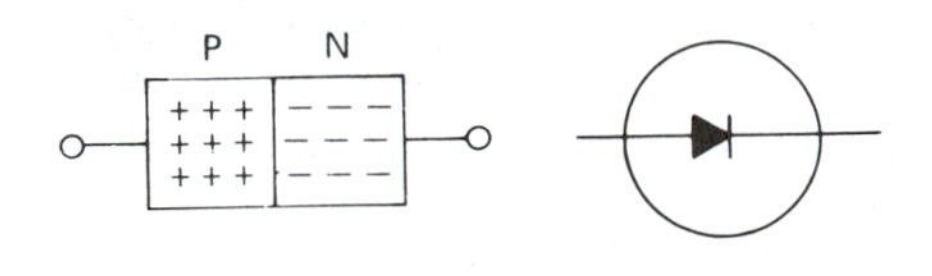

Figure 23-5B
PNP transistor and its schematic symbol

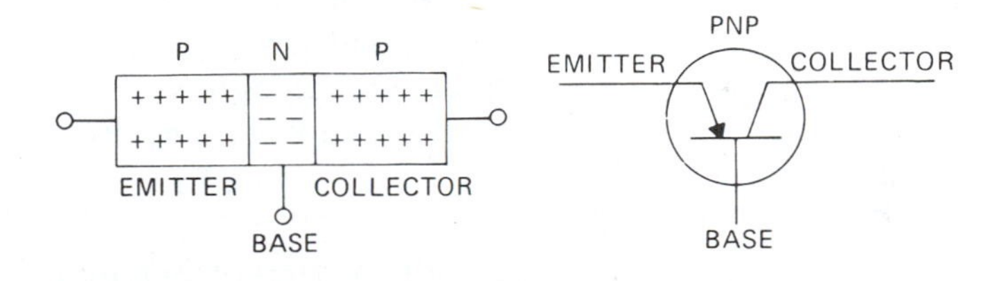

Figure 23-5C
NPN transistor and its schematic symbol

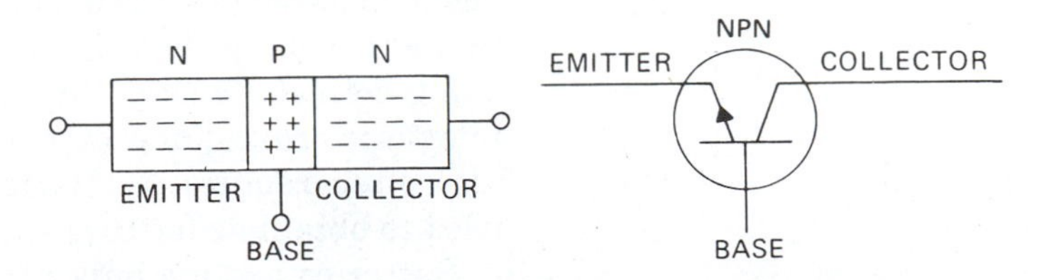

Figure 23-5D
Transistor

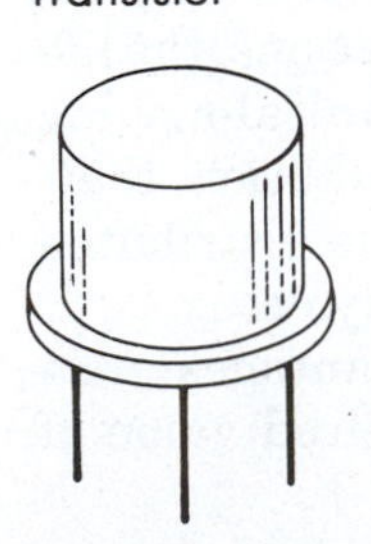

Figures 23-5A, 23-5B, and 23-5C illustrate the various methods of construction and the appropriate symbols. Figure 23-5D shows a typical transistor.

Diffusion occurs when P and N materials are joined together (Figure 23-5A). Some electrons in the N material, near the junction, are attracted to the holes in the P material, thus leaving holes in the N material. This diffusion of electrical charges produces a

Figure 23-6A
PN diode connected in a circuit (e indicates electron flow). Connections are for forward bias.

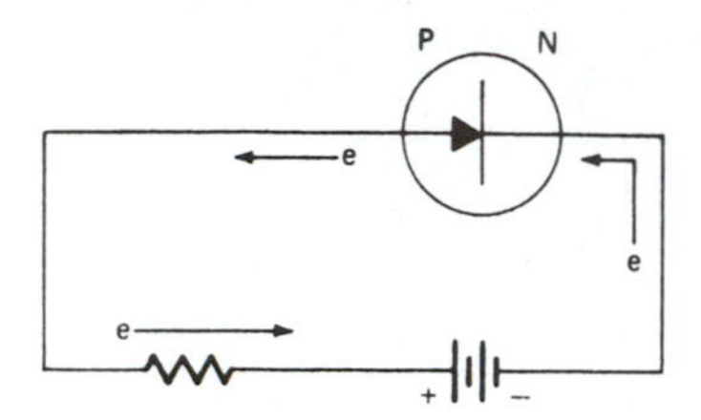

Figure 23-6B
PN diode connected for reverse bias blocks current flow.

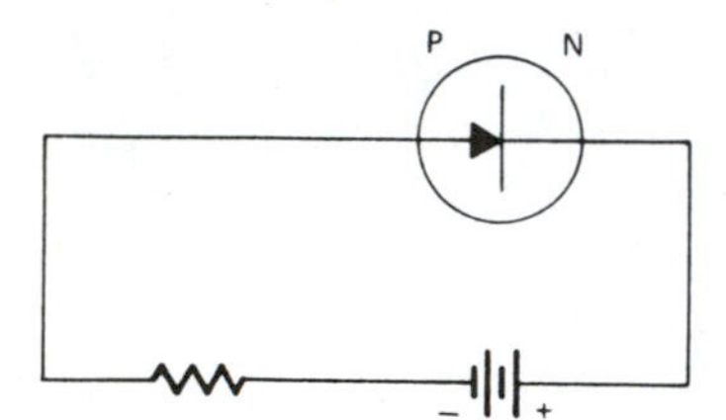

potential difference in a small area near the junction. As a result, the material will conduct in one direction but not in the opposite direction. For this reason, the area in which this emf exists is called a *barrier.*

A PN device is known as a *diode.* This type of construction permits current to flow in one direction but not in the opposite direction. If this device is connected into a circuit as shown in Figure 23-6A, current will flow. This is referred to as a *forward-bias connection.* The positive terminal of the battery attracts electrons from P material, leaving an excess of holes. Because electrons are drifting away from the junction, the excess holes tend to accumulate near the junction. At the same instant, electrons from the negative terminal of the battery are attracted to the less negative N material of the diode. This action overcomes the barrier at the junction and allows electrons to move into the excess holes of the P material. The result is a continuous flow of electrons in one direction.

Figure 23-6B shows the battery connections reversed (reverse bias). At the instant a reverse bias is applied, some electrons move away from the PN junction toward the positive terminal of the battery. At the same instant, a shift in electrons in the P material causes positive holes to appear farther away from the junction near the end of the diode, which is connected to the negative terminal of the battery. This action produces a wider barrier at the PN junction through which the electrons cannot flow. (A very small current leakage may occur.)

RECTIFIERS

Direct current has many uses in industry. Chemical applications that require dc are charging batteries, electroplating, and separating metal from ore. Applications that require a constant electric charge include dust and smoke precipitation, spray painting, and voltage testing of insulation. Mechanical applications using dc include some types of speed control, magnetic chucks, magnetic brakes, and magnetic clutches. Electromagnetic hoists also often require dc. Batteries, generators, and rectifiers are used to obtain dc for these applications.

Batteries are too bulky for certain applications, and they require maintenance and recharging. For some applications, the life of the battery is too limited. Generators are mechanical-moving equipment that require frequent maintenance. In addition, their noise during operation may be objectionable. If dc is distributed over long distances, losses in voltage and power may occur.

Rectifiers for converting ac to dc have many advantages. As a power source, ac is universally available. Any desired value of

Figure 23-7
Rectifier circuit using a single diode

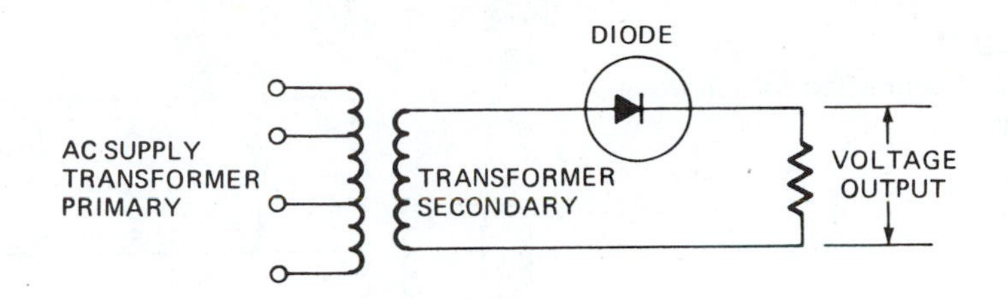

direct voltage can be obtained from an ac source by using transformers in combination with rectifiers. Further, electrical and electronic rectifiers are quiet and require minimum maintenance.

Types of Rectifiers

Rectifiers are classified as mechanical (generators and rotary convertors), electrical (dry metal, crystal, and electrolytic types), and electronic (semiconductor and tube types).

Half-Wave, Single-Phase Rectifiers

Half-wave rectifiers are devices that change alternating current to a pulsating direct current. They are used in such equipment as inexpensive battery chargers.

Figure 23-7 shows a simple type of rectifier circuit. On the supply side, a transformer can be used to change the value of voltage input to the rectifier. The rectifier acts as a single-pole switch open to one polarity and closed to the opposite polarity. Figure 23-8A shows the input voltage waveform. Figure 23-8B illustrates the waveform of the secondary voltage of the transformer. Figure 23-8C shows the output voltage waveform of the rectifier.

Figure 23-8A
Input voltage waveform to the rectifier circuit in Figure 23-7

Figure 23-8B
Waveform of secondary voltage of the transformer in Figure 23-7

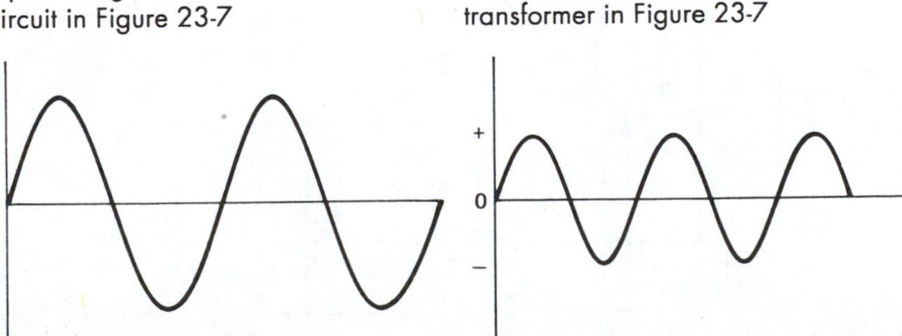

Figure 23-8C
Output voltage waveform of the rectifier circuit in Figure 23-7

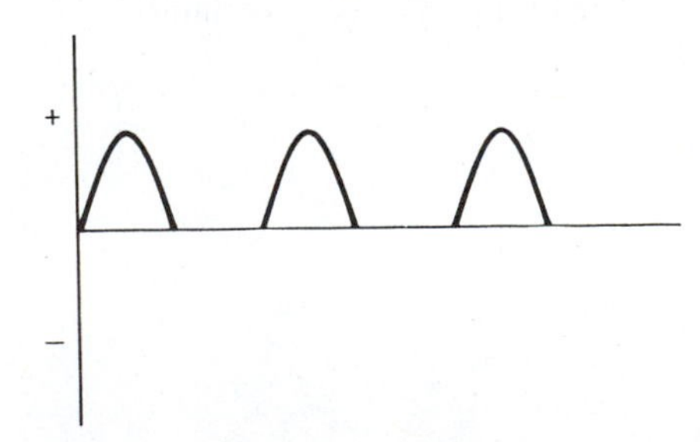

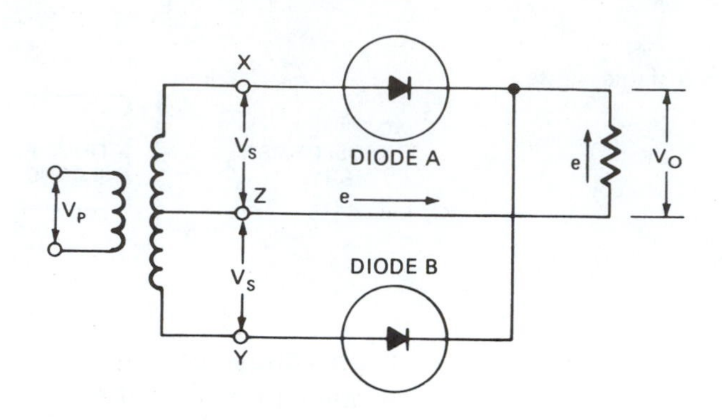

Figure 23-9
Diode circuit connected for full-wave rectification

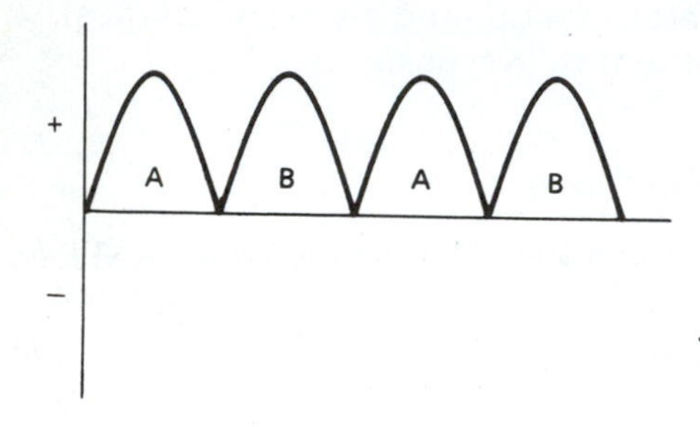

Figure 23-10
Output voltage waveform of a full-wave rectifier

Full-Wave, Single-Phase Rectifiers

Full-wave rectifiers are devices that convert alternating current to direct current using both halves of the ac wave. These rectifiers are used to provide direct current power for the operation of spray painting and electroplating equipment.

If two diodes and a transformer with a center tap are connected, as shown in Figure 23-9, it is possible to use both halves of the ac voltage wave. During each half cycle, the diodes conduct alternately. When point X is positive with respect to center tap Z, diode A conducts. Current flows from Z, through the load and diode A, to point X. During the next half cycle, Y is positive with respect to Z, and diode B conducts. The current flows from Z, through the load and diode B, to point Y. This arrangement provides a more constant voltage for the load. Figure 23-10 shows the rectifier output voltage waveform. This type of rectifier uses only one half of the transformer secondary voltage, requiring a transformer with a center tap. It is normally not suitable for voltages greater than 1000 volts.

The *bridge rectifier* is a full-wave rectifier that uses the entire transformer secondary voltage. The bridge rectifier, however, requires four diodes. Figure 23-11 is a schematic diagram of a bridge rectifier. When point X is positive, diodes A and B conduct.

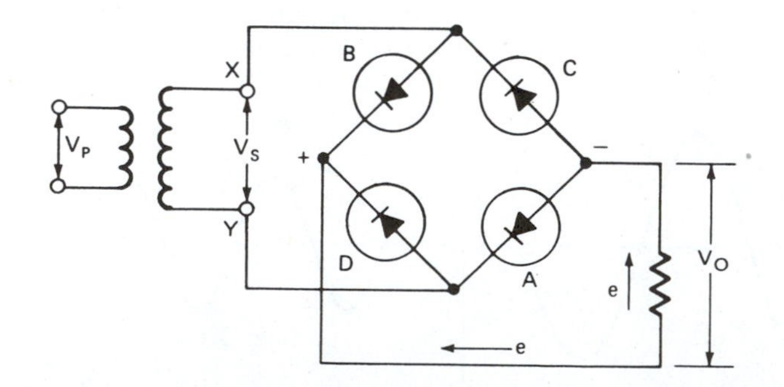

Figure 23-11
Diode bridge, full-wave rectifier

When point Y is positive, diodes C and D conduct. When point X is positive, the current flows from Y, through diode A, the load, and diode B, to point X. When Y is positive, the current flows from point X, through diode C, the load, and diode D, to point Y.

Copper Oxide Rectifiers

The copper oxide rectifier is an electrical rectifier made of copper. One element consists of a copper disk with a copper oxide layer formed on one side. A lead disk is pressed onto the oxide to form a good electrical connection, Figure 23-12A. Electrons can flow readily from the copper to the copper oxide but have great difficulty flowing in the opposite direction. The disks are stacked on an insulated rod. The number of disks required depends on the value of voltage. Figure 23-12B shows a typical copper oxide rectifier used in industry. Figure 23-12C shows the connections for half-wave rectification. Figure 23-12D shows the connections for full-wave rectification.

Figure 23-12A
Section of copper oxide rectifier

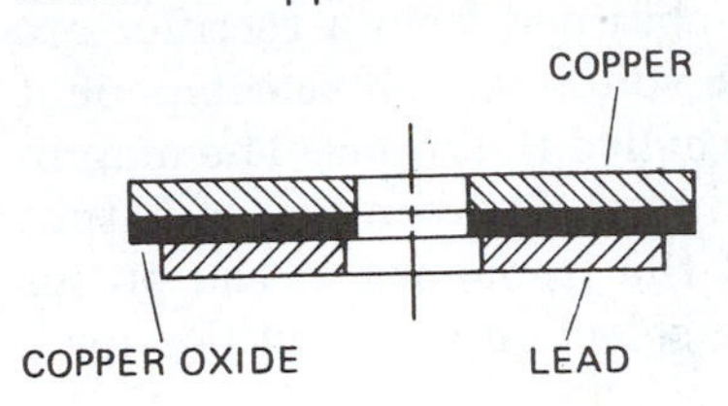

Figure 23-12B
Copper oxide rectifier

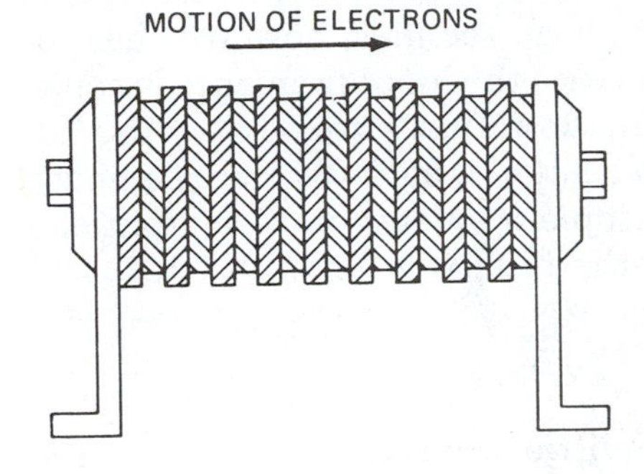

Figure 23-12C
Copper oxide rectifier connected for half-wave rectification

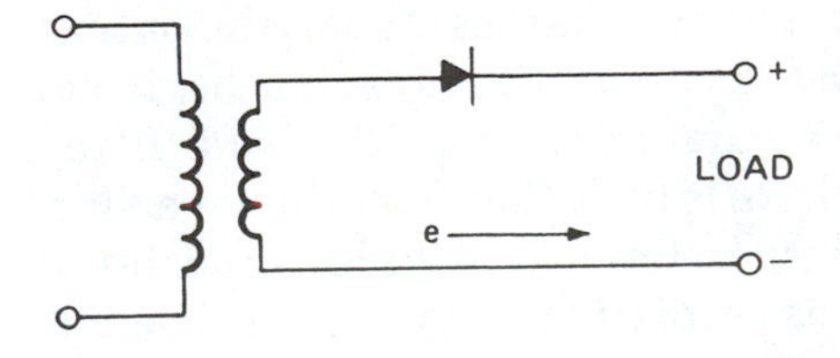

Figure 23-12D
Two copper oxide rectifiers connected for full-wave rectification

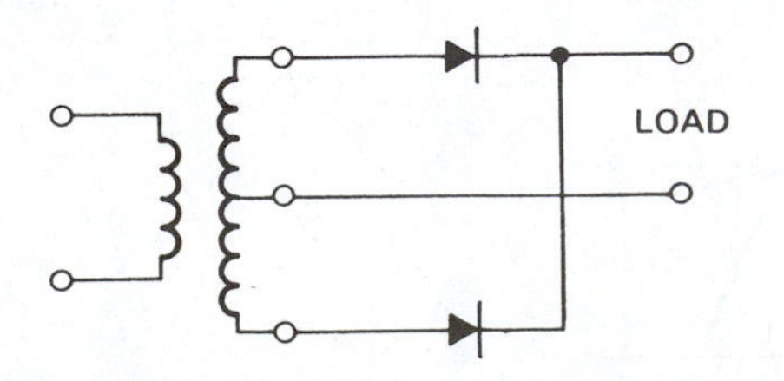

Selenium Rectifier

The selenium rectifier is used to control the speed of electric motors. Composed of an aluminum disk coated with selenium, the rectifier is sprayed with an alloy that melts at low temperatures. When a voltage is applied, current flows freely from the alloy to the selenium but has great difficulty flowing in the opposite direction.

The rectifying property of both the copper oxide and the selenium rectifiers is destroyed if they are subject to a voltage that is too high or to a current that is so great that it causes overheating. Either of these conditions will cause the resistance of the rectifier to decrease to a very low value in either direction.

An ohmmeter may be used to measure the resistance of a rectifier. The value obtained depends upon the voltage supplied by the ohmmeter. The best method for testing the rectifier is to measure the resistance of a similar rectifier that is in good condition. The value obtained when measuring the good rectifier can then be compared with the value of the rectifier in question. If these values are nearly equal, the questionable rectifier is not faulty.

Copper oxide rectifiers are generally used for low voltages and high currents. They are also used when intermittent or instantaneous operations are required. Selenium rectifiers are used for high voltages and low currents or when space is limited or weight must be kept to a minimum.

Ripple

The fluctuating output voltages obtained from a rectifier are actually a combination of ac and dc voltages. The ac component that causes the pulsating voltage is called the *ripple*. The magnitude of ripple voltages is measured in the percentage of output voltage. The percent of ripple and the frequency of the pulses are two factors that determine the selection of a rectifier for a specific job.

Two cycles of input voltage to a rectifier are shown in Figure 23-13A. Figure 23-13B shows that the output waveform of a half-wave, single-phase rectifier has the same ripple frequency as the input voltage because the distance from one peak to the next is the same. A full-wave, single-phase rectifier, however, has an output ripple frequency that is twice the input frequency because there are twice as many peaks in the output wave. Thus with a 60-hertz input, the ripple in the output of a half-wave, single-phase rectifier is also 60 hertz. The ripple frequency in the output of a full-wave, single-phase rectifier is 120 hertz, Figure 23-13C.

Figure 23-13A
Two cycles of input voltage to a rectifier

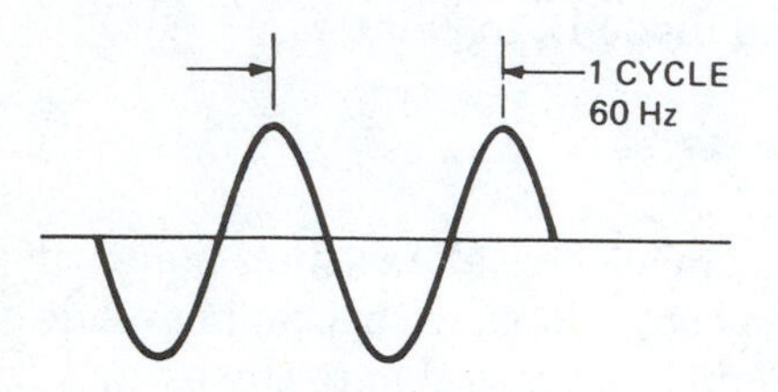

Figure 23-13B
Output voltage waveform of a half-wave, single-phase rectifier

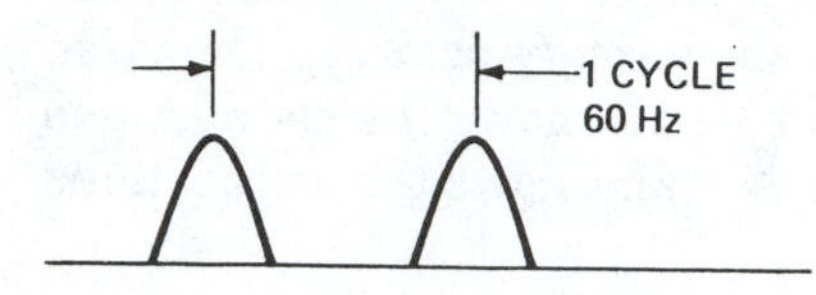

Figure 23-13C
Output voltage waveform of a full-wave, single-phase rectifier

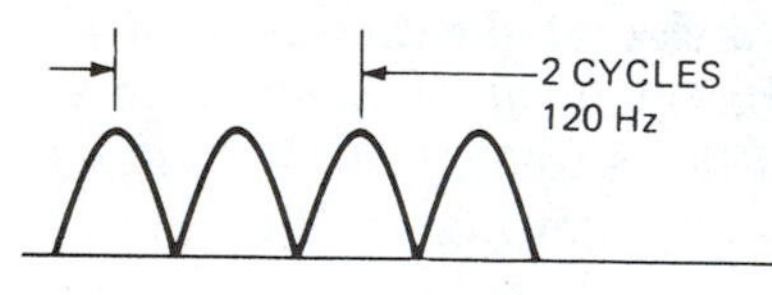

Filters

When dc voltages must have little or no ripple, a filter is used to smooth out the pulse. A filter usually consists of capacitance and inductance or resistance. The number of capacitors and inductors depends upon the degree of filtration required.

Figure 23-14A shows a filtering capacitor connected to the output of a rectifier. As illustrated in Figure 23-14B, when the secondary voltage rises, the capacitor charges. When the output voltage of the rectifier starts to decrease, the capacitor begins to discharge, thus maintaining a current through the load. If the resistance of the load is high, there is little flow from the capacitor, and it maintains almost full charge. However, if the load resistance is relatively low, there is a greater discharge from the capacitor, resulting in a pulsing voltage. The curves in Figures 23-14C and

Figure 23-14A
Filtering capacitor connected to the output of a rectifier

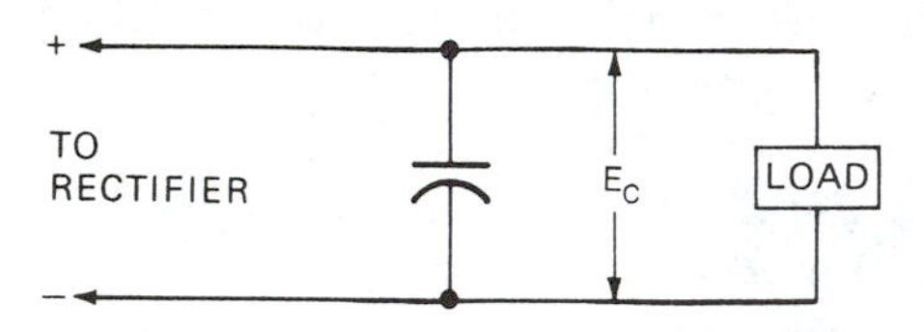

Figure 23-14B
Illustration of a filtering capacitor charge and discharge

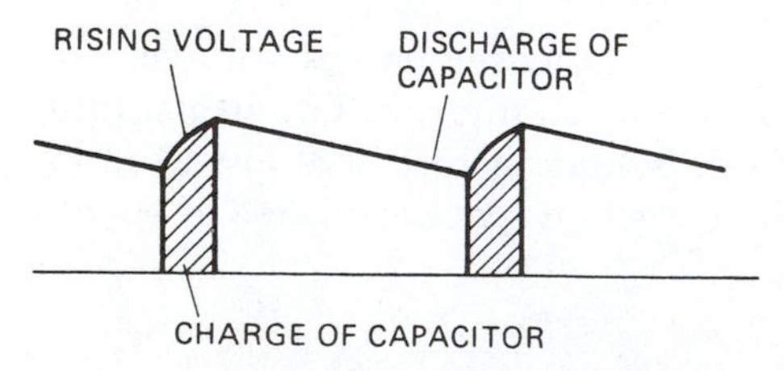

Figure 23-14C
Waveform of the output voltage of a half-wave rectifier with a single filtering capacitor

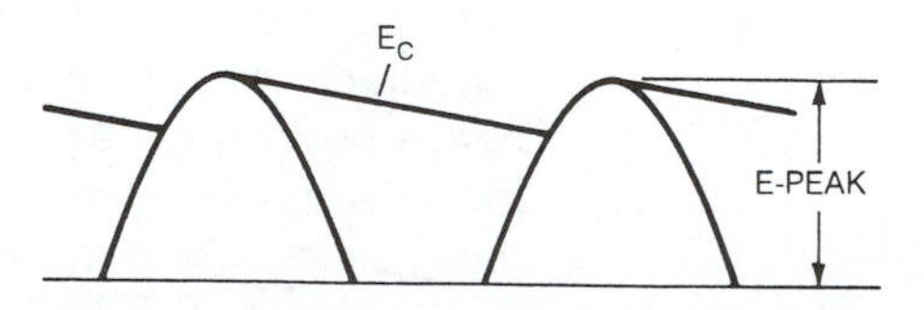

23-14D show waveforms from half-wave and full-wave rectifiers with a single filtering capacitor. The output obtained from the full-wave rectifier is steadier and, therefore, more desirable for most applications than the output from a half-wave rectifier.

The use of a capacitor and choke, Figure 23-15, further decreases the pulsing and produces a smoother load current. The addition of still more inductance and capacitance (LC) increases the filtering effect to the point where the voltage and current have practically no ripple. Figures 23-16A and 23-16B are circuit diagrams of capacitor input filters. Figures 23-15 and 23-16C are circuit diagrams of choke input filters. Comparative characteristics of the output voltage from the choke (inductive) and the capacitor (capacitive) filters are shown in Figure 23-17. With little or no load current, the output voltage is practically the peak voltage. When the load is high, the choke-input-type filter has better regulation.

Figure 23-14D
Waveform of the output voltage of a full-wave rectifier with a single filtering capacitor

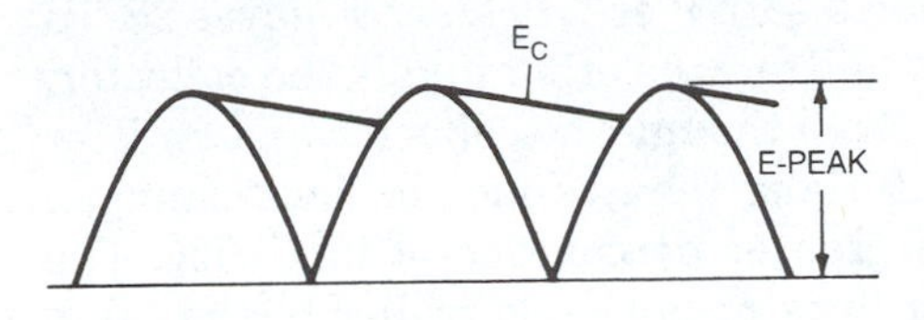

Figure 23-15
Single LC filter circuit supplied from a rectifier (choke input)

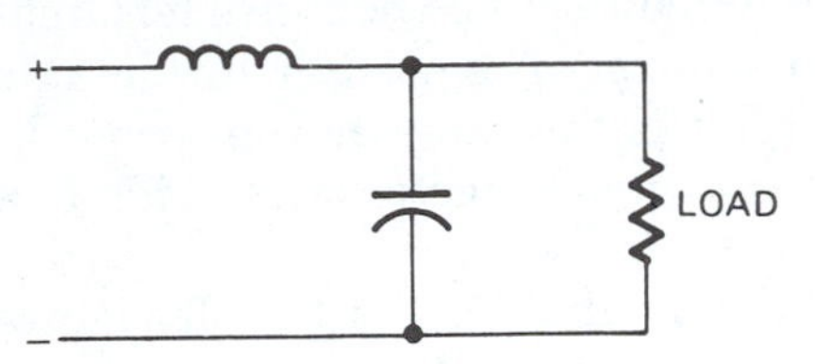

Figure 23-16A
Double-capacitor, single-choke filter circuit supplied from a rectifier (capacitor input)

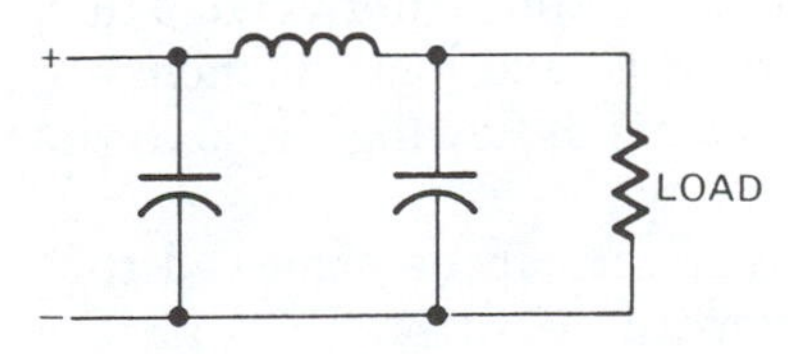

Figure 23-16B
Triple-capacitor, double-choke filter circuit supplied from a rectifier (capacitor input)

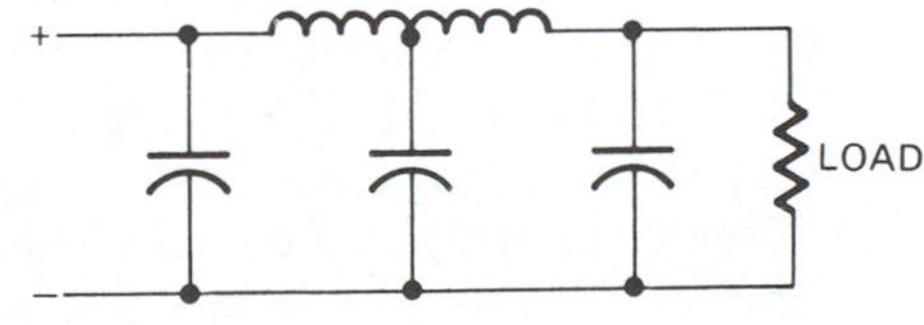

Figure 23-16C
Double-capacitor, double-choke filter circuit supplied from a rectifier (choke input)

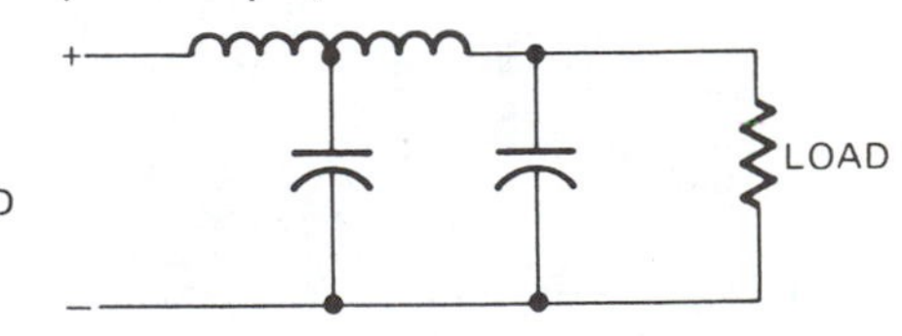

Figure 23-17
Characteristic curves of the output voltage for choke and capacitor filter circuits

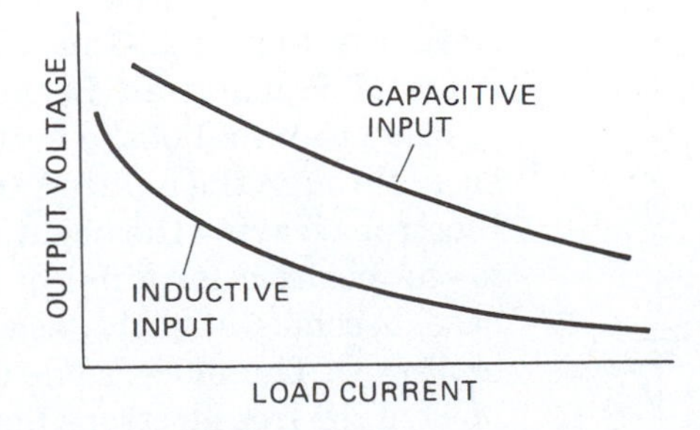

BIPOLAR TRANSISTORS

Because of their miniature size, long life, and cool operating properties, bipolar transistors offer a myriad of uses in industrial electricity, such as control circuits and computers. Bipolar transistors are made of three parts—PNP or NPN (see Figures 23-5B and 23-5C). One end is the emitter, the other end is the collector, and the center is the base. Even though the collector and emitter are made of the same materials, they cannot be interchanged because the emitter has a greater percentage of impurity. The electron flow through this transistor can be controlled by the value of voltage between the emitter and the base. Figure 23-18 shows a basic circuit for an NPN transistor.

Figure 23-18
Basic circuit for NPN transistor
e→ major electron flow
e⋯> minor electron flow

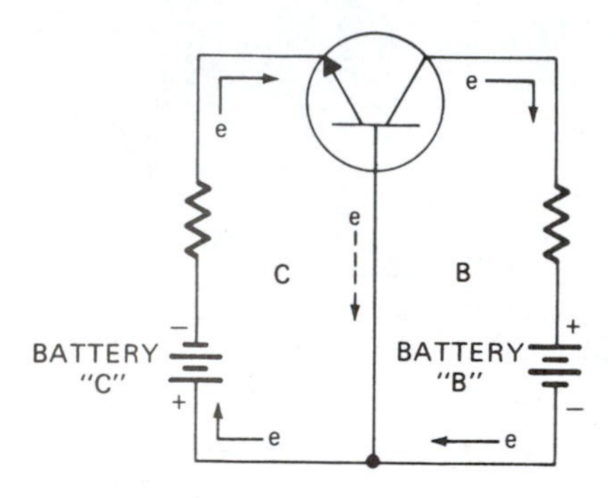

To understand the flow of current through this circuit, begin with the emitter-to-base circuit (circuit C). The negative terminal of battery C is connected to the emitter. The positive terminal is connected to the base. This circuit is the same as the circuit in Figure 23-6A and will have the same characteristics. This is a forward-bias diode circuit.

Circuit B is the same as C except that the battery leads are reversed. Circuit B, therefore, is connected for reverse bias. Normally, a reverse-bias diode blocks current flow.

From this observation, it appears that the only current flow is in circuit C, and no current flows through the transistor from the emitter to the collector. If this were true, circuit B would be useless. However, the construction of the transistor makes the difference. The P material is very thin compared to the N material. Thus, only a few free electrons flow in circuit C, and most of them are attracted to the positive charge of battery B flowing through the base and collector.

Figure 23-19
Basic circuit for PNP transistor
e→ major electron flow
e⋯> minor electron flow

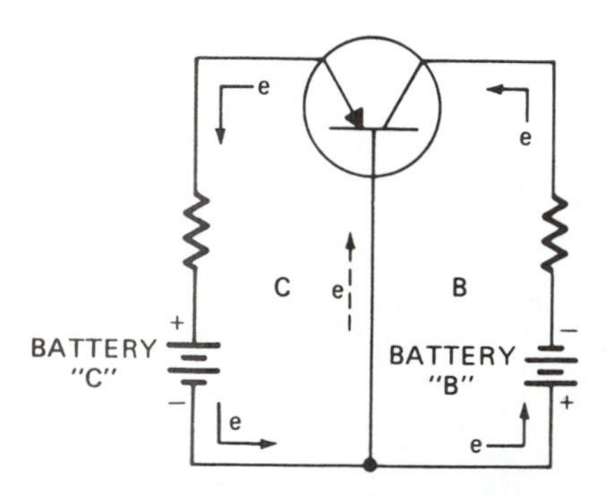

Reconsidering this combined circuit, it can be observed that most of the free electrons flow from the emitter, through the collector, battery B, and battery C, and back to the emitter. Only a few free electrons flow in circuit C.

The PNP transistor is wired as shown in Figure 23-19. This is similar to Figure 23-18 except that the collector and emitter are made of P material. Because the emitter is connected to the positive terminal of the battery, electrons are attracted from the base, through the emitter, to the positive terminal of the battery. Electrons leaving the negative terminal of battery C are attracted to the positive terminal of battery B. At the same instant, the base becomes slightly less negative than the negative side of battery C. Therefore, some of the electrons flow back to the base. Most of the free electrons flow from the base, through the emitter, battery C, battery B, and the collector, and back to the base.

AMPLIFIERS

An *amplifier* is a device that receives an input signal and draws from another power source to produce an output signal greater than the input. The signal may be electrical, mechanical, audio, light, heat, or other physical input. The amplifier may be a magnetic relay, an electron tube, a transistor, a speaker, or a generator. In general, amplifiers are used to control a large current with a small signal. In some installations, the output voltages are also large compared with the input voltages.

Figure 23-20 illustrates a magnetic amplifier. Figure 23-21 is a schematic diagram of a very basic one-stage electronic amplifier. It is often necessary to use more sophisticated arrangements to obtain the required amplification. For this purpose, multistage amplifiers are used.

Figure 23-20
Magnetic relay (power amplifier)

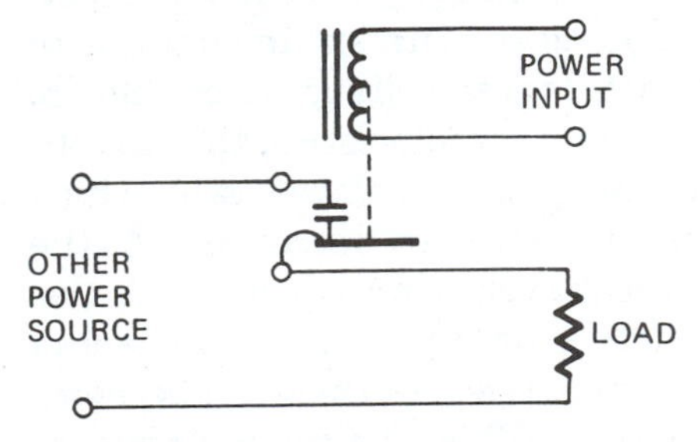

Figure 23-21
Transistor amplifier circuit

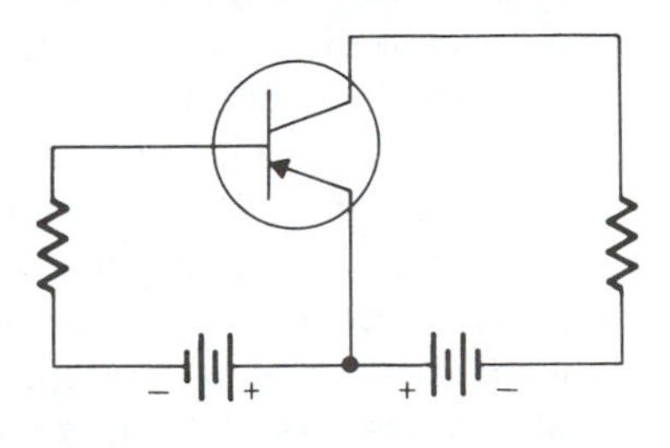

SPECIAL ELECTRONIC CIRCUITS

In some situations, a high-frequency ac must be obtained from a dc source, or the value of an ac frequency must be changed. To do this, an *oscillator circuit* may be used. A simplified oscillator circuit is shown in Figure 23-22. When the button is pressed, the capacitor is charged to the same value as the battery. When the button is released, the capacitor discharges through the coil. Because of inductance, the coil current lags the voltage by nearly 90 degrees. Therefore, the maximum current is reached at the instant the capacitor charge is zero. At this instant, the coil has maximum current and a strong magnetic field. As the field collapses, a voltage is induced in the coil that charges the capacitor. When the magnetic field has completely collapsed, the capacitor is fully charged. The capacitor will again discharge through the coil, producing a strong magnetic flux around the coil. This flux then collapses and again charges the capacitor. The new charge on the capacitor will be less than the original charge

Figure 23-22
Basic oscillator circuit

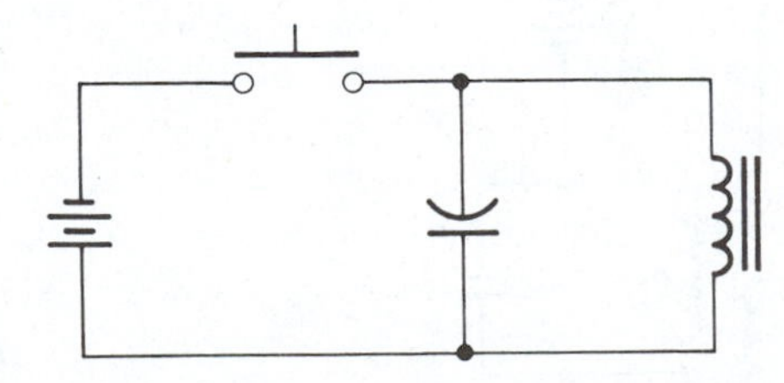

Figure 23-23
Waveform of the current in the circuit in Figure 23-21

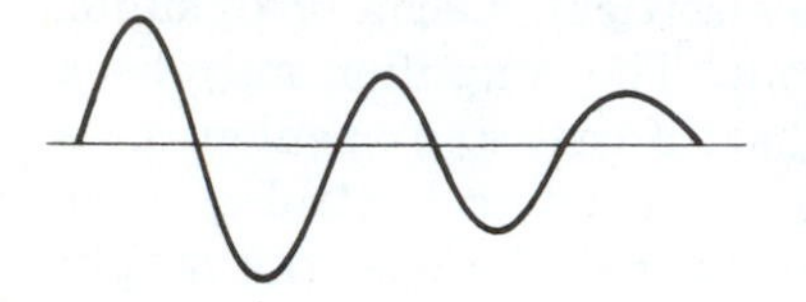

because of IR drop in the circuit. As this action continues, the oscillations reduce in size, Figure 23-23. The frequency, however, remains constant and may be determined by the following equation:

$$f = \frac{1}{2\pi\sqrt{LC}}$$

where f = the frequency, in hertz (Hz)
L = the inductance of the coil, in henrys (H)
C = the capacitance of the capacitor, in farads (F)

ELECTRONIC OSCILLATORS

Adding a transistor to the circuit in Figure 23-22 produces an oscillation. An elementary diagram of this arrangement is shown in Figure 23-24. An instrument designed for this purpose is an *electronic oscillator.*

When the switch in Figure 23-24 is closed, current begins to flow in circuit A. As the current increases through coil P, the expanding magnetic field induces a voltage across coil S. E_s will cause the base to become more positive, thus causing the current in circuit A to increase further and induce a still larger voltage across coil S. Because of the circuit and transistor characteristics, the current through circuit A now begins to increase at a slower rate, which causes the voltage induced across coil S to start decreasing. As the voltage across coil S decreases, the current through circuit A also decreases. Thus, the magnetic field produced by P contracts and causes the base to become more negative. As the base becomes negative, circuit A decreases further. When the collapsing magnetic field of P approaches zero, it decreases at a slower rate, causing less induced voltage across S. The base then becomes less negative, allowing the current through circuit A to increase.

While this action is occurring, another important process is also taking place. As the current through P is increasing, capacitor C is being charged. It discharges through P when the current

Figure 23-24
Transistor oscillator

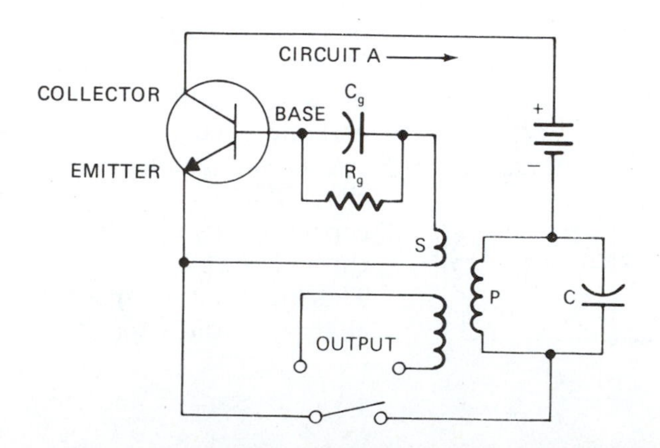

through circuit A begins to decrease. Thus, an oscillating current is produced in P and C. This arrangement of capacitor and inductor is frequently called a *tank circuit.*

The oscillating current would diminish if not for the fact that the current through circuit A adds to the oscillating current once during each half cycle. The frequency of the tank circuit depends upon the values of inductance and capacitance.

The current flowing through P induces an alternating voltage into the output coil. The frequency of this voltage is the same as that of P. Thus, the output frequency can be varied by varying the inductance and/or capacitance of the tank circuit.

Oscillator circuits can be designed to produce ac at a wide range of frequencies and waveforms.

SPECIAL ELECTRONIC DEVICES

Many transistors and tubes are designed for special functions, or they are combined with other devices to meet various requirements. Some solid-state devices are the diac, the triac, and various switching and rectifying devices. Some light-sensitive devices are the phototube, the photovoltaic cell, and the photoconductive cell.

Thyristors

A *thyristor* is an arrangement of semiconductor components designed for switching purposes. It is frequently used in switching control circuits to vary the speed of industrial motors. As solid-state switches, thyristors are fast acting and extremely accurate. Only a very small pulse current is required to trigger the thyristor into action.

Among the most commonly used thyristors are the *silicon-controlled rectifier* (SCR), which can be used to rectify as well as control the average dc output; the *diac,* which is a bidirectional diode; and the *triac,* which is a bidirectional triode transistor.

Thyristors are frequently used to control the speed and torque of ac and dc motors. They are used in heating, ventilating, and air-conditioning control systems, computer systems, and communication systems and for converting dc to ac.

Diacs

Diacs are generally used in phase control circuits. They accomplish such operations as speed control of motors, temperature control, and dimming of lights.

Figure 23-25
Schematic symbol for a diac

The diac is a bidirectional diode. It is a three-layer transistor that can block current flow in either direction. When the voltage across it exceeds a specific value (the breakdown voltage), the diac conducts. Because of their characteristics, diacs are useful in control circuits. The symbol for the diac is shown in Figure 23-25.

Triacs

The triac is similar to the diac in that it is a bidirectional transistor. It is, however, a three-element device that can block current flow in either direction or be triggered to conduct in either direction. A pulse applied to the gate starts conduction. This pulse may be either positive or negative. Once the conduction starts, it will continue to conduct until the circuit is opened.

Triacs are used primarily to control alternating current. The triac can conduct both alternations and regulate the average current by its triggering time. The symbol for the triac is shown in Figure 23-26.

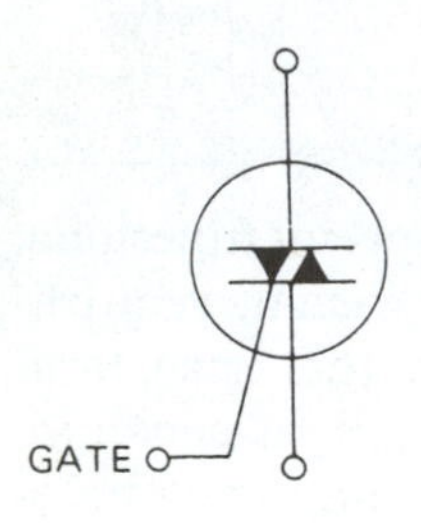

Figure 23-26
Schematic symbol for a triac

Silicon-Controlled Rectifiers

The SRC is a transistor device for rectifying and/or switching electrical power. It is capable of blocking current in either direction as well as controlling the dc output. It is made from a single, multilayer transistor containing four layers (PNPN), as illustrated in Figure 23-27A. Figure 23-27B shows the schematic symbol.

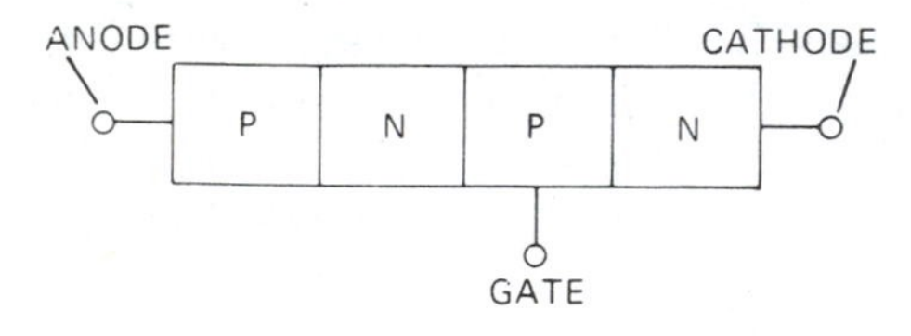

Figure 23-27A
Construction of an SCR transistor

This arrangement provides an anode, cathode, and gate (collector, emitter, and base). The left and right PN junctions conduct only under forward bias. The voltage applied across the transistor is of a polarity to cause the current to flow in the direction of low resistance. The middle junction (NP) blocks current when the entire unit is forward biased. If, under these conditions, a positive pulse is applied to the gate, the SCR will begin to conduct. Conduction will continue until the anode circuit is opened.

When the SCR is used to vary the average dc output, the phase-shift control method is generally used to determine when the gate is triggered. By this method, ac can be applied to both the anode and the gate. Figure 23-28 is a schematic diagram of an SCR circuit using phase-shift control. Because ac is supplied to the circuit, the gate will regain control when the current drops to zero.

Generally, SCRs can withstand large overcurrents for short periods of time. The gate circuit, however, is very sensitive and must be protected against overcurrent and/or overvoltage. Either condition may cause a short circuit from the cathode to the anode.

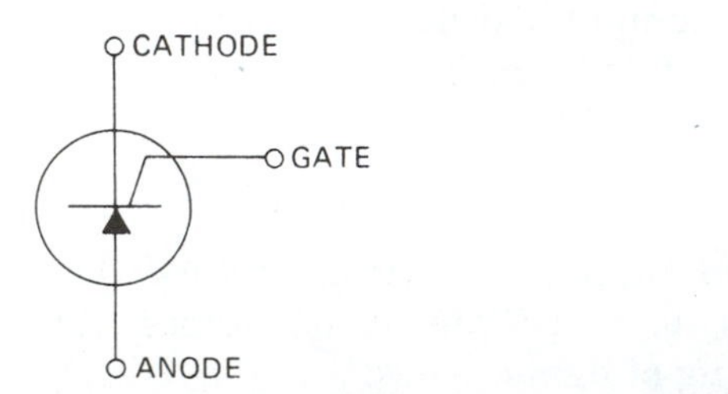

Figure 23-27B
Schematic symbol for an SCR transistor

The SCR is capable of controlling large amounts of power with high efficiency. It is small in size, has long life expectancy, and

Figure 23-28
Basic diagram of a SCR using phase-shift control

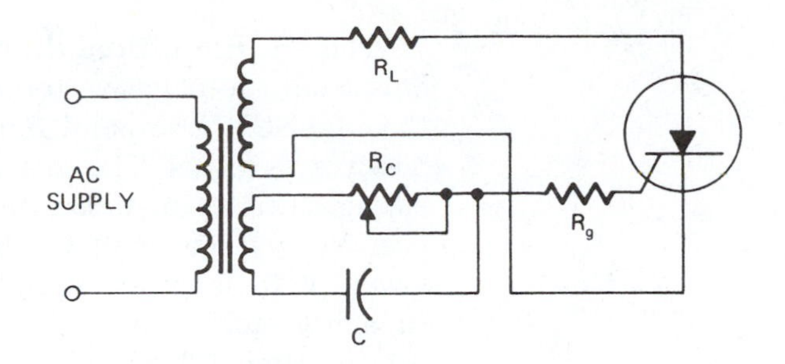

requires no maintenance. SCR voltage ratings range as high as 1500 volts, with current ratings as high as 800 amperes. These features make the SCR a very useful electronic control for industrial equipment.

SOLID-STATE INVERTERS

An inverter is a device that changes dc to ac and thereby controls industrial ac motors by varying the frequency. A very simple, but rather crude, type is shown in Figure 23-29. When switch S is closed, the current increases from zero to a maximum value. The expanding magnetic field produced by this current moves across the secondary of the transformer, inducing a voltage into the secondary coil. At the instant the current reaches a maximum value, relay TR opens contacts TR, and the primary current decreases to zero. The decreasing current causes the magnetic field around the secondary to collapse, inducing a voltage into the secondary winding in a direction opposite to the original voltage. Contacts TR close, and the cycle repeats. Although the operation of this inverter is relatively easy to understand, it is not practical. A much more useful type can be constructed with SCRs and solid-state circuitry. A basic SCR inverter is shown in figure 23-30.

Figure 23-29
Inverter consisting of a time-delay relay and a transformer

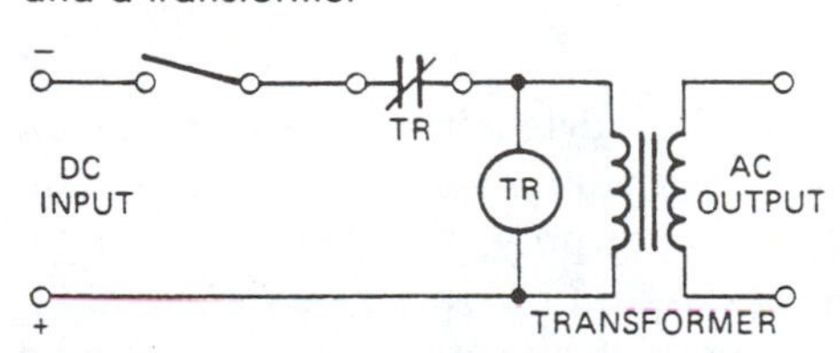

Figure 23-30
SCR inverter connected through a step-up transformer to an ac load

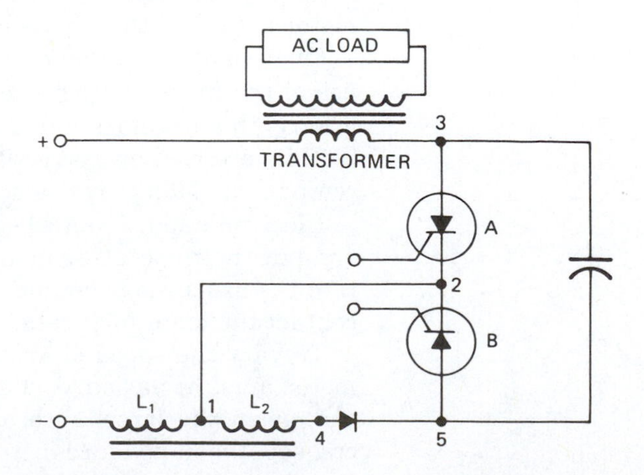

When SCR A is fired through the gate by a pulse, current will flow from the negative terminal, through L_1 to points 1 and 2, through SCR A to point 3, and through the transformer to the positive terminal. The increasing current through L_1 induces a voltage into L_2 in such a direction that point 4 is positive. This positive voltage attracts electrons from the capacitor, so the lower plate is positive and the upper plate is negative. The capacitor stores this charge, making the anode of SCR B more positive than its cathode. To turn off SCR A, a gate pulse is brought into SCR B. When SCR B fires, it causes the capacitor to discharge, producing an emf across SCR A. This emf reduces the current to a value below the point where it will maintain conductivity, and SCR A stops conducting. The voltage across the capacitor is now reversed, thereby producing a counter emf across SCR B and stopping current flow. At this point, a gate pulse is again applied to SCR A, and the cycle repeats. The frequency of the ac depends upon the timing of the gate pulses, the capacitance of C, and the inductance of L_1 and L_2. Any of these can be varied in order to obtain the required frequency.

ADJUSTABLE-SPEED MOTOR DRIVES

The speed of a dc motor may be varied in many ways, such as varying the current through the armature or field circuits. The dc motor may be operated from an ac supply, and its speed may be varied through a wide range above and below its basic speed by using electronic equipment. The motor speed and torque may be held constant as loads vary. The armature and/or field currents can be regulated through the use of SCRs.

AC motors can also be controlled through electronic equipment. The simplest method for controlling the speed of a standard ac motor is to vary the frequency.

One popular method of obtaining a variable frequency is the use of the inverter frequency speed controller. A simplified diagram of this system is shown in Figure 23-31. AC is supplied to the SCR inverter. The ac is rectified before being fed into the inverter component. This provides ac at a variable frequency.

Electronic speed control systems vary widely with manufacturers, but their operating principles are basically the same. To obtain the exact diagram and variations of operations, it is best to contact the manufacturers.

To vary the speed of an ac motor, the input frequency to the motor must be varied. A change in the frequency will result in a change in the impedance of the motor. If the frequency is increased, the impedance increases; a decrease in frequency results in a decrease in the impedance. To maintain the proper value of

Figure 23-31
Simplified SCR variable frequency speed controller

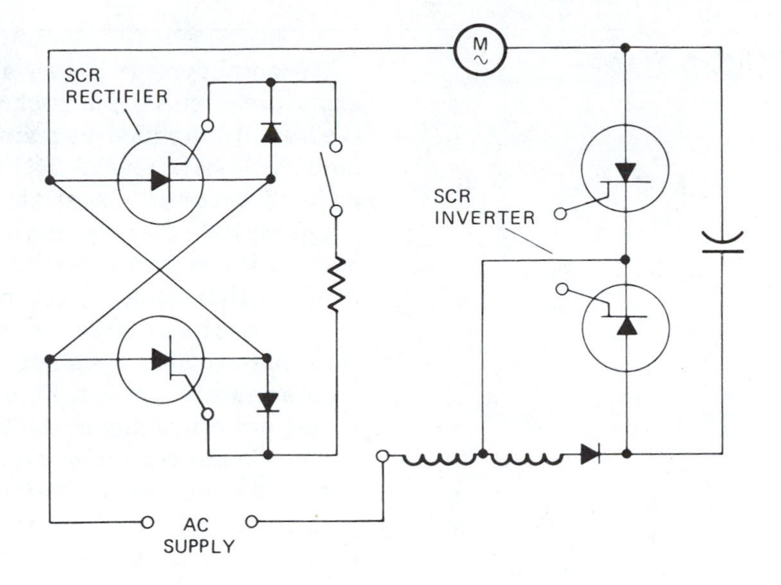

current, it is necessary to change the voltage with each change in the frequency. The SCRs provide this control of the voltage.

It is also necessary to consider the waveform of the voltage input to the motor. A change in waveform can also result in a change in the impedance. Squirrel-cage induction motors are generally designed to operate best when the input voltage form is in the shape of a sine wave.

Some electronic circuits generate a square wave as illustrated in Figure 23-32. Although many induction motors operate satisfactorily from a square wave input, a step wave is preferred and is usually more efficient. Figure 23-33 is a graph of a step wave.

To produce waveforms that have more exact characteristics to the sine wave, one must use logic and/or digital systems.

Figure 23-32
One cycle of the voltage output of an inverter frequency speed controller

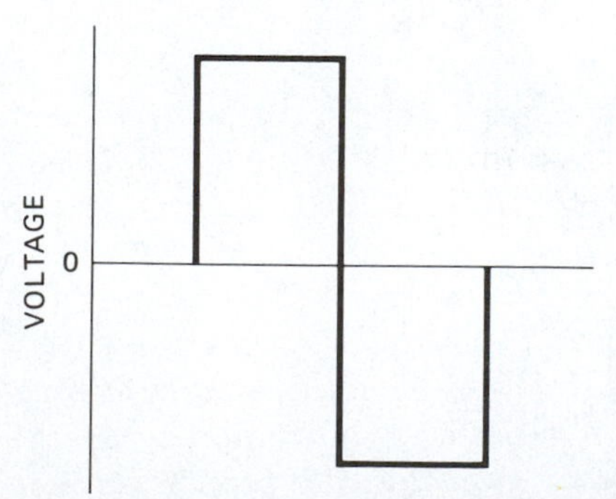

Figure 23-33
One cycle of voltage in step waveform

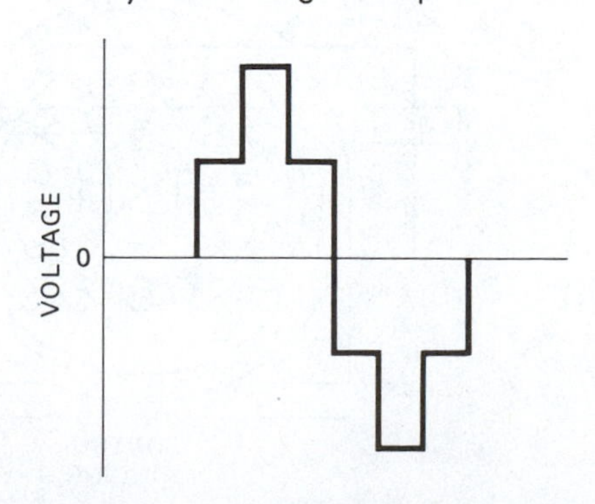

AMPLIDYNE

An amplidyne is a special type of dc generator. It has two armature circuits: one is short-circuited and the other supplies the load. A simplified diagram of an amplidyne is shown in Figures 23-34A, 23-34B, and 23-34C. If it is driven at its rated speed and a small current is sent through its control winding, a weak magnetic field develops in the armature, Figures 23-34A and 23-34B. As the armature winding moves through the flux, a voltage develops that causes the current to flow through the short circuit and across the brushes and armature. Because the resistance of the armature coils is negligible, a very weak field establishes a very high armature current, Figure 23-34A. This current produces a strong armature flux in the vertical direction, Figure 23-34C.

The armature is now cutting a very strong flux, producing a high voltage between the load brushes. This voltage supplies a high current through the armature windings and to the load, as indicated in Figure 23-34C. The low voltage supplied to the control windings governs the output voltage and current. A small change in the control current causes a large change in the output voltage.

It might be said that the amplidyne is a rotating amplifier. If the polarity of the control winding is reversed, the polarity of the load voltage is also reversed. This action is practically instantaneous. A change in the control current produces an immediate change in the output voltage.

The load current flowing through the armature produces a flux that opposes the flux produced by the control current, Figure 23-34C. This undesirable condition is prevented by adding a compensating winding that produces just enough flux to neutralize the flux produced by the load current.

Amplidynes are in use throughout the electrical industry as voltage regulators/exciters. They are especially useful when accurate voltage control is required (amplification of up to 10,000 to 1).

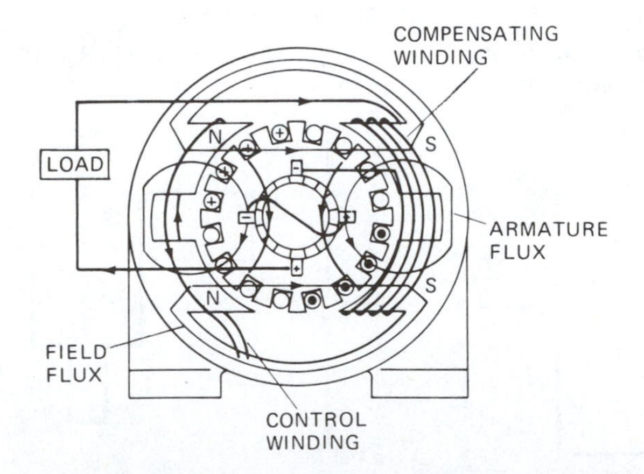

Figure 23-34A
Amplidyne generator

Figure 23-34B
High armature current in an amplidyne produced by a weak control circuit current

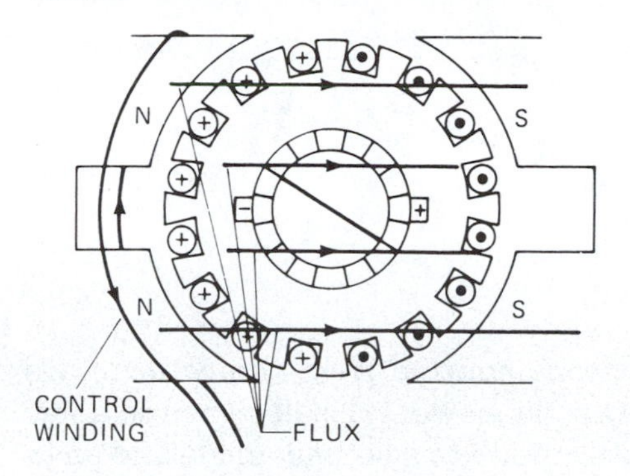

Figure 23-34C
Strong armature flux produced by the armature current in an amplidyne

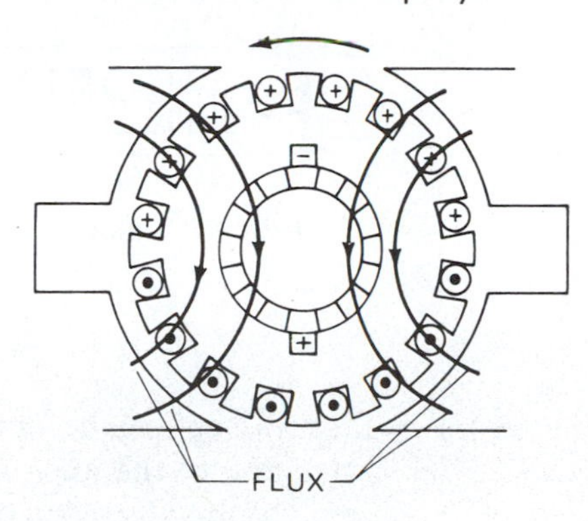

ELECTRON TUBES

The electron tube was the first device developed to provide electronic control of current flow. It performs much the same functions as solid-state, semiconductor devices. It is generally not as fast acting and much bulkier. It consists of a bulb containing a gas or vacuum and conducting materials called *elements*. The two primary elements are the *cathode* and *anode* (plate). Other elements contained within the bulb are called *grids*. The tube may have one or several grids, depending upon its purpose. The cathode, under certain conditions, emits electrons into the space within the bulb. The plate attracts the emitted electrons, and the grid controls the flow of the electrons between the cathode and the plate.

Cathode Ray Tube

Because most tubes have been replaced by semiconductor devices, it is not necessary to make a thorough study of all types of tubes that were in common use in the electrical industry. There is one tube, however, that is still in very common use. It is the cathode ray tube.

The cathode ray tube is used to produce a picture or graphic illustration. It is used to determine values of current and/or voltage. Various wave shapes and frequencies can also be determined with the tube. Practical applications are found in radar, sonar, and television (picture tube), and in the tuning of automobile engines.

There are two general types of cathode ray tubes: *electrostatic* and *electromagnetic*. The tubes operate on the principle of

Figure 23-35
Cathode ray tube

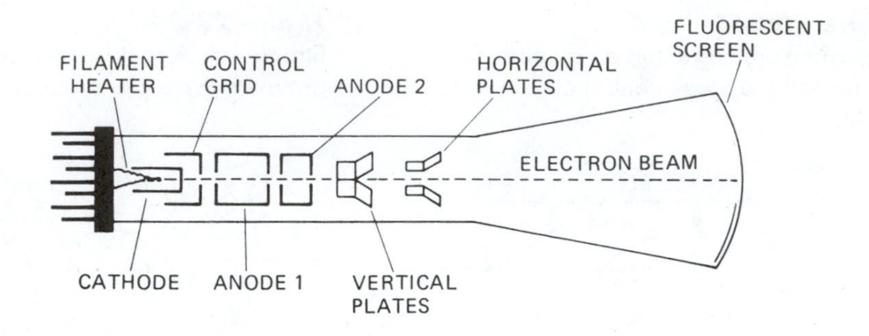

thermal emission. The electrons emitted from the cathode are attracted to the anode. The anode, however, is designed to allow some of the electrons to pass through to a second anode and on to the fluorescent screen, Figure 23-35. This flow of electrons is called an *electron beam*. The primary purpose of anodes 1 and 2 is to bring the electrons to a sharp focus on the screen. The screen emits light where the electrons strike it. The number of electrons striking the screen is controlled by the grid, which is kept negative with respect to the cathode.

The beam is deflected by a set of horizontal plates placed at right angles to each other, Figure 23-35. A variable voltage is applied to the plates. The voltage varies as illustrated by the curve in Figure 23-36. It is produced by an oscillator built within the case.

The purpose of the variable voltage is to cause the beam to move slowly from left to right and then to return instantly to the left. This cycle continues as long as the tube is energized. Applying an alternating voltage of the same frequency to the vertical plate produces a stationary trace on the screen, showing the variation of voltage applied to the vertical plates, Figure 23-37.

Figure 23-36
Waveform of voltage applied to the horizontal plates of a cathode ray tube

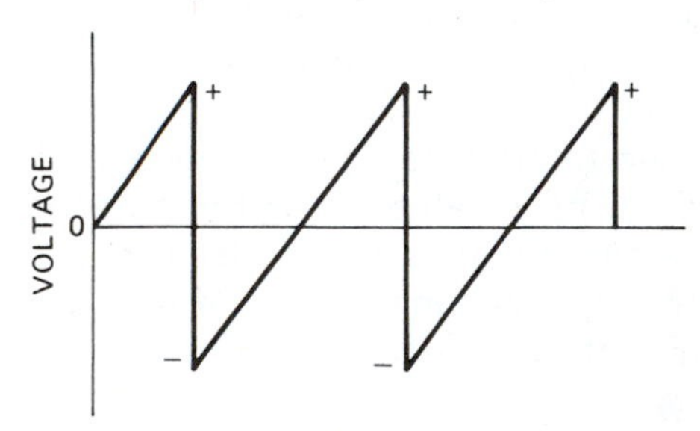

Figure 23-37
Waveform of voltage applied to the vertical plates of a cathode ray tube

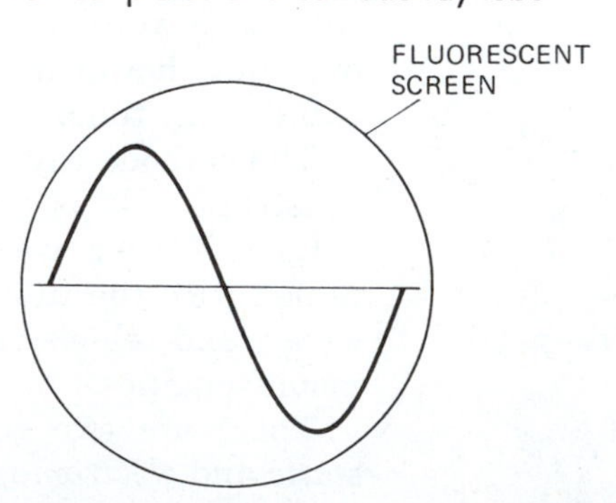

Only a single fluorescent spot occurs on the screen, but the spot moves so rapidly that, to the observer, it appears to be a continuous line.

The electrons that reach the screen must have a return path to the cathode. This is accomplished by metal deposits on the inside of the glass tube. A positive potential is applied to the metal coating. As the electrons bounce off the screen, they are attracted to the metal and return to the cathode.

OSCILLOSCOPE

One of the main uses of the cathode ray tube in the electrical industry is within the oscilloscope. An oscilloscope is an instrument used to obtain accurate measurements of current, voltage, and frequency. Instantaneous values can be determined easily from this instrument, which is also used to determine wave shapes and phase relationships between values of current and/or voltage.

The face of an oscilloscope is similar to a television screen. A graphic illustration of current or voltage is produced on the screen. An important feature of this instrument is that all measurements are taken within a specific time frame. To better understand how this is accomplished, refer to Figure 23-38. The face of the screen is divided into equal spaces both vertically and horizontally.

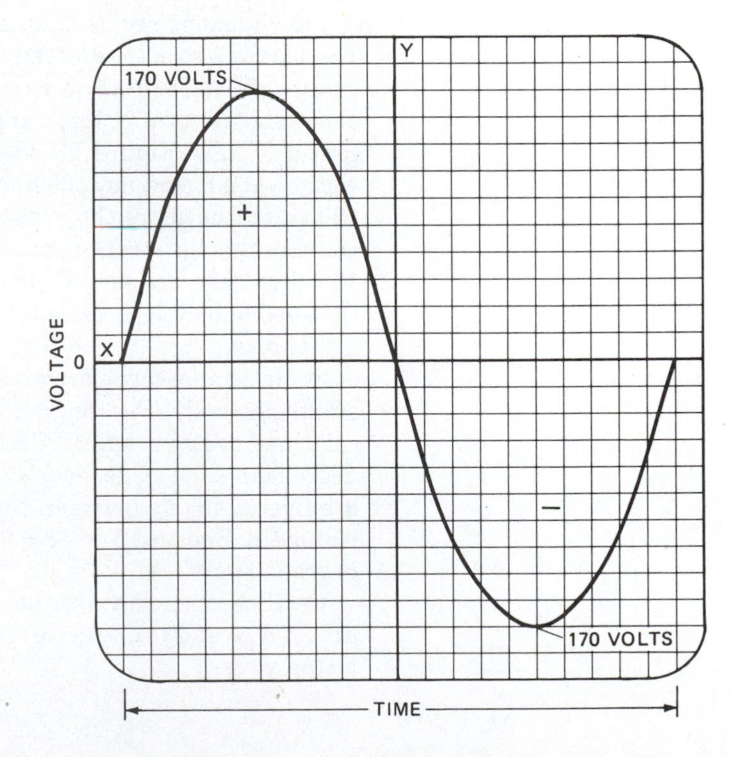

Figure 23-38
One cycle of alternating voltage

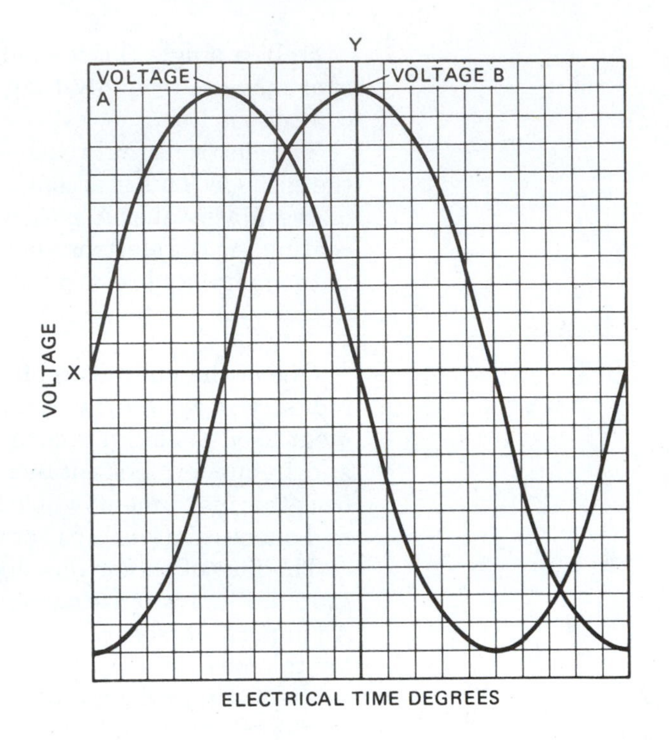

Figure 23-39
Two voltage curves 90 degrees out of phase with one another

The divisions are used for graphing the values to be measured. The vertical axis is referred to as the Y axis, and the horizontal axis is referred to as the X axis. Voltage and/or current is measured along the vertical Y axis, and time is measured along the horizontal X axis. Figure 23-38 illustrates one cycle of alternating voltage at a frequency of 60 hertz. The RMS value of the voltage is 120 volts; therefore the graph indicates a maximum value of 170 volts. In this illustration, each division along the Y axis represents 17 volts. Each division along the X axis represents 1/1200 second. Because one cycle is completed in 20/1200 second, the frequency is 60 hertz.

Some oscilloscopes are arranged to graph two or more voltages simultaneously. Figure 23-39 represents two voltages 90 electrical time degrees apart. Because there are 360 electrical time degrees in each cycle, each division represents 18 degrees. There are five divisions between the peak of voltage A and the peak of voltage B. Because $5 \times 18 = 90$, the voltages are 90 electrical time degrees apart.

Oscilloscopes can also be used to determine waveforms. Figure 23-40A shows one cycle of a square wave; Figure 23-40B shows a step wave.

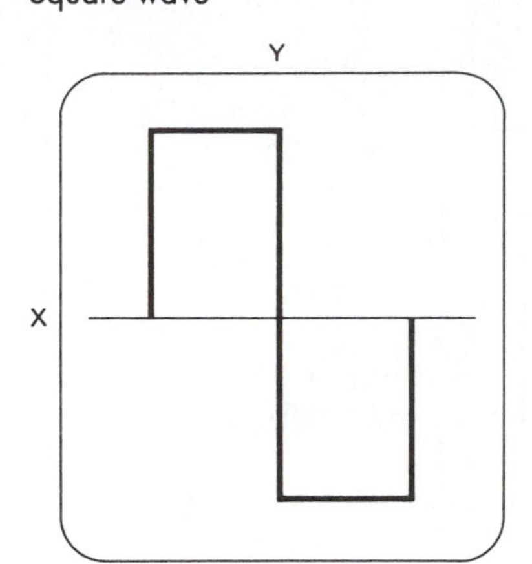

Figure 23-40A
Square wave

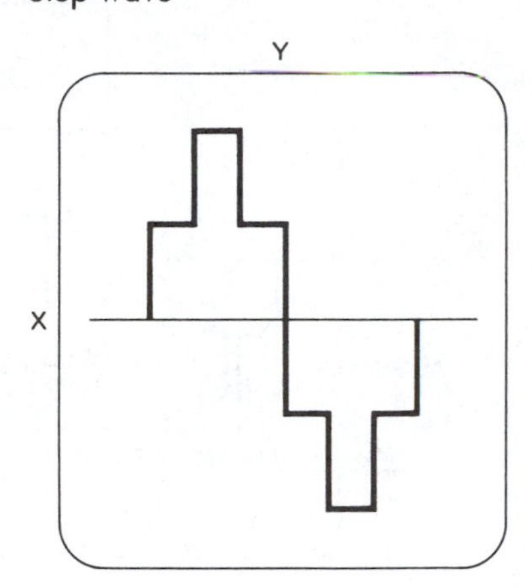

Figure 23-40B
Step wave

LIGHT-SENSITIVE DEVICES

There are three general types of light-sensitive devices: photoconductive, photoemissive, and photovoltaic. These devices are manufactured in solid-state form and in tube form.

Photoconductive Cell

The photoconductive cell is a solid-state device in which the value of resistance varies with the amount of light striking the light-sensitive material. Photoconductive cells are frequently used in control circuits to energize or deenergize the circuits. Examples include the controls of night-lights, alarm systems, floodlights, and oil-fired heating systems. These devices are made of cadium sulfide, lead sulfide, selenium, or germanium. They are generally impregnated with small amounts of gold, silver, or antimony. Photoconductive cells have a high resistance when they are not exposed to light. As the light intensity increases, the resistance of the cell decreases. The devices are capable of conducting enough current to operate small relays.

The cadium cell is commonly used to control the operating relay of residential oil-fired heating systems. The flame produces light that reduces the resistance of the cell, thereby allowing more current to flow in the relay coil. The relay contacts close, completing the circuit to the oil pump. The contacts remain closed as long as there is flame. If a malfunction occurs and the flame is extinguished, the cell's resistance increases, reducing the value of

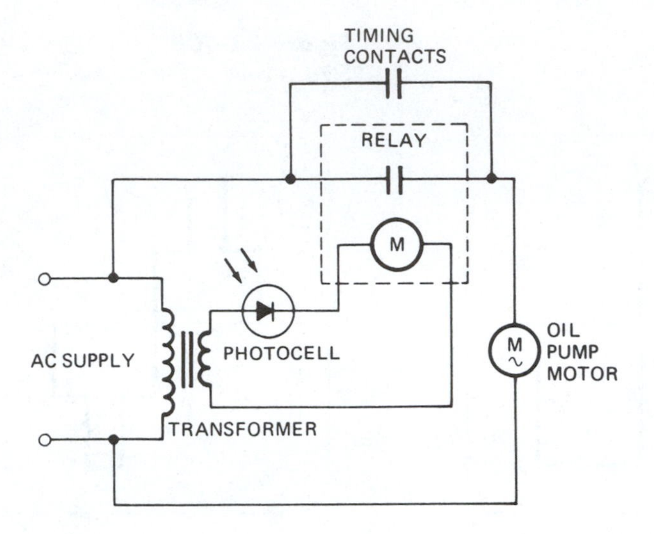

Figure 23-41
Simplified diagram of an oil pump controlled by a photoconductive cell

current flowing in the relay coil. The relay contacts open, and the pump no longer pumps oil into the hot fire chamber. Figure 23-41 is a simplified diagram of this operation.

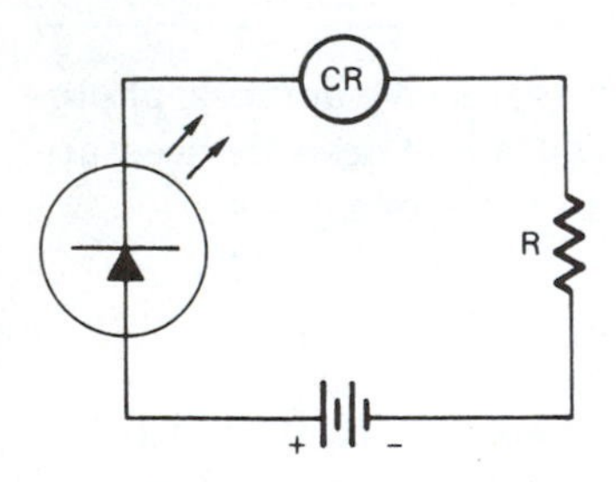

Figure 23-42
Circuit for a photoemissive tube

Photoemissive Cell

Photoemissive devices are manufactured in the form of electron tubes. The cathode consists of a semicylindrical plate coated on the inner surface with a material that emits electrons. The anode consists of a single wire. Both elements are enclosed in a glass tube that is evacuated or filled with an inert gas at low pressure. The circuit for the photoemissive tube is shown in Figure 23-42. The basic construction is shown in Figure 23-43. The tube does not produce electricity but will conduct when light strikes the sensitive material.

To make the cathode more sensitive to light, it is coated with lithium, sodium, potassium, rubidium, or cesium. The type of coating governs the luminous sensitivity and the spectral response. *Luminous sensitivity* is the amount of current for a given amount of light. *Spectral response* defines the amount of current obtained for the same amount of light of different colors. The amount of current a tube can conduct is generally in the range of microamperes (μA). The amount of light must be limited to a low value. approximately 2 lumens.

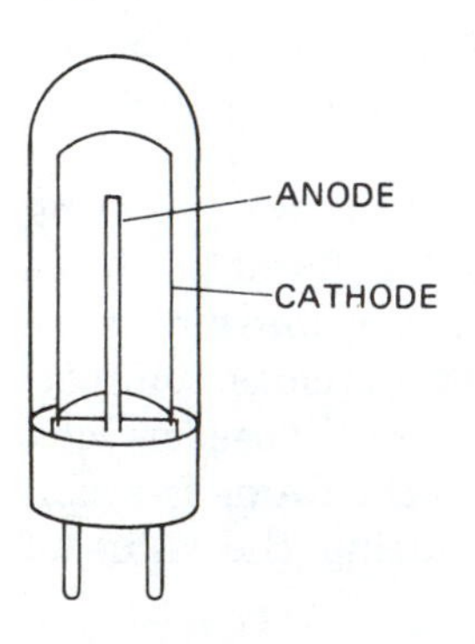

Figure 23-43
Basic construction of a photoemissive tube

With the vacuum-type photoemissive tube, a definite amount of light striking the cathode will cause the emission of a definite number of electrons. If the voltage applied between the cathode and the anode is gradually increased, more and more electrons will be attracted to the anode. Finally, at some predetermined voltage, usually 25 volts, all the electrons emitted from the cathode will be attracted to the anode. When this occurs, the saturation point has been reached. If the voltage is further increased, the current will not increase.

For the average illumination, the vacuum phototube conducts only a few microamperes. To increase the amount of current, an inert gas is added. When more than 25 volts are applied to the gas-filled phototube, the electrons leaving the cathode acquire sufficient velocity to knock electrons from the gas molecules, causing ionization. This action produces much higher current. Precautions must be taken not to exceed the voltage ratings. The value of voltage depends upon the type of tube and the amount of illumination. In general, the current-carrying capacity of a phototube is so small that it cannot be used even to actuate the most sensitive relay. However it can be used with amplifiers.

Photovoltaic Cell

The photovoltaic cell produces a voltage when light strikes it. There are several types available. One common type is made of copper and copper oxide with a barrier between them. When light strikes the copper oxide, some of the free electrons acquire enough energy to move through the barrier and accumulate in the copper. This action produces an excess of electrons in the copper and a shortage of electrons in the copper oxide. A potential difference then exists between the two ends of the cell.

These cells supply sufficient voltage to operate sensitive relays and other devices directly. They are slower acting than other light-sensitive devices. Therefore, they are not satisfactory for applications requiring extremely quick response.

Solar Cell

The solar cell is an improvement on the photovoltaic cell. It is a PN type of solid-state device and is very sensitive to light. Light striking the junction causes the residual electrons to develop enough energy to move from the P material into the N material. This produces an emf across the P and N terminals. The most common use for the solar cell is most likely in communication systems.

REVIEW

A. Multiple choice.
Select the best answer.

1. The primary function of electronic equipment is to
   a. improve power factor.
   b. relay signals.
   c. operate incandescent lights.
   d. provide field excitation for motors.
2. Electronic controllers have many functions similar to
   a. instrument transformers.
   b. pyrometers.
   c. electromagnetic relays.
   d. electric meters.
3. A semiconductor is
   a. a good conductor.
   b. a poor conductor.
   c. neither a good conductor nor a good insulator.
   d. a type of resistor.
4. Two of the most common materials used to make semiconductors are
   a. copper and aluminum.
   b. copper and magnesium.
   c. lead and silicon.
   d. germanium and silicon.
5. Covalent bonding is
   a. electrons moving from atom to atom.
   b. valance electrons sharing outer shells with adjacent atoms.
   c. the electrons in the atom that are bonded together.
   d. electrons bonded to protons.
6. The mixing of impurities (arsenic, indium, or gallium) with silicon or germanium is a chemical process called
   a. doping.
   b. bonding.
   c. building.
   d. purifying.
7. Semiconductors are divided into three groups called
   a. polar, nonpolar, and triode.
   b. unipolar, bipolar, and diode.
   c. unipolar, bipolar, and triode.
   d. polar, bipolar, and tripolar.

REVIEW *(continued)*

A. Multiple choice.
Select the best answer (continued).

8. A semiconductor material that has an excess of electrons is called
   a. P material.
   b. N material.
   c. V material.
   d. E material.
9. Diffusion occurs when P and N materials are
   a. separated.
   b. joined together.
   c. mixed.
   d. bonded.
10. A PN device is known as a
    a. diode.
    b. triode.
    c. biode.
    d. pentode.
11. A diode
    a. causes current flow.
    b. blocks current flow.
    c. increases current flow.
    d. blocks current flow in one direction.
12. Diodes conduct when they are connected in
    a. reverse bias.
    b. forward bias.
    c. neutral bias.
    d. none of the above.
13. Three classifications of rectifiers are
    a. mechanical, electrical, and electronic.
    b. unidirectional, bidirectional, and multidirectional.
    c. alternating current, direct current, and pulsating current.
    d. magnetic, mechanical, and electrical.
14. A rectifier is a device that changes
    a. ac to dc.
    b. dc to ac.
    c. the value of current.
    d. the value of voltage.

REVIEW *(continued)*

A. Multiple choice.
Select the best answer (continued).

15. Two types of rectifiers are
    a. copper oxide and aluminum.
    b. copper oxide and selenium.
    c. selenium and aluminum.
    d. aluminum and brass.

16. A half-wave rectifier produces a
    a. smooth dc.
    b. pulsating dc.
    c. square-wave ac.
    d. sawtooth-wave dc.

17. A filtering capacitor is used to
    a. change ac to dc.
    b. produce a square wave.
    c. change dc to ac.
    d. reduce ripple.

18. A three-element transistor has an
    a. emitter, collector, and base.
    b. emitter, collector, and grid.
    c. collector, grid, and base.
    d. emitter, collector, and resistor.

19. An amplifier is a device that
    a. produces sound.
    b. converts ac to dc.
    c. converts dc to ac.
    d. increases power.

20. An oscillator circuit is used to
    a. control dc motors.
    b. vary the frequency of an ac.
    c. change ac to dc.
    d. change dc to ac.

21. Oscillator circuits use
    a. resistance and capacitance.
    b. resistance and inductance.
    c. inductance and capacitance.
    d. reactance and resistance.

22. An arrangement of semiconductor components used for switching purposes is called a
    a. transistor switch.
    b. solid-state switch.
    c. thyristor.
    d. byristor.

REVIEW *(continued)*

A. Multiple choice.
Select the best answer (continued).

23. A transistor that performs such operations as speed control of motors, temperature control, and dimming of lights is called a
    a. triac.
    b. diac.
    c. quadac.
    d. diode.

24. A three-element device that can block current flow in either direction or be triggered to conduct in either direction is called a
    a. diac.
    b. triac.
    c. quadac.
    d. diode.

25. Diacs are generally used for
    a. voltage control.
    b. phase control.
    c. power factor control.
    d. capacitors.

26. Triacs conduct in
    a. one direction only.
    b. both directions.
    c. neither direction.
    d. specific periods of time.

27. SCR stands for
    a. solid-state control rectifier.
    b. silicon-controlled rectifier.
    c. solid-state current rectifier.
    d. silicon-controlled relay.

28. Solid-state inverters are used to
    a. change ac to dc.
    b. change dc to ac.
    c. produce an inverted waveform.
    d. invert two phases.

29. AC squirrel-cage induction motors can have their speed controlled through electronic equipment that controls the
    a. rotor current.
    b. armature current.
    c. frequency.
    d. rotor voltage.

REVIEW *(continued)*

A. Multiple choice.
Select the best answer (continued).

30. SCRs are frequently used to control the speed of
    a. dc motors.
    b. ac motors.
    c. neither a nor b.
    d. both a and b.

31. Inverters are frequently used to control the speed of
    a. shunt motors.
    b. wound-rotor motors.
    c. squirrel-cage induction motors.
    d. series motors.

32. The SCR is capable of controlling
    a. large amounts of power with high efficiency.
    b. small amounts of power with high efficiency.
    c. large amounts of power with poor efficiency.
    d. small amounts of power with poor efficiency.

33. An amplidyne is a special type of
    a. dc generator.
    b. ac generator.
    c. rotory converter.
    d. motor-generator set.

34. An amplidyne has
    a. one armature circuit.
    b. two armature circuits.
    c. three armature circuits.
    d. four armature circuits.

35. An amplidyne can be compared to
    a. a rectifier.
    b. an amplifier.
    c. an inverter.
    d. an inductor.

36. A weak magnetic field produced by the control circuit of an amplidyne will
    a. produce a weak armature current.
    b. produce a high value of armature current.
    c. have no effect on the armature current.
    d. produce a lagging armature current.

REVIEW *(continued)*

A. Multiple choice.
Select the best answer (continued).

37. A small change in the control current of an amplidyne will produce
    a. a small change in the output voltage.
    b. a large change in the output voltage.
    c. no change in the output voltage.
    d. a pulsating output voltage.
38. The output voltage of an amplidyne is governed by the
    a. control current.
    b. power factor of the load.
    c. both a and b.
    d. neither a nor b.
39. One use for the amplidyne in the electrical industry is to
    a. control the load current.
    b. change ac to dc.
    c. change dc to ac.
    d. control the load voltage.
40. The first device developed to produce electronic control of current was the
    a. transistor.
    b. electron tube.
    c. semiconductor.
    d. capacitor.
41. The cathode ray tube is used to
    a. rectify.
    b. amplify.
    c. produce a sine wave.
    d. produce a picture.
42. The oscilloscope is
    a. a type of capacitor.
    b. part of an oscillator circuit.
    c. used to determine wave shapes.
    d. used to control the speed of ac motors.
43. Instantaneous values of voltage can be measured with
    a. an oscilloscope.
    b. an SCR.
    c. a standard voltmeter.
    d. a synchroscope.

REVIEW *(continued)*

A. Multiple choice. Select the best answer (continued).

44. A phototube reacts to
    a. light.
    b. heat.
    c. voltage.
    d. current.
45. The most common type of photovoltaic cell is
    a. copper and lead.
    b. copper and copper oxide.
    c. lead and lead oxide.
    d. copper and lead oxide.
46. The cadmium cell is sensitive to
    a. heat.
    b. light.
    c. sound.
    d. motion.

B. Give complete answers.

1. Define *electronics.*
2. What is the primary function of electronic equipment?
3. List three advantages of electronic controls over electromagnetic controls.
4. What is a transistor?
5. What are semiconductors?
6. Describe the action that occurs when an impurity is added to silicon to produce N material.
7. How is P material developed?
8. What is meant by *covalent bonding?*
9. List three groups of semiconductors.
10. List three classifications of semiconductors according to materials.
11. What is a diode?
12. Describe the action that takes place when P and N materials are joined together.
13. List six types of rectifiers.
14. Name two types of electronic rectifiers.
15. Describe a half-wave rectifier.
16. Describe a full-wave rectifier.

REVIEW *(continued)*

B. Give complete answers (continued).

17. Describe a copper oxide rectifier.
18. Describe a selenium rectifier.
19. Define the term *ripple* with reference to dc.
20. Describe a PNP transistor.
21. Draw a diagram of a capacitor choke filter circuit.
22. Describe an NPN transistor.
23. Draw a diagram of a transistor amplifier.
24. Draw a transistor oscillator circuit.
25. Describe a tank circuit.
26. How can the output voltage frequency of an oscillator be varied?
27. What is a thyristor?
28. Describe a diac.
29. Describe a triac.
30. What is the purpose of the SCR?
31. What is an electronic inverter?
32. What type of equipment can be used to obtain a variable frequency ac?
33. What is an amplidyne?
34. What is a cathode ray tube?
35. Describe the operation of the cathode ray tube.
36. What is the purpose of the variable voltage in the cathode ray tube?
37. What is an oscilloscope?
38. Describe the difference between photoemissive, photoconductive, and photovoltaic.
39. Describe the construction and operation of the photovoltaic cell.
40. What is a cadium cell?
41. Describe one common use for a cadmium cell.

24
PROGRAMMABLE CONTROLLERS*

Objectives

After studying this chapter, the student will be able to:

- Identify and list the principal parts of a programmable controller.
- Explain why and under what conditions a programmable controller is used in preference to other types of controllers.
- Describe the relationship between the central processing unit, the programming device, and the input/output unit.
- Draw and interpret basic diagrams showing how the input/output module functions.
- Define various terms used in conjunction with programmable controllers.

BASIC COMPONENTS AND APPLICATION

A programmable controller (PC) is a special type of computer. It is designed to perform specific control functions in a logical order.

Differences Between the PC and the Common Computer

Some differences between a PC and a home and business computer are:

1. The PC is designed to be operated in an industrial environment. Any computer used in industry must be able to operate

* From Industrial Motor Control by *Stephen L. Herman and Walter N. Alerich. Copyright 1985 by Delmar Publishers Inc.* The information is used with permission from the author and the publisher.

in extremes of temperature; ignore voltage spikes and drops on the incoming power line; survive in an atmosphere that often contains corrosive vapors, oil, and dirt; and withstand shock and vibration.

2. Most PCs are designed to be programmed with relay schematic or ladder diagrams instead of the common computer languages such as BASIC or FORTRAN. An electrician who is familiar with relay logic diagrams can generally be trained to program a PC in a few hours, whereas it generally takes several months to train someone to program a standard computer.

Application

The PC is rapidly replacing other types of control equipment. The increasing need for faster-acting, reliable controls and the ability to meet system changes quickly and at minimum cost have brought about this change.

Electromagnetic relays and pneumatic, and motor-driven timers are slower acting, require more space and maintenance, and must be hard wired in. The first electronic controller met most of the industrial requirements but also brought with it a language unfamiliar to many electrical personnel.

The design of the PC has met the requirements of industry, and the language it uses is such that the electrical worker who installs and maintains the equipment need not have a strong background in computer technology.

Basic Components

PCs can be divided into four basic parts:

1. the power supply
2. the central processing unit
3. the program loader or terminal
4. the input/output, or I/O (pronounced eye-oh), track

The Power Supply. The power supply is used to lower the incoming ac voltage to the desired level, rectify it to direct current, and then filter and regulate it. The internal logic circuits of PCs operate on 5 to 15 volts dc depending on the type of controller. This voltage must be free of voltage spikes and other electrical noise. It must also be regulated to within 5 percent of the required voltage value. Some manufacturers of PCs use a separate power supply, and others build the power supply into the central processor.

The CPU. The central processing unit, or CPU, is the brain of the PC. It contains the microprocessor chip and related integrated

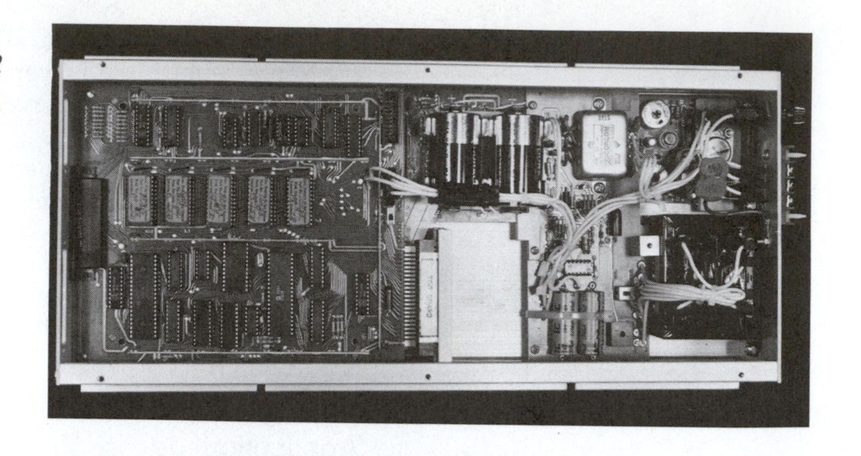

Figure 24-1
Inside the processor *(Courtesy of UTICOR Technology, Inc.)*

circuits to perform all the logic functions. The microprocessor chip used in most PCs is the same as the common computer chip used in many home and business machines, Figure 24-1.

The CPU generally has a key switch located on the front panel. This switch must be turned on before the CPU can be programmed. This is done to prevent the circuit from being changed accidentally. Plug connections mounted on the central processor are used to provide connections for the programming terminal and the I/O tracks, Figures 24-2A and 24-2B. Most CPUs are designed so that once the program has been tested, it can be stored on tape or disc. In this way, if a CPU fails and has to be replaced, the new unit can be reprogrammed from the tape or disc. This eliminates the time-consuming process of having to reprogram by hand.

Figure 24-2A
Central processing unit *(Courtesy of UTICOR Technology, Inc.)*

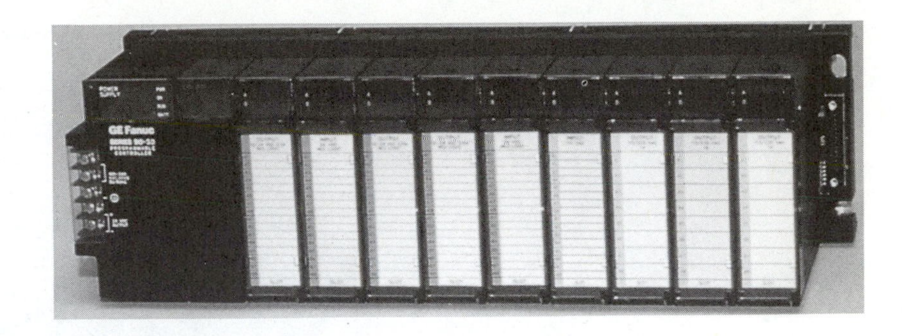

Figure 24-2B
Central processing unit *(Courtesy of GE Fanuc Automation North America, Inc.)*

The Programming Terminal. The programming terminal or loading terminal is a device used to program the CPU. Most terminals are one of two types. One type is a small, hand-held device that uses a liquid crystal display (LCD) to show the program, Figure 24-3. This terminal, however, displays only one line of the program at a time.

The other type of terminal uses a cathode ray tube (CRT) to show the program. This terminal looks similar to a portable television set with a keyboard attached, Figures 24-4A, 24-4B, and 24-5. It displays from four to six lines of the program at a time, depending on the manufacturer.

The terminal is used not only to program the controller but also to troubleshoot the circuit. When the terminal is connected to the CPU, the circuit can be examined while it is in operation. Figure 24-6 illustrates a circuit typical of those that are seen on the display. Notice that this schematic diagram is a little different from the typical ladder diagram. All of the line components are shown as normally open or normally closed contacts. There are no NEMA symbols for push buttons, float switches, limit switches, etc. The PC recognizes only open or closed contacts. It does not

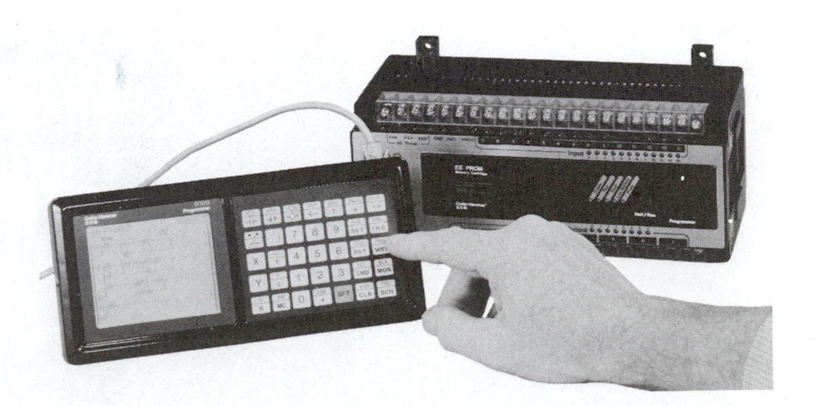

Figure 24-3
Small programmable controller and hand-held programming terminal *(Courtesy of EATON Corp., Cutler-Hammer Products)*

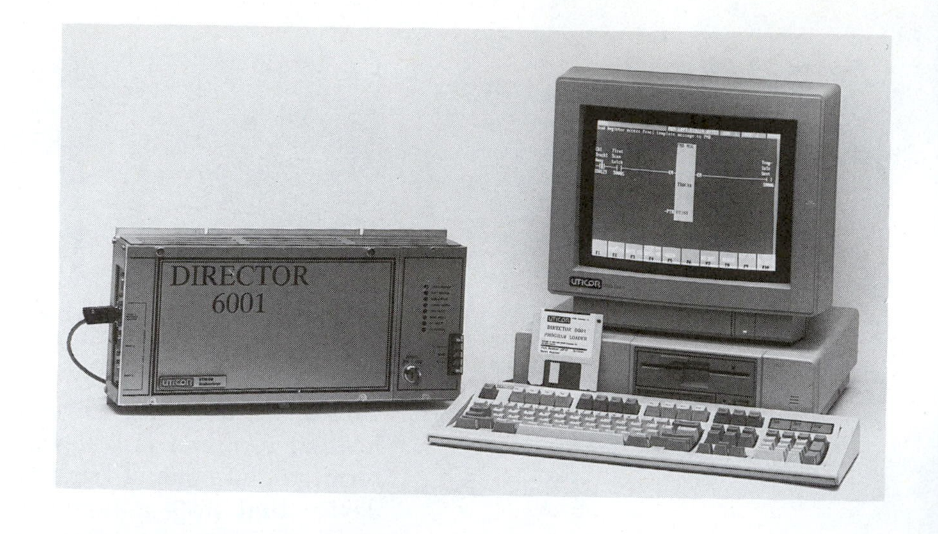

Figure 24-4A
Programming terminal *(Courtesy of UTICOR Technology, Inc.)*

Figure 24-4B
Programming terminal with key pad *(Courtesy of Allen-Bradley Co., Systems Division)*

Figure 24-5
Central processor unit, I/O track, and programming terminal *(Courtesy of General Electric Co.)*

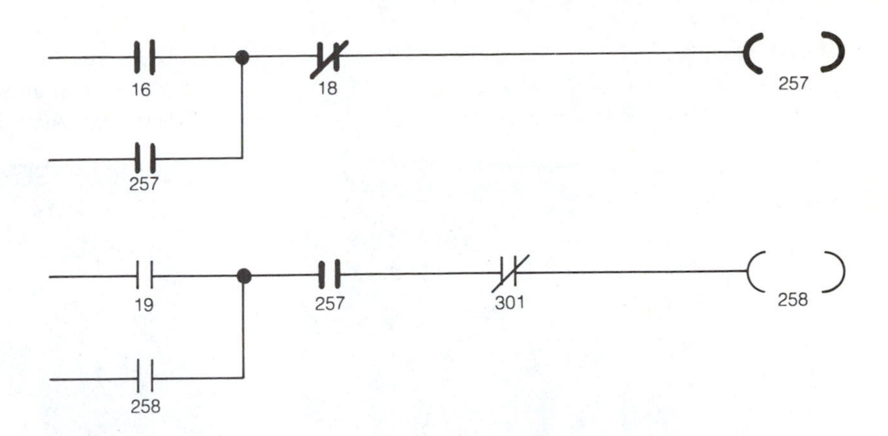

Figure 24-6
Analyzing circuit operation with the terminal *(From Herman & Alerich, Industrial Motor Control, copyright 1985 by Delmar Publishers Inc.)*

know if a contact is controlled by a push button, a limit switch, or a float switch. Each contact, however, does have a number. The number is used to distinguish one contact from another.

The coil symbols look like a set of parentheses instead of a circle as shown on most ladder diagrams. Each line ends with a coil, and each coil has a number. When a contact symbol has the same number as a coil, it means the contact is controlled by that coil. Figure 24-6 shows a coil numbered 257 and two contacts numbered 257. When relay coil 257 is energized, the controller interprets both of these contacts to be closed.

Notice that the 257 contacts, contacts 16 and 18, and coil 257 are drawn with dark, heavy lines. When a contact has a complete circuit through it, or a coil is energized, the terminal will illuminate that contact or coil. Contact 16 is illuminated, which means that it is closed, providing a current path. Contact 18 is closed, providing a current path to coil 257. Because coil 257 is energized, both 257 contacts are closed and providing current paths.

Contacts 19, 258, and 301 are not illuminated. This means that there is no complete circuit to coil 258. Coil 258 is not energized. A voltage does exist, however, at contact 301. This contact is shown as normally closed. Because it is not illuminated, it is open and no current path exists through it. Notice that the illumination of a contact does not mean that the contact has changed position; it means that there is a complete path for current flow.

When the terminal is used to load a program into the CPU, contact and coil symbols on the keyboard are used. These symbol keys are used to load a ladder diagram similar to the one shown in Figure 24-6 into the CPU. This program will show on the screen.

The I/O Track. The I/O track is used to connect the CPU to the outside world. It contains input modules that carry information to the CPU and output modules that carry information from the CPU. An I/O track with input and output modules is shown in

Figure 24-7A
I/O track with high-speed interface
(Courtesy of UTICOR Technology, Inc.)

Figure 24-7B
I/O track with input and output modules
(Courtesy of Allen-Bradley Co., Systems Division)

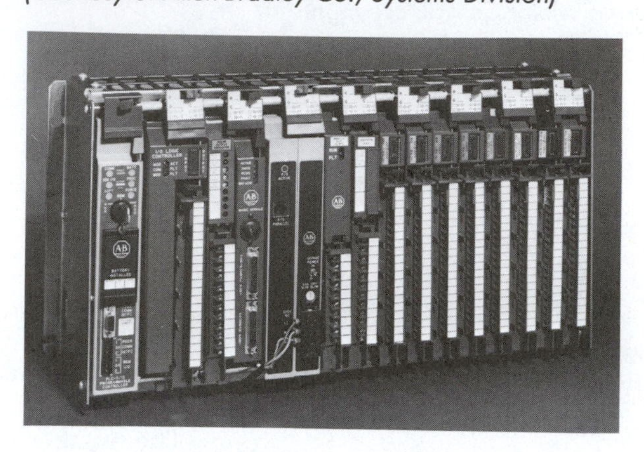

Figures 24-7A and B. Most modules contain more than one input or output. Any number from two to eight is common depending on the manufacturer. The modules shown in Figure 24-7A can each handle four connections. This means that each input module can handle four different inputs from pilot devices such as push buttons, float switches, or limit switches. Each output module can control four external devices such as pilot lights, solenoids, or motor starter coils. The operating voltage of modules can be alternating current or direct current and is generally either 120 or 24 volts. The I/O track in Figure 24-7A can handle eight modules. Because each module can accommodate four devices, this I/O track can control 32 inputs or outputs.

I/O Capacity. One factor that determines the size and cost of a PC is its I/O capacity. Many small units are designed to handle only 32 inputs or outputs. Large units can handle several hundred. The controller shown in Figure 24-2A is designed to handle eight I/O tracks. Because each I/O track has 32 inputs or outputs, the controller has an I/O capacity of 256. The I/O track in Figure 24-8 has a capacity for 80 inputs and outputs.

Figure 24-8
I/O track with input and output modules
(Courtesy of GE Fanuc Automation North America, Inc.)

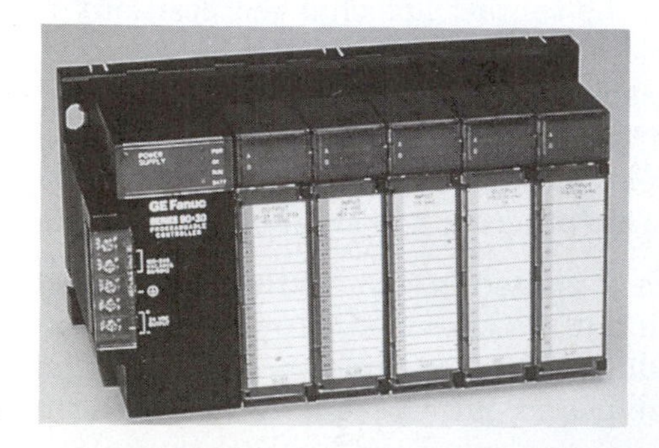

The Input Module. The CPU of a PC is extremely sensitive to voltage spikes and electrical noise. For this reason, the input I/O uses optoisolation to electrically separate the incoming signal from the CPU. Another job performed by the input I/O is debouncing any switch contacts connected to it.

Figure 24-9 shows a typical circuit used for the input. The bridge rectifier changes the ac voltage into dc voltage. A resistor is used to limit current to the light-emitting diode (LED). When the

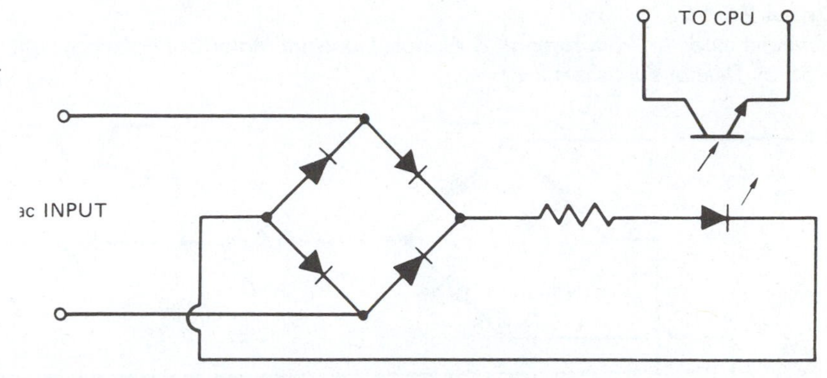

Figure 24-9
Input circuit *(From Herman & Alerich, Industrial Motor Control, copyright 1985 by Delmar Publishers Inc.)*

LED turns on, the light is detected by the phototransistor, which signals the CPU that there is a voltage present at the input terminal.

When the module has more than one input, the bridge rectifiers are connected together on one side to form a common terminal. On the other side, the rectifiers are labeled 1, 2, 3, and 4. Figure 24-10 shows four bridge rectifiers connected together to form a common terminal.

Figure 24-11 shows a limit switch connected to the input. Notice that the limit switch completes a circuit from the ac line to the bridge rectifier. When the limit switch closes, 120 volts ac is applied to the rectifier, causing the LED to turn on.

The Output Module. The output module is used to connect the CPU to the load. The output is an optoisolated, solid-state relay. The current rating can range from 0.5 to 3 amps, depending on the manufacturer. Voltage ratings are generally 24 or 120 volts and can be ac or dc.

If the output is designed to control a dc voltage, a power transistor is used to control the load, Figure 24-12. The transistor is a phototransistor that is operated by an LED. The LED is operated by the CPU.

If the output is designed to control an ac load, then a triac, rather than a power transistor, is used as the control device, Figure 24-13. A photodetector connected to the gate of the triac is used to control the output. When the LED is turned on by the CPU, the photodetector permits current to flow through the gate of the triac and turn it on.

If more than one output is contained in a module, the control devices are connected together on one side to form a common terminal. Figure 24-14 shows an output module that contains four outputs. Notice that one side of each triac has been connected to form a common terminal. On the other side, the triacs are labeled

Figure 24-10
Four-input module *(From Herman & Alerich, Industrial Motor Control, copyright 1985 by Delmar Publishers Inc.)*

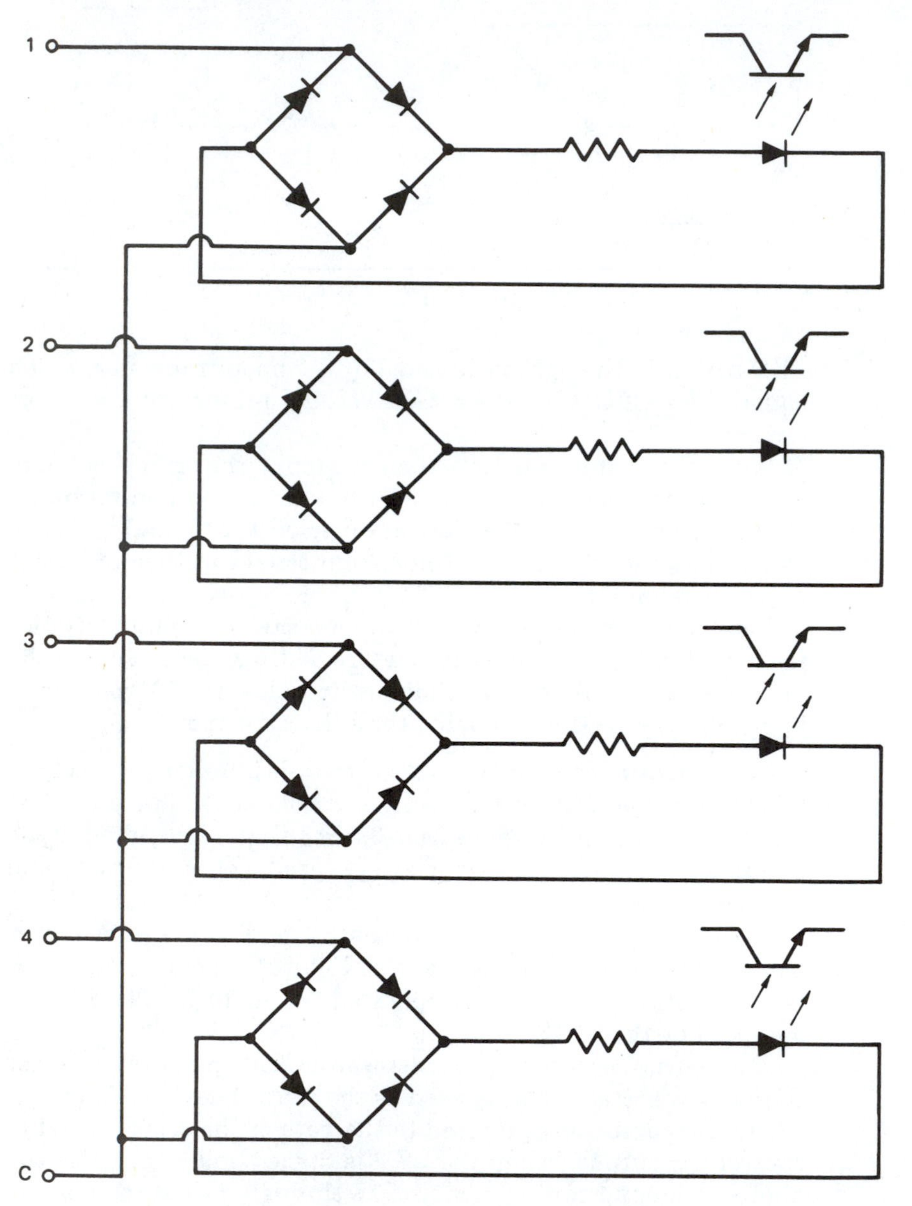

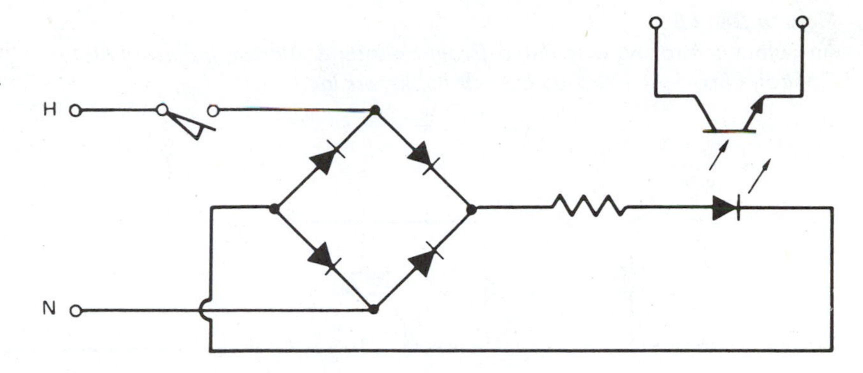

Figure 24-11
Limit switch completes circuit to rectifier
(From Herman & Alerich, Industrial Motor Control, copyright 1985 by Delmar Publishers Inc.)

Figure 24-12
Output module used to control a dc module
(From Herman & Alerich, Industrial Motor Control, copyright 1985 by Delmar Publishers Inc.)

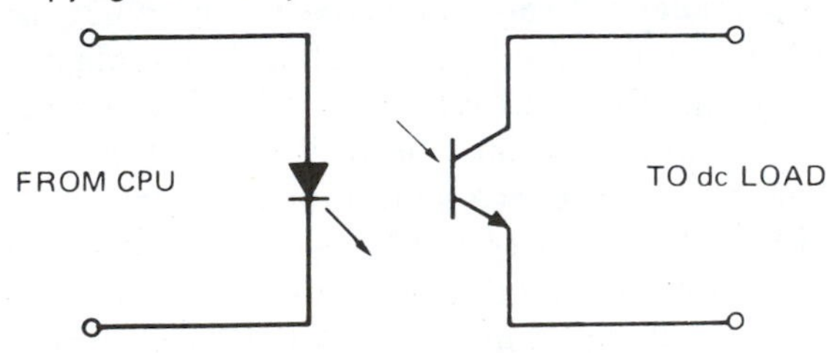

Figure 24-13
Output module used to control an ac module
(From Herman & Alerich, Industrial Motor Control, copyright 1985 by Delmar Publishers Inc.)

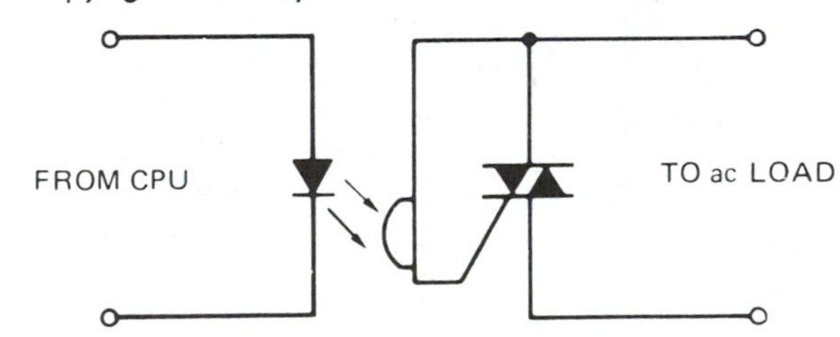

Figure 24-14
Four ac outputs in one module
(From Herman & Alerich, Industrial Motor Control, copyright 1985 by Delmar Publishers Inc.)

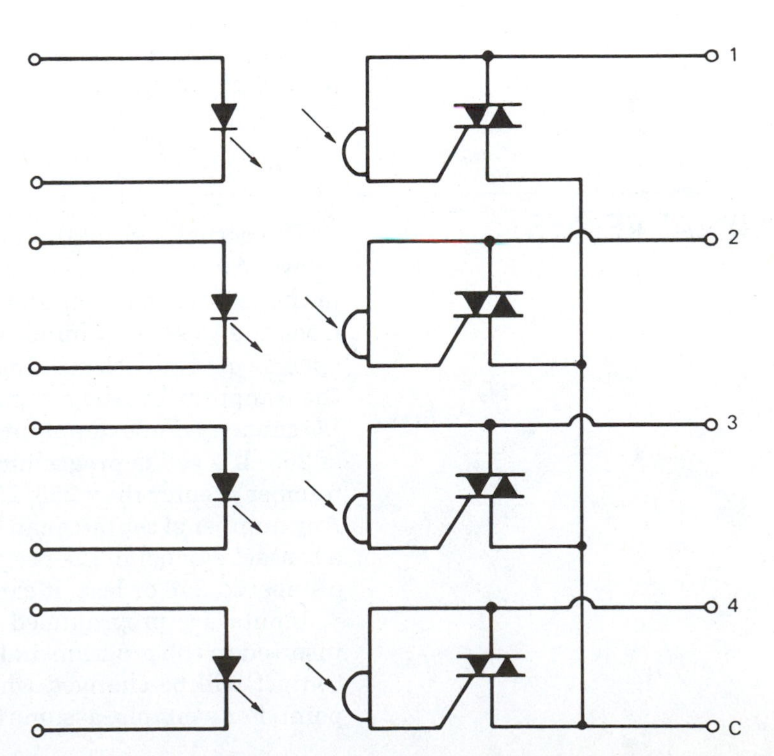

Figure 24-15
An output controlling a solenoid *(From Herman & Alerich, Industrial Motor Control, copyright 1985 by Delmar Publishers Inc.)*

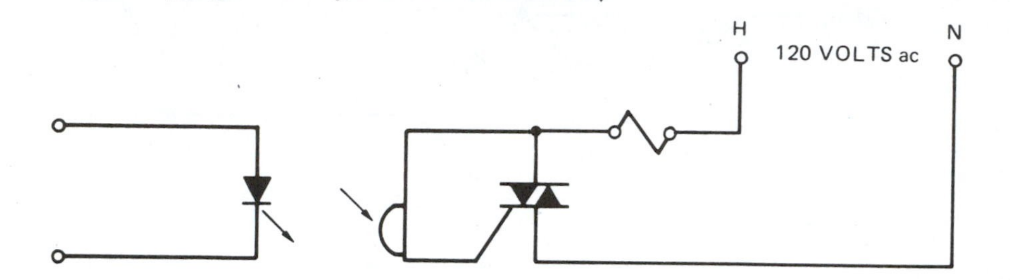

1, 2, 3, and 4. If power transistors are used as the control devices, then the emitters or the collectors can be connected to form a common terminal.

Figure 24-15 shows a solenoid coil connected to an ac output module. Notice that the triac is used as a switch to complete a circuit so that current can flow through the coil. The output module does not provide power to operate the load. The power must be provided by an external power source. The amount of current an output can control is limited. Small current loads, such as solenoid coils and pilot lights, can be controlled directly by the I/O output, but large current loads, such as motors, cannot. When a large amount of current must be controlled, the output is used to operate the coil of a motor starter or contactor, which can be used to control almost anything.

INTERNAL RELAYS

The actual logic of the control circuit is performed by *internal relays*. An internal relay is an imaginary device that exists only in the logic of the computer. It can have any number of contacts, from one to several hundred, and the contacts can be normally open or normally closed. Internal relays can be programmed into the computer by assigning a coil some number greater than the I/O capacity. For example, assume that the PC has an I/O capacity of 256. If a coil is programmed into the computer and assigned a number greater than 256, 257 for instance, it is an internal relay. Any number of contacts can be controlled by relay 257 by inserting a contact symbol in the program and numbering it 257. If a coil is numbered 256 or less, it can turn on an output when energized.

Inputs are programmed in a similar manner. If a contact is inserted in the program and assigned the number 256 or less, the contact will be changed when a voltage is sensed at that input point. For example, assume that a normally open contact has been programmed into the circuit and assigned number 22. When voltage is applied to input number 22, the contact will close. Because 22 is used as an input in this circuit, care must be taken not to assign number 22 to a coil. Terminal number 22 cannot be used as both an input and output at the same time.

Counters and Timers. The internal relays of a PC can be used as counters and timers. When timers are used, most of them are programmed in 0.1 second time intervals. For example, assume that a timer is to be used to provide a delay of 10 seconds. When the delay time is assigned to the timer, the number 00100 is used. This means the timer has been set for 100 tenths of a second, which is 10 seconds.

Off-Delay Circuit. The internal timers of a PC function as on-delay relays. A simple circuit can be used, however, to change the sense of the on-delay timer to make it perform as an off-delay timer. Figure 24-16 is this type of circuit. The desired operation of the circuit is as follows. When contact 350 closes, relay coil 12 energizes immediately and turns on a solenoid valve. When contact 350 opens, coil 12 remains energized for 10 seconds before it deenergizes and turns off the solenoid.

This logic is accomplished as follows:

1. When contact 350 closes, internal relay 400 energizes.
2. When coil 400 energizes, normally open contact 400 closes and completes a circuit to coil 12, and normally closed contact 400 connected in series with timer TO-1 opens.
3. When relay coil 12 energizes, both normally open 12 contacts close, and the I/O output at terminal 12 connects the solenoid coil to the power line.
4. When contact 350 opens, internal relay 400 deenergizes.

Figure 24-16
Off-delay circuit *(From Herman & Alerich, Industrial Motor Control, copyright 1985 by Delmar Publishers Inc.)*

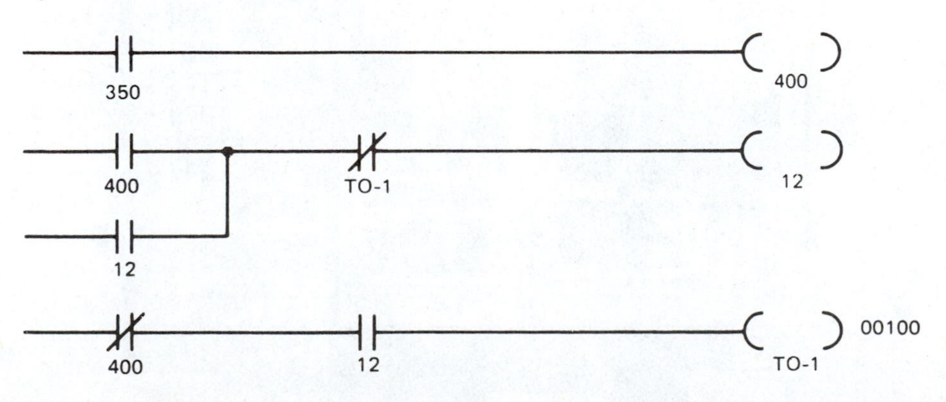

5. This causes both 400 contacts to change back to their original positions.
6. When normally open contact 400 returns to its open state, a continued current path to coil 12 is maintained by the now closed contact 12 connected parallel to it.
7. When normally closed contact 400 returns to its closed position, a circuit is completed through the now closed contact 12 to coil TO-1.
8. When coil TO-1 energizes, a 10-second timer starts. At the end of this time period, contact TO-1 opens and deenergizes coil 12.
9. When coil 12 deenergizes, both 12 contacts return to the open position and the I/O output turns the solenoid off.
10. Timer TO-1 deenergizes when contact 12 opens and the circuit is back in its original start condition.

The number of internal relays and timers contained in a PC is determined by the memory capacity of the computer. As a general rule, PCs that have a large I/O capacity have a large memory, and machines that have less I/O capacity have less memory.

The use of programmable controllers has steadily increased since their invention in the late 1960s. A PC can replace hundreds of relays and occupy only a fraction of the space. The circuit logic can be changed easily and quickly without requiring extensive hand rewiring. They have no moving parts or contacts to wear out, and their down time is less than an equivalent relay circuit. A programmable controller used to control a dc drive unit is shown in Figure 24-17.

Figure 24-17
DC drive unit controlled by a programmable controller *(Courtesy of Allen-Bradley Co., Drives Division)*

Electromagnetic Relays and PC Circuits

Ladder logic diagrams have been used in previous chapters to illustrate various motor control operations. They are called logic because they show specific operations in a logical sequence. The diagrams can be converted into schematics for use with PCs.

Figure 24-18A shows a standard stop/start station controlling a three-phase motor through an across-the-line starter. Figure 24-18B shows only the control circuit for this operation. A ladder logic diagram for this circuit is shown in Figure 24-19A.

Assume the circuit in Figure 24-19A is to be programmed into a PC. If the I/O unit has 24 terminals, half will be used for the input and half for the output. The stop, start, auxiliary, and overload contacts are all part of the input information. The normally closed stop button will be connected to terminal number

Figure 24-18A
Stop/start station connected to an across-the-line motor starter

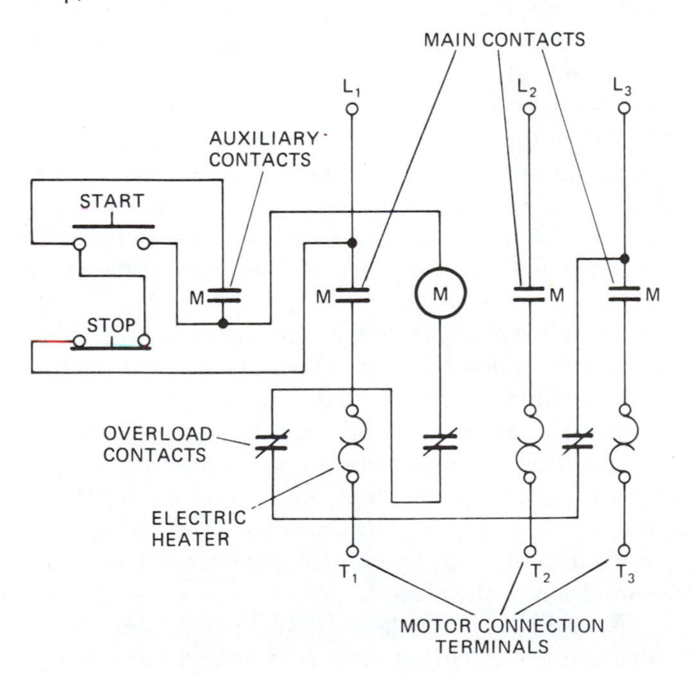

Figure 24-18B
Schematic diagram of the control circuit for a three-phase, across-the-line motor starter

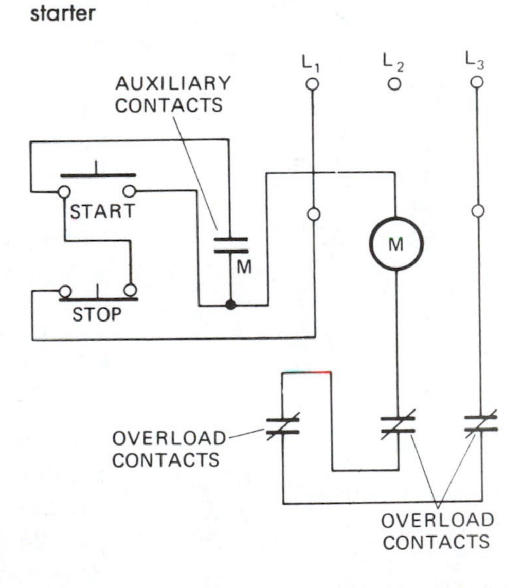

Figure 24-19A
Stop/start station connected to an across-the-line motor starter

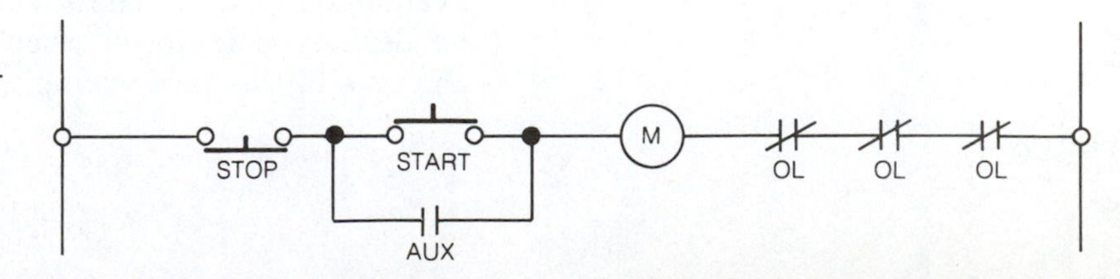

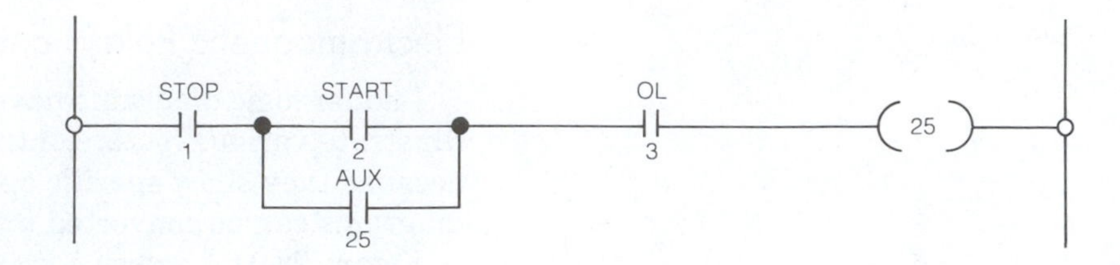

Figure 24-19B
Stop/start station connections for a programmable controller

1 of the I/O unit. The normally open start button is connected to number 2, and the normally closed overload contacts are connected to number 3. Notice that there are three sets of overload contacts, but because they are all normally closed and all perform the same function (to open the control circuit in case of overload), the series group is connected to one terminal of the I/O unit.

Figure 24-19 B illustrates the PC circuitry. Notice that all the contacts are programmed open. Their normal position is determined by supplying power to specific terminals. When power is supplied to an input terminal of the I/O unit, the CPU interprets this as a signal to change the position of the contacts. In Figure 24-19B, power is supplied to terminals 1 and 3 during normal circuit conditions. The CPU interprets these contacts to be closed and shows them illuminated on the screen.

Contact 2 is the start button and is connected to terminal number 2. Because the start button is normally open, no power is supplied to terminal number 2 during normal circuit conditions. Contacts 25 are normally open and are connected in parallel with contacts 2. Contacts 2 and 25 will not be illuminated when the motor is not in operation. Coil 25 will not be illuminated when it is not energized.

Diagrams for PCs always show coils at the end of the line. The number 25 has been assigned to the coil because it is not part of the I/O unit but is considered an internal relay.

When the start button is pressed, coil 25 is energized, causing contacts 25 to close. Releasing the start button does not break the circuit to the coil because contacts 25 are closed. Under these conditions, contacts 1, 3, and 25 will be illuminated on the screen. Coil 25 will also be illuminated. By checking the screen, one can determine the condition of the circuit.

Figures 19-10A, 19-10B, 19-10C, and 19-10D in Chapter 19 are diagrams illustrating the circuitry for a sequence, interlocking, electromagnetic controller. This same circuitry can be programmed into a PC. The advantage is that a separate program can be developed for each circuit—B, C, and D—without making changes in the hard wiring.

Figures 24-20A, 24-20B, and 24-20C show the same circuits diagrammed for a PC. With the PC, once the program has been developed, it can be stored and called into operation very quickly. With an electromagnetic controller, much of the hard wiring would have to be changed for the different operating sequences, or much extra hard wiring must take place initially. In either case, the PC is the simplest and quickest means of changing the sequence.

The programming methods presented in this text are for one type of programmable controller. Although there are about as many different methods of programming a programmable controller as there are manufacturers, the concepts presented here are basic to all controllers. It will be necessary, however, to consult the instruction manual when using a particular brand of controller.

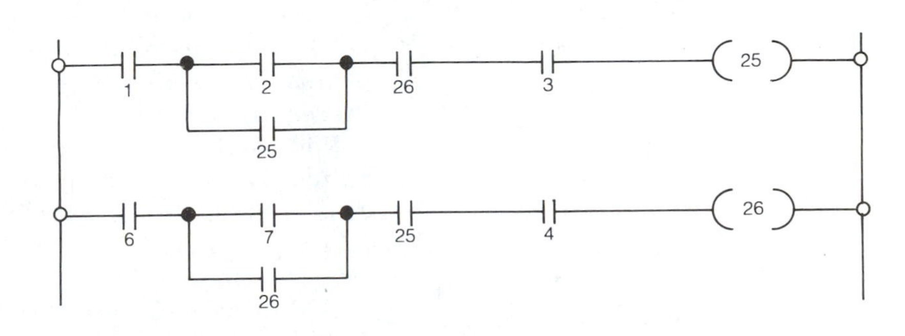

Figure 24-20A
PC circuit for interlocking motors. Motors cannot operate simultaneously.

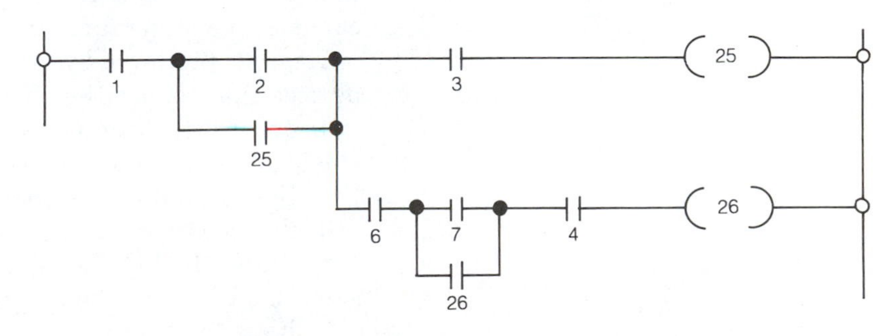

Figure 24-20B
PC circuit for interlocking motors. Second motor cannot start unless first motor is in operation.

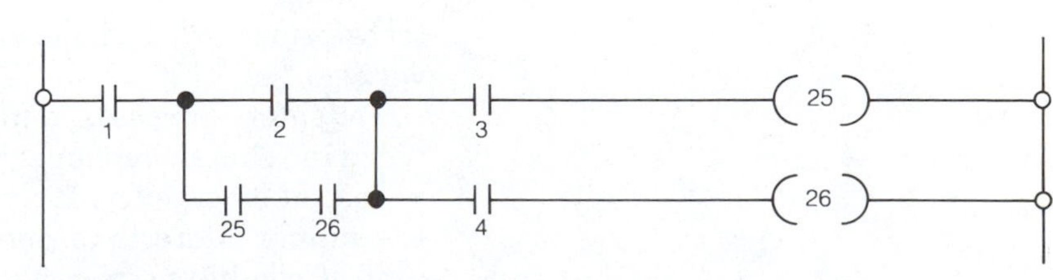

Figure 24-20C
PC circuit for interlocking motors. Both motors must operate simultaneously.

SELECTION AND INSTALLATION OF PC SYSTEMS

In the selection of a PC system, one must consider the following:

1. specific operations to be performed
2. location and space for equipment
3. power supply available
4. environmental conditions
5. availability of equipment
6. compatibility of equipment
7. flexibility of equipment

The amount of equipment to be controlled and the number and type of operations to be performed by each piece of equipment will be the determining factors in sizing the PC. Other factors to consider are memory capacity and compact design.

Some systems must provide control of both ac and dc equipment. Also, voltage fluctuations and electrical noise must be considered. Electrical noise is caused by electromagnetic and electrostatic fields, which are often present in industrial establishments. Suppliers must have knowledge of such conditions when determining PC requirements.

Voltage fluctuations are often caused by heavy loads that are switched on and off the system. Large-horse power motors tend to cause voltage drops, especially during the starting period. Certain types of electrical equipment tend to produce signals that cause voltage spikes.

Both electrical noise and voltage fluctuations can cause malfunctions in PC operations. To reduce problems caused by electrical noise, most manufacturers build into the PC a means to filter out signals produced by electrostatic fields. They also incorporate devices that reduce the effect of contact bounce.

In order to further protect against the effect of these conditions, shielded cable should be used in the installation of hard wiring. Effective grounding is also very important. Cable routes should be well planned to provide the shortest distance between the input or output and the PC. Power conductors should never be run in the same conduit with the control conductors.

Voltage fluctuations can be reduced by installing an isolation transformer, which is designed to provide a constant secondary voltage.

For safety purposes, a master disconnect switch must be hard wired in. This switch must be capable of being operated manually in case of emergency. It must disconnect the entire system from the supply. More than one switch, connected in series, may be used if conditions warrant. Other disconnect switches may be installed that isolate sections of the system if deemed necessary. All switches should be clearly identified.

All installations should be installed according to the manufacturer's specifications and national and local codes.

Once installation has been completed, a test of the entire system should be performed. The test should start with checking the installation against the specifications and code regulations. The PC should then be made to perform every operation for which it is designed. During the operation test, the emergency switches should be opened to ensure they function properly.

As a final step, before turning the system over to the consumer, the supplier should offer a brief training program for the operators and the maintenance personnel.

REVIEW

A. Multiple choice.
Select the best answer.

1. A PC is designed
   a. the same as an office computer.
   b. to operate in an industrial environment.
   c. the same as a home computer.
   d. to withstand heat.
2. Most PCs are designed to be programmed using
   a. BASIC computer language.
   b. FORTRAN computer language.
   c. ladder diagrams.
   d. special computer language.
3. The basic components of a PC are the
   a. power supply and central processing unit.
   b. program loader or terminal.
   c. input/output track.
   d. all of the above.
4. The internal logic circuits of a PC operate on
   a. 5–15 volts.
   b. 10–20 volts.
   c. 15–25 volts.
   d. 20–30 volts.
5. The supply voltage for a PC must be regulated to within
   a. 2% of the required voltage value.
   b. 3% of the required voltage value.
   c. 4% of the required voltage value.
   d. 5% of the required voltage value.

REVIEW *(continued)*

A. Multiple choice. Select the best answer (continued).

6. The CPU performs
   a. only special functions.
   b. all logic functions.
   c. only input functions.
   d. only output functions.
7. The programming terminal is a device used to program the
   a. power supply.
   b. input unit.
   c. output unit.
   d. central processing unit.
8. The I/O track is used to connect the
   a. input to the output.
   b. program terminal to the power supply.
   c. program terminal to the central processing unit.
   d. central processing unit to the outside world.
9. When the output of the PC is used to control an ac load, the control device is a
   a. power transistor.
   b. diode.
   c. triac.
   d. diac.
10. An internal relay is a
    a. solid-state relay built into the CPU.
    b. magnetic relay built into the CPU.
    c. relay outside of the PC that is assigned a number.
    d. relay built into the power supply.

B. Give complete answers.

1. What is a programmable controller?
2. List three advantages of a PC over electromagnetic control systems.
3. List two differences between PCs and home or business computers.
4. List four basic components of PCs.
5. In what section of a PC is the actual circuit logic performed?
6. What device is used to program a PC?
7. What device separates a PC from the outside circuits?
8. What two functions are performed by an input I/O?

REVIEW *(continued)*

B. Give complete answers (continued).

9. If an output I/O controls dc voltage, what electronic device controls the output circuit?
10. What is an internal relay?
11. What function does the power supply for a PC perform?
12. What is the purpose of the central processing unit in a PC?
13. Why is a key switch frequently used to control a CPU?
14. List one advantage of the CRT-type programming terminal over the LCD type.
15. How is a programming terminal used for troubleshooting?
16. What is the purpose of the I/O track?
17. How does the I/O track interact with the CPU?
18. In Figure 24-19B, what parts of the circuit are illuminated when coil 25 is energized?
19. In Figure 24-20A, what parts of the circuit are illuminated when neither coil is energized?
20. In Figure 24-20A, what parts of the circuit are illuminated when coil 25 is energized?
21. In Figure 24-20A, are contacts 2 ever illuminated? Under what conditions?
22. List one factor that determines the size and cost of a PC.
23. List three uses for internal relays.
24. List seven factors that must be considered in the selection of a PC system.
25. List three methods used to reduce electrical noise when installing a PC system.
26. List one method used to reduce voltage fluctuations when installing a PC system.
27. What factors in an industrial installation are most likely to be the cause of PC malfunctions?
28. What is electrical noise?
29. What are voltage spikes?
30. How are master switches installed?
31. Describe the testing procedure to be performed once an installation has been completed.
32. What final step should be taken before turning the system over to the consumer?

APPENDIXES

APPENDIX A
USA STANDARD SYMBOLS

Symbols Listed in Alphabetical Order

Symbol	Quantity	Symbol	Quantity
A	Area	n_1	Synchronous speed (of rotation)
B	Magnetic flux density	p	Number of poles
C	Capacitance	P	Power
d	Diameter	P_i	Input power
E	Voltage (see V)	P_o	Output power
f	Frequency	P_q	Reactive power
f_r	Resonance frequency	P_s	Apparent power
F	Force	Q	Electric charge
F_m	Magnetomotive force	R	Resistance
F_p	Power factor	R_m	Reluctance
F_q	Reactive factor	s	Slip
G	Conductance	t	Time
H	Magnetic field strength	t	Temperature
I	Electric current	T	Time constant
k	Thermal conductivity	V	Voltage
k_h	Hysteresis coefficient	W	Work
l	Length	X	Reactance
M	Torque (see T)	X_C	Capacitive reactance
n	Speed of rotation	X_L	Inductive reactance
n	Number of turns (in a winding)	Z	Impedance
n	Turns ratio		

APPENDIX B
METRIC AND CONVERSION TABLES

TABLE I STANDARD TABLES OF METRIC UNITS OF MEASURE

	Unit	Value in Meters	Symbol or Abbreviation
Linear Measure	micron	0.000 001	μ
	millimeter	0.001	mm
	centimeter	0.01	cm
	decimeter	0.1	dm
	meter (unit)	1.0	m
	dekameter	10.0	dam
	hectometer	100.0	hm
	kilometer	1 000.00	km
	myriameter	10 000.00	mym
	megameter	1 000 000.00	Mm
	Unit	**Value in Square Meters**	**Symbol or Abbreviation**
Surface Measure	square millimeter	0.000 001	mm^2
	square centimeter	0.000 1	cm^2
	square decimeter	0.01	dm^2
	square meter (centiare)	1.0	m^2
	square dekameter (are)	100.0	dam^2
	hectare	10 000.0	ha^2
	square kilometer	1 000 000.0	km^2
	Unit	**Value in Liters**	**Symbol or Abbreviation**
Volume	milliliter	0.001	mℓ
	centiliter	0.01	cℓ
	deciliter	0.1	dℓ
	liter (unit)	1.0	ℓ
	dekaliter	10.0	daℓ
	hectoliter	100.0	hℓ
	kiloliter	1 000.0	kℓ
	Unit	**Value in Grams**	**Symbol or Abbreviation**
Mass	microgram	0.000 001	μg
	milligram	0.001	mg
	centigram	0.01	cg
	decigram	0.1	dg
	gram (unit)	1.0	g
	dekagram	10.0	dag
	hectogram	100.0	hg
	kilogram	1 000.0	kg
	myriagram	10 000.0	myg
	quintal	100 000.0	q
	ton	1 000 000.0	t
	Unit	**Value in Cubic Meters**	**Symbol or Abbreviation**
Cubic Measure	cubic micron	10^{-18}	μ^3
	cubic millimeter	10^{-9}	mm^3
	cubic centimeter	10^{-6}	cm^3
	cubic decimeter	10^{-3}	dm^3
	cubic meter	1	m^3
	cubic dekameter	10^3	dam^3
	cubic hectometer	10^6	hm^3
	cubic kilometer	10^9	km^3

TABLE II CONVERSION OF ENGLISH AND METRIC UNITS OF MEASURE

Linear Measure

Unit	Inches to millimeters	Millimeters to inches	Feet to meters	Meters to feet	Yards to meters	Meters to yards	Miles to kilometers	Kilometers to miles
1	25.40	0.03937	0.3048	3.281	0.9144	1.094	1.609	0.6214
2	50.80	0.07874	0.6096	6.562	1.829	2.187	3.219	1.243
3	76.20	0.1181	0.9144	9.842	2.743	3.281	4.828	1.864
4	101.60	0.1575	1.219	13.12	3.658	4.374	6.437	2.485
5	127.00	0.1968	1.524	16.40	4.572	5.468	8.047	3.107
6	152.40	0.2362	1.829	19.68	5.486	6.562	9.656	3.728
7	177.80	0.2756	2.134	22.97	6.401	7.655	11.27	4.350
8	203.20	0.3150	2.438	26.25	7.315	8.749	12.87	4.971
9	228.60	0.3543	2.743	29.53	8.230	9.842	14.48	5.592

Example 1 in. = 25.40 mm, 1 m = 3.281 ft., 1 km = 0.6214 mi.

Surface Measure

Unit	Square inches to square centimeters	Square centimeters to square inches	Square feet to square meters	Square meters to square feet	Square yards to square meters	Square meters to square yards	Acres to hectares	Hectares to acres	Square miles to square kilometers	Square kilometers to square miles
1	6.452	0.1550	0.0929	10.76	0.8361	1.196	0.4047	2.471	2.59	0.3861
2	12.90	0.31	0.1859	21.53	1.672	2.392	0.8094	4.942	5.18	0.7722
3	19.356	0.465	0.2787	32.29	2.508	3.588	1.214	7.413	7.77	1.158
4	25.81	0.62	0.3716	43.06	3.345	4.784	1.619	9.884	10.36	1.544
5	32.26	0.775	0.4645	53.82	4.181	5.98	2.023	12.355	12.95	1.931
6	38.71	0.93	0.5574	64.58	5.017	7.176	2.428	14.826	15.54	2.317
7	45.16	1.085	0.6503	75.35	5.853	8.372	2.833	17.297	18.13	2.703
8	51.61	1.24	0.7432	86.11	6.689	9.568	3.237	19.768	20.72	3.089
9	58.08	1.395	0.8361	96.87	7.525	10.764	3.642	22.239	23.31	3.475

Example 1 sq. in. = 6.452 sq. cm, 1 sq. m = 1.196 sq. yd., 1 sq. mi. = 2.59 sq. km

Cubic Measure

Unit	Cubic inches to cubic centimeters	Cubic centimeters to cubic inches	Cubic feet to cubic meters	Cubic meters to cubic feet	Cubic yards to cubic meters	Cubic meters to cubic yards	Gallons to cubic feet	Cubic feet to gallons
1	16.39	0.06102	0.02832	35.31	0.7646	1.308	0.1337	7.481
2	32.77	0.1220	0.05663	70.63	1.529	2.616	0.2674	14.96
3	49.16	0.1831	0.08495	105.9	2.294	3.924	0.4010	22.44
4	65.55	0.2441	0.1133	141.3	3.058	5.232	0.5347	29.92
5	81.94	0.3051	0.1416	176.6	3.823	6.540	0.6684	37.40
6	98.32	0.3661	0.1699	211.9	4.587	7.848	0.8021	44.88
7	114.7	0.4272	0.1982	247.2	5.352	9.156	0.9358	52.36
8	131.1	0.4882	0.2265	282.5	6.116	10.46	1.069	59.84
9	147.5	0.5492	0.2549	371.8	6.881	11.77	1.203	67.32

Example 1 cu. cm = 0.06102 cu. in., 1 gal. = 0.1337 cu. ft.

Volume or Capacity Measure

Unit	Liquid ounces to cubic centimeters	Cubic centimeters to liquid ounces	Pints to liters	Liters to pints	Quarts to liters	Liters to quarts	Gallons to liters	Liters to gallons	Bushels to hectoliters	Hectoliters to bushels
1	29.57	0.03381	0.4732	2.113	0.9463	1.057	3.785	0.2642	0.3524	2.838
2	59.15	0.06763	0.9463	4.227	1.893	2.113	7.571	0.5284	0.7048	5.676
3	88.72	0.1014	1.420	6.340	2.839	3.785	11.36	0.7925	1.057	8.513
4	118.3	0.1353	1.893	8.454	3.170	4.227	15.14	1.057	1.410	11.35
5	147.9	0.1691	2.366	10.57	4.732	5.284	18.93	1.321	1.762	14.19
6	177.4	0.2029	2.839	12.68	5.678	6.340	22.71	1.585	2.114	17.03
7	207.0	0.2367	3.312	14.79	6.624	7.397	26.50	1.849	2.467	19.86
8	236.6	0.2705	3.785	16.91	7.571	8.454	30.28	2.113	2.819	22.70
9	266.2	0.3043	4.259	19.02	8.517	9.510	34.07	2.378	3.171	25.54

Example 1 ℓ = 2.113 pt., 1 gal. = 3.785 ℓ

TABLE III BASIC SI UNITS

Physical Quantity	SI Unit	SI Unit Symbol
Length (l)	meter	m
Mass (M)	kilogram	kg
Time (t)	second	s
Current (I)	ampere	A

TABLE IV DERIVED SI UNITS

Physical Quantity	SI Unit	SI Unit Symbol
Capacitance (C)	farad	F
Charge (Q)	coulomb	C
Conductance (G)	siemens	S
Energy (W)	joule	J
Frequency (f)	hertz	Hz
Force (F)	newton	N
Inductance (L)	henry	H
Power (P)	watt	W
Resistance (R)	ohm	Ω
Voltage (V or E)	volt	V

APPENDIX C
TABLE OF TRIGONOMETRIC FUNCTIONS

Degree	Sin	Cos	Tan	Cot	Sec	Csc	Degree	Sin	Cos	Tan	Cot	Sec	Csc
0.0	0.0000	1.0000	0.0000	∞	1.000	∞	45.5	0.7133	0.7009	1.018	0.9827	1.427	1.402
0.5	0.0087	1.0000	0.0087	114.6	1.000	114.6	46.0	0.7193	0.6947	1.036	0.9657	1.440	1.390
1.0	0.0175	0.9998	0.0175	57.29	1.000	57.30	46.5	0.7254	0.6884	1.054	0.9490	1.453	1.379
1.5	0.0262	0.9997	0.0262	38.19	1.000	38.20	47.0	0.7314	0.6820	1.072	0.9325	1.466	1.367
2.0	0.0349	0.9994	0.0349	28.64	1.001	28.65	47.5	0.7373	0.6756	1.091	0.9163	1.480	1.356
2.5	0.0436	0.9990	0.0437	22.90	1.001	22.93	48.0	0.7436	0.6691	1.111	0.9004	1.494	1.346
3.0	0.0523	0.9986	0.0524	19.08	1.001	19.11	48.5	0.7490	0.6626	1.130	0.8847	1.509	1.335
3.5	0.0610	0.9981	0.0612	16.35	1.002	16.38	49.0	0.7547	0.6561	1.150	0.8693	1.524	1.325
4.0	0.0698	0.9976	0.0699	14.30	1.002	14.34	49.5	0.7604	0.6494	1.171	0.8541	1.540	1.315
4.5	0.0785	0.9969	0.0787	12.71	1.003	12.75	50.0	0.7660	0.6428	1.192	0.8391	1.556	1.305
5.0	0.0872	0.9962	0.0875	11.43	1.004	11.47	50.5	0.7716	0.6361	1.213	0.8243	1.572	1.296
5.5	0.0958	0.9954	0.0963	10.39	1.005	10.43	51.0	0.7771	0.6293	1.235	0.8098	1.589	1.287
6.0	0.1045	0.9945	0.1051	9.514	1.006	9.567	51.5	0.7826	0.6225	1.257	0.7954	1.606	1.278
6.5	0.1132	0.9936	0.1139	8.777	1.006	8.834	52.0	0.7880	0.6157	1.280	0.7813	1.624	1.269
7.0	0.1219	0.9925	0.1228	8.144	1.008	8.206	52.5	0.7934	0.6088	1.303	0.7673	1.643	1.260
7.5	0.1305	0.9914	0.1317	7.596	1.009	7.661	53.0	0.7986	0.6081	1.327	0.7536	1.662	1.252
8.0	0.1392	0.9903	0.1405	7.115	1.010	7.185	53.5	0.8039	0.5948	1.351	0.7400	1.681	1.244
8.5	0.1478	0.9890	0.1495	6.691	1.011	6.765	54.0	0.8090	0.5878	1.376	0.7265	1.701	1.236
9.0	0.1564	0.9877	0.1584	6.314	1.012	6.392	54.5	0.8141	0.5807	1.402	0.7133	1.722	1.228
9.5	0.1650	0.9863	0.1673	5.976	1.014	6.059	55.0	0.8192	0.5736	1.428	0.7002	1.743	1.221
10.0	0.1736	0.9848	0.1763	5.671	1.015	5.759	55.5	0.8241	0.5664	1.455	0.6873	1.766	1.213
10.5	0.1822	0.9833	0.1853	5.396	1.017	5.487	56.0	0.8290	0.5592	1.483	0.6745	1.788	1.206
11.0	0.1908	0.9816	0.1944	5.145	1.019	5.241	56.5	0.8339	0.5519	1.511	0.6619	1.812	1.199
11.5	0.1994	0.9799	0.2035	4.915	1.020	5.016	57.0	0.8387	0.5446	1.540	0.6494	1.836	1.192
12.0	0.2079	0.9781	0.2126	4.705	1.022	4.810	57.5	0.8434	0.5373	1.570	0.6371	1.861	1.186
12.5	0.2164	0.9763	0.2217	4.511	1.024	4.620	58.0	0.8480	0.5299	1.600	0.6249	1.887	1.179
13.0	0.2250	0.9744	0.2309	4.331	1.026	4.445	58.5	0.8526	0.5225	1.632	0.6128	1.914	1.173
13.5	0.2334	0.9724	0.2401	4.165	1.028	4.284	59.0	0.8572	0.5150	1.664	0.6009	1.942	1.167
14.0	0.2419	0.9703	0.2493	4.011	1.031	4.134	59.5	0.8616	0.5075	1.698	0.5890	1.970	1.161
14.5	0.2504	0.9681	0.2586	3.867	1.033	3.994	60.0	0.8660	0.5000	1.732	0.5774	2.000	1.155
15.0	0.2588	0.9659	0.2679	3.732	1.035	3.864	60.5	0.8704	0.4924	1.767	0.5658	2.031	1.149
15.5	0.2672	0.9636	0.2773	3.606	1.038	3.742	61.0	0.8746	0.4848	1.804	0.5543	2.063	1.143
16.0	0.2756	0.9613	0.2867	3.487	1.040	3.628	61.5	0.8788	0.4772	1.842	0.5430	2.096	1.138
16.5	0.2840	0.9588	0.2962	3.376	1.043	3.521	62.0	0.8829	0.4695	1.881	0.5317	2.130	1.133
17.0	0.2924	0.9563	0.3057	3.271	1.046	3.420	62.5	0.8870	0.4617	1.921	0.5206	2.166	1.127
17.5	0.3007	0.9537	0.3153	3.172	1.048	3.326	63.0	0.8910	0.4540	1.963	0.5095	2.203	1.122
18.0	0.3090	0.9511	0.3249	3.078	1.051	3.236	63.5	0.8949	0.4462	2.006	0.4986	2.241	1.117
18.5	0.3173	0.9483	0.3346	2.989	1.054	3.152	64.0	0.8988	0.4384	2.050	0.4877	2.281	1.113
19.0	0.3256	0.9455	0.3443	2.904	1.058	3.072	64.5	0.9026	0.4305	2.097	0.4770	2.323	1.108
19.5	0.3338	0.9426	0.3541	2.824	1.061	2.996	65.0	0.9063	0.4226	2.145	0.4663	2.366	1.103
20.0	0.3420	0.9397	0.3640	2.747	1.064	2.924	65.5	0.9100	0.4147	2.194	0.4557	2.411	1.099
20.5	0.3502	0.9367	0.3739	2.675	1.068	2.855	66.0	0.9135	0.4067	2.246	0.4452	2.459	1.095
21.0	0.3584	0.9336	0.3839	2.605	1.071	2.790	66.5	0.9171	0.3987	2.300	0.4348	2.508	1.090
21.5	0.3665	0.9304	0.3939	2.539	1.075	2.729	67.0	0.9205	0.3907	2.356	0.4245	2.559	1.086
22.0	0.3746	0.9272	0.4040	2.475	1.079	2.669	67.5	0.9239	0.3827	2.414	0.4142	2.613	1.082
22.5	0.3827	0.9239	0.4142	2.414	1.082	2.613	68.0	0.9272	0.3746	2.475	0.4040	2.669	1.079
23.0	0.3907	0.9205	0.4245	2.356	1.086	2.559	68.5	0.9304	0.3665	2.539	0.3939	2.729	1.075
23.5	0.3987	0.9171	0.4348	2.300	1.090	2.508	69.0	0.9336	0.3584	2.605	0.3839	2.790	1.071
24.0	0.4067	0.9135	0.4452	2.246	1.095	2.459	69.5	0.9367	0.3502	2.675	0.3739	2.885	1.068
24.5	0.4147	0.9100	0.4557	2.194	1.099	2.411	70.0	0.9397	0.3420	2.747	0.3640	2.924	1.064
25.0	0.4226	0.9063	0.4663	2.145	1.103	2.366	70.5	0.9426	0.3338	2.824	0.3541	2.996	1.061
25.5	0.4305	0.9026	0.4770	2.097	1.108	2.323	71.0	0.9455	0.3256	2.904	0.3443	3.072	1.058
26.0	0.4384	0.8988	0.4877	2.050	1.113	2.281	71.5	0.9483	0.3173	2.989	0.3346	3.152	1.054
26.5	0.4462	0.8949	0.4986	2.006	1.117	2.241	72.0	0.9511	0.3090	3.078	0.3249	3.236	1.051
27.0	0.4540	0.8910	0.5095	1.963	1.122	2.203	72.5	0.9537	0.3007	3.172	0.3153	3.326	1.048
27.5	0.4617	0.8870	0.5206	1.921	1.127	2.166	73.0	0.9563	0.2924	3.271	0.3057	3.420	1.046
28.0	0.4695	0.8829	0.5317	1.881	1.133	2.130	73.5	0.9588	0.2840	3.376	0.2962	3.521	1.043
28.5	0.4772	0.8788	0.5430	1.842	1.138	2.096	74.0	0.9612	0.2756	3.487	0.2867	3.628	1.040
29.0	0.4848	0.8746	0.5543	1.804	1.143	2.063	74.5	0.9636	0.2672	3.606	0.2773	3.742	1.039
29.5	0.4924	0.8704	0.5658	1.767	1.149	2.031	75.0	0.9659	0.2588	3.732	0.2679	3.864	1.035
30.0	0.5000	0.8660	0.5774	1.732	1.155	2.000	75.5	0.9681	0.2504	3.867	0.2586	3.994	1.033
30.5	0.5075	0.8616	0.5890	1.698	1.161	1.970	76.0	0.9703	0.2419	4.011	0.2403	4.134	1.031
31.0	0.5150	0.8572	0.6009	1.664	1.167	1.942	76.5	0.9714	0.2334	4.165	0.2401	4.284	1.028
31.5	0.5225	0.8526	0.6128	1.632	1.173	1.914	77.0	0.9744	0.2250	4.331	0.2309	4.445	1.026
32.0	0.5299	0.8480	0.6249	1.600	1.179	1.887	77.5	0.9763	0.2164	4.511	0.2217	4.620	1.024
32.5	0.5373	0.8434	0.6371	1.570	1.186	1.861	78.0	0.9781	0.2079	4.705	0.2126	4.810	1.022
33.0	0.5446	0.8387	0.6494	1.540	1.192	1.836	78.5	0.9799	0.1994	4.915	0.2035	5.016	1.020
33.5	0.5519	0.8339	0.6619	1.511	1.199	1.812	79.0	0.9816	0.1908	5.145	0.1944	5.241	1.019
34.0	0.5592	0.8290	0.6745	1.483	1.206	1.788	79.5	0.9833	0.1822	5.396	0.1853	5.487	1.017
34.5	0.5664	0.8241	0.6873	1.455	1.213	1.766	80.0	0.9848	0.1736	5.671	0.1763	5.759	1.015
35.0	0.5736	0.8192	0.7002	1.428	1.221	1.743	80.5	0.9863	0.1650	5.976	0.1673	6.059	1.014
35.5	0.5807	0.8141	0.7133	1.402	1.228	1.722	81.0	0.9877	0.1564	6.314	0.1584	6.392	1.012
36.0	0.5878	0.8090	0.7265	1.376	1.236	1.701	81.5	0.9890	0.1478	6.691	0.1495	6.765	1.011
36.5	0.5948	0.8039	0.7400	1.351	1.244	1.681	82.0	0.9903	0.1392	7.115	0.1405	7.185	1.010
37.0	0.6018	0.7986	0.7536	1.327	1.252	1.662	82.5	0.9914	0.1305	7.596	0.1317	7.661	1.009
37.5	0.6088	0.7934	0.7673	1.303	1.260	1.643	83.0	0.9925	0.1219	8.144	0.1228	8.206	1.008
38.0	0.6157	0.7880	0.7813	1.280	1.269	1.624	83.5	0.9936	0.1132	8.777	0.1138	8.834	1.006
38.5	0.6225	0.7826	0.7954	1.257	1.278	1.606	84.0	0.9945	0.1045	9.514	0.1051	9.567	1.006
39.0	0.6293	0.7771	0.8098	1.235	1.287	1.589	84.5	0.9954	0.0958	10.39	0.0963	10.43	1.005
39.5	0.6361	0.7716	0.8243	1.213	1.296	1.572	85.0	0.9962	0.0872	11.43	0.0875	11.47	1.004
40.0	0.6428	0.7660	0.8391	1.192	1.305	1.556	85.5	0.9969	0.0785	12.71	0.0787	12.75	1.003
40.5	0.6494	0.7604	0.8541	1.171	1.315	1.540	86.0	0.9976	0.0698	14.30	0.0699	14.34	1.002
41.0	0.6561	0.7547	0.8693	1.150	1.325	1.524	86.5	0.9981	0.0610	16.35	0.0612	16.38	1.002
41.5	0.6626	0.7490	0.8847	1.130	1.335	1.509	87.0	0.9986	0.0523	19.08	0.0524	19.11	1.001
42.0	0.6691	0.7431	0.9004	1.111	1.346	1.494	87.5	0.9990	0.0436	22.90	0.0437	22.93	1.001
42.5	0.6756	0.7373	0.9163	1.091	1.356	1.480	88.0	0.9994	0.0349	28.64	0.0349	28.65	1.001
43.0	0.6820	0.7314	0.9325	1.072	1.367	1.466	88.5	0.9997	0.0262	38.19	0.0262	38.20	1.000
43.5	0.6884	0.7254	0.9490	1.054	1.379	1.453	89.0	0.9998	0.0175	57.29	0.0175	57.30	1.000
44.0	0.6947	0.7193	0.9657	1.036	1.390	1.440	89.5	1.0000	0.0087	114.6	0.0087	114.6	1.000
44.5	0.7009	0.7133	0.9827	1.018	1.402	1.427	90.0	1.0000	0.0000	—	0.0000	—	1.000
45.0	0.7071	0.7071	1.0000	1.000	1.414	1.414							

GLOSSARY

Across-the-line-starter. An electromagnetic controller designed to start motors at full voltage. It generally contains an overload protective mechanism, a main circuit, and a control circuit. The relay is usually controlled from a remote switch.

Air gap. With reference to a motor, the distance between the rotating component and the stationary component within the magnetic field.

Alloy. A mixture of two or more metals.

Alnico. A mixture of aluminum, nickel, cobalt, and iron used in the manufacturing of permanent magnets.

Alternating current (ac). The flow of electrons first in one direction and then in the opposite direction, in equal periods of time.

Alternator. An alternating current generator.

Ambient. Surrounding; for example, the air surrounding an electrical conductor.

American Wire Gauge (AWG). A standard table or scale used for the measurement of the most common wire conductors in use in electrical work. This scale ranges from No. 50, which is equal to 1 circular mil, to No. 0000, which is equal to 211,660 circular mils.

Ammeter. A meter used to measure the amount of current flowing in an electrical system, circuit, or component.

Ampacity. The current-carrying capacity of electrical conductors, expressed in amperes.

Ampere. The rate of flow of electrons through a circuit. One ampere is equal to the flow of 1 coulomb (628×10^{16} electrons) per second.

Amplification factor. The numerical maximum increase in signal strength that an electron tube can provide.

Amplifier. An electronic device used to increase power, voltage, current, and/or sound signals.

Anode. The element in an electron tube, sometimes called the plate, that attracts the electrons emitted by the cathode.

Apparent power. The product of the current and voltage in an ac system or part of a system. Its unit of measurement is the voltampere.

Arc heating. The act of producing heat by an electric arc.

Arc shield. Insulating walls between which the movable contacts of a switch pass. It is designed to quickly extinguish any arc that may form when breaking a circuit.

Armature. In general, a piece of magnetic material, sometimes surrounded by coiled conductors, arranged to be acted upon by a magnetic field established by current flow. On a generator, the coil(s) into which the voltage is induced is generally referred to as the armature. On a dc motor the rotating part is generally called the armature. It is the moving part of relays, buzzers, or loudspeakers.

Armature reactance. The result of self-induction in the armature coils of an ac motor or generator, which tends to reduce the amount of current flowing in the coils.

Armature reaction. The phenomenon that takes place in a motor or generator. Current flowing in the armature conductors produces a magnetic field that weakens and distorts the main field flux.

Atom. The smallest unit of any chemical element. Atoms consist of protons, neutrons, and electrons.

Automatic controller. A controller that is operated by one or more sensing devices such as a thermostat or pressurestat. This type of controller does not require human intervention.

Autotransformer. A transformer in which one winding is common to both the primary and secondary.

Average value. With reference to ac, the average of the instantaneous values of current or voltage for one half cycle.

Balanced system. A multiwire system in which the resultant current in each ungrounded conductor is equal and the current in the grounded neutral conductor is zero.

Ballast. A component used in fluorescent fixtures to change the value of voltage and to limit the amount of current.

Battery. Two or more cells connected to produce a specific voltage.

Bipolar transistor. A transistor consisting of a material sandwiched by a second material (NPN or PNP).

Black light. Ultraviolet light that is a form of radiant energy not visible to the human eye.

Breakdown torque. The maximum torque of a motor. This usually occurs at about 25 percent slip. Loading a motor beyond the breakdown torque will stall the rotor.

Brightness. A measurement of the amount of light reflected from an object or a light source.

British thermal unit. The amount of heat required to raise the temperature of one pound of water one degree Fahrenheit.

Brush. A device, usually made of carbon, used to complete the connection from a stationary circuit to a rotating circuit. Brushes are generally used on motors and generators.

Bus. A conducting bar, usually made of copper.

Bus (distribution). A conducting bar or combination of bars connected together for the purpose of distributing electrical energy to various points within an installation.

Bypass circuit. A subcircuit designed to shunt a component or part of a circuit.

Candlepower. A measurement of the intensity of light expressed in standard candles.

Capacitance. The ability to store an electrostatic charge.

Capacitor. A device consisting of conductors and insulators that is capable of storing an electric charge.

Cathode. The element of an electron tube that emits electrons.

Cell. A single chemical device capable of producing a direct voltage.

Central processing unit. A component of a programmable controller that performs all the logic functions.

Choke. An inductor used to limit the flow of ac.

Circuit breaker. A device designed to open and close a circuit by nonautomatic means. It opens the circuit automatically on a predetermined overcurrent, without injury to itself, when properly applied within its rating (NEC 1993). These devices can operate by thermal means or magnetic means or a combination of both.

Circular mil. The area of a circle 1 mil in diameter.

Coefficient of utilization. With reference to lighting systems, the ratio between the light reaching the working plane and the light produced by the luminaire.

Color (sensitivity). A term used to describe an imbalance of light energy reaching the eyes from sources and/or objects.

Combination circuit. A circuit containing groups of loads connected in series and in parallel.

Common ground. The grounding conductor that connects both the wiring system and the equipment to the grounding electrode.

Commutator. Copper bars mounted side by side on the shaft of a dc generator or motor. They are insulated from one another and from the shaft. They are arranged and connected to form a type of rotating switch. On a generator, the commutator serves to change the connections to the load at the same instant that the emf reverses in the armature, thus providing a unidirectional voltage to the load. On a motor, it serves to reverse the direction of the current in the armature at the precise instant to provide a unidirectional torque.

Compensating winding. Windings placed in the main pole faces of a generator or motor and connected in series with the armature. The purpose of these windings is to eliminate the effect of armature reaction.

Component. A part of a system, circuit, or apparatus.

Condenser. A term sometimes used for a capacitor. A device consisting of conductors and insulators that is capable of storing an electric charge.

Conductance. The ability of a material to permit current to flow. A material that has a conductance of one siemen will conduct one ampere of current when one volt of pressure is applied.

Conduction heating. The transfer of heat energy from one object to another by contact.

Conductor. Any material that offers very little resistance to electron flow.

Conduit. A tube, pipe, or duct for enclosing electrical conductors.

Consumer distribution system. A system of transmitting electrical energy to the consumer's premises.

Continuity. A continuous electrical connection between two points.

Continuity tester. An electrical instrument used to determine if there is a continuous electrical path between two points.

Contrast. With reference to light, either the difference in the amount of illumination from two sources or the difference in the amount of light reflected from various surfaces.

Controller. A device or group of devices that serve to govern, in some predetermined manner, the electrical power to the apparatus to which it is connected (NEC 1993). As used with a motor, a device used to adjust or vary the rotor speed.

Convection heating. The transmission of heat energy through a fluid or air.

Copper losses. Power dissipated as heat, caused by current flowing through the resistance of the conductors.

Coulomb. A measurement of a quantity of electrons. One coulomb is equal to 628×10^{16} electrons.

Counter electromotive force (cemf). A voltage induced into a coil that acts in the opposite direction to the applied voltage.

Covalent bonding. An arrangement of the atomic structure of a material that allows adjacent atoms to share valance electrons.

Current correction factor. A numerical value that is used to obtain the correct ampacity of a conductor installed in an area where the ambient temperature exceeds that used for the ampacity table. The ampacity from the table is multiplied by the correction factor to obtain the safe current-carrying capacity of the conductor(s).

Current interrupting capacity. The magnitude of current a device can interrupt without injury to itself or the equipment to which it is connected.

Cycle. A term used to indicate a complete change in values in both directions.

Dashpot. A cylindrically shaped container into which a plunger is inserted. The plunger is drawn up through air or liquid by a solenoid. The size of the hole in the plunger determines the rate of speed at which the plunger travels.

DC ripple. The pulse produced by a fluctuating direct current.

Dehumidifier. A machine that removes moisture from the air.

Demand factor. The ratio of the maximum demand of a system, or part of a system, to the total connected load of the system or the part of the system under consideration (NEC 1993).

Dielectric heating. The heating of electrically nonconductive materials by a high-frequency alternating current.

Diode tube. The simplest type of electron tube. It contains only two elements: a cathode and an anode.

Direct current (dc). The flow of electrons in one direction only.

Dynamic breaking. A method of stopping a motor quickly. With dc motors, the armature is disconnected from the supply and connected across a resistance load. With ac motors, the supply voltage is reversed and then the motor is disconnected from the source.

Eddy currents. Circulating currents caused by voltages induced into a conducting material.

Effective value. The effective value of an alternating current is that value that will produce the same heating effect as a specific value of a steady direct current. It is equal to the maximum value multiplied by 0.707.

Efficiency. The ratio of the power output to the power input.

Electrical network. A system of interconnecting loads and/or power sources.

Electric charge. An accumulation of electrons on an object or a lack of electrons on an object. When an object has an excessive number of electrons, it is negatively charged. If an object has lost some electrons so that it contains more protons than electrons, it is positively charged.

Electric current. The movement of electrons from atom to atom through a conductor.

Electrodynamometer. An electrical instrument that develops a torque. It is frequently used for meter movements.

Electromagnet. A magnet produced by an electric current flowing through a coil of wire.

Electromagnetic induction. The production of a voltage in a coil as a result of a magnetic field's moving across the coil or the coil's moving across the magnetic field.

Electromotive force (emf). The electrical pressure produced by a battery, generator, or other apparatus designed to produce a force to cause current flow. Electromotive force is usually called voltage.

Electron. The smallest part of an atom; the negative charge of an atom that orbits around the nucleus.

Electronic emission. Electrons being emitted from a material as a result of heat, light, impact, or electric charges.

Electronics. A branch of electrical science that pertains to the use and control of electrons in circuits containing gas or vacuum tubes, transistors, and amplifiers.

Electron tube. A tube used to control the flow of electrons in a circuit.

Energy. The ability to do work.

Equalizer bus. A conductor used to connect the series fields of two or more compound generators together when the generators are operating in parallel. The purpose of this connection is to equalize the currents through the series fields.

Equivalent circuit. A simplified circuit whose characteristics are equivalent to the original circuit.

Farad. A measurement of capacitance. A capacitor has a capacitance of 1 farad when it can be charged to 1 coulomb in 1 second by 1 volt of pressure. The farad is too large a measurement for practical use; for most purposes the microfarad is used.

Field excitation. Current flowing in the field coils of a generator. This current produces the flux necessary to induce a voltage into the armature coils.

Field rheostat. A variable resistor used to control the amount of current flowing through the field circuit of a motor or generator.

Filament. The element (usually a wire) through which current flows in an electron tube or electric lamp. The current flow produces heat and light.

Fitting. An accessory used in an electrical installation. A fitting is intended primarily to perform a mechanical rather than an electrical function.

Fluorescent. A material that glows when exposed to ultraviolet light.

Fluorescent lamp. A form of electric discharge lighting source. When a voltage is applied across the lamp, the gas within the tube ionizes, causing a radiation within the tube. The radiation activates a fluorescent material, producing a visible light.

Flux. Magnetic field.

Flux (alternating). Magnetic flux that varies in strength and direction at regular intervals; the magnetic field produced by ac.

Flux density. The number of lines of force for a given area, usually per square centimeter or square inch.

Flywheel. A wheel having a heavy outer perimeter. It is generally used to maintain a steady speed in machines whose load and/or driving force varies widely.

Flywheel effect. Characteristic of a machine that causes it to act as if it were attached to a flywheel.

Foot-candle. The intensity of illumination on a plane. One foot-candle is the intensity of illumination on a plane 1 foot away from a lighting source of 1 candlepower and at a right angle to the rays from the source.

Footlambert. A measurement of brightness. One footlambert is equal to 1 lumen of reflection or emission per square foot.

Frequency. In reference to ac, the number of cycles completed per unit of time, usually 1 second.

Fuse. An overcurrent protective device that contains a melting element.

Gas tube. An electron tube from which the air has been removed and gas filled.

Generator. A machine that converts mechanical energy into electrical energy; usually a machine that causes a coil to rotate through a magnetic field or a rotating magnetic field to move across a coil.

Gilbert. The unit of measurement of the magnetomotive force of a magnet.

Glare. An intense and/or bright light that tends to cause eye strain.

Glow switch. A type of fluorescent starter.

Grid bias. The voltage applied between the cathode and grid of an electron tube.

Ground conductor. A system or circuit conductor that is intentionally grounded.

Ground fault circuit interrupter. A device that senses small currents to ground and opens the circuit when the current exceeds a predetermined value.

Ground fault protector. A device that senses ground faults and opens the circuit when the current to ground reaches a predetermined value.

Grounding conductor. A conductor used to connect another conductor or equipment to the grounding electrode.

Harmonic. In an ac system, a voltage or current waveform that is a multiple of the initial frequency.
Heat pump. A device used to heat or cool dwellings powered by an electrically driven compressor.
Heat transfer. The exchange of heat energy from one object to another.
Henry. The unit of measurement of inductance. A coil has an inductance of 1 henry when an emf of 1 volt will cause the current to change at a rate of 1 ampere per second.
Hertz. The measurement of the number of cycles of an alternating current or voltage completed in 1 second.
Horsepower. A unit of measurement of mechanical power. One horsepower is equivalent to work produced at the rate of 33,000 foot-pounds per minute.
Humidifier. A machine that adds moisture to the air.
Humidistat. A device for controlling the amount of moisture in the air. It is a switch that opens and/or closes an electric circuit as the humidity changes.
Hypotenuse. The side of a right triangle opposite to the 90° angle. The longest side of a right triangle.
Hysteresis. A lagging effect. In magnetic circuits, hysteresis occurs when a ferromagnetic material is placed under the influence of a varying magnetic field. It is sometimes referred to as molecular friction.
Illumination. A measurement of the density of light projected on a surface. One lumen of light striking an area of 1 square foot will illuminate it at an intensity of 1 foot-candle.
Impedance. The total opposition to the flow of a pulsating or alternating current. Impedance is composed of reactance and resistance.
Impurity. A substance added to germanium or silicon to produce a semiconductor.
Incandescent. Light produced by heating an object to a high temperature.
Incandescent lamp. A lighting source, sometimes referred to as a light bulb, that emits light as a result of intense heat.
Inductance effect. The characteristic of an electromagnet that tends to prevent any change in the value of current flowing in the coil.
Induction heating. The heating of an electrically conductive material by a high-frequency alternating current.
Induction motor. A motor whose operation depends upon the principle of mutual induction. AC flowing in the stator windings induces a voltage into the rotor windings, developing two magnetic fields that interact to produce torque.
Inductor. A device that produces an inductive effect in a circuit.
In phase. Alternating currents or voltages are said to be in phase when they reach their maximum and zero values at the same time and in the same direction.
Input/Output (I/O) track. A component of a PC that connects the central processing unit to the devices outside the PC.
Instantaneous value. The value of an alternating current or voltage at a specific instant during the cycle.
Insulator. A device or material that has high electrical resistance. Insulators are used to separate or support conductors and prevent current flow between conductors and/or between conductors and ground.
Internal relay. A relay that is assigned a number greater than the input-output capacity of a programmable controller. It can then be programmed into a programmable controller.
Interpoles. Small poles placed midway between the main poles of a generator or motor and connected in series with the armature. The purpose of these poles is to compensate for armature reaction, thus reducing arcing at the brushes.
Inverter. A device that changes dc to ac.
Kirchhoff's Laws. Mathematical current and voltage laws frequently used to solve complex electrical circuit problems.
Lamp. The part of an electrical fixture that emits the light.
Light. A form of radiant energy that stimulates the visual senses.
Line drop. The voltage required to force a specific value of current through the resistance of the conductors between the supply and the load. Line drop may refer to the entire system or to a small part of the system.
Line loss. The power dissipated in the form of heat caused by current flowing through the conductors between the supply and the load. Line loss may refer to the entire system or to a small part of the system.
Liquidtight. Constructed and installed so as to prevent the entrance of liquids.
Load shifting. Removing some or all of the load from one voltage source and applying it to another source.
Loop distribution system. A system for transmitting electrical energy through conductors and equipment. The energy begins at a specific point, loops through a specified area, and returns to the point of origin.
Lumen. The unit for total light emitted from a lighting source.
Luminarie. A lighting source usually in the form of an electric lighting fixture.
Magnet. An object that attracts iron and steel.
Magnetic blowout coil. A coil connected across the contacts of a switch so that at the instant the switch is opened, the coil is energized. The magnetic field from the coil reacts with the magnetic field caused by the current arcing across the contacts; the result is a quick extinguishing of the arc.
Magnetic circuit. The path through which the magnetic lines of force flow.
Magnetic field. The area surrounding a magnet in which the magnetic force exists.
Magnetic flux. The total number of lines of force.
Magnetic lines of force. Imaginary lines used to indicate the direction of force within the magnetic field.
Magnetic saturation. The condition of a material when it is fully magnetized.
Magnetizing current. The current necessary to produce a magnetic field in a machine or in equipment.
Magneto. An ac generator that produces an electromotive force by rotating a coil of wire between the poles of a permanent magnet.
Magnetomotive force. The force that produces the flux emitting from a magnet.
Magnet wire. An insulated wire generally made of copper and used in the construction of electromagnets. The

wire is generally manufactured in sizes No. 36 AWG to No. 12 AWG, and has an exact and uniform diameter.

Maintenance factor. With reference to lighting systems, the maintenance factor is the percentage of light to be expected with the usual cleaning and/or painting of the luminaries and reflecting surfaces.

Manual controller. A controller that depends upon human intervention for its operation.

Matter. Anything that occupies space and has weight, such as air, wood, metal, and water.

Maximum value. With reference to ac, the greatest value reached during one cycle.

Maxwell. A unit of magnetic flux. One maxwell equals 1 gauss per square centimeter, or 1 magnetic line of force.

Megohmmeter. A high-range ohmmeter. Its power is provided by a small generator, often hand driven. It is frequently used to measure the resistance of insulation.

Mercury-vapor lamp. A type of electric discharge lamp, consisting of an arc tube sealed inside a glass bulb. Current flowing through mercury causes it to vaporize. The mercury vapor regulates the value of current.

Meter movement. The movable coil within a meter.

Meter range. The value of measurement covered by a scale on a meter; for example, 0-50. Some meters have several different ranges determined by the scale selected, such as 0–50, 0–500, and 0–5000.

Meter scale. The markings on the face of a meter that indicate the quantity of the value being measured.

Meter shunt. A conductor of specific resistance placed in parallel with the current coil of the meter. The resistance of the shunt determines the range of the meter.

Micrometer. A tool for precisely measuring small thicknesses such as the diameter of a round wire.

Mil. A unit of length equal to 1/1000 inch. It is frequently used to express the diameter of a wire.

Mil-foot. A unit of measurement consisting of an object 1 foot long and 1 mil in diameter.

Molecule. The smallest particle of matter without producing chemical change.

Momentary contact push button. A spring-loaded switching device containing one or more sets of contacts. The contacts are made only when the button is pressed.

Momentary push-button station. A spring-loaded switching device usually containing one or more sets of normally closed contacts and one or more sets of normally open contacts.

Motor. A machine that converts electrical energy into mechanical energy. A machine that produces a rotating motion.

Motor-generator set. An electric motor mechanically coupled to a generator. The motor is used to drive the generator.

Motor starter. A device used to regulate the current to a motor during the starting period. It may be used to make and break the circuit and/or limit the starting current. It is usually equipped with an overload protection device.

Multimeter. An electrical instrument designed to measure more than one unit; for example, voltage, current, and resistance.

Mutliwire circuit. A circuit consisting of two or more ungrounded conductors having a potential difference between them. It also has a grounded conductor with equal potential difference between it and each ungrounded conductor of the circuit, and is connected to the neutral conductor of the system.

Multiwire distribution system. A system containing two or more ungrounded conductors having a potential difference between them; also having a grounded conductor with equal potential difference between it and each ungrounded conductor of the system.

Mutual induction. The phenomenon by which a fluctuating current flowing in one coil induces an emf in an adjacent coil.

National Electrical Code (NEC). A set of recommended standards developed for the purpose of the practical safeguarding of persons and property from hazards arising from the use of electricity.

National Fire Protection Association (NFPA). A nonprofit organization established to aid the public in fire prevention. NFPA makes recommendations for manufacturing and installing equipment, devices, electrical conductors, etc. It has no power or authority to enforce compliance with their recommendations. The organization works in conjunction with insurance companies and fire departments to develop high standards of fire prevention.

Neutral conductor. The conductor of a multiwire circuit or system that is grounded and has an equal voltage between it and each ungrounded conductor of the circuit or system.

Neutral plane. An imaginary plane in a motor or generator forming right angles to the main field flux.

Neutron. The part of an atom that is electrically neutral; it is found in the nucleus of the atom.

No-field release coil. An electromagnet incorporated into some controllers. It is connected in series with the field on dc motors and allows the main circuit to open when the field current drops below a safe value.

No-load release coil. An electromagnet incorporated into some motor controllers, usually used with series motors. Its function is to release the controller switching mechanism, disconnecting the motor from the line when the load is removed from the motor.

No-voltage release coil. An electromagnet incorporated into some controllers. It is connected across the line and will allow the circuit to open when the voltage drops below a predetermined value.

Ohm. The unit of electrical resistance. One ohm of resistance will allow 1 ampere to flow when 1 volt is applied.

Ohmmeter. A meter used to measure the value of electrical resistance.

Ohm's Law. A mathematical law that expresses the relationship between current, voltage, and resistance in a dc circuit. It also expresses the relationship between current, voltage, and impedance in an ac circuit.

Orthographic drawing. A drawing in which only one side of an object is shown in each view. A two-dimensional drawing of an object formed by perpendicular intersecting lines.

Oscillate. To vary, fluctuate, or reverse direction in periodic intervals.

Oscillator. With reference to electron tubes, a tube circuit that converts dc to ac at a specific frequency.

Oscilloscope. An instrument that uses the cathode ray tube. It is used to determine wave shapes, frequency, values of current and voltage, to compare these values; and to determine phase relationships.

Out-of-phase lagging. A condition in an ac circuit in which the current reaches its peak value sometime later than the voltage. The current is lagging the voltage.

Out-of-phase leading. A condition in an ac circuit in which the current reaches its peak value sometime before the voltage. The current is leading the voltage.

Overload. An electrical condition in which an excessive amount of current flows. This condition arises when too much utilization equipment is connected into a circuit or system. When a motor is driving a load that requires more horsepower than its rating, it is overloaded, and excessive current flows.

Overload protector. With reference to motors, the overload protector is a device designed to open the circuit when the motor is overloaded. It does not necessarily protect against short circuits or grounds.

Overload relay. A device designed to protect a motor against excessive momentary surges, normal overloads existing for long periods of time, and high currents caused by an open phase.

Panel. A surface or cabinet on which electrical components and conductors are mounted.

Parallel circuit. A circuit that contains more than one path through which the electrons can flow.

Parallelogram. A four-sided flat figure having opposite sides parallel to each other.

Parallel-series circuit. A circuit in which the loads are grouped and connected in series; each series group is then connected in parallel with the other groups.

Peak value. With reference to ac, the highest value reached during one cycle.

Pentode tube. A five-element electron tube.

Permanent magnet. A magnet made of steel, rare earth, or alloys, which will retain its magnetism for an extremely long period of time.

Permeability. The measurement of the ability of a material to conduct magnetic lines of force. A material with high permeability is easily magnetized.

Phase. A position on an alternating current or voltage curve. The term three-phase indicates three separate voltages, which reach their maximum and zero values at three different but equidistant positions along the curve.

Phase angle. The angle of lead or lag between the current and the voltage in an ac circuit. In multiphase circuits, phase angle can refer to the angle between two voltages.

Phase sequence. A term referring to the order in which one voltage reaches maximum value in reference to the other.

Phase sequence meter. An instrument that indicates the order in which each voltage of a multiphase system reaches its maximum value.

Photocell. A material, generally placed within a tube, that produces an emf when light strikes it.

Phototube. An electron tube that emits electrons from the cathode when light strikes the cathode.

Polarity. An electrical condition indicating the direction in which dc is flowing through a circuit, instrument, or component.

Potential difference. The difference in electrical pressure between two points. Potential difference is measured in volts.

Power. The rate of doing work.

Power factor. The ratio of the true power to the apparent power. It is equal to the cosine of the phase angle between the current and voltage in an ac circuit.

Power factor (lagging). An ac circuit is said to contain a lagging power factor when it contains more inductance than capacitance. A lagging power factor occurs when the current is lagging the voltage.

Power factor (leading). An ac circuit is said to contain a leading power factor when it contains more capacitance then inductance. A leading power factor occurs when the current is leading the voltage.

Primary distribution system. The electrical conductors and equipment between the generator station and the final distribution point.

Prime mover. The machine that drives the rotating member of a generator.

Programmable controller (PC). A special type of computer designed to perform specific control functions in a logical order. It is used to control industrial machinery.

Prony brake. An instrument used to measure the torque of a motor.

Proton. A part of an atom containing a positive charge. Protons are found in the nucleus of the atom.

Pulsating direct current. An electric current that flows in one direction, but varies in strength.

Raceway. An enclosed channel designed to hold electrical conductors.

Radial distance. The distance from a center to a circular distance.

Radial distribution system. A system for distributing electrical energy. The energy is transmitted in all directions from one central point.

Radiation. With reference to heat transfer, the transfer of heat energy by way of electromagnetic waves from an infrared source to the object being heated.

Radiation heating. The transmission of heat energy through space. The rays are absorbed by the objects they contact.

Range selector. The switch or terminals on a meter that determine the desired scale.

Ratio. A mathematical relationship between two quantities usually expressed as a quotient (one quantity divided by the other).

Reactance (capacitive). The opposition to current flow caused by a capacitor or other device that has the characteristics of a capacitor.

Reactance (inductive). The opposition to the flow of a pulsating or alternating current caused by inductance in the circuit.

Reactive power. The power circulating between the source and the load. This is a result of reactance in the circuit. Reactive power is measured in vars (voltamperes reactive). If plotted on a graph, it has a negative value.

Rectifier. A device used to convert ac to dc.

Reduced voltage starter. A motor starter designed to lower the voltage applied to a motor during the starting period.

Relative humidity. The percentage of moisture in the air.

Relay. An electromagnetic switch. It consists of a coil and an armature. A small current through the coil causes the armature to move, opening or closing a switch.

Relay chatter. The noise caused by relays opening and closing rapidly as the ac passes through the zero values.

Reluctance. The opposition a material presents to magnetic lines of force. A material with high reluctance is difficult to magnetize.

Residual magnetism. The magnetism that remains in a material after it has been removed from the magnetic influence.

Resistance. The opposition to current flow caused by the material from which the circuit is made.

Resistance (effective). The total resistance to current flow in an ac circuit. Effective resistance is composed of dc resistance, skin effect, eddy current effect, and dielectric stress effect.

Resistance heating. The act of producing heat by current flowing through a resistance.

Resistance (pure). An ac circuit is said to be of pure resistance when it has no inductive or capacitive characteristics.

Resistivity. As used in the electrical industry, the resistivity of a conducting material is the resistance of the material per mil-foot at a temperature of 68°F (20°C).

Resistor. A device designed to offer a specific amount of opposition to electron flow. It is constructed of materials that have few free electrons.

Resonance. The characteristic of a circuit containing resistance, inductance, and capacitance that results in unity power factor. In a resonant circuit, $X_L = X_C$.

Resonant circuit. A circuit in which the inductive reactance is equal to the capacitive reactance. The result is zero reactance.

Resonant frequency. The frequency at which the inductive reactance is equal to the capacitive reactance. The frequency at which the resistance of a circuit is equal to the impedance.

Retentivity. The ability of a material to retain its magnetism.

Rheostat. A variable resistance. The resistance can be varied by turning a knob or moving a handle.

Right triangle. A triangle containing one 90° angle.

Rms value. Rms stands for root-mean-square. The value is the same as the effective value. It is calculated by determining the square root of the average of the squares of the instantaneous values for one cycle.

Rotary converter. A motor generator set housed in one unit. The motor drives the generator and is used for converting one type of electrical energy to another type. For example, it can change ac to dc, or 60 Hz to 180 Hz.

Rotor. The rotating part of a motor or generator.

Running current. The value of current required by a motor when it has reached its rated speed. The term usually refers to the value of current when the motor is running at full load and full speed.

Schematic diagram. An electrical diagram that shows the electrical connections of circuits and/or equipment in a simplified manner.

Secondary ties. Secondary conductors of a loop system to which load conductors are connected between transformers.

Self-induction. The inducing of an emf in a coil. It is caused by a varying current flowing in the coil.

Semiautomatic controller. A controller that depends upon human intervention for the initial operation after which it continues to operate through sensing devices.

Semiconductor. Materials that are neither good conductors nor good insulators. Certain combinations of these materials allow current to flow in one direction but not in the opposite direction. Semiconductor devices are rapidly replacing electron tubes.

Sensitivity (current). Generally used when referring to ammeters. It is the value of current required to produce full-scale deflection.

Sensitivity (voltage). Generally used with references to a voltmeter. It is expressed in the number of ohms of resistance in the meter circuit per volt of scale (ohms per volt). The greater the value of ohms per volt, the more sensitive the meter and the less effect it will have on the circuit characteristics.

Series circuit. A circuit that contains only one path over which the electrons can flow.

Series-parallel circuit. A circuit in which loads are connected to parallel groups; parallel groups are then connected in series.

Short circuit. A condition that exists in an electric circuit when conductors of opposite polarity come into contact with each other. This condition results in excessive current flow.

Siemens. The unit of electrical conductance. The reciprocal of electrical resistance. A resistance of 1 ohm has a conductance of 1 mho. A resistance of 3 ohms has a conductance of 1/3 siemens.

Single phase. A term applied to a single alternating current or voltage. A single waveform is produced within 360 electrical time degrees.

Skin effect. The effect produced by ac that tends to cause the current to flow along the outside of the conductor as opposed to flowing through the entire conductor.

Slip. With reference to an induction motor, slip is the difference between the synchronous speed and the rotor speed. It may be stated in revolutions per minute or percentage of the synchronous speed.

Slip rings. Copper or brass rings mounted on, and insulated from, the shaft of an alternator. On large alternators, they are used to make the connections from the dc source to the revolving field of the alternator. On small alternators, they are used to complete the connections from the alternator to the external circuit. Slip rings are sometimes used on motors to complete connections from a stationary circuit to a revolving circuit.

Solenoid. An electromagnet used to cause mechanical movement of an armature; for example, a solenoid valve.

Solid state. As used in electrical and electronic circuits, refers to the use of solid materials as opposed to gases. It usually refers to circuits using semiconductors.

Speed control. The ability to vary the speed of a motor by means of a device connected outside of the motor.
Speed regulation. The variation in the speed of a motor caused by a change in the load. A motor whose speed remains practically constant from no load to full load is considered to have good speed regulation.
Splice. A point in the wiring system where two or more conductors are joined both electrically and mechanically.
Square mil. The area of a square measuring 1 mil on a side.
Starter (fluorescent). An automatic switch installed in some types of fluorescent fixtures.
Starting current. The amount of current required by a motor at the instant of starting.
Starting torque. The torque developed on the rotor at the instant the motor is connected to the line.
Static control. Control devices that have no moving parts. They are made of semiconducting materials.
Stator. The stationary part of a motor or generator; the part of the machine that is secured to the frame.
Switch. A device for making or breaking the circuit or changing the current path.
Synchronized. With reference to alternators, all conditions are met so the machines can be connected in parallel.
Synchronizing. With reference to alternators, adjusting the phase sequence, phase relationship, frequency, and voltage of two or more alternators so they can be connected in parallel.
Synchronous speed. The speed at which the electromagnetic field revolves around the stator of an induction motor. The synchronous speed is determined by the frequency of the supply voltage and the number of poles on the motor stator.
Synchroscope. An instrument designed to indicate when the emf of two voltage sources are in phase and of the same frequency.
Temperature coefficient. The amount by which the resistance of a material changes per degree change in temperature for each ohm of resistance.
Terminal. A fitting for connecting electrical conductors to devices and/or equipment.
Thermal effect. The effect of temperature change on the resistance of a conducting material.
Thermionic emission. Electron emission caused by heat.
Thermocouple. Two dissimilar metals in contact with each other at one point. When the junction is heated, an emf is developed across the opposite ends of the metals.
Thermostat. A device for controlling temperature. It is a switch that opens and/or closes an electric circuit as the temperature changes.
Three-phase. A term applied to three alternating currents or voltages of the same frequency, type of wave, and amplitude. The currents and/or voltages are one third of a cycle (120 electrical time degrees) apart.
Thyristor. A transistor that has characteristics similar to a thyraton tube. It carries a high current and is triggered by a low impulse. It is fast acting.
Time-delay relay. A relay in which the armature movement is arranged to close or open contacts at a predetermined time after the signal has been received.
Torque. A turning or twisting force. A force that produces a rotation about an axis.
Transconductance. With reference to electron tubes, a comparison of the input and output. It is equal to the amplification factor divided by the plate resistance.
Transformer. A device that generally contains no moving parts and is designed to increase or decrease the voltage and/or isolate a part of the system from the supply.
Transformer (current). An isolation transformer designed to lower the value of current for measuring instruments and other devices without affecting the load current.
Transformer (isolation). A transformer designed to isolate a part of a system from the source.
Transformer (potential). A transformer designed to reduce the voltage for measuring purposes or to be used with relays and similar devices.
Transformer primary. The part of a transformer containing the winding to which the supply is connected.
Transformer ratio. The ratio of the number of turns of wire on the primary of a transformer to the number of turns on the secondary winding.
Transformer secondary. The part of a transformer containing the winding to which the load is connected.
Transformer (step-down). A transformer used to lower the supply voltage to a value required for the load.
Transformer (step-up). A transformer used to increase the supply voltage to a value required for transmission or for the load.
Transistor. A solid-state device made of certain semiconductor materials; generally used to control electron flow.
Transistor junction. The point at which the P material joins the N material.
Triangle. A three-sided figure formed by connecting three vectors end to end to form a closed pattern.
Trigonometry. The study of mathematics pertaining to the function of angles as ratios of the sides of a triangle. The use of vectors, angles, and trigonometry functions.
Triode tube. A three-element electron tube consisting of a cathode, anode, and grid.
True power. The power, in watts, utilized by a circuit. If plotted on a graph, the true power is the positive power, sometimes called the effective power.
Turbine. A revolving machine in which the energy of a moving fluid, gas or vapor is converted into rotary motion by the pressure of the fluid, gas, or vapor on blades, paddle wheels, or similar equipment.
Turbo-type alternator. An alternator driven by a turbine.
Two-phase. A term applied to two alternating currents or voltages of the same frequency, type of wave, and amplitude. There is a 90° phase displacement between each current and/or voltage.
Unipolar transistor. A transistor with only one junction. It consists of one P material and one N material.
Utility company. A company that supplies electrical energy to consumers.
Vacuum tube. An electronic device with physical characteristics similar to an incandescent lamp. The tube has the air removed and two or more elements are placed within the glass enclosure.

Valance. The outer shell of an atom that contains electrons.

Vector. A line segment that has a definite length and direction.

Vector diagram. Two or more vectors joined together to convey information.

Voltage. The electrical pressure that forces electrons to flow. Voltage is sometimes called electromotive force.

Voltage control. The ability to vary the output voltage of a generator by means of a device connected outside of the generator.

Voltage curve. A graph indicating the various values of voltage as produced by a generator or other voltage source.

Voltage regulation. The variation in output voltage of a generator, which changes with load. A generator whose voltage remains practically constant from no load to full load is said to have good voltage regulation.

Voltampere. The measurement of the apparent or reactive power of an ac circuit or system; the product of the current and voltage in an ac circuit.

Voltmeter. A meter used to measure the electrical pressure between two points.

Watt. A unit of measurement of electrical power; 746 watts is equal to 1 horsepower.

Watthour meter. An electrical instrument designed to measure the amount of energy (electrical power expended times the amount of time).

Wattmeter. An electrical instrument designed to measure the amount of true power utilized by a component, circuit, or system.

Work. The overcoming of resistance through a distance.

INDEX